physica status solidi (b)

ISSN 0370-1972 · SUBJECT AND AUTHOR INDEX (b) VOLUMES 121 TO 126 · 1984

ISSN 0031-8965 · SUBJECT AND AUTHOR INDEX (a) VOLUMES 81 TO 86 · 1984

physica status solidi (a)

AKADEMIE-VERLAG · BERLIN

Classification Scheme

1. Structure of Crystalline Solids
   1.1 Perfectly Periodic Structure
   1.2 Solid-State Phase Transformations
   1.3 Alloys. Metallurgy
   1.4 Microstructure (Magnetic Domains See 18; Ferroelectric Domains See 14.4.1)
   1.5 Films
   1.6 Surfaces
2. Non-Crystalline State
3. Crystal Growth
4. Bonding Properties
5. Mössbauer Spectroscopy
6. Lattice Dynamics. Phonons
7. Acoustic Properties
8. Thermal Properties
9. Diffusion
10. Defect Properties (Irradiation Defects See 11)
    10.1 Metals
    10.2 Non-Metals
11. Irradiation Effects (X-Ray Diffraction Investigations See 1 and 10)
12. Mechanical Properties (Plastic Deformations See 10)
    12.1 Metals
    12.2 Non-Metals
13. Electron States
    13.1 Band Structure
    13.2 Fermi Surfaces
    13.3 Surface and Interface States
    13.4 Impurity and Defect States
    13.5 Elementary Excitations (Phonons See 6)
        13.5.1 Excitons
        13.5.2 Plasmons
        13.5.3 Polarons
        13.5.4 Magnons
14. Electrical Properties. Transport Phenomena
    14.1 Metals. Semi-Metals
    14.2 Superconductivity. Superconducting Materials and Devices
    14.3 Semiconductors
        14.3.1 Films
        14.3.2 Surfaces and Interfaces
        14.3.3 Devices. Junctions (Contact Problems See 14.3.4)
        14.3.4 High-Field Phenomena, Space-Charge Effects, Inhomogeneities, Injected Carriers (Electroluminescence See 20.3; Junctions See 14.3.3)
    14.4 Dielectrics
        14.4.1 Ferroelectrics
15. Thermoelectric and Thermomagnetic Properties
16. Photoconductivity. Photovoltaic Effects
17. Emission of Electrons and Ions
    17.1 Field Emission Microscope Investigations
18. Magnetic Properties
    18.1 Paramagnetic Properties
    18.2 Ferromagnetic Properties
        18.2.1 Ferromagnetic Films
    18.3 Ferrimagnetic Properties
    18.4 Antiferromagnetic Properties

(Continued on cover three)

physica status solidi (b)

basic research

physica status solidi (a)

applied research

SUBJECT AND AUTHOR INDEX

Volumes 121 to 126, 1984

Volumes 81 to 86, 1984

ISSN 0370-1972/0031-8965

Edited by

E. BORCHARDT

AKADEMIE-VERLAG · BERLIN

Subscriptions and orders for this Subject and Author Index
or single copies of physica status solidi (b) and physica status solidi (a) should be directed
in the GDR:

to the Postzeitungsvertrieb or
to the Akademie-Verlag, DDR-1086 Berlin, Leipziger Str. 3—4;

in the other socialist countries:

to a book-shop for foreign language literature or
to the competent news-distributing agency;

in the FRG and BERLIN (WEST):

to a book-shop or
to the wholesale distributing agency Kunst und Wissen, Erich Bieber OHG, Wilhelmstr. 4—6, D-7000 Stuttgart 1;

in the other Western European countries:

to Kunst und Wissen Erich Bieber GmbH, Dufourstr. 51, CH-8008 Zürich;

in USA and CANADA:

to VCH Publishers, Inc., 303 N.W. 12th Avenue, Deerfield Beach, FL 33442-1705, USA;

in other countries:

to the international book- and journal-selling trade,
to Buchexport, Volkseigener Außenhandelsbetrieb der Deutschen Demokratischen Republik, DDR-7010 Leipzig, Postfach 160, or
to the Akademie-Verlag, DDR-1086 Berlin, Leipziger Straße 3—4.

Schriftleiter und verantwortlich für den Inhalt der Zeitschriften:

Professor Dr. Dr. h. c. P. Görlich, DDR-1086 Berlin, Leipziger Str. 3—4, Postfach 1233 bzw. DDR-6900 Jena, Schillbachstr. 24.

Herausgeber und verantwortlich für den Inhalt des Registers:

Dipl.-Phys. E. Borchardt, DDR-1086 Berlin, Leipziger Str. 3—4, Postfach 1233.

Verlag:

Akademie-Verlag, DDR-1086 Berlin, Leipziger Str. 3—4; Fernruf: 2236221 und 2236229; Telex-Nr.: 114420; Bank: Staatsbank der DDR, Berlin, Kto.-Nr.: 6836-26-20712.

Veröffentlicht unter der Lizenznummer 1310 und 1620 des Presseamtes beim Vorsitzenden des Ministerrates der Deutschen Demokratischen Republik.

Gesamtherstellung:

VEB Druckerei „Thomas Müntzer", DDR-5820 Bad Langensalza.

Bestellnummer des Registers 1068/85 R17.

Printed in the German Democratic Republic.

AN (EDV) 20831/20735

Preface

The seventeenth index volume contains volumes 121 to 126 of "physica status solidi (b) — basic research" as well as volumes 81 to 86 of "physica status solidi (a) — applied research", i.e. all volumes having appeared in 1984 containing 1512 papers.

The general scheme of last index volume was maintained. The basis for quoting is the subject classification given every paper in the respective issue. Nevertheless, before taking over these subject classifications into the index volume they were once more carefully revised and in some cases improved and completed in order to get a most uniform and clear classification for all papers. The average number of quotations per paper in this index volume is nearly the same as in the last one, namely about 3.0.

Again I have to express my thanks to the readers for their interest, and I will always readily take up any critical remarks and proposals for improvements. I am especially obliged to Prof. Dr. P. Görlich, Prof. Dr. E. Gutsche, Dr. H.-J. Hänsch, Dr. H. Lange, Dr. S. Oberländer, and Dipl. Phys. K. Müller for giving me advice and unselfish help. Furthermore I have to thank Mrs. E. Schwan for her exact and reliable secretary's work at the manuscript.

Berlin, October 1985

Eva Borchardt

Classification Scheme

1. Structure of Crystalline Solids
    1.1 Perfectly Periodic Structure
    1.2 Solid-State Phase Transformations
    1.3 Alloys. Metallurgy
    1.4 Microstructure (Magnetic Domains See 18; Ferroelectric Domains See 14.4.1)
    1.5 Films
    1.6 Surfaces
2. Non-Crystalline State
3. Crystal Growth
4. Bonding Properties
5. Mössbauer Spectroscopy
6. Lattice Dynamics. Phonons
7. Acoustic Properties
8. Thermal Properties
9. Diffusion
10. Defect Properties (Irradiation Defects See 11)
    10.1 Metals
    10.2 Non-Metals
11. Irradiation Effects (X-Ray Diffraction Investigations See 1 and 10)
12. Mechanical Properties (Plastic Deformations See 10)
    12.1 Metals
    12.2 Non-Metals
13. Electron States
    13.1 Band Structure
    13.2 Fermi Surfaces
    13.3 Surface and Interface States
    13.4 Impurity and Defect States
    13.5 Elementary Excitations (Phonons See 6)
        13.5.1 Excitons
        13.5.2 Plasmons
        13.5.3 Polarons
        13.5.4 Magnons
14. Electrical Properties. Transport Phenomena
    14.1 Metals. Semi-Metals
    14.2 Superconductivity. Superconducting Materials and Devices
    14.3 Semiconductors
        14.3.1 Films
        14.3.2 Surfaces and Interfaces
        14.3.3 Devices Junctions (Contact Problems See 14.3.4)
        14.3.4 High-Field Phenomena, Space-Charge Effects, Inhomogeneities, Injected Carriers (Electroluminescence See 20.3; Junctions See 14.3.3)
    14.4 Dielectrics
        14.4.1 Ferroelectrics
15. Thermoelectric and Thermomagnetic Properties
16. Photoconductivity. Photovoltaic Effects
17. Emission of Electrons and Ions
    17.1 Field Emission Microscope Investigations

18. Magnetic Properties
    - 18.1 Paramagnetic Properties
    - 18.2 Ferromagnetic Properties
        - 18.2.1 Ferromagnetic Films
    - 18.3 Ferrimagnetic Properties
    - 18.4 Antiferromagnetic Properties

19. Magnetic Resonance

20. Optical Properties
    - 20.1 Spectra. Optical Constants (X-Ray Spectra See 20)
    - 20.2 Lasers
    - 20.3 Luminescence (X-Ray Spectra See 20)

21. Metals. Alloys
    - 21.1 Transition Metals (Group Sc to Zn) and Their Alloys
        - 21.1.1 Fe and Fe Alloys
    - 21.2 Alkali-Metals
    - 21.3 Alkali-Earth Metals
    - 21.4 Rare-Earth Metals
    - 21.5 Actinides
    - 21.6 Noble Metals
    - 21.7 Semi-Metals (Semiconducting Alloys See 22.7)

22. Semiconductors and Ionic Crystals
    - 22.1 Elements
        - 22.1.1 Germanium
        - 22.1.2 Silicon
        - 22.1.3 Group VI Elements
    - 22.2 III–V Compounds
        - 22.2.1 Arsenides
        - 22.2.2 Phosphides
        - 22.2.3 Antimonides
        - 22.2.4 Mixed Crystals
    - 22.3 IV–IV, II–V, and III–VI Compounds
        - 22.3.1 Mixed Crystals
    - 22.4 II–VI Compounds
        - 22.4.1 Sulphides
        - 22.4.2 Selenides
        - 22.4.3 Tellurides
        - 22.4.4 Mixed Crystals
    - 22.5 Halides
        - 22.5.1 Silver Halides
        - 22.5.2 Alkali Halides
        - 22.5.3 Alkali-Earth Halides
        - 22.5.4 Mixed Crystals
    - 22.6 Simple Oxides (CdO, ZnO See 22.4)
        - 22.6.1 Mixed Crystals
    - 22.7 Semiconducting Intermetallic Compounds
        - 22.7.1 Mixed Crystals
    - 22.8 Tertiary and Higher Compounds
        - 22.8.1 Salts of Oxy-Acids (Carbonates, Phosphates, Silicates Including Mica etc).
        - 22.8.2 Spinels,.Garnets, Ferrites
    - 22.9 Organic Semiconductors

23. Solidified Gases

For comments on the use of the classification scheme see p. 6.

Some Comments on the Use of the Subject and Author Index

The classification system[1]) comprises 20 broad sections each devoted to a major field of solid state physics and 3 broad sections devoted to the main groups of substances. A section may contain a number of subsections dealing with aspects of this field of special substance, respectively, which tend to receive particular attention. A paper dealing with this field will either be classified under one or more of the subsections or, if none of these are relevant, it will just appear under the general section heading (i.e. sections and subsections are equivalent in this sense). For example, a paper dealing with problems of paramagnetism would appear under 18.1 but not 18, whereas a paper on diamagnetism would be quoted under 18; a paper dealing with GaP would appear under 22.2.2, not under 22 or 22.2, whereas a paper on AlN would be quoted under 22.2, and a paper on a II–IV or a IV–VI compound would be quoted under 22.

All papers will be quoted under each relevant section. There is made a distinction between one or two principal entries (refering to the main subject matter of the paper) and secondary entries (refering to subsidiary subject matter and substance, respectively).In the principal entries the title of the paper is given in its original language and in English. In addition, every principal entry is marked by * in front of the title of the paper. In the secondary entries the title is only given in English.

On the line following the title the subject classification is given (in parantheses) in the same manner as in the volumes. Each entry gives the volume, the number of the issue, the initial and final page numbers, and the year in which the paper appeared. A paper of physica status solidi (b) or physica status solidi (a) is marked by (b) or (a) in front of the line, respectively. The letter "K" before the page numbers indicates a short note published in the offset part of the journal.

Review articles appear, at the head of the list of entries under a given subject and are given the subtitle "Review Articles". These articles are listed in the chronological order of their publication, irrespective of being a paper of physica status solidi (b) or physica status solidi (a). The subject classification for a review article consisting of several parts is based on the article as a whole rather than on the individual parts.

The entries for original papers, short notes, and errata are made together in a single list for each subject and this follows the entries for review articles. These are again ordered according to data of publication, irrespective of being a paper of physica status solidi (b) or physica status solidi (a). Original papers are sometimes published in several parts under the same principal title (in general with special subtitles). The various parts of papers of this type are treated separately and each part is classified according to its particular contents.

In quoting errata the following rule is obeyed. If the erratum refers to a paper quoted in the preceding index, it is listed in the chronological order of its publication otherwise the erratum is quoted together with the paper to which it refers.

---

[1]) See cover two and three, and pages 4 and 5. For the case of losing the cover by bookbinding we repeat the classification scheme on pages 4 and 5.

In using the subject index attention should be given to the following details:

subject	quoted in	remarks
representations of space groups	1 and 13.1	
Alfen waves, helicon waves	13 and 20	
hot electrons	13 and 14.3.4	
transport phenomena	14 or subsection of 14	if not clearly to be assigned to 8 or 15
galvanomagnetic effects	14 or subsection of 14	not in 18
electro-optical effects	14 and 20 (or subsection of 14 and 20)	
piezoelectricity, electrostriction	14 or subsection of 14	not in 12 or subsection of 12; different electromechanical effects are quoted in 12 or subsection of 12
ionic conduction	14.4	not in 14.3
magneto-optical effects	18 and 20 (or subsection of 18 and 20)	possibly in additional sections, e.g. Faraday effect in 13
ferromagnetic films	18.2.1	not additionally in 1.5
X-ray spectra	20	but not X-ray studies of structure
MIS and MOS structures	the section assigned to the semiconductor	not in the section assigned to the oxide or the metal
MIM structures	the section assigned to the insulator	generally not in the section assigned to the metal
epitaxial layers	the section assigned to the layer	not in the section assigned to the substrate

The author index gives the names of the authors in alphabetical order. Each name is followed by the bibliographic details of the paper. A paper of physica status solidi (b) or physica status solidi (a) is again marked by (b) or (a) in front of the number of the volume, respectively. In the last column the subject classification of the paper is presented in the same manner as in the volumes. Behind the principal subject classification(s) the page number is given (in parantheses) on which the paper appears in the subject index. When several papers have been published by the same author they are arranged according to the dates of publication, irrespective of being a paper of physica status solidi (b) of physica status solidi (a).

Subject Index

1. Structure of Crystalline Solids

Original Papers and Short Notes

E. A. Tikhonova
*Non-Linear Effects of the Dynamical Scattering Theory in Slightly Distorted Crystals
(1; 10)
(a) Vol. **81**, No. 1, p. 69—75 (1984).

L. J. Balk, D. G. Davies, and N. Kultscher
*The Dependence of Scanning Electron Acoustic Microscopy (SEAM) Imaging on Chopping and Detection Frequency for Metal Samples
(1 and 7; 21.1)
(a) Vol. **82**, No. 1, p. 23—33 (1984).

H. Cerva and W. Graeff
*Contrast Investigations of Surface Acoustic Waves by Stroboscopic Topography (I): Orientation Contrast
(1 and 7; 14.4.1; 22.8.1)
(a) Vol. **82**, No. 1, p. 35—45 (1984).

D. G. Chambaere and E. De Grave
*A Study of the Non-Stoichiometrical Halogen and Water Content of β-FeOOH
(1; 5; 22)
(a) Vol. **83**, No. 1, p. 93—102 (1984).

A. M. Afanasev, R. M. Imamov, A. V. Maslov, and E. M. Pashaev
*Photoeffect in X-Ray Grazing Incidence Diffraction
(1 and 17; 22.1.2)
(a) Vol. **84**, No. 1, p. 73—78 (1984).

R. S. Vardanian
*On the Sign Choice in the Darwin Solution for X-Ray Diffraction in a Semi-Infinite Crystal
(1)
(a) Vol. **84**, No. 1, p. K7—K8 (1984).

V. P. GUPTA and O. P. SINGH
*Phillips Ionicity and Structural Classification of Layered Transition Metal Dichalcogenides
(1 and 4; 22)
(b) Vol. **125**, No. 1, p. K1—K5 (1984).

T. SATOW, O. UEMURA, K. HOSHINO, and T. WATANABE
*The Neutron Diffraction of Molten TlCl, TlBr, and TlI
(1 and 2; 22.5.2)
(a) Vol. **85**, No. 1, p. K5—K7 (1984).

M. DAVID, W. HERREMANS, and R. GEVERS
*Fast-Electron Diffraction near a Translation Interface on Edge. The Generalized Dispersion Relation
(1 and 10)
(a) Vol. **85**, No. 2, p. 375—388 (1984).

C. MALGRANGE and J. GRONKOWSKI
*Fraunhofer Diffraction of X-Ray Beams at Bragg Incidence in Distorted Crystals
(1)
(a) Vol. **85**, No. 2, p. 389—397 (1984).

S. CHATTOPADHYAY and C. MANDE
*XANES and EXAFS Study of the Compound CoAsSe
(1; 20; 21.1)
(b) Vol. **126**, No. 1, p. 53—57 (1984).

H. M. VAN NOORT, D. B. DE MOOIJ, and K. H. J. BUSCHOW
*Crystal Structure, Magnetic Properties, and ^{57}Fe Mössbauer Effect of PtFeSn
(1 and 18.2; 5; 22.8)
(a) Vol. **86**, No. 2, p. 655—662 (1984).

1.1 Perfectly Periodic Structure

Original Papers and Short Notes

G. NIHOUL and C. CESARI
*High Resolution Images for Large Cell Crystals Exhibiting Some Limits of the Projected Potential Approximation
(1.1 and 1.4; 22.6)
(a) Vol. **81**, No. 1, p. 87—98 (1984).

G. J. ASHWELL
(2,2-DHPE $Cl)_x(2,2\text{-}DPE)_{1-x}$(TCNQ), a Pauli Paramagnetic Organic Conductor
(14.3; 1.1; 18.1; 22.9)
(a) Vol. **81**, No. 1, p. 361—366 (1984).

G. J. ASHWELL and I. W. NOWELL
*4-Cyano-1,3,6-triazacycl[3.3.3]azine — A Homomolecular Organic Semiconductor
(1.1; 14.3; 22.9)
(a) Vol. **81**, No. 2, p. K97—K101 (1984).

G. J. ASHWELL, J. G. ALLEN, E. P. GOODINGS, and I. W. NOWELL
*Crystal Structure and Electrical Conductivity of [1,1′-bis(p-fluorophenyl)-4,4′-bipyridinium]$^{2+}$(TCNQ)$_2^{2-}$ and Electrical Characterisation of 51 TCNQ Simple Salts
(1.1 and 14.3; 22.9)
(a) Vol. **82**, No. 1, p. 301—306 (1984).

B. LAMBERT-ANDRON, P. CHAUDOUET, R. MADAR et R. FRUCHART
*Détermination de la structure cristallographique de la phase de type G $Ni_{16}Mn_6Aa_7$
(Determination of the Crystallographic Structure of the $Ni_{16}Mn_6As_2$ G-Type Phase)
(1.1; 22.8)
(a) Vol. **82**, No. 2, p. K133—K136 (1984).

U. PIETSCH, J. BAK-MISIUK, and V. GOTTSCHALCH
The Linear Thermal Expansion Coefficient of a $Ga_xIn_{1-x}As_yP_{1-y}$ Layer on InP:Sn Substrate
(8; 1.1; 22.2.4)
(a) Vol. **82**, No. 2, p. K137—K140 (1984).

G. J. ASHWELL and I. W. NOWELL
*Electrical Magnetic and Structural Characterisation of [1,2-Bis(1-ethyl-4-pyridinium)ethane]$^{2+}$ (7,7,8,8-Tetracyano-p-quinodimethanide)$_4^{2-}$
(1.1; 14.3; 18.1; 22.9)
(a) Vol. **83**, No. 1, p. 39—45 (1984).

R. PÉREZ, S. S. SHEININ, and K. Z. BOTROS
*Effects of Systematic Reflections on Diffraction Contrast in H.C.P. Cobalt
(1.1; 21.1)
(a) Vol. **83**, No. 2, p. 445—453 (1984).

YU. A. ALEKSANDROV, B. CHALUPA, J. KULDA, T. A. MACHEKHINA, R. MICHALEC, P. MIKULA, L. N. SEDLÁKOVÁ, J. VÁVRA, and M. VRÁNA
*Backscattering Diffraction of a Pulsed Neutron Beam on an Elastically Bent Single Crystal
(1.1; 22.1.2)
(a) Vol. **83**, No. 2, p. 455—460 (1984).

M. HERVIEU, N. NGUYEN, V. CAIGNAERT, and B. RAVEAU
*Calcium Mangano-Ferrites $(CaMO_3)_mCaMO_2 \square$, M = Fe, Mn, with an Oxygen-Defect Perovskite Structure Related to Brownmillerite. Electron Microscopy Investigation
(1.1 and 1.4; 22.8.2)
(a) Vol. **83**, No. 2, p. 473—483 (1984).

G. J. ASHWELL
Electrical, Magnetic, and Structural Characterisation of [1,2-Bis(1-methyl-4-pyridinium)ethylene]$^{2+}$ $(TCNQ)_4^{2-}$ (H_2O) and the Properties of Two Isostoichiometric TCNQ Salts
(14.3 and 18.1; 1.1; 15; 22.9)
(a) Vol. **83**, No. 2, p. 669—673 (1984).

Z. GALDECKI, P. GROCHULSKI, Z. WAWRZAK, and J. STEPIEŃ-DAMM
*Some Properties of Cyclotrimethylene–Trinitramine Single Crystals
(1.1; 22.9)
(a) Vol. **83**, No. 2, p. K113—K115 (1984).

M. SHIOJIRI, S. SEKIMOTO, T. MAEDA, Y. IKEDA, and K. IWAUCHI
*Crystal Structure of Fe_2TiO_5
(1.1; 22.8.1)
(a) Vol. **84**, No. 1, p. 55—64 (1984).

G. VAN TENDELOO and S. AMELINCKX
*A New Long-Period Derivative of the Au_5Mn_2 Structure
(1.1 and 1.4; 21.1; 21.6)
(a) Vol. **84**, No. 1, p. 185—192 (1984).

P. GROCHULSKI and Z. WAWRZAK
*Thermal Expansion of Sodium–Cadmium Formate. Powder Data
(1.1 and 8; 22.9)
(a) Vol. **84**, No. 1, p. K5—K6 (1984).

M. MESSAOUDI, A. KUHN, J. TOULOUSE, and J. DURAN
*An Isomer of the Organic Semiconductor Cs_2TCNQ_3. X-Ray, Optic, and EPR Analysis
(1.1; 19; 20,1; 22.9)
(a) Vol. **84**, No. 2, p. 411—416 (1984).

YU. N. MOSKVICH, A. A. SUKHOVSKII, and O. V. ROZANOV
*Deuteron Magnetic Resonance of Highly Deuterated Ammonium Hydrogen Selenate
(1.1 and 14.4.1; 19; 22.8.1)
(a) Vol. **84**, No. 2, p. K109—K113 (1984).

W. KULLMANN, J. GEURTS, W. RICHTER, N. LEHNER, H. RAUH, U. STEIGENBERGER, G. EICHHORN, and R. GEICK

Effect of Hydrostatic and Uniaxial Pressure on Structural Properties and Raman Active Lattice Vibrations in Bi_2Te_3
(4; 1.1; 6; 20.1; 22.7)
(b) Vol. **125**, No. 1, p. 131—138 (1984).

Z. JANÁČEK

*Coherence Properties of Radiation Diffracted by an Elastically Bent Crystal
(1.1; 22.1.2)
(a) Vol. **85**, No. 1, p. 77—82 (1984).

T. OHKAWA and H. HASHIMOTO

*On the Polarization Mixing of X-Rays. Quantum Theory of Interference of White X-Rays
(1.1)
(a) Vol. **85**, No. 2, p. 335—348 (1984).

P. A. BESIRGANYAN, R. TS. GABRIELYAN, V. G. KOHN, and A. H. TONEYAN

*Six-Beam X-Ray Diffraction in Ge Single Crystals
(1.1; 22.1.1)
(a) Vol. **85**, No. 2, p. 349—358 (1984).

P. A. ALEKSANDROV, A. M. AFANASIEV, and S. A. STEPANOV

*Bragg-Laue Diffraction in Inclined Geometry
(1.1; 1.6; 22.1.2)
(a) Vol. **86**, No. 1, p. 143—154 (1984).

A. A. KAMINSKII, E. L. BELOKONEVA, B. V. MILL, YU. V. PISAREVSKII, S. E. SARKISOV, I. M. SILVESTROVA, A. V. BUTASHIN, and G. G. KHODZHABAGYAN

Pure and Nd^{3+} Doped $Ca_3Ga_2Ge_4O_{14}$ and $Sr_3Ga_2Ge_4O_{14}$ Single Crystals, Their Structure, Optical, Spectral Luminescence, Electromechanical Properties, and Stimulated Emission
(14.4 and 20.2; 1.1; 12.2; 20.1; 20.3; 22.8.1)
(a) Vol. **86**, No. 1, p. 345—362 (1984).

N. KAMESWARI, K. V. RAMANUJACHARI, A. MEENAKSHI SUNDARAM, and C. S. SWAMY

*Synthesis and Structural Characteristics of $La_2Cu_{1-x}Co_xO_4$ ($x = 0.2, 0.4$, and 0.5)
(1.1 and 3; 22.6.1)
(a) Vol. **86**, No. 1, p. K15—K17 (1984).

1.2 Solid-State Phase Transformations

Review Article

C. BOULESTEIX
*A Survey of Domains and Domain Walls Generated by Crystallographic Phase Transitions Causing a Change of the Lattice
(1.2 and 1.4; 14.4.1; 22; 22.6; 22.8)
(a) Vol. **86**, No. 1, p. 11—42 (1984).

Original Papers and Short Notes

P. ENDERS
*Structure and Bonding in Cubic IV–VI Crystals (II): Instability of the TO(Γ) Mode: Unified Universal-Parameter Treatment of the Group-V Semimetals and Cubic IV–VI Semiconductors
(1.2; 4; 13; 22)
(b) Vol. **121**, No. 1, p. 39—46 (1984).

L. V. VASSILEV
Influence of the Phase Transition in $Pb_{1-x}Ge_xTe$ on the Electronic Dispersion Relation
(13.1; 1.2; 22.8)
(b) Vol. **121**, No. 1, p. 203—207 (1984).

T. SOMA and H.-M. KAGAYA
*NaCl-Type Lattice of GaAs and InSb under Pressure
(1.2; 22.2.1; 22.2.3)
(b) Vol. **121**, No. 1, p. K1—K5 (1984).

S. JURGA
*Molecular Motions and Polymorphic Phase Transitions in $(CH_3)_4$ $NClO_4$ and $(CH_3)_3NHClO_4$ as Studied by NMR
(1.2; 19; 22.9)
(a) Vol. **81**, No. 1, p. 77—85 (1984).

S. BANERJEE and K. URBAN
*Kinetics of Order–Disorder Transformation in Alloys under Electron Irradiation
(1.2; 1.3; 11; 21.1)
(a) Vol. **81**, No. 1, p. 145—162 (1984).

M. HIDAKA, S. YAMASHITA, and Y. OKAMOTO
*Study of Structural Phase Transitions of $KCaF_3$
(1.2; 6; 8; 22.5.4)
(a) Vol. **81**, No. 1, p. 177—183 (1984).

V. K. YANOVSKII
*Phase Transitions and Properties of Doped $KNbO_3$ Crystals
(1.2 and 14.4.1; 14.3; 15; 20.1; 22.8.1)
(a) Vol. **81**, No. 1, p. 399—406 (1984).

P. Enders
*Structure and Bonding in Cubic IV–VI Crystals (III): On the Cubic to Rhombohedral Phase Transition
(1.2; 4; 6; 22)
(b) Vol. **121**, No. 2, p. 461–470 (1984).

I. P. Makarova, S. V. Misyul, L. A. Muradyan, A. F. Bovina, V. I. Simonov, and K. S. Aleksandrov
*Anharmonic Thermal Atomic Vibrations in the Cubic Phase of $Cs_2NaNdCl_6$ Single Crystals
(1.2 and 6; 22.8)
(b) Vol. **121**, No. 2, p. 481–486 (1984).

V. E. Antonov, T. E. Antonova, I. T. Belash, V. Yu. Malyshev, and V. I. Rashupkin
Superconductivity and Atomic Ordering of Pd–Cu–H Solid Solutions
(14.2; 1.2; 21.6)
(a) Vol. **81**, No. 2, p. K185–K190 (1984).

N. K. Gobran, M. M. Danial, and R. Kamel
*Kinetics of the Amorphous–Crystalline Transition in the Metallic Glass $Fe_{40}Ni_{40}B_{20}$
(1.2; 2; 21.1; 21.1.1)
(a) Vol. **82**, No. 1, p. 63–66 (1984).

M. B. Zapart, W. Zapart, and A. I. Zvyagin
*Phase Transitions in Ferroelastic $RbIn(MoO_4)_2$ Crystals by Electron Paramagnetic Resonance of Cr^{3+} Ions
(1.2 and 19; 13.4; 22.8.1)
(a) Vol. **82**, No. 1, p. 67–73 (1984).

O. Bostanjoglo and E. Endruschat
*Time-Resolved TEM of Laser-Annealing
(1.2 and 1.4; 2; 11; 22.1.1)
(a) Vol. **82**, No. 1, p. K1–K3 (1984).

D. Roy, Devendra Kumar, and D. Bahadur
*Crystallisation Behaviour and Magnetic Properties of $Al_2O_3 \cdot Fe_2O_3 \cdot MnO \cdot SiO_2$ Glass
(1.2 and 18; 22.8.2)
(a) Vol. **82**, No. 1, p. K11–K15 (1984).

F. Smutný and M. Połomska
*Effects of Electric Field on the II–III Phase Transition in Lithium Ammonium Sulphate and Its Deuterated Analogue
(1.2 and 14.4.1; 22.8.1)
(a) Vol. **82**, No. 1, p. K33–K36 (1984).

B. Žekš, B. B. Lavrenčič, and R. Blinc
*Microscopic Model for the Ferroelastic Transition in $KLiSO_4$
(1.2; 22.8.1)
(b) Vol. **122**, No. 2, p. 399–404 (1984).

A. F. Barabanov, L. B. Litinskii, and A. V. Mikheenkov
*On the Atomic Phase of Crystalline Hydrogen
(1.2 and 13; 18.4; 23)
(b) Vol. **122**, No. 2, p. K109—K114 (1984).

L. G. Khvostantsev and V. A. Sidorov
*Phase Transitions in Antimony at Hydrostatic Pressure up to 9 GPa
(1.2; 14.1; 15; 21.7)
(a) Vol. **82**, No. 2, p. 389—398 (1984).

H. Broda
*Commensurate–Commensurate Phase Transitions in $(NH_4)_2CoCl_4$
(1.2; 22.8)
(a) Vol. **82**, No. 2, p. K117—K120 (1984).

A. Ioanid, M. Popescu, N. Vlahovici, and I. Bunget
*On the Phase Transition of KH_2PO_4 at $T \approx 110$ °C
(1.2; 14.4.1; 22.8.1)
(a) Vol. **82**, No. 2, p. K125—K128 (1984).

P. Orlandi and A. Rigamonti
*On the Phase Transition and Crystal Field Effects in $FeSiF_6 \cdot 6\,H_2O$
(1.2 and 13.4; 5; 22.8)
(a) Vol. **82**, No. 2, p. K129—K132 (1984).

Z. A. Matysina, D. R. Rizdvyanetskii, and S. Y. Zaginaichenko
*The Structural Diagrams of Magnetic Decompositive Alloys
(1.2 and 18; 21; 21.1)
(a) Vol. **82**, No. 2, p. K163—K168 (1984).

M. Wagner and G. Götz
*Crystallization of High-Dose Antimony-Implanted Silicon by Millisecond Pulse Laser Irradiation
(1.2 and 3; 22.1.2)
(a) Vol. **83**, No. 1, p. 69—76 (1984).

M. Kato and H.-R. Pak
*Thermodynamics of Stress-Induced First-Order Phase Transformations in Solids
(1.2; 8; 12)
(b) Vol. **123**, No. 2, p. 415—424 (1984).

S. Ghosh, S. Chatterjee, and A. N. Basu
*Phase Transition and Soft Mode Behaviour in Superionic Copper Halides
(1.2 and 6; 22.5)
(b) Vol. **123**, No. 2, p. 445—451 (1984).

A. V. Belushkin, I. Natkaniec, J. Wąsicki, and T. Zaleski
*The Phonon Density of States near the Structural Phase Transition of Solid Biphenyl
(1.2 and 6; 22.9)
(b) Vol. **123**, No. 2, p. K115—K118 (1984).

I. O. Bashkin, M. E. Kost, and E. G. Ponyatovskii
*Phase Transitions in Non-Stoichiometric Cerium Hydrides
(1.2; 21.4)
(a) Vol. **83**, No. 2, p. 461—465 (1984).

S. Tsunekawa, H. F. J. Watanabe, and H. Takei
*Linear Thermal Expansion of $SrTiO_3$
(1.2; 12.2; 22.8.1)
(a) Vol. **83**, No. 2, p. 467—472 (1984).
Erratum
(a) Vol. **86**, No. 2, p. 841 (1984).

Z. Bojarski, H. Morawiec, and T. J. Panek
*Positron Annihilation Study of Martensitic Transformation in Cu–Zn–Al Alloy
1.2 and 10.1; 21; 21.1)
(a) Vol. **83**, No. 2, p. K97—K101 (1984).

A. M. Balagurov, B. N. Savenko, M. Dlouhá, S. Vratislav, and Z. Jirák
*A Neutron Diffraction Study of Crystal and Domain Structure in $LiKSO_4$
(1.2; 22.8.1)
(a) Vol. **83**, No. 2, p. K117—K121 (1984).

A. N. Bogdanov, A. V. Zhuravlev, and V. T. Telepa
First-Order Phase Transition from Antiferromagnetic to Paramagnetic Phase in $(C_2H_5NH_3)_2CuCl_4$
(18.1 and 18.4; 1.2; 22.8)
(a) Vol. **83**, No. 2, p. K135—K140 (1984).

T. Soma, Y. Takahashi, and H.-Matsuo Kagaya
*Equation of State and Pressure Derivative of the Bulk Modulus for III-V Covalent Crystals
(1.2 and 4; 12.2; 22.2.1; 22.2.3)
(b) Vol. **124**, No. 1, p. K11—K14 (1984).

B. Chenevier, D. Fruchart, M. Bacmann, J. P. Senateur, P. Chaudouet, and L. Lundgren
Magnetic Properties of MnRhAs. Study of the Antiferromagnetic–Ferromagnetic Transition
(18.2 and 18.4; 1.2; 14.1; 21.1)
(a) Vol. **84**, No. 1, p. 199—206 (1984).

R. Zach, R. Duraj, and A. Szytuła
Pressure Effect on Magnetic Transformations in the $Co_xNi_{1-x}MnGe$ System
(18.2 and 18.4; 1.2; 22.8)
(a) Vol. **84**, No. 1, p. 229—236 (1984).

A. Ribeiro Filho, D. R. Tilley, E. L. Albuquerque, N. S. Almeida, and M. C. Oliveros
Light Scattering from Surfaces near Phase Transition Points
(1.6; 1.2)
(b) Vol. **124**, No. 2, p. 457—464 (1984).

F. Prokert and K. S. Aleksandrov
*Neutron Scattering Studies on Phase Transition and Phonon Dispersion in $Cs_2NaBiCl_6$
(1.2 and 6; 22.5)
(b) Vol. **124**, No. 2, p. 503—513 (1984).

V. I. Zametin
Absorption Edge Anomalies in Polar Semiconductors and Dielectrics at Phase Transitions
(20.1; 1.2; 14.4.1; 22; 22.5.2; 22.8.1; 22.9)
(b) Vol. **124**, No. 2, p. 625—640 (1984).

R. L. Davidovich, P. S. Gordienko, J. Grigas, T. A. Kaidalova, V. Urbonavičius, and L. A. Zemnukhova
*Phase Transitions in $KSbF_4$
(1.2 and 14.4.1; 8; 19; 22.8)
(a) Vol. **84**, No. 2, p. 387—392 (1984).

F. Reynaud
*Application of the Grand Canonical Ensemble in Classical Statistical Physics to the Study of Order–Disorder Transitions in Binary Substitutional Alloys
(1.2 and 1.3; 21)
(a) Vol. **84**, No. 2, p. 393—400 (1984).

M. T. Sebastian and P. Krishna
*X-Ray Diffraction from a SiC Crystal Undergoing the 3C–6H Solid State Transformation by Non-Random Microtwinning
(1.2; 10.2; 22.3)
(a) Vol. **84**, No. 2, p. 401—409 (1984).

V. I. Kononenko, A. L. Sukhman, S. L. Gruverman, and V. V. Torokin
Density and Surface Tension of Liquid Rare Earth Metals, Scandium and Yttrium
(2 and 12.1; 1.2; 21; 21.4)
(a) Vol. **84**, No. 2. p. 423—432 (1984).

A. V. Parygin, G. V. Chikvaidze, G. K. Semin, B. V. Lokshin, and Ya. A. Eiduss
*^{75}As-NQR Study of Yellow Crystalline Arsenic
(1.2; 19; 21.7)
(a) Vol. **84**, No. 2, p. K101—K103 (1984).

A. V. Salker, E. B. Mirza, V. V. Deshpande, and H. V. Keer
*Thermodynamic Analysis of the Phase Transition in V_4O_7
(1.2; 8; 14.1; 14.3; 18.1; 22.6)
(a) Vol. **84**, No. 2, p. K105—K108 (1984).

A. Ribeiro Filho and D. R. Tilley
*Thick Films near Phase Transitions. Green Functions, Fluctuation Spectra, and Light Scattering Theory
(1.2 and 20.1; 1.5)
(b) Vol. **125**, No. 1, p. 45—54 (1984).

B. Lorenz
Charge Ordering in the Anti-Th_3P_4 Structure
(4; 1.2; 22)
(b) Vol. **125**, No. 1, p. 375—379 (1984).

V. G. Chudinov and V. I. Protasov
*Kinetics of Vacancy Mechanism of the Solid–Liquid Transition in F.C.C. Metals
(1.2; 10.1; 21.1)
(a) Vol. **85**, No. 1, p. 105—112 (1984).

B. Wyncke, F. Brehat, M. El Sherif, and G. V. Kozlov
Far-Infrared and Submillimeter Dielectric Response of Sodium Nitrite
(6 and 20.1; 1.2; 14.4.1; 22.8.1)
(b) Vol. **125**, No. 2, p. 493—498 (1984).

S. K. Plachkova and L. A. Tersiev
*Phase Transitions and Intervalley Jahn-Teller Effect in the System $(GeTe)_{1-x}(AgSbTe_2)_x$
(1.2 and 14.4.1; 13.1; 22.8)
(b) Vol. **125**, No. 2, p. 659—664 (1984).

H. Werheit and R. Franz
*States of Thermal Equilibrium in β-Rhombohedral Boron and Their Alteration by Optical Excitation
(1.2; 13.4; 20.1; 22.1)
(b) Vol. **125**, No. 2, p. 779—784 (1984).

T. Soma, Y. Kitani, and H.-Matsuo Kagaya
*Mode Grüneisen Parameter and Pressure-Induced Phase Transition of Si–Ge Solid Solution
(1.2 and 6; 8; 22.1.1; 22.1.2)
(b) Vol. **125**, No. 2, p. K99—K101 (1984).

L. V. Vassilev and St. K. Dimitrova
Observation of Anderson-Like Transition in the System $Pb_{1-x}Ge_xTe$
(14.3 and 15; 1.2; 22.3.1)
(b) Vol. **125**, No. 2, p. K177—K180 (1984).

P. L. Loyzance and M. Couzi
*Group-to-Subgroup Relations in Space Groups. Application to the Structural Phase Transitions in the Perovskite-Type Layer Compounds $RbAlF_4$ and $RbVF_4$
(1.2; 22.8)
(a) Vol. **85**, No. 2, p. 359—374 (1984).

B. Topič, R. Blinc, and L. A. Shuvalov
*^{87}Rb NMR in Monoclinic RbD_2PO_4
(1.2; 14.4.1; 19; 22.8.1)
(a) Vol. **85**, No. 2, p. 409—415 (1984).

K. N. Jog, R. K. Singh, and S. P. Sanyal
*The Relative Stability and Phase Transition Studies of MgO and CaO Crystals
(1.2 and 12.2; 22.6)
(a) Vol. **85**, No. 2, p. 417—421 (1984).

H.-G. Unruh
*Dielectric Dispersion and Dynamics of Nitrite Ions in $K_2Ba(NO_2)_4$
(1.2 and 14.4; 6; 22.8.1)
(b) Vol. **126**, No. 1, p. 115—120 (1984).

W. Böhm, H. G. Kahle, and W. Wüchner
*Spectroscopic Study of the Crystal Field Splittings in $TbPO_4$
(1.2 and 13.4; 18.4; 20.1; 22.8.1)
(b) Vol. **126**, No. 1, p. 381—392 (1984).

E. R. Korutcheva and D. I. Uzunov
*On the Long-Range Random Critical Behaviour
(1.2 and 10)
(b) Vol. **126**, No. 1, p. K19—K22 (1984).

S. Trimper
Anisotropic Interactions in 2 + ε Dimensions
(18.2; 1.2)
(b) Vol. **126**, No. 1, p. K23—K27 (1984).

M. Hidaka, H. Fujii, B. J. Garrard, and B. M. Wanklyn
*Structural Phase Transitions of $RbVF_4$
(1.2; 22.5.4)
(a) Vol. **86**, No. 1, p. 75—83 (1984).

G. Sorge, J. Janich, L. A. Shuvalov, and N. M. Shchagina
*The Influence of an Electric DC Field on the New Intermediate Phase between the α- and β-Phases of $NaH_3(SeO_3)_2$ Crystals
(1.2 and 14.4.1; 12.2; 22.8)
(a) Vol. **86**, No. 1, p. 111—115 (1984).

A. A. Rempel, S. Z. Nazarova, and A. I. Gusev
*Effect of Atomic Ordering on the Heat Capacity of Non-Stoichiometric Niobium Carbide
(1.2 and 8; 6; 22)
(a) Vol. **86**, No. 1, p. K11—K14 (1984).

V. E. Arkhipov, V. I. Voronin, B. N. Goshchitskii, A. E. Karkin, V. N. Kozhanov, A. V. Mirmelshtein, and E. P. Romanov
The Structure, Superconducting Properties, and Electrical Rsistivity of V_2Zr and V_2Hf Irradiated with Fast Neutrons
(14.1 and 14.2; 1.2; 21; 21.1)
(a) Vol. **86**, No. 1, p. K59—K62 (1984).

V. I. Torgashev, V. Dvořák, and F. Smutný
*On Phase Transitions in $LiNH_4SO_4$
(1.2 and 14.4.1; 22.8.1)
(b) Vol. **126**, No. 2, p. 459—466 (1984).

D. M. Vanderwalker
The Nucleation of Ni Silicides on Dislocations and Twins in Silicon
(1.6 and 10.2; 1.2; 21.1; 22.1.2)
(a) Vol. **86**, No. 2, p. 507—516 (1984).

A. U. Sheleg and V. V. Zaretskii
*X-Ray Study of the Commensurate–Incommensurate Phase Transitions in α-ZnP_2
(1.2; 22.3)
(a) Vol. **86**, No. 2, p. 517—523 (1984).

S. N. VAIDYA
*Theory of Melting (I): Energy Balance Principle
(1.2 and 6; 8; 23)
(a) Vol. **86**, No. 2, p. 565—571 (1984).

TH. FRAUENHEIM, C. HAMANN, and M. MÜLLER
Electric Field-Induced Disorder–Order Transition in Organic Polycrystalline Films of Quasi-One-dimensional Lead-Phthalocyanine
(14.3.1; 1.2; 22.9)
(a) Vol. **86**, No. 2, p. 735—747 (1984).

H. GRUBER, E. KRAUTZ, H. P. FRITZER, K. GATTERER, and A. POPITSCH
Electronic Property Changes of η-Mo_4O_{11} by Doping with Tungsten, Rhenium, and Vanadium
(14.1; 1.2; 13.1; 18; 20.1; 22.6.1)
(a) Vol. **86**, No. 2, p. 749—757 (1984).

1.3 Alloys. Metallurgy

Original Papers and Short Notes

S. SIMOZAR and J. A. ALONSO
*On the Factors Controlling Glass Forming Ability of Metallic Alloys Formed by Fast Liquid Quenching
(1.3 and 2; 8; 21; 21.1)
(a) Vol. **81**, No. 1, p. 55–61 (1984).

G. GEVERS, A. S. BARRIÈRE, J. GRANNEC, L. LOZANO, and B. BLANCHARD
*Analysis of FeF_3 and NiF_2 Thin Films by Using RBS, (α X) Reactions, and SIMS Measurements
(1.3; 14.3.1; 22.5)
(a) Vol. **81**, No. 1, p. 105—122 (1984).

N. TANAKA and K. OHSHIMA
*High Resolution Electron Microscopic Observations of Disordered Cu–Pd Alloys
(1.3; 21.1)
(a) Vol. **81**, No. 1, p. 129—138 (1984).

S. BANERJEE and K. URBAN
Kinetics of Order–Disorder Transformation in Alloys under Electron Irradiation
(1.2; 1.3; 11; 21.1)
(a) Vol. **81**. No. 1, p. 145—162 (1984).

K. MUKHERJEE, TSING-WONG HSU, and M. KATO
*Fiber Morphology Resulting from Solid State Coalescence of Spherical Precipitates
(1.3 and 1.4)
(a) Vol. **81**, No. 2, p. 439—444 (1984).

K. SUMIYAMA and Y. NAKAMURA
*High Concentration Fe–Cu and Fe–Ag Alloys Produced by Vapor Quenching
(1.3; 5; 18.2; 21.1; 21.1.1; 21.6)
(a) Vol. **81**, No. 2, p. K109—K112 (1984).
Erratum
(a) Vol. **83**, No. 2, p. K219 (1984).

K. FELDMANN und L. FUENTES
*On the Possibility of Determining the Odd Part of the Orientation Distribution Function of Texturized Samples for Quartz-Like Structures Using Anomalous Scattering
(1.3 and 1.4; 22.6; 22.8.1)
(b) Vol. **122**, No. 1, p. K5—K9 (1984).

S. CHAKRABORTY, A. MANNA, and A. K. GHOSH
*Charge Transfer and Energy of Formation in Binary Alloys: CsCl Intermetallic Compounds
(1.3 and 4; 21.1; 21.1.1; 21.6)
(b) Vol. **122**, No. 2, p. 469—480 (1984).

J. Molenda
Electronic Processes in Electrode Materials of A_xMX_2-Type
(13.1; 1.3; 21.1; 21.2)
(b) Vol. **122**, No. 2, p. 591—598 (1984).

G. N. Gaidukov, A. A. Podrezov, and J. P. Hirth
*A Volterra-Type Dislocation Model of a Low-Angle [001] Twist Boundary in an F.C.C. Crystal
(1.3; 10.1; 21)
(a)Vol. **82**, No. 2, p. 355—363 (1984).

L. C. Otero-Diaz, J. Soria, and M. A. Alario-Franco
*Extended Defects in Vanadium Doped Rutile. An EPR and TEM Study
(1.3 and 10.2; 19; 22.6.1)
(a) Vol. **82**, No. 2, p. 379—388 (1984).

L. Uray
*The Effect of Void Size on the Resistivity of Sintered Tungsten
(1.3 and 14.1; 15; 21)
(a) Vol. **82**, No. 2, p. 545—552 (1984).

A. Baldan
*The Microstructural Properties of (β + $(Ni, Al)_7Hf_2$)Eutectics
(1.3; 10.1; 21; 21.1)
(a) Vol. **83**, No. 1, p. 47—57 (1984).

H. Q. Ye, Z. Zhang, and K. H. Kuo
*A High Resolution Electron Microscopic Study of Ordered Structures in $Ni_3(Ti, V)$ Alloys
(1.3 and 1.4; 21.1)
(a) Vol. **83**, No. 1, p. 77—84 (1984).

K. S. Vahvaselkä
*An X-Ray Investigation of Structural Properties and Electrical Resistivities of Liquid Al, Liquid Cu, and Liquid Al–Cu Alloys
(1.3 and 2; 14.1; 21; 21.1)
(a) Vol. **83**, No. 1, p. 103—112 (1984).

H. Schaeben
*Mathematical Modeling of the Orientation Distribution Function by the Vector Method of Quantitative Texture Analysis
(1.3 and 1.4)
(b) Vol. **123**, No. 2, p. 425—434 (1984).

J. A. Alonso, D. J. González, and M. P. Iñiguez
Role of the Excess Volume of Formation on Alloying
(4; 1.3; 21.2; 21.3)
(b) Vol. **123**, No. 2, p. 485—489 (1984).

A. Baldan
An Investigation of the Mechanism of Thermally Activated Deformation in Certain Ni–Al–Ta and Ni–Al–Hf Superalloys
(10.1; 1.3; 21; 21.1)
(a) Vol. **83**, No. 2, p. 507—511 (1984).

H. Kido, T. Hoshikawa, M. Shimada, and M. Koizumi
Crystal Structure and Magnetic and Electrical Properties of Pseudo-Ternary $GdCu_{1-x}Zn_xSi$ Compounds
(14.1 and 18.1; 1.3; 18.2; 22.8)
(a) Vol. **83**, No. 2, p. 561—565 (1984).

W. Hoyer, A. Müller, W. Matz, and M. Wobst
Neutron Scattering Investigation on Molten Ga–Te Alloys
(2; 1.3; 22.7)
(a) Vol. **84**, No. 1, p. 11—18 (1984).

S. A. Fysh and P. E. Clark
*On Mössbauer Analysis of Mineral Mixtures Having Environmentally Broadened Spectral Lines. Application to Bauxite
(1.3; 5; 22.6)
(a) Vol. **84**, No, 1, p. 31—38 (1984).

P. Mrozek, M. Menyhárd, J. Wernisch, and A. Jabłoński
*Surface Composition of the Ordered Al50–Co50 Alloy
(1.3 and 1.6; 21; 21.1)
(a) Vol. **84**, No. 1, p. 39—48 (1984).

C. A. Hewett, S. S. Lau, R. Fastow, and J. W. Mayer
*Pulsed Ion Beam Annealing of Ti–Au Alloys
(1.3 and 11; 1.5; 2; 21.1; 21.6)
(a) Vol. **84**, No. 1, p. 49—54 (1984).

L. Uray and M. Menyhárd
*The Segregation of Iron in Tungsten
(1.3; 1.4; 9; 14.1; 15; 21; 21.1.1)
(a) Vol. **84**, No. 1, p. 65—72 (1984).

R. Khanna
*Separation of Thermal Component in Diffuse Scattering from Disordered Binary Alloys
(1.3; 6; 21)
(a) Vol. **84**, No. 1, p. 95—101 (1984).

F. Reynaud
*Application of the Grand Canonical Ensemble in Classical Statistical Physics to the Study of Order–Disorder Transitions in Binary Substitutional Alloys
(1.2 and 1.3; 21)
(a) Vol. **84**, No. 2, p. 393—400 (1984).

A. I. Gusev and A. A. Rempel
*A Study of the Atomic Ordering in the Niobium Carbide Using the Magnetic Susceptibility Method
(1.3; 18.1; 22)
(a) Vol. **84**, No. 2, p. 527—534 (1984).

A. Mueller, W. Hoyer, E. Thomas, and M. Wobst
*X-Ray Scattering Investigation on Molten Ga–Te
(1.3 and 2; 22.7)
(a) Vol. **84**, No. 1, p. K97—K100 (1984).

Š. A. Alterovitz, G. H. Bu-Abbud, J. A. Woollam, and Ď. C. Liu
*Structural and Chemical Analysis of a Silicon Nitride Film on GaAs by Null Ellipsometry
(1.3 and 1.5; 22)
(a) Vol. **85**, No. 1, p. 69—76 (1984).

G. Schubert and J. Richter
*The Electrical Resistivity of Clustering Alloys
(1.3 and 14.1; 21)
(b) Vol. **125**, No. 2, p. 839—847 (1984).

J. M. López and J. A. Alonso
*The Atomic Size-Mismatch Contribution to the Enthalpy of Formation of Concentrated Substitutional Metallic Solid Solutions
(1.3 and 8; 21; 21.1)
(a) Vol. **85**, No. 2, p. 423—428 (1984).

B. Rauschenbach, G. Blasek, and R. Dietsch
*Modification of Concentration Profiles in Iron and Aluminium by High-Fluence Implantation of Nitrogen and Boron Ions
(1.3 and 11; 21; 21.1.1)
(a) Vol. **85**, No. 2, p. 473—480 (1984).

R. A. Dunlap, K. Dini, and P. Hargraves
*Hydrogen Absorption and Crystallisation in Amorphous $Co_{90}Zr_{10}$
(1.3; 2; 21.1)
(a) Vol. **85**, No. 2, p. K101—K103 (1984).

K. Helming and S. Matthies
*On the Interpretation of Orientation Distributions and Qualitative Ghost Corrections of Hexagonal-Orthorhombic Textures
(1.3 and 1.4; 21)
(b) Vol. **126**, No. 1, p. 43—52 (1984).

P. Malétras et P. Duval
*Evolution comparée d'alliages Cu–Al trempés revenus avant et après amincissement
(Comparison of Temper-Hardened Cu–Al Alloys before and after Thinning-Down)
(1.3; 21; 21.1)
(a) Vol. **86**, No. 1, p. 85—92 (1984).

M. G. Blanchin and L. A. Bursill
Extended versus Small Defect Equilibria in Non-Stoichiometric Rutile (I): Dissolution and Re-Precipitation of Extended Defects Observed by In-Situ Transmission Electron Microscopy
(10.2; 1.3; 22.6)
(a) Vol. **86**, No. 1, p. 101—109 (1984).

E. Zschech, W. Blau, H. Vega, K. Kleinstück, S. Mager, M. A. Kozlov, and M. A. Sheromov
*EXAFS and X-Ray Diffractional Investigation of the Heusler-Type Alloys Co_2MnSi and $Fe_{2,4}Mn_{0,6}Al$. Determination of the Ordering Probabilities
(1.3; 18; 21.1; 21.1.1)
(a) Vol. **86**, No. 1, p. 117—124 (1984).

Shu Zhen and G. J. Davies
*Calculation of Potential Energy Parameters for Some Intermetallic Compounds
(1.3 and 4; 22.7)
(a) Vol. **86**, No. 1, p. 155—161 (1984).

A. O. O. Mekhrabov and M. Doyama
*Electronic Theory of Atomic Short-Range Order for Ternary Alloys Using the Pseudopotential Approximation and Its Comparison with Experiments
(1.3 and 13; 21.1; 21.1.1)
(b) Vol. **126**, No. 2, p. 453—458 (1984).

M. G. Blanchin and L. A. Bursill
Extended Versus Small Defect Equilibria in Non-Stoichiometric Rutile (II): Structural Mechanisms for Dissolution and Reprecipitation of Extended Defects
(10.2; 1.3; 22.6)
(a) Vol. **86**, No. 2, p. 491—498 (1984).

H. J. Engelmann, K. Berndt, and F. Storbeck
*Quantitative Chemical Analysis of 3d Transition Metal Alloys by SXAPS
(1.3 and 13.3; 20; 21.1)
(a) Vol. **86**, No. 2, p. 499—506 (1984).

P. Siebert, M. Werner, and H. Löffler
*Heterogeneous Nucleation of Precipitates in an Al–12 at% Zn Alloy at Dislocation Loops
(1.3; 10.1; 21; 21.1)
(a) Vol. **86**, No. 2, p. 543—551 (1984).

F. C. Lovey, G. van Tendeloo, J. van Landuyt, L. Delaey, and S. Amelinckx
*On the Nature of Various Stacking Defects in 18R Martensite in Cu–Al Alloys. A Study by High Resolution Electron Microscopy
(1.3; 10.1; 21; 21.1)
(a) Vol. **86**, No. 2, p. 553—564 (1984).

Gyanesh Chandra and O. P. Katyal
Applicability of the Mayadas-Shatzkes Model on Electrical Resistivity of Polycrystalline Tin–Lead Alloy Films
(14.1; 1.3; 21)
(a) Vol. **86**, No. 2, p. 759—764 (1984).

1.4. Microstructure (Magnetic Domains See 18; Ferroelectric Domains See 14.4.1)

Review Article

C. Boulesteix
*A Survey of Domains and Domain Walls Generated by Crystallographic Phase Transitions Causing a Change of the Lattice
(1.2 and 1.4; 14.4.1; 22; 22.6; 22.8)
(a) Vol. **86**, No. 1, p. 11—42 (1984).

Original Papers and Short Notes

G. Nihoul and C. Cesari
*High Resolution Images for Large Cell Crystals Exhibiting Some Limits of the Projected Potential Approximation
(1.1 and 1.4; 22.6)
(a) Vol. **81**, No. 1, p. 87—98 (1984).

M. Pasemann, W. Bergholz, and W. Schröter
*Mössbauer Spectroscopical and Electron Microscopical Investigations of the Behaviour of Cobalt in Silicon
(1.4 and 10.2; 5; 22.1.2)
(a) Vol. **81**, No. 1, p. 273—280 (1984).

S. V. Tsivinskii
The Theory of Strain Hardening in Single and Polycrystals
(10.1; 1.4; 21; 21.1; 21.1.1; 22.1.1; 22.1.2)
(a) Vol. **81**, No. 1, p. 299—306 (1984).

S. Simon and Al. Nicula
*EPR Study of Gd^{3+} and Cu^{2+} Ions from Lithium-Borate Vitroceramics
(1.4 and 19; 22.8.1)
(a) Vol. **81**, No. 1, p. K1—K5 (1984).

A. A. Sitnikova, L. M. Sorokin, I. E. Talanin, E. G. Sheikhet, and E. S. Falkevich
*Electron-Microscopic Study of Microdefects in Silicon Single Crystals Grown at High Speed
(1.4; 10.2; 22.1.2)
(a) Vol. **81**, No. 2, p. 433—438 (1984).

K. Mukherjee, Tsing-Wong Hsu, and M. Kato
*Fiber Morphology Resulting from Solid State Coalescence of Spherical Precipitates
(1.3 and 1.4)
(a) Vol. **81**, No. 2, p. 439—444 (1984).

A. Baldan
*Microstructural Properties of the β-NiAl Phase in the Ternary Ni-Based Ni–Al–Ta Superalloys
(1.4; 21; 21.1)
(a) Vol. **81**, No. 2, p. 451—462 (1984)

T. Gessner, M. Pasemann, and B. Schmidt
Erratum to: Calculation of Neutron-Induced Defect Clusters in Silicon and Comparison with TEM Investigations
(11; 1.4; 22.1.2)
(a) Vol. **81**, No. 2, p. 751 (1984).

K. S. Raju and E. Pippel
*High Voltage Electron Microscopic (HVEM) Observations of Dislocations in Monocrystalline Gypsum
(1.4; 10.2; 22.8.1)
(a) Vol. **81**, No. 2, p. K93—K95 (1984).

M. Kittler, K.-W. Schröder, E. Bugiel, and C. Becker
*Quantitative EBIC Investigations on Bulk Stacking Faults in Silicon
(1.4 and 10.2; 14.3; 22.1.2)
(a) Vol. **81**, No. 2, p. K131—K135 (1984).

K. Feldmann and L. Fuentes
*On the Possibility of Determining the Odd Part of the Orientation Distribution Function of Texturized Samples for Quartz-Like Structures Using Anomalous Scattering
(1.3 and 1.4; 22.6; 22.8.1)
(b) Vol. **122**, No. 1, p. K5—K9 (1984).

G. van Tendeloo, S. Amelinckx, A. F. Marshall, and R. G. Walmsley
*A Complex High-Temperature Phase in Cu_9Zr_2
(1.4; 21.1)
(a) Vol. **82**, No. 1, p. 55—61 (1984).

G. K. Schmidt
Method for Obtaining the Spectrum of Plastic Strains of Mesoscopic Volume Elements in Fatigued Materials Investigated by X-Rays (I): Plastic and Elastic Strains in Fatigued Material
(10.1 and 12.1; 1.4; 21)
(a) Vol. **82**, No. 1, p. 75—84 (1984).

O. Bostanjoglo and E. Endruschat
*Time-Resolved TEM of Laser-Annealing
(1.2 and 1.4; 2; 11; 22.1.1)
(a) Vol. **82**, No. 1, p. K1—K3 (1984).

O. Simmich
*Influence of Non-Linear Terms in the Flux Equation on the Particle Growth
(1.4 and 9; 3; 21)
(a) Vol. **82**, No. 1, p. K21—K25 (1984).

V. I. Osinskii, F. M. Katsapov, and E. A. Tyavlovskaya
*Structural Perfection of Selective GaAs Regions in Si Substrate Windows
(1.4 and 1.6; 22.2.1)
(a) Vol. **82**, No. 2, p. 399—403 (1984).

G. K. Schmidt
Method for Obtaining the Spectrum of Plastic Strains of Mesoscopic Volume Elements in Fatigued Materials Investigated by X-Rays (II): Distributions of Elastic Strains and X-Ray Reflex Profiles
(10.1 and 12.1; 1.4; 21)
(a) Vol. **82**, No. 2, p. 413—420 (1984).

D. J. H. Cockayne, P. Pirouz, Z. Liu, G. R. Anstis, and P. Karnthaler
*A Quantitative Analysis of Image Contrast from Extrinsic Stacking Faults
(1.4 and 10.2; 22.1.2)
(a) Vol. **82**, No. 2, p. 425—439 (1984).

A. N. Pilyankevich and V. F. Britun
*Motion of Partial Dislocation in Silicon Carbide
(1.4 and 10.2; 22.3)
(a) Vol. **82**, No. 2, p. 449—457 (1984).

M. J. Marcinkowski
*The Mechanism of Cell Wall Formation
(1.4 and 10)
(a) Vol. **83**, No. 1, p. 59—68 (1984).
Erratum
(a) Vol. **86**, No. 1, p. 460 (1984).

H. Q. Ye, Z. Zhang, and K. H. Kuo
*A High Resolution Electron Microscopic Study of Ordered Structures in $Ni_3(Ti, V)$ Alloys
(1.3 and 1.4; 21.1)
(a) Vol. **83**, No. 1, p. 77—84 (1984).

M. Jergel and P. Mrafko
The Temperature Dependence of the Prepeak in the Diffraction Pattern of the Amorphous $Ti_{61}Cu_{16}Ni_{23}$ and $Ti_{62.5}Cu_{12}Ni_{23}Si_{2.5}$ Alloys
(2; 1.4; 21.1)
(a) Vol. **83**, No. 1, p. 113—121 (1984).

V. L. Indenbom, V. M. Kaganer, W. Möhling, and E. V. Suvorov
*Plane-Wave X-Ray Images of Dislocations Parallel to the Diffraction Vector
(1.4 and 10.2; 22.1.2)
(a) Vol. **83**, No. 1, p. 195—205 (1984).

A. I. Belyaeva, M. M. Kotlyarskii, O. V. Miloslavskaya, Yu. N. Stelmakhov, and V. P. Yurev
Investigations of the Crystal and Magnetic Structure Heterogeneities of $RbMnCl_3$
(18.4; 1.4; 22.5)
(a) Vol. **83**, No. 1, p. 245—253 (1984).

H. Schaeben
*Mathematical Modeling of the Orientation Distribution Function by the Vector Method of Quantitative Texture Analysis
(1.3 and 1.4)
(b) Vol. **123**, No. 2, p. 425—434 (1984).

M. Hervieu, N. Nguyen, V. Caignaert, and B. Raveau
*Calcium Mangano-Ferrites $(CaMO_3)_mCaMO_2$ □, M = Fe, Mn, with an Oxygen-Defect Perovskite Structure Related to Brownmillerite. Electron Microscopy Investigation
(1.1 and 1.4; 22.8.2)
(a) Vol. **83**, No. 2, p. 473—483 (1984).

Th. Karakostas and P. Delavignette
*Annealing Structures of Grain Boundaries in Zirconium
(1.4; 21)
(a) Vol. **83**, No. 2, p. 485—497 (1984)

H. Bartsch, D. Hoehl, and G. Kästner
*Radiation-Induced Rod-Like Defects in Silicon and Germanium
(1.4 and 11; 22.1.1; 22.1.2)
(a) Vol. **83**, No. 2, p. 543—551 (1984).

C. Hidalgo, N. de Diego, and M. A. Ochando
*A Positron Study of Iron Alloys
(1.4 and 10.1; 21.1.1)
(a) Vol. **83**, No. 2, p. K93–K96 (1984).

J. D. Stephenson
*Double Laue Pattern Topography of Top-Seeded $BaTiO_3$ Single Crystals Using White Synchrotron X-Radiation
(1.4 and 1.6; 14.4.1; 22.8.1)
(a) Vol. **84**, No. 1, p. 19—30 (1984).

L. Uray and M. Menyhárd
The Segregation of Iron in Tungsten
(1.3; 1.4; 9; 14.1; 15; 21; 21.1.1)
(a) Vol. **84**, No. 1, p. 65—72 (1984).

S. K. Maksimov, M. Ziegler, I. I. Khodos, I. I. Snighiryova, and M. Sh. Shikhsaidov
High Resolution Electron Microscopic Investigations of Dislocations in Deformed GaAs Single Crystals Doped with Te
(10.2; 1.4; 22.2.1)
(a) Vol. **84**, No. 1, p. 79—86 (1984).

L. Pasemann
*Theoretical Study of the Temperature Dependence of EBIC Contrast from Individual, Surface-Parallel Dislocations in a Schottky Diode
(1.4 and 10.2; 14.3.3; 22.1.2)
(a) Vol. **84**, No. 1, p. 133—142 (1984).

G. van Tendeloo and S. Amelinckx
*A New Long-Period Derivative of the Au_5Mn_2 Structure
(1.1 and 1.4; 21.1; 21.6)
(a) Vol. **84**, No. 1, p. 185—192 (1984).

D. P. Joshi and R. S. Srivastava
Effective Diffusion Length of Minority Carriers in Polycrystalline Silicon
(14.3.3 and 16; 1.4; 22.1.2)
(a) Vol. **84**, No. 1, p. 311—317 (1984).

N. Pashov, G. Kästner, M. Kalitzova, S. Simov, M. Mikhailov, and K. Kolev
*Electron Microscope and Electron Diffraction Studies of Amorphization Phenomena in Ge Implanted with Te Ions
(1.4 and 11; 2; 22.1.1)
(a) Vol. **84**, No. 1, p. K23—K27 (1984).

V. V. Aristov, I. E. Bondarenko, N. N. Dryomova, V. V. Kazmiruk, and E. B. Yakimov
*The EBIC Contrast of Dislocation Slip Planes in Silicon
(1.4 and 10.2; 14.3.4; 22.1.2)
(a) Vol. **84**, No. 1, p. K43—K45 (1984).

S. Kh. Milshtein, D. C. Joy, S. D. Ferris, and L. C. Kimerling
*Defect Characterization Using SEM–CCM. Relative Contrast Measurements
(1.4 and 10.2; 14.3.3; 22.1.2)
(a) Vol. **84**, No. 2, p. 363—369 (1984).

F. C. Lovey, G. van Tendeloo, and S. Amelinckx
*The Nature of Some Planar Defects in 2H Martensite of Cu–Al Alloys as Determined by HREM
(1.4; 10.1; 21; 21.1)
(a) Vol. **85**, No. 1, p. 29—37 (1984).

B. A. Men and M. L. Levitan
*Analysis of the Ground State of a Solid Solution with Interactions up to *k*-th Coordination Spheres of the Crystal Lattice. Selection of the Subdivision of the Lattice into Sublattices
(1.4)
(a) Vol. **85**, No. 1, p. 51—60 (1984).

D. M. Vanderwalker
*A Dark-Field TEM Method for Crystal Structure Determination of Cu-Rich Phases on Twins in Silicon
(1.4; 22.1.2)
(a) Vol. **85**, No. 1, p. 61—67 (1984).

E. Mizera, M. Sundberg, and P. Werner
*Defect Structure of ZnSe Crystals Investigated by Electron Microscopy
(1.4; 10.2; 22.4.2)
(a) Vol. **85**, No. 1, p. 83—88 (1984).

R. Pérez
*On the Identification of the Nature of Stacking Faults in H.C.P. Materials
(1.4; 10.1)
(a) Vol. **85**, No. 1, p. 113—119 (1984).

P. Gaworzewski, E. Hild, F.-G. Kirscht, and L. Vecsernyés
*Infrared Spectroscopical and TEM Investigations of Oxygen Precipitation in Silicon Crystals with Medium and High Oxygen Concentrations
(1.4 and 10.2; 20.1; 22.1.2)
(a) Vol. **85**, No. 1, p. 133—147 (1984).

I. Baker and E. M. Schulson
On Intrinsic Stacking Faults in Polycrystalline Ni_3Al
(10.1; 1.4; 21; 21.1)
(a) Vol. **85**, No. 2, p. 481—490 (1984).

K. Helming and S. Matthies
*On the Interpretation of Orientation Distributions and Qualitative Ghost Corrections for Hexagonal-Orthorhombic Textures
(1.3 and 1.4; 21)
(b) Vol. **126**, No. 1, p. 43—52 (1984).

N. I. Borgardt, S. K. Maksimov, and D. I. Piskunov
*Phase Information in Transmission Electron-Microscope Investigation of Defects. Dynamic Images with a Weak Amplitude Contrast
(1.4 and 10)
(a) Vol. **86**, No. 1, p. 55—65 (1984).

M. Takeda and H. Hashimoto
*A High Resolution Electron Microscopy Study of Domain Structure in CuAuI
(1.4; 21.1; 21.6)
(a) Vol. **86**, No. 1, p. 67—73 (1984).

H. Bender
Investigation of the Oxygen-Related Lattice Defects in Czochralski Silicon by Means of Electron Microscopy Techniques
(10.2; 1.4; 22.1.2)
(a) Vol. **86**, No. 1, p. 245—261 (1984).

K. Benhaddane and P. Beauchamp
*A Computation of (110) Intrinsic Stacking Fault Energies in B2 Ordered Alloys
(1.4; 10.1; 21; 21.1.1)
(a) Vol. **86**, No. 2, p. 613—621 (1984).

D. Wruck, M. Seifert, and W. Ulrici
*Microinhomogeneities of Donor and Acceptor Distributions in n-Type LEC-GaAs from Free-Carrier Infrared Absorption
(1.4 and 13.4; 20.1; 22.2.1)
(a) Vol. **86**, No. 2, p. 691—702 (1984).

1.5 Films

Original Papers and Short Notes

V. Yu. Balandin, A. V. Dvurechenskii, and L. N. Aleksandrov
Dynamics of Self-Sustaining Crystallization of Amorphous Silicon Layers
(2 and 3; 1.5; 22.1.2)
(a) Vol. **81**, No. 1, p. 63—68 (1984).

N. Kaiser
*Nucleation of Antimony on Amorphous Carbon
(1.5 and 3; 2; 22)
(a) Vol. **81**, No. 1, p. 99—103 (1984).

C.-D. Lien, M.-A. Nicolet, and S. S. Lau
*Low Temperature Formation of $NiSi_2$ from Evaporated Silicon
(1.5 and 3; 21)
(a) Vol. **81**, No. 1, p. 123—128 (1984).

D. Theis
*Selected Analytical Tools Yield a Better Insight into Electroluminescent Thin Films
(1.5 and 14.3.4; 20.3; 22.4.1)
(a) Vol. **81**, No. 2, p. 647—655 (1984).

I. G. Kaverina, V. V. Korobtsov, V. G. Zavodinskii, and A. V. Zotov
*Solid Phase Epitaxial Growth Anisotropy of Vacuum-Deposited Amorphous Silicon
(1.5 and 3; 2; 22.1.2)
(a) Vol. **82**, No. 2, p. 345—353 (1984).

E. M. Zub, E. V. Kuchis, and S. V. Sukhvalo
Electrical and Galvanomagnetic Properties of Films in the $Cu_xFe_{1-x}Cr_2S_4$ System
(14.3.1; 1.5; 22.8.2)
(a) Vol. **82**, No. 2, p. 569—574 (1984).

R. Klabes, J. Matthäi, M. Voelskow, E. Wieser, J.-W. Erben, W. Scharff, and C. Weissmantel
*Epitaxial Regrowth of Amorphous or Polycrystalline Silicon Layers on Silicon Single Crystals and Bridging Epitaxy by Flash Lamp Irradiation
(1.5 and 3; 2; 11; 22.1.2)
(a) Vol. **82**, No. 2, p. K121—K123 (1984).

M. Aoki, T. Nishikawa, and S. Okabe
*Pattern Variation of Evaporated Thin Films and Their Non-Pre-Excited TSEE
(1.5 and 17; 20.3; 22.5.2; 22.5.3)
(a) Vol. **83**, No. 1, p. 85—91 (1984).

K. Ots and M. Seilenthal
*On the Structual Model of CVD a-Si:O:H Films
(1.5 and 2; 20.1; 22.1.2; 22.6)
(a) Vol. **83**, No. 2, p. K109—K111 (1984).

C. A. Hewett, S. S. Lau, R. Fastow, and J. W. Mayer
Pulsed Ion Beam Annealing of Ti–Au Alloys
(1.3 and 11; 1.5; 2; 21.1; 21.6)
(a) Vol. **84**, No. 1, p. 49—54 (1984).

C. H. Ling, C. Y. Kwok, and K. Prasad
*Observation of an Anomalously High Etch Rate in Plasma-Enhanced CVD Silicon Nitride Films
(1.5; 22)
(a) Vol. **84**, No. 1, p. K1—K4 (1984).

W. Brodkorb, J. Salm, and E. Steinbeiss
*Discussion of Possible Surface Processes during the Reactive SiO_x Film Deposition in a Planar Magnetron Sputtering Equipment
(1.5 and 2; 22.6)
(a) Vol. **84**, No. 2, p. 379—386 (1984).

J. Baszyński, L. Smardz, and F. Stobiecki
*Magnetic Properties of Compositionally Modulated Ni/Cu Layers
(1.5 and 18.2.1; 19; 21.1)
(a) Vol. **84**, No. 2, p. K129—K131 (1984).

A. Ribeiro Filho and D. R. Tilley
Thick Films near Phase Transitions. Green Functions, Fluctuation Spectra, and Light Scattering Theory
(1.2 and 20.1; 1.5)
(b) Vol. **125**, No. 1, p. 45—54 (1984).

H. Berger, Yu. A. Kulyupin, S. A. Nepijko, and N. N. Sedov
*Quantitative Interference Electron Microscopy for Spherical Objects
(1.5 and 13.3; 21.7)
(a) Vol. **85**, No. 1, p. 45—50 (1984).

S. A. Alterovitz, G. H. Bu-Abbud, J. A. Woollam, and D. C. Liu
*Structural and Chemical Analysis of a Silicon Nitride Film on GaAs by Null Ellipsometry
(1.3 and 1.5; 22)
(a) Vol. **85**, No. 1, p. 69—76 (1984).

G. Berger and V. Atzrodt
*In Vitro Characterization of Bioactivity of Glassy or Glass-Crystalline Implant Materials Using Auger Electron Spectroscopy
(1.5; 2; 17; 22.6.1)
(a) Vol. **85**, No. 1, p. K9—K13 (1984).

T. Jung and L. Roth
*DC Plasma Nitridation of Thin Aluminium Films
(1.5; 11; 22)
(a) Vol. **85**, No. 1, p. K15—K18 (1984).

K. UHLMANN and M. SCHMIDT
*Influence of Work Function Change Due to Oxygen Chemisorption on the Secondary-Ion Emission Probability
(1.5 and 13.3; 17; 21.1.1)
(a) Vol. **85**, No. 1, p. K19—K22 (1984).

P. MICHON, H. ATMANI, and C. VAUTIER
*Crystallization Study of Evaporated a-Se Films by Means of DTA Technique
(1.5 and 3; 2; 22.1.3)
(a) Vol. **85**, No. 2, p. 399—407 (1984).

H. BOOYENS and J. H. BASSON
*On the Optimal Choice of Substrate for $Hg_{0.8}Cd_{0.2}Te$ Epitaxial Layers
(1.5 and 10.2; 3; 22.4.4)
(a) Vol. **85**, No. 2, p. 449—454 (1984).

YU. P. KHAPACHEV, A. A. DYSHEKOV, and D. S. KISELEV
*The Theory of X-Ray Diffraction Analysis of Elastic-Strain States in Epitaxial Films
(1.5; 12)
(b) Vol. **126**, No. 1, p. 37—42 (1984).

N. V. RUMAK, V. V. KHATKO, and V. N. PLOTNIKOV
*Structure and Properties of Silicon Dioxide Thermal Films (I): SiO_2 Films of up to 50 nm Thickness
(1.5 and 2; 3; 4; 22.6)
(a) Vol. **86**, No. 1, p. 93—100 (1984).

P. KOŠTIAL
*The Influence of Heating on the Attenuation of Surface Waves in Electrographic Se Layer
(1.5 and 7; 2; 22.1.3)
(a) Vol. **86**, No. 1, p. K7—K10 (1984).

M. F. ZHU, F. C. T. SO, E. T.-S. PAN, and M.-A. NICOLET
*Investigation of Amorphous $W_{60}Zr_{40}$ Film as a Diffusion Barrier in Metallization Schemes
(1.5; 2; 21)
(a) Vol. **86**, No. 2, p. 471—476 (1984).

N. V. RUMAK and V. V. KHATKO
*Structure and Properties of Silicon Dioxide Thermal Films (II): 110 nm Thick SiO_2 Films
(1.5 and 2; 3; 22.6)
(a) Vol. **86**, No. 2, p. 477—484 (1984).

1.6. Surfaces

Original Papers and Short Notes

P. A. ALEKSANDROV, A. M. AFANASIEV, M. K. MELKONYAN, and S. A. STEPANOV
*X-Ray Diffraction under Specular Reflection Conditions on Crystals with an Amorphous Surface Film
(1.6 and 2; 10.2; 22.1.2)
(a) Vol. **81**, No. 1, p. 47—53 (1984).

V. O. ESIN, V. I. DANILYUK, and V. N. POROZKOV
The Anisotropy of Interface Mobility in F.C.C. Crystals
(3; 1.6)
(a) Vol. **81**, No. 1, p. 163—169 (1984).

P. WEISSBRODT, F. STORBECK, and W. HAUFFE
*Sputtering Behaviour of Vanadium Silicide Single Crystals under 10 keV Kr^+ Ion Bombardment
(1.6 and 11; 21.1)
(a) Vol. **81**, No. 1, p. 259—265 (1984).

M. V. KOVALCHUK and E. KH. MUKHAMEDZHANOV
*The Use of a Gas-Flow Proportional Counter for the Energetical Analysis of Photoelectrons under X-Ray Diffraction Conditions
(1.6 and 17; 22.1.1)
(a) Vol. **81**, No. 2, p. 427—432 (1984).

T. JUNG, W. BRAUNE, M. SCHNÜRER, and J. SCHULZE
*Film and Interface Analysis of InSb MOS Structures
(1.6 and 14.3.2; 22.2.3)
(a) Vol. **81**, No. 2, p. 463—468 (1984).

P. ČAPKOVÁ and V. VALVODA
*The Surface Roughness Effect in Texture Measurements by the Schulz X-Ray Reflection Method
(1.6; 21.1)
(a) Vol. **81**, No. 2, p. K113—K115 (1984).

R. BASS
*A New Size Effect in Crystalline Slabs
(1.6)
(b) Vol. **122**, No. 1, p. K1—K4 (1984).

N. KEDDAR, D. L. KIRK, and S G. JOHNSON
*The Surface Degradation of the β''-Al_2O_3 Solid Electrolyte Utilised in Na–S Battery Cells
(1.6 and 14.3.2; 22.6)
(a) Vol. **82**, No. 1, p. 327—334 (1984).

R. TOGNATO
*Some Considerations on the Interfacial Free Energy between Dislocation Core and Bulk Material at High Temperatures
(1.6; 10.1; 21)
(a) Vol. **82**, No. 1, p. K17—K19 (1984).

V. Atzrodt and H. Lange
*Principal Component Analysis for Refractory Metal Silicide Investigations with Auger Electron Spectroscopy
(1.6; 17; 21; 22.1.2)
(a) Vol. **82**, No. 2, p. 373—378 (1984).

V. I. Osinskii, F. M. Katsapov, and E. A. Tyavlovskaya
*Structural Perfection of Selective GaAs Regions in Si Substrate Windows
(1.4 and 1.6; 22.2.1)
(a) Vol. **82**, No. 2, p. 399—403 (1984).

Th. Wirth, V. Atzrodt, and H. Lange
*Dynamic Alterations of the Surface Composition during Sputtering of Silicides
(1.6 and 11; 21)
(a) Vol. **82**, No. 2, p. 459—466 (1984).

A. M. Afanasev, R. M. Imamov, and E. Kh. Mukhamedzhanov
*The Yield of Photoelectrons of Different Energies in the X-Ray Laue Diffraction
(1.6; 17; 22.1.2)
(a) Vol. **83**, No. 1, p. K5—K9 (1984).

G. Dharmadurai
Analysis of Photoacoustically Investigated Gaseous Heat Transfer between Solid Surfaces
(6 and 8; 1.6)
(a) Vol. **83**, No. 1, p. K27—K30 (1984).

A. E. Smith
*A Calculation of Absorption Potentials for Low Energy Electrons in MgO, 2H-MoS_2, and 2H-$NbSe_2$
(1.6 and 13; 22; 22.6)
(b) Vol. **123**, No. 2, p. 619—626 (1984).

J. D. Stephenson
*Double Laue Pattern Topography of Top-Seeded $BaTiO_3$ Single Crystals Using White Synchrotron X-Radiation
(1.4 and 1.6; 14.4.1; 22.8.1)
(a) Vol. **84**, No. 1, p. 19—30 (1984).

P. Mrozek, M. Menyhárd, J. Wernisch, and A. Jabłoński
*Surface Composition of the Ordered Al50–Co50 Alloy
(1.3 and 1.6; 21;21.1)
(a) Vol. **84**, No. 1, p. 39—48 (1984).

Yu. B. Bolkhovityanov, Yu. D. Vaulin, B. Z. Olshanetskii, S. I. Stenin, and S. A. Teys
*Auger Analysis of a GaAs (111) A Surface after Its Contact with the In–Ga–As–P Saturated Liquid
(1.6 and 2; 22.2.1; 22.2.4)
(a) Vol. **84**, No. 1, p. K13—K16 (1984).

A. Ribeiro Filho, D. R. Tilley, E. L. Albuquerque, N. S. Almeida, and M. C. Oliveros
*Light Scattering from Surfaces near Phase Transition Points
(1.6; 1.2)
(b) Vol. **124**, No. 2, p. 457—464 (1984).

A. E. SMITH and D. F. LYNCH
*Low-Energy Electron Reflection Intensities from $2H\text{-}MoS_2$ as Derived by Scattering Matrix Calculations
(1.6 and 13; 22)
(b) Vol. **124**, No. 2, p. 649—655 (1984).

P. MICHEL, B. WEBER, and G. GÖTZ
*Formation of Pt Silicides by a Millisecond Laser Pulse
(1.6 and 11; 21)
(a) Vol. **85**, No. 1, p. K1—K4 (1984).

S. A. MAKSIMENKO
*On X-Ray Surface Diffraction
(1.6)
(a) Vol. **85**, No. 1, p. K23—K25 (1984).

M. A. ANDREEVA and R. N. KUZMIN
Mössbauer Conversion Electron Spectra at Total Reflection
(5; 1.6; 13.3; 21.1)
(b) Vol. **125**, No. 2, p. 461—465 (1984).

P. A. ALEKSANDROV, A. M. AFANASIEV, and S. A. STEPANOV
Bragg-Laue Diffraction in Inclined Geometry
(1.1; 1.6; 22.1.2)
(a) Vol. **86**, No. 1, p. 143—154 (1984).

A. M. AFANASIEV, S. M. AFANASIEV, P. A. ALEKSANDROV, R. M. IMAMOV, and E. M. PASHAEV
*Grazing Bragg-Laue Diffraction for Studying the Crystal Structure of Thin Films
(1.6 and 10.2; 22.2.1; 22.2.2)
(a) Vol. **86**, No. 1, p. K1—K5 (1984).

D. M. VANDERWALKER
*The Nucleation of Ni Silicides on Dislocations and Twins in Silicon
(1.6 and 10.2; 1.2; 21.1; 22.1.2)
(a) Vol. **86**, No. 2, p. 507—516 (1984).

2. Non-Crystalline State

Original Papers and Short Notes

L. Pajasová, P. Pajas, O. A. Makarov, and V. M. Zakharov
*Optical Properties of Amorphous GeS in the Spectral Region of 18 to 38 eV
(2 and 20.1; 13.1; 22)
(b) Vol. **121**, No. 1, p. 293—300 (1984).

H.-J. Unger
*Localization of Electronic States in Disordered Tight-Binding Systems (II): Mobility Edges
(2 and 13.4; 22)
(b) Vol. **121**, No. 1, p. 371—383 (1984).

H. Böttger and D. Wegener
*On the Percolation Approach to the Hopping Conduction in Strong Electrical Fields
(2 and 14.3; 6; 22)
(b) Vol. **121**, No. 1, p. 413—419 (1984).

J. Pozo and R. Ferrer
Specific Heat of Amorphous Magnets with Mixed Exchange Interactions
(8 and 18; 2)
(b) Vol. **121**, No. 1, p. K39—K41 (1984).

P. A. Aleksandrov, A. M. Afanasiev, M. K. Melkonyan, and S. A. Stepanov
*X-Ray Diffraction under Specular Reflection Conditions on Crystals with an Amorphous Surface Film
(1.6 and 2; 10.2; 22.1.2)
(a) Vol. **81**, No. 1, p. 47—53 (1984).

S. Simozar and J. A. Alonso
*On the Factors Controlling Glass Forming Ability of Metallic Alloys Formed by Fast Liquid Quenching
(1.3 and 2; 8; 21; 21.1)
(a) Vol. **81**, No. 1, p. 55—61 (1984).

V. Yu. Balandin, A. V. Dvurechenskii, and L. N. Aleksandrov
*Dynamics of Self-Sustaining Crystallization of Amorphous Silicon Layers
(2 and 3; 1.5; 22.1.2)
(a) Vol. **81**, No. 1, p. 63—68 (1984).

N. Kaiser
Nucleation of Antimony on Amorphous Carbon
(1.5 and 3; 2; 22)
(a) Vol. **81**, No. 1, p. 99—103 (1984).

E. M. Raspopova, V. T. Kalinnikov, T. N. Lyubovskaya, A. P. Chernov, and T. N. Fesenko
*Glass Formation and Some Properties of the $(Cd_2GeAs_4)_{1-x}(TlAs)_x$ System
(2; 14.3; 15; 22.8)
(a) Vol. **81**, No. 1, p. 139—143 (1984).

V. Z. Bengus, E. D. Tabachnikova, and V. I. Startsev
*Mechanical Behaviour of Some Metallic Glasses at 4.2 to 300 K
(2 and 12.1; 10.1; 21)
(a) Vol. **81**, No. 1, p. K11—K13 (1984).

A. F. Khokhlov, V. A. Panteleev, E. V. Dobrokhotov, G. A. Maksimov, and V. A. Sidorov
*Phosphorus Diffusion in Amorphous Silicon
(2 and 9; 22.1.2)
(a) Vol. **81**, No. 1, p. K15—K18 (1984).

A. U. Grinfelds, A. R. Silin, L. N. Skuya, and V. G. Plekhanov
Low Temperature Recombination Luminesence in Amorphous and Crystalline SiO_2
(20.3; 2; 13.5.3; 22.6)
(a) Vol. **81**, No. 1, p. K23—K26 (1984).

F. Momo, A. Sotgiu, M. Bettinelli, and A. Montenero
*Distribution of Magnetic Parameters in Some Copper Containing Glasses
(2 and 19)
(a) Vol. **81**, No. 1, p. K27—K30 (1984).

J. P. Schaffer, H. Park, J. H. Lind, and P. L. Jones
*Hydrogenated Amorphous Boron: Transient and Steady State Photoconductivity
(2 and 16; 22.1)
(a) Vol. **81**, No. 1, p. K51—K55 (1984).

N. V. Ershov, A. L. Ageev, A. V. Serikov, Yu. A. Babanov, and V. V. Vasin
*Regular Method in X-Ray Scattering Analysis of Amorphous and Liquid Metals
(2; 21; 21.1.1)
(b) Vol. **121**, No. 2, p. 451—460 (1984).

A. Lehmann, L. Schumann, and K. Hübner
Optical Phonons in Amorphous Silicon Oxides (II): Calculation of Phonon Spectra and Interpretation of the IR Transmission of SiO_x
(6 and 20.1; 2; 22.6)
(b) Vol. **121**, No. 2, p. 505—511 (1984).

T. Datta, S. D. LeVine, D. Thornberry, and E. R. Jones, Jr.
*Experimental Determination of the Edwards-Anderson Order Parameter in a High Temperature Spin Glass
(2 and 18; 21.1.1)
(b) Vol. **121**, No. 2, p. K125—K130 (1984).

E. Zedler and G. Lehmann
*Influence of Annealing Treatments on the Density of Amorphous $Fe_{40}Ni_{40}P_{14}B_6$ and $Fe_5Co_{70}Si_{15}B_{10}$ Alloys
(2; 21.1.1)
(a) Vol. **81**, No. 2, p. 445—449 (1984).

C. Chiang
*An Equation for Dynamic Switching in Amorphous Thin Films
(2 and 14.3.1)
(a) Vol. **81**, No. 2, p. 563—567 (1984).

S. Orzeszko, W. Bała, K. Fabisiak, and F. Rozpłoch
*DC Conductivity and ESR of Hydrogenated Amorphous Carbon Films
(2 and 14.3.1; 19; 22.1)
(a) Vol. **81**, No. 2, p. 579—584 (1984).

D. J. Easton and J. T. Edmond
*Glass-Crystal Transitions in Materials in the System As_2S_3–As_2Te_3
(2 and 8; 22)
(a) Vol. **81**, No. 2, p. K103—K107 (1984).

H. J. Goldsmid, H. Hora, and G. L. Paul
*Anomalous Heat Conduction of Ion-Implanted Amorphous Layers in Silicon Crystals Using a Laser-Probe Technique
(2 and 11; 8; 22.1.2)
(a) Vol. **81**, No. 2, p. K127—K130 (1984).

S. R. Rao, M. Nagabhooshanam, and V. Hari Babu
Electrical Properties of Annealed InSb Thin Films
(14.3.1; 2; 22.2.3)
(a) Vol. **81**, No. 2, p. K195—K197 (1984).

A. Richter, A. E. Romanov, W. Pompe, and V. I. Vladimirov
*Geometry and Energy of Disclinations in Topologically Disordered Systems
(2; 10.2; 12.2)
(b) Vol. **122**, No. 1, p. 35—45 (1984).

R. F. Konopleva, V. A. Evseev, B. A. Borisov, I. V. Nazarkin, and V. A. Chekanov
*Electrophysical Properties of Disordered Systems Based on Crystalline Germanium Compensated by Fast Neutron Irradiation
(2 and 14.3; 14.3.4; 22.1.1)
(b) Vol. **122**, No. 1, p. 349—359 (1984).

J. E. Griffiths, J. C. Phillips, G. P. Espinosa, J. P. Remeika, and P. M. Bridenbaugh
Assignment of the Companion A_1^c Line in $Ge_x(S, Se)_{1-x}$ Glasses
(6 and 20.1; 2; 22)
(b) Vol. **122**, No. 1, p. K11—K15 (1984).

V. Hinkel and D. Ploumbidis
*Temperature Dependence of the Magnetic Part of the Hyperfine Field at ^{123}Sb in the Liquid Sb–Te System
(2 and 13; 19; 21.7; 22.1)
(b) Vol. **122**, No. 1, p. K45—K48 (1984).

L. Lejček, M. Glogarová, and J. Pavel
Twist Disclinations in Thin SmC Liquid Crystal
(2 and 10)
(a) Vol. **82**, No. 1, p. 47—54 (1984).

N. K. Gobran, M. M. Danial, and R. Kamel
Kinetics of the Amorphous-Crystalline Transition in the Metallic Glass $Fe_{40}Ni_{40}B_{20}$
(1.2; 2; 21.1; 21.1.1)
(a) Vol. **82**, No. 1, p. 63—66 (1984).

D. Nesheva
Studies of Low-Energy Optical Absorption and Gap State Density of a-CdS by the Constant Photocurrent Method
(13.1 and 20.1; 2; 14.3.1; 22.4.1)
(a) Vol. **82**, No. 1, p. 243—248 (1984).

O. Bostanjoglo and E. Endruschat
Time-Resolved TEM of Laser-Annealing
(1.2 and 1.4; 2; 11; 22.1.1)
(a) Vol. **82**, No. 1, p. K1—K3 (1984).

W. Scharff, J.-W. Erben, A. Wolf, M. Heber, C. Hamann, C. Weissmantel, M. Voelskov, J. Matthäi, R. Kögler, R. Klabes, and E. Wieser
*Growth of Single-Crystalline Regions on Amorphous Insulating Substrates by Zone-Melting Recrystallization
(2 and 3; 11; 22.1.2)
(a) Vol. **82**, No. 1, p. K5—K9 (1984).

V. D. Sedykh and A. V. Serebryakov
*The Effect of Thermal Treatment on the Magnetic Anisotropy in Amorphous Fe-B Alloys
(2 and 18.2; 5; 21.1.1)
(a) Vol. **82**, No. 1, p. K63—K66 (1984).

Z. Černošek and E. Černošková
*Short-Range Order Model of Ge_xS_{1-x} Glasses and Utilization of This Model in Calculation of the Optical Gap and the IR 367 cm^{-1} Band Intensity
(2 and 20.1; 22)
(b) Vol. **122**, No. 2, p. 405—410 (1984).

M. Celasco, P. Mazzetti, and A. Stepanescu
*Theory of the Power Spectrum and Scatting Properties of Three-Dimensional Disordered Systems Having a Variable Degree of Short-Range Order (I): Continuous Spectrum
(2)
(b) Vol. **122**, No. 2, p. 411—421 (1984).

B. A. Vaid and K. C. Sharma
*Structural Disorder Model of Amorphous Semiconductors: Ge and Si
(2; 22.1.1; 22.1.2)
(b) Vol. **122**, No. 2, p. 423—429 (1984).

P. P. LOTTICI
On the Optic Vibrational Modes in Amorphous Arsenic
(6 and 20.1; 2; 22.1)
(b) Vol. **122**, No. 2, p. 431—434 (1984).

A. Z. MAKSYMOWICZ
*Short-Range Order and Magnetic Properties in Amorphous Heisenberg Ferromagnets
(2 and 18.2)
(b) Vol. **122**, No. 2, p. 519—526 (1984).

G. KNUYT and A. KEUPERS
*Calculation of the Resistivity of Liquid and Amorphous $Au_{1-x}Si_x$ Using Ziman Theory
(2 and 14.1; 21.6)
(b) Vol. **122**, No. 2, p. 781—790 (1984).

I. G. KAVERINA, V. V. KOROBTSOV, V. G. ZAVODINSKII, and A. V. ZOTOV
Solid Phase Epitaxial Growth Anisotropy of Vacuum-Deposited Amorphous Silicon
(1.5 and 3; 2; 22.1.2)
(a) Vol. **82**, No. 2, p. 345—353 (1984).

C. MOŢOC, R. MITROI, and C. ROŞU
Photovoltaic Effects in Nematic and Cholesteric Mesophases Containing Nonmesogenic Compounds
(16; 2; 22.9)
(a) Vol. **82**, No. 2, p. 575—579 (1984).

R. KLABES, J. MATTHÄI, M. VOELSKOW, E. WIESER, J.-W. ERBEN, W. SCHARFF, and C. WEISSMANTEL
Epitaxial Regrowth of Amorphous or Polycrystalline Silicon Layers on Silicon Single Crystals and Bridging Epitaxy by Flash Lamp Irradiation
(1.5 and 3; 2; 11; 22.1.2)
(a) Vol. **82**, No. 2, p. K121—K123 (1984).

V. HAJKO, A. ZENTKO, M. TIMKO, and V. HAJKO, JR.
Isotopic Effect in Neutron-Irradiated $Fe_{40}Ni_{40}B_{20}$ Alloys
(11 and 18.2; 2; 21.1.1)
(a) Vol. **82**, No. 2, p. K159—K162 (1984).

T. SOUMURA, S. SAITO, T. KURAISHI, and T. MAEDA
*Magnetic Moment and Its Temperature Variation of Amorphous $(Fe_{0.5}Ni_{0.5})_{1-x}B_x$ Alloys
(2 and 18.2; 21.1; 21.1.1)
(a) Vol. **82**, No. 2, p. K169—K172 (1984).

R. KRISHNAN
*Magnetic Moment of Amorphous $Co_{1-x}Nb_x$ Alloys and Friedel's Model
(2 and 18.2; 21; 21.1)
(a) Vol. **82**, No. 2, p. K173—K175 (1984).

R. ŻUBEREK, F. STOBIECKI, and J. WOSIK
*Magnetoelastic Properties of $Fe_{1-x}B_x$ Amorphous Thin Films
(2 and 18.2.1; 21.1.1)
(a) Vol. **82**, No. 2, p. K177—K179 (1984).

A. I. ONIPKO
*Application of the Cluster Size Distribution in Binary Randomly Disordered Finite Chains. Exact Solution of the Random Walk Problem for a Finite Chain with an Arbitary Concentration of Pure Absorbers
(2 and 10)
(b) Vol. **123**, No. 1, p. 37—44 (1984).

R. RICHTER and F. GOEDSCHE
*On the Thermoelectric Power of Amorphous Ferromagnets
(2 and 15; 14.1; 18.2; 21)
(b) Vol. **123**, No. 1, p. 143—154 (1984).

H. SOLBRIG
*Asymptotic Medium Concept Applied to Correlated Point Scatterers in the Weak-Scattering Limit
(2 and 13)
(b) Vol. **123**, No. 1, p. 237—246 (1984).

K. S. VAHVASELKÄ
*An X-Ray Investigation of Structural Properties and Electrical Resistivities of Liquid Al, Liquid Cu, and Liquid Al–Cu Alloys
(1.3 and 2; 14.1; 21; 21.1)
(a) Vol. **83**, No. 1, p. 103—112 (1984).

M. JERGEL and P. MRAFKO
*The Temperature Dependence of the Prepeak in the Diffraction Pattern of the Amorphous $Ti_{61}Cu_{16}Ni_{23}$ and $Ti_{62.5}Cu_{12}Ni_{23}Si_{2.5}$ Alloys
(2; 1.4; 21.1)
(a) Vol. **83**, No. 1, p. 113—121 (1984).

B. V. KOBRIN and V. P. SHANTAROVICH
*Study of Ge_2S_3 Glass Undoped and Doped with Metals by Positron Annihilation Technique
(2 and 10.2; 22)
(a) Vol. **83**, No. 1, p. 159—164 (1984).

H. REUTHER
*Contribution to the Thermotransport in Silicate Glasses
(2 and 9; 22.6.1)
(a) Vol. **83**, No. 1, p. 173—178 (1984).

K. MAEDA, S. KOBAYASHI, S. TAKEUCHI, T. CHIBA, and K. OSAMURA
*Absence of Dilatational Defects in Deformed $Cu_{57}Zr_{43}$ Amorphous Alloy
(2 and 10.1; 21.1)
(a) Vol. **83**, No. 1, p. 219—226 (1984).

Y. SHIMADA
Amorphous Co-Metal Films Produced by Sputtering
(18.2.1; 2; 21.1)
(a) Vol. **83**, No. 1, p. 255—261 (1984).

F. S. Nasredinov, A. A. Andreev, A. N. Kurmantaev, A. R. Regel, and P. P. Seregin
Tin Impurity Atoms in Vacuum Deposited Amorphous Silicon
(13.4 and 14.3.1; 2; 20.1; 22.1.2)
(a) Vol. **83**, No. 1, p. 285—289 (1984).

R. M. Mehra, Hemant Kumar, Surinder Koul, and P. C. Mathur
Thickness Dependence of DC Conductivity of Amorphous Se and Binary Amorphous Se–Te, Se–Ge, and Se–Sb Films
(14.3.1; 2; 22.1.3)
(a) Vol. **83**, No. 1, p. 341—347 (1984).

L. Červinka, V. Cháb, and S. Koc
*Erratum to: XPS Investigation of the Amorphous $TlSe_x$ System
(2 and 4; 17; 22.3)
(a) Vol. **83**, No. 1, p. 418 (1984).

E. Culea, Al. Nicula, and I. Bratu
An Infrared Study of $xV_2O_5 \cdot (1 - x) B_2O_3$ Glasses
(6 and 20.1; 2; 22.6.1)
(a) Vol. **83**, No. 1, p. K15—K18 (1984).

J. Teteris
Photoinduced Optical Absorption in As–Se
(20.1; 2; 22)
(a) Vol. **83**, No. 1, p. K47—K50 (1984).

S. Chaudhuri, S. K. Biswas, and A. Choudhury
*Effects of Heat Treatment on the Optical Absorption of Amorphous $In_{30}Se_{70}$ Films
(2 and 20.1; 22.3)
(a) Vol. **83**, No. 1, p. K57—K60 (1984).

S. A. Azimov, P. M. Karageorgy-Alkalaev, A. Yu. Leiderman, and V. S. Rubin
*Computer Simulation of Generation–Recombination Currents in Amorphous Silicon p–n Diode Structures
(2 and 14.3; 14.3.4; 22.1.2)
(a) Vol. **83**, No. 1, p. K81—K86 (1984).

M. Celasco, P. Mazzetti, and A. Stepanescu
*Theory of the Power Spectrum and Scattering Properties of Three-Dimensional Disordered Systems Having a Variable Degree of Short-Range Order (II): Line Spectrum and Cross-Relation Technique
(2)
(b) Vol. **123**, No. 2, p. 403—413 (1984).

T. Kaneyoshi and I. Tamura
*The Distribution of Magnetic Moments and the Temperature Dependence of Reduced Hyperfine Fields in Amorphous Ferromagnets
(2 and 18.2; 21.1.1)
(b) Vol. **123**, No. 2, p. 525—531 (1984).

N. M. Ravindra, J. F. Morhange, P. A. Thomas, and C. Ance
*Polarizability Studies of Amorphous Silicon
(2 and 14.4; 22.1.2)
(b) Vol. **123**, No. 2, p. 611—618 (1984).

S. K. J. Al-Ani, C. A. Hogarth, and D. N. Waters
The Effect of Temperature on the Optical Absorption Edge of Amorphous Thin Films of Silicon Monoxide
(20.1; 2; 13; 22.6)
(b) Vol. **123**, No. 2, p. 653—658 (1984).

G. Knuyt
*Electrical Resistivity Calculations for Amorphous Cu_xAl_{1-x}
(2 and 14.1; 21; 21.1)
(b) Vol. **123**, No. 2, p. 763—771 (1984).

J. Šlechta
On the Magnetic Part of C_v of Amorphous Materials
(8 and 18; 2)
(b) Vol. **123**, No. 2, p. K133—K137 (1984).

H.-U. Tschirner and K. Uhlig
*Some Remarks on Activation Energies Observed in the Liquid Semiconductor Ga_2Te_3
(2 and 14.3; 13.4; 22)
(b) Vol. **123**, No. 2, p. K177—K181 (1984).

R. V. Gopala Rao and Sobhen Kumar Pal
*Molten Rubidium Bromide — A Rigorous Study of Its Partial Structure and Allied Functions
(2 and 4; 22.5.2)
(a) Vol. **83**, No. 2, p. 429—436 (1984).

A. Turos, H. Frey, O. Meyer, W. Müller, and J. M. Pirrung
*Ion Beam Analysis of Amorphous Silicon Films Produced by Magnetron Sputtering
(2 and 14.3.1; 3; 22.1.2)
(a) Vol. **83**, No. 2, p. 437—443 (1984).

M. Doi, Y. Itoh, D.-Y. Chang, and T. Imura
SIMS Studies on the Impurity Diffusion of Boron in Amorphous Alloys
(9; 2; 21.1; 21.1.1)
(a) Vol. **83**, No. 2, p. 529—536 (1984).

V. N. Solovev
*Atomic Particle Delocalisation Effect in Disordered Media
(2 and 9; 22.1.2)
(a) Vol. **83**, No. 2, p. 553—559 (1984).

M. Vaněček, A. Abrahám, O. Štika, J. Stuchlík, and J. Kočka
Gap States Density in a-Si: H Deduced from Subgap Optical Absorption Measurement on Schottky Solar Cells
(13.1 and 20.1; 2; 22.1.2)
(a) Vol. **83**, No. 2, p. 617—623 (1984).

K. MUKHERJEE, M. CHANDRASEKARAN, and M. KATO
Electrical Resistivity of an Equiatomic Solid and Liquid Au–Cd Alloy
(14.1; 2; 8; 21; 21.6)
(a) Vol. **83**, No. 2, p. 657—667 (1984).

G. FLEURY, A. HAMOU, C. LHERMITTE, and C. VIGER
*Bismuth Influence on a-Se Film Crystallization
(2 and 14.3.1; 22.1.3)
(a) Vol. **83**, No. 2, p. K103—K108 (1984).

K. OTS and M. SEILENTHAL
*On the Structural Model of CVD a-Si:O:H Films
(1.5 and 2; 20.1; 22.1.2; 22.6)
(a) Vol. **83**, No. 2, p. K109—K111 (1984).

T. SOUMURA, S. SAITO, and T. MAEDA
*Electrical Resistivity and Structural Stability for Amorphous $(Fe_{0.5}Ni_{0.5})_{1-x}B_x$ Alloys
(2 and 14.1; 21.1; 21.1.1)
(a) Vol. **83**, No. 2, p. K195—K198 (1984).

V. B. GOHEL, C. K. ACHARYA, and A. R. JANI
On the Determination of Pseudopotential Parameters and the Comprehensive Study of Metallic Properties
(13.1; 2; 14.1; 15; 21)
(b) Vol. **124**, No. 1, p. 137—145 (1984).

P. V. ELYUTIN, B. HICKEY, G. J. MORGAN, and G. F. WEIR
Diffusion and Localisation in Random Systems
(13; 2)
(b) Vol. **124**, No. 1, p. 279—286 (1984).

R. JANKOWIAK, B. RIES, and H. BÄSSLER
Spectral Diffusion and Triplet Exciton Localization in an Organic Glass
(13.5.1 and 20.3; 2; 22.9)
(b) Vol. **124**, No. 1, p. 363—371 (1984).

G. JUNGK and A. RÖSELER
LO–TO-Splitting and Many-Particle Effects
(6 and 20.1; 2; 22.6)
(b) Vol. **124**, No. 1, p. K1—K4 (1984).

W. HOYER, A. MÜLLER, W. MATZ, and M. WOBST
*Neutron Scattering Investigation on Molten Ga–Te Alloys
(2; 1.3; 22.7)
(a) Vol. **84**, No. 1, p. 11—18 (1984).

C. A. HEWETT, S. S. LAU, R. FASTOW, and J. W. MAYER
Pulsed Ion Beam Annealing of Ti–Au Alloys
(1.3 and 11; 1.5; 2; 21.1; 21.6)
(a) Vol. **84**, No. 1, p. 49—54 (1984).

A. A. Simashkevich and S. D. Shutov
Evidence of Schottky Barrier Formation at Contact of Metal with Chalcogenide Glassy Semiconductor
(14.3.4; 2; 16; 22)
(a) Vol. **84**, No. 1, p. 343—352 (1984).

Yu. B. Bolkhovityanov, Yu. D. Vaulin, B. Z. Olshanetskii, S. I. Stenin, and S. A. Teys
*Auger Analysis of a GaAs (111) A Surface after Its Contact with the In–Ga–As–P Saturated Liquid
(1.6 and 2; 22.2.1; 22.2.4)
(a) Vol. **84**, No. 1, p. K13—K16 (1984).

E. Culea, Al. Nicula, and I. Bratu
An Infrared Study of $V_2O_5 \cdot As_2O_3$ Glasses
(6 and 20.1; 2; 22.6.1)
(a) Vol. **84**, No. 1, p. K17—K21 (1984).

N. Pashov, G. Kästner, M. Kalitzova, S. Simov, M. Mikhailov, and K. Kolev
Electron Microscope and Electron Diffraction Studies of Amorphization Phenomena in Ge Implanted with Te Ions
(1.4 and 11; 2; 22.1.1)
(a) Vol. **84**, No. 1, p. K23—K27 (1984).

Y. Shimada
*Initial Permeability of Amorphous Films with Multilayered Structure
(2 and 18.2.1; 21)
(a) Vol. **84**, No. 1, p. K55—K58 (1984).

E. Canessa, D. F. Mariani, and J. Vignolo
*Structure of Non-Simple Liquid Metals
(2; 21; 21.7)
(b) Vol. **124**, No. 2, p. 465—472 (1984).

K. Winer and F. Wooten
*Structure and Vibrational Spectra of a Model of a-Si:H with Periodic Boundary Conditions
(2 and 6; 22.1.2)
(b) Vol. **124**, No. 2, p. 473—480 (1984).

A. Schönhals and E. Donth
*A Dynamic Percolation Approach to the Andrade Creep
(2; 12.2)
(b) Vol. **124**, No. 2, p. 515—524 (1984).

N. P. Kovalenko and L. M. Kuzmina
*The Influence of Many-Body Interactions on the Speed of Sound in Liquid Metals
(2 and 12.1; 7; 21.2)
(b) Vol. **124**, No. 2, p. 537—544 (1984).

B. Vasvári
A Simple Muffin-Tin Model for Liquid Metals
(13; 2; 21.1)
(b) Vol. **124**, No. 2, p. 683—692 (1984).

E. Burzo, M. Chipară, D. Ungur, and I. Ardelean
*Electron Paramagnetic Resonance Study of $xFe_2O_3(1-x)$ [$B_2O_3 \cdot PbO$] Glasses
(2 and 19; 22.8)
(b) Vol. **124**, No. 2, p. K117—K120 (1984).

K. Hricovíni and J. Krempaský
On the Electronic Structure of Iron–Boron Alloys
(13.1; 2; 17; 21.1.1)
(b) Vol. **124**, No. 2, p. K147—K150 (1984).

B. G. Mulimani and R. Harris
A Note on the Electron–Phonon Enhancement of the Thermopower in Amorphous Alloys
(15; 2; 21)
(b) Vol. **124**, No. 2, p. K181—K185 (1984).

G. Cocco, S. Enzo, C. Antonione, L. Battezzati, and G. Riontino
*Structure Features in the Palladium–Beryllium Amorphous System
(2; 21; 21.3)
(a) Vol. **84**, No. 2, p. 371—377 (1984).

W. Brodkorb, J. Salm, and E. Steinbeiss
*Discussion of Possible Surface Processes during the Reactive SiO_x Film Deposition in a Planar Magnetron Sputtering Equipment
(1.5 and 2; 22.6)
(a) Vol. **84**, No. 2, p. 379—386 (1984).

V. I. Kononenko, A. L. Sukhman, S. L. Gruverman, and V. V. Torokin
*Density and Surface Tension of Liquid Rare Earth Metals, Scandium and Yttrium
(2 and 12.1; 1.2; 21; 21.4)
(a) Vol. **84**, No. 2, p. 423—432 (1984).

L. S. Grigoryan, M. V. Simonyan, and E. G. Sharoyan
*Electrical, Magnetic, and Spectral Properties of the New Class of Charge Transfer Salts Based on Phthalocyanines and Iodine
(2 and 14.3.1; 19; 20.1; 22.9)
(a) Vol. **84**, No. 2, p. 597—606 (1984).

T. Luciński and J. Baszyński
Influence of Composition and Structural Transformation on Transport Properties in Amorphous $Fe_{1-x}Si_x$ Films
(14.1; 2; 18.2.1; 21.1.1)
(a) Vol. **84**, No. 2, p. 607—612 (1984).

A. Mueller, W. Hoyer, E. Thomas, and M. Wobst
*X-Ray Scattering Investigation on Molten Ga–Te
(1.3 and 2; 22.7)
(a) Vol. **84**, No. 2, p. K97—K100 (1984).

L. N. Skuja, A. N. Trukhin, and A. E. Plaudis
Luminescence in Germanium-Doped Glassy SiO_2
(20.3; 2; 10.2; 13.4; 20.1; 22.6)
(a) Vol. **84**, No. 2, p. K153—K157 (1984).

C. Chiang
*Photo-Illumination and Current Excitation of Switching in Amorphous Thin Films
(2 and 14.3.1; 16)
(a) Vol. **84**, No. 2, p. K189—K191 (1984).

B. R. Gelchinskii and R. M. Ryskulbekova
*On Correlation between Structural Features of Liquid Metals and the Effective Ion–Ion Potential
(2 and 4; 21.2; 21.7)
(b) Vol. **125**, No. 1, p. 81—86 (1984).

A. Alegría, J. M. Barandiarán, and J. Colmenero
*Dielectric Relaxation at the Glass Transition as a Free Volume Process (II): A Continuous Distribution of Relaxation Times
(2 and 14.4; 22.9)
(b) Vol. **125**, No. 1, p. 409—419 (1984).

R. N. Joarder, S. Palchaudhuri, and R. V. Gopala Rao
*Entropy and Heat Capacity of Liquid Metals
(2 and 8; 21.2)
(b) Vol. **125**, No. 1, p. K33—K36 (1984).

J. Kočka and K. Čermák
*Drift Mobility of a-Si: H
(2 and 14.3.1; 14.3.4; 22.1.2)
(b) Vol. **125**, No. 1, p. K89—K91 (1984).

R. V. Sharma
*Partial Higher-Order Structure Factors in the Long-Wavelength Limit for Molten Cuprous Chloride
(2; 22.5)
(a) Vol. **85**, No. 1, p. 39—43 (1984).

D. S. Misra, A. Kumar, and S. C. Agarwal
Intensity and Temperature Dependence of Steady-State Photoconductivity Down to 8 K and DOS Distribution Obtained from These Measurements in a-Si:H
(16; 2; 22.1.2)
(a) Vol. **85**, No. 1, p. 297—303 (1984).

T. Satow, O. Uemura, K. Hoshino, and T. Watanabe
*The Neutron Diffration of Molten TlCl, TlBr, and TlI
(1 and 2; 22.5.2)
(a) Vol. **85**, No. 1, p. K5—K7 (1984).

G. Berger and V. Atzrodt
In Vitro Characterization of Bioactivity of Glassy or Glass–Crystalline Implant Materials Using Auger Electron Spectroscopy
(1.5; 2; 17; 22.6.1)
(a) Vol. **85**, No. 1, p. K9—K13 (1984).

B. A. Vaid and K. C. Sharma
*Ultrasonic Attenuation in Liquid Metals
(2 and 7; 21)
(b) Vol. **125**, No. 2, p. 505—512 (1984).

C. Pisani and E. Marchese
*Critical Analysis and Simplified Formulation of the Travelling Cluster Approximation for Disordered Systems
(2 and 13)
(b) Vol. **125**, No. 2, p. 721—733 (1984).

A. Paja and T. Stobiecki
Electrical Resistivity of Amorphous $Fe_{1-x}B_x$ Alloys
(14.1; 2; 21.1.1)
(b) Vol. **125**, No. 2, p. K171—K176 (1984).

P. Michon, H. Atmani, and C. Vautier
Crystallization Study of Evaporated a-Se Films by Means of DTA Technique
(1.5 and 3; 2; 22.1.3)
(a) Vol. **85**, No. 2, p. 399—407 (1984).

V. G. Chudinov, N. V. Moseev, B. N. Goshchitskii, and V. I. Protasov
Possible Amorphisation and Phase Separation of Intermetallic Mo_3Si Compound under Radiation. Molecular Dynamics Study
(11; 2; 21.1)
(a) Vol. **85**, No. 2, p. 435—440 (1984).

F. Walz
Transition-Induced Magnetic After-Effect in the Amorphous Alloy $Fe_{20}Ni_{56}B_{24}$ at the Critical Temperature
(18.2; 2; 10.1; 21.1; 21.1.1)
(a) Vol. **85**, No. 2, p. 503—509 (1984).

J. M. Alameda, M. C. Contreras and H. Rubio
Magnetic Properties of Amorphous Fe_xSi_{1-x} Films (I): Transverse Susceptibility, Structure Constant, and Local Anisotropy
(18.2.1; 2; 21.1.1)
(a) Vol. **85**, No. 2, p. 511—522 (1984).

G. Micocci, A. Rizzo, and A. Tepore
Photoconductivity in Halogenated and Hydrogenated Amorphous Silicon Films
(16; 2; 22.1.2)
(a) Vol. **85**, No. 2, p. 609—614 (1984).

R. A. Dunlap, K. Dini, and P. Hargraves
Hydrogen Absorption and Crystallisation in Amorphous $Co_{90}Zr_{10}$
(1.3; 2; 21.1)
(a) Vol. **85**, No. 2, p. K101—K103 (1984).

J. Málek, L. Tichý, and J. Klikorka
*Exponential Absorption Edge in $Ge_{20}Sb_{15}S_{65}$ Glass Depending on Sample Preparation
(2 and 20.1; 22.8)
(a) Vol. **85**, No. 2, p. K123—K127 (1984).

A. Bărbulescu and Lucia Sincan
Absorption Edge Shift in the $PbO-SiO_2$ System
(20.1; 2; 22.6.1)
(a) Vol. **85**, No. 2, p. K129—K131 (1984).

Y. Fukuda, A. Tomita, and N. Takeuchi
Thermoluminescence and Thermally Stimulated Exoelectron Emission in Glass and Sintered $CaB_4O_7:CuCl_2$
(17 and 20.3; 2; 22.6)
(a) Vol. **85**, No. 2, p. K141—K144 (1984).

O. Kohmoto
Electrical Resistivity of Amorphous Co-Rich FeCo–SiB Alloys
(14.1; 2; 18.3; 21.1; 21.1.1)
(a) Vol. **85**, No. 2, p. K155—K157 (1984).

G. Parthasarathy and E. S. R. Gopal
*Unusual Effect of Tellurium Doping on the Electrical Properties of Bulk Amorphous Selenium
(2 and 14.3; 22.1.3)
(a) Vol. **85**, No. 2, p. K165—K167 (1984).

B. A. Vaid, K. C. Sharma, and V. K. Syal
*A Disorder Model of Melting (II): Alkali Halides
(2; 8; 22.5.2)
(b) Vol. **126**, No. 1, p. 59—62 (1984).

D. Della Sala, C. Giovannella, and F. Evangelisti
*Study of Bonding Configurations in Amorphous Ge_xSi_{1-x}:H Alloys
(2; 4; 20.1; 22.1.1; 22.1.2)
(b) Vol. **126**, No. 1, p. 125—131 (1984).

A. Z. Maksymowicz, T. Stobiecki, E. Jarocki, and W. Karaś
Charge Transfer and Concentration Dependence of Magnetization in Amorphous Fe–B Alloys
(13 and 18; 2; 21.1.1)
(b) Vol. **126**, No. 1, p. 191—195 (1984).

N. M. KHAMIDULLINA
*Calculation of Some Characteristics of an Intrinsic Random Field in Hydrogenated Amorphous Silicon
(2; 20.1; 22.1.2)
(b) Vol. **126**, No. 1, p. 253—260 (1984).

S. K. J. AL-ANI and C. A. HOGARTH
*The Spin Density and Optical Properties of Amorphous Thin Arsenic Silicate Films
(2 and 13.1; 19; 20.1; 22.6)
(b) Vol. **126**, No. 1, p. 293—304 (1984).

R. HAYN and W. JOHN
*Non-Exponential Localization in an Inhomogeneously Disordered Chain
(2 and 13)
(b) Vol. **126**, No. 1, p. 335—341 (1984).

M. SUHRKE and B. ESSER
Time-Resolved Luminescence of Charge Carriers in Localized States in Amorphous Semiconductors
(20.3; 2; 13.4; 22)
(b) Vol. **126**, No. 1, p. 393—406 (1984).

D. KORN
*Electrical Resistivity of Amorphous SnMn
(2 and 14.1; 21)
(b) Vol. **126**, No. 1, p. 425—432 (1984).

E. CANESSA, D. F. MARIANI, and J. VIGNOLO
*Liquid Structure Factor of Sb: Application of the Canessa-Mariani-Vignolo Model
(2; 21.7)
(b) Vol. **126**, No. 1, p. K1—K4 (1984).

R. V. G. RAO and S. KUMAR PAL
*A New Method of Evaluation of the Structure of Molten Sodium Chloride
(2; 4; 22.5.2)
(a) Vol. **86**, No. 1, p. 43—54 (1984).

N. V. RUMAK, V. V. KHATKO, and V. N. PLOTNIKOV
*Structure and Properties of Silicon Dioxide Thermal Films (I): SiO_2 Films of up to 50 nm Thickness
(1.5 and 2; 3; 4; 22.6)
(a) Vol. **86**, No. 1, p. 93—100 (1984).

I. P. ZVYAGIN
*Planar Conduction in Inhomogeneous Films. Application to Amorphous Silicon
(2 and 14.3.2; 22.1.2)
(a) Vol. **86**, No. 1, p. 375—381 (1984).

B. T. KOLOMIETS, V. M. LYUBIN, M. A. TAGUYRDZHANOV, and G. ZENTAI
A Comparative Study of Photoconductivity in Vitreous As_2Se_3 and a-Si:H Thin Films
(16; 2; 14.3.1; 22; 22.1.2)
(a) Vol. **86**, No. 1, p. 383—387 (1984).

P. Koštial
The Influence of Heating on the Attenuation of Surface Waves in Electrographic Se Layer
(1.5 and 7; 2; 22.1.3)
(a) Vol. **86**, No. 1, p. K7–K10 (1984).

S. Palchaudhuri, R. N. Joarder, and R. V. Gopala Rao
*Partial Structure Factors and Self-Diffusion in Liquid Copper–Lead Alloys
(2 and 9; 21; 21.1)
(b) Vol. **126**, No. 2, p. 443—452 (1984).

Juh Tzeng Lue
Time-Resolved Reflectivity Measurements on RF Sputtered a-Si:H Alloy
(13.4 and 20.1; 2; 16; 22.1.2)
(b) Vol. **126**, No. 2, p. 713—720 (1984).

M. A. Alam and M. Tomak
*Electrical Resistivity of Liquid Noble Metal Alloys
(2 and 14.1; 21.1; 21.6)
(b) Vol. **126**, No. 2, p. 729—732 (1984).

M. F. Zhu, F. C. T. So, E. T.-S. Pan, and M.-A. Nicolet
Investigation of Amorphous $W_{60}Zr_{40}$ Film as a Diffusion Barrier in Metallization Schemes
(1.5; 2; 21)
(a) Vol. **86**, No. 2, p. 471—476 (1984).

N. V. Rumak and V. V. Khatko
*Structure and Properties of Silicon Dioxide Thermal Films (II): 110 nm Thick SiO_2 Films
(1.5 and 2; 3; 22.6)
(a) Vol. **86**, No. 2, p. 477—484 (1984).

C. I. Nistor, V. Spânu, S. Mandache, G. Filoti, and V. Florescu
Hyperfine Field Distribution in Metglasses
(5; 2; 21.1.1)
(a) Vol. **86**, No. 2, p. 525–529 (1984).

G. Berger and M. Giehler
*Characterization of in Vitro Corroded Surfaces of Bioactive Glasses (P_2O_5–CaO–SiO_2) with Infrared Reflection Spectroscopy
(2 and 20.1; 4; 6; 22.6.1)
(a) Vol. **86**, No. 2, p. 531—542 (1984).

3. Crystal Growth

Original Papers and Short Notes

V. Yu. Balandin, A. V. Dvurechenskii, and L. N. Aleksandrov
*Dynamics of Self-Sustaining Crystallization of Amorphous Silicon Layers
(2 and 3; 1.5; 22.1.2)
(a) Vol. **81**, No. 1, p. 63—68 (1984).

N. Kaiser
*Nucleation of Antimony on Amorphous Carbon
(1.5 and 3; 2; 22)
(a) Vol. **81**, No. 1, p. 99—103 (1984).

C.-D. Lien, M.-A. Nicolet, and S. S. Lau
*Low Temperature Formation of $NiSi_2$ from Evaporated Silicon
(1.5 and 3; 21)
(a) Vol. **81**, No. 1, p. 123—128 (1984).

V. O. Esin, V. I. Danilyuk, and V. N. Porozkov
*The Anisotropy of Interface Mobility in F.C.C. Crystals
(3; 1.6)
(a) Vol. **81**, No. 1, p. 163—169 (1984).

P. J. Dean
Comparison of MOCVD-Grown with Conventional II-VI Materials Parameters for EL Thin Films
(14.3.4 and 20.3; 3; 13.4; 22.4.1; 22.4.2)
(a) Vol. **81**, No. 2, p. 625—646 (1984).

H. Kerkow, G. Kreysch, and B. Lukasch
The Influence of Foreign Atoms on the Epitaxial Annealing of Ion-Implanted Silicon
(11; 3; 22.1.2)
(a) Vol. **82**, No. 1, p. 125—133 (1984).

W. Scharff, J.-W. Erben, A. Wolf, M. Heber, C. Hamann, C. Weissmantel, M. Voelskov, J. Matthäi, R. Kögler, R. Klabes, and E. Wieser
*Growth of Single-Crystalline Regions on Amorphous Insulating Substrates by Zone-Melting Recrystallization
(2 and 3; 11; 22.1.2)
(a) Vol. **82**, No. 1, p. K5—K9 (1984).

O. Simmich
Influence of Non-Linear Terms in the Flux Equation on the Particle Growth
(1.4 and 9; 3; 21)
(a) Vol. **82**, No. 1, p. K21—K25 (1984).

I. G. Kaverina, V. V. Korobtsov, V. G. Zavodinskii, and A. V. Zotov
*Solid Phase Epitaxial Growth Anisotropy of Vacuum-Deposited Amorphous Silicon
(1.5 and 3; 2; 22.1.2)
(a) Vol. **82**, No. 2, p. 345—353 (1984).

R. Klabes, J. Matthäi, M. Voelskow, E. Wieser, J.-W. Erben, W. Scharff, and C. Weissmantel
*Epitaxial Regrowth of Amorphous or Polycrystalline Silicon Layers on Silicon Single Crystals and Bridging Epitaxy by Flash Lamp Irradiation
(1.5 and 3; 2; 11; 22.1.2)
(a) Vol. **82**, No. 2, p. K121—K123 (1984).

M. Wagner and G. Götz
*Crystallization of High-Dose Antimony-Implanted Silicon by Millisecond Pulse Laser Irradiation
(1.2 and 3; 22.1.2)
(a) Vol. **83**, No. 1, p. 69—76 (1984).

M. Wagner, H.-D. Geiler, G. Andrä, and G. Götz
*Control of Explosive Liquid-Phase Crystallization of Ion-Implanted Silicon Layers by Double Pulse Laser Irradiation
(3 and 11; 22.1.2)
(a) Vol. **83**, No. 1, p. K1—K3 (1984).

A. Turos, H. Frey, O. Meyer, W. Müller, and J. M. Pirrung
Ion Beam Analysis of Amorphous Silicon Films Produced by Magnetron Sputtering
(2 and 14.3.1; 3; 22.1.2)
(a) Vol. **83**, No. 2, p. 437—443 (1984).

A. Tanaka, T. Sukegawa, and T. Sugiura
*Phosphorus Pressure Effects on GaP Crystal Quality in the Solution Growth
(3; 10.2; 22.2.2)
(a) Vol. **84**, No. 1, p. K9—K12 (1984).

P. Michon, H. Atmani, and C. Vautier
*Crystallization Study of Evaporated a-Se Films by Means of DTA Technique
(1.5 and 3; 2; 22.1.3)
(a) Vol. **85**, No. 2, p. 399—407 (1984).

H. Booyens and J. H. Basson
On the Optimal Choice of Substrate for $Hg_{0.8}Cd_{0.2}Te$ Epitaxial Layers
(1.5 and 10.2; 3; 22.4.4)
(a) Vol. **85**, No. 2, p. 449—454 (1984).

D. Baltrūnas
*A Study of the Possibility of Restoration of Composition of SnTe Oxidized Epitaxial Layers by Means of NGR Spectroscopy
(3 and 5; 22)
(a) Vol. **85**, No. 2, p. K97—K99 (1984).

N. V. Rumak, V. V. Khatko, and V. N. Plotnikov
Structure and Properties of Silicon Dioxide Thermal Films (I): SiO_2 Films of up to 50 nm Thickness
(1.5 and 2; 3; 4; 22.6)
(a) Vol. **86**, No. 1, p. 93—100 (1984).

N. Kameswari, K. V. Ramanujachari, A. Meenakshi Sundaram, and C. S. Swamy
*Synthesis and Structural Characteristics of $La_2Cu_{1-x}Co_xO_4$ ($x = 0.2, 0.4$, and 0.5)
(1.1 and 3; 22.6.1)
(a) Vol. **86**, No. 1, p. K15—K17 (1984).

N. V. Rumak and V. V. Khatko
Structure and Properties of Silicon Dioxide Thermal Films (II): 110 nm Thick SiO_2 Films
(1.5 and 2; 3; 22.6)
(a) Vol. **86**, No. 2, p. 477—484 (1984).

4. Bonding Properties

Review Article

J. Shanker and G. G. Agrawal
*Van der Waals Potentials in Ionic Crystals
(4; 6; 12.2; 22.5.2)
(b) Vol. **123**, No. 1, p. 11—26 (1984).

Original Papers and Short Notes

P. Enders
Structure and Bonding in Cubic IV–VI Crystals (II): Instability of the TO(Γ) Mode: Unified Universal–Parameter Treatment of the Group-V Semimetals and Cubic IV–VI Semiconductors
(1.2; 4; 13; 22)
(b) Vol. **121**, No. 1, p. 39—46 (1984).

K. C. Gupta and R. P. S. Rathore
*Lattice Equilibrium and Phonons in Molybdenum and Tungsten
(4 and 6; 12.1; 21)
(b) Vol. **121**, No. 1, p. 73—79 (1984).

M. C. Böhm
Symmetry Breaking in Organometallic Polymers Due to Interchain Interactions
(13.1; 4; 22.9)
(b) Vol. **121**, No. 1, p. 255—264 (1984).

M. S. Islam and C. Mande
*X-Ray Spectroscopic Study of Re_3W and Re_3Mo Systems
(4; 20; 21)
(a) Vol. **81**, No. 1, p. 197—201 (1984).

V. Valvoda and P. Čapková
*X-Ray Diffraction Study of Charge Distribution in Titanium Carbide
(4; 21.1)
(a) Vol. **81**, No. 1, p. 203—208 (1984).

P. Enders
Structure and Bonding in Cubic IV–VI Crystals (III): On the Cubic to Rhombohedral Phase Transition
(1.2; 4; 6; 22)
(b) Vol. **121**, No. 2, p. 461—470 (1984).

R. P. Singh, J. P. S. Rana, and G. B. Varshney
*The Contribution of Zero-Point Vibration toward the Static and Thermodynamical Properties of Potassium Halides
(4; 12.2; 22.5.2)
(b) Vol. **121**, No. 2, p. 493—500 (1984).

J. SHANKER, J. NARAIN, and A. K. RAJAURIA
*Theory of Compression in Ionic Crystals
(4; 22.5.2)
(b) Vol. **122**, No. 2, p. 435—440 (1984).

S. N. MAITI and D. ROY
*Lattice Dilatation in High-Concentration Binary Solid Solutions
(4; 12.2; 22.5.4)
(b) Vol. **122**, No. 2, p. 449—460 (1984).

S. CHAKRABORTY, A. MANNA, and A. K. GHOSH
*Charge Transfer and Energy of Formation in Binary Alloys: CsCl Intermetallic Compounds
(1.3 and 4; 21.1; 21.1.1; 21.6)
(b) Vol. **122**, No. 2, p. 469—480 (1984).

V. G. TSIRELSON, O. V. KOROLKOVA, I. S. REZ, and R. P. OZEROV
*Extended Model of Bond Charges and Its Application in Calculation of Optical Properties of Crystals with Different Types of Chemical Bonds
(4; 20.1; 22; 22.5.2; 22.6; 22.8.1)
(b) Vol. **122**, No. 2, p. 599—612 (1984).

W. LORENZ and C. ENGLER
Quantum-Theoretical Model of Adiabatic Rate Processes in Condensed Systems or at Interfaces
(13.3; 4; 6; 8; 22)
(b) Vol. **122**, No. 2, p. 745—755 (1984)

J. MINGE and S. WAPLAK
Successive Phase Transitions in $(NH_4)_3H(SO_4)_2$ and $(ND_4)_3D(SO_4)_2$ Crystals Studied by EPR of the VO^{2+} Ion and SeO_3^- Radical
(14.4.1; 4; 19; 22.8.1)
(b) Vol. **123**, No. 1, p. 27—36 (1984).

A. J. KAUR, W. N. BHENDE, P. S. BAKHSHI, and J. SHANKER
*Analysis of the Short-Range Repulsive Potential in Alkali Halides
(4; 6; 8; 22.5.2)
(b) Vol. **123**, No. 1, p. 71—78 (1984).

B. PRASAD
Phonon Frequencies and Binding Energy of F.C.C. Calcium
(6; 4; 21.3)
(b) Vol. **123**, No. 1, p. K9—K11 (1984).

K. B. GARG, J. LEIRO, M. HEINONEN, and M. RICHARDSON
*X-Ray Emission Study of Bonding in SmT_2 Compounds (T ≡ Fe, Co, Ni)
(4 and 13; 20; 21.1; 21.1.1; 21.4)
(a) Vol. **83**, No. 1, p. 279—284 (1984).

L. ČERVINKA, V. CHÁB, and S. KOC
*Erratum to: XPS Investigation of the Amorphous $TlSe_x$ System
(2 and 4; 17; 22.3)
(a) Vol. **83**, No. 1, p. 418 (1984).

Chia-Nan-Chang
The Lattice Dynamics of Solid HCl and HBr in Their Low-Temperature Phase
(6; 4; 22.5)
(b) Vol. **123**, No. 2, p. 435—444 (1984).

R. K. Singh, Neeta Agnihotri, S. V. Suryanarayana, and S. P. Sanyal
*Study of Cohesion, Harmonic and Anharmonic Elastic Properties of CaF_2–SrF_2 Mixed Crystals
(4; 12.2; 22.5.4)
(b) Vol. **123**, No. 2, p. 453—462 (1984).

D. W. Field
*Electron Transfer and Thermal Vibration Parameters in Titanium Nitride: An X-Ray Diffraction Study
(4; 6; 22)
(b) Vol. **123**, No. 2, p. 479—483 (1984).

J. A. Alonso, D. J. González, and M. P. Iñiguez
*Role of the Excess Volume of Formation on Alloying
(4; 1.3; 21.2; 21.3)
(b) Vol. **123**, No. 2, p. 485—489 (1984).

S. K. Agrawal, J. Narain, and J. Shanker
*Investigation of the Equation of State and the Grüneisen Parameters for Transition Metal Oxides
(4; 6; 12.2; 22.6)
(b) Vol. **123**, No. 2, p. 497—501 (1984).

R. P. Goyal and B. R. K. Gupta
*Evaluation of Ionic Compressibilities and Static Polarizabilities of Cations and Anions of Ammonium Halides
(4; 22.5)
(b) Vol. **123**, No. 2, p. K109—K114 (1984).

R. V. Gopala Rao and Sobhen Kumar Pal
*Molten Rubidium Bromide — A Rigorous Study of Its Partial Structure and Allied Functions
(2 and 4; 22.5.2)
(a) Vol. **83**, No. 2, p. 429—436 (1984).

R. K. Singh and P. Khare
*Analysis of Anharmonic Properties of Some Partially Covalent Crystals
(4 and 12.2; 22.5; 22.5.2)
(b) Vol. **124**, No. 1, p. 29—36 (1984).

T. Soma, Y. Takahashi, and H.-Matsuo Kagaya
*Equation of State and Pressure Derivative of the Bulk Modulus for III–V Covalent Crystals
(1.2 and 4; 12.2; 22.2.1; 22.2.3)
(b) Vol. **124**, No. 1, p. K11—K14 (1984).

P. Enders
*On the Dielectric Theory: Does the Static Ionic Charge Scale with $f_i^{1/2}$?
(4; 22)
(b) Vol. **124**, No. 1, p. K21—K23 (1984).

U. WALZER
*Elastic and Bonding Properties of Transition Metals
(4; 12.1; 13.1; 21.1)
(b) Vol. **125**, No. 1, p. 55—64 (1984).
Erratum
(b) Vol. **126**, No. 2, p. 752 (1984).

B. R. GELCHINSKII and R. M. RYSKULBEKOVA
*On Correlation between Structural Features of Liquid Metals and the Effective Ion–Ion Potential
(2 and 4; 21.2; 21.7)
(b) Vol. **125**, No. 1, p. 81—86 (1984).

S. CHAKRABORTY, A. MANNA, and A. K. GHOSH
*Charge Transfer and Energy of Formation in Ordered Binary Alloys
(4 and 13.1; 21.1; 21.6)
(b) Vol. **125**, No. 1, p. 97—100 (1984).

A. PASTUREL, C. COLINET, C. ALLIBERT, P. HICTER, A. PERCHERON-GUEGAN, and J. C. ACHARD
*A Theoretical and Experimental Study of the Enthalpies of Formation of $LaNi_5$-Type Compounds
(4 and 13.1; 21.1; 21.4)
(b) Vol. **125**, No. 1, p. 101—106 (1984).

N. AGNIHOTRI, R. K. SINGH, and S. P. SANYAL
*Interionic Forces and Elastic Properties of Na_2S
(4; 6; 12.2; 22; 22.5.3)
(b) Vol. **125**, No. 1, p. 113—119 (1984).

W. KULLMANN, J. GEURTS, W. RICHTER, N. LEHNER, H. RAUH, U. STEIGENBERGER, G. EICHHORN, and R. GEICK
*Effect of Hydrostatic and Uniaxial Pressure on Structural Properties and Raman Active Lattice Vibrations in Bi_2Te_3
(4; 1.1; 6; 20.1; 22.7)
(b) Vol. **125**, No. 1, p. 131—138 (1984).

B. LORENZ
*Charge Ordering in the Anti-Th_3P_4 Structure
(4; 1.2; 22)
(b) Vol. **125**, No. 1, p. 375—379 (1984).

V. P. GUPTA and O. P. SINGH
*Phillips Ionicity and Structural Classification of Layered Transition Metal Dichalcogenides
(1 and 4; 22)
(b) Vol. **125**, No. 1, p. K1—K5 (1984).

J. SHANKER, W. N. BHENDE, and P. S. BAKHSHI
*Analysis of the Equation of State for LiF Crystal
(4; 12.2; 22.5.2)
(b) Vol. **125**, No. 1, p. K21—K26 (1984).

I. M. Antonova
*On the Effective Potential for the Anion in MgO and CaO Crystals
(4 and 13; 22.6)
(b) Vol. **125**, No. 1, p. K55—K58 (1984).

W. Burckhardt, P. Bussemer, and U. Dietrich
*Structure and Molar Refraction and Its Wavelength Dependence at Different Alkali and Ammonium Halides
(4 and 20.1; 22.5; 22.5.2)
(a) Vol. **85**, No. 1, p. 97—103 (1984).

Ch. Opitz, H. Müller, and H. Miessner
*Bonding Mechanism for Interstitial Hydrogen in Nickel
(4; 10.1; 13.4; 21.1)
(b) Vol. **125**, No. 2, p. K93—K97 (1984).

J. Narain, G. G. Agrawal, and J. Shanker
*Compressibility of Rare-Gas Solids under the Effect of High Pressure
(4 and 12.2; 23)
(b) Vol. **126**, No. 1, p. 83—86 (1984).

D. Della Sala, C. Giovannella, and F. Evangelisti
Study of Bonding Configurations in Amorphous Ge_xSi_{1-x}:H Alloys
(2; 4; 20.1; 22.1.1; 22.1.2)
(b) Vol. **126**, No. 1, p. 125—131 (1984).

K. Čermák
*The Width of the Energy Gap of $Tl_3^IB^VC_4^{VI}$-Type Compounds
(4 and 13.1; 20.1; 22.8)
(b) Vol. **126**, No. 1, p. 273—277 (1984).

R. K. Singh and D. K. Neb
*Mean Square Displacements of Rare Gas Solids
(4 and 6; 8; 23)
(b) Vol. **126**, No. 1, p. K15—K18 (1984).

R. V. G. Rao and S. Kumar Pal
A New Method of Evaluation of the Structure of Molten Sodium Chloride
(2; 4; 22.5.2)
(a) Vol. **86**, No. 1, p. 43—54 (1983).

N. V. Rumak, V. V. Khatko, and V. N. Plotnikov
Structure and Properties of Silicon Dioxide Thermal Films (I): SiO_2 Films of up to 50 nm Thickness
(1.5 and 2; 3; 4; 22.6)
(a) Vol. **86**, No. 1, p. 93—100 (1984).

Shu Zhen and G. J. Davies
*Calculation of Potential Energy Parameters for Some Intermetallic Compounds
(1.3 and 4; 22.7)
(a) Vol. **86**, No. 1, p. 155—161 (1984).

R. P. Singh, J. P. S. Rana, A. Daud, and G. B. Varshney
*The Role of the Zero-Point Vibration towards Crystal Properties of Rubidium Halide Crystals
(4; 12.2; 22.5.2)
(b) Vol. **126**, No. 2, p. 495—499 (1984).

B. Holland, H. S. Greenside, and M. Schlüter
*Cohesive Properties of Ge, Si, and Diamond Calculated with Minimum Basis Sets
(4 and 13; 12.2; 22.1; 22.1.1; 22.1.2)
(b) Vol. **126**, No. 2, p. 511—515 (1984).

U. Pietsch
*Simulation of Dynamical Bond Charge Transfer Properties in Elemental Semiconductors by Means of a Simple Quantum-Mechanical Model
(4; 13; 22.1.2)
(b) Vol. **126** , No. 2, p. 595—605 (1984).

G. Berger and M. Giehler
Characterization of in Vitro Corroded Surfaces of Bioactive Glasses (P_2O_5–CaO–SiO_2) with Infrared Reflection Spectroscopy
(2 and 20.1; 4; 6; 22.6.1)
(a) Vol. **86**, No. 2, p. 531—542 (1984).

5. Mössbauer Spectroscopy

Original Papers and Short Notes

K. K. P. Srivastava and S. N. Mishra
*Spin Relaxation Consideration in Mössbauer Spectra of $FeSiF_6 \cdot 6\,H_2O$
(5; 22.8)
(b) Vol. **121**, No. 1, p. K83–K86 (1984).

W. Bergholz, S. Damgaard, J. W. Petersen, and G. Weyer
Reactions of Cobalt in Silicon with Electron-Irradiation-Induced Defects
(10.2; 5; 22.1.2)
(a) Vol. **81**, No. 1, p. 239–246 (1984).

M. Pasemann, W. Bergholz, and W. Schröter
Mössbauer Spectroscopical and Electron Microscopical Investigations of the Behaviour of Cobalt in Silicon
(1.4 and 10.2; 5; 22.1.2)
(a) Vol. **81**, No. 1, p. 273–280 (1984).

V. P. Gupta, A. K. Singh, K. Chandra, and S. K. Jaireth
Mössbauer Investigations on Pyrrhotite
(10.2; 5; 18.3; 18.4; 22.8)
(a) Vol. **81**, No. 1, p. 281–291 (1984).

K. Sumiyama and Y. Nakamura
High Concentration Fe–Cu and Fe–Ag Alloys Produced by Vapor Quenching
(1.3; 5; 18.2; 21.1; 21.1.1; 21.6)
(a) Vol. **81**, No. 2, p. K109–K112 (1984).
Erratum
(a) Vol. **83**, No. 2, p. K219 (1984).

T. Tsuji, K. Naito, and K. Ishigure
*Effect of Particle Size on Mössbauer Parameters of α-Fe_2O_3
(5 and 18; 22.6)
(a) Vol. **82**, No. 1, p. K57–K61 (1984).

V. D. Sedykh and A. V. Serebryakov
The Effect of Thermal Treatment on the Magnetic Anisotropy in Amorphous Fe–B Alloys
(2 and 18.2; 5; 21.1.1)
(a) Vol. **82**, No. 1, p. K63–K66 (1984).

P. Orlandi and A. Rigamonti
On the Phase Transition and Crystal Field Effects in $FeSiF_6 \cdot 6\,H_2O$
(1.2 and 13.4; 5; 22.8)
(a) Vol. **82**, No. 2, p. K129–K132 (1984).

S. K. Roy and D. L. Bhattacharya
Debye-Waller Factor for Small Concentration of Mössbauer Impurities in Metallic Solids
(6; 5; 10.1; 21)
(b) Vol. **123**, No. 1, p. 45–51 (1984).

A. YA. DZYUBLIK
*Effect of Forced Vibrations on the Scattering of X-Rays and Mössbauer Radiation by a Crystal (I)
(5 and 6)
(b) Vol. **123**, No. 1, p. 53—64 (1984).

I. N. GROZNOV, S. P. IONOV, I. B. KEVDINA, V. S. LUBIMOV, E. F. MAKAROV, O. V. MARCHENKO, N. N. MIKHAILOV, V. A. ONISHCHUK, and V. P. SHANTAROVICH
Mössbauer and Positron Studies of Superconduction Oxides $BaPb_{1-x}Bi_xO_3$
(14.2; 5; 22.6.1)
(b) Vol. **123**, No. 1, p. 183—190 (1984).

D. G. CHAMBAERE and E. DE GRAVE
A Study of the Non-Stoichiometrical Halogen and Water Content of β-FeOOH
(1; 5; 22)
(a) Vol. **83**, No. 1, p. 93—102 (1984).

E. K. SADYKOV
*The Theory of Double Gamma Resonance in Paramagnetics
(5 and 19; 18.1)
(b) Vol. **123**, No. 2, p. 703—709 (1984).

L. CSER, I. GLADKIH, and C. HARGITAI
Impurity Charge Screening in 3d Alloys
(13.4; 5; 21.1.1)
(b) Vol. **124**, No. 1, p. 271—278 (1984).

S. A. FYSH and P. E. CLARK
On Mössbauer Analysis of Mineral Mixtures Having Environmentally Broadened Spectral Lines. Application to Bauxite
(1.3; 5; 22.6)
(a) Vol. **84**, No. 1, p. 31—38 (1984).

W. WINKLER
Mössbauer Studies of Small Magnetite Particles of a Magnetic Fluid
(18.2; 5; 22.6)
(a) Vol. **84**, No. 1, p. 193—198 (1984).

B. G. TAGIEV, G. D. SULTANOV, G. M. NIFTIEV, I. M. ASKEROV, and G. K. ASLANOV
*The Mössbauer Effect and Magnetic Susceptibilities in $Ga_2Se_3:{}^{57}Fe$ Single Crystals
(5 and 18; 22.3)
(a) Vol. **84**, No. 1, p. K59—K62 (1984).

R. DOERFLER, D. L. NAGY, H.-D. PFANNES, A. PUTZKA, G. RITTER, and N. ZEMAN
Non-Equilibrium Population of Fe^{3+} Electronic States after the Electron Capture (I): ^{57}Co in $LiNbO_3$ and $LiTaO_3$
(13.4; 5; 22.8.1)
(b) Vol. **124**, No. 2, p. 767—772 (1984).

PRAN KISHAN, CHANDRA PRAKASH, J. S. BAIJAL, and K. K. LAROIA
*Mössbauer Studies on Hyperfine Interactions in Titanium Substituted Lithium Ferrites
(5; 22.8.2)
(a) Vol. **84**, No. 2, p. 535—540 (1984).

A. R. MKRTCHYAN, A. R. ARAKELYAN, R. G. GABRIELYAN, L. A. KOCHARYAN, G. R. GRIGORYAN, and M. M. SLAVINSKII
*Time Dependence of Resonance γ-Radiation Modulated by Acoustic Excitations. Experiment
(5)
(b) Vol. **125**, No. 1, p. K67—K70 (1984).

M. A. ANDREEVA and R. N. KUZMIN
*Mössbauer Conversion Electron Spectra at Total Reflection
(5; 1.6; 13.3; 21.1)
(b) Vol. **125**, No. 2, p. 461—465 (1984).

F. SCHNEIDER and S. UNTERRICKER
Radiation Damage and Its Annealing Behaviour in InP after Recoil Implantation of ^{118}Sb and ^{111}In Observed by TDPAC
(10.2 and 11; 5; 22.2.2)
(a) Vol. **85**, No. 2, p. 455—461 (1984).

D. BALTRŪNAS
*A Study of the Possibility of Restoration of Composition of SnTe Oxidized Epitaxial Layers by Means of NGR Spectroscopy
(3 and 5; 22)
(a) Vol. **85**, No. 2, p. K97—K99 (1984).

C. I. NISTOR, V. SPÂNU, S. MANDACHE, G. FILOTI, and V. FLORESCU
*Hyperfine Field Distribution in Metglasses
(5; 2; 21.1.1)
(a) Vol. **86**, No. 2, p. 525—529 (1984).

H. M. VAN NOORT, D. B. DE MOOIJ, and K. H. J. BUSCHOW
Crystal Structure, Magnetic Properties, and ^{57}Fe Mössbauer Effect of PtFeSn
(1 and 18.2; 5; 22.8)
(a) Vol. **86**, No. 2, p. 655—662 (1984).

T. PANNAPARAYIL, M. OSKOOIE-TABRIZI, C. LO, and L. N. MULAY
*Infrared Spectroscopic, Magnetic, and Mössbauer Studies of Aluminosilicate Cage Structures (Fe/H-Mordenites) for Syngas Conversion
(5; 18.1; 20.1; 22.8.1)
(a) Vol. **86**, No. 2, p. 663—671 (1984).

6. Lattice Dynamics. Phonons

Review Articles

B. M. ASKEROV, B. I. KULIEV, and S. R. FIGAROVA
Electron Transport Phenomena in Conductive Films
(14.3.1; 6; 13.1; 15; 22)
(b) Vol. **121**, No. 1, p. 11—37 (1984).

R. RAMJI RAO and A. RAMANAND
Thermal and Elastic Properties of Rare-Earth Metals
(8 and 12.1; 6; 21.4)
(b) Vol. **122**, No. 1, p. 11—27 (1984).

J. SHANKER and G. G. AGRAWAL
Van der Waals Potentials in Ionic Crystals
(4; 6; 12.2; 22.5.2)
(b) Vol. **123**, No. 1, p. 11—26 (1984).

K. N. SHRIVASTAVA
*Origin of Superexchange Interactions in 3d Magnetic Solids (I)
(6 and 18.2; 13; 22.5)
(b) Vol. **125**, No. 1, p. 11—43 (1984).

K. N. SHRIVASTAVA
*Origin of Superexchange Interactions in 3d Magnetic Solids (II)
(6 and 18.2; 13; 22.5)
(b) Vol. **125**, No. 2, p. 441—460 (1984).

A. BLACHA, H. PRESTING, and M. CARDONA
*Deformation Potentials of $\boldsymbol{k} = \boldsymbol{0}$ States of Tetrahedral Semiconductors
(6 and 13.1; 22.1; 22.2; 22.4)
(b) Vol. **126**, No. 1, p. 11—36 (1984).

Original Papers and Short Notes

B. GAVEAU, J. ŁAWRYNOWICZ, and L. WOJTCZAK
*Statistical Mechanics of Anharmonic Crystals
(6)
(b) Vol. **121**, No. 1, p. 47—58 (1984).

K. C. GUPTA and R. P. S. RATHORE
*Lattice Equilibrium and Phonons in Molybdenum and Tungsten
(4 and 6; 12.1; 21)
(b) Vol. **121**, No. 1, p. 73—79 (1984).

J. OTTO and F. FISCHER
*FIR Studies of OCN^-, SCN^-, and $SeCN^-$ Centre Gap Modes in KI Single Crystals
(6 and 20.1; 22.5.2)
(b) Vol. **121**, No. 1, p. 81—90 (1984).

H. BÖTTGER and D. WEGENER
On the Percolation Approach to the Hopping Conduction in Strong Electrical Fields
(2 and 14.3; 6; 22)
(b) Vol. **121**, No. 1, p. 413—419 (1984).

V. ČÁPEK
*Initial Condition Term and Method of Artificial Temperature for Generalized Master Equations
(6 and 13; 22)
(b) Vol. **121**, No. 1, p. K7—K12 (1984).

V. A. HAISLER, V. M. ZALETIN, A. F. KRAVCHENKO, and G. Y. YASHIN
*Raman Scattering of First and Second Order in Red Mercury Iodide
(6 and 20.1; 22.5)
(b) Vol. **121**, No. 1, p. K13—K17 (1984).

M. HIDAKA, S. YAMASHITA, and Y. OKAMOTO
Study of Structural Phase Transitions of $KCaF_3$
(1.2; 6; 8; 22.5.4)
(a) Vol. **81**, No. 1, p. 177—183 (1984).

C. JÄGER and K. HEIDE
*Temperature Dependence of the ^{23}Na NMR Central Lineshape in $Na_2Ca(CO_3)_2$
(6 and 19; 22.8.1)
(a) Vol. **81**, No. 1, p. K7—K10 (1984).

P. ENDERS
Structure and Bonding in Cubic IV-VI Crystals (III): On the Cubic to Rhombohedral Phase Transition
(1.2; 4; 6; 22)
(b) Vol. **121**, No. 2, p. 461—470 (1984).

L. MERTEN and W. GOLTSCHE
*Stimulated Raman Effect on Plasmons and Longitudinal Plasmon-Phonons in Simple Cubic Crystals
(6 and 20; 13.5.2; 22)
(b) Vol. **121**, No. 2, p. 471—480 (1984).

I. P. MAKAROVA, S. V. MISYUL, L. A. MURADYAN, A. F. BOVINA, V. I. SIMONOV, and K. S. ALEKSANDROV
*Anharmonic Thermal Atomic Vibrations in the Cubic Phase of $Cs_2NaNdCl_6$ Single Crystals
(1.2 and 6; 22.8)
(b) Vol. **121**, No. 2, p. 481—486 (1984).

T. SOMA, Y. KITANI, and H.-MATSUO KAGAYA
*Lattice Dynamics and Thermal Properties of Si–Ge Solid Solution
(6; 8; 22.3)
(b) Vol. **121**, No. 2, p. 501—504 (1984).

A. LEHMANN, L. SCHUMANN, and K. HÜBNER
*Optical Phonons in Amorphous Silicon Oxides (II): Calculation of Phonon Spectra and Interpretation of the IR Transmission of SiO_x
(6 and 20.1; 2; 22.6)
(b) Vol. **121**, No. 2, p. 505—511 (1984).

S. SIMONS
*Lattice Thermal Conduction in the Absence of Three-Phonon Interactions
(6 and 8)
(b) Vol. **121**, No. 2, p. K109—K112 (1984).

H.-M. KAGAYA and T. SOMA
*Lattice Dynamics of InP and ZnTe
(6; 22.2.2; 22.4.3)
(b) Vol. **121**, No. 2, p. K113—K118 (1984).

L. WENDLER and E. JÄGER
*Surface Polaritons in a SiO_2/Si Bilayer System
(6 and 13.5; 22.1.2; 22.6)
(b) Vol. **121**, No. 2, p. K119—K124 (1984).

A. A. ALYBAKOV, V. A. GUBANOVA, M. ZHANIBEKOV, and N. TOICHEV
Formation and Stability of Paramagnetic Centers in KBr:CrO_4^{2-}, Ca^{2+} Crystals
(19 and 20.1; 6; 13.4; 22.5.2)
(b) Vol. **121**, No. 2, p. K205—K208 (1984).

G. J. ASHWELL and C. M. CARE
*Optical Phonon Assisted Hopping in (DPPE) $(TCNQ)_5(H_2O)$
(6 and 14; 22.9)
(a) Vol. **81**, No. 2, p. 557—561 (1984).

F. CERDEIRA, V. LEMOS, F. E. A. MELO, and M. CARDONA
*Effective Charges and Their Pressure Dependence in α-$LiIO_3$
(6 and 20.1; 22.8.1)
(b) Vol. **122**, No. 1, p. 53—64 (1984).

B. DAUDIN, W. MUTSCHELLER, and M. WAGNER
*Heat Conductivity of the Cooperative Vibronic System $TmVO_4$
(6 and 8; 22.8.1)
(b) Vol. **122**, No. 1, p. 73—79 (1984).

P. KLEINERT
*Coherent-Potential Approximation for the Lattice Vibrations of Mixed III–V Crystals
(6; 20.1; 22.2.4)
(b) Vol. **122**, No. 1, p. 81—89 (1984).

U. SLIWCZUK, H. STOLZ, and W. VON DER OSTEN
Indirect-Forbidden Exciton Transitions in AgBr
(13.5.1; 6; 20.1; 22.5.1)
(b) Vol. **122**, No. 1, p. 203—209 (1984).

YU. B. GAIDIDEI and A. A. VAKHNENKO
Nonequilibrium Kinetics of Exciton Wave Packets in Crystals
(13.5.1; 6; 22)
(b) Vol. **122**, No. 1, p. 239—247 (1984).

J. E. GRIFFITHS, J. C. PHILLIPS, G. P. ESPINOSA, J. P. REMEIKA, and P. M. BRIDENBAUGH
*Assignment of the Companion A_1^c Line in $Ge_x(S, Se)_{1-x}$ Glasses
(6 and 20.1; 2; 22)
(b) Vol. **122**, No. 1, p. K11—K15 (1984).

A. KRÓL, W. NAZAREWICZ, and M. J. KOZIELSKI
Al_{Zn}–V_{Zn} Pairs in Wurtzite-Type ZnS
(10.2 and 20.1; 6; 22.4.1)
(b) Vol. **122**, No 1, p. K23—K25 (1984).

P. P. LOTTICI
*On the Optic Vibrational Modes in Amorphous Arsenic
(6 and 20.1; 2; 22.1)
(b) Vol. **122**, No. 2, p. 431—434 (1984).

K. AOKI and M. CARDONA
*Effect of Pressure on the Raman Scattering Efficiency of GaP
(6 and 20.1; 22.2.2)
(b) Vol. **122**, No. 2, p. 441—447 (1984).

C. PATEL, T. J. PARKER, H. JAMSHIDI, and W. F. SHERMAN
*Phonon Frequencies in GaAs
(6 and 20.1; 22.2.1)
(b) Vol. **122**, No. 2, p. 461—467 (1984).

V. I. BELITSKY, A. V. GOLTSEV, I. G. LANG, and S. T. PAVLOV
*Two-Phonon Resonance Raman Scattering and Quantum Interference Effects in Polar Semiconductors in High Magnetic Fields
(6 and 20.1; 22)
(b) Vol. **122**, No. 2, p. 581—590 (1984).

B. GIESE
Electron Scattering Interaction and Elementary Excitations in Lead Chalcogenides
(13.1 and 13.5; 6; 22)
(b) Vol. **122**, No. 2, p. 641—650 (1984).

W. LORENZ and C. ENGLER
Quantum-Theoretical Model of Adiabatic Rate Processes in Condensed Systems or at Interfaces
(13.3; 4; 6; 8; 22)
(b) Vol. **122**, No. 2, p. 745—755 (1984).

V. M. POLYANOVSKII
*Stark Cyclotron-Phonon Oscillations of Sound Absorption in Semiconductors with Superlattice
(6 and 7; 19; 22)
(b) Vol. **122**, No. 2, p. K115—K118 (1984).

H. C. SAXENA and R. P. S. RATHORE
*Lattice Vibrations in Hafnium
(6; 21)
(b) Vol. **122**, No. 2, p. K119—K124 (1984).

T. Soma, Y. Kitani, and H.-Matsuo Kagaya
*Local and Band Mode Frequencies of $Si_{1-x}Ge_x$ Solid Solution
(6; 22.1.1; 22.1.2)
(b) Vol. **122**, No. 2, p. K125—K126 (1984).

F. Kh. Abdullaev and A. A. Abdumalikov
Dynamics of Magnetic Solitons in Planar Ferromagnets with Random Anisotropy
(13.5 and 18.2; 6; 13.5.1)
(b) Vol. **122**, No. 2, p. K141—K146 (1984).

M. Krzesińska and T. Szuta-Buchacz
Hypersonic Wave Attenuation in Yttrium Aluminium Garnet and Gadolinium Gallium Garnet
(7; 6; 12.2; 22.8.2)
(a) Vol. **82**, No. 2, p. 421—424 (1984).

S. K. Roy and D. L. Bhattacharya
*Debye-Waller Factor for Small Concentration of Mössbauer Impurities in Metallic Solids
(6; 5; 10.1; 21)
(b) Vol. **123**, No. 1, p. 45—51 (1984).

A. Ya. Dzyublik
*Effect of Forced Vibrations on the Scattering of X-Rays and Mössbauer Radiation by a Crystal (I)
(5 and 6)
(b) Vol. **123**, No. 1, p. 53—64 (1984).

G. Jungk and A. Röseler
*Ellipsometric Investigation of Lattice Vibrations in Anisotropic Systems. α-Quartz
(6 and 20.1; 22.6)
(b) Vol. **123**, No. 1, p. 65—70 (1984).

A. J. Kaur, W. N. Bhende, P. S. Bakhshi, and J. Shanker
Analysis of the Short-Range Repulsive Potential in Alkali Halides
(4; 6; 8; 22.5.2)
(b) Vol. **123**, No. 1, p. 71—78 (1984).

V. V. Krasnogolovets and P. M. Tomchuk
On the Proton Conductivity Mechanism in Systems Containing Hydrogen-Bonded Chains
(14.4; 6; 22.8.1)
(b) Vol. **123**, No. 2, p. 365—376 (1984).

S. A. Mazen, N. A. Hakeem, and B. A. Sabrah
*The IR Absorption Spectra of $Cu_{1-x}Cd_xFe_2O_4$
(6 and 20.1; 22.8.2)
(b) Vol. **123**, No. 1, p. K1—K4 (1984).

B. Prasad
*Phonon Frequencies and Binding Energy of F.C.C. Calcium
(6; 4; 21.3)
(b) Vol. **123**, No. 1, p. K9—K11 (1984).

M. Pagannone, B. Fornari, G. Mattei, and L. Mattioli
*Radiative (Brewster-Type) Surface Mode Dispersion in Sodium Chlorate
(6 and 20.1; 13.5; 22.8.1)
(b) Vol. **123**, No. 1, p. K13—K16 (1984).

A. Król and R. Zielińska
*A New Defect Complex in CdTe: In, Li
(6 and 10.2; 20.1; 22.4.3)
(b) Vol. **123**, No. 1, p. K17—K20 (1984).

K. R. Zhdanov, A. V. Mishenko, F. S. Rakhmenkulov, and V. E. Fedorov
*Structural Anisotropy and Heat Capacity of NbS_3
(6 and 8; 22)
(a) Vol. **83**, No. 1, p. 147—152 (1984).

A. L. Shabalov and M. S. Feldman
*Optical Properties of Thin GeO_x Films
(6 and 20.1; 22.6)
(a) Vol. **83**, No. 1, p. K11—K14 (1984).

E. Culea, Al. Nicula, and I. Bratu
*An Infrared Study of $xV_2O_5 \cdot (1 - x) B_2O_3$ Glasses
(6 and 20.1; 2; 22.6.1)
(a) Vol. **83**, No. 1, p. K15—K18 (1984).

G. Dharmadurai
*Analysis of Photoacoustically Investigated Gaseous Heat Transfer between Solid Surfaces
(6 and 8; 1.6)
(a) Vol. **83**, No. 1, p. K27—K30 (1984).

Chia-Nan Chang
*The Lattice Dynamics of Solid HCl and HBr in Their Low-Temperature Phase
(6; 4; 22.5)
(b) Vol. **122**, No. 2, p. 435—444 (1984).

S. Ghosh, S. Chatterjee, and A. N. Basu
*Phase Transition and Soft Mode Behaviour in Superionic Copper Halides
(1.2 and 6; 22.5)
(b) Vol. **123**, No. 2, p. 445—451 (1984).

T. Soma, T. Itoh, and H.-Matsuo Kagaya
*Lattice Dynamics of Al
(6; 8; 21)
(b) Vol. **123**, No. 2, p. 463—467 (1984).

L. Wendler
*Phonon-Polaritons in Bilayer Systems. p-Polarized Guided Wave Phonon-Polaritons and Guided Wave Surface Phonon-Polaritons
(6 and 20.1; 13.5; 22.2.1; 22.2.2; 22.5.3)
(b) Vol. **123**, No. 2, p. 469—477 (1984).

D. W. Field
Electron Transfer and Thermal Vibration Parameters in Titanium Nitride: An X-Ray Diffraction Study
(4; 6; 22)
(b) Vol. **123**, No. 2, p. 479—483 (1984).

B. B. S. Negi and P. N. Ram
First-Order Raman Scattering in RbCl Containing F-Centres
(10.2; 6; 20.1; 22.5.2)
(b) Vol. **123**, No. 2, p. 491—495 (1984).

S. K. Agrawal, J. Narain, and J. Shanker
Investigation of the Equation of State and the Grüneisen Parameters for Transition Metal Oxides
(4; 6; 12.2; 22.6)
(b) Vol. **123**, No. 2, p. 497—501 (1984).

S. P. Gwet, Y. Mathey, and C. Sourisseau
*The Infrared, Raman, Resonance Raman Spectra, and the Valence Force Field of the HfS_3 Layer-Type Compound
(6 and 20.1; 22)
(b) Vol. **123**, No. 2, p. 503—517 (1984).

J. M. Wesselinowa
Electron–Phonon Interaction in Ferromagnetic Semiconductors
(18.2; 6; 13; 22)
(b) Vol. **123**, No. 2, p. 585—593 (1984).

J. Loos
*On the Fluctuations and Phase Transitions in Dicke-Like Models
(6 and 18.3)
(b) Vol. **123**, No. 2, p. 595—603 (1984).

N. I. Likholit, H.-E. Ponath, V. L. Strizhevskii, and Yu. N. Yashkir
Transient Coherent Anti-Stokes Raman Spectroscopy of Surface Polaritons
(13.5 and 20.1; 6; 22)
(b) Vol. **123**, No. 2, p. 691—701 (1984).

W. Wojdowski
*Infrared Spectra of Thin Films of $Bi_{12}GeO_{20}$ and $Bi_{12}SiO_{20}$
(6 and 20.1; 22.8)
(b) Vol. **123**, No. 2, p. K101—104 (1984).

Sh. M. Efendiev
*Raman Spectra in $Bi_{12}TiO_{20}$ Single Crystals
(6 and 20.1; 22.8)
(b) Vol. **123**, No. 2, p. K105—K108 (1984).

A. V. Belushkin, I. Natkaniec, J. Wąsicki, and T. Zaleski
*The Phonon Density of States near the Structural Phase Transition of Solid Biphenyl
(1.2 and 6; 22.9)
(b) Vol. **123**, No. 2, p. K115—K118 (1984).

V. Riede, H. Neumann, H. Sobotta, W. Kissinger, M. S. Omar, and B. R. Pamplin
*Infrared Study of Lattice and Free Carrier Effects in p-Type $CuGe_2P_3$
(6 and 20.1; 13.1; 22.8)
(b) Vol. **123**, No. 2, p. K119—K123 (1984).

V. Valvoda
*X-Ray Debye Temperatures of HfC and α-W_2C
(6 and 8; 21)
(a) Vol. **83**, No. 2, p. K123—K125 (1984).

H.-Matsuo Kagaya and T. Soma
*Lattice Dynamics of Tetrahedral Compounds with the Local Heine-Abarenkov Model Potential
(6; 22.2; 22.4)
(b) Vol. **124**, No. 1, p. 37—44 (1984).

H. Reimer and F. Fischer
*SERS from Ag Colloids in Alkali Halide Crystals
(6 and 20.1; 10.2; 22.5.2)
(b) Vol. **124**, No. 1, p. 61—67 (1984).

C. M. Weinert and U. Scherz
*Calculation of the Jahn-Teller Coupling Forces for $ZnS:Cu^{2+}$ and $GaP:Ni^{+}$
(6 and 20.1; 13.4; 22.2.2; 22.4.1)
(b) Vol. **124**, No. 1, p. 287—296 (1984).

G. Jungk and A. Röseler
*LO–TO-Splitting and Many-Particle Effects
(6 and 20.1; 2; 22.6)
(b) Vol. **124**, No. 1, p. K1—K4 (1984).

H. Englisch
*The Density of Low Frequencies in a Disordered Chain
(6)
(b) Vol. **124**, No. 1, p. K15—K19 (1984).

H. Neumann, P. Deus, R. D. Tomlinson, G. Kühn, and B. Hintze
*Thermal Expansion of $CuInTe_2$ from 30 to 300 K
(6 and 8; 22.8)
(a) Vol. **84**, No. 1, p. 87—93 (1984).

R. Khanna
Separation of Thermal Component in Diffuse Scattering from Disordered Binary Alloys
(1.3; 6; 21)
(a) Vol. **84**, No. 1, p. 95—101 (1984).

D. Senulienė, G. Babonas, E. I. Leonov, I. Muminov, and V. M. Orlov
*Optical Properties of $Bi_{12}Si_{1-x}Ti_xO_{20}$ Single Crystals
(6 and 20.1; 22.8)
(a) Vol. **84**, No. 1, p. 113—117 (1984).

E. CULEA, AL. NICULA, and I. BRATU
*An Infrared Study of $V_2O_5 \cdot As_2O_3$ Glasses
(6 and 20.1; 2; 22.6.1)
(a) Vol. **84**, No. 1, p. K17—K21 (1984).

K. WINER and F. WOOTEN
*Structure and Vibrational Spectra of a Model of a-Si:H with Periodic Boundary Conditions
(2 and 6; 22.1.2)
(b) Vol. **124**, No. 2, p. 473—480 (1984).

R. SUDHARSANAN, T. K. K. SRINIVASAN, and S. RADHAKRISHNA
*Raman and Infrared Studies on Ag_2HgI_4 and Cu_2HgI_4 Superionic Compounds
(6 and 20.1; 22.5.4)
(b) Vol. **124**, No. 2, p. 481—487 (1984)

M. S. PATHAK, S. KHAN, and S. GHOSH
*Longitudinal Phonon–Plasmon Interactions in Semiconducting Plasmas
(6 and 13.5.2; 7; 22.2.3)
(b) Vol. **124**, No. 2, p. 489—496 (1984).

B. PRASAD
*First-Principles Thermal Expansion in Rb and Cs
(6 and 8; 21.2)
(b) Vol. **124**, No. 2, p. 497—501 (1984).

F. PROKERT and K. S. ALEKSANDROV
*Neutron Scattering Studies on Phase Transition and Phonon Dispersion in $Cs_2NaBiCl_6$
(1.2 and 6; 22.5)
(b) Vol. **124**, No. 2, p. 503—513 (1984).

T. SOMA, H. OHSUGI, and H.-MATSUO KAGAYA
*Phonon Dispersion Curves of Cs–K, Cs–Rb, and Rb–K Alloys
(6; 8; 21.2)
(b) Vol. **124**, No. 2, p. 525—529 (1984).

B. A. LUKIYANETS, K. D. TOVSTYUK, V. V. BOICHUK, and K. K. DANILEVICH-TOVSTYUK
*The Jahn-Teller Effect in Layer Crystals
(6 and 13; 22.3)
(b) Vol. **124**, No. 2, p. 693—698 (1984).

K. KAMBAS, C. JULIEN, M. JOUANNE, A. LIKFORMAN, and M. GUITTARD
*Raman Spectra of α- and γ-In_2Se_3
(6 and 20.1; 22.3)
(b) Vol. **124**, No. 2, p. K105—K108 (1984).

M. ISHII, M. SAEKI, and I. KAWADA
*Raman Study of Non-Stoichiometric Titanium Sulfides
(6 and 20.1; 22)
(b) Vol. **124**, No. 2, p. K109—K112 (1984).

R. P. GAIROLA
*Low-Temperature Lattice Thermal Conductivity of Nonmetallic Solids with Isotopic Impurities
(6 and 8; 10.2; 22; 22.1.1)
(b) Vol. 125, No. 1, p. 65—74 (1984).

SH. M. EFENDIEV, V. M. NAGIEV, and V. M. BURLAKOV
*Anharmonicity Effects in IR-Reflection Spectra of Sheelites
(6 and 20.1; 22.8.1)
(b) Vol. 125, No. 1, p. 75—79 (1984).

M. H. BEN GHOZLEN, Y. MLIK, and A. DAOUD
*Raman Study of Mono-, Di-, and Trimethyl Ammonium Hexachlorostannates
(6 and 20.1; 22.9)
(b) Vol. 125, No. 1, p. 87—96 (1984).

T. SOMA, T. ITOH, and H.-MATSUO KAGAYA
*Pressure Effect on Phonon Frequencies and Elastic Stiffness Constants of Al
(6; 12.1; 21)
(b) Vol. 125, No. 1, p. 107—112 (1984).

N. AGNIHOTRI, R. K. SINGH, and S. P. SANYAL
Interionic Forces and Elastic Properties of Na_2S
(4; 6; 12.2; 22; 22.5.3)
(b) Vol. 125, No. 1, p. 113—119 (1984).

V. FREI and P. DEUS
*A Frequency Sum Rule and Spectral Moments in Lattice Dynamics
(6; 22.1; 22.1.1; 22.2.2; 22.5.2; 22.8)
(b) Vol. 125, No. 1, p. 121—130 (1984).

W. KULLMANN, J. GEURTS, W. RICHTER, N. LEHNER, H. RAUH, U. STEIGENBERGER, G. EICHHORN, and R. GEICK
Effect of Hydrostatic and Uniaxial Pressure on Structural Properties and Raman Active Lattice Vibrations in Bi_2Te_3
(4; 1.1; 6; 20.1; 22.7)
(b) Vol. 125, No. 1, p. 131—138 (1984).

SHI-WEI GU and YU-KUI ZHOU
*Properties of the Exciton-Phonon System in a Magnetic Field
(6 and 13.5.1)
(b) Vol. 125, No. 1, p. 323—331 (1984).

G. I. BERSUKER and V. Z. POLINGER
*Multimode Pseudo Jahn-Teller Effect for Off-Centre Impurities in Crystals
(6 and 13.4; 10.2; 22.5.2)
(b) Vol. 125, No. 1, p. 401—408 (1984).

M. M. BABAEV, T. M. GASYMOV, and A. A. KATANOV
Thermopower and Thermomagnetic Effects of Hot Carriers under Conditions of Mutual Electron–Phonon Drag
(15; 6; 13; 22)
(b) Vol. 125, No. 1, p. 421—429 (1984).

A. PARISINI, P. P. LOTTICI, and RAZZETTI
*Comments on the Optic Vibrational Modes in Defect Chalcopyrites
(6 and 20.1; 22.8)
(b) Vol. **125**, No. 1, p. K7—K11 (1984).

G. CORRADI, E. RZEPKA et K. POLGÁR
*Influence des impuretés cationiques (Cu, Cr, Ni, Er) sur les spectres Raman du niobate de lithium
(Influence of Cationic Impurities (Cu, Cr, Ni, Er) on the Raman Spectra of Lithium Nitrate)
(6 and 20.1; 22.8.1)
(b) Vol. **125**, No. 1, p. K13—K16 (1984).

H. SOBOTTA, H. NEUMANN, N. N. SYRBU, and V. RIEDE
*Infrared Study of the Lattice Vibrations in Tetragonal CdP_2
(6 and 20.1; 22.3)
(b) Vol. **125**, No. 1, p. K17—K20 (1984).

L. WENDLER
*Coupled Surface Plasmon–Phonon Polaritons in Bilayer Systems
(6 and 20; 13.5.2; 21.6; 22.2.2)
(b) Vol. **125**, No. 1, p. K27—K31 (1984).

V. M. NAGIEV, SH. M. EFENDIEV, and V. M. BURLAKOV
*Vibrational Spectra of Crystals with Sheelite Structure and the Solid Solutions on Their Basis
(6 and 20.1; 22.8.1)
(b) Vol. **125**, No. 2, p. 467—475 (1984).

A. N. VASILEV and V. ČÁPEK
*Vibrational Relaxation in a Localized Excited Electronic State by the GME Method
(6; 13.4)
(b) Vol. **125**, No. 2, p. 477—482 (1984).

B. WYNCKE, F. BREHAT, M. EL SHERIF, and G. V. KOZLOV
*Far-Infrared and Submillimeter Dielectric Response of Sodium Nitrite
(6 and 20.1; 1.2; 14.4.1; 22.8.1)
(b) Vol. **125**, No. 2, p. 493—498 (1984).

K. K. CHOPRA and S. BOUARAB
*Off-Symmetry Phonon Frequencies of Copper
(6; 21.1)
(b) Vol. **125**, No. 2, p. 499—504 (1984).

B. S. TOŠIĆ and J. P. ŠETRAJČIĆ
*Exciton Drops in One-Dimensional Molecular Chains. New Method of Solving of Gap Equation
(6 and 13.5.1)
(b) Vol. **125**, No. 2, p. 743—749 (1984).

T. SOMA, Y. KITANI, and H.-MATSUO KAGAYA
*Mode Grüneisen Parameter and Pressure-Induced Phase Transition of Si–Ge Solid Solution
(1.2 and 6; 8; 22.1.1; 22.1.2)
(b) Vol. **125**, No. 2, p. K99—K101 (1984).

V. A. Haisler, V. M. Zaletin, and A. F. Kravchenko
*Raman Study of Anharmonic Effects in Red Mercury Iodide
(6 and 20.1; 22.5)
(b) Vol. **125**, No. 2, p. K103—K108 (1984).

I. J. Lalov
*Temperature Vibronic Spectra of Molecular Crystals
(6 and 13.5.1; 20.1; 22.9)
(b) Vol. **125**, No. 2, p. K147—K150 (1984).

M. J. Škrinjar, D. V. Kapor, Ž. M. Škrbić, and J. P. Šetrajčić
*On the Problem of Exciton Coupling in Molecular Crystals
(6 and 13.5.1; 22.9)
(b) Vol. **125**, No. 2, p. K151—K154 (1984).

U. F. Kozmidis-Luburić, J. P. Šetrajčić, Ž. M. Škrbić, and B. S. Tošić
*Exciton–Phonon Interaction and Bound States
(6 and 13.5.1)
(b) Vol. **125**, No. 2, p. K155—K159 (1984).

A. R. Vasconcellos and R. Luzzi
On Laser-Induced Lattice Heating in a Polar Semiconductor
(8 and 11; 6; 22; 22.2.1)
(b) Vol. **126**, No. 1, p. 63—70 (1984).

T. Soma, T. Itoh, and H.-Matsuo Kagaya
*Pressure Dependence of the Melting Point of Al
(6; 8; 21)
(b) Vol. **126**, No. 1, p. 71—75 (1984).

S. P. Singh and R. P. S. Rathore
*Phonons in Face-Centred Cubic Calcium and Strontium
(6; 12.1; 21.3)
(b) Vol. **126**, No. 1, p. 87—90 (1984).

A. P. Mayer and R. K. Wehner
*Calculation of Grüneisen Constants in Si
(6; 22.1.2)
(b) Vol. **126**, No. 1, p. 91—103 (1984).

S. Mochizuki
Radiative Optical Modes of Vibration in NiO Thin Film
(20.1; 6; 13.5; 22.6)
(b) Vol. **126**, No. 1, p. 105—114 (1984).

H.-G. Unruh
Dielectric Dispersion and Dynamics of Nitrite Ions in $K_2Ba(NO_2)_4$
(1.2 and 14.4; 6; 22.8.1)
(b) Vol. **126**, No. 1, p. 115—120 (1984).

J. Estel, J. Kalus, and L. Pintschovius
*Measurement of Phonon Energies in Silicon under Uniaxial Stress
(6; 22.1.2)
(b) Vol. **126**, No. 1, p. 121—124 (1984).

T. Soma and H.-Matsuo Kagaya
*Compression Effect on Mode Grüneisen Parameters of Alkali Metals
(6 and 8; 12.1; 21.2)
(b) Vol. **126**, No. 1, p. K5—K7 (1984).

H.-Matsuo Kagaya and T. Soma
Dynamical Elastic Constants of Covalent Compounds
(12.2; 6; 22.2; 22.4)
(b) Vol. **126**, No. 1, p. K9—K14 (1984).

R. K. Singh and D. K. Neb
*Mean Square Displacements of Rare Gas Solids
(4 and 6; 8; 23)
(b) Vol. **126**, No. 1, p. K15—K18 (1984).

G. M. Shmelev, Nguyen Hong Shon, and G. I. Tsurkan
*Magnetic Oscillations of Acoustoelectrical Current of Photoelectrons in p-GaAs Type Semiconductors
(6 and 16; 7; 22.2.1)
(b) Vol. **126**, No. 1, p. K105—K108 (1984).

A. A. Rempel, S. Z. Nazarova, and A. I. Gusev
Effect of Atomic Ordering on the Heat Capacity of Non-Stoichiometric Niobium Carbide
(1.2 and 8; 6; 22)
(a) Vol. **86**, No. 1, p. K11—K14 (1984).

W. S. Tse and C. C. Chen
*Lattice Vibrations of Crystalline Silicon and Germanium Tetrachlorides
(6 and 20.1; 22)
(a) Vol. **86**, No. 1, p. K23—K25 (1984).

M. D. Blokh and M. V. Entin
*The Photoexcited Flow of Nonequilibrium Phonons along a Crystal Surface
(6)
(b) Vol. **126**, No. 2, p. 487—493 (1984).

V. M. Yartsev
Charge Transfer and Electron–Molecular Vibration Coupling in Tetramerized Quasi-1D Semiconductors
(14.3; 6; 20.1; 22.9)
(b) Vol. **126**, No. 2, p. 501—510 (1984).

E. Kaden and H.-L. Günter
*The Thermoelectric Power of p-Ge at Low Temperatures
(6 and 15; 14.3; 22.1.1)
(b) Vol. **126**, No. 2, p. 733—740 (1984).

T. Soma and H.-Matsuo Kagaya
*Mode Grüneisen Parameter for Local and Band Modes of Rb–K, Cs–K, and Cs–Rb Alloy Systems
(6; 8; 12.1; 21.2)
(b) Vol. **126**, No. 2, p. K109—K112 (1984).

M. P. Dos Santos, C. Hirlimann, and M. Balkanski
*Raman Detection of the Nitrogen Localized Mode in GaP:N
(6 and 20.1; 22.2.2)
(b) Vol. **126**, No. 2, p. K113—K116 (1984).

G. Berger and M. Giehler
Characterization of in Vitro Corroded Surfaces of Bioactive Glasses (P_2O_5–CaO–SiO_2) with Infrared Reflection Spectroscopy
(2 and 20.1; 4; 6; 22.6.1)
(a) Vol. **86**, No. 2, p. 531—542 (1984).

S. N. Vaidya
*Theory of Melting (I): Energy Balance Principle
(1.2 and 6; 8; 23)
(a) Vol. **86**, No. 2, p. 565—571 (1984).

B. R. Nag, S. R. Ahmed, and M. Deb Roy
Submicron Electron Transport in Silicon at 300 and 77 K
(14.3.4; 6; 13.1; 22.1.2)
(a) Vol. **86**, No. 2, p. 773—779 (1984).

7. Acoustic Properties

Original Papers and Short Notes

V. V. Gudkov and V. D. Fil
*Ultrasonic Wave Field Transport by Conduction Electrons in Indium
(7 and 14.1; 13.2; 21)
(b) Vol. **121**, No. 1, p. 433—440 (1984).

S. Ghosh and R. B. Saxena
Parametric Conversion of an Electromagnetic Wave into an Acoustic Wave in Magnetised Cubic Piezoelectric Semiconducting Plasmas
(13.5.2; 7; 22.2.3)
(b) Vol. **121**, No. 2, p. 661—672 (1984).

G. M. Shmelev, Nguyen Quoc Anh, G. I. Tsurkan, and S. Y. Mensah
*"Currentless" Amplification of Hypersound in a Planar Configuration by Inelastic Scattering of Electrons
(7 and 14.3; 22)
(b) Vol. **121**, No. 2, p. K209—K213 (1984).

S. N. Barilo, A. P. Ges, L. E. Soshnikov, and T. V. Smirnova
Thermal Conductivity of Rare-Earth Orthoferrites at Reorientation Phase Transitions
(8 and 18.4; 7; 22.8.2)
(b) Vol. **122**, No. 1, p. 47—52 (1984).

H. Szymczak
*Acoustical Relaxation Due to the Bound Small Polaron in Cubic Crystals
(7 and 13.5.3; 22.6)
(b) Vol. **122**, No. 1, p. K37—K40 (1984).

L. J. Balk, D. G. Davies, and N. Kultscher
*The Dependence of Scanning Electron Acoustic Microscopy (SEAM) Imaging on Chopping and Detection Frequency for Metal Samples
(1 and 7; 21.1)
(a) Vol. **82**, No. 1, p. 23—33 (1984).

H. Cerva and W. Graeff
*Contrast Investigations of Surface Acoustic Waves by Stroboscopic Topography (I): Orientation Contrast
(1 and 7; 14.4.1; 22.8.1)
(a) Vol. **82**, No. 1, p. 35—45 (1984).

V. M. Polyanovskii
*Stark Cyclotron-Phonon Oscillations of Sound Absorption in Semiconductors with Superlattice
(6 and 7; 19; 22)
(b) Vol. **122**, No. 2, p. K115—K118 (1984).

M. Krzesińska and T. Szuta-Buchacz
*Hypersonic Wave Attenutation in Yttrium Aluminium Garnet and Gadolinium Gallium Garnet
(7; 6; 12.2; 22.8.2)
(a) Vol. **82**, No. 2, p. 421—424 (1984).

J. Dominec, K. Misek, W. Kraak, and R. Herrmann
*Ultrasound Attenuation and Quantum Oscillations in Bi–Sb Alloys
(7 and 13; 21.7)
(b) Vol. **123**, No. 2, p. 635—640 (1984).

R. G. Gabrielyan and H. A. Aslanyan
*Modulation of γ- and X-Rays by Ultrasonic Vibrations in Crystals
(7)
(b) Vol. **123**, No. 2, p. K97—K99 (1984).

S. Guha, C. Basu, and M. Bose
*Parametric Excitation of Electron-Acoustic Waves in Magnetised Semiconductor Plasmas
(7 and 14.3; 22.2.3)
(a) Vol. **83**, No. 2, p. 675—680 (1984).

V. L. Aksenov, A. Yu. Didyk, and N. M. Plakida
*Ultrasonic Attenuation in Quasi-One-Dimensional Ferroelectrics
(7 and 14.4.1; 22.8.1)
(b) Vol. **124**, No. 1, p. 45—54 (1984).

S. Ghosh and S. Dixit
Parametric Decay of a High-Power Helicon Wave into a Transverse Acoustic and Another Helicon Waves in Longitudinally Magnetised Cubic Piezoelectric Semiconducting Plasmas
(13.5.2; 7; 22.2.3)
(b) Vol. **124**, No. 1, p. 395—402 (1984).

M. S. Pathak, S. Khan, and S. Ghosh
Longitudinal Phonon–Plasmon Interactions in Semiconducting Plasmas
(6 and 13.5.2; 7; 22.2.3)
(b) Vol. **124**, No. 2, p. 489—496 (1984).

N. P. Kovalenko and L. M. Kuzmina
The Influence of Many-Body Interactions on the Speed of Sound in Liquid Metals
(2 and 12.1; 7; 21.2)
(b) Vol. **124**, No. 2, p. 537—544 (1984).

A. K. Sharma and D. P. Tewari
*Stimulated Brillouin Scattering of an Electromagnetic Wave in CdS. Thermal Nonlinearity
(7; 22.4.1)
(a) Vol. **84**, No. 2, p. 439—442 (1984).

Nguyen Van Hieu and Ha Vinh Tan
Polariton Effect in Non-Linear Acousto-Optical Phenomena
(13.5 and 20; 7)
(b) Vol. **125**, No. 1, p. 245—250 (1984).

N. M. L. Köche, P. C. Morais, and K. Skeff Neto
*The Usefulness of the Photoacoustic Cell for Magnetic Measurements
(7 and 18; 19; 22.8.2)
(a) Vol. **85**, No. 1, p. K65—K67 (1984).

M. Adolf, J. Rosenzweig, and U. Birkholz
*Bipolar Photoconductivity and Hall Effect of the Acoustoelectric Current in CdTe
(7 and 16; 14.3; 22.4.3)
(a) Vol. **85**, No. 1, p. K81—K84 (1984).

B. A. Vaid and K. C. Sharma
*Ultrasonic Attenuation in Liquid Metals
(2 and 7; 21)
(b) Vol. **125**, No. 2, p. 505—512 (1984).

J. Dominec
*Semiconductor–Semimetal Transition in Bi–Sb Alloys Observed in Ultrasound Attenuation
(7 and 14.3; 14.1; 22.7)
(b) Vol. **126**, No. 1, p. K87—K90 (1984).

G. M. Shmelev, Nguyen Hong Shon, and G. I. Tsurkan
Magnetic Oscillations of Acoustoelectrical Current of Photoelectrons in p-GaAs Type Semiconductors
(6 and 16; 7; 22.2.1)
(b) Vol. **126**, No. 1, p. K105—K108 (1984).

P. C. Morais, A. L. Tronconi, and K. Skeff Neto
*Photoacoustic Signal Behavior near the Ferromagnetic Resonance Field
(7; 19; 22.8.2)
(a) Vol. **86**, No. 1, p. 163—168 (1984).

P. Koštial
*The Influence of Heating on the Attenuation of Surface Waves in Electrographic Se Layer
(1.5 and 7; 2; 22.1.3)
(a) Vol. **86**, No. 1, p. K7—K10 (1984).

M. E. Abu-Zeid, M. M. El-Oker, A. I. Abu-Shady, and M. S. Adili
*Photoacoustic and Ellipsometric Study of Black Nickel Coatings on Copper Substrates
(7 and 20.1; 21.1)
(a) Vol. **86**, No. 2, p. 805—811 (1984).

I. A. Ryzhkin
*Propagation of Elastic Waves in Ice
(7; 22)
(a) Vol. **86**, No. 2, p. K87—K90 (1984).

8. Thermal Properties

Review Article

R. RAMJI RAO and A. RAMANAND
*Thermal and Elastic Properties of Rare-Earth Metals
(8 and 12.1; 6; 21.4)
(b) Vol. **122**, No. 1, p. 11—27 (1984).

Original Papers and Short Notes

J. POZO and R. FERRER
*Specific Heat of Amorphous Magnets with Mixed Exchange Interactions
(8 and 18; 2)
(b) Vol. **121**, No. 1, p. K39—K41 (1984).

S. SIMOZAR and J. A. ALONSO
On the Factors Controlling Glass Forming Ability of Metallic Alloys Formed by Fast Liquid Quenching
(1.3 and 2; 8; 21; 21.1)
(a) Vol. **81**, No. 1, p. 55—61 (1984).

V. B. KOKSHENEV, K. BALCEREK, T. TYC, and A. JEŻOWSKI
*Discussion of the Heat Flux Rectification in the Solid–Solid System in the Acoustic-Mismatch Theory Framework
(8; 22.6)
(a) Vol. **81**, No. 1, p. 171—176 (1984).

M. HIDAKA, S. YAMASHITA, and Y. OKAMOTO
Study of Structural Phase Transitions of $KCaF_3$
(1.2; 6; 8; 22.5.4)
(a) Vol. **81**, No. 1, p. 177—183 (1984).

Cz. MARUCHA and J. RAFAŁOWICZ
*Deviation from Matthiessen's Rule for Aluminium Thermal Conductivity
(8; 21)
(a) Vol. **81**, No. 1, p. 185—189 (1984).

T. SOMA, Y. KITANI, and H.-MATSUO KAGAYA
Lattice Dynamics and Thermal Properties of Si–Ge Solid Solution
(6; 8; 22.3)
(b) Vol. **121**, No. 2, p. 501—504 (1984).

S. SIMONS
*Lattice Thermal Conduction in the Absence of Three-Phonon Interactions
(6 and 8)
(b) Vol. **121**, No. 2, p. K109—K112 (1984).

F. Völklein and E. Kessler
A Method for the Measurement of Thermal Conductivity, Thermal Diffusivity, and Other Transport Coefficients of Thin Films
(14.1 and 14.3.1; 8; 15; 21.7)
(a) Vol. **81**, No. 2, p. 585—596 (1984).

D. J. Easton and J. T. Edmond
*Glass–Crystal Transitions in Materials in the System As_2S_3–As_2Te_3
(2 and 8; 22)
(a) Vol. **81**, No. 2, p. K103—K107 (1984).

H. J. Goldsmid, H. Hora, and G. L. Paul
Anomalous Heat Conduction of Ion-Implanted Amorphous Layers in Silicon Crystals Using a Laser-Probe Technique
(2 and 11; 8; 22.1.2)
(a) Vol. **81**, No. 2, p. K127—K130 (1984).

S. N. Barilo, A. P. Ges, L. E. Soshnikov, and T. V. Smirnova
*Thermal Conductivity of Rare-Earth Orthoferrites at Reorientation Phase Transitions
(8 and 18.4; 7; 22.8.2)
(b) Vol. **122**, No. 1, p. 47—52 (1984).

B. Daudin, W. Mutscheller, and M. Wagner
*Heat Conductivity of the Cooperative Vibronic System $TmVO_4$
(6 and 8; 22.8.1)
(b) Vol. **122**, No. 1, p. 73—79 (1984).

B. Prasad
*Thermal Expansion and Mode Grüneisen Parameters in Lithium
(8; 21.2)
(b) Vol. **122**, No. 1, p. 91—93 (1984).

M. G. Abramishvili, V. I. Altukhov, T. L. Kalabegishvili, and V. G. Kvachadze
*Thermal Conductivity and Optical and Paramagnetic Absorption Spectra of Reactor Irradiated MgO Crystals
(8 and 11; 20.1; 22.6)
(b) Vol. **122**, No. 1, p. 95—104 (1984).

D. Coquillat, A. Pradel, G. Brun, and J. C. Tedenac
Growth and Physical Properties of $TlBiTe_2$ Crystals
(13.1 and 14.3; 8; 15; 20.1; 22.8)
(a) Vol. **82**, No. 1, p. 295—299 (1984).

G. Spörl, D. D. Chat, and E. Hegenbarth
*Thermal Conductivity of Pseudo-One-Dimensional Ferroelectric CsH_2PO_4
(8 and 14.4.1; 22.8.1)
(a) Vol. **82**, No. 1, p. K27—K31 (1984).

W. Lorenz and C. Engler
Quantum-Theoretical Model of Adiabatic Rate Processes in Condensed Systems or at Interfaces
(13.3; 4; 6; 8; 22)
(b) Vol. **122**, No. 2, p. 745—755 (1984).

S. M. Wasim and A. Noguera
Transport Properties of n-Type $CuInSe_2$
(13.4 and 14.3; 8; 22.8)
(a) Vol. **82**, No. 2, p. 553—559 (1984).

U. Pietsch, J. Bak-Misiuk, and V. Gottschalch
*The Linear Thermal Expansion Coefficient of a $Ga_xIn_{1-x}As_yP_{1-y}$ Layer on InP:Sn Substrate
(8; 1.1; 22.2.4)
(a) Vol. **82**, No. 2, p. K137—K140 (1984).

A. J. Kaur, W. N. Bhende, P. S. Bakhshi, and J. Shanker
Analysis of the Short-Range Repulsive Potential in Alkali Halides
(4; 6; 8; 22.5.2)
(b) Vol. **123**, No. 1, p. 71—78 (1984).

A. A. El-Sharkawy, A. S. Hilal, and S. R. Atalla
*Modified Plane, Radial, and Angström Temperature-Wave Techniques for the Measurement of Thermal Properties of Non-Conducting Materials
(8; 22.9)
(a) Vol. **83**, No. 1, p. 123—129 (1984).

S. R. Bakhchieva, N. P. Kekelidze, and M. G. Kekua
*Thermal Conductivity of Germanium Doped with Silicon, Tin, and Aluminium
(8; 22.1.1)
(a) Vol. **83**, No. 1, p. 139—145 (1984).

K. R. Zhdanov, A. V. Mishenko, F. S. Rakhmenkulov, and V. E. Fedorov
*Structural Anisotropy and Heat Capacity of NbS_3
(6 and 8; 22)
(a) Vol. **83**, No. 1, p. 147—152 (1984).

S. K. Plachkova
Thermoelectric Figure of Merit of the System $(GeTe)_{1-x}(AgSbTe_2)_x$
(14.3 and 15; 8; 22.8)
(a) Vol. **83**, No. 1, p. 349—355 (1984).

I. Henning and E. Hegenbarth
*Specific Heat of $SrTiO_3$ in the Range of 0.5 to 6 K
(8 and 14.4.1; 22.8.1)
(a) Vol. **83**, No. 1, p. K23—K26 (1984).

G. Dharmadurai
*Analysis of Photoacoustically Investigated Gaseous Heat Transfer between Solid Surfaces
(6 and 8; 1.6)
(a) Vol. **83**, No. 1, p. K27—K30 (1984).

M. KATO and H.-R. PAK
Thermodynamics of Stress-Induced First-Order Phase Transformations in Solids
(1.2; 8; 12)
(b) Vol. **123**, No. 2, p. 415—424 (1984).

T. SOMA, T. ITOH, and H.-MATSUO KAGAYA
Lattice Dynamics of Al
(6; 8; 21)
(b) Vol. **123**, No. 2, p. 463—467 (1984).

K. BÄRNER, C. SANTANDREA, U. NEITZEL, and E. GMELIN
*Thermal and Magnetic Properties of $Cr_xMn_{1-x}As$ Compounds
(8 and 18; 21.1)
(b) Vol. **123**, No. 2, p. 541—552 (1984).

J. ŠLECHTA
*On the Magnetic Part of C_V of Amorphous Materials
(8 and 18; 2)
(b) Vol. **123**, No. 2, p. K133—K137 (1984).

K. MUKHERJEE, M. CHANDRASEKARAN, and M. KATO
Electrical Resistivity of an Equiatomic Solid and Liquid Au–Cd Alloy
(14.1; 2; 8; 21; 21.6)
(a) Vol. **83**, No. 2, p. 657—667 (1984).

V. VALVODA
*X-Ray Debye Temperatures of HfC and α-W_2C
(6 and 8; 21)
(a) Vol. **83**, No. 2, p. K123—K125 (1984).

T. J. SINGH and G. S. VERMA
*Green's Functions Study of the Thermal Conductivity of Solids
(8)
(b) Vol. **124**, No. 1, p. K5—K9 (1984).

H. NEUMANN, P. DEUS, R. D. TOMLINSON, G. KÜHN, and B. HINTZE
*Thermal Expansion of $CuInTe_2$ from 30 to 300 K
(6 and 8; 22.8)
(a) Vol. **84**, No. 1, p. 87—93 (1984).

P. GROCHULSKI and Z. WAWRZAK
*Thermal Expansion of Sodium–Cadmium Formate. Powder Data
(1.1 and 8; 22.9)
(a) Vol. **84**, No. 1, p. K5—K6 (1984).

B. PRASAD
*First-Principles Thermal Expansion in Rb and Cs
(6 and 8; 21.2)
(b) Vol. **124**, No. 2, p. 497—501 (1984).

T. Soma, H. Ohsugi, and H.-Matsuo Kagaya
Phonon Dispersion Curves of Cs–K, Cs–Rb, and Rb–K Alloys
(6; 8; 21.2)
(b) Vol. **124**, No. 2, p. 525—529 (1984).

N. Dass and M. Kumari
*On the Isothermal Anderson-Grüneisen Parameter in Solids
(8; 22.5.2)
(b) Vol. **124**, No. 2, p. 531–536 (1984).

Do Yi-Jing, Chen Li-Rong, and Yen Tzu-Tung
*The Relation between the Lindemann Coefficient and the Temperature for the Alkali Metals
(8; 21.2)
(b) Vol. **124**, No. 2, p. K113—K116 (1984).

R. L. Davidovich, P. S. Gordienko, J. Grigas, T. A. Kaidalova, V. Urbonavičius, and L. A. Zemnukhova
Phase Transitions in $KSbF_4$
(1.2 and 14.4.1; 8; 19; 22.8)
(a) Vol. **84**, No. 2, p. 387—392 (1984).

A. V. Salker, E. B. Mirza, V. V. Deshpande, and H. V. Keer
Thermodynamic Analysis of the Phase Transition in V_4O_7
(1.2; 8; 14.1; 14.3; 18.1; 22.6)
(a) Vol. **84**, No. 2, p. K105—K108 (1984).

R. P. Gairola
*Low-Temperature Lattice Thermal Conductivity of Nonmetallic Solids with Isotopic Impurities
(6 and 8; 10.2; 22; 22.1.1)
(b) Vol. **125**, No. 1, p. 65—74 (1984).

R. N. Joarder, S. Palchaudhuri, and R. V. Gopala Rao
*Entropy and Heat Capacity of Liquid Metals
(2 and 8; 21.2)
(b) Vol. **125**, No. 1, p. K33—K36 (1984).

A. I. Gusev
Structural Vacancies in Nonstoichiometric Compounds at High Pressure. Thermodynamic Model
(10.1; 8; 12.1; 21.1)
(a) Vol. **85**, No. 1, p. 159—166 (1984).

P. Středa
Universal Transport Properties of Inversion Layers in Quantizing Magnetic Field
(13.3; 8; 14.1; 15; 18; 21)
(b) Vol. **125**, No. 2, p. 849—860 (1984).

T. Soma, Y. Kitani, and H.-Matsuo Kagaya
Mode Grüneisen Parameter and Pressure-Induced Phase Transition of Si–Ge Solid Solution
(1.2 and 6; 8; 22.1.1; 22.1.2)
(b) Vol. **125**, No. 2, p. K99—K101 (1984).

J. M. López and J. A. Alonso
*The Atomic Size-Mismatch Contribution to the Enthalpy of Formation of Concentrated Substitutional Metallic Solid Solutions
(1.3 and 8; 21; 21.1)
(a) Vol. **85**, No. 2, p. 423—428 (1984).

A. A. El-Sharkawy, I. H. Rashed, M. S. Zaghloul, and M. H. Ghoniem
*Study of Thermal Properties of Polycrystalline KCl, KBr, and KI in the Temperature Range 300 to 600 K
(8; 22.5.2)
(a) Vol. **85**, No. 2, p. 429—434 (1984).

B. A. Vaid, K. C. Sharma, and V. K. Syal
A Disorder Model of Melting (II): Alkali Halides
(2; 8; 22.5.2)
(b) Vol. **126**, No. 1, p. 59—62 (1984).

A. R. Vasconcellos and R. Luzzi
*On Laser-Induced Lattice Heating in a Polar Semiconductor
(8 and 11; 6; 22; 22.2.1)
(b) Vol. **126**, No. 1, p. 63—70 (1984).

T. Soma, T. Itoh, and H.-Matsuo Kagaya
Pressure Dependence of the Melting Point of Al
(6; 8; 21)
(b) Vol. **126**, No. 1, p. 71—75 (1984).

B. Lorenz and H. Fehske
Critical Study of the Static Functional Integral Method in the Hubbard Model
(13 and 18; 8)
(b) Vol. **126**, No. 1, p. 235—245 (1984).

T. Soma and H. Matsuo Kagaya
*Compression Effect on Mode Grüneisen Parameters of Alkali Metals
(6 and 8; 12.1; 21.2)
(b) Vol. **126**, No. 1, p. K5—K7 (1984).

R. K. Singh and D. K. Neb
Mean Square Displacements of Rare Gas Solids
(4 and 6; 8; 23)
(b) Vol. **126**, No. 1, p. K15—K18 (1984).

A. A. Rempel, S. Z. Nazarova, and A. I. Gusev
*Effect of Atomic Ordering on the Heat Capacity of Non-Stoichiometric Niobium Carbide
(1.2 and 8; 6; 22)
(a) Vol. **86**, No. 1, p. K11—K14 (1984).

M. A. Aldzhanov, D. A. Guseinov, and R. K. Veliev
*Low-Temperature Heat Capacity of $CdGa_2S_4$
(8; 22.8)
(a) Vol. **86**, No. 1, p. K19—K21 (1984).

S. Mahajan and S. Prakash
*Heat of Solution of Hydrogen in Aluminum
(8; 10.1; 21)
(b) Vol. **126**, No. 2, p. 467—480 (1984).

T. Soma and H.-Matsuo Kagaya
Mode Grüneisen Parameter for Local and Band Modes of Rb–K, Cs–K, and Cs–Rb Alloy Systems
(6; 8; 12.1; 21.2)
(b) Vol. **126**, No. 2, p. K109—K112 (1984).

S. N. Vaidya
Theory of Melting (I): Energy Balance Principle
(1.2 and 6; 8; 23)
(a) Vol. **86**, No. 2, p. 565—571 (1984).

V. E. Borisenko, V. V. Gribkovskii, V. A. Labunov, and S. G. Yudin
*Pulsed Heating of Semiconductors
(8; 11; 22.1; 22.2)
(a) Vol. **86**, No. 2, p. 573—583 (1984).

9. Diffusion

Review Article

V. A. Pavlov, V. I. Shalaev, and V. M. Alyabiev
Effect of Dislocation Structure on Creep and Fracture of Metals and Alloys (I): Normal Deformation Conditions
(10.1; 9; 21)
(a) Vol. **85**, No. 1, p. 11—27 (1984).

Original Papers and Short Notes

K. D. Becker
*The Encounter Model in Impurity Diffusion. Correlation Effects in NMR
(9 and 10.2; 19; 22.5.1; 22.5.2)
(b) Vol. **121**, No. 1, p. 91—100 (1984).

R. Fuentes-Samaniego and J. P. Hirth
*Asymptotic Solutions for the Differential Equation Describing Solute Drag by Dislocations
(9 and 10)
(b) Vol. **121**, No. 1, p. 101—109 (1984).

E. Wieser, H. Syhre, F. G. Rüdenauer, and W. Steiger
*Influence of Flash-Lamp Annealing on the Diffusion Behaviour of Ion Implanted Boron Profiles
(9; 11; 22.1.2)
(a) Vol. **81**, No. 1, p. 247—252 (1984).

A. F. Khokhlov, V. A. Panteleev, E. V. Dobrokhotov, G. A. Maksimov, and V. A. Sidorov
*Phosphorus Diffusion in Amorphous Silicon
(2 and 9; 22.1.2)
(a) Vol. **81**, No. 1, p. K15—K18 (1984).

E. J. Savino and A. M. Monti
Calculation of Lattice Defect Saddle-Point Configurations
(10.1; 9; 21.3)
(b) Vol. **121**, No. 2, p. 513—524 (1984).

J. Kučera, L. Kozák, and H. Mehrer
*Magnetic Anomalies of Self-Diffusion and Co Heterodiffusion in α-Fe
(9; 10.1; 18.2; 21.1.1)
(a) Vol. **81**, No. 2, p. 497—505 (1984).

J. J. Kleperis, S. J. Takeris, A. R. Lusis, and J. P. Stradins
*The Investigation of Hydrogen Diffusion in Palladium by the Chemichromic Effect
(9; 21.6)
(a) Vol. **81**, No. 2, p. K121—K125 (1984).

P. SÜPTITZ and A. FISCHER
*Lateral Diffusion of Photodoped Silver in Amorphous $Ge_{25}Se_{75}$ Layers
(9; 20.1; 22)
(a) Vol. **82**, No. 1, p. 157—161 (1984).

O. SIMMICH
*Influence of Non-Linear Terms in the Flux Equation on the Particle Growth
(1.4 and 9; 3; 21)
(a) Vol. **82**, No. 1, p. K21—K25 (1984).

DAO KHAC AN
*An Approximate Method for Determining the Segregation Coefficient of the Boron Drive Diffusion in Oxidizing Ambients
(9; 22.1.2; 22.6)
(a) Vol. **82**, No. 2, p. K145—K148 (1984).

O. J. ŻOGAŁ, CH. JÄGER, H. DÖHLER, and B. SCHNABEL
*Multipulse ^1H NMR Study of Proton Self-Diffusion in Scandium and Lutetium Dihydride
(9 and 19; 21.4)
(a) Vol. **82**, No. 2, p. K153—K157 (1984).

U. POTZEL, J. VÖLKL, H. WIPF, and A. MAGERL
*Tracer Diffusion and Chemical Diffusion of Hydrogen in TaH_x
(9; 21)
(b) Vol. **123**, No. 1, p. 85—92 (1984).

R. BLINC, J. DOLINŠEK, G. LAHAJNAR, I. ZUPANČIČ, L. A. SHUVALOV, and A. I. BARANOV
*Spin–Lattice Relaxation and Self-Diffusion Study of the Protonic Superionic Conductors $CsHSeO_4$ and $CsHSO_4$
(9 and 14.4; 19; 22.8.1)
(b) Vol. **123**, No. 1, p. K83—K87 (1984).

H. REUTHER
*Contribution to the Thermotransport in Silicate Glasses
(2 and 9; 22.6.1)
(a) Vol. **83**, No. 1, p. 173—178 (1984).

G. OELGART and G. ARNOLD
Internal Electric Field-Enhanced Impurity Diffusion in $GaAs_{1-x}P_x$
(13.4 and 14.3.3; 9; 22.2.4)
(a) Vol. **83**, No. 1, p. 393—402 (1984).

N. P. TISHCHENKO
*Kinetics of Impurity Particles in Low-Temperature Matrices
(9; 10.2; 23)
(a) Vol. **83**, No. 2, p. 513—519 (1984).

M. Doi, Y. Itoh, D.-Y. Chang, and T. Imura
*SIMS Studies on the Impurity Diffusion of Boron in Amorphous Alloys
(9; 2; 21.1; 21.1.1)
(a) Vol. **83**, No. 2, p. 529—536 (1984).

V. N. Solovev
*Atomic Particle Delocalisation Effect in Disordered Media
(2 and 9; 22.1.2)
(a) Vol. **83**, No. 2, p. 553—559 (1984).

S. K. Maksimov
Correlation between Thermodynamic Instability of $A^{III}B^{V}$ Heterocompositions and Degradation of Devices Based on These Materials
(10.2 and 14.3.2; 9; 22.2.4)
(a) Vol. **83**, No. 2, p. 685—692 (1984).

L. Uray and M. Menyhárd
The Segregation of Iron in Tungsten
(1.3; 1.4; 9; 14.1; 15; 21; 21.1.1)
(a) Vol. **84**, No. 1, p. 65—72 (1984).

A. V. Dvurechenskii, R. Grötzschel, N. M. Igonina, B. A. Koval, and N. I. Lebedeva
*Diffusion of Impurities in Undercooled Melt of Pulse Heated Ion-Implanted Silicon
(9; 11; 22.1.2)
(a) Vol. **84**, No. 1, p. 171—177 (1984).

E. Nebauer and M. Trapp
*Ge and Au Profiles in GaAs Produced from AuGe Contacts and Studied by SIMS
(9; 14.3.2; 22.2.1)
(a) Vol. **84**, No. 1, p. K39—K42 (1984).

A. J. Szadkowski and A. Lubomirska-Wittlin
Compensation of p-Type $Cd_{1-x}Mn_xTe$ by Diffusion of Ag
(14.3 and 16; 9; 13.4; 22.4.4)
(a) Vol. **84**, No. 2, p. K181—K184 (1984).

R. Fuentes-Samaniego and J. P. Hirth
*Excess Solute Atmosphere Carried by a Potential Well Characteristic of a Moving Dislocation Core
(9 and 10)
(b) Vol. **125**, No. 1, p. 147—154 (1984).

V. Suriyanarayanan, Y. Madhavan, and K. Ramachandran
*Isotope Diffusion in Cadmium Telluride
(9; 22.4.3)
(b) Vol. **125**, No. 2, p. 513—516 (1984).

P. Varotsos, K. Alexopoulos, and M. Lazaridou
On the Proportionality of the Migration Volume and the Migration Enthalpy in Fluorides
(10.2; 9; 22.5.3)
(b) Vol **125**, No. 2, p. K109—K112 (1984).

E. Sandow, W. Wesch, and E. Nebauer
Ohmic AuGe Contacts on n-GaAs Produced by Nanosecond Laser Pulse Irradiation
(14.3.4; 9; 11; 22.2.1)
(a) Vol. **85**, No. 2, p. K169—K173 (1984).

V. V. Kirsanov, S. B. Kislitsin, and E. M. Kislitsina
*Interstitial Migration in a Stress Gradient
(9 and 10.1; 21.1.1)
(a) Vol. **86**, No. 1, p. 199—206 (1984).

S. Palchaudhuri, R. N. Joarder, and R. V. Gopala Rao
*Partial Structure Factors and Self-Diffusion in Liquid Copper–Lead Alloys
(2 and 9; 21; 21.1)
(b) Vol. **126**, No. 2, p. 443—452 (1984).

R. Döhl, M.-P. Macht, and V. Naundorf
*Measurement of the Diffusion Coefficient of Cobalt in Copper
(9; 10.1; 21.1)
(a) Vol. **86**, No. 2, p. 603—612 (1984).

D. Shaw
*Alternative Mechanisms for the Diffusion of Sn and Zn in GaAs
(9; 10.2; 22.2.1)
(a) Vol. **86**, No. 2, p. 629—635 (1984).

10. Defect Properties (Irradiation Defects See 11)

Original Papers and Short Notes

R. FUENTES-SAMANIEGO and J. P. HIRTH
*Asymptotic Solutions for the Differential Equation Describing Solute Drag by Dislocations
(9 and 10)
(b) Vol. **121**, No. 1, p. 101—109 (1984).

V. V. ANTONOV-ROMANOVSKII
*On the Formation of Large Clusters of Recombining Particles of One Kind in Solids under Excitation (One-Dimensional Continuous and Discrete Models)
(10)
(b) Vol. **121**, No. 1, p. 133—144 (1984).

E. A. TIKHONOVA
Non-Linear Effects of the Dynamical Scattering Theory in Slightly Distorted Crystals
(1; 10)
(a) Vol. **81**, No. 1, p. 69—75 (1984).

L. LEJČEK, M. GLOGAROVÁ, and J. PAVEL
Twist Disclinations in Thin SmC Liquid Crystal
(2 and 10)
(a) Vol. **82**, No. 1, p. 47—54 (1984).

A. I. ONIPKO
*Application of the Cluster Size Distribution in Binary Randomly Disordered Finite Chains. Exact Solution of the Random Walk Problem for a Finite Chain with an Arbitrary Concentration of Pure Absorbers
(2 and 10)
(b) Vol. **123**, No. 1, p. 37—44 (1984).

E. NEMBACH
*Critical Resolved Shear Stress of Materials Which Simultaneously Contain Various Types of Obstacles Impeding the Glide of Dislocations
(10)
(b) Vol. **123**, No. 1, p. K21—K24 (1984).

M. J. MARCINKOWSKI
*The Mechanism of Cell Wall Formation
(1.4 and 10)
(a) Vol. **83**, No. 1, p. 59—68 (1984).
Erratum
(a) Vol. **86**, No. 1, p. 460 (1984).

R. FUENTES-SAMANIEGO and J. P. HIRTH
*Excess Solute Atmosphere Carried by a Potential Well Characteristic of a Moving Dislocation Core
(9 and 10)
(b) Vol. **125**, No. 1, p. 147—154 (1984).

V. V. Antonov-Romanovskii
*On the Formation of Large Clusters of Recombining Particles of One Kind in Solids during Reaction A + B = 0. Two- and Three-Dimensional Discrete Models
(10)
(b) Vol. **125**, No. 2, p. 517–521 (1984).

M. David, W. Herremans, and R. Gevers
*Fast-Electron Diffraction near a Translation Interface on Edge. The Generalized Dispersion Relation
(1 and 10)
(a) Vol. **85**, No. 2, p. 375–388 (1984).

C. W. Lung
*Interaction of a Dislocation with a Crack
(10)
(a) Vol. **85**, No. 2, p. K113–K116 (1984).

P. T. Landsberg and S. G. Canagaratna
*The Grand Partition Function in Defect Statistics
(10)
(b) Vol. **126**, No. 1, p. 141–156 (1984).

E. R. Korutcheva and D. I. Uzunov
*On the Long-Range Random Critical Behaviour
(1.2 and 10)
(b) Vol. **126**, No. 1, p. K19–K22 (1984).

N. I. Borgardt, S. K. Maksimov, and D. I. Piskunov
*Phase Information in Transmission Electron-Microscope Investigation of Defects. Dynamic Images with a Weak Amplitude Contrast
(1.4 and 10)
(a) Vol. **86**, No. 1, p. 55–65 (1984).

M. J. Marcinkowski
*The Role of Surface Dislocations in the Continuum Theory of Lattice Defects
(10)
(b) Vol. **126**, No. 2, p. 527–535 (1984).

R. C. Ecob
*The Measurement of Dislocation Link Length Distributions in a Tilted TEM Foil
(10)
(a) Vol. **86**, No. 2, p. 585–592 (1984).

10.1 Metals

Review Articles

V. A. Pavlov, V. I. Shalaev, and V. M. Alyabiev
*Effect of Dislocation Structure on Creep and Fracture of Metals and Alloys (I): Normal Deformation Conditions
(10.1; 9; 21)
(a) Vol. **85**, No. 1, p. 11—27 (1984).

V. A. Pavlov, V. I. Shalaev, and V. M. Alyabiev
*Effect of Dislocation Structure on Creep and Fracture of Metals and Alloys (II): Irradiation
(10.1; 11; 21)
(a) Vol. **85**, No. 2, p. 315—333 (1984).

Original Papers and Short Notes

D. Segers, P. van Mourik, M. H. van Wijngaarden, and B. M. Rao
*Precipitation of Silicon in a Solid Quenched Aluminium–Silicon (1.3 at%) Alloy Studied by Positron Annihilation
(10.1; 21)
(a) Vol. **81**, No. 1, p. 209—216 (1984).

H. J. Blythe, F. Walz, and H. Kronmüller
*Magnetic Investigation of the Annealing Behaviour of Some Dilute Iron–Nickel Alloys after Low-Temperature Electron Irradiation
(10.1 and 18.2; 11; 21.1; 21.1.1)
(a) Vol. **81**, No. 1, p. 227—238 (1984).

S. V. Tsivinskii
*The Theory of Strain Hardening in Single and Polycrystals
(10.1; 1.4; 21; 21.1; 21.1.1; 22.1.1; 22.1.2)
(a) Vol. **81**, No. 1, p. 299—306 (1984).

V. M. Chernov and M. M. Savin
*Erratum to: Elastic Interaction of Edge and Screw Dislocations with Point Defects in Cubic Crystals
(10.1; 21)
(a) Vol. **81**, No. 1, p. 415 (1984).

V. Z. Bengus, E. D. Tabachnikova, and V. I. Startsev
Mechanical Behaviour of Some Metallic Glasses at 4.2 to 300 K
(2 and 12.1; 10.1; 21)
(a) Vol. **81**, No. 1, p. K11—K13 (1984).

E. J. Savino and A. M. Monti
*Calculation of Lattice Defect Saddle-Point Configurations
(10.1; 9; 21.3)
(b) Vol. **121**, No. 2, p. 513—524 (1984).

P. M. HAZZLEDINE, H. P. KARNTHALER, and A. KORNER
*The Application of the Wulff Construction to Dislocation Problems
(10.1; 12.1; 21.1; 21.6)
(a) Vol. **81**, No. 2, p. 473—484 (1984).

K. IWASAKI
*High-Temperature Internal Friction Peaks of Pure Aluminium Single-, Bi-, and Polycrystals Measured with an Inverted Flexure Pendulum
(10.1 and 12.1; 21)
(a) Vol. **81**, No. 2, p. 485—496 (1984).

J. KUČERA, L. KOZÁK, and H. MEHRER
Magnetic Anomalies of Self-Diffusion and Co Heterodiffusion in α-Fe
(9; 10.1; 18.2; 21.1.1)
(a) Vol. **81**, No. 2, p. 497—505 (1984).

G. K. SCHMIDT
*Method for Obtaining the Spectrum of Plastic Strains of Mesoscopic Volume Elements in Fatigued Materials Investigated by X-Rays (I): Plastic and Elastic Strains in Fatigued Material
(10.1 and 12.1; 1.4; 21)
(a) Vol. **82**, No. 1, p. 75—84 (1984).

A. NOHARA, M. IZUMI, H. SAKA, and T. IMURA
*Plastic Deformation Behavior of β-CuZn Single Crystals at Low and High Temperatures
(10.1; 21.1)
(a) Vol. **82**, No. 1, p. 163—170 (1984).

R. TOGNATO
Some Considerations on the Interfacial Free Energy between Dislocation Core and Bulk Material at High Temperatures
(1.6; 10.1; 21)
(a) Vol. **82**, No. 1, p. K17—K19 (1984).

G. N. GAIDUKOV, A. A. PODREZOV, and J. P. HIRTH
A Volterra-Type Dislocation Model of a Low-Angle [001] Twist Boundary in an F.C.C. Crystal
(1.3; 10.1; 21)
(a) Vol. **82**, No. 2, p. 355—363 (1984).

G. K. SCHMIDT
*Method for Obtaining the Spectrum of Plastic Strains of Mesoscopic Volume Elements in Fatigued Materials Investigated by X-Rays (II): Distributions of Elastic Strains and X-Ray Reflex Profiles
(10.1 and 12.1; 1.4; 21)
(a) Vol. **82**, No. 2, p. 413—420 (1984).

S. K. ROY and D. L. BHATTACHARYA
Debye-Waller Factor for Small Concentration of Mössbauer Impurities in Metallic Solids
(6; 5; 10.1; 21)
(b) Vol. **123**, No. 1, p. 45—51 (1984).

A. Baldan
The Microstructural Properties of (β + (Ni, Al)$_7$Hf$_2$) Eutectics
(1.3; 10.1; 21; 21.1)
(a) Vol. **83**, No. 1, p. 47—57 (1984).

K. Maeda, S. Kobayashi, S. Takeuchi, T. Chiba, and K. Osamura
*Absence of Dilatational Defects in Deformed $Cu_{57}Zr_{43}$ Amorphous Alloy
(2 and 10.1; 21.1)
(a) Vol. **83**, No. 1, p. 219—226 (1984).

D. O. Northwood and I. O. Smith
*Internal and Friction Stresses for Al–Mg Class I High Temperature Creep Alloys
(10.1 and 12.1; 21)
(a) Vol. **83**, No. 1, p. 227—234 (1984).

P. Delobelle, D. Varchon, and C. Oytana
*Short-Range Reordering of Heavy Interstitials in Ta, Nb, and Fe during Relaxation and Static Strain Aging
(10.1; 12.1; 21; 21.1.1)
(a) Vol. **83**, No. 1, p. 235—243 (1984).

A. Baldan
*An Investigation of the Mechanism of Thermally Activated Deformation in Certain Ni–Al–Ta and Ni–Al–Hf Superalloys
(10.1; 1.3; 21; 21.1)
(a) Vol. **83**, No. 2, p. 507—511 (1984).

C. Hidalgo, N. de Diego, and M. A. Ochando
*A Positron Study of Iron Alloys
(1.4 and 10.1; 21.1.1)
(a) Vol. **83**, No. 2, p. K93—K96 (1984).

Z. Bojarski, H. Morawiec, and T. J. Panek
*Positron Annihilation Study of Martensitic Transformation in Cu–Zn–Al Alloy
(1.2 and 10.1; 21; 21.1)
(a) Vol. **83**, No. 2, p. K97—K101 (1984).

A. M. Gómez and J. Serna
*Positron Lifetime in Polycrystalline Gadolinium
(10.1; 21.4)
(a) Vol. **83**, No. 2, p. K131—K133 (1984).

W. Kienle and W. Frank
*Calculation of the Intensity of the Small-Angle X-Ray Scattering by Small Clusters of Intrinsic Point Defects
(10.1; 21.1)
(b) Vol. **124**, No. 1, p. 69—77 (1984).

U. H. Gläser
*On the Calculation of the Residual Resistivity of Precipitates
(10.1 and 14.1; 21; 21.1)
(b) Vol. **124**, No. 1, p. 423—429 (1984).

T. S. Kê, P. Cui, and C. M. Su
*Internal Friction in High-Purity Aluminium Single Crystals
(10.1 and 12.1; 21)
(a) Vol. **84**, No. 1, p. 157—164 (1984).

R. Basu
*A Mechanism for Drag on Dislocations in Zinc
(10.1; 21.1)
(a) Vol. **84**, No. 1, p. K35—K38 (1984).

A. R. Ali, S. A. Mahmoud, and G. Said
*On the Role of Cu Atoms on the Behaviour of Domain Wall–Dislocation Interaction in Ni–Cu Alloy
(10.1 and 18.2; 21.1)
(a) Vol. **84**, No. 1, p. K67—K70 (1984).

T. S. Kê, L. D. Zhang, P. Cui, Q. Huang, and B. S. Zhang
*Internal-Friction Peak in Polygonized High-Purity Aluminium and In-Situ Observation by Transmission Electron Microscope
(10.1 and 12.1; 21)
(a) Vol. **84**, No. 2, p. 465—474 (1984).

R. Zürcher, V. Gröger, and F. Stangler
*Discontinuous Flow in Cu–Be2 between 4.2 and 12.5 K
(10.1; 21.1)
(a) Vol. **84**, No. 2, p. 475—480 (1984).

A. Nohara and T. Imura
*Yield-Point Phenomenon in a β-CuZn Single Crystal at High Temperatures
(10.1; 12.1; 21.1)
(a) Vol. **84**, No. 2, p. 501—507 (1984).

D. O. Northwood, L. Moerner, and I. O. Smith
*The Breakdown in Creep by Viscous Glide in Al–Mg Solid Solution Alloys at High Stress Levels
(10.1; 21)
(a) Vol. **84**, No. 2, p. 509—515 (1984).

O. L. de Lange and C. J. Young
*Interaction between Dislocation Arrays
(10.1; 21)
(a) Vol. **84**, No. 2, p. 517—526 (1984).

M. Z. Butt, P. Feltham, and S. M. Raza
*Stress Relaxation in Nickel between 15 and 300 K
(10.1; 21.1)
(a) Vol. **84**, No. 2, p. K125—K127 (1984).

F. C. Lovey, G. van Tendeloo, and S. Amelinckx
The Nature of Some Planar Defects in 2H Martensite of Cu–Al Alloys as Determined by HREM
(1.4; 10.1; 21; 21.1)
(a) Vol. **85**, No. 1, p. 29—37 (1984).

V. G. Chudinov and V. I. Protasov
Kinetics of Vacancy Mechanism of the Solid-Liquid Transition in F. C. C. Metals
(1.2; 10.1; 21.1)
(a) Vol. **85**, No. 1, p. 105—112 (1984).

R. Pérez
On the Identification of the Nature of Stacking Faults in H.C.P. Materials
(1.4; 10.1)
(a) Vol. **85**, No. 1, p. 113—119 (1984).

N. Kuramochi, H. Mizubayashi, and S. Okuda
*Dislocation Pinning at Low Temperatures in Nb
(10.1 and 12.1; 21)
(a) Vol. **85**, No. 1, p. 121—131 (1984).

D. O. Northwood and I. O. Smith
*Steady-State Creep and Strain Transients for Stress-Dip Tests in Polycrystalline Magnesium at 300 °C
(10.1 and 12.1; 21.3)
(a) Vol. **85**, No. 1, p. 149—158 (1984).

A. I. Gusev
*Structural Vacancies in Nonstoichiometric Compounds at High Pressure. Thermodynamic Model
(10.1; 8; 12.1; 21.1)
(a) Vol. **85**, No. 1, p. 159—166 (1984).

R. Tognato
*Influence of Dislocation Core Density on Overheating Absence
(10.1; 21)
(a) Vol. **85**, No. 1, p. K47—K50 (1984).

I. G. Ratishvili
*Annihilation of Frenkel Pairs in B.C.C. Metals (α-Iron) (I): The Case of the Unstressed Lattice (General Consideration)
(10.1; 21.1.1)
(b) Vol. **125**, No. 2, p. 533—542 (1984).

Y. Matsuoka, S. Terakado, and S. Tanigawa
*Positron Age–Momentum Correlation in Ultrafine Particles of Silver and Gold
(10.1; 13.3; 21.6)
(b) Vol. **125**, No. 2, p. 823—829 (1984).

Ch. Opitz, H. Müller, and H. Miessner
Bonding Mechanism for Interstitial Hydrogen in Nickel
(4; 10.1; 13.4; 21.1)
(b) Vol. **125**, No. 2, p. K93—K97 (1984).

G. Fantozzi, C. Esnouf, J. Lauzier, and C. Minier
*Internal Friction at Low Temperature in High Purity Magnesium
(10.1 and 12.1; 21.3)
(a) Vol. **85**, No. 2, p. 463—468 (1984).

I. Baker and E. M. Schulson
*On Intrinsic Stacking Faults in Polycrystalline Ni_3Al
(10.1; 1.4; 21; 21.1)
(a) Vol. **85**, No. 2, p. 481—490 (1984).

F. Walz
Transition-Induced Magnetic After-Effect in the Amorphous Alloy $Fe_{20}Ni_{56}B_{24}$ at the Critical Temperature
(18.2; 2; 10.1; 21.1; 21.1.1)
(a) Vol. **85**, No. 2, p. 503—509 (1984).

A. Pawełek
*A Simple Model of Small Yield Point Due to Strain Rate Change Based on the Dynamics of the Source Generated Dislocation Group
(10.1 and 12.1; 21)
(a) Vol. **85**, No. 2, p. K117—K121 (1984).

A. M. Monti
*Divacancy in H.C.P. Metals
(10.1; 21)
(b) Vol. **126**, No. 1, p. 163—175 (1984).

I. N. Karnaukhov and A. F. Zhuravlev
*Magnetic State of Crystals with Dislocations
(10.1 and 18; 19)
(b) Vol. **126**, No. 1, p. 207—212 (1984).

H. C. González
*Hardening of Single Crystals of Magnesium by Low Neutron Doses at 77 K
(10.1 and 11; 21.3)
(a) Vol. **86**, No. 1, p. 169—175 (1984).

S. V. Sokolskii, F. F. Lavrentev, and O. P. Salita
*Jump-Like Deformation and Stability of the Hardened and Structural States of Zinc Single Crystals with Forest Dislocations in the Temperature Range 293 to 4.2 K
(10.1; 21.1)
(a) Vol. **86**, No. 1, p. 177—183 (1984).

V. V. Kirsanov, S. B. Kislitsin, and E. M. Kislitsina
*Interstitial Migration in a Stress Gradient
(9 and 10.1; 21.1.1)
(a) Vol. **86**, No. 1, p. 199—206 (1984).

H. M. Eissa and M. M. El-Oker
*The Interaction of Straight Dislocations with Dislocation Loops in an Anisotropic Material (Zinc)
(10.1; 21.1)
(a) Vol. **86**, No. 1, p. 207—218 (1984).

H. Yamaguchi, T. Tsuchida, I. Hashimoto, Y. Ohtani, and S. Ohtaki
*Stage II Recovery of Defects in Cold-Worked Aluminium
(10.1; 14.1; 21)
(a) Vol. **86**, No. 1, p. 237—244 (1984).

M. B. Shah and M. S. C. Bose
*Magnetic NDT Technique to Evaluate Fatigue Damage
(10.1 and 18.2; 21)
(a) Vol. **86**, No. 1, p. 275—281 (1984).

A. Pawełek
*On the Thermodynamic Criterion for the Unstable Motion of a Source Generated Dislocation Group
(10.1 and 12.1)
(a) Vol. **86**, No. 1, p. K27—K30 (1984).

S. Mahajan and S. Prakash
Heat of Solution of Hydrogen in Aluminum
(8; 10.1; 21)
(b) Vol. **126**, No. 2, p. 467—480 (1984).

A. I. Melker and S. N. Romanov
Athermal Vacancy Cluster Formation in F. C. C. Metals by High Energy PKA
(11; 10.1; 21.1; 21.6)
(b) Vol. **126**, No. 2, p. 517—525 (1984).

P. Siebert, M. Werner, and H. Löffler
Heterogeneous Nucleation of Precipitates in an Al–12 at% Zn Alloy at Dislocation Loops
(1.3; 10.1; 21; 21.1)
(a) Vol. **86**, No. 2, p. 543—551 (1984).

F. C. Lovey, G. van Tendeloo, J. van Landuyt, L. Delaey, and S. Amelinckx
On the Nature of Various Stacking Defects in 18R Martensite in Cu–Al Alloys. A Study by High Resolution Electron Microscopy
(1.3; 10.1; 21; 21.1)
(a) Vol. **86**, No. 2, p. 553—564 (1984).

T. S. Kê, P. Cui, S. C. Yan, and Q. Huang
*Internal Friction Peaks of Super-High-Purity Aluminium at Medium Temperatures
(10.1 and 12.1; 21)
(a) Vol. **86**, No. 2, p. 593—601 (1984).

R. Döhl, M.-P. Macht, and V. Naundorf
Measurement of the Diffusion Coefficient of Cobalt in Copper
(9; 10.1; 21.1)
(a) Vol. **86**, No. 2, p. 603—612 (1984).

K. Benhaddane and P. Beauchamp
A Computation of (110) Intrinsic Stacking Fault Energies in B2 Ordered Alloys
(1.4; 10.1; 21; 21.1.1)
(a) Vol. **86**, No. 2, p. 613—621 (1984).

K. Iwasaki
*High-Temperature Internal Friction Peaks of Dilute Aluminium Alloys
(10.1 and 12.1; 21)
(a) Vol. **86**, No. 2, p. 637—648 (1984).

10.2 Non-Metals

Original Papers and Short Notes

K. D. Becker
*The Encounter Model in Impurity Diffusion. Correlation Effects in NMR
(9 and 10.2; 19; 22.5.1; 22.5.2)
(b) Vol. **121**, No. 1, p. 91—100 (1984).

A. Kessler and P. Kizler
*Investigation of Z Center Conversions in KCl:$SrCl_2$ (I): Optical Measurement of the Kinetics of the Thermal $Z_1 \rightarrow F, Z_2$ Decay
(10.2; 20.1; 22.5.2)
(b) Vol. **121**, No. 1, p. 111—116 (1984).

A. Watterich, I. Földvári, G. Corradi, and H. Opyrchał
*Comment on the Identification of Z Centres in LiF:Mg and LiF:Mg, Ti Single Crystals
(10.2 and 20.1; 22.5.2)
(b) Vol. **121**, No. 1, p. 117—125 (1984).

J. Prakash
*Photo-Chemical Reaction Studies in Selenium-Doped Potassium Iodide
(10.2 and 20.1; 22.5.2)
(b) Vol. **121**, No. 1, p. 127—132 (1984).

J. Manam and V. V. Ratnam
V_3 and F_A Centers in KCl:Na Crystals. A Thermoluminescence Study
(13.4 and 20.3; 10.2; 22.5.2)
(b) Vol. **121**, No. 1, p. 393—399 (1984).

V. N. Salomatov and I. A. Parfianovich
The F_H-Centre Energy Structure in KCl and KBr
(13.4 and 20.1; 10.2; 22.5.2)
(b) Vol. **121**, No. 1, p. 401—406 (1984).

P. A. Aleksandrov, A. M. Afanasiev, M. K. Melkonyan, and S. A. Stepanov
X-Ray Diffraction under Specular Reflection Conditions on Crystals with an Amorphous Surface Film
(1.6 and 2; 10.2; 22.1.2)
(a) Vol. **81**, No. 1, p. 47—53 (1984).

W. Bergholz, S. Damgaard, J. W. Petersen, and G. Weyer
*Reactions of Cobalt in Silicon with Electron-Irradiation-Induced Defects
(10.2; 5; 22.1.2)
(a) Vol. **81**, No. 1, p. 239—246 (1984).

C. Y. Chen, K. L. Sweeney, and L. E. Halliburton
*Reduction and Radiation Effects in Lithium Tantalate
(10.2; 11; 19; 20.1; 22.8.1)
(a) Vol. **81**, No. 1, p. 253—257 (1984).

M. Pasemann, W. Bergholz, and W. Schröter
*Mössbauer Spectroscopical and Electron Microscopical Investigations of the Behaviour of Cobalt in Silicon
(1.4 and 10.2; 5; 22.1.2)
(a) Vol. **81**, No. 1, p. 273—280 (1984).

V. P. Gupta, A. K. Singh, K. Chandra, and S. K. Jaireth
*Mössbauer Investigations on Pyrrhotite
(10.2; 5; 18.3; 18.4; 22.8)
(a) Vol. **81**, No. 1, p. 281—291 (1984).

P. Kindl
Positron Lifetime Studies on γ-Irradiated PTFE of Different Initial Crystallinity
(11; 10.2; 22.9)
(a) Vol. **81**, No. 1, p. 293—298 (1984).

V. D. Tkachev, A. V. Mudryi, and N. S. Minaev
*Noble Gas Atoms as Chemical Impurities in Silicon
(10.2 and 13.4; 20.3; 22.1.2)
(a) Vol. **81**, No. 1, p. 313—321 (1984).

H. W. Helberg
*Dielectric Relaxation Behaviour of Dislocations in Semiconductors
(10.2 and 14.3; 14.4; 22)
(a) Vol. **81**, No. 1, p. 381—390 (1984).

A. A. Sitnikova, L. M. Sorokin, I. E. Talanin, E. G. Sheikhet, and E. S. Falkevich
Electron-Microscopic Study of Microdefects in Silicon Single Crystals Grown at High Speed
(1.4; 10.2; 22.1.2)
(a) Vol. **81**, No. 2, p. 433—438 (1984).

K. S. Raju and E. Pippel
High Voltage Electron Microscopic (HVEM) Observations of Dislocations in Monocrystalline Gypsum
(1.4; 10.2; 22.8.1)
(a) Vol. **81**, No. 2, p. K93—K95 (1984).

A. Fousková, M. Pavel, and B. Březina
*Light Scattering by Defects in TGS
(10.2; 14.4.1; 22.9)
(a) Vol. **81**, No. 2, p. K117—K119 (1984).

M. Kittler, K.-W. Schröder, E. Bugiel, and C. Becker
*Quantitative EBIC Investigations on Bulk Stacking Faults in Silicon
(1.4 and 10.2; 14.3; 22.1.2)
(a) Vol. **81**, No. 2, p. K131—K135 (1984).

S. Ramana Murthy and T. Seshagiri Rao
*Internal Friction in Ni–Zn Ferrites
(10.2 and 12.2; 22.8.2)
(a) Vol. **81**, No. 2, p. K137—K140 (1984).

D. N. POPOV, N. M. NANCHEVA, M. R. MINEV, and N. S. BAKALOVA
*On the Nature of the X-Centers in KCl, Studied by Positron Angular Correlation
(10.2; 20.1; 22.5.2)
(a) Vol. **81**, No. 2, p. K141—K144 (1984).

K. SCHMALZ, P. GAWORZEWSKI, and F.-G. KIRSCHT
*Deep Levels in Czochralski p-Si Due to Heat Treatment at 600 to 900 °C
(10.2 and 13.4; 14.3; 22.1.2)
(a) Vol. **81**, No. 2, p. K165—K169 (1984).

A. RICHTER, A. E. ROMANOV, W. POMPE, and V. I. VLADIMIROV
Geometry and Energy of Disclinations in Topologically Disordered Systems
(2; 10.2; 12.2)
(b) Vol. **122**, No. 1, p. 35—45 (1984).

A. KESSLER and P. KIZLER
*Investigation of Z Center Conversions in KCl:$SrCl_2$ (II): Dielectric Relaxation Caused by Mutual Conversions of Z_1, Z_2, and F Centers
(10.2; 14.4; 22.5.2)
(b) Vol. **122**, No. 1, p. 105—112 (1984).

S. BANDYOPADHYAY, C. NEOGY, and S. K. DEB
*Activation Energy of Substitutional Defects in Alkali Chlorides
(10.2; 22.5.2)
(b) Vol. **122**, No. 1, p. 113—123 (1984).

T. NAKAMURA and K. KAWAMURA
*The Frustration of Antiferromagnetic Bonds in the Screw Dislocation
(10.2 and 18.4)
(b) Vol. **122**, No. 1, p. 141—150 (1984).

A. V. EGRANOV and A. I. NEPOMNYACHIKH
*Magnesium Colour Centres in NaF
(10.2 and 13.4; 20.1; 20.3; 22.5.2)
(b) Vol. **122**, No. 1, p. 249—254 (1984).

S. SHIMANUKI and N. NISHIMAKI
*A Correlation between the Spectral Location of the Absorption Bands Due to the Localized Centers and the Lattice Constant in Alkali Halide Crystals. Anionic Impurity Centers with s^2-Configuration and F Center which Have Cubic Symmetry
(10.2 and 13.4; 20.1; 22.5.2)
(b) Vol. **122**, No. 1, p. 269—283 (1984).

Y. WATANABE
ESR and Magnetic Investigation of Lithium Intercalated VSe_2
(13.4; 10.2; 18.1; 19; 22.8)
(b) Vol. **122**, No. 1, p. 309—316 (1984).

FAT DUEN HO
*The Improved Point-Ion Potentials for the F^+-Centers in $LiNbO_3$ and $LiTaO_3$
(10.2 and 13.4; 20.1; 22.8.1)
(b) Vol. **122**, No. 1, p. 337—344 (1984).

A. KRÓL, W. NAZAREWICZ, and M. J. KOZIELSKI
*Al_{Zn}–V_{Zn} Pairs in Wurtzite-Type ZnS
(10.2 and 20.1; 6; 22.4.1)
(b) Vol. **122**, No. 1, p. K23—K25 (1984).

G. BALDACCHINI, G. P. GALLERANO, U. M. GRASSANO, M. MEUCCI, A. SCACCO, F. SOMMA, A. TIOURINE, and M. TONELLI
*Optical Pumping and Reorientation of the F_A Centers in KCl: Li^+
(10.2 and 13.4; 20.1; 22.5.2)
(b) Vol. **122**, No. 1, p. K83—K86 (1984).

J. A. GORRI, A. PAZ, and F. GUIU
*On the Stress Dependence of the Dislocation Velocity in Alkali Halide Crystals
(10.2; 22.5.2)
(a) Vol. **82**, No. 1, p. 85—91 (1984).

D. M. VANDERWALKER
*The Influence of Cu on Dislocation Motion in Si
(10.2; 22.1.2)
(a) Vol. **82**, No. 1, p. 93—99 (1984).

R. D. BAËTA
*Steady-State Creep in Quartz
(10.2; 22.6)
(a) Vol. **82**, No. 1, p. 101—109 (1984).

E. MACHO, J. LLOPIS, A. REMÓN, C. BALLESTEROS, and J. PIQUERAS
*Optical Study of Defects in Quenched MgO Crystals
(10.2 and 20.3; 20.1; 22.6)
(a) Vol. **82**, No. 1, p. 111—117 (1984).

Y. KAMIURA, F. HASHIMOTO, S. YONEYAMA, and T. NOBUSADA
*Reduction of Copper Contamination in Germanium on Heat Treatments above 700 °C
(10.2; 13.4; 22.1.1)
(a) Vol. **82**, No. 1, p. 135—141 (1984).

G. BRAUER, F. KERBE, ZS. KAJCSOS, and A. ASHRY
*Positron Annihilation Studies on BeO Powders
(10.2; 22.6)
(a) Vol. **82**, No. 1, p. 143—148 (1984).

K. NARASIMHA REDDY, U. V. SUBBA RAO, and M. LAKSHMIPATHI RAO
*Pre- and Post-Irradiation Deformation Effects on Thermoluminescence Studies in Barium-Doped NaCl Crystals
(10.2 and 20.3; 22.5.2)
(a) Vol. **82**, No. 1, p. 229—233 (1984).

M. Jolivet, M. Cadene, J. M. Abadie, and M. Rolland
*Morphology, Dopant Concentration, and Electrical Properties of H_2SO_4-and $BiCl_3$-Doped Polyacetylene Films
(10.2 and 14.3.1; 22.9)
(a) Vol. **82**, No. 1, p. 257—263 (1984).

G. A. Bielig
*Room Temperature Effects of γ-Rays on the Dislocation Damping Coefficient in LiF Crystals
(10.2 and 11; 22.5.2)
(a) Vol. **82**, No. 1, p. K43—K47 (1984).

K. S. Jones and S. J. Pearton
*Grain Boundaries in Germanium: Effects of Exposure to Plasmas
(10.2 and 14.3; 22.1.1)
(a) Vol. **82**, No. 1, p. K101—K105 (1984).

S. Dzhumanov and P. K. Khabibulaev
*The Coexistence of Unstable, Metastable, and Separated Frenkel Pair Defects in Solids (I): Basic Characteristics
(10.2; 13.4; 22.5.2)
(b) Vol. **122**, No. 2, p. 481—488 (1984).

T. Eshita, K. Tanimura, and N. Itoh
*Photo-Induced Transformation of Close Frenkel Pairs in Strontium Fluorite
(10.2 and 20.1; 20.3; 22.5.3)
(b) Vol. **122**, No. 2, p. 489—500 (1984).

A. Tóth, T. Keszthelyi, P. Kálmán, and J. Sárközi
*Diffusion Model for Charge Transport by Moving Dislocations in Simple Ionic Crystals
(10.2; 22.5.2)
(b) Vol. **122**, No. 2, p. 501—508 (1984).

R. K. Chaudhary, A. V. Krishna Rao, and K. V. Rao
*Effect of Magnetic Field on the F-Band Absorption in NaCl and KCl Single Crystals Irradiated with X-Rays under DC Fields
(10.2 and 20.1; 22.5.2)
(b) Vol. **122**, No. 2, p. 709—713 (1984).

A. Akilbekov and A. Elango
*Low-Temperature Pair Associates of H Centres in KBr
(10.2 and 13.4; 20.1; 22.5.2)
(b) Vol. **122**, No. 2, p. 715—723 (1984).

E. N. S. Muccillo and R. Muccillo
*Thermally Stimulated Depolarization Currents in γ-Irradiated (Mrad Range) $KBr:Sr^{++}$ Crystals
(10.2 and 11; 14.4; 22.5.2)
(b) Vol. **122**, No. 2, p. K127—K132 (1984).

V. N. Salomatov, N. N. Kristoffel, and I. A. Parfianovich
On the Nature of Metastable Color Centers in KBr
(13.4; 10.2; 22.5.2)
(b) Vol. **122**, No. 2, p. K183—K185 (1984).

N. M. Olekhnovich and A. L. Karpei
*Dynamical Effects of X-Ray Scattering in Laue Geometry for Si Crystals with Structure Defects
(10.2; 22.1.2)
(a) Vol. **82**, No. 2, p. 365—371 (1984).

L. C. Otero-Diaz, J. Soria, and M. A. Alario-Franco
*Extended Defects in Vanadium Doped Rutile. An EPR and TEM Study
(1.3 and 10.2; 19; 22.6.1)
(a) Vol. **82**, No. 2, p. 379—388 (1984).

D. J. H. Cockayne, P. Pirouz, Z. Liu, G. R. Anstis, and P. Karnthaler
*A Quantitative Analysis of Image Contrast from Extrinsic Stacking Faults
(1.4 and 10.2; 22.1.2)
(a) Vol. **82**, No. 2, p. 425—439 (1984).

B. I. Smirnov, T. N. Snezhkova, and O. L. Volkova
*Fracture in LiF Bicrystals at Plastic Deformation
(10.2; 22.5.2)
(a) Vol. **82**, No. 2, p. 441—447 (1984).

A. N. Pilyankevich and V. F. Britun
*Motion of Partial Dislocation in Silicon Carbide
(1.4 and 10.2; 22.3)
(a) Vol. **82**, No. 2, p. 449—457 (1984).

L. A. Kazakevich, P. F. Lugakov, and I. M. Filippov
*The Effect of Dislocations on the Charge-Carrier Recombination Processes in Irradiated Silicon
(10.2 and 13.4; 11; 14.3; 22.1.2)
(a) Vol. **82**, No. 2, p. 511—518 (1984).

B. Sorkin and H. Käämbre
*V_K Centres in Thermo-Exoemission from CsBr:Tl
(10.2 and 17; 13.4; 22.5.2)
(a) Vol. **82**, No. 2, p. K149—K152 (1984).

C. Ortiz, C. N. Afonso, R. M. Shelby, A. C. Tropper, and G. C. Bjorklund
*Photochemical Hole Burning in Polycrystalline and Neutron Irradiated NaF
(10.2 and 20.3; 22.5.2)
(b) Vol. **123**, No. 1, p. 79—84 (1984).

R. Rauch
*Photochemical Conversion of M_A Centres and Identification of M_A^+ Centres in CaF_2:Na
(10.2; 20.1; 20.3; 22.5.3)
(b) Vol. **123**, No. 1, p. 93--98 (1984).

M. J. Springis and J. A. Valbis
*Visible Luminescence of Colour Centres in Sapphire
(10.2 and 20.3; 22.6)
(b) Vol. **123**, No. 1, p. 335—343 (1984).

A. KRÓL and R. ZIELIŃSKA
*A New Defect Complex in CdTe: In, Li
(6 and 10.2; 20.1; 22.4.3)
(b) Vol. **123**, No. 1, p. K17—K20 (1984).

E. RADZHABOV
Impurity Centers on LiF: Li_2O
(13.4 and 20.3; 10.2; 22.5.2)
(b) Vol. **123**, No. 1, p. K79—K82 (1984).

L. A. SHUVALOV, S. A. GRIDNEV, V. I. KUDRYASH, and B. N. PRASOLOV
Ferroelastic Domain Structure Dynamics and Internal Friction of $KH_3(SeO_3)_2$ Single Crystals
(14.4.1; 10.2; 12.2; 22.8.1)
(a) Vol. **83**, No. 1, p. 131—137 (1984).

A. YA. NEIMAN, A. A. AFANASIEV, L. M. FEODOROVA, V. T. GABRIELIAN, and S. M. KARAGEZIAN
*Deviation from Stoichiometry and Electron Transfer in $PbMoO_4$
(10.2 and 14.3; 22.8.1)
(a) Vol. **83**, No. 1, p. 153—158 (1984).

B. V. KOBRIN and V. P. SHANTAROVICH
*Study of Ge_2S_3 Glass Undoped and Doped with Metals by Positron Annihilation Technique
(2 and 10.2; 22)
(a) Vol. **83**, No. 1, p. 159—164 (1984).

F. APPEL, M. BARTSCH, U. MESSERSCHMIDT, E. M. NADGORNYI, and S. N. VALKOVSKII
*Dislocation Motion and Plasticity in MgO Single Crystals
(10.2; 22.6)
(a) Vol. **83**, No. 1, p. 179—194 (1984).

V. L. INDENBOM, V. M. KAGANER, W. MÖHLING, and E. V. SUVOROV
*Plane-Wave X-Ray Images of Dislocations Parallel to the Diffraction Vector
(1.4 and 10.2; 22.1.2)
(a) Vol. **83**, No. 1, p. 195—205 (1984).

K. YASUTAKE, M. UMENO, and H. KAWABE
*Oxygen Precipitation and Microdefects in Czochralski-Grown Silicon Crystals
(10.2; 22.1.2)
(a) Vol. **83**, No. 1, p. 207—217 (1984).

K. LÖHNERT and E. KUBALEK
*The Cathodoluminescence Contrast Formation of Localized Non-Radiative Defects in Semiconductors
(10.2 and 20.3; 13.4; 22.2)
(a) Vol. **83**, No. 1, p. 307—314 (1984).

B. P. CHANDRA, R. D. VERMA, B. MAJUMDAR, and V. P. BHATT
Dependence of Mechanoluminescence on the Microhardness and Dislocation Density in Crystals
(12.2 and 20.3; 10.2; 22.5.2; 22.8.1)
(a) Vol. **83**, No. 1, p. K19—K22 (1984).

A. M. Evstigneev, L. F. Linnik, L. G. Linnik, E. A. Salkov, A. N. Krasiko, and O. V. Snitko
*Observation of Anisotropic Internal Stresses by Electroreflectance and Transverse Photovoltaic Effect Measurements
(10.2; 14.3.4; 16; 20.1; 22.1.1)
(a) Vol. **83**, No. 1, p. K31—K33 (1984).

V. N. Brudnyi, V. A. Novikov, A. D. Pogrebnyak, and Yu P. Surov
*Positron Annihilation in Electron-Irradiated p-$ZnGeP_2$ Compound
(10.2 and 11; 14.3; 22.8)
(a) Vol. **83**, No. 1, p. K35—K38 (1984).

B. Sorkin and H. Käämbre
*A New Evidence for Hole-Induced Thermo-Exoemission
(10.2; 17; 20.3; 22.5.2)
(a) Vol. **83**, No. 1, p. K39—K42 (1984).

B. B. S. Negi and P. N. Ram
*First-Order Raman Scattering in RbCl Containing F-Centres
(10.2; 6; 20.1; 22.5.2)
(b) Vol. **123**, No. 2, p. 491—495 (1984).

L. Bosi, P. Gagliardelli, and M. Nimis
*New Results on the Decay Properties of Perturbed and Unperturbed M-Centers in NaCl:$CaCl_2$
(10.2; 20.1; 20.3; 22.5.2)
(b) Vol. **123**, No. 2, p. 519—524 (1984).

N. P. Tishchenko
Kinetics of Impurity Particles in Low-Temperature Matrices
(9; 10.2; 23)
(a) Vol. **83**, No. 2, p. 513—519 (1984).

M. Böhm and A. Scharmann
*Entropy Change in Order–Disorder Transitions of Dipolar Relaxations
(10.2; 14.4; 22.8.1)
(a) Vol. **83**, No. 2, p. 537—541 (1984).

C. Ballesteros, J. Piqueras, J. Llopis, and R. González
Cathodoluminescence from MgO Single Crystals Containing a High Concentration of Anion Vacancies
(20.3; 10.2; 13.4; 22.6)
(a) Vol. **83**, No. 2, p. 645—649 (1984).

S. K. Maksimov
*Correlation between Thermodynamic Instability of $A^{III}B^{V}$ Heterocompositions and Degradation of Devices Based on These Materials
(10.2 and 14.3.2; 9; 22.2.4)
(a) Vol. **83**, No. 2, p. 685—692 (1984).

T. Kurobori and N. Takeuchi
Room Temperature, Stable Laser Action Using F_2 Centers in LiF:Mg, OH Crystals
(20.2; 10.2; 22.5.2)
(a) Vol. **83**, No. 2, p. K179—K182 (1984).

H. Reimer and F. Fischer
SERS from Ag Colloids in Alkali Halide Crystals
(6 and 20.1; 10.2; 22.5.2)
(b) Vol. **124**, No. 1, p. 61—67 (1984).

A. A. Alybakov, V. A. Gubanova, G. S. Denisov, and B. S. Umurzakov
Influence of Oxygen Centers on the Thermoluminescence of NaF Crystals
(13.4 and 20.3; 10.2; 22.5.2)
(b) Vol. **124**, No. 1, p. K75—K78 (1984).

M. Moreno and M. T. Barriuso
*An Insight into the $10Dq$ Value Found in $CaZnF_4:Mn^{2+}$
(10.2 and 13.4; 20.3; 22.5)
(b) Vol. **124**, No. 1, p. K83—K86 (1984).

S. K. Maksimov, M. Ziegler, I. I. Khodos, I. I. Snighiryova, and M. Sh. Shikhsaidov
*High Resolution Electron Microscopic Investigations of Dislocations in Deformed GaAs Single Crystals Doped with Te
(10.2; 1.4; 22.2.1)
(a) Vol. **84**, No. 1, p. 79—86 (1984).

G. Kowalski
*X-Ray Diffraction Contrast and Ray-Paths in the Case of a Dislocation in a Particular Position
(10.2; 22.1.2)
(a) Vol. **84**, No. 1, p. 119—132 (1984).

L. Pasemann
*Theoretical Study of the Temperature Dependence of EBIC Contrast from Individual, Surface-Parallel Dislocations in a Schottky Diode
(1.4 and 10.2; 14.3.3; 22.1.2)
(a) Vol. **84**, No. 1, p. 133—142 (1984).

E. Bugiel, M. Kittler, A. Borchardt, and H. Richter
*On the Gettering Efficiency of Crystal Defects in Silicon
(10.2; 14.3.3; 22.1.2)
(a) Vol. **84**, No. 1, p. 143—147 (1984).

V. V. Kveder, R. Labusch, and Yu. A. Ossipyan
*The Exodiffusion of Hydrogen in Dislocated Crystalline Silicon
(10.2 and 13.4; 22.1.2)
(a) Vol. **84**, No. 1, p. 149—156 (1984).

A. Watterich, E. Szontagh, and É. Kocsárdy
*Paramagnetic Centres in Aluminium-Oxide Layers Prepared by Anodic Oxidation
(10.2; 19; 20.1; 22.6)
(a) Vol. **84**, No. 1, p. 165—169 (1984).

R. Pareja, R. González, and M. A. Pedrosa
*Study of Thermochemically Reduced and Electron-Irradiated $LiNbO_3$ Single Crystals by Positron Annihilation and Optical Absorption Measurements
(10.2 and 11; 20.1; 22.8.1)
(a) Vol. **84**, No. 1, p. 179—183 (1984).

A. Tanaka, T. Sukegawa, and T. Sugiura
Phosphorus Pressure Effects on GaP Crystal Quality in the Solution Growth
(3; 10.2; 22.2.2)
(a) Vol. **84**, No. 1, p. K9—K12 (1984).

J. Muto and Y. Kobayashi
*Photobleaching Behaviours of Rhodamine 6G Diffused in Polymethyl Methacrylate and in the Copolymer of Methyl Methacrylate with Methacrylic Acid
(10.2 and 20; 22.9)
(a) Vol. **84**, No. 1, p. K29—K33 (1984).

V. V. Aristov, I. E. Bondarenko, N. N. Dryomova, V. V. Kazmiruk, and E. B. Yakimov
*The EBIC Contrast of Dislocation Slip Planes in Silicon
(1.4 and 10.2; 14.3.4; 22.1.2)
(a) Vol. **84**, No. 1, p. K43—K45 (1984).

S. Bandyopadhyay and S. K. Deb
*Solution Enthalpies of Divalent Defects in Alkali Fluorides and Chlorides
(10.2; 22.5.2)
(b) Vol. **124**, No. 2, p. 545—557 (1984).

P. Kleinert
*Cluster-Bethe-Lattice Treatment of the Interstitial Boron Centre in Irradiated Gallium Arsenide
(10.2; 11; 22.2.1)
(b) Vol. **124**, No. 2, p. 559—564 (1984).

S. Minagawa
*Mechanical, Electrical, and Thermal Fields Produced by Moving Dislocations and Disclinations in Thermo-Piezoelectric Continua
(10.2 and 14.4)
(b) Vol. **124**, No. 2, p. 565—571 (1984).

L. O. Schwan, R. Florian, D. Schmid, E. Betz, W. Schrof, B. Walker, and H. Port
*Dynamics of the Optical Excitation–Deexcitation Cycle of O_2^- Centers in KCl
(10.2 and 13.4; 20.3; 22.5.2)
(b) Vol. **124**, No. 2, p. 741—745 (1984).

S. Kh. Milshtein, D. C. Joy, S. D. Ferris, and L. C. Kimerling
*Defect Characterization Using SEM–CCM. Relative Contrast Measurements
(1.4 and 10.2; 14.3.3; 22.1.2)
(a) Vol. **84**, No. 2, p. 363—369 (1984).

M. T. Sebastian and P. Krishna
X-Ray Diffraction from a SiC Crystal Undergoing the 3C–6H Solid State Transformation by Non-Random Microtwinning
(1.2; 10.2; 22.3)
(a) Vol. **84**, No. 2, p. 401—409 (1984).

E. B. YAKIMOV, N. A. YARYKIN, and V. I. NIKITENKO
*The Influence of Thermal Treatment on Thermally Stimulated Depolarization Spectrum of the Electret State in Plastically Deformed Silicon
(10.2 and 13.4; 14.4; 22.1.2)
(a) Vol. **84**, No. 2, p. 443—450 (1984).

G. BRAUER, F. KERBE, ZS. KAJCSOS, and A. ASHRY
*Positron Annihilation Studies on Al_2O_3 Powders
(10.2; 13.3; 22.6)
(a) Vol. **84**, No. 2, p. 451—455 (1984).

P. F. LUGAKOV and V. V. LUKYANITSA
*Peculiarities of Radiation Defect Formation and Annealing in n-Si Due to Their Interaction with Each Other and Defect Clusters
(10.2 and 11; 14.3; 22.1.2)
(a) Vol. **84**, No. 2, p. 457—464 (1984).

P. GROSBRAS, J. L. DEMENET, H. GAREM et J. C. DESOYER
*Equilibre statique entre dislocations partielles dans du silicium déformé sous contrainte uniaxiale (Static Equilibrium between Partial Dislocations in Silicon Deformed by Uniaxial Stress)
(10.2; 22.1.2)
(a) Vol. **84**, No. 2, p. 481—492 (1984).

P. GILLE and M. SCHENK
*Structural Incorporation and Electrical Efficiency of Tl in PbTe
(10.2; 14.3; 22)
(a) Vol. **84**, No. 2, p. K121—K124 (1984).

L. N. SKUJA, A. N. TRUKHIN, and A. E. PLAUDIS
Luminescence in Germanium-Doped Glassy SiO_2
(20.3; 2; 10.2; 13.4; 20.1; 22.6)
(a) Vol. **84**, No. 2, p. K153—K157 (1984).

N. N. KOREN, V. I. LEVCHENKO, V. V. DIKAREVA, and A. V. KAZUSHCHIK
Physical Properties of Germanium Telluride Crystals Grown from Gas Phase
(14.3 and 15; 10.2; 22)
(a) Vol. **84**, No. 2, p. K173—K176 (1984).

R. P. GAIROLA
Low-Temperature Lattice Thermal Conductivity of Nonmetallic Solids with Isotopic Impurities
(6 and 8; 10.2; 22; 22.1.1)
(b) Vol. **125**, No. 1, p. 65—74 (1984).

V. NAGIRNYI, T. SOOVIK, and S. ZAZUBOVICH
*Photostimulated Reoriention of $H_z^{[001]}(Pb^{2+})$ Centres in a $KCl:PbCl_2$ Crystal
(10.2 and 20.3; 22.5.2)
(b) Vol. **125**, No. 1, p. 139—145 (1984).

G. R. Augst and Yu. I. Vladislavskii
*A Theoretical Study of Simple Complexes of Point Defects in Alkali Halides by the Lattice Static Method
(10.2; 22.5.2)
(b) Vol. **125**, No. 1, p. 155—160 (1984).

E. Mugeński and R. Cywiński
Low-Temperature Photoluminescence of the Aggregate Centres in Eu^{2+}-Doped KCl Crystals
(20.3; 10.2; 22.5.2)
(b) Vol. **125**, No. 1, p. 381—386 (1984).

G. I. Bersuker and V. Z. Polinger
Multimode Pseudo Jahn-Teller Effect for Off-Centre Impurities in Crystals
(6 and 13.4; 10.2; 22.5.2)
(b) Vol. **125**, No. 1, p. 401—408 (1984).

Yu. M. Annenkov, V. F. Stolarenko, and T. S. Frangulian
*Effect of Powerful Pulsed Electron Irradiation on Creation of Hole Colour Centres in KCl at 300 K
(10.2 and 20.1; 11; 22.5.2)
(b) Vol. **125**, No. 1, p. K37—K40 (1984).

V. A. Chepurnoi, S. N. Mysovskii, I. A. Parfianovich, V. N. Salomatov, and A. P. Shkadarevich
*Influence of Oxygen Defects on the Position and Form of the F_2^+ Center Zero-Phonon Lines in LiF
(10.2 and 13.4; 20.3; 22.5.2)
(b) Vol. **125**, No 1, p. K41—K44 (1984).

A. V. Mudryi, A. L. Pushkarchuk, V. D. Tkachev, and A. G. Ulyashin
*Noble Gas Atoms as Interstitials in Silicon and Diamond
(10.2 and 13.4; 22.1; 22.1.2)
(b) Vol. **125**, No. 1, p. K75—K78 (1984).

V. Bethkenhagen and K. Hübner
*Bethe-Lattice Approach to the Electronic Structure of Intrinsic Defects in v-SiO_2
(10.2 and 13.4; 22.6)
(b) Vol. **125**, No. 1, p. K79—K83 (1984).

E. Mizera, M. Sundberg, and P. Werner
Defect Structure of ZnSe Crystals Investigated by Electron Microscopy
(1.4; 10.2; 22.4.2)
(a) Vol. **85**, No. 1, p. 83—88 (1984).

P. Gaworzewski, E. Hild, F.-G. Kirscht, and L. Vecsernyés
*Infrared Spectroscopical and TEM Investigations of Oxygen Precipitation in Silicon Crystals with Medium and High Oxygen Concentrations
(1.4 and 10.2; 20.1; 22.1.2)
(a) Vol. **85**, No. 1, p. 133—147 (1984).

V. G. Artyushenko, V. V. Voitsekhovskii, L. S. Kornienko, I. S. Lisitskii, and A. O. Rybaltovskii
Effect of Ionizing Radiation on Optical Properties of Thallium Halides in the 0.36 to 15 μm Range
(11 and 20.1; 10.2; 22.5)
(a) Vol. **85**, No. 1, p. 167—171 (1984).

Y. Kamiura and F. Hashimoto
*Origin of a Shallow Acceptor in Quenched Germanium
(10.2 and 13.4; 14.3; 22.1.1)
(a) Vol. **85**, No. 1, p. 227—234 (1984).

H. Booyens and J. H. Basson
*Piezoelectrically Induced Charge Distribution around Dislocations in CdTe and HgCdTe
(10.2 and 13.4; 14.4; 22.4.3; 22.4.4)
(a) Vol. **85**, No. 1, p. 243—248 (1984).

K. Somaiah and V. Hari Babu
*Two Types of F-Centres and Thermoluminescence in BaFCl Crystals
(10.2; 20.1; 20.3; 22.5.4)
(a) Vol. **85**, No. 1, p. 265—271 (1984).

D. Hildebrandt, H. Strusny, R. Groetzschel, E. Kotai, and F. Paszti
*Damage and Trapping Behaviour of Crystalline Silicon at Low Energy Deuterium Implantation
(10.2 and 11; 22.1.2)
(a) Vol. **85**, No. 1, p. K35—K38 (1984).

A. V. Dvurechenskii, B. P. Kashnikov, G. P. Pokhil, V. P. Popov, A. V. Tulinov, and A. A. Turinge
*Defect Structure Study with Planar Channeling in Pulse-Annealed Ion-Implanted Silicon
(10.2 and 11; 22.1.2)
(a) Vol. **85**, No. 1, p. K39—K42 (1984).

V. D. Tkachev, G. Hölzer, and A. R. Chelyadinskii
*Damage Profiles in Ion Implanted Silicon
(10.2 and 11; 22.1.2)
(a) Vol. **85**, No. 1, p. K43—K46 (1984).

G. Heder and J. M. Spaeth
*ENDOR Investigation of H_i^0 (Li) Centres in KCl
(10.2 and 19; 13.4; 22.5.2)
(b) Vol. **125**, No. 2, p. 523—531 (1984).

D. Vignaud and J. L. Farvacque
Optical Absorption Studies of Plastically Deformed InSb
(13.4 and 20.1; 10.2; 22.2.3)
(b) Vol. **125**, No. 2, p. 785—794 (1984).

P. Varotsos, K. Alexopoulos, and M. Lazaridou
*On the Proportionality of the Migration Volume and the Migration Enthalpy in Fluorides
(10.2; 9; 22.5.3)
(b) Vol. **125**, No. 2, p. K109—K112 (1984).

A. Dominguez-Rodriguez, J. Cabrera-Caño, R. Marquez, and J. Castaing
*High-Temperature Creep of Equimolecular NiO–CoO Solid Solution
(10.2 and 12.2; 22.6.1)
(a) Vol. **85**, No. 2, p. 445—448 (1984).

H. BOOYENS and J. H. BASSON
*On the Optimal Choice of Substrate for $Hg_{0.8}Cd_{0.2}Te$ Epitaxial Layers
(1.5 and 10.2; 3; 22.4.4)
(a) Vol. **85**, No. 2, p. 449—454 (1984).

F. SCHNEIDER and S. UNTERRICKER
*Radiation Damage and Its Annealing Behaviour in InP after Recoil Implantation of ^{118}Sb and ^{111}In Observed by TDPAC
(10.2 and 11; 5; 22.2.2)
(a) Vol. **85**, No. 2, p. 455—461 (1984).

M. SUEZAWA and K. SUMINO
*On the Annihilation of Thermal Donors in Silicon Crystals
(10.2 and 13.4; 20.1; 22.1.2)
(a) Vol. **85**, No. 2, p. 469—472 (1984).

A. K. ARORA
*Isochronal Annealing Study of Aggregation in Doped Alkali Halide Crystals
(10.2; 22.5.2)
(a) Vol. **85**, No. 2, p. 491—495 (1984).

W. HERGERT and L. PASEMANN
Theoretical Study of the Information Depth of the Cathodoluminescence Signal in Semiconductor Materials
(20.3; 10.2; 14.3; 22)
(a) Vol. **85**, No. 2, p. 641—648 (1984).

J. L. LINDSTRÖM, B. G. SVENSSON, J. W. CORBETT, and G. S. OEHRLEIN
*On the Complex of the Oxygen Interstitial and the Silicon Interstitial in Silicon
(10.2; 22.1.2)
(a) Vol. **85**, No. 2, p. K109—K111 (1984).

A. R. KRASNAYA, L. N. OSTER, and V. YA. YASKOLKO
Exoelectron Emission from Calcium Sulfate Excited by Slow Electrons
(17; 10.2; 22.8.1)
(a) Vol. **85**, No. 2, p. K145—K148 (1984).

K. WESSER, S. N. SAHU, J. P. KARINS, and J. W. CORBETT
*Comparison of an Extended Interstitial with a Split Interstitial in Silicon
(10.2 and 13.4; 22.1.2)
(b) Vol. **126**, No. 1, p. 157—161 (1984).

I. S. SMIRNOVA
*Dislocations in $A^{II}B^{VI}$ and $A^{III}B^{V}$ Piezoelectric Crystals
(10.2 and 14.3; 22.2; 22.4)
(b) Vol. **126**, No. 1, p. 177—189 (1984).

D. VIGNAUD and J. L. FARVACQUE
Free-Carrier Optical Absorption Induced by Dislocation Scattering Mechanisms in III–V-Compounds
(13.1 and 20.1; 10.2; 22.2)
(b) Vol. **126**, No. 1, p. 351—359 (1984).

V. Bethkenhagen and K. Hübner
Cluster-Bethe-Lattice Approach to the Electronic Structure of a Twofold Coordinated Silicon Atom as a New Intrinsic Defect in v-SiO_2
(13.4; 10.2; 22.6)
(b) Vol. **126**, No. 1, p. K71—K76 (1984).

R. A. Vardanian and G. G. Kirakosian
*Electron Scattering by a Charged Dislocation
(10.2 and 14.3; 22.1.1)
(b) Vol. **126**, No. 1, p. K83—K86 (1984).

M. G. Blanchin and L. A. Bursill
*Extended versus Small Defect Equilibria in Non-Stoichiometric Rutile (I): Dissolution and Re-Precipitation of Extended Defects Observed by In-Situ Transmission Electron Microscopy
(10.2; 1.3; 22.6)
(a) Vol. **86**, No. 1, p. 101—109 (1984).

V. V. Kalinin and N. N. Gerasimenko
*On the Effect of Doping Impurities on the Formation of Rod-Like Defects in Silicon
(10.2 and 11; 22.1.2)
(a) Vol. **86**, No. 1, p. 185—190 (1984).

Ch. Lushchik, J. Kolk, A. Lushchik, and N. Lushchik
*Radiational Creation of Frenkel Defects in KCl–Tl
(10.2; 19; 20.1; 20.3; 22.5.2)
(a) Vol. **86**, No. 1, p. 219—227 (1984).

H.-R. Höche and J. Schreiber
*Anisotropic Deformation Behaviour of GaAs
(10.2; 22.2.1)
(a) Vol. **86**, No. 1, p. 229—236 (1984).

H. Bender
*Investigation of the Oxygen-Related Lattice Defects in Czochralski Silicon by Means of Electron Microscopy Techniques
(10.2; 1.4; 22.1.2)
(a) Vol. **86**, No. 1, p. 245—261 (1984).

A. Kozanecki and H. Rzewuski
Distant DAP Bands in Electron-Irradiated and Ion-Implanted CdSe
(11 and 20.3; 10.2; 22.4.2)
(a) Vol. **86**, No. 1, p. 263—270 (1984).

E. Hartmann, L. Kovács, and J. Paitz
*Electrical Conductivity of Gadolinium–Gallium Garnet (GGG) Crystals
(10.2 and 14.4; 22.8.2)
(a) Vol. **86**, No. 1, p. 401—405 (1984).

A. M. Afanasiev, S. M. Afanasiev, P. A. Aleksandrov, R. M. Imamov, and E. M. Pashaev
*Grazing Bragg-Laue Diffraction for Studying the Crystal Structure of Thin Films
(1.6 and 10.2; 22.2.1; 22.2.2)
(a) Vol. **86**, No. 1, p. K1—K5 (1984).

A. A. Evtushenko, V. F. Petrenko, and I. A. Ryzhkin
*Electric Polarization of Ice at Nonuniform Elastic Strains
(10.2 and 14.4; 12.2; 22)
(a) Vol. **86**, No. 1, p. K31—K34 (1984).

R. Tognato
*Overheating in Semiconductors with a Diamond-Like Structure
(10.2; 22)
(a) Vol. **86**, No. 1, p. K35—K37 (1984).

M. S. Jahan and D. W. Cooke
*Thermal Evolution of Low-Temperature Manganese Centers in X-Irradiated CaF_2:Mn
(10.2 and 19; 11; 20.3; 22.5.3)
(b) Vol. **126**, No. 2, p. 687—692 (1984).

K. Iwahana
*Magnetic Field Dependence of the Spin Polarization of the F Center
(10.2; 18.1; 22.5.2)
(b) Vol. **126**, No. 2, p. 693—702 (1984).

S. Bandyopadhyay and S. K. Deb
*Solution Enthalpies of Trivalent Defects in Alkali Chlorides
(10.2; 22.5.2)
(b) Vol. **126**, No. 2, p. K117—K121 (1984).

M. Lazaridou
*Connection of the Defect Volumes to the Bulk Properties in CaF_2
(10.2; 22.5.3)
(b) Vol. **126**, No. 2, p. K123—K124 (1984).

M.G. Blanchin and L. A. Bursill
*Extended Versus Small Defect Equilibria in Non-Stoichiometric Rutile (II): Structural Mechanisms for Dissolution and Reprecipitation of Extended Defects
(10.2; 1.3; 22.6)
(a) Vol. **86**, No. 2, p. 491—498 (1984).

D. M. Vanderwalker
*The Nucleation of Ni Silicides on Dislocations and Twins in Silicon
(1.6 and 10.2; 1.2; 21.1; 22.1.2)
(a) Vol. **86**, No. 2, p. 507—516 (1984).

D. Shaw
Alternative Mechanisms for the Diffusion of Sn and Zn in GaAs
(9; 10.2; 22.2.1)
(a) Vol. **86**, No. 2, p. 629—635 (1984).

S. Ihara, Y. Warita, and K. Suzuki
*Ionic Conductivity in $AgI_{1-x}Cl_x$
(10.2 and 14.4; 22.5.4)
(a) Vol. **86**, No. 2, p. 729—734 (1984).

V. M. Babich, Yu. P. Dotsenko, and V. B. Kovalchuk
*Some Characteristic Properties of Thermal Donor Formation in Oxygen-Containing Silicon at 450 °C
(10.2; 22.1.2)
(a) Vol. **86**, No. 2, p. K91—K94 (1984).

Y. Kamiura and F. Hashimoto
*Annealing Behaviour of Shallow Acceptors in Quenched Germanium
(10.2; 13.4; 14.3; 22.1.1)
(a) Vol. **86**, No. 2, p. K95—K97 (1984).

11. Irradiation Effects (X-Ray Diffraction Investigations See 1 and 10)

Review Article

V. A. Pavlov, V. I. Shalaev, and V. M. Alyabiev
Effect of Dislocation Structure on Creep and Fracture of Metals and Alloys (II): Irradiation
(10.1; 11; 21)
(a) Vol. **85**, No. 2, p. 315—333 (1984).

Original Papers and Short Notes

H. Stachowiak
*Screening of a Proton Moving through the Electron Gas
(11 and 13.4; 21)
(b) Vol. **121**, No. 1, p. 307—315 (1984).

S. Banerjee and K. Urban
Kinetics of Order–Discorder Transformation in Alloys under Electron Irradiation
(1.2; 1.3; 11; 21.1)
(a) Vol. **81**, No. 1, p. 145—162 (1984).

A. D. Pogrebnyak, E. Yu. Boyarko, Yu. Yu. Kryuchkov, and S. A. Vorobev
*Positron Annihilation and Profiles of Radiation Damages in GaAs and Si Crystals Irradiated by Supercurrent Proton or Electron Beams
(11; 14.3; 22.1.2; 22.2.1)
(a) Vol. **81**, No. 1, p. 217—225 (1984).

H. J. Blythe, F. Walz, and H. Kronmüller
Magnetic Investigation of the Annealing Behaviour of Some Dilute Iron–Nickel Alloys after Low-Temperature Electron Irradiation
(10.1 and 18.2; 11; 21.1; 21.1.1)
(a) Vol. **81**, No. 1, p. 227—238 (1984).

E. Wieser, H. Syhre, F. G. Rüdenauer, and W. Steiger
Influence of Flash-Lamp Annealing on the Diffusion Behaviour of Ion Implanted Boron Profiles
(9; 11; 22.1.2)
(a) Vol. **81**, No. 1, p. 247—252 (1984).

C. Y. Chen, K. L. Sweeney, and L. E. Halliburton
Reduction and Radiation Effects in Lithium Tantalate
(10.2; 11; 19; 20.1; 22.8.1)
(a) Vol. **81**, No. 1, p. 253—257 (1984).

P. Weissbrodt, F. Storbeck, and W. Hauffe
*Sputtering Behaviour of Vanadium Silicide Single Crystals under 10 keV Kr^+ Ion Bombardment
(1.6 and 11; 21.1)
(a) Vol. **81**, No. 1, p. 259—265 (1984).

J. J. Jimenez-Rodriguez, A. Gras-Marti, and G. Carter
*Description of Damage Accumulation and Overlap in Ion Irradiated Solids
(11)
(a) Vol. **81**, No. 1, p. 267—272 (1984).

P. Kindl
*Positron Lifetime Studies on γ-Irradiated PTFE of Different Initial Crystallinity
(11; 10.2; 22.9)
(a) Vol. **81**, No. 1, p. 293—298 (1984).

A. Chandorkar, V. T. Karulkar, A. S. Vengurlekar, and K. V. Ramanathan
MOS Capacitance–Voltage Characteristics and Dielectric Properties of Ion Implanted Thermal Oxides on Silicon
(14.3.3; 11; 14.4; 22.1.2)
(a) Vol. **81**, No. 1, p. 407—414 (1984).

V. E. Gusakov and V. Yu. Yavid
*The Role of Nonelastic Nuclear Processes in the Formation of Defect Clusters in Semiconductors Irradiated with High-Energy Protons
(11; 22.1.1)
(a) Vol. **81**, No. 2, p. 469—472 (1984).

T. Gessner, M. Pasemann, and B. Schmidt
*Erratum to: Calculation of Neutron-Induced Defect Clusters in Silicon and Comparison with TEM Investigations
(11; 1.4; 22.1.2)
(a) Vol. **81**, No. 2, p. 751 (1984).

H. J. Goldsmid, H. Hora, and G. L. Paul
*Anomalous Heat Conduction of Ion-Implanted Amorphous Layers in Silicon Crystals Using a Laser-Probe Technique
(2 and 11; 8; 22.1.2)
(a) Vol. **81**, No. 2, p. K127—K130 (1984).

M. S. Yunusov, S. N. Abdurakhmanova, M. A. Zaikovskaya, and B. L. Oksengendler
*The Influence of Minority Carrier Injection on Subthreshold Defects in Silicon
(11 and 14.3.4; 22.1.2)
(a) Vol. **81**, No. 2, p. K145—K149 (1984).

Z. Januškevičius, V. Koval, A. Kudin, A. Sodeika, V. Tartachnik, and I. Tychina
*Radiation Defects in Single Crystals of Zinc Diphosphide
(11 and 14.3; 22.3)
(a) Vol. **81**, No. 2, p. K191—K193 (1984).

E. G. Vyatkin, A. M. Taratin, and S. A. Vorobiev
*Computer Simulation of Glancing Scattering of a Fast Proton Beam by a Crystal Surface
(11; 21)
(b) Vol. **122**, No. 1, p. 29—33 (1984).

M. G. Abramishvili, V. I. Altukhov, T. L. Kalabegishvili, and V. G. Kvachadze
*Thermal Conductivity and Optical and Paramagnetic Absorption Spectra of Reactor Irradiated MgO Crystals
(8 and 11; 20.1; 22.6)
(b) Vol. **122**, No. 1, p. 95—104 (1984).

A. P. Pathak
Position Dependence of Channeling Stopping Power
(11 and 13; 22.1.2)
(b) Vol. **122**, No. 1, p. 171—174 (1984).

A. I. Melker and S. N. Romanov
*Athermal Divacancy Formation in High Energy Displacement Cascades
(11)
(b) Vol. **122**, No. 1, p. K17—K21 (1984).

W. Hohenau
*The Influence of Silver and Copper Electrolysis on the Radiation-Induced Optical Absorption of Natural Quartz
(11 and 20.1; 22.6)
(a) Vol. **82**, No. 1, p. 119—124 (1984).

H. Kerkow, G. Kreysch, and B. Lukasch
*The Influence of Foreign Atoms on the Epitaxial Annealing of Ion-Implanted Silicon
(11; 3; 22.1.2)
(a) Vol. **82**, No. 1, p. 125—133 (1984).

C. Wagner and A. El-Sadek
*The Annealing Behaviour of Arsenic-Implanted Silicon between 600 and 800 °C Characterized by Carrier Density Profiles
(11 and 14.3; 22.1.2)
(a) Vol. **82**, No. 1, p. 149—155 (1984).

E. Wieser and D. Panknin
*Electrical Activation and Damage Annealing of Boron-Implanted Silicon by Flash-Lamp Irradiation
(11; 14.3; 22.1.2)
(a) Vol. **82**, No. 1, p. 171—177 (1984).

O. Bostanjoglo and E. Endruschat
Time-Resolved TEM of Laser-Annealing
(1.2 and 1.4; 2; 11; 22.1.1)
(a) Vol. **82**, No. 1, p. K1—K3 (1984).

W. Scharf, J.-W. Erben, A. Wolf, M. Heber, C. Hamann, C. Weissmantel, M. Voelskov, J. Matthäi, R. Kögler, R. Klabes, and E. Wieser
Growth of Single-Crystalline Regions on Amorphous Insulating Substrates by Zone-Melting Recrystallization
(2 and 3; 11; 22.1.2)
(a) Vol. **82**, No. 1, p. K5—K9 (1984).

G. A. BIELIG
*Room Temperature Effects of γ-Rays on the Dislocation Damping Coefficient in LiF Crystals
(10.2 and 11; 22.5.2)
(a) Vol. **82**, No. 1, p. K43—K47 (1984).

E. N. S. MUCCILLO and R. MUCCILLO
*Thermally Stimulated Depolarization Currents in γ-Irradiated (Mrad Range) KB:Sr^{++} Crystals
(10.2 and 11; 14.4; 22.5.2)
(b) Vol. **122**, No. 2, p. K127—K132 (1984).

TH. WIRTH, V. ATZRODT, and H. LANGE
*Dynamic Alternations of the Surface Composition during Sputtering of Silicides
(1.6 and 11; 21)
(a) Vol. **82**, No. 2, p. 459—466 (1984).

V. I. KUZNETSOV, P. F. LUGAKOV, and A. V. TSIKUNOV
*Interaction of Radiation Defects with the Surface of Silicon
(11 and 14.3.2; 13.4; 22.1.2)
(a) Vol. **82**, No. 2, p. 467—473 (1984).

G. A. KACHURIN, V. A. MAYER, S. I. ROMANOV, M. VOELSKOW, R. KLABES, and E. WIESER
*Low-Energy Implantation of Arsenic in Silicon
(11; 14.3; 22.1.2)
(a) Vol. **82**, No. 2, p. 475—480 (1984).

L. A. KAZAKEVICH, P. F. LUGAKOV, and I. M. FILIPPOV
The Effect of Dislocations on the Charge-Carrier Recombination Processes in Irradiated Silicon
(10.2 and 13.4; 11; 14.3; 22.1.2)
(a) Vol. **82**, No. 2, p. 511—518 (1984).

R. KLABES, J. MATTHÄI, M. VOELSKOW, E. WIESER, J.-W. ERBEN, W. SCHARFF, and C. WEISSMANTEL
Epitaxial Regrowth of Amorphous or Polycrystalline Silicon Layers on Silicon Single Crystals and Bridging Epitaxy by Flash Lamp Irradiation
(1.5 and 3; 2; 11; 22.1.2)
(a) Vol. **82**, No. 2, p. K121—K123 (1984).

F. EICHHORN, J. KULDA, and P. MIKULA
*Small Angle Neutron Scattering on a Reactor Beam Irradiated Silicon Single Crystal
(11; 22.1.2)
(a) Vol. **82**, No. 2, p. K141—K143 (1984).

V. HAJKO, A. ZENTKO, M. TIMKO, and V. HAJKO, JR.
*Isotopic Effect in Neutron-Irradiated $Fe_{40}Ni_{40}B_{20}$ Alloys
(11 and 18.2; 2; 21.1.1)
(a) Vol. **82**, No. 2, p. K159—K162 (1984).

V. N. BRUDNYI, M. A. KRIVOV, A. I. POTAPOV, and YU. V. RUD
*Electrical Properties of Electron- and Proton-Irradiated GaAs and $ZnGeAs_2$ Solid Solutions
(11 and 14.3; 22.2.1; 22.8)
(a) Vol. **82**, No. 2, p. K191—K194 (1984).

V. Rybka, V. Hnatowicz, J. Kvítek, J. Vacík, and B. Schmidt
*Determination of the Range Profiles of Boron Implanted into Si and SiO_2
(11; 22.1.2; 22.6)
(a) Vol. **83**, No. 1, p. 165—171 (1984).

I. I. Kolkovskii, P. F. Lugakov, and V. V. Shusha
*Charge-Carrier Recombination in Silicon Irradiated with γ-Rays of Different Energies
(11 and 13.4; 22.1.2)
(a) Vol. **83**, No. 1, p. 299—306 (1984).

A. P. Tyutnev, V. S. Saenko, A. I. Karpechin, G. S. Mingaleev, V. I. Arkhipov, A. I. Rudenko, and A. V. Vannikov
*Radiation-Induced Conductivity in Polymers under Continous Irradiation
(11 and 14.3; 22.9)
(a) Vol. **83**, No. 1, p. 365—373 (1984).

S. Kashchieva, P. Danesh, and A. Dyakov
*RF Annealing of the Implantation-Induced Defects in Silicon Using Hydrogen Plasma
(11 and 13.4; 14.3.3; 22.1.2)
(a) Vol. **83**, No. 1, p. 411—417 (1984).

M. Wagner, H.-D. Geiler, G. Andrä, and G. Götz
*Control of Explosive Liquid-Phase Crystallization of Ion-Implanted Silicon Layers by Double Pulse Laser Irradiation
(3 and 11; 22.1.2)
(a) Vol. **83**, No. 1, p. K1—K3 (1984).

N. Brudnyi, V. A. Novikov, A. D. Pogrebnyak, and Yu. P. Surov
*Positron Annihilation in Electron-Irradiated p-$ZnGeP_2$ Compound
(10.2 and 11; 14.3; 22.8)
(a) Vol. **83**, No. 1, p. K35—K38 (1984).

E. Ejimanya and D. V. Morgan
Damage Induced Degradation of Metal Contacts on Gallium Arsenide
(14.3.3; 11; 22.2.1)
(a) Vol. **83**, No. 1, p. K87—K92 (1984).

A. Golzené, B. Meyer, and C. Schwab
*Experimental Evidence for an Associated Defect Model for the Neutron Generated As_{Ga} Center in Gallium Arsenide
(11 and 19; 13.4; 22.2.1)
(b) Vol. **123**, No. 2, p. K125—K128 (1984).

P. F. Lugakov and V. V. Lukyanitsa
*Efficiency of Radiation Defect Formation in Silicon at Various Intensities of Electron and γ-Ray Irradiation
(11; 13.4; 22.1.2)
(a) Vol. **83**, No. 2, p. 521—528 (1984).

H. Bartsch, D. Hoehl, and G. Kästner
*Radiation-Induced Rod-Like Defects in Silicon and Germanium
(1.4 and 11; 22.1.1; 22.1.2)
(a) Vol. **83**, No. 2, p. 543–551 (1984).

V. G. Grygoriev, F. G. Neshov, A. A. Puzanov, and A. R. Urmanov
*Effects of Electronic Structure on Ion Stopping Cross-Section in Solids
(11; 22.1.2)
(a) Vol. **83**, No. 2, p. 573—580 (1984).

N. K. Bondareva, I. V. Krylova, and V. B. Golubev
Thermostimulated Exoemission and Radical Recombination in Polyethylene Irradiated with an Electron Beam
(13.3 and 17; 11; 19; 22.9)
(a) Vol. **83**, No. 2, p. 589—595 (1984).

R. L. Calvert and R. J. Danby
Thermoluminescence and Radiophotoluminescence from Eu- and Sm-Doped $CaSO_4$
(20.3; 11; 13.4; 22.8.1)
(a) Vol. **83**, No. 2, p. 597—604 (1984).

S. Bhushan, D. Diwan, and S. P. Kathuria
Thermoluminescence of ZnO:Pr and ZnO:Gd under UV, β- and γ-Ray Irradiations
(20.3; 11; 13.4; 22.4)
(a) Vol. **83**, No. 2, p. 605—611 (1984).

K. D. Glinchuk, A. V. Prokhorovich, and N. S. Zayats
Effect of 2.2 MeV Electron Irradiation and Annealing on the 1.37 eV Emission Band in Zinc-Doped p-Type GaAs
(13.4 and 20.3; 11; 22.2.1)
(a) Vol. **83**, No. 2, p. 625—630 (1984).

B. Sorkin and H. Käämbre
Thermo-Exoemission of NaCl:Ag Crystals, Cleaved before and after X-Irradiation
(17; 11; 13.3; 13.4; 22.5.2)
(a) Vol. **83**, No. 2, p. K171—K174 (1984).

K. S. V. Nambi
Comment on "The Rôle of Sulphur-Oxyradicals in Thermoluminescence and Exoelectron Emision of $CaSO_4$ Phosphors" by R. Huzimura and K. Atarashi
(17 and 20.3; 11; 13.4; 19; 22.8.1)
(a) Vol. **83**, No. 2, p. K175—K177 (1984).

N. V. Baramidze, V. L. Bonch-Bruevich, and N. I. Kurdiani
Optical Absorption in Neutron Irradiated GaAs
(13.4 and 20.1; 11; 22.2.1)
(b) Vol. **124**, No. 1, p. 241—247 (1984).

C. A. Hewett, S. S. Lau, R. Fastow, and J. W. Mayer
*Pulsed Ion Beam Annealing of Ti–Au Alloys
(1.3 and 11; 1.5; 2; 21.1; 21.6)
(a) Vol. **84**, No. 1, p. 49—54 (1984).

A. V. Dvurechenskii, R. Grötzschel, N. M. Igonina, B. A. Koval, and N. I. Lebedeva
Diffusion of Impurities in Undercooled Melt of Pulse Heated Ion-Implanted Silicon
(9; 11; 22.1.2)
(a) Vol. **84**, No. 1, p. 171—177 (1984).

R. Pareja, R. González, and M. A. Pedrosa
*Study of Thermochemically Reduced and Electron-Irradiated $LiNbO_3$ Single Crystals by Positron Annihilation and Optical Absorption Measurements
(10.2 and 11; 20.1; 22.8.1)
(a) Vol. **84**, No. 1, p. 179—183 (1984).

G. S. Mingaleev, A. P. Tyutnev, V. I. Arkhipov, A. I. Rudenko, A. V. Vannikov, and V. S. Saenko
Computer Simulation and Analytical Solutions in the Study of the Transient Radiation-Induced Conductivity in Polymers
(14.3; 11; 22.9)
(a) Vol. **84**, No. 1, p. 327—336 (1984).

N. Pashov, G. Kästner, M. Kalitzova, S. Simov, M. Mikhailov, and K. Kolev
*Electron Microscope and Electron Diffraction Studies of Amorphization Phenomena in Ge Implanted with Te Ions
(1.4 and 11; 2; 22.1.1)
(a) Vol. **84**, No. 1, p. K23—K27 (1984).

A. Subrahmanyam and K. V. S. Badarinath
Dielectric Properties of RbCl Single Crystals
(14.4; 11; 22.5.2)
(a) Vol. **84**, No. 1, p. K93—K96 (1984).

P. Kleinert
Cluster-Bethe-Lattice Treatment of the Interstitial Boron Centre in Irradiated Gallium Arsenide
(10.2; 11; 22.2.1)
(b) Vol. **124**, No. 2, p. 559—564 (1984).

A. M. Taratin and S. A. Vorobev
*Multiple Scattering of Relativistic Electrons under Axial Channeling in a Crystal. Model of the Lattice of the Atomic Rows
(11; 13; 22.1.2)
(b) Vol. **124**, No. 2, p. 641—648 (1984).

P. F. Lugakov and V. V. Lukyanitsa
*Peculiarities of Radiation Defect Formation and Annealing in n-Si Due to Their Interaction with Each Other and Defect Clusters
(10.2 and 11; 14.3; 22.1.2)
(a) Vol. **84**, No. 2, p. 457—464 (1984).

K. D. Glinchuk, A. V. Prokhorovich, and N. S. Zayats
Effect of 2.2 MeV Electron Irradiation and Annealing on the Non-Copper-Induced 1.35 eV Emission Band in n-Type GaAs
(13.4 and 20.3; 11; 22.2.1)
(a) Vol. **84**, No. 2, p. 567—572 (1984).

A. P. Tyutnev, P. I. Dubenskov, V. S. Saenko, and A. V. Vannikov
*Radiation-Induced Conductivity of Poly-(N-Vinylcarbazole) and Poly-(N-Epoxypropylcarbazole)
(11 and 14.3; 22.9)
(a) Vol. **84**, No. 2, p. 585—596 (1984).

S. Shindo
*The Recoil Effect on the Electronic Stopping Power of Solids for Slowly Moving Atoms
(11)
(b) Vol. **125**, No. 1, p. 161—167 (1984).

M. A. Chubisov and A. F. Akkerman
*Simulation of Z_1-Oscillations in Low-Energy Heavy Ion Ranges in Solids
(11; 22.1.2)
(b) Vol. **125**, No. 1, p. 169—175 (1984).

D. E. Popov, E. I. Rozum, S. R. Uglov, and S. A. Vorobiev
*Anomalous Radiation of Fast Channeled Electrons in Optically Transparent Crystals
(11 and 20; 13; 22.1.3; 22.5.2)
(b) Vol. **125**, No. 1, p. 251—258 (1984).

Yu. M. Annenkov, V. F. Stolarenko, and T. S. Frangulian
Effect of Powerful Pulsed Electron Irradiation on Creation of Hole Colour Centres in KCl at 300 K
(10.2 and 20.1; 11; 22.5.2)
(b) Vol. **125**, No. 1, p. K37—K40 (1984).

V. G. Artyushenko, V. V. Voitsekhovskii, L. S. Kornienko, I. S. Lisitskii, and A. O. Rybaltovskii
*Effect of Ionizing Radiation on Optical Properties of Thallium Halides in the 0.36 to 15 μm Range
(11 and 20.1; 10.2; 22.5)
(a) Vol. **85**, No. 1, p. 167—171 (1984).

M. Gaber and H.-J. Fitting
Energy–Depth Relation of Electrons in Bulk Targets by Monte-Carlo Calculations
(17; 11; 21)
(a) Vol. **85**, No. 1, p. 195—198 (1984).

G. Oelgart and U. Werner
*Kilovolt Electron Energy Loss Distribution in GaAsP
(11; 13; 22.2.4)
(a) Vol. **85**, No. 1, p. 205—213 (1984).

P. Michel, B. Weber, and G. Götz
*Formation of Pt Silicides by a Millisecond Laser Pulse
(1.6 and 11; 21)
(a) Vol. **85**, No. 1, p. K1—K4 (1984).

T. Jung and L. Roth
DC Plasma Nitridation of Thin Aluminium Films
(1.5; 11; 22)
(a) Vol. **85**, No. 1, p. K15—K18 (1984).

A. D. Pogrebnyak, Sh. R. Mastov, M. F. Kuznetsov, K. P. Arefiev, S. V. Rakitin, and S. A. Vorobiev
*Observation of RF Pulses from Solids during Laser Irradiation
(11 and 20; 22.1.2; 22.2.1)
(a) Vol. **85**, No. 1, p. K31—K33 (1984).

D. Hildebrandt, H. Strusny, R. Groetzschel, E. Kotai, and F. Paszti
*Damage and Trapping Behaviour of Crystalline Silicon at Low Energy Deuterium Implantation
(10.2 and 11; 22.1.2)
(a) Vol. **85**, No. 1, p. K35—K38 (1984).

A. V. Dvurechenskii, B. P. Kashnikov, G. P. Pokhil, V. P. Popov, A. V. Tulinov, and A. A. Turinge
*Defect Structure Study with Planar Channeling in Pulse-Annealed Ion-Implanted Silicon
(10.2 and 11; 22.1.2)
(a) Vol. **85**, No. 1, p. K39—K42 (1984).

V. D. Tkachev, G. Hölzer, and A. R. Chelyadinskii
*Damage Profiles in Ion Implanted Silicon
(10.2 and 11; 22.1.2)
(a) Vol. **85**, No. 1, p. K43—K46 (1984).

V. I. Gatalskaya, E. M. Gololobov, and L. A. Kurochkin
*Relaxation Time and Electron Scattering Cross Section in Irradiated p-Silicon at Temperatures of Liquid Helium from Cyclotron Resonance Data
(11; 19; 22.1.2)
(a) Vol. **85**, No. 1, p. K51—K55 (1984).

V. M. Lomako and P. Ya. Starostin
*Temperature Dependence of the Charge-Carrier Removal Rate in n-GaAs Containing Defect Clusters
(11 and 13.4; 22.2.1)
(a) Vol. **85**, No. 1, p. K57—K60 (1984).

V. A. Bazylev and A. V. Demura
*Atomic Particle Ionization in a Crystal in Random Orientation
(11; 13)
(b) Vol. **125**, No. 2, p. 583—591 (1984).

S. Sapru and K. L. Tan
Eu^{2+} Ions Doped in Alkali Halides Act as Donors on Irradiation
(20.3; 11; 13.4; 22.5.2)
(b) Vol. **125**, No. 2, p. K161—K163 (1984).

V. G. Chudinov, N. V. Moseev, B. N. Goshchitskii, and V. I. Protasov
*Possible Amorphisation and Phase Separation of Intermetallic Mo_3Si Compound under Radiation. Molecular Dynamics Study
(11; 2; 21.1)
(a) Vol. **85**, No. 2, p. 435—440 (1984).

P. F. LUGAKOV and T. A. LUKASHEVICH
*Formation and Parameters of Boron–Divacancy Complexes in Irradiated p-Si
(11; 13.4; 22.1.2)
(a) Vol. **85**, No. 2, p. 441—444 (1984).

F. SCHNEIDER and S. UNTERRICKER
*Radiation Damage and Its Annealing Behaviour in InP after Recoil Implantation of ^{118}Sb and ^{111}In Observed by TDPAC
(10.2 and 11; 5; 22.2.2)
(a) Vol. **85**, No. 2, p. 455—461 (1984).

B. RAUSCHENBACH, G. BLASEK, and R. DIETSCH
*Modification of Concentration Profiles in Iron and Aluminium by High-Fluence Implantation of Nitrogen and Boron Ions
(1.3 and 11; 21; 21.1.1)
(a) Vol. **85**, No. 2, p. 473—480 (1984).

V. I. GUBSKAYA, P. V. KUCHINSKII, and V. M. LOMAKO
*The Effect of the Vacancy Charge State on the Radiation Defect Formation in Silicon
(11 and 13.4; 22.1.2)
(a) Vol. **85**, No. 2, p. 585—590 (1984).

A. P. TYUTNEV, V. S. SAENKO, A. F. DUNAEV, E. D. POZHIDAEV, and A. V. VANNIKOV
Temperature Dependence of Transient Radiation- Induced Conductivity in Polymers
(14.3; 11; 22.9)
(a) Vol. **85**, No. 2, p. 591—602 (1984).

S. A. AZIMOV, D. A. ARONOV, V. M. RUBINOV, and M. TUICHIEV
On the Mechanism of Current Passage in Gallium Phosphide Films Excited by Light or a Beam of Accelerated Electrons
(14.3.1; 11; 22.2.2)
(a) Vol. **85**, No. 2, p. K159—K164 (1984).

E. SANDOW, W. WESCH, and E. NEBAUER
Ohmic AuGe Contacts on n-GaAs Produced by Nanosecond Laser Pulse Irradiation
(14.3.4; 9; 11; 22.2.1)
(a) Vol. **85**, No. 2, p. K169—K173 (1984).

A. R. VASCONCELLOS and R. LUZZI
*On Laser-Induced Lattice Heating in a Polar Semiconductor
(8 and 11; 6; 22; 22.2.1)
(b) Vol. **126**, No. 1, p. 63—70 (1984).

A. I. MELKER and S. N. ROMANOV
*Group Representation of the Formation Probabilities of Athermal Vacancy Clusters in Atomic Collision Cascades
(11)
(b) Vol. **126**, No. 1, p. 133—140 (1984).

H. C. González
*Hardening of Single Crystals of Magnesium by Low Neutron Doses at 77 K
(10.1 and 11; 21.3)
(a) Vol. **86**, No. 1, p. 169—175 (1984).

V. V. Kalinin and N. N. Gerasimenko
*On the Effect of Doping Impurities on the Formation of Rod-Like Defects in Silicon
(10.2 and 11; 22.1.2)
(a) Vol. **86**, No. 1, p. 185—190 (1984).

A. D. Pogrebnyak
*Positron and Positronium States in Semiconductors Irradiated by Supercurrent Beams of Charged Particles
(11; 22.1.2; 22.2.1)
(a) Vol. **86**, No. 1, p. 191—198 (1984).

A. Kozanecki and H. Rzewuski
*Distant DAP Bands in Electron-Irradiated and Ion-Implanted CdSe
(11 and 20.3; 10.2; 22.4.2)
(a) Vol. **86**, No. 1, p. 263—270 (1984).

A. V. Dvurechenskii, B. P. Kashnikov, and V. V. Suprunchik
New EPR Defects in Si⟨Al⟩
(13.4 and 19; 11; 22.1.2)
(a) Vol. **86**, No. 1, p. 313—318 (1984).

S. C. Sabharwal, S. P. Kathuria, and B. Ghosh
Kinetics of Thermoluminescence in γ-Irradiated NaI(Tl)
(20.3; 11; 13.4; 22.5.2)
(a) Vol. **86**, No. 1, p. 327—331 (1984).

A. P. Tyutnev, V. S. Saenko, A. V. Vannikov, and V. E. Oskin
*Radiation Induced Dielectric Effect in Polymers
(11 and 14.4; 22.9)
(a) Vol. **86**, No. 1, p. 363—374 (1984).

S. A. Grusha, A. M. Evstigneev, K. A. Ismailov, R. V. Konakova, A. N. Krasiko, I. K. Sinishchuk, Yu. A. Tkhorik, and V. I. Fainberg
*Electric Field Effect on Radiation-Induced Gettering of Defects in n^+–n–n^{++}-GaAs
(11 and 14.3.3; 22.2.1)
(a) Vol. **86**, No. 1, p. K83—K86 (1984).

A. I. Melker and S. N. Romanov
*Athermal Vacancy Cluster Formation in F.C.C. Metals by High Energy PKA
(11; 10.1; 21.1; 21.6)
(b) Vol. **126**, No. 2, p. 517—525 (1984).

V. V. Kaplin, S. B. Nurmagambetov, V. I. Gridnev, E. I. Rozum, S. Pak, S. R. Uglov, and S. A. Vorobiev
*Angular Distributions of Fast Electrons at Planar Channeling in Silicon Crystals
(11; 13; 22.1.2)
(b) Vol. **126**, No. 2, p. 565—574 (1984).

M. S. Jahan and D. W. Cooke
Thermal Evolution of Low-Temperature Manganese Centers in X-Irradiated CaF_2:Mn
(10.2 and 19; 11; 20.3; 22.5.3)
(b) Vol. **126**, No. 2, p. 687—692 (1984).

D. Fasold, K. Hehl, and S. Jetschke
Refractive Index and Absorption Measurement in a Thin Layer by Two Reflection Measurements
(20.1; 11; 22.8.1)
(a) Vol. **86**, No. 2, p. 485—490 (1984).

V. E. Borisenko, V. V. Gribkovskii, V. A. Labunov, and S. G. Yudin
Pulsed Heating of Semiconductors
(8; 11; 22.1; 22.2)
(a) Vol. **86**, No. 2, p. 573—583 (1984).

A. P. Tyutnev, V. S. Saenko, A. V. Vannikov, and E. D. Pozhidaev
Concerning the Radiation-Induced Surface Conductivity in Polymers
(14.3.2; 11; 22.9)
(a) Vol. **86**, No. 2, p. 709—716 (1984).

S. B. Sevastianov, N. N. Gerasimenko, and N. V. Vershinina
*Inhomogeneitis in Irradiated SiO_2–Si Structures
(11 and 14.3.3; 14.3.4; 22.1.2)
(a) Vol. **86**, No. 2, p. 717—727 (1984).

M. Voelskow, J. Matthäi, and R. Klabes
*Electrical Properties of Ion Implanted and Short Time Annealed Polycrystalline Silicon
(11 and 14.3.1; 22.1.2)
(a) Vol. **86**, No. 2, p. 781—788 (1984).

H. Terletzki, A. Boden, F. Wulf, and W. R. Fahrner
*Laterally Inhomogeneous Charge Build-Up in CMOS Inverters during Ionizing Irradiation
(11 and 14.3.3; 14.3.2; 22.1.2)
(a) Vol. **86**, No. 2, p. 789—794 (1984).

V. Rybka, F. Deml, and P. Hampl
*Dual Implantation of Si and P into GaAs
(11 and 14.3; 22.2.1)
(a) Vol. **86**, No. 2, p. K117—K119 (1984).

12. Mechanical Properties (Plastic Deformations See 10)

Original Papers

M. KATO and H.-R. PAK
Thermodynamics of Stress-Induced First-Order Phase Transformations in Solids
(1.2; 8; 12)
(b) Vol. **123**, No. 2, p. 415—424 (1984).

YU. P. KHAPACHEV, A. A. DYSHEKOV, and D. S. KISELEV
The Theory of X-Ray Diffraction Analysis of Elastic-Strain States in Epitaxial Films
(1.5; 12)
(b) Vol. **126**, No. 1, p. 37—42 (1984).

12.1 Metals

Review Article

R. Ramji Rao and A. Ramanand
*Thermal and Elastic Properties of Rare-Earth Metals
(8 and 12.1; 6; 21.4)
(b) Vol. **122**, No. 1, p. 11—27 (1984).

Original Papers and Short Notes

K. C. Gupta and R. P. S. Rathore
Lattice Equilibrium and Phonons in Molybdenum and Tungsten
(4 and 6; 12.1; 21)
(b) Vol. **121**, No. 1, p. 73—79 (1984).

V. Z. Bengus, E. D. Tabachnikova, and V. I. Startsev
*Mechanical Behaviour of Some Metallic Glasses at 4.2 to 300 K
(2 and 12.1; 10.1; 21)
(a) Vol. **81**, No. 1, p. K11—K13 (1984).

P. M. Hazzledine, H. P. Karnthaler, and A. Korner
The Application of the Wulff Construction to Dislocation Problems
(10.1; 12.1; 21.1; 21.6)
(a) Vol. **81**, No. 2, p. 473—484 (1984).

K. Iwasaki
*High-Temperature Internal Friction Peaks of Pure Aluminium Single-, Bi-, and Polycrystals Measured with an Inverted Flexure Pendulum
(10.1 and 12.1; 21)
(a) Vol. **81**, No. 2, p. 485—496 (1984).

G. K. Schmidt
*Method for Obtaining the Spectrum of Plastic Strains of Mesoscopic Volume Elements in Fatigued Materials Investigated by X-Rays (I): Plastic and Elastic Strains in Fatigued Material
(10.1 and 12.1; 1.4; 21)
(a) Vol. **82**, No. 1, p. 75—84 (1984).

A. D. Korotaev, Yu. I. Chumlyakov, V. F. Esipenko, and L. S. Bushnev
*Superelasticity Effects in Alloy Single Crystals of Cu–15% Al–2% Co with Non-Coherent Particles Due to Twinning
(12.1; 21.1)
(a) Vol. **82**, No. 2, p. 405—412 (1984).

G. K. Schmidt
*Method for Obtaining the Spectrum of Plastic Strains of Mesoscopic Volume Elements in Fatigued Materials Investigated by X-Rays (II): Distributions of Elastic Strains and X-Ray Reflex Profiles
(10.1 and 12.1; 1.4; 21)
(a) Vol. **82**, No. 2, p. 413—420 (1984).

D. GERLICH
*Fourth-Order Elastic Constants of Solids
(12.1 and 12.2; 21; 21.6; 22.5.2)
(b) Vol. **123**, No. 1, p. K5—K7 (1984).

D. O. NORTHWOOD and I. O. SMITH
*Internal and Friction Stresses for Al–Mg Class I High Temperature Creep Alloys
(10.1 and 12.1; 21)
(a) Vol. **83**, No. 1, p. 227—234 (1984).

P. DELOBELLE, D. VARCHON, and C. OYTANA
Short-Range Reordering of Heavy Interstitials in Ta, Nb, and Fe during Relaxation and Static Strain Aging
(10.1; 12.1; 21; 21.1.1)
(a) Vol. **83**, No. 1, p. 235—243 (1984).

T. S. KÊ, P. CUI, and C. M. SU
*Internal Friction in High-Purity Aluminium Single Crystals
(10.1 and 12.1; 21)
(a) Vol. **84**, No. 1, p. 157-164 (1984).

N. P. KOVALENKO and L. M. KUZMINA
*The Influence of Many-Body Interactions on the Speed of Sound in Liquid Metals
(2 and 12.1; 7; 21.2)
(b) Vol. **124**, No. 2, p. 537—544 (1984).

V. I. KONONENKO, A. L. SUKHMAN, S. L. GRUVERMAN, and V. V. TOROKIN
*Density and Surface Tension of Liquid Rare Earth Metals, Scandium and Yttrium
(2 and 12.1; 1.2; 21; 21.4)
(a) Vol. **84**, No. 2, p. 423—432 (1984).

T. S. KÊ, L. D. ZHANG, P. CUI, Q. HUANG, and B. S. ZHANG
*Internal-Friction Peak in Polygonized High-Purity Aluminium and In-Situ Observation by Transmission Electron Microscope
(10.1 and 12.1; 21)
(a) Vol. **84**, No. 2, p. 465—474 (1984).

A. NOHARA and T. IMURA
Yield-Point Phenomenon in a β-CuZn Single Crystal at High Temperatures
(10.1; 12.1; 21.1)
(a) Vol. **84**, No. 2, p. 501—507 (1984).

U. WALZER
Elastic and Bonding Properties of Transition Metals
(4; 12.1; 13.1; 21.1)
(b) Vol. **125**, No. 1, p. 55—64 (1984).
Erratum
(b) Vol. **126**, No. 2, p. 752 (1984).

T. Soma, T. Itoh, and H.-Matsuo Kagaya
Pressure Effect on Phonon Frequencies and Elastic Stiffness Constants of Al
(6; 12.1; 21)
(b) Vol. **125**, No. 1, p. 107—112 (1984).

A. C. Gupta and A. N. Nigam
Effect of Cold Work on Hydrogenation of Nickel as Seen in the Thermoelectric Behaviour
(13.2; 12.1; 15; 21.1)
(b) Vol. **125**, No. 1, p. 207—209 (1984).

H. M. Ledbetter
*Monocrystal–Polycrystal Elastic Constants of a Stainless Steel
(12.1; 21.1.1)
(a) Vol. **85**, No. 1, 89—96 (1984).

N. Kuramochi, H. Mizubayashi, and S. Okuda
*Dislocation Pinning at Low Temperatures in Nb
(10.1 and 12.1; 21)
(a) Vol. **85**, No. 1, p. 121—131 (1984).

D. O. Northwood and I. O. Smith
*Steady-State Creep and Strain Transients for Stress-Dip Tests in Polycrystalline Magnesium at 300 °C
(10.1 and 12.1; 21.3)
(a) Vol. **85**, No. 1, p. 149—158 (1984).

A. I. Gusev
Structural Vacancies in Nonstoichiometric Compounds at High Pressure, Thermodynamic Model
(10.1; 8; 12.1; 21.1)
(a) Vol. **85**, No. 1, p. 159—166 (1984).

G. Fantozzi, C. Esnouf, J. Lauzier, and C. Minier
*Internal Friction at Low Temperature in a High Purity Magnesium
(10.1 and 12.1; 21.3)
(a) Vol. **85**, No. 2, p. 463—468 (1984).

A. Pawełek
*A Simple Model of Small Yield Point Due to Strain Rate Change Based on the Dynamics of the Source Generated Dislocation Group
(10.1 and 12.1; 21)
(a) Vol. **85**, No. 2, p. K117—K121 (1984).

S. P. Singh and R. P. S. Rathore
Phonons in Face-Centred Cubic Calcium and Strontium
(6; 12.1; 21.3)
(b) Vol. **126**, No. 1, p. 87—90 (1984).

T. Soma and H.-Matsuo Kagaya
Compression Effect on Mode Grüneisen Parameters of Alkali Metals
(6 and 8; 12.1; 21.2)
(b) Vol. **126**, No. 1, p. K5—K7 (1984).

A. Pawelek
*On the Thermodynamic Criterion for the Unstable Motion of a Source Generated Dislocation Group
(10.1 and 12.1)
(a) Vol. **86**, No. 1, p. K27—K30 (1984).

T. Soma and H.-Matsuo Kagaya
Mode Grüneisen Parameter for Local and Band Modes of Rb–K, Cs–K, and Cs–Rb Alloy Systems
(6; 8; 12.1; 21.2)
(b) Vol. **126**, No. 2, p. K109—K112 (1984).

T. S. Kê, P. Cui, S. C. Yan, and Q. Huang
*Internal Friction Peaks of Super-High-Purity Aluminium at Medium Temperatures
(10.1 and 12.1; 21)
(a) Vol. **86**, No. 2, p. 593—601 (1984).

K. Iwasaki
*High-Temperature Internal Friction Peaks of Dilute Aluminium Alloys
(10.1 and 12.1; 21)
(a) Vol. **86**, No. 2, p. 637—648 (1984).

M. S. C. Bose
*Effect of Saturated Magnetic Field on Fatigue Life of Carbon Steel
(12.1 and 18.2; 21.1.1)
(a) Vol. **86**, No. 2, p. 649—654 (1984).

12.2 Non-Metals

Review Article

J. Shanker and G. G. Agrawal
Van der Waals Potentials in Ionic Crystals
(4; 6; 12.2; 22.5.2)
(b) Vol. **123**, No. 1, p. 11–26 (1984).

Original Papers and Short Notes

M. Gatulle and M. Fischer
*Elastic Constants of the Layered Compounds GaS, GaSe, InSe, and Their Pressure Dependence (II): Theoretical Part
(12.2; 22.3)
(b) Vol. **121**, No. 1, p. 59–71 (1984).

T. S. Verma, V. Kumar, and J. Shanker
*Evaluation of Photoelastic Constants and Strain Derivatives of the Static Dielectric Constant of Silver and Thallium Halides
(12.2; 14.4; 22.5)
(b) Vol. **121**, No. 2, p. 487—491 (1984).

R. P. Singh, J. P. S. Rana, and G. B. Varshney
The Contribution of Zero-Point Vibration toward the Static and Thermodynamical Properties of Potassium Halides
(4; 12.2; 22.5.2)
(b) Vol. **121**, No. 2, p. 493—500 (1984).

S. Motogi and G. A. Maugin
Effects of Magnetostriction on Vibrations of Bloch and Néel Walls
(18.2; 12.2; 22.1.2; 22.6)
(a) Vol. **81**, No. 2, p. 519—532 (1984).

S. Ramana Murthy and T. Seshagiri Rao
*Internal Friction in Ni–Zn Ferrites
(10.2 and 12.2; 22.8.2)
(a) Vol. **81**, No. 2, p. K137—K140 (1984).

A. Richter, A. E. Romanov, W. Pompe, and V. I. Vladimirov
Geometry and Energy of Disclinations in Topologically Disordered Systems
(2; 10.2; 12.2)
(b) Vol. **122**, No. 1, p. 35—45 (1984).

R. K. Varshney and J. Shanker
*Temperature Dependence of Third-Order Elastic Constants of NaCl-Structure Alkali Halide Crystals
(12.2; 22.5.2)
(b) Vol. **122**, No. 1, p. 65—72 (1984).

U. Schaller, W. Kraak, and R. Herrmann
Metal–Insulator-Transition in n-InSb under High Hydrostatic Pressure
(14.1 and 14.3; 12.2; 22.2.3)
(b) Vol. **122**, No. 1, p. K87—K91 (1984).

P. V. Reddy and T. S. Rao
*Elastic Behaviour of Mixed Manganese–Magnesium Ferrites at Room Temperature
(12.2; 22.8.2)
(a) Vol. **82**, No. 1, p. K37—K41 (1984).

S. N. Maiti and D. Roy
Lattice Dilatation in High-Concentration Binary Solid Solutions
(4; 12.2; 22.5.4)
(b) Vol. **122**, No. 2, p. 449—460 (1984).

M. Krzesińska and T. Szuta-Buchacz
*Hypersonic Wave Attenuation in Yttrium Aluminium Garnet and Gadolinium Gallium Garnet
(7; 6; 12.2; 22.8.2)
(a) Vol. **82**, No. 2, p. 421—424 (1984).

D. Gerlich
*Fourth-Order Elastic Constants of Solids
(12.1 and 12.2; 21; 21.6; 22.5.2)
(b) Vol. **123**, No. 1, p. K5—K7 (1984).

L. A. Shuvalov, S. A. Gridnev, V. I. Kudryash, and B. N. Prasolov
Ferroelastic Domain Structure Dynamics and Internal Friction of $KH_3(SeO_3)_2$ Single Crystals
(14.4.1; 10.2; 12.2; 22.8.1)
(a) Vol. **83**, No. 1, p. 131—137 (1984).

B. P. Chandra R. D. Verma, B. Majumdar, and V. P. Bhatt
*Dependence of Mechanoluminescence on the Microhardness and Dislocation Density in Crystals
(12.2 and 20.3; 10.2; 22.5.2; 22.8.1)
(a) Vol. **83**, No. 1, p. K19—K22 (1984).

R. K. Singh, Neeta Agnihotri, S. V. Suryanarayana, and S. P. Sanyal
Study of Cohesion, Harmonic and Anharmonic Elastic Properties of CaF_2–SrF_2 Mixed Crystals
(4; 12.2; 22.5.4)
(b) Vol. **123**, No. 2, p. 453—462 (1984).

S. K. Agrawal, J. Narain, and J. Shanker
Investigation of the Equation of State and the Grüneisen Parameters for Transition Metal Oxides
(4; 6; 12.2; 22.6)
(b) Vol. **123**, No. 2, p. 497—501 (1984).

S. Tsunekawa, H. F. J. Watanabe, and H. Takei
Linear Thermal Expansion of $SrTiO_3$
(1.2; 12.2; 22.8.1)
(a) Vol. **83**, No. 2, p. 467—472 (1984).
Erratum
(a) Vol. **86**, No. 2, p. 841 (1984).

B. Müller, P. Grau, and G. Kluge
*The Hardness of Chromium- and Titanium-Doped Sapphire Crystals
(12.2; 22.6)
(a) Vol. **83**, No. 2, p. 499—506 (1984).

R. K. Singh and P. Khare
*Analysis of Anharmonic Properties of Some Partially Covalent Crystals
(4 and 12.2; 22.5; 22.5.2)
(b) Vol. **124**, No. 1, p. 29—36 (1984).

R. Piotrzkowski
Pressure Study of Photoconductivity in Indium Antimonide (I): Low-Concentration n-Type at 77 K. Pressure-Dependent Capture Coefficient of a Hole-Trapping Center
(13.4 and 16; 12.2; 22.2.3)
(b) Vol. **124**, No. 1, p. 411—421 (1984).

T. Soma, Y. Takahashi, and H.-Matsuo Kagaya
Equation of State and Pressure Derivative of the Bulk Modulus for III–V Covalent Crystals
(1.2 and 4; 12.2; 22.2.1; 22.2.3)
(b) Vol. **124**, No. 1, p. K11—K14 (1984).

A. Schönhals and E. Donth
A Dynamic Percolation Approach to the Andrade Creep
(2; 12.2)
(b) Vol. **124**, No. 2, p. 515—524 (1984).

A. I. Melker and A. V. Ivanov
*Computer Simulation of the Fracture of Two Coupled Anharmonic Chains with a Strong Interaction
(12.2; 22.9)
(a) Vol. **84**, No. 2, p. 417—422 (1984).

N. Agnihotri, R. K. Singh, and S. P. Sanyal
Interionic Forces and Elastic Properties of Na_2S
(4; 6; 12.2; 22; 22.5.3)
(b) Vol. **125**, No. 1, p. 113—119 (1984).

I. Gorczyca
Calculation of the Effect of Hydrostatic Pressure on the Band Structure of GaP and GaSb
(13.1; 12.2; 22.2.2; 22.2.3)
(b) Vol. **125**, No. 1, p. 229—235 (1984).

J. Shanker, W. N. Bhende, and P. S. Bakhshi
Analysis of the Equation of State for LiF Crystal
(4; 12.2; 22.5.2)
(b) Vol. **125**, No. 1, p. K21—K26 (1984).

G. S. Kumar, S. Sudhakar, R. Ethiraj, and V. G. Krishna Murthy
*Strain-Optical Constant (p_{44}) of Mixed Crystals of KCl–KBr of Equimolar Concentration
(12.2 and 20; 22.5.4)
(a) Vol. **85**, No. 1, p. K27—K30 (1984).

K. N. Jog, R. K. Singh, and S. P. Sanyal
*The Relative Stability and Phase Transition Studies of MgO and CaO Crystals
(1.2 and 12.2; 22.6)
(a) Vol. **85**, No. 2, p. 417—421 (1984).

A. Dominguez-Rodriguez, J. Cabrera-Caño, R. Marquez, and J. Castaing
*High-Temperature Creep of Equimolecular NiO–CoO Solid Solution
(10.2 and 12.2; 22.6.1)
(a) Vol. **85**, No. 2, p. 445—448 (1984).

A. Sasaki and M. Iwata
*Anisotropy of Microhardness in Anthracene Single Crystals
(12.2; 22.9)
(a) Vol. **85**, No. 2, p. K105—K107 (1984).

W. Kraak, U. Schaller, and R. Herrmann
*Sensitive Pressure Gauge on the Basis of Pure n-InSb
(12.2 and 14.3; 22.2.3)
(a) Vol. **85**, No. 2, p. K183—K187 (1984).

R. K. Varshney and J. Shanker
*Temperature Dependence of Third-Order Elastic Constants of CsCl, CsBr, and CsI Crystals
(12.2; 22.5.2)
(b) Vol. **126**, No. 1, p. 77—82 (1984).

J. Narain, G. G. Agrawal, and J. Shanker
*Compressibility of Rare-Gas Solids under the Effect of High Pressure
(4 and 12.2; 23)
(b) Vol. **126**, No. 1, p. 83—86 (1984).

H.-Matsuo Kagaya and T. Soma
*Dynamical Elastic Constants of Covalent Compounds
(12.2; 6; 22.2; 22.4)
(b) Vol. **126**, No. 1, p. K9—K14 (1984).

G. Sorge, J. Janich, L. A. Shuvalov, and N. M. Shchagina
The Influence of an Electric DC Field on the New Intermediate Phase between the α- and β-Phases of $NaH_3(SeO_3)_2$ Crystals
(1.2 and 14.4.1; 12.2; 22.8)
(a) Vol. **86**, No. 1, p. 111—115 (1984).

A. A. Kaminskii, E. L. Belokoneva, B. V. Mill, Yu. V. Pisarevskii, S. E. Sarkisov, I. M. Silvestrova, A. V. Butashin, and G. G. Khodzhabagyan
Pure and Nd^{3+} Doped $Ca_3Ga_2Ge_4O_{14}$ and $Sr_3Ga_2Ge_4O_{14}$ Single Crystals, Their Structure, Optical, Spectral Luminescence, Electrochemical Properties, and Stimulated Emission
(14.4 and 20.2; 1.1; 12.2; 20.1; 20.3; 22.8.1)
(a) Vol. **86**, No. 1, p. 345—362 (1984).

A. A. Evtushenko, V. F. Petrenko, and I. A. Ryzhkin
Electric Polarization of Ice at Nonuniform Elastic Strains
(10.2 and 14.4; 12.2; 22)
(a) Vol. **86**, No. 1, p. K31—K34 (1984).

R. P. Singh, J. P. S. Rana, A. Daud, and G. B. Varshney
The Role of the Zero-Point Vibration towards Crystal Properties of Rubidium Halide Crystals
(4; 12.2; 22.5.2)
(b) Vol. **126**, No. 2, p. 495—499 (1984).

B. Holland, H. S. Greenside, and M. Schlüter
Cohesive Properties of Ge, Si, and Diamond Calculated with Minimum Basis Sets
(4 and 13; 12.2; 22.1; 22.1.1; 22.1.2)
(b) Vol. **126**, No. 2, p. 511—515 (1984).

K. R. Allakhverdiev, S. S. Babaev, N. A. Bakhyshov, T. G. Mamedov, E. Yu. Salaev, and I. K. Efendieva
The Influence of Hydrostatic Pressure on the Fundamental Absorption Edge of Crystals with TlSe Type Structure
(13.1 and 20.1; 12.2; 22; 22.8)
(b) Vol. **126**, No. 2, p. K139—K143 (1984).

H. Nakanishi and T. Irie
Pressure Effect on the Absorption Edges of $CdIn_2S_4$ and Related Compounds
(13.1 and 20.1; 12.2; 22.8.2)
(b) Vol. **126**, No. 2, p. K145—K148 (1984).

13. Electron States

Review Articles

K. N. Shrivastava
Origin of Superexchange Interactions in 3d Magnetic Solids (I)
(6 and 18.2; 13; 22.5)
(b) Vol. **125**, No. 1, p. 11—43 (1984).

K. N. Shrivastava
Origin of Superexchange Interactions in 3d Magnetic Solids (II)
(6 and 18.2; 13; 22.5)
(b) Vol. **125**, No. 2, p. 441—460 (1984).

Original Papers and Short Notes

P. Enders
Structure and Bonding in Cubic IV–VI Crystals (II): Instability of the TO(Γ) Mode: Unified Universal-Parameter Treatment of the Group-V Semimetals and Cubic IV–VI Semiconductors
(1.2; 4; 13; 22)
(b) Vol. **121**, No. 1, p. 39—46 (1984).

E. V. Chulkov, I. Yu. Sklyadneva, and V. E. Panin
*The First-Principles Calculation of the Norm-Conserving Non-Local Singular Atomic Pseudopotentials
(13; 21; 22)
(b) Vol. **121**, No. 1, p. 265—274 (1984)

J. M. Wesselinowa
*Damping of the Transverse and Longitudinal Excitations of the Ising Model in a Transverse Field
(13 and 14.4.1; 22.8.1)
(b) Vol. **121**, No. 1, p. 317—327 (1984).

V. Čápek
*Initial Condition Term and Method of Artificial Temperature for Generalized Master Equations
(6 and 13; 22)
(b) Vol. **121**, No. 1, p. K7—K12 (1984).

H. E. Borges and P. R. Silva
*Bidimensional Spin — 1/2 Ising Model in a Random Field
(13 and 18.2)
(b) Vol. **121**, No. 1, p. K19—K24 (1984).

H. E. Borges and P. R. Silva
*Tridimensional Spin — 1/2 Ising Model in a Random Field
(13 and 18.2)
(b) Vol. **121**, No. 1, p. K25—K28 (1984).

H. Stachowiak
*On the Positron Schrödinger Equation in a Metal
(13; 21)
(b) Vol. **121**, No. 1, p. K69—K73 (1984).

G. H. Wannier
*Wannier Functions from Landau Functions?
(13)
(b) Vol. **121**, No. 2, p. 603—607 (1984).

W. Richert and W. Ebeling
*Thermodynamic Functions of the Electron Fluid for a Wide Density–Temperature Range
(13; 13.5; 22)
(b) Vol. **121**, No. 2, p. 633—639 (1984).

G. Seifert, H.-G. Fritsche, P. Ziesche, and V. Heera
*On the Electronic Structure of Palladium–Hydrogen and Platinum–Hydrogen Systems
(13; 21.6)
(b) Vol. **121**, No. 2, p. 705—715 (1984).

P. R. Silva, F. C. Sá Barreto, and H. E. Borges
*Simulation of the Site Diluted Ising Model through an Infinite Random Field
(13 and 18.2)
(b) Vol. **121**, No. 2, p. K151—K155 (1984).

J. M. Wesselinowa
*Relaxation Time for H-Bonded Ferroelectrics
(13 and 14.4.1)
(b) Vol. **121**, No. 2, p. K157—K160 (1984).

B. Dusowski and G. Heber
*Magnetic Properties of a Random Ensemble of Two-Spin Clusters. Molecular-Field Approach
(13 and 18)
(b) Vol. **122**, No. 1, p. 133—140 (1984).

P. S. Riseborough
*De Haas-Van Alphen Effect in Nearly Magnetic Transition Metals
(13; 18.1; 21; 21.4)
(b) Vol. **122**, No. 1, p. 161—170 (1984).

A. P. Pathak
*Position Dependence of Channeling Stopping Power
(11 and 13; 22.1.2)
(b) Vol. **122**, No. 1, p. 171—174 (1984).

J. B. Radunović
*Conductivity of an Electron Gas in a Low-Frequency Electric Field Due to Large-Scale Fluctuations
(13 and 14)
(b) Vol. **122**, No. 1, p. 383—387 (1984).

V. Hinkel and D. Ploumbidis
*Temperature Dependence of the Magnetic Part of the Hyperfine Field at ^{123}Sb in the Liquid Sb–Te System
(2 and 13; 19; 21.7; 22.1)
(b) Vol. **122**, No. 1, p. K45—K48 (1984).

R. Rünger and D. Ploumbidis
*High Temperature NMR in the V_3Au Compound
(13 and 19; 21.1)
(b) Vol. **122**, No. 1, p. K49—K52 (1984).

R. S. Gekht
*Low-Temperature Properties of One-Dimensional Heisenberg Systems with Random Anisotropy
(13 and 18.2)
(b) Vol. **122**, No. 2, p. 509—517 (1984).

O. P. Aggarwal and N. N. Saxena
*K-Absorption Spectra of Copper Oxides
(13 and 20; 22.6)
(b) Vol. **122**, No. 2, p. 669—673 (1984).

S. Guha and C. Basu
*Stimulation Brillouin Scattering of Laser Radiation by an Electron-Acoustic Wave in Magnetised Semiconductor Plasmas
(13 and 20; 18)
(b) Vol. **122**, No. 2, p. 799—804 (1984).

A. F. Barabanov, L. B. Litinskii, and A. V. Mikheenkov
*On the Atomic Phase of Crystalline Hydrogen
(1.2 and 13; 18.4; 23)
(b) Vol. **122**, No. 2, p. K109—K114 (1984).

M. J. Škrinjar, D. V. Kapor, J. P. Šetrajčić, and Z. Ivić
*New Approach to the Theory of Hybrid Excitations in Magnetic Dielectrics
(13 and 18.4; 20.1)
(b) Vol. **123**, No. 1, p. 135—141 (1984).

W. Nolting and M. Matlak
*Complete Analytical Solution for the Zero Bandwidth s–f Model
(13 and 18)
(b) Vol. **123**, No. 1, p. 155—168 (1984).

H. Solbrig
*Asymptotic Medium Concept Applied to Correlated Point Scatterers in the Weak-Scattering Limit
(2 and 13)
(b) Vol. **123**, No. 1, p. 237—246 (1984).

K. B. GARG, J. LEIRO, M. HEINONEN, and M. RICHARDSON
*X-Ray Emission Study of Bonding in SmT_2 Compounds (T ≡ Fe, Co, Ni)
(4 and 13; 20; 21.1; 21.1.1; 21.4)
(a) Vol. **83**, No. 1, p. 279—284 (1984).

H. FEHSKE, E. KOLLEY, and W. KOLLEY
Spin-Glass Behaviour in Disordered Hubbard Alloys
(18; 13; 21)
(b) Vol. **123**, No. 2, p. 533—540 (1984).

J. M. WESSELINOWA
Electron–Phonon Interaction in Ferromagnetic Semiconductors
(18.2; 6; 13; 22)
(b) Vol. **123**, No. 2, p. 585—593 (1984).

A. E. SMITH
*A Calculation of Absorption Potentials for Low Energy Electrons in MgO, 2H-MoS_2, and 2H-$NbSe_2$
(1.6 and 13; 22; 22.6)
(b) Vol. **123**, No. 2, p. 619—626 (1984).

J. DOMINEC, K. MISEK, W. KRAAK, and R. HERRMANN
*Ultrasound Attenuation and Quantum Oscillations in Bi–Sb Alloys
(7 and 13; 21.7)
(b) Vol. **123**, No. 2, p. 635—640 (1984).

M. ŠOB, S. SZUSZKIEWICZ, and M. SZUSZKIEWICZ
*Polarized Positron Annihilation and Enhancement Effects in Ferromagnetic Iron
(13; 18.2; 21.1.1)
(b) Vol. **123**, No. 2, p. 649—652 (1984).

S. K. J. AL-ANI, C. A. HOGARTH, and D. N. WATERS
The Effect of Temperature on the Optical Absorption Edge of Amorphous Thin Films of Silicon Monoxide
(20.1; 2; 13; 22.6)
(b) Vol. **123**, No. 2, p. 653—658 (1984).

H. ENGLISCH and R. ENGLISCH
*Exact Density Functionals for Ground-State Energies (I): General Results
(13)
(b) Vol. **123**, No. 2, p. 711—721 (1984).

J. M. WESSELINOWA
*Relaxation Time for Ferromagnetic Semiconductors
(13 and 18.2; 22)
(b) Vol. **123**, No. 2, p. K129—K132 (1984).

V. G. KOHN and H. ESCHRIG
*On the Theory of Exchange and Correlation in the Electron Gas at Metallic Densities
(13)
(b) Vol. **123**, No. 2, p. K149—K153 (1984).

H. Englisch
*Remark on a Paper by Stephen about Localization in a Wire
(13 and 14)
(b) Vol. **123**, No. 2, p. K169—K171 (1984).

V. L. Bonch-Bruevich and Le Vu Ky
*Stochastic Self-Oscillations in a Two-Temperature Electron–Hole Plasma in a Semiconductor
(13 and 14.3.4)
(b) Vol. **124**, No. 1, p. 111—116 (1984).

A. Rubaszek and H. Stachowiak
*Application of the Quadratic Response Theory to the Screening of a Positron in an Electron Gas of Metallic Density
(13; 13.2; 21)
(b) Vol. **124**, No. 1, p. 159—166 (1984).

P. V. Elyutin, B. Hickey, G. J. Morgan, and G. F. Weir
*Diffusion and Localisation in Random Systems
(13; 2)
(b) Vol. **124**, No. 1, p. 279—286 (1984).

H. Englisch and R. Englisch
*Exact Density Functionals for Ground-State Energies (II): Details and Remarks
(13)
(b) Vol. **124**, No. 1, p. 373—379 (1984).

R. B. Prabhu and M. N. Ghatikar
*Isomer Shift and Susceptibility of Mix-Valent $EuCu_2Si_2$
(13 and 18; 22. 8)
(b) Vol. **124**, No. 1, p. K25—K29 (1984).

V. G. Baryshevskii
*Positronium Spin Rotation and Time Oscillation of the Normal to the Three-Photon Positronium Decay Plane in Magnetic Field
(13; 19)
(b) Vol. **124**, No. 2, p. 619—623 (1984).

A. M. Taratin and S. A. Vorobev
Multiple Scattering of Relativistic Electrons under Axial Channeling in a Crystal. Model of the Lattice of the Atomic Rows
(11; 13; 22.1.2)
(b) Vol. **124**, No. 2, p. 641—648 (1984).

A. E. Smith and D. F. Lynch
*Low-Energy Electron Reflection Intensities from 2H-MoS_2 as Derived by Scattering Matrix Calculations
(1.6 and 13; 22)
(b) Vol. **124**, No. 2, p. 649—655 (1984).

M. Okusawa
*Splitting of Core-Level XPS Lines of Transition-Metal Ions in Compounds and Moment of the Lines
(13 and 17; 22.5)
(b) Vol. **124**, No. 2, p. 673—681 (1984).

B. Vasvári
*A Simple Muffin-Tin Model for Liquid Metals
(13; 2; 21.1)
(b) Vol. **124**, No. 2, p. 683—692 (1984).

B. A. Lukiyanets, K. D. Tovstyuk, V. V. Boichuk, and K. K. Danilevich-Tovstyuk
*The Jahn-Teller Effect in Layer Crystals
(6 and 13; 22.3)
(b) Vol. **124**, No. 2, p. 693—698 (1984).

H. Berndt and H.-J. Hunger
*Experimental Determination of the M-Shell Ionization Cross Section
(13; 21.6)
(a) Vol. **84**, No. 2, p. K149—K152 (1984).

S. N. Gupta and S. Shah
*Evidence of Open 4f Shells in Rare Earth Compounds
(13; 20; 21.4)
(b) Vol. **125**, No. 1, p. 211—214 (1984).

D. E. Popov, E. I. Rozum, S. R. Uglov, and S. A. Vorobiev
Anomalous Radiation of Fast Channeled Electrons in Optically Transparent Crystals
(11 and 20; 13; 22.1.3; 22.5.2)
(b) Vol. **125**, No. 1, p. 251—258 (1984).

A. Fortini
*Quantum Corrections to the Screening Length and Anisotropy in Magnetic Field
(13)
(b) Vol. **125**, No. 1, p. 259—266 (1984).

J. A. Leiro
The First Born Approximation in Estimates of the Threshold Anomalies in the X-Ray Spectra of Simple Metals
(20; 13; 21)
(b) Vol. **125**, No. 1, p. 305—310 (1984).

H. Englisch
*Instabilities and δ-Distributions in Solid State Physics
(13; 20.3; 22)
(b) Vol. **125**, No. 1, p. 393—400 (1984).

M. M. Babaev, T. M. Gasymov, and A. A. Katanov
Thermopower and Thermomagnetic Effects of Hot Carriers under Conditions of Mutual Electron–Phonon Drag
(15; 6; 13; 22)
(b) Vol. **125**, No. 1, p. 421—429 (1984).

I. M. Antonova
*On the Effective Potential for the Anion in MgO and CaO Crystals
(4 and 13; 22.6)
(b) Vol. **125**, No. 1, p. K55—K58 (1984).

G. Oelgart and U. Werner
Kilovolt Electron Energy Loss Distribution in GaAsP
(11; 13; 22.2.4)
(a) Vol. **85**, No. 1, p. 205—213 (1984).

V. Andreev and M. Pope
*Generation of Hot Free Electrons Due to the Interaction between Charge-Transfer Excitons and Thermalized Free Electrons in Molecular Crystals
(13 and 13.5.1; 22.9)
(b) Vol. **125**, No. 2, p. 573—581 (1984).

V. A. Bazylev and A. V. Demura
Atomic Particle Ionization in a Crystal in Random Orientation
(11; 13)
(b) Vol. **125**, No. 2, p. 583—591 (1984).

O. V. Boev and K. P. Arefiev
*Delocalized Positronium in Alkali Halides
(13; 22.5.2)
(b) Vol. **125**, No. 2, p. 619—627 (1984).

P. Streitenberger
*Effective Exchange and Correlation Interaction and Dielectric Screening in the Electron Gas. A Functional Approach to the Local-Field Correction
(13; 14.4; 21)
(b) Vol. **125**, No. 2, p. 681—692 (1984).

C. Pisani and E. Marchese
*Critical Analysis and Simplified Formulation of the Travelling Cluster Approximation for Disordered Systems
(2 and 13)
(b) Vol. **125**, No. 2, p. 721—733 (1984).

D. Hajduković and I. M. Savić
*Ground-State Degeneracy of the One-Dimensional Potts and Ising Magnets with Competing Interactions
(13 and 18)
(b) Vol. **125**, No. 2, p. K113—K116 (1984).

V. S. Rukhlov and G. B. Teitelbaum
*Effect of Spin–Spin Reservoir on the Magnetic Bistability
(13 and 18)
(b) Vol. **125**, No. 2, p. K121—K124 (1984).

A. V. Soldatov and A. N. Gusatinskii
*Energy Delocalization of the Rare-Earth f-States in Some Compounds
(13 and 20; 22)
(b) Vol. **125**, No. 2, p. K129—K132 (1984).

S. J. Mali and D. G. Kanhere
*Calculation of X-Ray Photoemission Shifts for Some Metallic Elements Using Slater's Transition State Theory
(13 and 17; 21)
(b) Vol. **125**, No. 2, p. K143—K145 (1984).

A. Z. Maksymowicz, T. Stobiecki, E. Jarocki, and W. Karaś
*Charge Transfer and Concentration Dependence of Magnetization in Amorphous Fe–B Alloys
(13 and 18; 2; 21.1.1)
(b) Vol. **126**, No. 1, p. 191—195 (1984).

B. Lorenz and H. Fehske
*Critical Study of the Static Functional Integral Method in the Hubbard Model
(13 and 18; 8)
(b) Vol. **126**, No. 1, p. 235—245 (1984).

R. Hayn and W. John
*Non-Exponential Localization in an Inhomogeneously Disordered Chain
(2 and 13)
(b) Vol. **126**, No. 1, p. 335—341 (1984).

M. Mondal and K. P. Ghatak
*Magnetic Susceptibility of Electrons in Ultrathin Films of Semiconductors in the Presence of a Quantizing Magnetic Field
(13 and 18; 22.2.1)
(b) Vol. **126**, No. 1, p. K41—K45 (1984).

A. O. O. Mekhrabov and M. Doyama
*Electronic Theory of Atomic Short-Range Order for Ternary Alloys Using the Pseudopotential Approximation and Its Comparison with Experiments
(1.3 and 13; 21.1; 21.1.1)
(b) Vol. **126**, No. 2, p. 453—458 (1984).

B. Holland, H. S. Greenside, and M. Schlüter
*Cohesive Properties of Ge, Si, and Diamond Calculated with Minimum Basis Sets
(4 and 13; 12.2; 22.1; 22.1.1; 22.1.2)
(b) Vol. **126**, No. 2, p. 511—515 (1984).

V. V. Kaplin, S. B. Nurmagambetov, V. I. Gridnev, E. I. Rozum, S. Pak, S. R. Uglov, and S. A. Vorobiev
Angular Distributions of Fast Electrons at Planar Channeling in Silicon Crystals
(11; 13; 22.1.2)
(b) Vol. **126**, No. 2, p. 565—574 (1984).

U. PIETSCH
Simulation of Dynamical Bond Charge Transfer Properties in Elemental Semiconductors by Means of a Simple Quantum-Mechanical Model
(4; 13; 22.1.2)
(b) Vol. **126**, No. 2, p. 595—605 (1984).

M. ASCHE and O. G. SARBEI
*The Role of Electron–Hole Interaction in the Cooling Process of Highly Excited Carriers
(13 and 20; 22.2.1)
(b) Vol. **126**, No. 2, p. 607—616 (1984).

U. H. GLÄSER and H. LÖFFLER
Effective Medium Approach for the Residual Resistivity of Alloys Containing Guinier-Preston Zones
(14.1; 13; 21; 21.1)
(b) Vol. **126**, No. 2, p. 741—751 (1984).

13.1 Band Structure

Review Articles

B. M. Askerov, B. I. Kuliev, and S. R. Figarova
Electron Transport Phenomena in Conductive Films
(14.3.1; 6; 13.1; 15; 22)
(b) Vol. **121**, No. 1, p. 11—37 (1984).

L. B. Dubrovskaya and S. Z. Nazarova
Some Regularities in the Behaviour of the Magnetic, Electrokinetic, and Superconducting Properties of Transition Metal Monocarbide and Mononitride Solid Solutions
(14.3 and 18; 13.1; 14.2; 15; 22)
(a) Vol. **83**, No. 1, p. 11—38 (1984).

G. Dräger and O. Brümmer
*Polarized X-Ray Emission Spectra of Single Crystals
(13.1 and 20; 21; 22)
(b) Vol. **124**, No. 1, p. 11—28 (1984).

A. Blacha, H. Presting, and M. Cardona
*Deformation Potentials of $\boldsymbol{k} = \boldsymbol{0}$ States of Tetrahedral Semiconductors
(6 and 13.1; 22.1; 22.2; 22.4)
(b) Vol. **126**, No. 1, p. 11—36 (1984).

Original Papers and Short Notes

L. V. Vassilev
*Influence of the Phase Transition in $Pb_{1-x}Ge_xTe$ on the Electronic Dispersion Relation
(13.1; 1.2; 22.8)
(b) Vol. **121**, No. 1, p. 203—207 (1984).

S. B. Shrivastava and V. K. Ojha
*Band Structure Calculation of the Momentum Density in F.C.C. Manganese
(13.1; 21.1)
(b) Vol. **121**, No. 1, p. 225—231 (1984).

W. John, V. V. Nemoshkalenko, V. N. Antonov, and Vl. N. Antonov
*Electron–Phonon Interaction in Transition Metals. Results of Relativistic APW Band Structure Calculations
(13.1; 21)
(b) Vol. **121**, No. 1, p. 233—239 (1984).

O. V. Farberovich, G. P. Nizhnikova, S. V. Vlasov, and E. P. Domashevskaya
*Local-Density-Functional Approximation to the Energy Band Structure of TmS Using the Self-Consistent Relativistic Linearized-Augmented-Plane-Wave Method
(13.1; 21.4)
(b) Vol. **121**, No. 1, p. 241—253 (1984).

M. C. Böhm
*Symmetry Breaking in Organometallic Polymers Due to Interchain Interactions
(13.1; 4; 22.9)
(b) Vol. **121**, No. 1, p. 255—264 (1984).

L. A. Manakova and S. P. Ionov
*The Ordering Types in the Extended Hubbard Model with Electron–Phonon Interaction
(13.1; 13.5.3)
(b) Vol. **121**, No. 1, p. 275—287 (1984).

J. Cisowski, J. C. Portal, E. K. Arushanov, J. M. Broto, S. Huant, and L. C. Brunel
*Shubnikov-De Haas Effect and Cyclotron Resonance in p-CdSb
(13.1; 22.3)
(b) Vol. **121**, No. 1, p. 289—292 (1984).

L. Pajasová, P. Pajas, O. A. Makarov, and V. M. Zakharov
Optical Properties of Amorphous GeS in the Spectral Region of 18 to 38 eV
(2 and 20.1; 13.1; 22)
(b) Vol. **121**, No. 1, p. 293—300 (1984).

V. Cháb and I. Bartoš
*Band Structure and Lifetime Determinations for GeS from Angle-Resolved Photoemission
(13.1 and 17; 22)
(b) Vol. **121**, No. 1, p. 301—306 (1984).

J. Monecke
*Dielectric Function and Conductivity of p-Type Semiconductors
(13.1 and 14.3; 22)
(b) Vol. **121**, No. 1, p. 329—338 (1984).

A. K. Ray, S. B. Trickey, and A. B. Kunz
Optical Absorption of Solid Xenon at High Pressure
(20.1; 13.1; 23)
(b) Vol. **121**, No. 1, p. K47—K50 (1984).

K. Žičkus and A. Audzijonis
The Exponential Optical Absorption Edge Tail of SbSeI and BiSeI
(20.1; 13.1; 22.8)
(b) Vol. **121**, No. 1, p. K51—K53 (1984).

W. Hörig, H. Neumann, and G. Kühn
The Fundamental Absorption Edge of $LiInSe_2$
(20.1; 13.1; 22.8)
(b) Vol. **121**, No. 1, p. K55—K58 (1984).

V. I. Ivashchenko, A. A. Lisenko, and E. A. Zhurakovskii
*Electronic Structure of YC_x
(13.1; 22)
(b) Vol. **121**, No. 2, p. 583—588 (1984).

G. L. GUTSEV, YU. M. SHULGA, and YU. G. BORODKO

*The Discrete Variational X_α-Method Interpretation of the C KVV and Ti $L_{2,3}M_{2,3}V$ Auger Spectra of TiC
(13.1; 17; 21.1)
(b) Vol. **121**, No. 2, p. 595—601 (1984).

H. NEUMANN, D. W. BULLETT, H. SOMMER, R. D. TOMLINSON, and W. JOHN

*Valence Bands of the Cu–III–VI_2 Chalcopyrites Studied by Photoemission Spectra, X-Ray Emission Spectra, and Electronic Structure Calculations
(13.1; 17; 20; 22.8)
(b) Vol. **121**, No. 2, p. 641—647 (1984).

P. RENNERT, H. SCHELLE, and U.-H. GLÄSER

*Calculation of Polarized K and L X-Ray-Emission Spectra and of Partial Densities of States of the H.C.P. Metals Be, Mg, and Zn
(13.1; 20; 21.3)
(b) Vol. **121**, No. 2, p. 673—684 (1984).

Z. PAWŁOWSKA

*Positron Annihilation in Zinc. Theory and Experiment
(13.1; 21.1)
(b) Vol. **121**, No. 2, p. 695—703 (1984).

M. STORDEUR and H. T. LANGHAMMER

*Erratum to: Anisotropy of Transport Properties Due to Direction-Dependent Band Nonparabolicity
(13.1; 14.3; 15; 22)
(b) Vol. **121**, No. 2, p. 765 (1984).

XIAN-WU ZOU

*Narrowing and Splitting of the 2S Band in Lithium with Increasing Atomic Separation
(13.1; 21.2)
(b) Vol. **121**, No. 2, p. K171—K174 (1984).

C. ALBERS and TH. MORITZ

*Comments on the Applicability of Energy Gap versus Refractive Index Relations to Lead Chalcogenides
(13.1; 20.1; 22)
(b) Vol. **121**, No. 2, p. K175—K180 (1984).

SANG DON CHOI and OK HEE CHUNG

*Theory of Cyclotron Resonance Lineshape in an Electron–Phonon System
(13.1 and 19; 22.2.3)
(b) Vol. **121**, No. 2, p. K181—K185 (1984).

W. JOHN

*Rigid Muffin-Tin Approximation for the Electron–Phonon Matrix Element. Nonlocal and Nonspherical Potentials
(13.1; 21)
(b) Vol. **122**, No. 1, p. 197—201 (1984).

G. ANGONOA, R. DOVESI, C. PISANI, and C. ROETTI
* A New Technique for the Evaluation of Densities of States in ab initio Calculations of Periodic Systems
(13.1; 13.2; 21.2; 21.3)
(b) Vol. **122**, No. 1, p. 211—220 (1984).

R. A. EVARSTOV and V. P. SMIRNOV
*Application of the Band Representations of Space Groups in the Theory of Electronic States of Crystalline Solids (I): General Consideration of the Band Representations
(13.1)
(b) Vol. **122**, No. 1, p. 231—238 (1984).

M. NYLÉN
*Electronic Structure of an Unreconstructed $(10\bar{1}0)$ α-Quartz Surface
(13.1 and 13.3; 22.6)
(b) Vol. **122**, No. 1, p. 301—308 (1984).

S. K. PLACHKOVA
*Estimation of Some Band-Edge Structure Parameters of the System $(GeTe)_{1-x}(AgSbTe_2)_x$
(13.1; 14.3; 15; 22)
(b) Vol. **122**, No. 1, p. K53—K58 (1984).

D. NESHEVA
*Studies of Low-Energy Optical Absorption and Gap State Density of a-CdS by the Constant Photocurrent Method
(13.1 and 20.1; 2; 14.3.1; 22.4.1)
(a) Vol. **82**, No. 1, p. 243—248 (1984).

D. COQUILLAT, A. PRADEL, G. BRUN, and J. C. TEDENAC
*Growth and Physical Properties of $TlBiTe_2$ Crystals
(13.1 and 14.3; 8; 15; 20.1; 22.8)
(a) Vol. **82**, No. 1, p. 295—299 (1984).

R. A. EVARESTOV and V. P. SMIRNOV
*Application of the Band Representations of Space Groups in the Theory of Electronic States of Crystalline Solids (II): Band Representations for Some Space Groups
(13.1; 22)
(b) Vol. **122**, No. 2, p. 559—567 (1984).

J. MOLENDA
*Electronic Processes in Electrode Materials of A_xMX_2-Type
(13.1; 1.3; 21.1; 21.2)
(b) Vol. **122**, No. 2, p. 591—598 (1984).

S. LOGOTHETIDIS, S. VES, and J. SPYRIDELIS
Ellipsometric Determination of the Optical Constants of $ZnIn_2S_4$
(20.1; 13.1; 22.8)
(b) Vol. **122**, No. 2, p. 613—622 (1984).

G. Valiukonis, F. M. Gashimzade, D. A. Guseinova, G. Krivaitė, M. M. Mamedov, and A. Šileika
*Electroreflectance and Thermoreflectance Spectra of SnSe Crystals
(13.1 and 20.1; 22)
(b) Vol. **122**, No. 2, p. 623—627 (1984).

B. Giese
*Electron Scattering Interaction and Elementary Excitations in Lead Chalcogenides
(13.1 and 13.5; 6; 22)
(b) Vol. **122**, No, 2, p. 641—650 (1984).

W. Kautek, F. Decker, and H. Gerischer
Optical Anisotropy of Transition Metal Dichalcogenides. A Photoelectrochemical Determination
(20.1; 13.1; 16; 22)
(b) Vol. **122**, No. 2, p. 651—659 (1984).

S. N. Sahu, J. T. Borenstein, V. A. Singh, and J. W. Corbett
*Semi-Empirical Tight Binding Calculations for the Energy Bands of the Diamond and Zincblende Type Semiconductors
(13.1; 22.1; 22.2)
(b) Vol. **122**, No. 2, p. 661—667 (1984).

F. Lukeš, E. Schmidt, J. Humlíček, and P. Dub
*Optical Properties of GeS
(13.1 and 20.1; 22)
(b) Vol. **122**, No. 2, p. 675—686 (1984).

W. Knap, J. Kossut, and J. Mycielski
*Photoelectromagnetic Effect and Photoconductivity in Quantizing Magnetic Fields
(13.1; 16; 22)
(b) Vol. **122**, No. 2, p. 761—773 (1984).

M. Hanke, D. Hennig, and R. Schuchardt
*Cluster-Size Dependence of the Electronic Density of States within the Bethe-Lattice Approach
(13.1; 22)
(b) Vol. **122**, No. 2, p. K151—K154 (1984).

I. Bartoš
*On Inhomogeneous Damping of Electrons in Crystals
(13.1; 22)
(b) Vol. **122**, No. 2, p. K159—K162 (1984).

K. Bhattacharyya and P. K. Basu
*Variational Calculation of Ground Subband Energy of Two-Dimensional Electron Gas in p–n^+–p Structure
(13.1 and 14.3.3; 22.2.1)
(b) Vol. **122**, No. 2, p. K191—K194 (1984).

Yu. S. Bulyshev, I. M. Kashirskii, and V. V. Sinitskii
Photoconductivity and Energy Level Structure of 2,4,7-Trinitro-9-Fluorenone Thin Films
(14.3.1 and 16; 13.1; 22.9)
(a) Vol. **82**, No. 2, p. 537—544 (1984).

MUKESH JAIN, A. V. R. WARRIER, and H. K. SEHGAL
Composition Dependence of the Optical Energy Gap in $Pb_{1-x}Hg_xTe$ Alloy Thin Films
(20.1; 13.1; 22.8)
(a) Vol. **82**, No. 2, p. K181—K184 (1984).

S. YU. POTAPENKO and A. M. SATANIN
*Electron States and Conductivity of Size Quantized Systems with Rough Boundaries
(13.1; 13.3; 14.3.1; 22)
(b) Vol. **123**, No. 1, p. 191—199 (1984).

J. NAUDTS, A. WERNER, and H. D. HOCHHEIMER
*Pressure–Volume Relation in Intermediate Valence Compounds
(13.1; 21.4)
(b) Vol. **123**, No. 1, p. 209—217 (1984).

F. LOPEZ-AGUILAR and J. COSTA-QUINTANA
*A First Principles Pseudopotential Model for the Strong Intrasite Interaction Applied to the $4f^{13}$ Configuration (Yb_2O_3)
(13.1; 22.6)
(b) Vol. **123**, No. 1, p. 219—228 (1984).

U. BRÜSTEL and K. UNGER
*Recursion Method with Plane Wave Basis for Band Structure Calculations
(13.1)
(b) Vol. **123**, No. 1, p. 229—236 (1984).

M. BEILER, M. MOCKER, and O. ZIEP
*Theory of Second-Order Auger Recombination in Strong Degenerate Small-Gap Semiconductors. Application to PbSSe
(13.1 and 14.3; 20.3; 22.8)
(b) Vol. **123**, No. 1, p. 247—255 (1984).

N. E. CHRISTENSEN
*First-Principles Calculation of Deformation Potentials of Copper Halides
(13.1; 22.5)
(b) Vol. **123**, No. 1, p. 281—288 (1984).

M. HANKE, W. KÜHN, and R. STREHLOW
*Correctly Weighting Tetrahedron Method for ***k***-Space Integration of F.C.C. Structures
(13.1)
(b) Vol. **123**, No. 1, p. K39—K42 (1984).

J. LEOTIN, M. GOIRAN, and S. ASKENAZY
*Cyclotron Resonance of Holes in Cadmium Antimonide
(13.1; 22.3)
(b) Vol. **123**, No. 1, p. K43—K46 (1984).

H. T. LANGHAMMER, M. STORDEUR, H. SOBOTTA, and V. RIEDE
IR Transmission Investigations of Sb_2Te_3 Single Crystals
(20.1; 13.1; 22.7)
(b) Vol. **123**, No. 1, p. K47—K51 (1984).

V. CRIŞAN, I. POP, V. TOŞA, and N. RUSU
*Band Structure of $Gd_2Ni_2Al_{15}$ Intermetallic Compound
(13.1 and 18.3; 21.1; 21.4)
(b) Vol. **123**, No. 1, p. K53—K56 (1984).

M. ZAŁUŻNY
*On the Influence of the Depolarization Field on the Intersubband Adsorption in Thin Semiconductor Films
(13.1 and 20.1; 22; 22.2.1)
(b) Vol. **123**, No. 1, p. K57—K61 (1984).

W. HENRION and H. LANGE
Optical Properties of $CrSi_2$
(20.1; 13.1; 21)
(b) Vol. **123**, No. 1, p. K69—K73 (1984).

M. M. EL-OKER, F. A. SULTAN, S. A. YOUSEF, and S. A. EL-SAHHAR
Surface Plasmon Modes and Energy Loss Function of Indium Thin Films on Mica Substrates
(13.5.2 and 20.1; 13.1; 21)
(a) Vol. **83**, No. 1, p. 263—267 (1984).

G. DEBIAIS
Optical and Electrical Characterization of Mercury Sulfide Mixed Films
(14.3.1 and 20.1; 13.1; 22.4.4)
(a) Vol. **83**, No. 1, p. 269—278 (1984).

S. B. MASLENKOV, A. I. KOZLENKOV, S. A. FILIN, and A. I. SHULGIN
*Soft X-Ray Emission Spectra from Some Ni–Al Alloys
(13.1 and 20; 21; 21.1)
(b) Vol. **123**, No. 2, p. 605—609 (1984).

K. ŁAWNICZAK-JABŁOŃSKA, E. MINNI, J. PEŁKA, E. SUONINEN, and J. AULEYTNER
*XPS and XES Study of the Valence Band of Cr–Al B.C.C. Solid Solutions
(13.1; 20; 21.1)
(b) Vol. **123**, No. 2, p. 627—634 (1984).

H.-G. BARTEL, G. SCHOLZ, R. NEUMANN, and W. SCHMIDT
*The Electronic Structure of Octahedral Platinum and Iridium Clusters (Pt_6, Ir_6). A Comparison between Xα-SW and EHT Calculations
(13.1; 21)
(b) Vol. **123**, No. 2, p. 641—648 (1984).

T. GRAFE and R. STREHLOW
*Influence of Si–Si Bonds on the Local Electronic Density of States of a-SiO_2
(13.1 and 13.4; 22.6)
(b) Vol. **123**, No. 2, p. 663—669 (1984).

B. GIESE and M. MOCKER
*Carrier Heating in Lead Chalcogenides
(13.1 and 14.3.4; 22)
(b) Vol. **123**, No. 2, p. 755—762 (1984).

V. V. NEMOSHKALENKO, A. E. KRASOVSKII, V. N. ANTONOV, VL. N. ANTONOV, U. FLECK, H. WONN, and P. ZIESCHE
*Addendum to: The Relativistic Linear Muffin-Tin Orbital Method. Application to Au
(13.1; 21.6)
(b) Vol. **123**, No. 2, p. 787 (1984).

V. RIEDE, H. NEUMANN, H. SOBOTTA, W. KISSINGER, M. S. OMAR, and B. R. PAMPLIN
Infrared Study of Lattice and Free Carrier Effects in p-Type $CuGe_2P_3$
(6 and 20.1; 13.1; 22.8)
(b) Vol. **123**, No. 2, p. K119—K123 (1984).

M. MONDAL and K. P. GHATAK
*Influence of Magnetic Quantization on the Effective Electron Mass in Zero-Gap Semiconductors
(13.1 and 18; 22)
(b) Vol. **123**, No. 2, p. K143—K148 (1984).

A. ANAGNOSTOPOULOS, K. KAMBAS, and B. PLOSS
*Optical Measurements and Temperature Dependence of the Energy Gap in $ZnIn_2S_4$ Layered Compound
(13.1 and 20.1; 22.8)
(b) Vol. **123**, No. 2, p. K155—K159 (1984).

N. S. BUINOV and S. R. SYRTSOV
*Ferroelectric Phase Transition in the Field of Nonresonance Electromagentic Waves
(13.1 and 14.4.1; 22)
(b) Vol. **123**, No. 2, p. K165—K167 (1984).

M. VANĚČEK, A. ABRAHÁM, O. ŠTIKA, J. STUCHLÍK, and J. KOČKA
*Gap States Density in a-Si:H Deduced from Subgap Optical Absorption Measurement on Schottky Solar Cells
(13.1 and 20.1; 2; 22.1.2)
(a) Vol. **83**, No. 2, p. 617—623 (1984).

R. HERRMANN, G. NACHTWEI, and W. KRAAK
On the Barrier Height in Grain Boundaries of InSb Bicrystals
(14.3; 13.1; 22.2.3)
(a) Vol. **83**, No. 2, p. K207—K210 (1984).

V. B. GOHEL, C. K. ACHARYA, and A. R. JANI
*On the Determination of Pseudopotential Parameters and the Comprehensive Study of Metallic Properties
(13.1; 2; 14.1; 15; 21)
(b) Vol. **124**, No. 1, p. 137—145 (1984).

H. S. BRANDI, G. JALBERT, and O. L. MALTA
*Multiphoton Absorption in Solids. Non-Perturbative Approach in the Weak-Field Limit
(13.1 and 20.1; 22)
(b) Vol. **124**, No. 1, p. 147—157 (1984).

W. HUHN and A. STAHL
*Self-Consistent Field Theory Applied to the Semiconductor Band Edge
(13.1 and 13.5; 14.3; 20; 22)
(b) Vol. **124**, No. 1, p. 167—177 (1984).

V. DRCHAL and J. KUDRNOVSKÝ
*The Auger Spectra from Alloys. Effect of Partial Long-Range Order
(13.1; 17; 21)
(b) Vol. **124**, No. 1, p. 179—189 (1984).

W. GIRIAT and J. STANKIEWICZ
Absorption Edge in $Mn_xZn_{1-x}Te$
(20.1; 13.1; 22.4.4)
(b) Vol. **124**, No. 1, p. K53—K57 (1984).

T. A. MATVEEVA
*4f States in Spectra of Rare-Earth Metals
(13.1; 20; 21.4)
(b) Vol. **124**, No. 1, p. K59—K62 (1984).

N. B. MUSTAFAEV and M. G. SHAKHTAKHTINSKII
Thermopower of a Thin Bi Film in a Classical Strong Magnetic Field
(15; 13.1; 21.7)
(b) Vol. **124**, No. 1, p. K93—K96 (1984).

H. NAKANISHI, S. ENDO, and T. IRIE
Piezoresistance in $CdIn_2S_4$
(14.3; 13.1; 22.8)
(b) Vol. **124**, No. 1, p. K101—K104 (1984).

T. M. RAZYKOV and B. KH. KADIROV
*Energy Band Diagrams of Cu_2S–$Zn_xCd_{1-x}S$ ($0 \leqq x \leqq 1$) Heterojunctions
(13.1 and 14.3.3; 22.4.4)
(a) Vol. **84**, No. 1, p. K71—K74 (1984).

A. L. PUSHKARCHUK, P. P. MARDILOVICH, A. I. TROKHIMETS, G. M. ZHIDOMIROV, and S. G. GAGARIN
*The Energy Band Structure of the Corundum (0001) Surface with Different Extent of Hydroxylation
(13.1 and 13.3; 22.6)
(b) Vol. **124**, No. 2, p. 699—705 (1984).

M. STORDEUR and G. SIMON
Investigation of the Weak-Field Charge Transport in Semiconducting V_2–VI_3 Compounds with Trigonal Symmetry (II): Interpretation of the Weak-Field Charge Transport in Sb_2Te_3 Single Crystals
(14.3; 13.1; 15; 22.7)
(b) Vol. **124**, No. 2, p. 799—806 (1984).

K. ČERMÁK
The Exponential Optical Absorption Edge Tail of Tl_3VS_4 and Tl_3TaS_4
(20.1; 13.1; 22.8)
(b) Vol. **124**, No. 2, p. K143—K145 (1984).

K. Hricovíni and J. Krempaský
*On the Electronic Structure of Iron–Boron Alloys
(13.1; 2; 17; 21.1.1)
(b) Vol. **124**, No. 2, p. K147—K150 (1984).

N. B. Mustafaev and M. G. Shakhtakhtinskii
*Optical Absorption in Size-Quantized Bi Films
(13.1 and 20.1; 21.7)
(b) Vol. **124**, No. 2, p. K151—K154 (1984).

U. Walzer
Elastic and Bonding Properties of Transition Metals
(4; 12.1; 13.1; 21.1)
(b) Vol. **125**, No. 1, p. 55—64 (1984).
Erratum
(b) Vol. **126**, No. 2, p. 752 (1984).

S. Chakraborty, A. Manna, and A. K. Ghosh
*Charge Transfer and Energy of Formation in Ordered Binary Alloys
(4 and 13.1; 21.1; 21.6)
(b) Vol. **125**, No. 1, p. 97—100 (1984).

A. Pasturel, C. Colinet, C. Allibert, P. Hicter, A. Percheron-Guegan, and J. C. Achard
*A Theoretical and Experimental Study of the Enthalpies of Formation of $LaNi_5$-Type Compounds
(4 and 13.1; 21.1; 21.4)
(b) Vol. **125**, No. 1, p. 101—106 (1984).

S. Huant, L. Dmowski, M. Baj, and L. C. Brunel
*Intraconduction Band Magneto-Optical Study of InSb under Hydrostatic Pressure
(13.1 and 19; 16; 22.2.3)
(b) Vol. **125**, No. 1, p. 215—219 (1984).

M. De Dios Leyva, R. P. Alvarez, and J. L. Gondar
*Fundamental Zero Energy Gap States in Superlattices
(13.1; 22.2.1; 22.2.3)
(b) Vol. **125**, No. 1, p. 221—228 (1984).

I. Gorczyca
*Calculation of the Effect of Hydrostatic Pressure on the Band Structure of GaP and GaSb
(13.1; 12.2; 22.2.2; 22.2.3)
(b) Vol. **125**, No. 1, p. 229—235 (1984).

T. Chassé, U. Berg, and O. Brümmer
*Angular-Resolved X-Ray Photoelectron Spectra of Valence and Core Levels from GaS
(13.1 and 17; 22.3)
(b) Vol. **125**, No. 1, p. 267—274 (1984).

H. J. Fischbeck
*Kink Lattice Solution in a Quasi-One-Dimensional Peierls-Fröhlich System with a Nearly One-Third-Filled Band (II): Thermodynamic Considerations
(13.1)
(b) Vol. **125**, No. 1, p. 275—292 (1984).

M. C. Böhm
*A Crystal Orbital (CO) Model for Low-Dimensional Segregated Donor and Acceptor Stacks
(13.1 and 13.4; 22.9)
(b) Vol. **125**, No. 1, p. 311—322 (1984).

N. E. Christensen
*Deformation Potentials of the Valence Band Edge in CuBr
(13.1; 22.5)
(b) Vol. **125**, No. 1, p. K59—K61 (1984).

C. Prasad and J. D. Dubey
*Electronic Structure and Properties of Cubic Boron Nitride
(13.1; 22.2)
(b) Vol. **125**, No. 2, p. 629—638 (1984).

D. Fröhlich, A. Nöthe, and K. Reimann
Determination of Valence Band Parameters in ZnTe
(13.5.1 and 20.1; 13.1; 22.4.3)
(b) Vol. **125**, No. 2, p. 653—657 (1984).

S. K. Plachkova and L. A. Tersiev
Phase Transitions and Intervalley Jahn-Teller Effect in the System $(Ge\text{-}Te)_{1-x}(AgSbTe_2)_x$
(1.2 and 14.4.1; 13.1; 22.8)
(b) Vol. **125**, No. 2, p. 659—664 (1984).

H. J. Fischbeck
*The Incommensurate CDW Phase of a Quasi-One-Dimensional Peierls-Fröhlich System with a Nearly Quarter-Filled Band (I): Analytical Treatment
(13.1)
(b) Vol. **125**, No. 2, p. 665—680 (1984).

M. Scrocco
*X-Ray Photoelectron Spectra of Ag, Ga, In, and β-Sn Metals. Multielectron Excitations
(13.1 and 17; 21; 21.6)
(b) Vol. **125**, No. 2, p. 713—720 (1984).

G. Milchberg, K. Saminadayar, E. Molva, and H. R. Zelsmann
Optical Determination of the Valence-Band Parameters in CdTe
(13.4 and 20.3; 13.1; 22.4.3)
(b) Vol. **125**, No. 2, p. 795—803 (1984).

C. Fau, J. F. Dame, M. De Carvalho, J. Calas, M. Averous, and B. A. Lombos
Three-Band Model Applied to Narrow-Gap HgCdTe
(14.3; 13.1; 22.4.4)
(b) Vol. **125**, No. 2, p. 831—838 (1984).

S. K. Ghatak and M. N. Ghatikar
*On the Valence State of Rare-Earth Ion (R) in $RuCu_2Si_2$ and RCu_2Ge_2
(13.1; 22.8)
(b) Vol. **125**, No. 2, p. K125—K127 (1984).

P. Enders
*Electronic Structure of the Group-V Semimetals: Tight-Binding Parameter Estimation from Chemical Trends
(13.1; 21.7)
(b) Vol. **125**, No. 2, p. K133—K135 (1984).
Erratum
(b) Vol. **126**, No. 2, p. K177 (1984).

K. Čermák
*The Width of the Energy Gap of $Tl^{I}_{3}B^{V}C^{VI}_{4}$-Type Compounds
(4 and 13.1; 20.1; 22.8)
(b) Vol. **126**, No. 1, p. 273—277 (1984).

B. A. Orłowski and P. Plenkiewicz
*Electronic Band Structure of CdF_2. Photoemission Experiment and Pseudopotential Calculations
(13.1; 22.5)
(b) Vol. **126**, No. 1, p. 285—292 (1984).

S. K. J. Al-Ani and C. A. Hogarth
*The Spin Density and Optical Properties of Amorphous Thin Arsenic Silicate Films
(2 and 13.1; 19; 20.1; 22.6)
(b) Vol. **126**, No. 1, p. 293—304 (1984).

A. Sh. Makhmudov, Z. M. Khakimov, A. A. Levin, M. S. Yunusov, and P. K. Khabibullaev
*Green's Function Method in the Equivalent Orbital Basis. Point Defects and Complexes of Two Point Defects in Silicon
(13.1 and 13.4; 22.1.2)
(b) Vol. **126**, No. 1, p. 311—321 (1984).

D. Vignaud and J. L. Farvacque
*Free-Carrier Optical Absorption Induced by Dislocation Scattering Mechanisms in III–V Compounds
(13.1 and 20.1; 10.2; 22.2)
(b) Vol. **126**, No. 1, p. 351—359 (1984).

M. Mondal and K. P. Ghatak
*On the Effective Mass in Tetragonal Semiconductors in the Presence of an Arbitrarily Oriented Quantizing Magnetic Field
(13.1; 22)
(b) Vol. **126**, No. 1, p. K47—K51 (1984).

A. V. Skripov and A. P. Stepanov
*Electronic Structure of V_3Si and V_3Ge. Comparison of NMR Data with Band-Structure Calculations
(13.1 and 19; 21.1)
(b) Vol. **126**, No. 2, p. 557—563 (1984).

F. Bechstedt, R. Enderlein, and O. Heinrich
*The Effect of Many-Electron Correlation on Photothresholds of Semiconductors and Valence Band Discontinuities at Heterojunctions
(13.1; 22)
(b) Vol. **126**, No. 2, p. 575—585 (1984).

A. M. El-Dib and H. F. Hassan
*Temperature Dependence of the Electronic Energy Bands of $NaNO_2$ Crystal from 20 to 245 °C
(13.1; 22.8)
(b) Vol. **126**, No. 2, p. 587—593 (1984).

K. R. Allakhverdiev, S. S. Babaev, N. A. Bakhyshov, T. G. Mamedov, E. Yu. Salaev, and I. K. Efendieva
*The Influence of Hydrostatic Pressure on the Fundamental Absorption Edge of Crystals with TlSe Type Structure
(13.1 and 20.1; 12.2; 22; 22.8)
(b) Vol. **126**, No. 2, p. K139—K143 (1984).

H. Nakanishi and T. Irie
*Pressure Effect on the Absorption Edges of $CdIn_2S_4$ and Related Compounds
(13.1 and 20.1; 12.2; 22.8.2)
(b) Vol. **126**, No. 2, p. K145—K148 (1984).

H. Boudriot, R. Gründler, K. Deus, T. Stöckert, and H. A. Schneider
*Electron Energy Loss Measurements on $CdSnP_2$: Plasmon Data, Dielectric Function, Band Structure Data
(13.1 and 13.5.2; 14.4; 22.8)
(b) Vol. **126**, No. 2, p. K149—K154 (1984).

L. Borisova
*Density-of-States Effective Mass in Solid Solutions of the $(Ag_{x/2}Pb_{1-x}Sb_{x/2})Te$ Type
(13.1; 14.3; 15; 22)
(b) Vol. **126**, No. 2, p. K155—K158 (1984)

H. Gruber, E. Krautz, H. P. Fritzer, K. Gatterer, and A. Popitsch
Electronic Property Changes of η-Mo_4O_{11} by Doping with Tungsten, Rhenium, and Vanadium
(14.1; 1.2; 13.1; 18; 20.1; 22.6.1)
(a) Vol. **86**, No. 2, p. 749—757 (1984).

B. R. Nag, S. R. Ahmed, and M. Deb Roy
Submicron Electron Transport in Silicon at 300 and 77 K
(14.3.4; 6; 13.1; 22.1.2)
(a) Vol. **86**, No. 2, p. 773—779 (1984).

L. G. Khvostantsev, V. A. Sidorov, and O. B. Tsiok
*Gapless State and Anomalous Behaviour of Thermopower of Black Phosphorus at Pressure up to 12 GPa
(13.1 and 15; 14.3; 22.1)
(a) Vol. **86**, No. 2, p. K99—K104 (1984).

13.2 Fermi Surfaces

Original Papers and Short Notes

V. V. Gudkov and V. D. Fil
Ultrasonic Wave Field Transport by Conduction Electrons in Indium
(7 and 14.1; 13.2; 21)
(b) Vol. **121**, No. 1, p. 433—440 (1984).

G. Angonoa, R. Dovesi, C. Pisani, and C. Roetti
A New Technique for the Evaluation of Densities of States in ab initio Calculations of Periodic Systems
(13.1; 13.2; 21.2; 21.3)
(b) Vol. **122**, No. 1, p. 211—220 (1984).

A. Rubaszek and H. Stachowiak
Application of the Quadratic Response Theory to the Screening of a Positron in an Electron Gas of Metallic Density
(13; 13.2; 21)
(b) Vol. **124**, No. 1, p. 159—166 (1984).

A. C. Guptá and A. N. Nigam
*Effect of Cold Work on Hydrogenation of Nickel as Seen in the Thermoelectric Behaviour
(13.2; 12.1; 15; 21.1)
(b) Vol. **125**, No. 1, p. 207—209 (1984).

13.3 Surface and Interface States

Original Papers and Short Notes

R. Honmura, E. F. Sarmento, I. P. Fittipaldi, and T. Kaneyoshi
Phase Diagram for the Ising Model with a Free Surface
(18.2; 13.3)
(b) Vol. **121**, No. 1, p. 197–202 (1984).

C. L. Roy and G. Roy
*A Study of Relativistic Interface States
(13.3)
(b) Vol. **121**, No. 2, p. 717–730 (1984).

V. E. Godwin and M. Tomak
*Impurity States in the Presence of an Interface
(13.3 and 13.4)
(b) Vol. **122**, No. 1, p. 295–299 (1984).

M. Nylén
*Electronic Structure of an Unreconstructed $(10\bar{1}0)$ α-Quartz Surface
(13.1 and 13.3; 22.6)
(b) Vol. **122**, No. 1, p. 301–308 (1984).

T. L. Reinecke and F. Crowne
*Electron–Phonon Coupling in Inhomogeneous Electronic Systems
(13.3)
(b) Vol. **122**, No. 1, p. K65–K68 (1984).

S. J. Lee and Chung-In Um
*Impurity and Thickness Effects on the Cyclotron Resonance Linewidth of Electrons in Si Inversion Layers
(13.3; 14.3.2; 19; 22.1.2)
(b) Vol. **122**, No. 1, p. K73–K78 (1984).

J. Kelber, Ch. Schnittler, and G. Paasch
*The Completeness Relation and the Local Density Approximation for Layer Structures in Microelectronics
(13.3; 22)
(b) Vol. **122**, No. 1, p. K99–K103 (1984).

W. Lorenz and C. Engler
*Quantum-Theoretical Model of Adiabatic Rate Processes in Condensed Systems or at Interfaces
(13.3; 4; 6; 8; 22)
(b) Vol. **122**, No. 2, p. 745–755 (1984).

S. Yu. Potapenko and A. M. Satanin
Electron States and Conductivity of Size Quantized Systems with Rough Boundaries
(13.1; 13.3; 14.3.1; 22)
(b) Vol. **123**, No. 1, p. 191–199 (1984).

E. M. Kogan, Yu. N. Gornostiriev, and V. V. Ustinov
*Conduction Electron Scattering by a Short-Range Ordered Metal Surface
(13.3; 21)
(b) Vol. **123**, No. 1, p. 289—293 (1984).

G. Kaden
Consideration of Surface Band Bending in Silicon in Case of Internal Photoemission in MOS Structures
(14.3.3 and 14.3.4; 13.3; 22.1.2)
(a) Vol. **83**, No. 1, p. 375—386 (1984).

N. K. Bondareva, I. V. Krylova, and V. B. Golubev
*Thermostimulated Exoemission and Radical Recombination in Polyethylene Irradiated with an Electron Beam
(13.3 and 17; 11; 19; 22.9)
(a) Vol. **83**, No. 2, p. 589—595 (1984).

S. N. Kozlov and N. L. Levshin
*Isotopic Effect in the Recombination of Nonequilibrium Charge Carriers on the Germanium Surface
(13.3; 22.1.1)
(a) Vol. **83**, No. 2, p. K149—K152 (1984).

B. Sorkin and H. Käämbre
Thermo-Exoemission of NaCl:Ag Crystals, Cleaved before and after X-Irradiation
(17; 11; 13.3; 13.4; 22.5.2)
(a) Vol. **83**, No. 2, p. K171—K174 (1984).

J. Lopez, J. C. Le Bossé, and R. Baudoing
*Critical Analysis of the Excitation Map Concept for Studying Threshold Effects in VLEED
(13.3; 21.1)
(b) Vol. **124**, No. 1, p. 381—388 (1984).

H. Solbrig
*Comment on "Critical Analysis of the Excitation Map Concept for Studying Threshold Effects in VLEED" by J. Lopez, J. C. Bossé, and R. Baudoing
(13.3; 21.1)
(b) Vol. **124**, No. 1, p. K63—K67 (1984).

W. Frentrup, M. Griepentrog, H. Klose, G. Kreysch, and U. Müller-Jahreis
The Influence of Alkali Atoms Implanted in Silicon on the Negative Secondary Ion Emission
(17; 13.3; 22.1.2)
(a) Vol. **84**, No. 1, p. 269—272 (1984).

M. Jirsa
*On the Angular Dependence of the Effective Spin-Pinning in Thin Insulator Film
(13.3 and 19)
(b) Vol. **124**, No. 2, p. 609—618 (1984).

A. L. Pushkarchuk, P. P. Mardilovich, A. I. Trokhimets, G. M. Zhidomirov, and S. G. Gagarin
*The Energy Band Structure of the Corundum (0001) Surface with Different Extent of Hydroxylation
(13.1 and 13.3; 22.6)
(b) Vol. **124**, No. 2, p. 699—705 (1984).

L. Wendler
*Surface Plasmon–Polaritons in Bilayer Systems
(13.3 and 13.5.2)
(b) Vol. **124**, No. 2, p. K161—K165 (1984).

A. Stern and K. Hübner
*Electron Density of States of an Si–SiO_2 Interface with a Gradual Chemical Transition
(13.3; 22.1.2; 22.6)
(b) Vol. **124**, No. 2, p. K171—K175 (1984).

G. Brauer, F. Kerbe, Zs. Kajcsos, and A. Ashry
Positron Annihilation Studies on Al_2O_3 Powders
(10.2; 13.3; 22.6)
(a) Vol. **84**, No. 2, p. 451—455 (1984).

S. Alexandrova, A. Szekeres, and L. Kassamakova
*RF Plasma Induced Lateral Nonuniformities in Si–SiO_2 Structures
(13.3 and 14.3.2; 22.1.2)
(a) Vol. **84**, No. 2, p. 561—565 (1984).

M. Jirsa
*Angular Dependence of SSWR in Thin Metallic Films. Approximate Analytical Solution
(13.3 and 19; 21)
(b) Vol. **125**, No. 1, p. 187—196 (1984).

R. Herrmann, W. Kraak, and M. Gliński
*Quantized Hall Effect in the n-Inversion Layer in InSb Grain Boundaries
(13.3 and 14.3; 22.2.3)
(b) Vol. **125**, No. 1, p. K85—K88 (1984).

H. Berger, Yu. A. Kulyupin, S. A. Nepijko, and N. N. Sedov
*Quantitative Interference Electron Microscopy for Spherical Objects
(1.5 and 13.3; 21.7)
(a) Vol. **85**, No. 1, p. 45—50 (1984).

K. Uhlmann and M. Schmidt
*Influence of Work Function Change Due to Oxygen Chemisorption on the Secondary-Ion Emission Probability
(1.5 and 13.3; 17; 21.1.1)
(a) Vol. **85**, No. 1, p. K19—K22 (1984).

V. A. Bespalov, V. F. Kiselev, G. S. Plotnikov, and A. M. Saletskii
*Fluorescence Decay of Adsorbed Dye Molecules on Solid Surfaces
(13.3 and 20.3; 22.1.1; 22.6)
(a) Vol. **85**, No. 1, p. K73—K76 (1984).

M. A. Andreeva and R. N. Kuzmin
Mössbauer Conversion Electron Spectra at Total Reflection
(5; 1.6; 13.3; 21.1)
(b) Vol. **125**, No. 2, p. 461—465 (1984).

M. S. Brodin, V. M. Bandura, and M. G. Matsko
Excitonic Surface Polaritons in Luminescence from ZnTe Crystals
(13.5.1 and 20.3; 13.3; 22.4.3)
(b) Vol. **125**, No. 2, p. 613—618 (1984).

V. N. Ermakov, D. V. Korbutyak, V. G. Litovchenko, O. Yu. Mikityuk, and V. V. Nitsovich
A Study of the Reflection Spectra of Surface Exciton–Impurity Complexes
(13.4 and 13.5.1; 13.3; 20.1; 22.4.1)
(b) Vol. **125**, No. 2, p. 815—822 (1984).

Y. Matsuoka, S. Terakado, and S. Tanigawa
Positron Age-Momentum Correlation in Ultrafine Particles of Silver and Gold
(10.1; 13.3; 21.6)
(b) Vol. **125**, No. 2, p. 823—829 (1984).

P. Středa
*Universal Transport Properties of Inversion Layers in Quantizing Magnetic Fields
(13.3; 8; 14.1; 15; 18; 21)
(b) Vol. **125**, No. 2, p. 849—860 (1984).

I. M. Nepomnyashi and V. N. Nishanov
*The Localization Energy of Wannier-Mott Excitons at the Interface
(13.3 and 13.5.1)
(b) Vol. **126**, No. 1, p. K53—K57 (1984).

G. Bambakidis and W. T. McKeown
*C_5H_9 Cluster as a Model for the (111) 1 × 1 Diamond Surface
(13.3; 22.1)
(b) Vol. **126**, No. 1, p. K67—K70 (1984).

W. Braune, M. Schnürer, N. Kubicki, and R. Herrmann
**C–U* Measurements of Anodized InSb MOS Structures at 77 and 4.2 K
(13.3 and 14.3.3; 22.2.3)
(a) Vol. **86**, No. 1, p. 427—431 (1984).

M. Jirsa and V. Kamberský
*Angular Dependence of Spin-Wave Resonance in Thin Films with Asymmetrical Boundary Conditions. Analytical Treatment
(13.3 and 19)
(b) Vol. **126**, No. 2, p. 547—555 (1984).

G. Thörner and G. Borstel
Relativistic Theory of Photoemission from Solids (I): Key Aspects of the Model
(17; 13.3)
(b) Vol. **126**, No. 2, p. 617—628 (1984).

S. Jalaj, I. Singh, and S. K. Joshi
*Tight-Binding Calculations for the Angle-Resolved Local Density of States of InP(110) Surface
(13.3; 22.2.2)
(b) Vol. **126**, No. 2, p. 669—676 (1984).

H. J. ENGELMANN, K. BERNDT, and F. STORBECK
*Quantitative Chemical Analysis of 3d Transition Metal Alloys by SXAPS
(1.3 and 13.3; 20; 21.1)
(a) Vol. **86**, No. 2, p. 499–506 (1984).

I. THURZO and E. PINČÍK
*Some Peculiar Features of Small-Signal Charge DLTS Response of GaAs MOS Capacitors
(13.3 and 14.3.3; 22.2.1)
(a) Vol. **86**, No. 2, p. 795—804 (1984).

13.4 Impurity and Defect States

Original Papers and Short Notes

A. Kasten, D. Weschenfelder, and G. Hess
Spectroscopic Investigation of the Magnetic Properties of $Ho_2O_2SO_4$
(18.4; 13.4; 18.3; 22.8.1)
(b) Vol. **121**, No. 1, p. 145—155 (1984).

H. Stachowiak
*Screening of a Proton Moving through the Electron Gas
(11 and 13.4; 21)
(b) Vol. **121**, No. 1, p. 307—315 (1984).

R. M. Bilbé, J. E. Nicholls, and J. J. Davies
*Electron Paramagnetic Resonance Investigation of Pb- and Ge-Doped CdTe
(13.4 and 19; 22.4.3)
(b) Vol. **121**, No. 1, p. 339—344 (1984).

M. Cahay and E. Kartheuser
*Polarizability of Donors in Polar Semiconductors: Effect of a Weak Magnetic Field
(13.4; 22.1.2; 22.2.2; 22.4.2)
(b) Vol. **121**, No. 1, p. 345—356 (1984).

M. T. Primatarowa
*Influence of Planar Defects and Disorder on the Energy Spectrum of Molecular Crystals
(13.4 and 13.5.1)
(b) Vol. **121**, No. 1, p. 357—369 (1984).

H.-J. Unger
*Localization of Electronic States in Disordered Tight-Binding Systems (II): Mobility Edges
(2 and 13.4; 22)
(b) Vol. **121**, No. 1, p. 371—383 (1984).

S. Žurauskas and A. Dargys
*Field Ionization of Phosphorus Atoms in Silicon
(13.4; 14.3.3; 22.1.2)
(b) Vol. **121**, No. 1, p. 385—391 (1984).

J. Manam and V. V. Ratnam
*V_3 and F_A Centers in KCl:Na Crystals. A Thermoluminescence Study
(13.4 and 20.3; 10.2; 22.5.2)
(b) Vol. **121**, No. 1, p. 393—399 (1984).

V. N. Salomatov and I. A. Parfianovich
*The F_H-Centre Energy Structure in KCl and KBr
(13.4 and 20.1; 10.2; 22.5.2)
(b) Vol. **121**, No. 1, p. 401—406 (1984).

Y. Ishihara, Y. Ohno, and I. Nakada
*Anisotropic Electrical Properties of GeSe
(13.4 and 14.3; 22)
(b) Vol. **121**, No. 1, p. 407—412 (1984).

S. M. Basha, K. Purandar, and Y. P. Reddy
*Electronic Absorption Spectrum of Fe^{3+} Doped in Calcium Tartrate Tetrahydrate
(13.4 and 20.1; 22.9)
(b) Vol. **121**, No. 1, p. K79—K81 (1984).

F. V. Kusmartsev and E. I. Rashba
*Symmetry Breaking in the Theory of Self-Trapping Barrier and in Allied Problems
(13.4)
(b) Vol. **121**, No. 1, p. K87—K90 (1984).

V. D. Tkachev, A. V. Mudryi, and N. S. Minaev
*Noble Gas Atoms as Chemical Impurities in Silicon
(10.2 and 13.4; 20.3; 22.1.2)
(a) Vol. **81**, No. 1, p. 313—321 (1984).

A. Abdel-Kader, F. J. Bryant, and J. H. C. Hogg
*Tellurium Doping and Implantation of Zinc Sulphide
(13.4; 20.3; 22.4.1)
(a) Vol. **81**, No. 1, p. 333—342 (1984).

J. Oswald, H. Feichtinger, and R. Czaputa
*A Useful Method to Improve Convergence in Least-Squares Fitting Procedures
(13.4; 14.3; 20.3; 22.8)
(a) Vol. **81**, No. 1, p. 343—351 (1984).

K. Žďánský and N. T. Thuc Hien
*Analysis of DLTS Curves Corresponding to Non-Exponential Transients
(13.4 and 14.3.3; 22.2.2)
(a) Vol. **81**, No. 1, p. 353—360 (1984).

B. Kaufmann
*Photoluminescence of a Deep Level in Undoped PbTe
(13.4 and 20.3; 22)
(a) Vol. **81**, No. 1, p. K31—K35 (1984).

C. Rincón and S. M. Wasim
*Influence of Intrinsic Defects on the Electrical Properties of $CuInSe_2$
(13.4 and 14.3; 22.8)
(a) Vol. **81**, No. 1, p. K77—K80 (1984).

M. A. Ivanov, V. Ya. Mitrofanov, and A. Ya. Fishman
Magnetic Anisotropy of Disordered Magnetics with Jahn-Teller Ions
(18.3; 13.4; 22.8.2)
(b) Vol. **121**, No. 2, p. 547—559 (1984).

E. A. GOREMYCHKIN, E. MÜHLE, P. G. IVANITSKII, V. T. KROTENKO, M. V. PASECHNIK, V. V. SLISENKO, A. A. VASILKEVICH, B. LIPPOLD, O. D. CHISTYAKOV, and E. M. SAVITSKII
*Crystal Electric Field Splitting in $TbNi_5$ and $ErNi_5$ Studied by Inelastic Neutron Scattering
(13.4; 21.4)
(b) Vol. **121**, No. 2, p. 623—631 (1984).

S. WYSOCKI
*Kinetics of the Decay of Surface γ-Conductivity of Magnesium Oxide
(13.4; 14.3.2; 22.6)
(b) Vol. **121**, No. 1, p. 731—736 (1984).

B. G. TAGIEV, G. M. NIFTIEV, and S. A. ABUSHOV
Photoluminescence of GaSe:Mn Single Crystals
(20.3; 13.4; 22.3)
(b) Vol. **121**, No. 2, p. K195—K199 (1984).

V. K. JAIN, V. P. SETH, and R. S. BANSAL
Electron Paramagnetic Resonance of Cr^{3+} in $Al_2(SO_4)_3 \cdot 18\ H_2O$ Single Crystals
(19; 13.4; 22.8.1)
(b) Vol. **121**, No. 2, p. K201—K204 (1984).

A. A. ALYBAKOV, V. A. GUBANOVA, M. ZHANIBEKOV, and N. TOICHEV
Formation and Stability of Paramagnetic Centers in $KBr:CrO_4^{2-}$, Ca^{2+} Crystals
(19 and 20.1; 6; 13.4; 22.5.2)
(b) Vol. **121**, No. 2, p. K205—K208 (1984).

B. S. ACHARYA, M. L. MUKHERJEE, C. M. SUNTA, and S. P. KATHURIA
*Thermoluminescence Emission and Decay Kinetics of $NaCl:SnCl_2$ Single Crystals
(13.4 and 20,3; 22.5.2)
(a) Vol. **81**, No. 2, p. 533—540 (1984).

S. DEMIAN, F. EL AKKAD, B. FARAG, H. MATHIEU, M. ROUZEYRE, and J. CHEVALLIER
*Photoluminescences and Optical Properties of Lithium-Doped $Mg_xZn_{1-x}Te$ Alloys
(13.4 and 20.3; 20.1; 22.4.4)
(a) Vol. **81**, No. 2, p. 549—555 (1984).

P. J. DEAN
Comparison of MOCVD-Grown with Conventional II–VI Materials Parameters for EL Thin Films
(14.3.4 and 20.3; 3; 13.4; 22.4.1; 22.4.2)
(a) Vol. **81**, No. 2, p. 625—646 (1984).

A. S. POPOV
*Degradation Model of Red GaP LEDs
(13.4 and 14.3.3; 20.3.; 22.2.2)
(a) Vol. **81**, No. 2, p. 669—674 (1984).

J. KLUGE and J. DONECKER
Transfer Effects in ODMR Spectra of ZnSe:Mn
(19 and 20.3; 13.4; 22.4.2)
(a) Vol. **81**, No. 2, p. 675—679 (1984).

W. E. Hagston
Mechanism of Excitation of Rare-Earth Complexes in Electroluminescence Devices
(14.3.4 and 20.3; 13.4; 22.4)
(a) Vol. **81**, No. 2, p. 687—693 (1984).

J. Kreissl and W. Gehlhoff
*EPR Investigations of ZnS:Mn and ZnSe:Mn. Single Crystals, Powders, Thin-Film Structures
(13.4 and 19; 22.4.1; 22.4.2)
(a) Vol. **81**, No. 2, p. 701—707 (1984).

K. Schmalz, P. Gaworzewski, and F.-G. Kirscht
*Deep Levels in Czochralski p-Si Due to Heat Treatment at 600 to 900 °C
(10.2 and 13.4; 14.3; 22.1.2)
(a) Vol. **81**, No. 2, p. K165—K169 (1984).

A. E. Aksenov, Yu. M. Fedorov, and A. A. Leksikov
Systems of Interacting Centers of Photoluminescence in $RbMnF_3$ and $KMnF_3$
(20.3; 13.4; 22.5)
(a) Vol. **81**, No. 2, p. K171—K174 (1984).

C. Sanchez, M. Henry, R. Morineau, and M. C. Leroy
*Small Polaron Mobility in α-$Li_xV_2O_5$
(13.4 and 13.5.3; 14.3; 19; 22.6.1)
(b) Vol. **122**, No. 1, p. 175—182 (1984).

A. V. Egranov and A. I. Nepomnyachikh
*Magnesium Colour Centres in NaF
(10.2 and 13.4; 20.1; 20.3; 22.5.2)
(b) Vol. **122**, No. 1, p. 249—254 (1984).

M. Korkmaz, R. Bessege, and A. Gavaix
*EPR Studies of Cu^{2+} in Anhydrous Lead Tartrate
(13.4 and 19; 22.8.1)
(b) Vol. **122**, No. 1, p. 255—262 (1984).

S. Shimanuki and N. Nishimaki
*A Correlation between the Spectral Location of the Absorption Bands Due to the Localized Centers and the Lattice Constant in Alkali Halide Crystals. Anionic Impurity Centers with s^2-Configuration and F Center which Have Cubic Symmetry
(10.2 and 13.4; 20.1; 22.5.2)
(b) Vol. **122**, No. 1, p. 269—283 (1984).

W. Schrittenlacher, H. Nelkowski, and H. Pradella
*ESR and ODMR of Zinc–Vacancy-Associated Defects in ZnSe
(13.4 and 19; 22.4.2)
(b) Vol. **122**, No. 1, p. 285—294 (1984).

V. E. Godwin and M. Tomak
*Impurity States in the Presence of an Interface
(13.3 and 13.4)
(b) Vol. **122**, No. 1, p. 295—299 (1984).

Y. WATANABE
*ESR and Magnetic Investigation of Lithium Intercalated VSe_2
(13.4; 10.2; 18.1; 19; 22.8)
(b) Vol. **122**, No. 1, p. 309—316 (1984).

H. J. A. KOOPMANS, M. M. A. PERIK, B. NIEUWENHUIJSE, and P. J. GELLINGS
*The Simulation and Interpretation of the EPR Powder Spectra of Gd^{3+}-Doped $LaAlO_3$
(13.4 and 19; 22.8.1)
(b) Vol. **122**, No. 1, p. 317—330 (1984).

J. M. BARANOWSKI and PHUONG AN
*The Ni Donor and Acceptor Levels in CdSe
(13.4 and 20.1; 22.4.2)
(b) Vol. **122**, No. 1, p. 331—336 (1984).

FAT DUEN HO
*The Improved Point-Ion Potentials for the F^+-Centers in $LiNbO_3$ and $LiTaO_3$
(10.2 and 13.4; 20.1; 22.8.1)
(b) Vol. **122**, No. 1, p. 337—344 (1984).

G. BALDACCHINI, G. P. GALLERANO, U. M. GRASSANO, M. MEUCCI, A. SCACCO, F. SOMMA, A. TIOURINE, and M. TONELLI
*Optical Pumping and Reorientation of the F_A Centers in KCl:Li^+
(10.2 and 13.4; 20.1; 22.5.2)
(b) Vol. **122** No. 1, p. K83—K86 (1984).

M. B. ZAPART, W. ZAPART, and A. I. ZVYAGIN
Phase Transitions in Ferroelastic $RbIn(MoO_4)_2$ Crystals by Electron Paramagnetic Resonance of Cr^{3+} Ions
(1.2 and 19; 13.4; 22.8.1)
(a) Vol. **82**, No. 1, p. 67—73 (1984).

Y. KAMIURA, F. HASHIMOTO, S. YONEYAMA, and T. NOBUSADA
Reduction of Copper Contamination in Germanium on Heat Treatments above 700 °C
(10.2; 13.4; 22.1.1)
(a) Vol. **82**, No. 1, p. 135—141 (1984).

K. SOMAIAH and V. HARI BABU
*Glow Curves and the Emission of Flux Grown BaFCl:Gd Crystals
(13.4 and 20.3; 20.1; 22.5.4)
(a) Vol. **82**, No. 1, p. 201—206 (1984).

A. N. GEORGOBIANI, A. N. GRUZINTSEV, Z. P. ILYUKHINA, V. E. TEZLEVAN, and I. M. TIGINYANU
*Photoconductivity and Non-Equilibrium Carrier-Recombination in $CdIn_2S_4$ Single Crystals
(13.4 and 16; 22.8)
(a) Vol. **82**, No. 1, p. 207—212 (1984).

H. L. OCZKOWSKI
*The Retrapping Process in a Thermoluminescence Model with an Intra-Pair Metastable State
(13.4 and 20.3; 22.4.2)
(a) Vol. **82**, No. 1, p. 213—219 (1984).

A. V. NABOK, B. A. NESTERENKO, YU. M. SHIRSHOV, YU. V. GOLTVIANSKII, and A. P. DUBCHAK
*Profile of Trapped Charge in Silicon Nitride Films in MNOS Structures
(13.4 and 14.3.4; 14.4; 22)
(a) Vol. **82**, No. 1, p. 221—227 (1984).

M. SUEZAWA and K. SUMINO
*Nature of Thermal Donors in Silicon Crystals
(13.4; 20.1; 22.1.2)
(a) Vol. **82**, No. 1, p. 235—242 (1984).

J.-P. DOUMERC, J. HEJTMANEK, J.-P. CHAMINADE, M. POUCHARD, and M. KRUSSANOVA
A Photoelectrochemical Study of $CuWO_4$ Single Crystals
(14.3; 13.4; 15; 16; 22.8.1)
(a) Vol. **82**, No. 1, p. 285—294 (1984).

T. WIKTORCZYK and C. WESOŁOWSKA
*Application of TSEE and TSD Methods to Investigations on Trap Levels in Yb_2O_3 Thin Films Deposited by Electron Beam Evaporation
(13.4; 22.6)
(a) Vol. **82**, No. 1, p. K67—K70 (1984).

G. MICOCCI and A. TEPORE
*Thermally Stimulated Current in PbI_2 at Low Electric Fields
(13.4; 14.4; 22.5)
(a) Vol. **82**, No. 1, p. K79—K82 (1984).

S. DZHUMANOV and P. K. KHABIBULLAEV
The Coexistence of Unstable, Metastable, and Separated Frenkel Pair Defects in Solids (I): Basic Characteristics
(10.2; 13.4; 22.5.2)
(b) Vol. **122**, No. 2, p. 481—488 (1984).

A. MANDELIS
*Self-Consistent Semi-Classical Dynamic Theory of Non-Radiative Capture and Emission Statistics in Defect Semiconductors
(13.4; 22)
(b) Vol. **122**, No. 2, p. 687—701 (1984).

A. AKILBEKOV and A. ELANGO
*Low-Temperature Pair Associates of H Centres in KBr
(10.2 and 13.4; 20.1; 22.5.2)
(b) Vol. **122**, No. 2, p. 715—723 (1984).

A. K. PETROSYAN, R. M. KHACHATRYAN, and E. G. SHAROYAN
*Jahn-Teller Effect in EPR and Optical Absorption Spectra of Cu^{2+} in $LiNbO_3$ Single Crystals
(13.4; 19; 20.1; 22.8.1)
(b) Vol. **122**, No. 2, p. 725—734 (1984).

W. Schrittenlacher
*Analysis of the Hyperfine Interactions of the Vacancy–Iodine Complex in ZnSe
(13.4; 19; 22.4.2)
(b) Vol. **122**, No. 2, p. 735—743 (1984)

E. A. Harris, J. H. Mellor, and S. Parke
*Electron Paramagnetic Resonance of Tetravalent Praseodymium in Zircon
(13.4 and 19; 22.8.1)
(b) Vol. **122**, No. 2, p. 757—760 (1984).

D. L. Alov, S. I. Gubarev, and V. B. Timofeev
*Determination of the s–d Exchange Interaction Sign in $Cd_{1-x}Mn_xS$ Crystals
(13.4 and 18.2; 22.4.4)
(b) Vol. **122**, No. 2, p. K133—K136 (1984).

V. V. Baidakov, V. N. Ermakov, N. N. Grigorev, V. V. Kolomoets, and T. A. Kudykina
*Breakdown of Impurity States of As and Sb in Germanium at Uniaxial Compression $\boldsymbol{P} \parallel [001] \parallel \boldsymbol{E}$
(13.4 and 14.3.4; 22.1.1)
(b) Vol. **122**, No. 2, p. K163—K167 (1984).

T. Lideikis, V. Noreika, A. Petraitiene, I. Šimkiene, and K. Valacka
Optical Absorption in Indium Doped Lead Tin Telluride
(20.1; 13.4; 22)
(b) Vol. **122**, No. 2, p. K179—K182 (1984).

V. N. Salomatov, N. N. Kristoffel, and I. A. Parfianovich
*On the Nature of Metastable Color Centers in KBr
(13.4; 10.2; 22.5.2)
(b) Vol. **122**, No. 2, p. K183—K185 (1984).

V. I. Kuznetsov, P. F. Lugakov, and A. V. Tsikunov
Interaction of Radiation Defects with the Surface of Silicon
(11 and 14.3.2; 13.4; 22.1.2)
(a) Vol. **82**, No. 2, p. 467—473 (1984).

K. D. Glinchuk, A. V. Prokhorovich, and N. S. Zayats
*A New (Non-Copper-Induced) 1.35 eV Emission Band in n-Type GaAs. Origin and Characteristics
(13.4 and 20.3; 22.2.1)
(a) Vol. **82**, No. 2, p. 503—510 (1984).

L. A. Kazakevich, P. F. Lugakov, and I. M. Filippov
*The Effect of Dislocations on the Charge-Carrier Recombination Processes in Irradiated Silicon
(10.2 and 13.4; 11; 14.3; 22.1.2)
(a) Vol. **82**, No. 2, p. 511—518 (1984).

E. Chimczak, W. S. Gordon, and M. Bertrandt-Żytkowiak
Model of the Electroluminescence in Short-Pulse-Excited Thin-Film Structures Based on ZnS:Mn
(14.3.3 and 20.3; 13.4; 22.4.1)
(a) Vol. **82**, No. 2, p. 527—531 (1984).

P. I. BARANSKII, V. M. BABICH, N. P. BARAN, A. A. BUGAY, YU. P. DOTSENKO, and V. B. KOVALCHUK
*The Effect of Heat Treatment on Compensated CZ Silicon
(13.4; 14.3; 19; 22.1.2)
(a) Vol. **82**, No. 2, p. 533—536 (1984).

S. M. WASIM and A. NOGUERA
*Transport Properties of n-Type $CuInSe_2$
(13.4 and 14.3; 8; 22.8)
(a) Vol. **82**, No. 2, p. 553—559 (1984).

P. ORLANDI and A. RIGAMONTI
*On the Phase Transition and Crystal Field Effects in $FeSiF_6 \cdot 6\,H_2O$
(1.2 and 13.4; 5; 22.8)
(a) Vol. **82**, No. 2, p. K129—K132 (1984).

B. SORKIN and H. KÄÄMBRE
V_K Centres in Thermo-Exoemission from CsBr:TI
(10.2 and 17; 13.4; 22.5.2)
(a) Vol. **82**, No. 2, p. K149—K152 (1984).

R. JAŠINSKAITĖ, A. SAKALAS, A. MARTINAITIS, and P. HÖSCHL
Native Acceptor Defects in Chlorine Doped Single Crystals of CdTe
(14.3; 13.4; 22.4.3)
(a) Vol. **82**, No. 2, p. K199—K204 (1984).

YU. V. VOROBEV, E. I. TOLPYGO, and M. K. SHEINKMAN
*The Auger Processes with Energy Transfer to the Bound Charge Carriers in Multi-Valley Semiconductors
(13.4 and 14.3; 22.2.1)
(b) Vol. **123**, No. 1, p. 295—308 (1984).

B. ČABRIĆ and LJ. M. RISTOVSKI
*Resonant Energy Transfer between Localized Excitations in a Linear Polymer Chain
(13.4 and 13.5; 22.9)
(b) Vol. **123**, No. 1, p. 309—317 (1984).

K. LISCHKA, R. DURSTBERGER, G. LINDEMANN, and H. STAUDINGER
*Defect States in $Pb_{1-x}Sn_xTe$
(13.4; 16; 22.8)
(b) Vol. **123**, No. 1, p. 319—324 (1984).

L. N. KANTOROVICH
*Multipole Theory of the Polarization of Solids by Point Defects (II); The Point Charge Approximation
(13.4 and 14.4; 22.5.2)
(b) Vol. **123**, No. 1, p. 325—334 (1984).

V. A. KOVARSKII, L. V. CHERNYSH, and A. V. BELOUSOV
*About the Low-Temperature Dependence of Activation for Capture Cross-Sections of Charge Carriers in Semiconductors with Persistent Photoconductivity
(13.4; 16; 22)
(b) Vol. **123**, No. 1, p. 345—351 (1984).

V. S. Grunin and I. B. Patrina
*Impurities and Charge Compensation in Some Molybdenium Oxides
(13.4 and 19; 22.6; 22.8.1)
(b) Vol. **123**, No. 1, p. 353—363 (1984).

A. S. Barrière and G. Gevers
Impurity Conduction along the Least Resistance Paths of Granular Insulating Thin Films
(14.3.1; 13.4; 22.5)
(b) Vol. **123**, No. 1, p. 377—391 (1984).

S. V. J. Lakshman and A. S. Jacob
*Electronic Absorption Spectrum of Cu^{2+} in Sodium Zinc Sulphate Tetrahydrate Single Crystals
(13.4 and 20.1; 22.8.1)
(b) Vol. **123**, No. 1, p. K75—K77 (1984).

E. Radzhabov
*Impurity Centers in LiF:Li_2O
(13.4 and 20.3; 10.2; 22.5.2)
(b) Vol. **123**, No. 1, p. K79—K82 (1984).

F. S. Nasredinov, A. A. Andreev, A. N. Kurmantaev, A. R. Regel, and P. P. Seregin
*Tin Impurity Atoms in Vacuum Deposited Amorphous Silicon
(13.4 and 14.3.1; 2; 20.1; 22.1.2)
(a) Vol. **83**, No. 1, p. 285—289 (1984).

J. J. Kleperis, P. D. Cikmach, and A. R. Lusis
*Colour Centres in Amorphous Tungsten Trioxide Thin Films
(13.4 and 20.1; 14.3.1; 19; 22.6)
(a) Vol. **83**, No. 1, p. 291—297 (1984).

I. I. Kolkovskii, P. F. Lugakov, and V. V. Shusha
*Charge-Carrier Recombination in Silicon Irradiated with γ-Rays of Different Energies
(11 and 13.4; 22.1.2)
(a) Vol. **83**, No. 1, p. 299—306 (1984).

K. Löhnert and E. Kubalek
The Cathodoluminescence Contrast Formation of Localized Non-Radiative Defects in Semiconductors
(10.2 and 20.3; 13.4; 22.2)
(a) Vol. **83**, No. 1, p. 307—314 (1984).

A. Krier and F. J. Bryant
The Hot Electron Energy Distribution Function in ZnS:Cu:Cl:Er Thin Films
(20.3; 13.4; 22.4.1)
(a) Vol. **83**, No. 1, p. 315—322 (1984).

G. Oelgart and G. Arnold
*Internal Electric Field-Enhanced Impurity Diffusion in $GaAs_{1-x}P_x$
(13.4 and 14.3.3; 9; 22.2.4)
(a) Vol. **83**, No. 1, p. 393—402 (1984).

S. Kashchieva, P. Danesh, and A. Dyakov
*RF Annealing of the Implantation-Induced Defects in Silicon Using Hydrogen Plasma
(11 and 13.4; 14.3.3; 22.1.2)
(a) Vol. **83**, No. 1, p. 411—417 (1984).

B. G. Tagiev, G. M. Niftiev, and S. A. Abushov
*Photoelectrical and Optical Properties of GaSe:Yb Single Crystals
(13.4; 16; 20.3; 22.3)
(a) Vol. **83**, No. 1, p. K61—K64 (1984).

A. A. Kaminskii, L. P. Kozeeva, and A. A. Pavlyuk
Stimulated Emission of Er^{3+} and Ho^{3+} Ions in $KLa(MoO_4)_2$ Crystals
(20.2 and 20.3; 13.4; 22.8)
(a) Vol. **83**, No. 1, p. K65—K67 (1984).

R. Brazis, P. Pipinys, and A. Rimeika
*Temperature Dependence of Currents in Thin Films of Silicon Oxide
(13.4 and 14.3.4; 22.6)
(a) Vol. **83**, No. 1, p. K69—K71 (1984).

M. Ferrari, M. Montagna, L. Gonzo, G. Viliani, O. Pilla, and G. Farace
*Resonant Energy Transfer in Ruby under Magnetic Field
(13.4; 20.3; 22.6)
(b) Vol. **123**, No. 2, p. 659—662 (1984).

T. Grafe and R. Strehlow
*Influence of Si–Si Bonds on the Local Electronic Density of States of a-SiO_2
(13.1 and 13.4; 22.6)
(b) Vol. **123**, No. 2, p. 663—669 (1984).

V. A. Kovarskii, E. P. Sinyavskii, and L. V. Chernysh
*Optical Properties of Bielectron–Impurity Complexes
(13.4; 20.1; 22.5)
(b) Vol. **123**, No. 2, p. 671—677 (1984).

R. Borromei and E. Cavalli
*Low-Temperature Absorption Spectrum of Ni^{2+} Ion in Single Crystals of Ni_2NbBO_6
(13.4 and 20.1; 22.8.1)
(b) Vol. **123**, No. 2, p. 679—689 (1984).

A. Golzené, B. Meyer, and C. Schwab
Experimental Evidence for an Associated Defect Model for the Neutron Generated As_{Ga} Center in Gallium Arsenide
(11 and 19; 13.4; 22.2.1)
(b) Vol. **123**, No. 2, p. K125—K128 (1984).

H.-U. Tschirner and K. Uhlig
Some Remarks on Activation Energies Observed in the Liquid Semiconductor Ga_2Te_3
(2 and 14.3; 13.4; 22)
(b) Vol. **123**, No. 2, p. K177—K181 (1984).

P. F. LUGAKOV and V. V. LUKYANITSA
Efficiency of Radiation Defect Formation in Silicon at Various Intensities of Electron and γ-Ray Irradiation
(11; 13.4; 22.1.2)
(a) Vol. **83**, No. 2, p. 521—528 (1984).

R. L. CALVERT and R. J. DANBY
Thermoluminescence and Radiophotoluminescence from Eu- and Sm-Doped $CaSO_4$
(20.3; 11; 13.4; 22.8.1)
(a) Vol. **83**, No. 2, p. 597—604 (1984).

S. BHUSHAN, D. DIWAN, and S. P. KATHURIA
Thermoluminescence of ZnO:Pr and ZnO:Gd under UV, β- and γ-Ray Irradiations
(20.3; 11; 13.4; 22.4)
(a) Vol. **83**, No. 2, p. 605—611 (1984).

K. D. GLINCHUK, A. V. PROKHOROVICH, and N. S. ZAYATS
*Effect of 2.2 MeV Electron Irradiation and Annealing on the 1.37 eV Emission Band in Zinc-Doped p-Type GaAs
(13.4 and 20.3; 11; 22.2.1)
(a) Vol. **83**, No. 2, p. 625—630 (1984).

H.-D. AUTENRIETH, S. KEMMLER-SACK und W. WISCHERT
*Lumineszenz und sensibilisierte Emission der dreiwertigen Seltenen Erden in $Ca_3La_2W_2O_{12}$
(Luminescence and Sensitized Emission of Trivalent Rare Earth Ions in $Ca_3La_2W_2O_{12}$)
(13.4 and 20.3; 22.8)
(a) Vol. **83**, No. 2, p. 631—636 (1984).

H. LEMKE
*Eigenschaften von substitutionell gelöstem Mangan in Silizium
(Properties of Substitutional Manganese in Silicon)
(13.4; 14.3; 22.1.2)
(a) Vol. **83**, No. 2, p. 637—643 (1984).

C. BALLESTEROS, J. PIQUERAS, J. LLOPIS, and R. GONZÁLEZ
Cathodoluminescence from MgO Single Crystals Containing a High Concentration of Anion Vacancies
(20.3; 10.2; 13.4; 22.6)
(a) Vol. **83**, No. 2, p. 645—649 (1984).

R. K. GARTIA and R. SINGHA
Analysis of Thermoluminescence of Topaz
(20.3; 13.4; 22.8.1)
(a) Vol. **83**, No. 2, p. 651—655 (1984).

N. G. FILONOV, N. K. MAKSIMOVA, A. P. VYATKIN, I. D. ROMANOVA, and A. M. MYSIK
Transport Mechanism in GaAs Schottky Diodes. Deep Centres Effects
(14.3.3; 13.4; 22.2.1)
(a) Vol. **83**, No. 2, p. 701—708 (1984).

N. YU. KONSTANTINOV, L. G. KARASEVA, V. V. GROMOV, and A. V. YAKOVLEV
*Interconfiguration Transitions of $4f^3$–$4f^2 5d^1$ of the Nd^{3+} Ion in YAG Single Crystals
(13.4 and 20.1; 22.8.2)
(a) Vol. **83**, No. 2, p. K153—K157 (1984).

B. SORKIN and H. KÄÄMBRE
Thermo-Exoemission of NaCl:Ag Crystals, Cleaved before and after X-Irradiation
(17; 11; 13.3; 13.4; 22.5.2)
(a) Vol. **83**, No. 2, p. K171—K174 (1984).

K. S. V. NAMBI
Comment on "The Role of Sulphur-Oxyradicals in Thermoluminescence and Exoelectron Emission of $CaSO_4$ Phosphors" by R. Huzimura and K. Atarashi
(17 and 20.3; 11; 13.4; 19; 22.8.1)
(a) Vol. **83**, No. 2, p. K175—K177 (1984).

N. V. BARAMIDZE, V. L. BONCH-BRUEVICH, and N. I. KURDIANI
*Optical Absorption in Neutron Irradiated GaAs
(13.4 and 20.1; 11; 22.2.1)
(b) Vol. **124**, No. 1, p. 241—247 (1984).

A. M. RATNER and E. I. TARASOVA
*Local Optical Centers in Rare Gas Crystals
(13.4; 13.5.1; 23)
(b) Vol. **124**, No. 1, p. 249—260 (1984).

L. CSER, I. GLADKIH, and C. HARGITAI
*Impurity Charge Screening in 3d Alloys
(13.4; 5; 21.1.1)
(b) Vol. **124**, No. 1, p. 271—278 (1984).

C. M. WEINERT and U. SCHERZ
Calculation of the Jahn-Teller Coupling Forces for $ZnS:Cu^{2+}$ and $GaP:Ni^+$
(6 and 20.1; 13.4; 22.2.2; 22.4.1)
(b) Vol. **124**, No. 1, p. 287—296 (1984).

P. MORETTI, A. RANFAGNI, and A. AGRESTI
*Ground and Excited States of Tl^+ Impurity in KCl Crystal by the Use of the Green's Function Method
(13.4; 22.5.2)
(b) Vol. **124**, No. 1, p. 297—306 (1984).

I. BALTOG and L. MIHUŢ
Luminescence Decay in NaI:Pb
(20.3; 13.4; 22.5.2)
(b) Vol. **124**, No. 1, p. 307—313 (1984).

M. HEMING and G. LEHMANN
*Correlation of Zero Field Splittings and Site Distortions. Interstitial Incorporation of Mn^{2+} in Urea-Ammoniumchloride $(H_2N)_2CO \cdot NH_4Cl$
(13.4 and 19; 22.8)
(b) Vol. **124**, No. 1, p. 315—323 (1984).

M. Kibler, G. Grenet, J. C. Gâcon, and B. Jacquier
*Energy Levels of Sm^{2+} in BaClF and SrClF Crystals under Magnetic Field (I): Theory
(13.4 and 20.3; 18; 22.5.4)
(b) Vol. **124**, No. 1, p. 325—334 (1984).

O. Goede and Dang Dinh Thong
*Energy Transfer Processes in (Zn, Mn)S Mixed Crystals
(13.4 and 20.3; 22.4.1)
(b) Vol. **124**, No. 1, p. 343—353 (1984).

M. Budzynski, E. A. Goremychkin, O. I. Kochetov, A. Latuszynski, P. Mikolajczak, E. Mühle, A. I. Muminov, and M. Subotowicz
*Electric Field Gradient at Gd in Gadolinium and Rare Earth Trifluoride Single Crystals
(13.4; 19; 22.5)
(b) Vol. **124**, No. 1, p. 355—362 (1984).

R. Piotrzkowski
*Pressure Study of Photoconductivity in Indium Antimonide (I): Low–Concentration n–Type at 77 K. Pressure- Dependent Capture Coefficient of a Hole-Trapping Center
(13.4 and 16; 12.2; 22.2.3)
(b) Vol. **124**, No. 1, p. 411—421 (1984).

V. F. Zolin, V. M. Markushev, M. N. Popova, R. A. Aliev, G. D. Guseinov, and R. M. Sardarly
*Neodymium Luminescence in the Ternary Semiconducting Chalcogenides $TlGaS_2$ and $TlGaSe_2$
(13.4 and 20.3; 22.8)
(b) Vol. **124**, No. 1, p. K69—K73 (1984).

A. A. Alybakov, V. A. Gubanova, G. S. Denisov, and B. S. Umurzakov
*Influence of Oxygen Centers on the Thermoluminescence of NaF Crystals
(13.4 and 20.3; 10.2; 22.5.2)
(b) Vol. **124**, No. 1, p. K75—K78 (1984).

E. M. Goldys
*The Donor Wave Functions in Gallium Phosphide and the Real Band Structure
(13.4 and 20.1; 18; 22.2.2)
(b) Vol. **124**, No. 1, p. K79—K82 (1984).

M. Moreno and M. T. Barriuso
*An Insight into the $10Dq$ Value Found in $CaZnF_4$:Mn^{2+}
(10.2 and 13.4; 20.3; 22.5)
(b) Vol. **124**, No. 1, p. K83—K86 (1984).

V. V. Kveder, R. Labusch, and Yu. A. Ossipyan
*The Exodiffusion of Hydrogen in Dislocated Crystalline Silicon
(10.2 and 13.4; 22.1.2)
(a) Vol. **84**, No. 1, p. 149—156 (1984).

V. V. Borimskii, K. D. Glinchuk, N. M. Litovchenko, and Z. A. Salnik
*New Oxygen-Induced Recombination Centres in 600 to 800 °C Heat-Treated Silicon
(13.4; 22.1.2)
(a) Vol. **84**, No. 1, p. 237—241 (1984).

W. ULRICI
*Evaluation of Chromium Concentration in Semi-Insulating GaAs:Cr from Optical Absorption
(13.4 and 20.1; 22.2.1)
(a) Vol. **84**, No. 1, p. 243—250 (1984).

W. SIEGEL and G. KÜHNEL
*Influence of the Temperature Dependence of the Thermal Impurity Activation Energy on the Analysis of Hall Measurements on GaP
(13.4 and 14.3; 22.2.2)
(a) Vol. **84**, No. 1, p. 251—262 (1984).

V. M. BABICH, N. P. BARAN, V. L. BORBLIK, YU. P. DOTSENKO, V. B. KOVALCHUK, and V. A. SHERSHEL
*Specific Features of Piezogalvanomagnetic Effects in Oxygen-Containing Silicon Crystals in the Presence of Thermal Donors-I and -II
(13.4; 14.3; 22.1.2)
(a) Vol. **84**, No. 1, p. 263—268 (1984).

A. S. BARRIÈRE and G. GEVERS
*Spatial Distribution of Localized Electronic States Responsible for the Polarization of Granular Insulating Thin Films
(13.4 and 14.3.1; 22.5)
(a) Vol. **84**, No. 1, p. 273—284 (1984).

O. YU. BORKOVSKAYA, N. L. DMITRUK, V. G. LITOVCHENKO, and O. I. MAEVA
The Influence of Deep Levels on Photomemory Effect in Structures with a Potential Barrier
(16; 13.4; 22.2.1)
(a) Vol. **84**, No. 1, p. 285—290 (1984).

M. BÖHM, O. ERB, and A. SCHARMANN
Electrical Conduction in $PbMoO_4$ Crystals
(16 and 20.1; 13.4; 22.8.1)
(a) Vol. **84**, No. 1, p. 291—299 (1984).

A. A. KAMINSKII, Z. I. ZHMUROVA, V. A. LOMONOV, and S. E. SARKISOV
Two Stimulated Emission $^4F_{3/2} \rightarrow {}^4I_{11/2,13/2}$ Channels of Nd^{3+} Ions in Crystals of the CaF_2–ScF_3 System
(20.2 and 20.3; 13.4; 22.8)
(a) Vol. **84**, No. 1, p. K81—K84 (1984).

G. MARTÍNEZ and R. RAMÍREZ
*Correlation Effects in the Rotationally Invariant Asymmetric Anderson Model
(13.4 and 18)
(b) Vol. **124**, No. 2, p. 573—586 (1984).

R. BORROMEI, L. OLEARI, and P. DAY
*The Absorption Spectrum of Manganate(VI) in $BaSO_4$. A Reinterpretation
(13.4 and 20.1; 22.8.1)
(b) Vol. **124**, No. 2, p. 707—718 (1984).

T. F. VEREMEICHIK
*Similarity of Ground- and Excited-State Absorption Spectra of Cr^{3+} Ions in a Strong Crystal Field
(13.4 and 20.1; 22.6)
(b) Vol. **124**, No. 2, p. 719—729 (1984).

M. KIBLER, J. C. GÂCON, G. GRENET, B. JACQUIER, and L. C. BRUNEL
*Energy Levels of Sm^{2+} in BaClF and SrClF Crystals under Magnetic Field (II): Results
(13.4 and 20.3; 18; 22.5.4)
(b) Vol. **124**, No. 2, p. 731—739 (1984).

L. O. SCHWAN, R. FLORIAN, D. SCHMID, E. BETZ, W. SCHROF, B. WALKER, and H. PORT
*Dynamics of the Optical Excitation–Deexcitation Cyle of O_2^- Centers in KCl
(10.2 and 13.4; 20.3; 22.5.2)
(b) Vol. **124**, No. 2, p. 741—745 (1984).

H.-J. MÜLLER and H.-J. WÜNSCHE
*Model Calculations for Bound Excitons and Bound Excitonic Molecules in $GaAs_{1-x}P_x$:N
(13.4 and 13.5.1; 20.3; 22.2.4)
(b) Vol. **124**, No. 2, p. 747—756 (1984).

M. MATOS and B. MAFFEO
*Molecular Cluster-INDO Calculations of Substitutional Defects in Alkali Halides
(13.4; 20.1; 22.5.2)
(b) Vol. **124**, No. 2, p. 757—765 (1984).

R. DOERFLER, D. L. NAGY, H.-D. PFANNES, A. PUTZKA, G. RITTER, and N. ZEMAN
*Non-Equilibrium Population of Fe^{3+} Electronic States after the Electron Capture (I): ^{57}Co in $LiNbO_3$ and $LiTaO_3$
(13.4; 5; 22.8.1)
(b) Vol. **124**, No. 2, p. 767—772 (1984).

E. B. YAKIMOV, N. A. YARYKIN, and V. I. NIKITENKO
*The Influence of Thermal Treatment on Thermally Stimulated Depolarization Spectrum of the Electret State in Plastically Deformed Silicon
(10.2 and 13.4; 14.4; 22.1.2)
(a) Vol. **84**, No. 2, p. 443—450 (1984).

M. PARDAVI-HORVÁTH, I. FÖLDVÁRI, I. FELLEGVÁRI, L. GOSZTONYI, and J. PAITZ
Spectroscopic Properties of Ca^{2+}-Doped GGG
(20.1; 13.4; 22.8.2)
(a) Vol. **84**, No. 2, p. 547—553 (1984).

S. S. DIMOV, L. I. PAVLOV, K. V. STAMENOV, and G. B. ALTSHULLER
*Dispersion of Laser-Induced Nonlinear Optical Susceptibility in CdS
(13.4; 20.2; 22.4.1)
(a) Vol. **84**, No. 2, p. 555—560 (1984).

K. D. GLINCHUK, A. V. PROKHOROVICH, and N. S. ZAYATS
*Effect of 2.2 MeV Electron Irradiation and Annealing on the Non-Copper-Induced 1.35 eV Emission Band in n-Type GaAs
(13.4 and 20.3; 11; 22.2.1)
(a) Vol. **84**, No. 2, p. 567—572 (1984).

P. KRISPIN and J. MAEGE
*V_P Related Defects in Diffused GaP:N Diodes
(13.4 and 14.3.3; 22.2.2)
(a) Vol. **84**, No. 2, p. 573—584 (1984).

B. OLEJNÍKOVÁ
*Non-Exponential Capacitance Transient Due to Field-Enhanced Emission of Electrons from Deep Traps
(13.4 and 14.3.4; 22.2.1)
(a) Vol. **84**, No. 2, p. 621—630 (1984).

M. ASLAM, R. SINGH, and P. BALK
Nature of Electron and Hole Traps in MOS Systems with Poly-Si-Electrode
(14.3.3; 13.4; 22.1.2)
(a) Vol. **84**, No. 2, p. 659—668 (1984).

P. LOŠŤÁK, J. HORÁK, and L. KOUDELKA
Some Physical Properties and Point Defects in $Bi_2Te_{3-x}S_x$ Mixed Crystals
(20.1; 13.4; 14.3; 22.7.1)
(a) Vol. **84**, No. 2, p. K143—K147 (1984).

L. N. SKUJA, A. N. TRUKHIN, and A. E. PLAUDIS
Luminescence in Germanium-Doped Glassy SiO_2
(20.3; 2; 10.2; 13.4; 20.1; 22.6)
(a) Vol. **84**, No. 2, p. K153—K157 (1984).

TRAN KIM ANH, NGUYEN DAI HUNG, NQUY HUU CHI, and NGUYEN MANH SON
Energy Transfer between Tb^{3+} and Eu^{3+} in Rare Earth Pentaphosphates
(20.1 and 20.3; 13.4; 22.8.1)
(a) Vol. **84**, No. 2, p. K159—K163 (1984).

A. J. SZADKOWSKI and A. LUBOMIRSKA-WITTLIN
Compensation of p-Type $Cd_{1-x}Mn_xTe$ by Diffusion of Ag
(14.3 and 16; 9; 13.4; 22.4.4)
(a) Vol. **84**, No. 2, p. K181—K184 (1984).

M. C. BÖHM
*A Crystal Orbital (CO) Model for Low-Dimensional Segregated Donor and Acceptor Stacks
(13.1 and 13.4; 22.9)
(b) Vol. **125**, No. 1, p. 311—322 (1984).

L. ABELLO, S. J. SCHWERDTFEGER, and C. F. SCHWERDTFEGER
*ESR Study of Cu^{2+} Diffused into Sodium β''-Alumina
(13.4 and 19; 22.6.1)
(b) Vol. **125**, No. 1, p. 333—340 (1984).

T. R. GAWRON and J. MYCIELSKI
*Shallow Impurity States in Strong Magnetic Field
(13.4; 20.1; 22)
(b) Vol. **125**, No. 1, p. 341—348 (1984).

R. Borromei
*The Electronic Spectrum of the CrO_4^{3-} Ion in the $Ca_2SiO_4 \cdot CaCl_2$ Host Lattice
(13.4 and 20.1; 22.8)
(b) Vol. **125**, No. 1, p. 349—355 (1984).

D. Thébault, J. Barrau, G. Armelles, N. Lauret, and J. P. Noguier
*The No-Phonon Luminescence at 0.793 eV from Si:Au. Uniaxial Stress and Zeeman Spectroscopy Experiments
(13.4 and 20.3; 22.1.2)
(b) Vol. **125**, No. 1, p. 357—364 (1984).

L. E. Kholodenkov and A. G. Makhanek
Two-Photon Luminescence Excitation Spectra of Eu^{3+} in Yttrium Aluminium Garnet
(20.1 and 20.3; 13.4; 22.8.2)
(b) Vol. **125**, No. 1, p. 365—374 (1984).

G. I. Bersuker and V. Z. Polinger
*Multimode Pseudo Jahn-Teller Effect for Off-Centre Impurities in Crystals
(6 and 13.4; 10.2; 22.5.2)
(b) Vol. **125**, No. 1, p. 401—408 (1984).

V. A. Chepurnoi, S. N. Mysovskii, I. A. Parfianovich, V. N. Salomatov, and A. P. Shkadarevich
*Influence of Oxygen Defects on the Position and Form of the F_2^+ Center Zero-Phonon Lines in LiF
(10.2 and 13.4; 20.3; 22.5.2)
(b) Vol. **125**, No. 1, p. K41—K44 (1984).

A. Dargys and A. Matulis
*Field Tunneling from Impurities in Case of Anisotropic Effective Masses
(13.4)
(b) Vol. **125**, No. 1, p. K71—K73 (1984).

A. V. Mudryi, A. L. Pushkarchuk, V. D. Tkachev, and A. G. Ulyashin
*Noble Gas Atoms as Interstitials in Silicon and Diamond
(10.2 and 13.4; 22.1; 22.1.2)
(b) Vol. **125**, No. 1, p. K75—K78 (1984).

V. Bethkenhagen and K. Hübner
*Bethe-Lattice Approach to the Electronic Structure of Intrinsic Defects in v-SiO_2
(10.2 and 13.4; 22.6)
(b) Vol. **125**, No. 1, p. K79—K83 (1984).

G. A. Adegboyega
*Electrical Measurements on Silver-Diffused GaAs
(13.4 and 14.3; 22.2.1)
(a) Vol. **85**, No. 1, p. 215—218 (1984).

K. Žďánský and N. T. Thuc Hien
*Analysis of DLTS Curves of Aggregated Deep Level Impurities
(13.4 and 14.3.3; 22.4.3)
(a) Vol. **85**, No. 1, p. 219—226 (1984).

Y. Kamiura and F. Hashimoto
*Origin of a Shallow Acceptor in Quenched Germanium
(10.2 and 13.4; 14.3; 22.1.1)
(a) Vol. **85**, No. 1, p. 227—234 (1984).

P. Lilley, M. R. Czerniak, and J. E. Nicholls
Photoluminescence and Electrical Properties of Vapour Phase Epitaxial ZnSe Grown on GaAs
(14.3.1 and 20.3; 13.4; 22.4.2)
(a) Vol. **85**, No. 1, p. 235—242 (1984).

H. Booyens and J. H. Basson
*Piezoelectrically Induced Charge Distributions around Dislocations in CdTe and HgCdTe
(10.2 and 13.4; 14.4; 22.4.3; 22.4.4)
(a) Vol. **85**, No. 1, p. 243—248 (1984).

S. V. Vintsents, V. F. Kiselev, and G. S. Plotnikov
*Energy Transfer between Excited Adsorbed Dye Molecules and Charged Defects in Insulator–Semiconductor Structures
(13.4; 19; 20.3; 22.1.1; 22.1.2)
(a) Vol. **85**, No. 1, p. 273—281 (1984).

V. M. Lomako and P. Ya. Starostin
*Temperature Dependence of the Charge-Carrier Removal Rate in n-GaAs Containing Defect Clusters
(11 and 13.4; 22.2.1)
(a) Vol. **85**, No. 1, p. K57—K60 (1984).

A. N. Vasilev and V. Čápek
Vibrational Relaxation in a Localized Excited Electronic State by the GME Method
(6; 13.4)
(b) Vol. **125**, No. 2, p. 477—482 (1984).

G. Heder and J. M. Spaeth
ENDOR Investigation of H_i^0(Li) Centres in KCl
(10.2 and 19; 13.4; 22.5.2)
(b) Vol. **125**, No. 2, p. 523—531 (1984).

V. N. Zabluda, A. V. Malakhovskii, and I. S. Edelman
Magnetooptical Effects and Optical Absorption of $FeBO_3$ in the Region of $^6A_{1g} \rightarrow {}^4T_{1g}(^4G), {}^4T_{2g}(^4G)$ Transitions
(18.2 and 20.1; 13.4; 22.8.1)
(b) Vol. **125**, No. 2, p. 751—757 (1984).

G. Ambrazevičius, G. Babonas, and Yu. V. Rud
*Excitation Spectra of Photoluminescence Bands in $Cd_{1-x}Mn_xTe$
(13.4 and 20.3; 22.4.4)
(b) Vol. **125**, No. 2, p. 759—764 (1984).

P. Petelenz
*Optical Detrapping in Molecular Crystals (II): Second Conduction Band
(13.4; 20.1; 22.9)
(b) Vol. **125**, No. 2, p. 765—771 (1984).

U. W. Pohl, H.-E. Gumlich, and W. Busse
*Zero-Phonom Lines in the Excitation Band Centred at 23200 cm^{-1} Due to Mn^{2+} in Polymorphic ZnS
(13.4; 20.1; 22.4.1)
(b) Vol. **125**, No. 2, p. 773—777 (1984).

H. Werheit and R. Franz
States of Thermal Equilibrium in β-Rhombohedral Boron and Their Alteration by Optical Excitation
(1.2; 13.4; 20.1; 22.1)
(b) Vol. **125**, No. 2, p. 779—784 (1984).

D. Vignaud and J. L. Farvacque
*Optical Absorption Studies of Plastically Deformed InSb
(13.4 and 20.1; 10.2; 22.2.3)
(b) Vol. **125**, No. 2, p. 785—794 (1984).

G. Milchberg, K. Saminadayar, E. Molva, and H. R. Zelsmann
*Optical Determination of the Valence-Band Parameters in CdTe
(13.4 and 20.3; 13.1; 22.4.3)
(b) Vol. **125**, No. 2, p. 795—803 (1984).

S. W. Biernacki
*Born-von Karman Cell Model for Deep Impurities
(13.4; 22.4.2)
(b) Vol. **125**, No. 2, p. 805—814 (1984).

V. N. Ermakov, D. V. Korbutyak, V. G. Litovchenko, O. Yu. Mikityuk, and V. V. Nitsovich
*A Study of the Reflection Spectra of Surface Exciton–Impurity Complexes
(13.4 and 13.5.1; 13.3; 20.1; 22.4.1)
(b) Vol. **125**, No. 2, p. 815—822 (1984).

V. A. Kasiyan, D. D. Nedeoglo, A. V. Simashkevich, and I. N. Timchenko
*Hopping Conductivity and Electron Localization in Heavily Doped ZnSe
(13.4 and 14.3; 22.4.2)
(b) Vol. **125**, No. 2, p. 869—878 (1984).

Ch. Opitz, H. Müller, and H. Miessner
Bonding Mechanism for Interstitial Hydrogen in Nickel
(4; 10.1; 13.4; 21.1)
(b) Vol. **125**, No. 2, p. K93—K97 (1984).

S. Sapru and K. L. Tan
Eu^{2+} Ions Doped in Alkali Halides Act as Donors on Irradiation
(20.3; 11; 13.4; 22.5.2)
(b) Vol. **125**, No. 2, p. K161—K163 (1984).

P. F. Lugakov and T. A. Lukashevich
Formation and Parameters of Boron–Divacancy Complexes in Irradiated p-Si
(11; 13.4; 22.1.2)
(a) Vol. **85**, No. 2, p. 441—444 (1984).

M. SUEZAWA and K. SUMINO
*On the Annihilation of Thermal Donors in Silicon Crystals
(10.2 and 13.4; 20.1; 22.1.2)
(a) Vol. **85**, No. 2, p. 469—472 (1984).

A. A. KAMINSKII, S. E. SARKISOV, G. A. DENISENKO, V. V. RYABCHENKOV, V. A. LOMONOV, YU. E. PERLIN, M. G. BLAZHA, D. SCHULTZE, B. HERMONEIT, and P. REICHE
Growth, Spectral and Luminescence Study of Cubic $Bi_4Ge_3O_{12}$:Pr^{3+} Crystals
(20.1 and 20.3; 13.4; 22.8)
(a) Vol. **85**, No. 2, p. 553—567 (1984).

H. KRAUSE, M. SCHELL, and H.-CHR. GOLINSKY
*Hopping Conduction in SiO_2 Films Thermally Grown on Phosphorus-Doped Polycrystalline Silicon
(13.4 and 14.3.1; 22.6)
(a) Vol. **85**, No. 2, p. 569—573 (1984).

V. V. EMTSEV, YU. N. DALUDA, P. GAWORZEWSKI, and K. SCHMALZ
*Electrical Properties of Thermal Donors Formed in CZ-Si during Heat Treatment at 450 °C
(13.4; 14.3; 20.1; 22.1.2)
(a) Vol. **85**, No. 2, p. 575—584 (1984).

V. I. GUBSKAYA, P. V. KUCHINSKII, and V. M. LOMAKO
*The Effect of the Vacancy Charge State on the Radiation Defect Formation in Silicon
(11 and 13.4; 22.1.2)
(a) Vol. **85**, No. 2, p. 585—590 (1984).

B. P. TYAGI and K. SEN
Grain Size and Temperature Dependence of the Electrical Behaviour of Polysilicon
(14.3.4; 13.4; 22.1.2)
(a) Vol. **85**, No. 2, p. 603—608 (1984).

S. NEŠPŮREK, O. ZMEŠKAL, and F. SCHAUER
An Ohmic Contact Test Using the Activation Energy of Steady-State Space-Charge-Limited Current
(14.3.4; 13.4; 22.1.2; 22.4.2)
(a) Vol. **85**, No. 2, p. 619—625 (1984).

R. PICKENHAIN, K. JACOBS, W. SEIFERT, V. CHERNYI, and S. BREHME
Green and Yellow Light Emitting Diodes Produces from Vapour Phase Epitaxial GaP:N (I): Deep Levels in Diodes Produced by either Diffusion or Implantation of Zinc
(14.3.3 and 20.3; 13.4; 22.2.2)
(a) Vol. **85**, No. 2, p. 627—639 (1984).

H. LEMKE
*Eigenschaften der Energieniveaus von Kobalt in Silizium
(Porperties of Energy Levels of Cobalt in Silicon)
(13.4; 14.3; 22.1.2)
(a) Vol. **85**, No. 2, p. K133—K136 (1984).

H. Lemke
*Eigenschaften der Energieniveaus von Rhodium und Iridium in Silizium
(Properties of Energy Levels of Rhodium and Iridium in Silicon)
(13.4; 14.3; 22.1.2)
(a) Vol. **85**, No. 2, p. K137—K140 (1984).

G. Rudlof and H. Glaefeke
*Remarks about a New, Constructive Method for the Analysis of Thermally Stimulated Relaxation Processes (TSRP)
(13.4 and 17; 22)
(a) Vol. **85**, No. 2, p. K149—K153 (1984).

K. Wesser, S. N. Sahu, J. P. Karins, and J. W. Corbett
*Comparison of an Extended Interstitial with a Split Interstitial in Silicon
(10.2 and 13.4; 22.1.2)
(b) Vol. **126**, No. 1, p. 157—161 (1984).

A. Sh. Makhmudov, Z. M. Khakimov, A. A. Levin, M. S. Yunusov, and P. K. Khabibullaev
*Green's Function Method in the Equivalent Orbital Basis. Point Defects and Complexes of Two Point Defects in Silicon
(13.1 and 13.4; 22.1.2)
(b) Vol. **126**, No. 1, p. 311—321 (1984).

M. Korkmaz, B. Aktas et M. Dupont
*Le spectre de RPE de l'ion Mn^{2+} dans le tartrate de cadmium ($CdC_4H_4O_6 \cdot 5\ H_2O$)
(EPR Spectra of the Mn^{2+} Ion in Cadmium Tartrate ($CdC_4H_4O_6 \cdot 5\ H_2O$))
(13.4 and 19; 22.8.1)
(b) Vol. **126**, No. 1, p. 323—327 (1984).

F. Dujardin, B. Stébé, and G. Munschy
*Neutral Bound Excitions in a Low Magnetic Field
(13.4 and 13.5.1; 22.2; 22.4)
(b) Vol. **126**, No. 1, p. 329—334 (1984).

D. Backs and A. Hertrampf
*Optically Induced EPR of Conduction and Donor Band Electrons in Zinc Sulphide
(13.4 and 19; 22.4.1)
(b) Vol. **126**, No. 1, p. 343—349 (1984).

R. Chen, V. K. Mathur, J. F. Rhodes, M. D. Brown, S. W. McKeever, and R. K. Bull
*Thermoluminescence Governed by Simultaneous Thermal Stimulation of Electrons and Holes
(13.4 and 20.3; 22)
(b) Vol. **126**, No. 1, p. 361—369 (1984).

J. Lech, A. Ślęzak, and I. Bójko
*EPR Powder Spectra of Low-Symmetric Mn^{2+} Complexes in Aragonite Lattice
(13.4 and 19; 22.8)
(b) Vol. **126**, No. 1, p. 371—380 (1984).

W. Böhm, H. G. Kahle, and W. Wüchner
*Spectroscopic Study of the Crystal Field Splittings in $TbPO_4$
(1.2 and 13.4; 18.4; 20.1; 22.8.1)
(b) Vol. **126**, No. 1, p. 381—392 (1984).

M. Suhrke and B. Esser
Time-Resolved Luminescence of Charge Carriers in Localized States in Amorphous Semiconductors
(20.3; 2; 13.4; 22)
(b) Vol. **126**, No. 1, p. 393—406 (1984).

Zh. Egemberdiev, V. Nagirnyi, T. Soovik, and S. Zazubovich
*Decay Kinetics and Polarization of the A_T Emission of Pb^{2+} Centres of Different Structure in KBr: : $PbBr_2$
(13.4 and 20.3; 22.5.4)
(b) Vol. **126**, No. 1, p. 407—414 (1984).

Zh. Egemberdiev, V. Nagirnyi, T. Soovik, and S. Zazubovich
*Slow Decay Component of the A_T Emission in KBr:Pb Crystals
(13.4 and 20.3; 22.5.2)
(b) Vol. **126**, No. 1, p. K63—K66 (1984).

V. Bethkenhagen and K. Hübner
*Cluster-Bethe-Lattice Approach to the Electronic Structure of a Twofold Coordinated Silicon Atom as a New Intrinsic Defect in v-SiO_2
(13.4; 10.2; 22.6)
(b) Vol. **126**, No. 1, p. K71—K76 (1984).

W. Pastusiak
*Dynamical Relativistic Crystal Field Contributions to the Zero Field Splitting of S-State $4f^7$ Ions
(13.4 and 19; 22.5.3)
(b) Vol. **126**, No. 1, p. K77—K81 (1984).

D. I. Aladashvili, L. Konczewicz, and S. Porowski
*Studies of the Deep Levels in p-Type InSb under Pressure
(13.4; 14.3; 22.2.3)
(a) Vol. **86**, No. 1, p. 301—308 (1984).

P. F. Lugakov and V. V. Shusha
A Model of Nonequilibrium Charge-Carrier Recombination in Semiconductors Containing Nonuniformities
(14.3.4; 13.4; 22.1.2)
(a) Vol. **86**, No. 1, p. 309—312 (1984).

A. V. Dvurechenskii, B. P. Kashnikov, and V. V. Suprunchik
*New EPR Defects in Si⟨Al⟩
(13.4 and 19; 11; 22.1.2)
(a) Vol. **86**, No. 1, p. 313—318 (1984).

I. Tale and J. Rosa
*Fractional Glow Technique Spectroscopy of Traps in Heavily Doped AlN:O
(13.4; 20.1; 22.3; 22.2)
(a) Vol. **86**, No. 1, p. 319—326 (1984).

S. C. Sabharwal, S. P. Kathuria, and B. Ghosh
Kinetics of Thermoluminescence in γ-Irradiated NaI(Tl)
(20.3; 11; 13.4; 22.5.2)
(a) Vol. **86**, No. 1, p. 327—331 (1984).

R. Capelletti and M. Manfredi
The Role of Eu^{++} Aggregation on Optical Absorption, Emission, Luminescence Time Decay, and ITC Plots in NaCl:Eu^{++}
(20.1 and 20.3; 13.4; 22.5.2)
(a) Vol. **86**, No. 1, p. 333—343 (1984).

U. Braatz and D. Zappe
Investigation of Space Charge Limited, Poole-Frenkel, and Isothermal Currents on HgI_2 Single Crystals
(14.3.4; 13.4; 22.5)
(a) Vol. **86**, No. 1, p. 407—411 (1984).

H. Lemke
*Über die mögliche Existenz „repulsiver" Zentren mit großen Einfangquerschnitten in Silizium
(On the Possible Existence of „Repulsive" Centres with Large Capture Cross Section in Silicon)
(13.4; 14.3; 22.1.2)
(a) Vol. **86**, No. 1, p. K39—K43 (1984).

S. Shigetomi, T. Ikari, Y. Koga, and S. Shigetomi
Annealing Behaviour of Electrical Properties of n-InSe Single Crystals
(14.3; 13.4; 22.3)
(a) Vol. **86**, No. 1, p. K69—K72 (1984).

K. Masuda-Jindo
*Tight-Binding Electronic Theory for Lattice Defects in Transition Metals. Correlation Effects
(13.4; 21; 21.1.1)
(b) Vol. **126**, No. 2, p. 643—652 (1984).

V. Nagirnyi, T. Soovik, and S. Zazubovich
*Decay Kinetics and Polarization of the A_x Emission of KI:In Crystal
(13.4 and 20.3; 22.5.2)
(b) Vol. **126**, No. 2, p. 653—658 (1984).

L. E. Kholodenkov, A. G. Makhanek, and A. A. Kaminskii
One- and Two-Photon Luminescence Excitation of Eu^{3+} in a LaF_3 Monocrystal
(20.1 and 20.3; 13.4; 22.5)
(b) Vol. **126**, No. 2, p. 659—667 (1984).

C. Tuncay and M. Tomak
*Ideal Vacancy States in Si. A Study of the Empirical Tight-Binding Method
(13.4; 22.1.2)
(b) Vol. **126**, No. 2, p. 677—685 (1984).

Juh Tzeng Lue
*Time-Resolved Reflectivity Measurements on RF Sputtered a-Si:H Alloy
(13.4 and 20.1; 2; 16; 22.1.2)
(b) Vol. **126**, No. 2, p. 713—720 (1984).

V. V. Borimskii, K. D. Glinchuk, N. M. Litovchenko, and Z. A. Salnik
*Influence of Carbon and Preannealing on the Formation of Oxygen-Induced Recombination Centres in Heat-Treated Silicon
(13.4; 22.1.2)
(a) Vol. **86**, No. 2, p. 623—627 (1984).

V. M. Babich, N. P. Baran, A. A. Bugai, Yu. P. Dotsenko, V. B. Kovalchuk, and V. A. Shershel
*On the Properties of Thermodonors-II in Cz-Si Crystals of High Carbon Content
(13.4; 14.3; 19; 22.1.2)
(a) Vol. **86**, No. 2, p. 679—683 (1984).

D. Wolf and S. Kemmler-Sack
*IR Emission and Sensitized Luminescence of the Trivalent Rare Earth Ions in $18R\text{-}Ba_6B_2W_3O_{18}$ (B = Y, Er, Yb)
(13.4 and 20.3; 22.8)
(a) Vol. **86**, No. 2, p. 685—690 (1984).

D. Wruck, M. Seifert, and W. Ulrici
*Microinhomogeneities of Donor and Acceptor Distributions in n-Type LEC-GaAs from Free-Carrier Infrared Absorption
(1.4 and 13.4; 20.1; 22.2.1)
(a) Vol. **86**, No. 2, p. 691—702 (1984).

R. M. Bilbé, B. Lunn, J. E. Nicholls, and D. H. J. Totterdell
*The Role of Copper as an Impurity in High Resistivity Càdmium Telluride and Its Influence on γ-Ray Spectrometer Performance
(13.4 and 14.3; 20.1; 22.4.3)
(a) Vol. **86**, No. 2, p. 821—826 (1984).

Y. Kamiura and F. Hashimoto
Annealing Behaviour of Shallow Acceptors in Quenched Germanium
(10.2; 13.4; 14.3; 22.1.1)
(a) Vol. **86**, No. 2, p. K95—K97 (1984).

13.5 Elementary Excitations (Phonons See 6)

Original Papers and Short Notes

D. Lj. Mirjanić, M. M. Marinković, G. Knežević, and B. S. Tošić
*Two-Particle Solitary Waves
(13.5)
(b) Vol. **121**, No. 2, p. 589—593 (1984).

M. Rösler, R. Zimmermann, and W. Richert
*The Electron–Hole Plasma at Finite Temperatures. Calculation of Exchange-Correlation Energies and Padé Approximants
(13.5; 22.1.1; 22.2.1)
(b) Vol. **121**, No. 2, p. 609—621 (1984).

W. Richert and W. Ebeling
Thermodynamic Functions of the Electron Fluid for a Wide Density–Temperature Range
(13; 13.5; 22)
(b) Vol. **121**, No. 2, p. 633—639 (1984).

L. Wendler and E. Jäger
*Surface Polaritons in a SiO_2/Si Bilayer System
(6 and 13.5; 22.1.2; 22.6)
(b) Vol. **121**, No. 2, p. K119—K124 (1984).

B. Giese
*Electron Scattering Interaction and Elementary Excitations in Lead Chalcogenides
(13.1 and 13.5; 6; 22)
(b) Vol. **122**, No. 2, p. 641—650 (1984).

F. Kh. Abdullaev and A. A. Abdumalikov
*Dynamics of Magnetic Solitons in Planar Ferromagnets with Random Anisotropy
(13.5 and 18.2; 6; 13.5.1)
(b) Vol. **122**, No. 2, p. K141—K146 (1984).

M. Nawrocki, R. Planel, F. Mollot, and M. J. Kozielski
Magnetization and Spin-Flip Raman Scattering in $Cd_{1-x}Mn_xS$
(18.4 and 20.1; 13.5; 22.8)
(b) Vol. **123**, No. 1, p. 99—103 (1984).

Th. Hayne and G. Weiser
*Dispersion of Exciton Polaritons in a Monoclinic Cyanine Dye Crystal. Evaluation of the Reflectivity on Arbitrary Faces
(13.5 and 20.1; 22.9)
(b) Vol. **123**, No. 1, p. 271—280 (1984).

B. Čabrić and Lj. M. Ristovski
*Resonant Energy Transfer between Localized Excitations in a Linear Polymer Chain
(13.4 and 13.5; 22.9)
(b) Vol. **123**, No. 1, p. 309—317 (1984).

13.5 Elementary Excitations

M. PAGANNONE, B. FORNARI, G. MATTEI, and L. MATTIOLI
Radiative (Brewster-Type) Surface Mode Dispersion in Sodium Chlorate
(6 and 20.1; 13.5; 22.8.1)
(b) Vol. **123**, No. 1, p. K13—K16 (1984).

L. WENDLER
Phonon-Polaritons in Bilayer Systems. p-Polarized Guided Wave Phonon-Polaritons and Guided Wave Surface Phonon-Polaritons
(6 and 20.1; 13.5; 22.2.1; 22.2.2; 22.5.3)
(b) Vol. **123**, No. 2, p. 469—477 (1984).

N. I. LIKHOLIT, H.-E. PONATH, V. L. STRIZHEVSKII, and YU. N. YASHKIR
*Transient Coherent Anti-Stokes Raman Spectroscopy of Surface Polaritons
(13.5 and 20.1; 6; 22)
(b) Vol. **123**, No. 2, p. 691—701 (1984).

V. I. KRUGLOV
*On the Theory of Solitary Wave Excitation in Molecular Crystals by a Coherent Electromagnetic Field
(13.5)
(b) Vol. **124**, No. 1, p. 127—136 (1984).

W. HUHN and A. STAHL
*Self-Consistent Field Theory Applied to the Semiconductor Band Edge
(13.1 and 13.5; 14.3; 20; 22)
(b) Vol. **124**, No. 1, p. 167—177 (1984).

H. STOLZ and R. ZIMMERMANN
*Screening in a System of Many Particles Composed of Two Charged Fermions. Application to Excitons
(13.5; 13.5.1; 22.2.1)
(b) Vol. **124**, No. 1, p. 201—212 (1984).

L. F. GUDYMENKO and E. G. GULE
*The Nature of Non-Linear Radiation of $A^{II}B^{VI}$ Single Crystals
(13.5 and 20.3; 22.4.1)
(b) Vol. **124**, No. 2, p. K139—K142 (1984).

F. THUSELT and K. UNGER
*High Carrier Densities and Gap Shrinkage in p–i–n-Type Structures
(13.5; 14.3.3; 22.2.2)
(a) Vol. **84**, No. 2, p. 677—691 (1984).

U. LINDNER and S. SCHERF
*Two-Soliton Solution in a Modified Hubbard Chain
(13.5 and 18.4)
(b) Vol. **125**, No. 1, p. 177—185 (1984).

NGUYEN VAN HIEU and HA VINH TAN
*Polariton Effect in Non-Linear Acousto-Optical Phenomena
(13.5 and 20; 7)
(b) Vol. **125**, No. 1, p. 245—250 (1984).

V. ČÁPEK
*To the Interplay of Vibrational Relaxation and Exciton Motion
(13.5)
(b) Vol. **125**, No. 2, p. 639—643 (1984).

S. MOCHIZUKI
Radiative Optical Modes of Vibration in NiO Thin Film
(20.1; 6; 13.5; 22.6)
(b) Vol. **126**, No. 1, p. 105—114 (1984).

HOANG NGOC CAM, NGUYEN VAN HIEU, and NGUYEN AI VIET
*On the Damping of the Excitonic Polaritons
(13.5)
(b) Vol. **126**, No. 1, p. 247—252 (1984).

13.5.1 Excitons

Original Papers and Short Notes

M. T. Primatarowa
*Influence of Planar Defects and Disorder on the Energy Spectrum of Molecular Crystals
(13.4 and 13.5.1)
(b) Vol. **121**, No. 1, p. 357—369 (1984).

V. N. Abramov, A. N. Ermoshkin, A. I. Kuznetsov, and V. V. Mürk
*Breakdown of Excitonic Absorption and Luminescence Induced by a d-Type Conduction Band in Sc_2O_3
(13.5.1 and 20.3; 20.1; 22.6)
(b) Vol. **121**, No. 1, p. K59—K63 (1984).

J. Y. Bigot and B. Hönerlage
*On the Dielectric Function of Three-Level Systems under High Excitation (I): Theoretical Approach
(13.5.1; 20.1; 22.5)
(b) Vol. **121**, No. 2, p. 649—659 (1984).

F. Henneberger and H. Rossmann
*Resonatorless Optical Bistability Based on Increasing Nonlinear Absorption
(13.5.1 and 20.1; 22.4.1)
(b) Vol. **121**, No. 2, p. 685—693 (1984).

F. Henneberger, J. Puls, H. Rossmann, and I. N. Uraltsev
*On the Refractive Index at the Exciton Resonance of Highly Excited CdS
(13.5.1 and 20.1; 22.4.1)
(b) Vol. **121**, No. 2, p. K191—K194 (1984).

R. Stegmann, G. Oelgart, W. Sauer, A. Scheifler, and M. Trapp
*Investigation of the Recombination Mechanism in $GaAs_{0.11}P_{0.89}$:N
(13.5.1 and 20.3; 22.2.4)
(a) Vol. **81**, No. 2, p. 541—548 (1984).

U. Sliwczuk, H. Stolz, and W. von der Osten
*Indirect-Forbidden Exciton Transitions in AgBr
(13.5.1; 6; 20.1; 22.5.1)
(b) Vol. **122**, No. 1, p. 203—209 (1984).

E. L. Ivchenko, V. P. Kochereshko, G. V. Mikhailov, and I. N. Uraltsev
Resonance Magneto-Spatial Dispersion of Crystals
(18 and 20.1; 13.5.1; 22.4.2)
(b) Vol. **122**, No. 1, p. 221—230 (1984).

Yu. B. Gaididei and A. A. Vakhnenko
*Nonequilibrium Kinetics of Exciton Wave Packets in Crystals
(13.5.1; 6; 22)
(b) Vol. **122**, No. 1, p. 239—247 (1984).

Q. Kim and D. W. Langer
*Effects of Excitation Intensity on Photoluminescence of Pure CdTe
(13.5.1 and 20.3; 22.4.3)
(b) Vol. **122**, No. 1, p. 263—268 (1984).

R. M. Vlasova, V. A. Kapustin, V. N. Semkin, G. Sekretarczyk, L. S. Agroskin, G. V. Papayan, and L. P. Rautian
Molecular Excitons in Quasi-One-Dimensional Conducting Crystals (Semiconductors and Metals)
(20.1; 13.5.1; 22.9)
(b) Vol. **122**, No. 2, p. 549—557 (1984).

E. Ya. Glushko
*Metallic Quenching of Exciton Luminescence in Anthracene
(13.5.1 and 20.3; 22.9)
(b) Vol. **122**, No. 2, p. 569—580 (1984).

R. Schwabe, J. L. Staehli, I. Streit, and F. Thuselt
*EHP Luminescence of GaP:N under Resonant Free-Exciton Excitation
(13.5.1 and 20.3; 22.2.2)
(b) Vol. **122**, No. 2, p. 629—639 (1984).

V. I. Sokolov and V. V. Chernyaev
*Detection of Manganese Photoionization in $Cd_{1-x}Mn_xTe$ Solid Solutions
(13.5.1; 16; 20.1; 22.4.4)
(b) Vol. **122**, No. 2, p. 703—708 (1984).

F. Kh. Abdullaev and A. A. Abdumalikov
Dynamics of Magnetic Solitons in Planar Ferromagnets with Random Anisotropy
(13.5 and 18.2; 6; 13.5.1)
(b) Vol. **122**, No. 2, p. K141—K146 (1984).

P. Świderski and A. Twardowski
*The Excited States of Free Exciton in $Zn_{0.972}Mn_{0.028}Te$
(13.5.1; 22.4.4)
(b) Vol. **122**, No. 2, p. K147—K150 (1984).

R. Bacewicz
*Excitonic Wavelength Modulated Reflectance of $CdGa_2Se_4$
(13.5.1 and 20.1; 22.8)
(b) Vol. **122**, No. 2, p. K155—K158 (1984).

A. I. Bobrysheva, V. V. Baltaga, and M. V. Grodetskii
*Biexciton Two-Photon Absorption and Raman Scattering in CuBr
(13.5.1; 20.1; 22.5)
(b) Vol. **123**, No. 1, p. 169—181 (1984).

B. Hönerlage and J. Y. Bigot
*On the Dielectric Function of Three-Level Systems under High Excitation (II): Exiton-Biexciton Transitions and Population Effects in CuCl
(13.5.1; 20.1; 22.5)
(b) Vol. **123**, No. 1, p. 201—208 (1984).

YU. P. GNATENKO, P. A. SKUBENKO, Z. D. KOVALYUK, and V. M. KAMINSKII
*An Isothermal Annealing Effect on the Luminescence Spectra of GaSe Single Crystals
(13.5.1 and 20.3; 22.3)
(b) Vol. **123**, No. 1, p. K63—K67 (1984).

A. VERVOITTE, J. C. CANIT, B. BRIAT, and U. KAMBLI
*Optical and Magnetic Circular Dichroism Study of the 2D Antiferromagnet Rb_2MnCl_4
(13.5.1 and 13.5.4; 18.4; 20.1; 22.5)
(b) Vol. **124**, No. 1, p. 87—102 (1984).

P. K. SEN
*Nonlinear Absorption in III—V Semiconductors
(13.5.1 and 20.1; 22.2)
(b) Vol. **124**, No. 1, p. 117—125 (1984).

HOANG XUAN NGUYEN and R. ZIMMERMANN
*Nonlinear Exciton Transmission in CdS. Non-Homogenous and Non Stationary Model Calculations
(13.5.1 and 20; 22.4.1)
(b) Vol. **124**, No. 1, p. 191—200 (1984).

H. STOLZ and R. ZIMMERMANN
Screening in a System of Many Particles Composed of Two Charged Fermions. Application to Excitons
(13.5; 13.5.1; 22.2.1)
(b) Vol. **124**, No. 1, p. 201—212 (1984).

V. P. KOCHERESHKO, J. RÖSELER, I. N. URALTSEV, and K. HENNEBERGER
*Dipole Forbidden Transmission of Magnetopolaritons in CdSe
(13.5.1 and 20.1; 18; 22.4.2)
(b) Vol. **124**, No. 1, p. 213—219 (1984).

B. HÖNERLAGE and J. Y. BIGOT
*Dispersion and Absorption Anomalies in Three-Level Systems
(13.5.1 and 20.1; 22.5)
(b) Vol. **124**, No. 1, p. 221—227 (1984).

J. P. ŠETRAJČIĆ, D. V. KAPOR, and D. MIRJANIĆ
*On the Possibility of Bose-Einstein Condensation and the Superfluidity of Frenkel Excitons
(13.5.1; 20.1)
(b) Vol. **124**, No. 1, p. 235—239 (1984).

A. M. RATNER and E. I. TARASOVA
Local Optical Centers in Rare Gas Crystals
(13.4; 13.5.1; 23)
(b) Vol. **124**, No. 1, p. 249—260 (1984).

R. GAETHKE, P. GÜRTLER, R. KINK, E. ROICK, and G. ZIMMERER
*Vibrational Relaxation and Hot Luminescence of Ne_2^* Centers in Solid Neon
(13.5.1 and 20.3; 23)
(b) Vol. **124**, No. 1, p. 335—342 (1984).

R. JANKOWIAK, B. RIES, and H. BÄSSLER
*Spectral Diffusion and Triplet Exciton Localization in an Organic Glass
(13.5.1 and 20.3; 2; 22.9)
(b) Vol. **124**, No. 1, p. 363—371 (1984).

J. P. ŠETRAJČIĆ, M. J. ŠKRINJAR, and D. MIRJANIĆ
*On the Possibility of Bose-Einstein Condensation and the Superfluidity of Spin Waves
(13.5.1 and 13.5.4; 18.2)
(b) Vol. **124**, No. 1, p. K45—K48 (1984).

T. IKARI and S. SHIGETOMI
Photoluminescence of n-Type InSe
(20.3; 13.5.1; 22.3)
(b) Vol. **124**, No. 1, p. K49—K51 (1984).

P. REINEKER and B. J. SCHMID
*The Mixing of Frenkel and Charge-Transfer Excited States (I): General Treatment of Singlet and Triplet Excitations and Analytic Solutions of Simple Models
(13.5.1; 22.9)
(b) Vol. **124**, No. 2, p. 657—671 (1984).

H.-J. MÜLLER and H.-J. WÜNSCHE
*Model Calculations for Bound Excitons and Bound Excitonic Molecules in $GaAs_{1-x}P_x$:N
(13.4 and 13.5.1; 20.3; 22.2.4)
(b) Vol. **124**, No. 2, p. 747—756 (1984).

V. I. SOKOLOV, A. N. MAMEDOV, T. P. SURKOVA, G. A. EMELCHENKO, and L. G. KOLINOVA
*Fine Structure of the Photoionization Band Edge in Ni-Doped ZnO
(13.5.1 and 20.1; 14.3.4; 22.4)
(b) Vol. **124**, No. 2, p. K155—K159 (1984).

R. R. GUSEINOV
*Coulomb Interaction and Excitons in a Superlattice
(13.5.1)
(b) Vol. **125**, No. 1, p. 237—243 (1984).

TH. KLEINEFELD and W. VON DER OSTEN
*Localized and Free Biexcitons in Silver Bromide
(13.5.1 and 20.3; 22.5.1)
(b) Vol. **125**, No. 1, p. 293—303 (1984).

SHI-WEI GU and YU-KUI ZHOU
*Properties of the Exciton–Phonon System in a Magnetic Field
(6 and 13.5.1)
(b) Vol. **125**, No. 1, p. 323—331 (1984).

V. ANDREEV and M. POPE
*Generation of Hot Free Electrons Due to the Interaction between Charge-Transfer Excitons and Thermalized Free Electrons in Molecular Crystals
(13 and 13.5.1; 22.9)
(b) Vol. **125**, No. 2, p. 573—581 (1984).

B. J. Schmid and P. Reineker
*The Mixing of Frenkel and Charge-Transfer Excited States (II): Numerical Treatment of Non-Symmetric Dimers
(13.5.1; 22.9)
(b) Vol. **125**, No. 2, p. 593—604 (1984).

M. S. Brodin, V. M. Bandura, and M. G. Matsko
*Excitonic Surface Polaritons in Luminescence from ZnTe Crystals
(13.5.1 and 20.3; 13.3; 22.4.3)
(b) Vol. **125**, No. 2, p. 613—618 (1984).

D. Fröhlich, A. Nöthe, and K. Reimann
*Determination of Valence Band Parameters in ZnTe
(13.5.1 and 20.1; 13.1; 22.4.3)
(b) Vol. **125**, No. 2, p. 653—657 (1984).

G. Manzke, V. May, and K. Henneberger
*Excited States and Collective Excitations in a Dense Gas of Excitons
(13.5.1; 22)
(b) Vol. **125**, No. 2, p. 693—704 (1984).

M. P. Lisitsa, A. V. Stolyarenko, S. F. Terekhova, and N. A. Onishchenko
*Refraction Index Variation in the Exciton Resonance of CdS and $TlInS_2$ Single Crystals under Laser Excitation
(13.5.1; 20.1; 22.3.1; 22.4.1)
(b) Vol. **125**, No. 2, p. 705—711 (1984).

Y. Unuma, Y. Abe, Y. Masumoto, and S. Shionoya
*Dynamical Behaviour of the High Density Electron–Hole System in CdSe
(13.5.1 and 20.3; 22.4.2)
(b) Vol. **125**, No. 2, p. 735—742 (1984).

B. S. Tošić and J. P. Šetrajčić
*Exciton Drops in One-Dimensional Molecular Chains. New Method of Solving of Gap Equation
(6 and 13.5.1)
(b) Vol. **125**, No. 2, p. 743—749 (1984).

V. N. Ermakov, D. V. Korbutyak, V. G. Litovchenko, O. Yu. Mikityuk, and V. V. Nitsovich
*A Study of the Reflection Spectra of Surface Exciton–Impurity Complexes
(13.4 and 13.5.1; 13.3; 20.1; 22.4.1)
(b) Vol. **125**, No. 2, p. 815—822 (1984).

V. Čápek and V. Szöcs
*Is the Sink Model of Exciton Trapping in Molecular Condensates Satisfactory?
(13.5.1; 22.9)
(b) Vol. **125**, No. 2, p. K137—K142 (1984).

L. J. Lalov
*Temperature Vibronic Spectra of Molecular Crystals
(6 and 13.5.1; 20.1; 22.9)
(b) Vol. **125**, No. 2, p. K147—K150 (1984).

M. J. Škrinjar, D. V. Kapor, Ž. M. Škrbić, and J. P. Šetrajčić
*On the Problem of Exciton Coupling in Molecular Crystals
(6 and 13.5.1; 22.9)
(b) Vol. **125**, No. 2, p. K151—K154 (1984).

U. F. Kozmidis-Luburić, J. P. Šetrajčić, Ž. M. Škrbić, and B. S. Tošić
*Exciton–Phonon Interaction and Bound States
(6 and 13.5.1)
(b) Vol. **125**, No. 2, p. K155—K159 (1984).

D. Fröhlich, H.-J. Weber, R. Wille, and R. Francini
*Temperature Dependence of the Exciton Lineshape of GeO_2
(13.5.1 and 20.1; 22.6)
(b) Vol. **126**, No. 1, p. 305—310 (1984).

F. Dujardin, B. Stébé, and G. Munschy
*Neutral Bound Excitons in a Low Magnetic Field
(13.4 and 13.5.1; 22.2; 22.4)
(b) Vol. **126**, No. 1, p. 329—334 (1984).

I. M. Nepomnyashi and V. N. Nishanov
*The Localization Energy of Wannier-Mott Excitons at the Interface
(13.3 and 13.5.1)
(b) Vol. **126**, No. 1, p. K53—K57 (1984).

V. V. Sobolev and A. I. Kozlov
*Free Exciton and Exciton–Impurity Complexes of Zinc Diphosphide and Zinc Diarsenide
(13.5.1 and 20.1; 22.3)
(b) Vol. **126**, No. 1, p. K59—K62 (1984).

R. Baltramiejūnas, I. Rückmann, G. Tamulaitis, and A. Žukauskas
*Sample-Dependent Luminescence Spectra in Highly Photoexcited CdS and CdSe
(13.5.1 and 20.1; 22.4.1; 22.4.2)
(b) Vol. **126**, No. 2, p. K135—K138 (1984).

13.5.2 Plasmons

Original Papers and Short Notes

L. Merten and W. Goltsche
Stimulated Raman Effect of Plasmons and Longitudinal Plasmon-Phonos in Simple Cubic Crystals
(6 and 20; 13.5.2; 22)
(b) Vol. **121**, No. 2, p. 471—480 (1984).

S. Gosh and R. B. Saxena
*Parametric Conversion of an Electromagnetic Wave into an Acoustic Wave in Magnetised Cubic Piezoelectric Semiconducting Plasmas
(13.5.2; 7; 22.2.3)
(b) Vol. **121**, No. 2, p. 661—672 (1984).

H. Zscheile, R. Gründler, U. Dahms, J. Fröhner, and G. Lehmann
*π-Plasmon Dispersion of Trans-$(CH)_x$ at Small Momentum Transfer by Electron Energy Loss Spectroscopy
(13.5.2; 14.4; 22.9)
(b) Vol. **121**, No. 2, p. K161—K164 (1984).

T. Chattopadhyay and A. R. Chetal
*Plasmon and X-Ray Chemical Shifts in Some Copper Systems
(13.5.2 and 20; 22; 22.5)
(b) Vol. **122**, No. 1, p. 193—196 (1984).

F. Thuselt
*Low-Density Approach to RPA Self-Energies
(13.5.2; 22.1.1; 22.1.2; 22.2.1; 22.2.2)
(b) Vol. **123**, No. 1, p. 257—269 (1984).

M. M. El-Oker, F. A. Sultan, S. A. Yousef, and S. A. El-Sahhar
*Surface Plasmon Modes and Energy Loss Function of Indium Thin Films on Mica Substrates
(13.5.2 and 20.1; 13.1; 21)
(a) Vol. **83**, No. 1, p. 263—267 (1984).

T. Stöckert, R. Gründler, H. Boudriot, and H. A. Schneider
*Volume Plasmon Dispersion of Polycrystalline GaP and InP
(13.5.2; 22.2.2)
(b) Vol. **123**, No. 2, p. K161—K164 (1984).

M. M. El-Oker
*Detection of Surface Plasmons in Some Indium Films by Attenuated Total Reflection
(13.5.2 and 20.1; 21)
(a) Vol. **83**, No. 2, p. 613—616 (1984).

K. M. Paraskevopoulos, D. Siapkas, and N. A. Economou
*Infrared Reflectivity Plasma Spectra of $TlBiTe_2$ Single Crystals
(13.5.2 and 20.1; 22.8)
(b) Vol. **124**, No. 1, p. 229—234 (1984).

S. Ghosh and S. Dixit
*Parametric Decay of a High-Power Helicon Wave into a Transverse Acoustic and Another Helicon Waves in Longitudinally Magnetised Cubic Piezoelectric Semiconducting Plasmas
(13.5.2; 7; 22.2.3)
(b) Vol. **124**, No. 1, p. 395—402 (1984).

M. S. Pathak, S. Khan, and S. Ghosh
*Longitudinal Phonon–Plasmon Interactions in Semiconducting Plasmas
(6 and 13.5.2; 7; 22.2.3)
(b) Vol. **124**, No. 2, p. 489—496 (1984).

L. Wendler
*Surface Plasmon–Polaritons in Bilayer Systems
(13.3 and 13.5.2)
(b) Vol. **124**, No. 2, p. K161—K165 (1984).

L. Wendler
Coupled Surface Plasmon–Phonon Polaritons in Bilayer Systems
(6 and 20; 13.5.2; 21.6; 22.2.2)
(b) Vol. **125**, No. 1, p. K27—K31 (1984).

M. Rösler and W. Brauer
On the Theory of Ion-Induced Electron Emission
(17; 13.5.2; 21)
(b) Vol. **126**, No. 2, p. 629—642 (1984).

H. Boudriot, R. Gründler, K. Deus, T. Stöckert, and H. A. Schneider
*Electron Energy Loss Measurements on $CdSnP_2$: Plasmon Data, Dielectric Function, Band Structure Data
(13.1 and 13.5.2; 14.4; 22.8)
(b) Vol. **126**, No. 2, p. K149—K154 (1984).

Y. Q. Sheng, R. Y. Li, and P. Munz
*The Study of Mode Characteristics of Metal Clad Dielectric Waveguides
(13.5.2 and 20; 21.6; 22.5.2)
(a) Vol. **86**, No. 2, p. 813—819 (1984).

13.5.3 Polarons

Original Papers and Short Notes

D. IHLE
*Polaronic Effects on the Verwey Transition in Fe_3O_4
(13.5.3; 22.6)
(b) Vol. **121**, No. 1, p. 217—224 (1984).

L. A. MANAKOVA and S. P. IONOV
The Ordering Types in the Extended Hubbard Model with Electron–Phonon Interaction
(13.1; 13.5.3)
(b) Vol. **121**, No. 1, p. 275—287 (1984).

A. U. GRINFELDS, A. R. SILIN, L. N. SKUYA, and V. G. PLEKHANOV
Low Temperature Recombination Luminescence in Amorphous and Crystalline SiO_2
(20.3; 2; 13.5.3; 22.6)
(a) Vol. **81**, No. 1, p. K23—K26 (1984).

S. W. GU and J. ZHANG
*Note on Strong Coupling Polarons in a Semi-Infinite Ionic Crystal
(13.5.3; 22)
(b) Vol. **121**, No. 2, p. K165—K170 (1984).

Y. LÉPINE and M. CHARBONNEAU
*Evaluation of the First Relaxed Excited State Energy of the Free Fröhlich Polaron
(13.5.3)
(b) Vol. **122**, No. 1, p. 151—160 (1984).

C. SANCHEZ, M. HENRY, R. MORINEAU, and M. C. LEROY
*Small Polaron Mobility in α-$Li_xV_2O_5$
(13.4 and 13.5.3; 14.3; 19; 22.6.1)
(b) Vol. **122**, No. 1, p. 175—182 (1984).

H. SZYMCZAK
*Acoustical Relaxation Due to the Bound Small Polaron in Cubic Crystals
(7 and 13.5.3; 22.6)
(b) Vol. **122**, No. 1, p. K37—K40 (1984).

S. W. GU and S. L. BAN
*The Ground State Effective Hamiltonian of Strong Coupling One-Dimensional Optical Exciton
(13.5.3; 22)
(b) Vol. **122**, No. 1, p. K41—K44 (1984).

N. B. IVANOV
*Generalized Dispersion Relations for Surface Polaritons in Uniaxial Antiferromagnetic Crystals
(13.5.3 and 18.4)
(b) Vol. **126**, No. 2, p. K131—K134 (1984).

13.5.4 Magnons

Original Papers and Short Notes

A. A. Kazakov
*Two Types of Low-Temperature Phase Transitions in Uniaxial Ferromagnets with High Single-Ion Anisotropy
(13.5.4; 18.2)
(b) Vol. **121**, No. 1, p. 177—183 (1984).

J. M. Wesselinowa
Magnon Damping of the s–d Model above T_C
(18.2; 13.5.4; 22)
(b) Vol. **121**, No. 2, p. K147—K149 (1984).

M. Sh. Erukhimov and S. G. Ovchinnikov
*Atomic Operators in the Theory of Narrow-Band Magnetic Semiconductors
(13.5.4; 18.2; 22)
(b) Vol. **123**, No. 1, p. 105—114 (1984).

Kh. I. Pushkarov and M. T. Primatarowa
*Solitary Clusters of Spin Deviations and Lattice Deformation in an Anharmonic Ferromagnetic Chain
(13.5.4 and 18.2)
(b) Vol. **123**, No. 2, p. 573—584 (1984).

I. Ya. Korenblit and B. G. Tankhilevich
*Fluctuations in a Nonequilibrium System of Electrons and Magnons in Ferromagnetic Semiconductors
(13.5.4 and 18.2; 22)
(b) Vol. **124**, No. 1, p. 79—86 (1984).

A. Vervoitte, J. C. Canit, B. Briat, and U. Kambli
*Optical and Magnetic Circular Dichroism Study of the 2D Antiferromagnet Rb_2MnCl_4
(13.5.1 and 13.5.4; 18.4; 20.1; 22.5)
(b) Vol. **124**, No. 1, p. 87—102 (1984).

U. Lindner
*Solitary Solutions of the Heisenberg Model
(13.5.4; 18.2)
(b) Vol. **124**, No. 1, p. K39—K44 (1984).

J. P. Šetrajčić, M. J. Škrinjar, and D. Mirjanić
*On the Possibility of Bose-Einstein Condensation and the Superfluidity of Spin Waves
(13.5.1 and 13.5.4; 18.2)
(b) Vol. **124**, No. 1, p. K45—K48 (1984).

V. L. Safonov
*Interactions of Magnons in the Six-Sublattice Antiferromagnet
(13.5.4 and 18.4; 22.5)
(b) Vol. **126**, No. 1, p. 197—206 (1984).

L. Merten, R. Kotzott, and A. Lehmeyer
*Dispersion Relations for Magnon-Polaritons in Uniaxial Ferrimagnetic Crystals
(13.5.4 and 20.1; 18.3)
(b) Vol. **126**, No. 1, p. 223—228 (1984).

14. Electrical Properties. Transport Phenomena

Original Papers and Short Notes

S. K. Zhilinskii, M. E. Itkis, and F. Ya. Nad
*Frequency Dependence of Conductivity in TaS_3 at Helium Temperatures
(14; 22)
(a) Vol. **81**, No. 1, p. 367—375 (1984).

G. J. Ashwell and C. M. Care
*Optical Phonon Assisted Hopping in (DPPE) $(TCNQ)_5(H_2O)$
(6 and 14; 22.9)
(a) Vol. **81**, No. 2, p. 557—561 (1984).

C. Bansal and K. Surendranath
*Thermoelectric Power and Electrical Resistivity of $Fe_2Nb_3Se_{15}$
(14 and 15; 22.8)
(b) Vol. **122**, No. 1, p. 345—348 (1984).

B. Lippold and H. Müller
*Kondo-Like Behaviour of the Electrical Resistivity of the Compound $Pr_{0.25}La_{0.75}Al_3$
(14; 21.4)
(b) Vol. **122**, No. 1, p. 377—382 (1984).

J. B. Radunović
*Conductivity of an Electron Gas in a Low-Frequency Electric Field Due to Large-Scale Fluctuations
(13 and 14)
(b) Vol. **122**, No. 1, p. 383—387 (1984).

G. De Mey
*Potential Distribution in Anisotropic Layers and Hall Plates with Arbitary Geometries
(14)
(a) Vol. **82**, No. 1, p. K91—K93 (1984).

H. Englisch
*Remark on a Paper by Stephen about Localization in a Wire
(13 and 14)
(b) Vol. **123**, No. 2, p. K169—K171 (1984).

S. Ghosh and S. Khan
*Amplitude Modulation and Demodulation of Electromagnetic Waves in Magnetised Semiconducting Plasmas
(14 and 20; 22)
(a) Vol. **85**, No. 2, p. 535—541 (1984).

K. Tennakone and P. de S. Kaviratne
*Activation of Li^+ Ion Transport in Lithium Ferricyanide by Interstitial Water
(14; 22.8)
(a) Vol. **86**, No. 2, p. K121—K123 (1984).

14.1 Metals. Semi-Metals

Original Papers and Short Notes

J. Kilmer, C. M. van Vliet, G. Bosman, A. van der Ziel, and P. H. Handel
*Evidence of Electromagnetic Quantum 1/*f* Noise Found in Gold Thin Films
(14.1; 21.6)
(b) Vol. **121**, No. 1, p. 429—432 (1984).

V. V. Gudkov and V. D. Fil
*Ultrasonic Wave Field Transport by Conduction Electrons in Indium
(7 and 14.1; 13.2; 21)
(b) Vol. **121**, No. 1, p. 433—440 (1984).

V. Christoph and H. Wegener
*Conductivity of Strongly Disordered Metallic Alloys
(14.1; 21)
(b) Vol. **121**, No. 1, p. K103—K107 (1984).

V. I. Pecherskaya, D. N. Bolshutkin, V. A. Desnenko, and V. Ya. Ilichev
Electrical and Magnetic Properties of the Fe18Cr(30–40)Ni Alloys at Low Temperatures
(18.1 and 18.2; 14.1; 21.1; 21.1.1)
(a) Vol. **81**, No. 2, p. 507—512 (1984).

F. Völklein and E. Kessler
*A Method for the Measurement of Thermal Conductivity, Thermal Diffusivity, and Other Transport Coefficients of Thin Films
(14.1 and 14.3.1; 8; 15; 21.7)
(a) Vol. **81**, No. 2, p. 585—596 (1984).

U. Schaller, W. Kraak, and R. Herrmann
*Metal–Insulator-Transition in n-InSb under High Hydrostatic Pressure
(14.1 and 14.3; 12.2; 22.2.3)
(b) Vol. **122**, No. 1, p. K87—K91 (1984).

G. Knuyt and A. Keupers
*Calculation of the Resistivity of Liquid and Amorphous $Au_{1-x}Si_x$ Using Ziman Theory
(2 and 14.1; 21.6)
(b) Vol. **122**, No. 2, p. 781—790 (1984).

L. G. Khvostantsev and V. A. Sidorov
Phase Transitions in Antimony at Hydrostatic Pressure up to 9 GPa
(1.2; 14.1; 15; 21.7)
(a) Vol. **82**, No. 2, p. 389—398 (1984).

H. Ratajczak and I. Gościańska
*Planar Hall Effect in Thin Amorphous RE–TM Films with Perpendicular Anisotropy
(14.1 and 18.2.1; 21.4)
(a) Vol. **82**, No. 2, p. 491—496 (1984).

L. Uray
*The Effect of Void Size on the Resistivity of Sintered Tungsten
(1.3 and 14.1; 15; 21)
(a) Vol. **82**, No. 2, p. 545—552 (1984).

R. Richter and F. Goedsche
On the Thermoelectric Power of Amorphous Ferromagnets
(2 and 15; 14.1; 18.2; 21)
(b) Vol. **123**, No. 1, p. 143—154 (1984).

K. S. Vahvaselkä
An X-Ray Investigation of Structural Properties and Electrical Resistivities of Liquid Al, Liquid Cu, and Liquid Al–Cu Alloys
(1.3 and 2; 14.1; 21; 21.1)
(a) Vol. **83**, No. 1, p. 103—112 (1984).

V. K. Arora and H. N. Spector
*Quantum Freeze-Out of Carriers in Semimetallic and Semiconducting Thin Wires
(14.1 and 14.3; 21.7)
(b) Vol. **123**, No. 2, p. 747—753 (1984).

G. Knuyt
*Electrical Resistivity Calculations for Amorphous Cu_xAl_{1-x}
(2 and 14.1; 21; 21.1)
(b) Vol. **123**, No. 2, p. 763—771 (1984).

H. Kido, T. Hoshikawa, M. Shimada, and M. Koizumi
*Crystal Structure and Magnetic and Electrical Properties of Pseudo-Ternary $GdCu_{1-x}Zn_xSi$ Compounds
(14.1 and 18.1; 1.3; 18.2; 22.8)
(a) Vol. **83**, No. 2, p. 561—565 (1984).

K. Mukherjee, M. Chandrasekaran, and M. Kato
*Electrical Resistivity of an Equiatomic Solid and Liquid Au–Cd Alloy
(14.1; 2; 8; 21; 21.6)
(a) Vol. **83**, No. 2, p. 657—667 (1984).

G. L. F. Fraga, J. V. Kunzler, F. Ogiba, and D. E. Brandão
*Electrical Resistivity of the Ni_2MnIn Heusler Alloy
(14.1; 21.1)
(a) Vol. **83**, No. 2, p. K187—K190 (1984).

T. Soumura, S. Saito, and T. Maeda
*Electrical Resistivity and Structural Stability for Amorphous $(Fe_{0.5}Ni_{0.5})_{1-x}B_x$ Alloys
(2 and 14.1; 21.1; 21.1.1)
(a) Vol. **83**, No. 2, p. K195—K198 (1984).

V. B. Gohel, C. K. Acharya, and A. R. Jani
On the Determination of Pseudopotential Parameters and the Comprehensive Study of Metallic Properties
(13.1; 2; 14.1; 15; 21)
(b) Vol. **124**, No. 1, p. 137—145 (1984).

U. H. Gläser
*On the Calculation of the Residual Resistivity of Precipitates
(10.1 and 14.1; 21; 21.1)
(b) Vol. **124**, No. 1, p. 423—429 (1984).

L. Uray and M. Menyhárd
The Segregation of Iron in Tungsten
(1.3; 1.4; 9; 14.1; 15; 21.1; 21.1.1)
(a) Vol. **84**, No. 1, p. 65—72 (1984).

B. Chenevier, D. Fruchart, M. Bacmann, J. P. Senateur, P. Chaudouet, and L. Lundgren
Magnetic Properties of MnRhAs. Study of the Antiferromagnetic–Ferromagnetic Transition
(18.2 and 18.4; 1.2; 14.1; 21.1)
(a) Vol. **84**, No. 1, p. 199—206 (1984).

R. Dimmich and F. Warkusz
*On the Electrical Transport Phenomena in Thin Wires
(14.1)
(a) Vol. **84**, No. 1, p. K85—K88 (1984).

T. Luciński and J. Baszyński
*Influence of Composition and Structural Transformation on Transport Properties in Amorphous $Fe_{1-x}Si_x$ Films
(14.1; 2; 18.2.1; 21.1.1)
(a) Vol. **84**, No. 2, p. 607—612 (1984).

A. V. Salker, E. B. Mirza, V. V. Deshpande, and H. V. Keer
Thermodynamic Analysis of the Phase Transition in V_4O_7
(1.2; 8; 14.1; 14.3; 18.1; 22.6)
(a) Vol. **84**, No. 2, p. K105—K108 (1984).

A. I. Romanenko, F. S. Rakhmenkulov, I. N. Kuropyatnik, V. E. Fedorov, and A. V. Mishchenko
*Electrical Resistance of the Mo-Chain Compound Mo_2S_3 in Equilibrium and Non-Equilibrium States
(14.1; 21)
(a) Vol. **84**, No. 2, p. K165—K167 (1984).

Yu. V. Zakharov and Yu. I. Mankov
*Magnetoresistance of Ferromagnets with Domain Structure
(14.1; 18.2; 21)
(b) Vol. **125**, No. 1, p. 197—205 (1984).

G. I. Makovetskii and V. P. Dymont
*Crystallographic, Magnetic, and Electrical Properties of Nickel-Substituted Chromium Telluride
(14.1 and 18.2; 22.8)
(a) Vol. **85**, No. 1, p. K69—K71 (1984).

J. V. Yakhmi, I. K. Gopalakrishnan, and A. K. Grover
*Electrical Resistivity Studies on the Heusler Alloys $CO_2T_{1-x}Al_{1+x}$(T = Ti or Zr)
(14.1; 21.1)
(a) Vol. **85**, No. 1, p. K89—K92 (1984).

G. Schubert and J. Richter
*The Electrical Resistivity of Clustering Alloys
(1.3 and 14.1; 21)
(b) Vol. **125**, No. 2, p. 839—847 (1984).

P. Středa
Universal Transport Properties of Inversion Layers in Quantizing Magnetic Field
(13.3; 8; 14.1; 15; 18; 21)
(b) Vol. **125**, No. 2, p. 849—860 (1984).

A. Paja and T. Stobiecki
*Electrical Resistivity of Amorphous $Fe_{1-x}B_x$ Alloys
(14.1; 2; 21.1.1)
(b) Vol. **125**, No. 2, p. K171—K176 (1984).

H. Taguchi, M. Shimada, and M. Koizumi
Magnetic Properties in the System $(Sr_{1-x}Ca_x)CoO_{3-\delta}$
(18.2; 14.1; 22.6.1)
(a) Vol. **85**, No. 2, p. 523—527 (1984).

O. Kohmoto
*Electrical Resistivity of Amorphous Co-Rich FeCo–SiB Alloys
(14.1; 2; 18.3; 21.1; 21.1.1)
(a) Vol. **85**, No. 2, p. K155—K157 (1984).

D. Korn
*Electrical Resistivity of Amorphous SnMn
(2 and 14.1; 21)
(b) Vol. **126**, No. 1, p. 425—432 (1984).

J. Dominec
Semiconductor–Semimetal Transition in Bi–Sb Alloys Observed in Ultrasound Attenuation
(7 and 14.3; 14.1; 22.7)
(b) Vol. **126**, No. 1, p. K87—K90 (1984).

E. Chilla and W. Schiller
*The Influence of Electron–Phonon Interaction on the Thermoelectric Power at High Temperatures
(14.1 and 15; 21)
(b) Vol. **126**, No. 1, p. K91—K96 (1984).

G. Hüber and D. Ploumbidis
*Electrical Resistivity and NMR Investigations of the V–Co System at High Temperatures
(14.1 and 19; 21.1)
(b) Vol. **126**, No. 1, p. K101—K104 (1984).

H. Yamaguchi, T. Tsuchida, I. Hashimoto, Y. Ohtani, and S. Ohtaki
Stage II Recovery of Defects in Cold-Worked Aluminium
(10.1; 14.1; 21)
(a) Vol. **86**, No. 1, p. 237—244 (1984).

V. E. Arkhipov, V. I. Voronin, B. N. Goshchitskii, A. E. Karkin, V. N. Kozhanov, A. V. Mirmelshtein, and E. P. Romanov
*The Structure, Superconducting Properties, and Electrical Resistivity of V_2Zr and V_2Hf Irradiated with Fast Neutrons
(14.1 and 14.2; 1.2; 21; 21.1)
(a) Vol. **86**, No. 1, p. K59—K62 (1984).

O. P. Hansen and I. F. I. Mikhail
*Galvanomagnetic Effects of Bismuth in Non-Quantising Magnetic Fields. Pseudo-Parabolic Model
(14.1; 15; 21.7)
(b) Vol. **126**, No. 2, p. 721—728 (1984).

M. A. Alam and M. Tomak
*Electrical Resistivity of Liquid Noble Metal Alloys
(2 and 14.1; 21.1; 21.6)
(b) Vol. **126**, No. 2, p. 729—732 (1984).

U. H. Gläser and H. Löffler
*Effective Medium Approach for the Residual Resistivity of Alloys Containing Guinier-Preston Zones
(14.1; 13; 21; 21.1)
(b) Vol. **126**, No. 2, p. 741—751 (1984).

H. Gruber, E. Krautz, H. P. Fritzer, K. Gatterer, and A. Popitsch
*Electronic Property Changes of η-Mo_4O_{11} by Doping with Tungsten, Rhenium, and Vanadium
(14.1; 1.2; 13.1; 18; 20.1; 22.6.1)
(a) Vol. **86**, No. 2, p. 749—757 (1984).

Gyanesh Chandra and O. P. Katyal
*Applicability of the Mayadas-Shatzkes Model on Electrical Resistivity of Polycrystalline Tin–Lead Alloy Films
(14.1; 1.3; 21)
(a) Vol. **86**, No. 2, p. 759—764 (1984).

Gyanesh Chandra and O. P. Katyal
*The Study of Electrical Properties of Polycrystalline Tin Films
(14.1; 21)
(a) Vol. **86**, No. 2, p. 765—772 (1984).

14.2 Superconductivity. Superconducting Materials and Devices

Review Article

L. B. Dubrovskaya and S. Z. Nazarova
Some Regularities in the Behaviour of the Magnetic, Electrokinetic, and Superconducting Properties of Transition Metal Monocarbide and Mononitride Solid Solutions
(14.3 and 18; 13.1; 14.2; 15; 22)
(a) Vol. **83**, No. 1, p. 11—38 (1984).

Original Papers and Short Notes

M. Surma
*Electron–Phonon Coupling and Electron–Electron Coulomb Interaction in Superconductors
(14.2)
(b) Vol. **121**, No. 1, p. 209—216 (1984).

P. Tekiel
*Perturbation of Homogeneity of Mixed State in Type II Superconductors
(14.2)
(b) Vol. **121**, No. 1, p. K91—K95 (1984).

L. N. Shehata and A. G. Saif
*Pinning of a Curved Flux Line by Macroscopic Inclusions in a Type II Superconductor
(14.2)
(b) Vol. **121**, No. 2, p. 749—756 (1984).

V. E. Antonov, T. E. Antonova, I. T. Belash, V. Yu. Malyshev, and V. I. Rashupkin
*Superconductivity and Atomic Ordering of Pd–Cu–H Solid Solutions
(14.2; 1.2; 21.6)
(a) Vol. **81**, No. 2, p. K185—K190 (1984).

M. A. Savchenko and A. V. Stefanovich
*Fluctuation Theory of Magnetic Superconductors
(14.2; 21.4)
(b) Vol. **122**, No. 1, p. 367—376 (1984).

P. Seidel and H.-G. Meyer
*Critical Josephson Current of Tunnel Junctions with Proximity Electrodes Containing Magnetic Impurities
(14.2; 21)
(b) Vol. **122**, No. 1, p. K105—K107 (1984).

Z. Przełożny and G. Kozłowski
*The Magnetic Field Distribution in Deformed FLL in Type II Superconductors
(14.2)
(b) Vol. **122**, No. 2, p. 775—780 (1984).

TH. ROATSCH and H.-G. MEYER
*The Double SQUID as a Digital Memory Cell
(14.2)
(a) Vol. **82**, No. 2, p. 581—590 (1984).

I. N. GROZNOV, S. P. IONOV, I. B. KEVDINA, V. S. LUBIMOV, E. F. MAKAROV, O. V. MARCHENKO, N. N. MIKHAILOV, V. A. ONISHCHUK, and V. P. SHANTAROVICH
*Mössbauer and Positron Studies of Superconducting Oxides $BaPb_{1-x}Bi_xO_3$
(14.2; 5; 22.6.1)
(b) Vol. **123**, No. 1, p. 183—190 (1984).

G. M. VUJIČIĆ
*Electron-Two-Level System Interaction and Superconductivity
(14.2)
(b) Vol. **123**, No. 1, p. K93—K95 (1984).

T. HÄDRICH
*Investigation of the Current Density Distribution of Small Josephson Tunnel Junctions by Fourier Transformation
(14.2)
(a) Vol. **83**, No. 1, p. 387—392 (1984).

E. BORCHI and S. DE GENNARO
*Superconducting Properties of Rare-Earth Amorphous Alloys
(14.2; 21.4)
(b) Vol. **123**, No. 2, p. 723—731 (1984).

G. J. CHASHOO and Y. SINGH
*Effect of Supercurrent on Charge Imbalance Relaxation in a Phase Slip Centre
(14.2)
(b) Vol. **123**, No. 2, p. 733—737 (1984).

G. VONES and E. SCHACHINGER
*Time-Dependent Ginzburg-Landau Equations for Anisotropic Superconducting Alloys
(14.2)
(b) Vol. **123**, No. 2, p. 773—786 (1984).

V. P. IORDATII, L. Z. KON, and V. P. TABAKAR
*The Critical Temperature of Superconductors in the „Jellium“ Model with Vertex Correction to Electron–Electron Interaction
(14.2)
(b) Vol. **123**, No. 2, p. K173—K175 (1984).

I. M. CHAPNIK
*Superconductivity and the Electronic Structure
(14.2; 21.1)
(b) Vol. **123**, No. 2, p. K183—K187 (1984).

C. SUŁKOWSKI and J. SZYMASZEK
*Normal State Resistance and Superconducting Properties of $LaMo_6S_8$ Films
(14.2)
(a) Vol. **83**, No. 2, p. K191—K194 (1984).

R. SOBOLEWSKI, C. V. STANCAMPIANO, G. MOUROU, and D. BUTLER
*Picosecond Pulse Generation Capability of Optically Driven Superconducting Thin Films
(14.2)
(a) Vol. **83**, No. 2, p. K203—K205 (1984).

P. CHATTERJEE
*Superconducting Properties of the Hexagonal-Close-Packed Alloy System TcZr
(14.2; 21)
(b) Vol. **124**, No. 2, p. 773—779 (1984).

H. LUCHIAN
*Magnetic Symmetry in the A15 Type of Structure
(14.2 and 18; 21)
(b) Vol. **124**, No. 2, p. K125—K128 (1984).

G. GIAQUINTA, C. DI MAURO, and R. FAZIO
*Vortex Pair Binding in 2D Superconductors and Surface Effects
(14.2)
(b) Vol. **124**, No. 2, p. K177—K180 (1984).

S. L. KAKANI and U. N. UPADHYAYA
*Superconducting and Magnetic Order Parameters in the Coexistence Region of Ternary Superconductors
(14.2 and 18.2; 22.8)
(b) Vol. **125**, No. 2, p. 861—867 (1984).

G. GIAQUINTA, C. DI MAURO, and R. FAZIO
*On the Vortex Pair Collapse in 2D Superconductors
(14.2)
(b) Vol. **125**, No. 2, p. K185—K187 (1984).

I. M. CHAPNIK
*Some Regularities in the Occurence of Superconductivity
(14.2; 21)
(b) Vol. **125**, No. 2, p. K189—K191 (1984).

H.-J. KÖHLER, M. RIEDEL, H.-G. MEYER, W. KRECH, P. SEIDEL, and K.-H. BERTHEL
*The Influence of Microwaves on the Current–Voltage Characteristic of Nb–NbO_x–Pb Josephson Tunnel Junctions
(14.2)
(a) Vol. **85**, No. 2, p. K175—K177 (1984).

N. E. ALEKSEEVSKII, A. V. MITIN, E. P. KHLYBOV, C. BAZAN, B. GREŃ, and A. GILEWSKI
*The Influence of Alloying on the Properties of Superconducting Molybdenum Sulphides
(14.2)
(a) Vol. **86**, No. 1, p. 389—400 (1984).

V. E. Arkhipov, V. I. Voronin, B. N. Goshchitskii, A. E. Karkin, V. N. Kozhanov, A. V. Mirmelshtein, and E. P. Romanov

*The Structure, Superconducting Properties, and Electrical Resistivity of V_2Zr and V_2Hf Irradiated with Fast Neutrons
(14.1 and 14.2; 1.2; 21; 21.1)
(a) Vol. **86**, No. 1, p. K59—K62 (1984).

V. N. Rudko and V. I. Sugakov

*Superconductivity of Metals with a Superlattice of Voids
(14.2)
(b) Vol. **126**, No. 2, p. 703—712 (1984).

14.3 Semiconductors

Review Article

L. B. Dubrovskaya and S. Z. Nazarova
*Some Regularities in the Behaviour of the Magnetic, Electrokinetic, and Superconducting Properties of Transition Metal Monocarbide and Mononitride Solid Solutions
(14.3 and 18; 13.1; 14.2; 15; 22)
(a) Vol. **83**, No. 1, p. 11—38 (1984).

Original Papers and Short Notes

J. Monecke
*Dielectric Function and Conductivity of p-Type Semiconductors
(13.1 and 14.3; 22)
(b) Vol. **121**, No. 1, p. 329—338 (1984).

Y. Ishihara, Y. Ohno, and I. Nakada
*Anisotropic Electrical Properties of GeSe
(13.4 and 14.3; 22)
(b) Vol. **121**, No. 1, p. 407—412 (1984).

Böttger and D. Wegener
*On the Percolation Approach to the Hopping Conduction in Strong Electrical Fields
(2 and 14.3; 6; 22)
(b) Vol. **121**, No. 1, p. 413—419 (1984).

E. M. Raspopova, V. T. Kalinnikov, T. N. Lyubovskaya, A. P. Chernov, and T. N. Fesenko
Glass Formation and Some Properties of the $(Cd_2GeAs_4)_{1-x}(TIAs)_x$ System
(2; 14.3; 15; 22.8)
(a) Vol. **81**, No. 1, p. 139—143 (1984).

A. D. Pogrebnyak, E. Yu. Boyarko, Yu. Yu. Kryuchkov, and S. A. Vorobev
Positron Annihilation and Profiles of Radiation Damages in GaAs and Si Crystals Irradiated by Supercurrent Proton or Electron Beams
(11; 14.3; 22.1.2; 22.2.1)
(a) Vol. **81**, No. 1, p. 217—225 (1984).

J. Oswald, H. Feichtinger, and R. Czaputa
A Useful Method to Improve Convergence in Least-Squares Fitting Procedures
(13.4; 14.3; 20.3; 22.8)
(a) Vol. **81**, No. 1, p. 343—351 (1984).

G. J. Ashwell
*(2,2-DHPE Cl)$_x$(2,2-DPE)$_{1-x}$(TCNQ), a Pauli Paramagnetic Organic Conductor
(14.3; 1.1; 18.1; 22.9)
(a) Vol. **81**, No. 1, p. 361—366 (1984).

G. F. Myachina, T. G. Ermakova, and V. A. Lopyrev
*Compensation Effect in the Electrical Conduction in Some Conjugated Polymers
(14.3; 22.9)
(a) Vol. **81**, No. 1, p. 377—380 (1984).

H. W. Helberg
*Dielectric Relaxation Behavior of Dislocations in Semiconductors
(10.2 and 14.3; 14.4; 22)
(a) Vol. **81**, No. 1, p. 381—390 (1984).

V. K. Yanovskii
Phase Transitions and Properties of Doped $KNbO_3$ Crystals
(1.2 and 14.4.1; 14.3; 15; 20.1; 22.8.1)
(a) Vol. **81**, No. 1, p. 399—406 (1984).

I. Hevesi, L. B. Kiss, M. I. Török, and L. Nánai
*Electron Concentration in V_2O_5 Single Crystals as Determined by $1/f$ Noise Measurements
(14.3; 22.6)
(a) Vol. **81**, No. 1, p. K67—K70 (1984).

C. Rincón and S. M. Wasim
*Influence of Intrinsic Defects on the Electrical Properties of $CuInSe_2$
(13.4 and 14.3; 22.8)
(a) Vol. **81**, No. 1, p. K77—K80 (1984).

M. Stordeur and H. T. Langhammer
Erratum to: Anisotropy of Transport Properties Due to Direction-Dependent Band Nonparabolicity
(13.1; 14.3; 15; 22)
(b) Vol. **121**, No. 2, p. 765 (1984).

G. M. Shmelev, Nguyen Quoc Anh, G. I. Tsurkan, and S. Y. Mensah
*"Currentless" Amplification of Hypersound in a Planar Configuration by Inelastic Scattering of Electrons
(7 and 14.3; 22)
(b) Vol. **121**, No. 2, p. K209—K213 (1984).

A. K. Sarychev and A. P. Vinogradov
*Erratum to: Effective Medium Theory for the Magnetoconductivity Tensor of Disordered Materials
(14.3; 22.3)
(b) Vol. **121**, No. 2, p. K215 (1984).

G. J. Ashwell and I. W. Nowell
4-Cyano-1,3,6-triazacycl[3.3.3]azine — A Homomolecular Organic Semiconductor
(1.1; 14.3; 22.9)
(a) Vol. **81**, No. 2, p. K97—K101 (1984).

M. Kittler, K.-W. Schröder, E. Bugiel, and C. Becker
Quantitative EBIC Investigations on Bulk Stacking Faults in Silicon
(1.4 and 10.2; 14.3; 22.1.2)
(a) Vol. **81**, No. 2, p. K131—K135 (1984).

A. A. Ghani Awad
*Thermopower and Resistivity in Correlation to B-Site Magnetic Ordering in Cu–Ni and Cu–Cd Ferrospinels
(14.3 and 15; 22.8.2)
(a) Vol. **81**, No. 2, p. K155—K158 (1984).

K. Schmalz, P. Gaworzewski, and F.-G. Kirscht
Deep Levels in Czochralski p-Si Due to Heat Treatment at 600 to 900 °C
(10.2 and 13.4; 14.3; 22.1.2)
(a) Vol. **81**, No. 2, p. K165—K169 (1984).

G. M. Niftiev, B. G. Tagiev, A. O. Khalilov, and S. A. Abushov
*Lasting Relaxations and Residual Conductivity in $InSe_{1-x}S_x$ Single Crystals
(14.3; 22.3.1)
(a) Vol. **81**, No. 2, p. K175—K178 (1984).

B. Karvaly, B. Mallik, G. Kemeny, and S. Trivedi
*Anomalous Effects of Adsorption on the Electric Conductivity of Some Organic Semiconductor Powders under Pressure
(14.3; 22.9)
(a) Vol. **81**, No. 2, p. K179—K184 (1984).

Z. Januškevičius, V. Koval, A. Kudin, A. Sodeika, V. Tartachnik, and I. Tychina
*Radiation Defects in Single Crystals of Zinc Diphosphide
(11 and 14.3; 22.3)
(a) Vol. **81**, No. 2, p. K191—K193 (1984).

C. Sanchez, M. Henry, R. Morineau, and M. C. Leroy
Small Polaron Mobility in α-$Li_xV_2O_5$
(13.4 and 13.5.3; 14.3; 19; 22.6.1)
(b) Vol. **122**, No. 1, p. 175—182 (1984).

R. F. Konopleva, V. A. Evseev, B. A. Borisov, I. V. Nazarkin, and V. A. Chekanov
*Electrophysical Properties of Disordered Systems Based on Crystalline Germanium Compensated by Fast Neutron Irradiation
(2 and 14.3; 14.3.4; 22.1.1)
(b) Vol. **122**, No. 1, p. 349—359 (1984).

G. I. Tsurkan, G. M. Shmelev, and Nguyen Hong Shon
*Photostimulated Uneven Magnetorestistance at Inelastic Scattering of Electrons in Semiconductors
(14.3; 18; 22)
(b) Vol. **122**, No. 1, p. 361—366 (1984).

S. K. Plachkova
Estimation of Some Band-Edge Structure Parameters of the System $(GeTe)_{1-x}(AgSbTe_2)_x$
(13.1; 14.3; 15; 22)
(b) Vol. **122**, No. 1, p. K53—K58 (1984).

U. SCHALLER, W. KRAAK, and R. HERRMANN
*Metal–Insulator-Transition in n-InSb under High Hydrostatic Pressure
(14.1 and 14.3; 12.2; 22.2.3)
(b) Vol. **122**, No. 1, p. K87—K91 (1984).

M. MONDAL and K. P. GHATAK
*On the Modification of the Einstein Relation for a Semiconductor Superlattice
(14.3; 22)
(b) Vol. **122**, No. 1, p. K93—K97 (1984).

C. WAGNER and A. EL-SADEK
*The Annealing Behaviour of Arsenic-Implanted Silicon between 600 and 800 °C Characterized by Carrier Density Profiles
(11 and 14.3; 22.1.2)
(a) Vol. **82**, No. 1, p. 149—155 (1984).

E. WIESER and D. PANKNIN
Electrical Activation and Damage Annealing of Boron-Implanted Silicon by Flash-Lamp Irradiation
(11; 14.3; 22.1.2)
(a) Vol. **82**, No. 1, p. 171—177 (1984).

J.-P. DOUMERC, J. HEJTMANEK, J.-P. CHAMINADE, M. POUCHARD, and M. KRUSSANOVA
*A Photoelectrochemical Study of $CuWO_4$ Single Crystals
(14.3; 13.4; 15; 16; 22.8.1)
(a) Vol. **82**, No. 1, p. 285—294 (1984).

D. COQUILLAT, A. PRADEL, G. BRUN, and J. C. TEDENAC
*Growth and Physical Properties of $TlBiTe_2$ Crystals
(13.1 and 14.3; 8; 15; 20.1; 22.8)
(a) Vol. **82**, No. 1, p. 295—299 (1984).

G. J. ASHWELL, J. G. ALLEN, E. P. GOODINGS, and I. W. NOWELL
*Crytal Structure and Electrical Conductivity of $[1,1'\text{-bis(p-fluorophenyl)-4,4'-bipyridinium}]^{2+}$ $(TCNQ)_2^{2-}$ and Electrical Characterisation of 51 TCNQ Simple Salts
(1.1 and 14.3; 22.9)
(a) Vol. **82**, No. 1, p. 301—306 (1984).

K. S. JONES and S. J. PEARTON
*Grain Boundaries in Germanium: Effects of Exposure to Plasmas
(10.2 and 14.3; 22.1.1)
(a) Vol. **82**, No. 1, p. K101—K105 (1984).

P. KUIVALAINEN, H. ISOTALO, and H. STUBB
*Thermoelectric Power in Inhomogeneous Metallic Polymers
(14.3 and 15; 22.9)
(b) Vol. **122**, No. 2, p. 791—797 (1984).

YOUNG HWAN CHOI, YOUN SOO LIM, CHUL KOO KIM, and KYUN NAHM
*Electrical Conductivity of Magnetite below the Verwey Transition Temperature
(14.3; 22.6)
(b) Vol. **122**, No. 2, p. K187—K190 (1984).

G. A. Kachurin, V. A. Mayer, S. I. Romanov, M. Voelskow, R. Klabes, and E. Wieser
Low-Energy Implantation of Arsenic in Silicon
(11; 14.3; 22.1.2)
(a) Vol. **82**, No. 2, p. 475—480 (1984).

L. A. Kazakevich, P. F. Lugakov, and I. M. Filippov
The Effect of Dislocations on the Charge-Carrier Recombination Processes in Irradiated Silicon
(10.2 and 13.4; 11; 14.3; 22.1.2)
(a) Vol. **82**, No. 2, p. 511—518 (1984).

P. I. Baranskii, V. M. Babich, N. P. Baran, A. A. Bugay, Yu. P. Dotsenko, and V. B. Kovalchuk
The Effect of Heat Treatment on Compensated CZ Silicon
(13.4; 14.3; 19; 22.1.2)
(a) Vol. **82**, No. 2, p. 533—536 (1984).

S. M. Wasim and A. Noguera
*Transport Properties of n-Type $CuInSe_2$
(13.4 and 14.3; 8; 22.8)
(a) Vol. **82**, No. 2, p. 553—559 (1984).

H. Süssmann, A. Priemuth, and U. Pröhl
*Doping Properties of Pb and Ge in Bi_2Te_3 and Sb_2Te_3
(14.3 and 15; 22.7)
(a) Vol. **82**, No. 2, p. 561—567 (1984).

V. N. Brudnyi, M. A. Krivov, A. I. Potapov, and Yu. V. Rud
*Electrical Properties of Electron- and Proton-Irradiated GaAs and $ZnGeAs_2$ Solid Solutions
(11 and 14.3; 22.2.1; 22.8)
(a) Vol. **82**, No. 2, p. K191—K194 (1984).

R. Jašinskaitė, A. Sakalas, A. Martinaitis, and P. Höschl
*Native Acceptor Defects in Chlorine Doped Single Crystals of CdTe
(14.3; 13.4; 22.4.3)
(a) Vol. **82**, No. 2, p. K199—K204 (1984).

M. Beiler, M. Mocker, and O. Ziep
*Theory of Second-Order Auger Recombination in Strong Degenerate Small-Gap Semiconductors. Application to PbSSe
(13.1 and 14.3; 20.3; 22.8)
(b) Vol. **123**, No. 1, p. 247—255 (1984).

Yu. V. Vorobev, E. I. Tolpygo, and M. K. Sheinkman
*The Auger Processes with Energy Transfer to the Bound Charge Carriers in Multi-Valley Semiconductors
(13.4 and 14.3; 22.2.1)
(b) Vol. **123**, No. 1, p. 295—308 (1984).

G. J. Ashwell and I. W. Nowell
Electrical, Magnetic, and Structural Characterisation of $[1,2\text{-Bis}(1\text{-ethyl-4-pyridinium})\text{ethane}]^{2+}$ $(7,7,8,8\text{-Tetracyano-p-quinodimethanide})_4^{2-}$
(1.1; 14.3; 18.1; 22.9)
(a) Vol. **83**, No. 1, p. 39—45 (1984).

A. YA. NEIMAN, A. A. AFANASIEV, L. M. FEODOROVA, V. T. GABRIELIAN, and S. M. KARAGEZIAN
*Deviation from Stoichiometry and Electron Transfer in $PbMoO_4$
(10.2 and 14.3; 22.8.1)
(a) Vol. **83**, No. 1, p. 153—158 (1984).

T. SŁUPKOWSKI
*Electrical Conductivity of Mixtures of Conducting and Insulating Particles
(14.3; 22.9)
(a) Vol. **83**, No. 1, p. 329—333 (1984).

S. K. PLACHKOVA
*Thermoelectric Figure of Merit of the System $(GeTe)_{1-x}(AgSbTe_2)_x$
(14.3 and 15; 8; 22.8)
(a) Vol. **83**, No. 1, p. 349—355 (1984).

J. ROSENZWEIG, J. ZHANG, and U. BIRKHOLZ
*Annealing Effects on Electrical Conductivity and Thermoelectric Power of Sintered PbTe
(14.3; 15; 22)
(a) Vol. **83**, No. 1, p. 357—363 (1984).

A. P. TYUTNEV, V. S. SAENKO, A. I. KARPECHIN, G. S. MINGALEEV, V. I. ARKHIPOV, A. I. RUDENKO, and A. V. VANNIKOV
*Radiation-Induced Conductivity in Polymers under Continuous Irradiation
(11 and 14.3; 22.9)
(a) Vol. **83**, No. 1, p. 365—373 (1984).

V. N. BRUDNYI, V. A. NOVIKOV, A. D. POGREBNYAK, and YU. P. SUROV
Positron Annihilation in Electron-Irradiated p-$ZnGeP_2$ Compound
(10.2 and 11; 14.3; 22.8)
(a) Vol. **83**, No. 1, p. K35—K38 (1984).

S. A. AZIMOV, P. M. KARAGEORGY-ALKALAEV, A. YU. LEIDERMAN, and V. S. RUBIN
*Computer Simulation of Generation–Recombination Currents in Amorphous Silicon p–n Diode Structures
(2 and 14.3; 14.3.4; 22.1.2)
(a) Vol. **83**, No. 1, p. K81—K86 (1984).

V. K. ARORA and H. N. SPECTOR
*Quantum Freeze-Out of Carriers in Semimetallic and Semiconducting Thin Wires
(14.1 and 14.3; 21.7)
(b) Vol. **123**, No. 2, p. 747—753 (1984).

H.-U. TSCHIRNER and K. UHLIG
*Some Remarks on Activation Energies Observed in the Liquid Semiconductor Ga_2Te_3
(2 and 14.3; 13.4; 22)
(b) Vol. **123**, No. 2, p. K177—K181 (1984).

H. LEMKE
Properties of Substitutional Manganese in Silicon
(13.4; 14.3; 22.1.2)
(a) Vol. **83**, No. 2, p. 637—643 (1984).

G. J. ASHWELL
*Electrical, Magnetic, and Structural Characterisation of [1,2-Bis(1-methyl-4-pyridinium)ethylene]$^{2+}$-(TCNQ)$_4^{2-}$ (H_2O) and the Properties of Two Isostoichiometric TCNQ Salts
(14.3 and 18.1; 1.1; 15; 22.9)
(a) Vol. **83**, No. 2, p. 669—673 (1984).

S. GUHA, C. BASU, and M. BOSE
*Parametric Excitation of Electron-Acoustic Waves in Magnetised Semiconductor Plasmas
(7 and 14.3; 22.2.3)
(a) Vol. **83**, No. 2, p. 675—680 (1984).

M. RICK, J. ROSENZWEIG, and U. BIRKHOLZ
*Anisotropy of Electrical Conductivity in Si_2Te_3
(14.3; 22)
(a) Vol. **83**, No. 2, p. K183—K186 (1984).

C. BUTTI, A. RAYMOND, and G. MASSÉ
*Conductivity Anisotropy in α-HgS
(14.3; 22.4.1)
(a) Vol. **83**, No. 2, p. K199—K202 (1984).

R. HERRMANN, G. NACHTWEI, and W. KRAAK
*On the Barrier Height in Grain Boundaries of InSb Bicrystals
(14.3; 13.1 22.2.3)
(a) Vol. **83**, No. 2, p. K207—K210 (1984).

W. HUHN and A. STAHL
Self-Consistent Field Theory Applied to the Semiconductor Band Edge
(13.1 and 13.5; 14.3; 20; 22)
(b) Vol. **124**, No. 1, p. 167—177 (1984).

A. CONAN, A. BONNET, M. ZOAETER, and D. RAMOUL
*Dependence of the Total Mobility in a One-Band Model. Application to n-Type $MoTe_2$
(14.3 and 15; 22)
(b) Vol. **124**, No. 1, p. 403—410 (1984).

M. STRORDEUR
*Investigation of the Weak-Field Charge Transport in Semiconducting V_2–VI_3 Compounds with Trigonal Symmetry (I): Expanded Model for the Transport Coefficients
(14.3; 15; 22.7)
(b) Vol. **124**, No. 1, p. 439—446 (1984).

D. V. GITSU and F. M. MUNTYANU
*Magnetoresistance of $Bi_{1-x}Sb_x$ Alloys ($0.08 \leqq x \leqq 0.15$) in Transversal Quantizing Magnetic Fields
(14.3; 18; 22.7)
(b) Vol. **124**, No. 1, p. K87—K91 (1984).

H. NAKANISHI, S. ENDO, and T. IRIE
*Piezoresistance in $CdIn_2S_4$
(14.3; 13.1; 22.8)
(b) Vol. **124**, No. 1, p. K101—K104 (1984).

W. SIEGEL and G. KÜHNEL
*Influence of the Temperature Dependence of the Thermal Impurity Activation Energy on the Analysis of Hall Measurements on GaP
(13.4 and 14.3; 22.2.2)
(a) Vol. **84**, No. 1, p. 251—262 (1984).

V. M. BABICH, N. P. BARAN, V. L. BORBLIK, YU. P. DOTSENKO, V. B. KOVALCHUK, and V. A. SHERSHEL
Specific Features of Piezogalvanomagnetic Effects in Oxygen-Containing Silicon Crystals in the Presence of Thermal Donors-I and -II
(13.4; 14.3; 22.1.2)
(a) Vol. **84**, No. 1, p. 263—268 (1984).

G. S. MINGALEEV, A. P. TYUTNEV, V. I. ARKHIPOV, A. I. RUDENKO, A. V. VANNIKOV, and V. S. SAENKO
*Computer Simulation and Analytical Solutions in the Study of the Transient Radiation-Induced Conductivity in Polymers
(14.3; 11; 22.9)
(a) Vol. **84**, No. 1, p. 327—336 (1984).

A. A. GHANI, S. A. MAZEN, and A. H. ASHOUR
*The Annealing Effect on the Conduction Mechanism of Copper Ferrite
(14.3 and 15; 22.8.2)
(a) Vol. **84**, No. 1, p. 337—341 (1984).

I. N. VORONIN
*Two Kinds of Self-Oscillations and the Flip-Flop Effect in a Semiconductor with Shallow Donors
(14.3)
(b) Vol. **124**, No. 2, p. 793—798 (1984).

M. STORDEUR and G. SIMON
*Investigation of the Weak-Field Charge Transport in Semiconducting V_2–VI_3 Compounds with Trigonal Symmetry (II): Interpretation of the Weak-Field Charge Transport in Sb_2Te_3 Single Crystals
(14.3; 13.1; 15; 22.7)
(b) Vol. **124**, No. 2, p. 799—806 (1984).

P. F. LUGAKOV and V. V. LUKYANITSA
Peculiarities of Radiation Defect Formation and Annealing in n-Si Due to Their Interaction with Each Other and Defect Clusters
(10.2 and 11, 14.3; 22.1.2)
(a) Vol. **84**, No. 2, p. 457—464 (1984).

A. P. TYUTNEV, P. I. DUBENSKOV, V. S. SAENKO, and A. V. VANNIKOV
*Radiation-Induced Conductivity of Poly-(N-Vinylcarbazole) and Poly-(N-Epoxypropylcarbazole)
(11 and 14.3; 22.9)
(a) Vol. **84**, No. 2, p. 585—596 (1984).

H. S. MAITI, S. RAJENDRAN, and V. SITAKARA RAO
*Electrical Conductivity and Thermoelectric Power Studies of Iron(III)–Nickel(II) Mixed Oxides
(14.3; 15; 22.6.1)
(a) Vol. **84**, No. 2, p. 631—638 (1984).

I. I. Zyatkov, V. P. Miroshkin, and Ya. I. Panova
*High-Frequency Conductivity of Manganese–Zinc Ferrites
(14.3; 22.8.2)
(a) Vol. **84**, No. 2, p. 645—650 (1984).

R. Satyanarayana and S. Ramana Murthy
*Electrical Conductivity of Barium–Strontium Ferrites
(14.3; 15; 22.8.2)
(a) Vol. **84**, No. 2, p. 655—658 (1984).

A. V. Salker, E. B. Mirza, V. V. Deshpande, and H. V. Keer
Thermodynamic Analysis of the Phase Transition in V_4O_7
(1.2; 8; 14.1; 14.3; 18.1; 22.6)
(a) Vol. **84**, No. 2, p. K105—K108 (1984).

P. Gille and M. Schenk
Structural Incorporation and Electrical Efficiency of Tl in PbTe
(10.2; 14.3; 22)
(a) Vol. **84**, No. 2, p. K121—K124 (1984).

P. Lošťák, J. Horák, and L. Koudelka
Some Physical Properties and Point Defects in $Bi_2Te_{3-x}S_x$ Mixed Crystals
(20.1; 13.4; 14.3; 22.7.1)
(a) Vol. **84**, No. 2, p. K143—K147 (1984).

N. N. Koren, V. I. Levchenko, V. V. Dikareva, and A. V. Kazushchik
*Physical Properties of Germanium Telluride Crystals Grown from Gas Phase
(14.3 and 15; 10.2; 22)
(a) Vol. **84**, No. 2, p. K173—K176 (1984).

A. J. Szadkowski and A. Lubomirska-Wittlin
*Compensation of p-Type $Cd_{1-x}Mn_xTe$ by Diffusion of Ag
(14.3 and 16; 9; 13.4; 22.4.4)
(a) Vol. **84**, No. 2, p. K181—K184 (1984).

E. E. Matyas and T. E. Borisenko
*Electrical Properties of PbTe–SnS Solid Solutions
(14.3; 22)
(a) Vol. **84**, No. 2, p. K193—K196 (1984).

R. Herrmann, W. Kraak, and M. Gliński
*Quantized Hall Effect in the n-Inversion Layer in InSb Grain Boundaries
(13.3 and 14.3; 22.2.3)
(b) Vol. **125**, No. 1, p. K85—K88 (1984).

G. A. Adegboyega
*Electrical Measurements on Silver-Diffused GaAs
(13.4 and 14.3; 22.2.1)
(a) Vol. **85**, No. 1, p. 215—218 (1984).

Y. Kamiura and F. Hashimoto
Origin of a Shallow Acceptor in Quenched Germanium
(10.2 and 13.4; 14.3; 22.1.1)
(a) Vol. **85**, No. 1, p. 227—234 (1984).

I. Diaconu, D. D. Eley, and M. R. Willis
*AC Conductivity in Single Crystals and Compactions of TCNQ Complex Salts
(14.3; 22.9)
(a) Vol. **85**, No. 1, p. 283—289 (1984).

M. Messaoudi and J. Duran
*Conductivity in One-Dimensional Disordered Cs_2TCNQ_3 Salts
(14.3; 22.9)
(a) Vol. **85**, No. 1, p. 291—296 (1984).

M. Adolf, J. Rosenzweig, and U. Birkholz
Bipolar Photoconductivity and Hall Effect of the Acoustoelectric Current in CdTe
(7 and 16; 14.3; 22.4.3)
(a) Vol. **85**, No. 1, p. K81—K84 (1984).

I. N. Grincheshen, N. S. Popovich, and A. A. Shtanov
*Relaxation Peculiarities in $TlSbSe_2$ Crystals
(14.3 and 16; 22.8)
(a) Vol. **85**, No. 1, p. K85—K88 (1984).

C. Fau, J. F. Dame, M. De Carvalho, J. Calas, M. Averous, and B. A. Lombos
*Three-Band Model Applied to Narrow-Gap HgCdTe
(14.3; 13.1; 22.4.4)
(b) Vol. **125**, No. 2, p. 831—838 (1984).

V. A. Kasiyan, D. D. Nedeoglo, A. V. Simashkevich, and I. N. Timchenko
*Hopping Conductivity and Electron Localization in Heavily Doped ZnSe
(13.4 and 14.3; 22.4.2)
(b) Vol. **125**, No. 2, p. 869—878 (1984).

L. V. Vassilev and St. K. Dimitrova
*Observation of Anderson-Like Transition in the System $Pb_{1-x}Ge_xTe$
(14.3 and 15; 1.2; 22.3.1)
(b) Vol. **125**, No. 2, p. K177—K180 (1984).

K. Tennakone and M. Kahanda
*Conductivity Transition in Cesium Doped Iodine
(14.3 and 14.4; 22.5)
(b) Vol. **125**, No. 2, p. K181—K183 (1984).

B. Lorenz
*Correct Low Temperature Expansion for the DC Conductivity of Magnetite
(14.3; 22.6)
(b) Vol. **125**, No. 2, p. K193—K195 (1984).

V. V. Emtsev, Yu. N. Daluda, P. Gaworzewski, and K. Schmalz
Electrical Properties of Thermal Donors Formed in CZ-Si during Heat Treatment at 450 °C
(13.4, 14.3; 20.1; 22.1.2)
(a) Vol. **85**, No. 2, p. 575—584 (1984).

A. P. Tyutnev, V. S. Saenko, A. F. Dunaev, E. D. Pozhidaev, and A. V. Vannikov
*Temperature Dependence of Transient Radiation-Induced Conductivity in Polymers
(14.3; 11; 22.9)
(a) Vol. **85**, No. 2, p. 591—602 (1984).

W. Hergert and L. Pasemann
Theoretical Study of the Information Depth of the Cathodoluminescence Signal in Semiconductor Materials
(20.3; 10.2; 14.3; 22)
(a) Vol. **85**, No. 2, p. 641—648 (1984).

H. Lemke
Properties of Energy Levels of Cobalt in Silicon
(13.4; 14.3; 22.1.2)
(a) Vol. **85**, No. 2, p. K133—K136 (1984).

H. Lemke
Properties of Energy Levels of Rhodium and Iridium in Silicon
(13.4; 14.3; 22.1.2)
(a) Vol. **85**, No. 2, p. K137—K140 (1984).

G. Parthasarathy and E. S. R. Gopal
*Unusual Effect of Tellurium Doping on the Electrical Properties of Bulk Amorphous Selenium
(2 and 14.3; 22.1.3)
(a) Vol. **85**, No. 2, p. K165—K167 (1984).

W. Kraak, U. Schaller, and R. Herrmann
*Sensitive Pressure Gauge on the Basis of Pure n-InSb
(12.2 and 14.3; 22.2.3)
(a) Vol. **85**, No. 2, p. K183—K187 (1984).

I. S. Smirnova
*Dislocations in $A^{II}B^{VI}$ and A^{III}/B^{V} Piezoelectric Crystals
(10.2 and 14.3; 22.2; 22.4)
(b) Vol. **126**, No. 1, p. 177—189 (1984).

R. A. Vardanian and G. G. Kirakosian
*Electron Scattering by a Charged Dislocation
(10.2 and 14.3; 22.1.1)
(b) Vol. **126**, No. 1, p. K83—K86 (1984).

J. Dominec
*Semiconductor–Semimetal Transition in Bi–Sb Alloys Observed in Ultrasound Attenuation
(7 and 14.3; 14.1; 22.7)
(b) Vol. **126**, No. 1, p. K87—K90 (1984).

A. A. Wafik and S. A. Mazen
Barkhausen Jumps and Resistivity of Cu–Cd Ferrites
(18.3; 14.3; 22.8.2)
(a) Vol. **86**, No. 1, p. 271—274 (1984).

D. I. Aladashvili, L. Konczewicz, and S. Porowski
Studies of the Deep Levels in p-Type InSb under Pressure
(13.4; 14.3; 22.2.3)
(a) Vol. **86**, No. 1, p. 301—308 (1984).

H. Lemke
On the Possible Existence of „Repulsive" Centres with Large Capture Cross Section in Silicon
(13.4; 14.3; 22.1.2)
(a) Vol. **86**, No. 1, p. K39—K43 (1984).

O. A. Golikova, N. E. Solovev, Ya. A. Ugai, and V. A. Feigelman
*Electrophysical Properties of α-Rhombohedral Boron
(14.3 and 15; 22.1)
(a) Vol. **86**, No. 1, p. K51—K54 (1984).

A. M. Konin and A. P. Saščiuk
*Thermogradient Magnetoconcentration Effect
(14.3; 22.1.1; 22.2.3)
(a) Vol. **86**, No. 1, p. K55—K58 (1984).

P. I. Baranskii, T. G. Grishchenko, V. V. Savyak, and Yu. V. Simonenko
*Temperature Dependence of the Thermoelectric Characteristics of n-Si Elastically Strained in the [100] Direction
(14.3 and 15; 22.1.2)
(a) Vol. **86**, No. 1, p. K63—K67 (1984).

S. Shigetomi, T. Ikari, Y. Koga, and S. Shigetomi
*Annealing Behaviour of Electrical Properties of n-InSe Single Crystals
(14.3; 13.4; 22.3)
(a) Vol. **86**, No. 1, p. K69—K72 (1984).

V. M. Yartsev
*Charge Transfer and Electron–Molecular Vibration Coupling in Tetramerized Quasi-1D Semiconductors
(14.3; 6; 20.1; 22.9)
(b) Vol. **126**, No. 2, p. 501—510 (1984).

E. Kaden and H.-L. Günter
The Thermoelectric Power of p-Ge at Low Temperatures
(6 and 15; 14.3; 22.1.1)
(b) Vol. **126**, No. 2, p. 733—740 (1984).

L. Borisova
Density-of-States Effective Mass in Solid Solutions of the $(Ag_{x/2}Pb_{1-x}Sb_{x/2})Te$ Type
(13.1; 14.3; 15; 22)
(b) Vol. **126**, No. 2, p. K155—K158 (1984).

V. M. BABICH, N. P. BARAN, A. A. BUGAI, YU. P. DOTSENKO, V. B. KOVALCHUK, and V. A. SHERSHEL
On the Properties of Thermodonors-II in Cz-Si Crystals of High Carbon Content
(13.4; 14.3; 19; 22.1.2)
(a) Vol. **86**, No. 2, p. 679—683 (1984).

R. M. BILBÉ, B. LUNN, J. E. NICHOLLS, and D. H. J. TOTTERDELL
*The Role of Copper as an Impurity in High Resistivity Cadmium Telluride and Its Influence on γ-Ray Spectrometer Performance
(13.4 and 14.3; 20.1; 22.4.3)
(a) Vol. **86**, No. 2, p. 821—826 (1984).

Y. KAMIURA and F. HASHIMOTO
Annealing Behaviour of Shallow Acceptors in Quenched Germanium
(10.2; 13.4; 14.3; 22.1.1)
(a) Vol. **86**, No. 2, p. K95—K97 (1984).

L. G. KHVOSTANTSEV, V. A. SIDOROV, and O. B. TSIOK
Gapless State and Anomalous Behaviour of Thermopower of Black Phosphorus at Pressure up to 12 GPa
(13.1 and 15; 14.3; 22.1)
(a) Vol. **86**, No. 2, p. K99—K104 (1984).

P. G. RUSTAMOV, F. M. SADIGOV, and CH. I. ABILOV
*Electrophysical Properties of $CeBiTe_3$
(14.3 and 15; 22.8)
(a) Vol. **86**, No. 2, p. K113—K115 (1984).

V. RYBKA, F. DEML, and P. HAMPL
*Dual Implantation of Si and P into GaAs
(11 and 14.3; 22.2.1)
(a) Vol. **86**, No. 2, p. K117—K119 (1984).

14.3.1 Films

Review Article

B. M. ASKEROV, B. I. KULIEV, and S. R. FIGAROVA
*Electron Transport Phenomena in Conductive Films
(14.3.1; 6; 13.1; 15; 22)
(b) Vol. **121**, No. 1, p. 11—37 (1984).

Original Papers and Short Notes

G. GEVERS, A. S. BARRIÈRE, J. GRANNEC, L. LOZANO, and B. BLANCHARD
Analysis of FeF_3 and NiF_2 Thin Films by Using RBS, (α, X) Reactions, and SIMS Measurements
(1.3; 14.3.1; 22.5)
(a) Vol. **81**, No. 1, p. 105-122 (1984).

M. P. MIKALKEVIČIUS and K. K. TUMKEVIČIUS
*A Mechanism of Current Percolation in a Heterogeneous Layer of PbTe
(14.3.1 and 16; 22)
(a) Vol. **81**, No. 1, p. K43—K45 (1984).

H. VON BERLEPSCH
*Trapping Model for the Surface Potential Kinetics in HDPE
(14.3.1 and 14.3.2; 22.9)
(a) Vol. **81**, No. 1, p. K63—K65 (1984).

E. GERLACH
*Scattering by Ions near Interfaces
(14.3.1; 22)
(b) Vol. **121**, No. 2, p. 757—764 (1984).

C. CHIANG
*An Equation for Dynamic Switching in Amorphous Thin Films
(2 and 14.3.1)
(a) Vol. **81**, No. 2, p. 563—567 (1984).

S. ORZESZKO, W. BAŁA, K. FABISIAK, and F. ROZPŁOCH
*DC Conductivity and ESR of Hydrogenated Amorphous Carbon Films
(2 and 14.3.1; 19; 22.1)
(a) Vol. **81**, No. 2, p. 579—584 (1984).

F. VÖLKLEIN and E. KESSLER
*A Method for the Measurement of Thermal Conductivity, Thermal Diffusivity, and Other Transport Coefficients of Thin Films
(14.1 and 14.3.1; 8; 15; 21.7)
(a) Vol. **81**, No. 2, p. 585—596 (1984).

N. A. Vlasenko, Z. L. Denisova, and V. S. Khomchenko
Electroluminescence of SiO_x–LnF_3 Thin Films
(14.3 4 and 20.3; 14.3.1; 22.6)
(a) Vol. **81**, No. 2, p. 657—660 (1984).

R. Mach, G. O. Müller, and G. Schulz
On the Time Behaviour of Electron Impact Excitation in Thin Film Electroluminescent Devices
(14.3.4 and 20.3; 14.3.1; 22.4.1)
(a) Vol. **81**, No. 2, p. 723—732 (1984).

S. R. Rao, M. Nagabhooshanam, and V. Hari Babu
*Electrical Properties of Annealed InSb Thin Films
(14.3.1; 2; 22.2.3)
(a) Vol. **81**, No. 2, p. K195—K197 (1984).

D. Nesheva
Studies of Low-Energy Optical Absorption and Gap State Density of a-CdS by the Constant Photocurrent Method
(13.1 and 20.1; 2; 14.3.1; 22.4.1)
(a) Vol. **82**, No. 1, p. 243—248 (1984).

M. Jolivet, M. Cadene, J. M. Abadie, and M. Rolland
*Morphology, Dopant Concentration, and Electrical Properties of H_2SO_4-and $BiCl_3$-Doped Polyacetylene Films
(10.2 and 14.3.1; 22.9)
(a) Vol. **82**, No. 1, p. 257—263 (1984).

M. Jałochowski
*Electron Tunnelling in Pb/Ag Ultrathing Layered Structures
(14.3.1; 14.3.3; 21; 21.6)
(a) Vol. **82**, No. 2, p. 497—502 (1984).

Yu. S. Bulyshev, I. M. Kashirskii, and V. V. Sinitskii
*Photoconductivity and Energy Level Structure of 2,4,7-Trinitro-9-Fluorenone Thin Films
(14.3.1 and 16; 13.1; 22.9)
(a) Vol. **82**, No. 2, p. 537—544 (1984).

E. M. Zub, E. V. Kuchis, and S. V. Sukhvalo
*Electrical and Galvanomagnetic Properties of Films in the $Cu_xFe_{1-x}Cr_2S_4$ System
(14.3.1; 1.5; 22.8.2)
(a) Vol. **82**, No. 2, p. 569—574 (1984).

S. H. Pawar, S. P. Tamhankar, and C. D. Lokhande
*Deposition of Bi_2CdS_4 Films by the Spray Pyrolysis Technique
(14.3.1; 20.1; 22.8)
(a) Vol. **82**, No. 2, p. K195—K198 (1984).

S. Yu. Potapenko and A. M. Satanin
Electron States and Conductivity of Size Quantized Systems with Rough Boundaries
(13.1; 13.3; 14.3.1; 22)
(b) Vol. **123**, No. 1, p. 191—199 (1984).

A. S. Barrière and G. Gevers
*Impurity Conduction along the Least Resistance Paths of Granular Insulating Thin Films
(14.3.1; 13.4; 22.5)
(b) Vol. **123**, No. 1, p. 377—391 (1984).

G. Debiais
*Caractérisation optique et électrique de couches mixtes de sulfure de mercure
(Optical and Electrical Characterization of Mercury Sulfide Mixed Films)
(14.3.1 and 20.1; 13.1; 22.4.4)
(a) Vol. **83**, No. 1, p. 269—278 (1984).

F. S. Nasredinov, A. A. Andreev, A. N. Kurmantaev, A. R. Regel, and P. P. Seregin
*Thin Impurity Atoms in Vacuum Deposited Amorphous Silicon
(13.4 and 14.3.1; 2; 20.1; 22.1.2)
(a) Vol. **83**, No. 1, p. 285—289 (1984).

J. J. Kleperis, P. D. Cikmach, and A. R. Lusis
Colour Centres in Amorphous Tungsten Trioxide Thin Films
(13.4 and 20.1; 14.3.1; 19; 22.6)
(a) Vol. **83**, No. 1, p. 291—297 (1984).

G. S. Nadkarni, S. Radhakrishnan, and N. Sankarraman
*Space Charge Limited Conduction in Thin Film V_2O_5–P_2O_5–ZnO Glasses
(14.3.1 and 14.3.4; 22.6.1)
(a) Vol. **83**, No. 1, p. 335—340 (1984).

R. M. Mehra, Hemant Kumar, Surinder Koul, and P. C. Mathur
*Thickness Dependence of DC Conductivity of Amorphous Se and Binary Amorphous Se–Te, Se–Ge, and Se–Sb Films
(14.3.1; 2; 22.1.3)
(a) Vol. **83**, No. 1, p. 341—347 (1984).

S. Santhanam and A. K. Chaudhuri
*Electrical Conduction in SnTe Thin Films
(14.3.1; 22.7)
(a) Vol. **83**, No. 1, p. K77—K80 (1984).

A. Turos, H. Frey, O. Meyer, W. Müller, and J. M. Pirrung
*Ion Beam Analysis of Amorphous Silicon Films Produced by Magnetron Sputtering
(2 and 14.3.1; 3; 22.1.2)
(a) Vol. **83**, No. 2, p. 437—443 (1984).

G. Fleury, A. Hamou, C. Lhermitte, and C. Viger
*Bismuth Influence on a-Se Film Crystallization
(2 and 14.3.1; 22.1.3)
(a) Vol. **83**, No. 2, p. K103—K108 (1984).

A. S. Barrière and G. Gevers
*Spatial Distribution of Localized Electronic States Responsible for the Polarization of Granular Insulating Thin Films
(13.4 and 14.3.1; 22.5)
(a) Vol. **84**, No. 1, p. 273—284 (1984).

S. El-Halawany, R. Bacewicz, J. Filipowicz, and R. Trykozko
Photoconductivity of Flash Evaporated $CuInSe_2$ Thin Films
(16; 14.3.1; 20.1; 22.8)
(a) Vol. **84**, No. 1, p. K89—K92 (1984).

L. S. Grigoryan, M. V. Simonyan, and E. G. Sharoyan
*Electrical, Magnetic, and Spectral Properties of the New Class of Charge Transfer Salts Based on Phthalocyanines and Iodine
(2 and 14.3.1; 19; 20.1; 22.9)
(a) Vol. **84**, No. 2, p. 597—606 (1984).

T. K. Bandyopadhyay and S. R. Chaudhuri
*Electronic Properties of Anodic n-CdS Films from Photoelectrochemical Measurements
(14.3.1; 16; 22.4.1)
(a) Vol. **84**, No. 2, p. 651—654 (1984).

G. I. Ibaev, S. I. Kulieva, and T. G. Magerramov
*Frequency Dependences of the Dynamic Current–Voltage Characteristic of the Se-CdSe Heterostructure Based on a Single-Crystal Selenium Film
(14.3.1 and 14.3.3; 22.1.3; 22.4.2)
(a) Vol. **84**, No. 2, p. K185—K188 (1984).

C. Chiang
*Photo-Illumination and Current Excitation of Switching in Amorphous Thin Films
(2 and 14.3.1; 16)
(a) Vol. **84**, No. 2, p. K189—K191 (1984).

J. Kočka and K. Čermák
*Drift Mobility of a-Si:H
(2 and 14.3.1; 14.3.4; 22.1.2)
(b) Vol. **125**, No. 1, p. K89—K91 (1984).

P. Lilley, M. R. Czerniak, and J. E. Nicholls
*Photoluminescence and Electrical Properties of Vapour Phase Epitaxial ZnSe Grown on GaAs
(14.3.1 and 20.3; 13.4; 22.4.2)
(a) Vol. **85**, No. 1, p. 235—242 (1984).

H. Krause, M. Schell, and H.-Chr. Golinsky
*Hopping Conduction in SiO_2 Films Thermally Grown on Phosphorus-Doped Polycrystalline Silicon
(13.4 and 14.3.1; 22.6)
(a) Vol. **85**, No. 2, p. 569—573 (1984).

S. A. Azimov, D. A. Aronov, V. M. Rubinov, and M. Tuichiev
*On the Mechanism of Current Passage in Gallium Phosphide Films Excited by Light or a Beam of Accelerated Electrons
(14.3.1; 11; 22.2.2)
(a) Vol. **85**, No. 2, p. K159—K164 (1984).

B. T. Kolomiets, V. M. Lyubin, M. A. Taguyrdzhanov, and G. Zentai
A Comparative Study of Photoconductivity in Vitreous As_2Se_3 and a-Si:H Thin Films
(16; 2; 14.3.1; 22; 22.1.2)
(a) Vol. **86**, No. 1, p. 383—387 (1984).

DANG TRAN QUAN
*Electrical Properties and Optical Absorption of SnSe Evaporated Thin Films
(14.3.1; 20.1; 22)
(a) Vol. **86**, No. 1, p. 421–426 (1984).

TH. FRAUENHEIM, C. HAMANN, and M. MÜLLER
*Electric Field-Induced Disorder–Order Transition in Organic Polycrystalline Films of Quasi-One-dimensional Lead-Phthalocyanine
(14.3.1; 1.2; 22.9)
(a) Vol. **86**, No. 2, p. 735–747 (1984).

M. VOELSKOW, J. MATTHÄI, and R. KLABES
*Electrical Properties of Ion Implanted and Short Time Annealed Polycrystalline Silicon
(11 and 14.3.1; 22.1.2)
(a) Vol. **86**, No. 2, p. 781–788 (1984).

14.3.2 Surfaces and Interfaces

Original Papers and Short Notes

H. von Berlepsch
*Trapping Model for the Surface Potential Kinetics in HDPE
(14.3.1 and 14.3.2; 22.9)
(a) Vol. **81**, No. 1, p. K63—K65 (1984).

S. Wysocki
Kinetics of the Decay of Surface γ-Conductivity of Magnesium Oxide
(13.4; 14.3.2; 22.6)
(b) Vol. **121**, No. 2, p. 731—736 (1984).

P. K. Basu and J. B. Roy
*Warm Electron Coefficient of Two-Dimensional Electron Gas in Silicon Inversion Layer at Low Temperatures
(14.3.2 and 14.3.4; 22.1.2)
(b) Vol. **121**, No. 2, p. 743—748 (1984).

T. Jung, W. Braune, M. Schnürer, and J. Schulze
*Film and Interface Analysis of InSb MOS Structures
(1.6 and 14.3.2; 22.2.3)
(a) Vol. **81**, No. 2, p. 463—468 (1984).

Chu Hao, J. Zimmermann, M. Charef, R. Fauquembergue, and E. Constant
*A Semi-Classical Model for Simulating Inversion Carrier Transport in Si MOS Devices
(14.3.2 and 14.3.3; 22.1.2)
(a) Vol. **81**, No. 2, p. 569—577 (1984).

S. J. Lee and Chung-In Um
Impurity and Thickness Effects on the Cyclotron Resonance Linewidth of Electrons in Si Inversion Layers
(13.3; 14.3.2; 19; 22.1.2)
(b) Vol. **122**, No. 1, p. K73—K78 (1984).

S. Sikorski
*Theory of *C–U* Profiling of a Nonuniformly Doped Semiconductor
(14.3.2; 14.3.3)
(a) Vol. **82**, No. 1, p. 265—274 (1984).

N. Keddar, D. L. Kirk, and S. G. Johnson
*The Surface Degradation of the β″-Al_2O_3 Solid Electrolyte Utilised in Na–S Battery Cells
(1.6 and 14.3.2; 22.6)
(a) Vol. **82**, No. 1, p. 327—334 (1984).

V. I. Kuznetsov, P. F. Lugakov, and A. V. Tsikunov
*Interaction of Radiation Defects with the Surface of Silicon
(11 and 14.3.2; 13.4; 22.1.2)
(a) Vol. **82**, No. 2, p. 467—473 (1984).

S. K. Maksimov
*Correlation between Thermodynamic Instability of $A^{III}B^{V}$ Heterocompositions and Degradation of Devices Based on These Materials
(10.2 and 14.3.2; 9; 22.2.4)
(a) Vol. **83**, No. 2, p. 685–692 (1984).

P. Středa
*Critical Current of the Quantum Hall Regime
(14.3.2 and 14.3.4; 22.2.1)
(b) Vol. **124**, No. 1, p. K97—K99 (1984).

W. Possart and A. Röder
*Measurement of Electrical Potential Distribution in a Polymer near the Contact to a Metal by Means of Scanning Electron Microscopy
(14.3.2; 22.9)
(a) Vol. **84**, No. 1, p. 319—325 (1984).

E. Nebauer and M. Trapp
Ge and Au Profiles in GaAs Produced from AuGe Contacts and Studied by SIMS
(9; 14.3.2; 22.2.1)
(a) Vol. **84**, No. 1, p. K39—K42 (1984).

S. Alexandrova, A. Szekeres, and L. Kassamakova
*RF Plasma Induced Lateral Nonuniformities in Si–SiO_2 Structures
(13.3 and 14.3.2; 22.1.2)
(a) Vol. **84**, No. 2, p. 561—565 (1984).

I. P. Zvyagin
*Planar Conduction in Inhomogeneous Films. Application to Amorphous Silicon
(2 and 14.3.2; 22.1.2)
(a) Vol. **86**, No. 1, p. 375—381 (1984).

A. J. Muhammad, D. V. Morgan, and A. E. Guile
*The Effect of Overlap of Ohmic Contacts on the Characteristics of a MIM Structure Having a Single Discrete Shallow Trap Level
(14.3.2 and 14.3.4)
(a) Vol. **86**, No. 1, p. 439—452 (1984).

D. Ray Chaudhuri, J. B. Roy, and P. K. Basu
Density of the 2D Electron Gas in Modulation Doped GaInAs/AlInAs Layers from a Charge Control Analysis
(14.3.3; 14.3.2; 22.2.4)
(a) Vol. **86**, No. 1, p. K79—K82 (1984).

H. N. Spector and V. K. Arora
*Scattering of Electrons by a Sheet of Impurities in a Quasi-Two-Dimensional Heterostructure
(14.3.2)
(b) Vol. **126**, No. 2, p. K171—K174 (1984).

A. Zoul
*Influence of Deep Surface Levels on *C–U* Schottky Doping Profils of n-GaAs VPE Layers
(14.3.2; 14.3.4; 22.2.1)
(a) Vol. **86**, No. 2, p. 673—678 (1984).

A. P. Tyutnev, V. S. Saenko, A. V. Vannikov, and E. D. Pozhidaev
*Concerning the Radiation-Induced Surface Conductivity in Polymers
(14.3.2; 11; 22.9)
(a) Vol. **86**, No. 2, p. 709—716 (1984).

H. Terletzki, A. Boden, F. Wulf, and W. R. Fahrner
Laterally Inhomogeneous Charge Build-Up in CMOS Inverters during Ionizing Irradiation
(11 and 14.3.3; 14.3.2; 22.1.2)
(a) Vol. **86**, No. 2, p. 789—794 (1984).

14.3.3 Devices. Junctions (Contact Problems See 14.3.4)

Review Articles

D. A. Aronov, P. I. Knigin, Yu. S. Korolev, and V. V. Rubinov
*Exclusion Effect in Semiconductors with Non-Injecting Contacts
(14.3.3 and 14.3.4; 16; 22.1.1; 22.1.2; 22.2.2)
(a) Vol. **81**, No. 1, p. 11—45 (1984).

J. Pułtorak
*The l–h Junction as an Element of the Semiconductor Device
(14.3.3; 22.1.1)
(a) Vol. **82**, No. 1, p. 11—22 (1984).

Original Papers and Short Notes

S. Žurauskas and A. Dargys
Field Ionization of Phosphorus Atoms in Silicon
(13.4; 14.3.3; 22.1.2)
(b) Vol. **121**, No. 1, p. 385—391 (1984).

K. Žďánský and N. T. Thuc Hien
*Analysis of DLTS Curves Corresponding to Non-Exponential Transients
(13.4 and 14.3.3; 22.2.2)
(a) Vol. **81**, No. 1, p. 353—360 (1984).

A. Chandorkar, V. T. Karulkar, A. S. Vengurlekar, and K. V. Ramanathan
*MOS Capacitance–Voltage Characteristics and Dielectric Properties of Ion Implanted Thermal Oxides on Silicon
(14.3.3; 11; 14.4; 22.1.2)
(a) Vol. **81**, No. 1, p. 407—414 (1984).

H. K. Phan, P. H. Binh, and L. H. Phu
*A Simple Method for Investigating Charge Storage Effect in MIS Switching Diodes
(14.3.3; 14.3.4; 22.1.2)
(a) Vol. **81**, No. 1, p. K81—K84 (1984).

J. Walachová and D. Nohavíca
*Erratum to: Determination of p–n Junctions in InP/GaInAsP Laser Heterostructures Using a Profilometer
(14.3.3; 22.2)
(a) Vol. **81**, No. 1, p. K91 (1984).

Chu Hao, J. Zimmermann, M. Charef, R. Fauquembergue, and E. Constant
*A Semi-Classical Model for Simulating Inversion Carrier Transport in Si MOS Devices
(14.3.2 and 14.3.3; 22.1.2)
(a) Vol. **81**, No. 2, p. 569—577 (1984).

A. S. Popov
*Degradation Model of Red GaP LEDs
(13.4 and 14.3.3; 20.3; 22.2.2)
(a) Vol. **81**, No. 2, p. 669—674 (1984).

S. Sikorski
Theory of *C–U* Profiling of a Nonuniformly Doped Semiconductor
(14.3.2; 14.3.3)
(a) Vol. **82**, No. 1, p. 265—274 (1984).

S. S. Simeonov and M. D. Ivanovich
*Recombination Current in Abrupt Semiconductor p–n Junctions
(14.3.3; 22)
(a) Vol. **82**, No. 1, p. 275—284 (1984).

G. Arnold and G. Oelgart
*Ionization Rates for Electrons and Holes in GaAsP
(14.3.3; 14.3.4; 22.2.4)
(a) Vol. **82**, No. 1, p. 319—325 (1984).

Ya. I. Verzimacha, A. V. Kovalshuk, M. V. Kurik, C. Hamann, and A. Mrwa
*Solar Cells of Schottky Type with Ni–PbPc Interface
(14.3.3 and 16; 22.9)
(a) Vol. **82**, No. 1, p. K111—K115 (1984).

K. Bhattacharyya and P. K. Basu
*Variational Calculation of Ground Subband Energy of Two-Dimensional Electron Gas in p–n^+–p Structure
(13.1 and 14.3.3; 22.2.1)
(b) Vol. **122**, No. 2, p. K191—K194 (1984).

M. Jałochowski
Electron Tunnelling in Pb/Ag Ultrathin Layered Structures
(14.3.1; 14.3.3; 21; 21.6)
(a) Vol. **82**, No. 2, p. 497—502 (1984).

M. Hrobár, M. Grendel, and I. Thurzo
*Current DLTS Spectra of MIS Structures Due to Dielectric Polarization of the Insulator
(14.3.3; 14.3.4)
(a) Vol. **82**, No. 2, p. 519—525 (1984).

E. Chimczak, W. S. Gordon, and M. Bertrandt-Żytkowiak
*Model of the Electroluminescence in Short-Pulse-Excited Thin-Film Structures Based on ZnS:Mn
(14.3.3 and 20.3; 13.4; 22.4.1)
(a) Vol. **82**, No. 2, p. 527—531 (1984).

U. Woggon, K. Wandel, and H. Weinert
*Temperature Dependence of Minority Carrier Lifetime in VPE–GaP:N,Te
(14.3.3; 22.2.4)
(a) Vol. **82**, No. 2, p. K205—K208 (1984).

I. Zólomy
*Negative Differential Resistance in the I–U Characteristics of Surface Oxide Transistors
(14.3.3; 14.3.4; 22.1.2)
(a) Vol. **82**, No. 2, p. K209—K211 (1984).

G. Kaden
*Zur Berücksichtigung der Oberflächenbandverbiegung im Silizium bei Photoinjektion in MOS-Strukturen
(Consideration of Surface Band Bending in Silicon in Case of Internal Photoemission in MOS Structures)
(14.3.3 and 14.3.4; 13.3; 22.1.2)
(a) Vol. **83**, No. 1, p. 375—386 (1984).

G. Oelgart and G. Arnold
*Internal Electric Field-Enhanced Impurity Diffusion in $GaAs_{1-x}P_x$
(13.4 and 14.3.3; 9; 22.2.4)
(a) Vol. **83**, No. 1, p. 393—402 (1984).

P. Mialhe, A. Khoury, and J. P. Charles
*A Review of Techniques to Determine the Series Resistance of Solar Cells
(14.3.3 and 16; 22.1.2)
(a) Vol. **83**, No. 1, p. 403—409 (1984).

S. Kashchieva, P. Danesh, and A. Dyakov
RF Annealing of the Implantation-Induced Defects in Silicon Using Hydrogen Plasma
(11 and 13.4; 14.3.3; 22.1.2)
(a) Vol. **83**, No. 1, p. 411—417 (1984).

E. Ejimanya and D. V. Morgan
*Damage Induced Degradation of Metal Contacts on Gallium Arsenide
(14.3.3; 11; 22.2.1)
(a) Vol. **83**, No. 1, p. K87—K92 (1984).

P. C. Dhanasekaran, P. R. Vaya, and B. S. V. Gopalam
*Spectral Sensitivity Calculations on n^+–p and n^+–p–p^+ Silicon Solar Cells
(14.3.3 and 16; 22.1.2)
(a) Vol. **83**, No. 2, p. 693—700 (1984).

N. G. Filonov, N. K. Maksimova, A. P. Vyatkin, I. D. Romanova, and A. M. Mysik
*Transport Mechanism in GaAs Schottky Diodes. Deep Centres Effects
(14.3.3; 13.4; 22.2.1)
(a) Vol. **83**, No. 2, p. 701—708 (1984).

P. J. Reuters, L. DoThanh, and P. Balk
*Determination of Carrier Type in MIS Structures with Pyrolytic AlN Insulator Films
(14.3.3; 14.4; 22; 22.1.2)
(a) Vol. **83**, No. 2, p. 719—727 (1984).

D. T. Rasulov and R. Akhmedov
*Thin-Film Solar Cell on the Basis of Polycrystalline Cadmium Telluride
(14.3.3 and 16; 22.4.1; 22.4.3; 22.4.4)
(a) Vol. **83**, No. 2, p. K211—K213 (1984).

E. A. Akopyan, A. Sh. Mekhtiev, and A. L. Poimanov
*Photon Storage Mode in a Forward Biased p–n Junction
(14.3.3; 22)
(a) Vol. **83**, No. 2, p. K215—K217 (1984).

L. Pasemann
Theoretical Study of the Temperature Dependence of EBIC Contrast from Individual, Surface-Parallel Dislocations in a Schottky Diode
(1.4 and 10.2; 14.3.3; 22.1.2)
(a) Vol. **84**, No. 1, p. 133—142 (1984).

E. Bugiel, M. Kittler, A. Borchardt, and H. Richter
On the Gettering Efficiency of Crystal Defects in Silicon
(10.2; 14.3.3; 22.1.2)
(a) Vol. **84**, No. 1, p. 143—147 (1984).

D. A. Aronov, P. I. Knigin, and B. R. Mamatkulov
*Exclusion Effect Quenching in Compensated Semiconductors under the Action of Impurity Illumination Generating Majority Current Carriers
(14.3.3 and 14.3.4; 16; 22)
(a) Vol. **84**, No. 1, p. 301—310 (1984).

D. P. Joshi and R. S. Srivastava
*Effective Diffusion Length of Minority Carriers in Polycrystalline Silicon
(14.3.3 and 16; 1.4; 22.1.2)
(a) Vol. **84**, No. 1, p. 311—317 (1984).

T. M. Razykov and B. Kh. Kadirov
*Energy Band Diagrams of Cu_2S–$Zn_xCd_{1-x}S$ ($0 \leqq x \leqq 1$) Heterojunctions
(13.1 and 14.3.3; 22.4.4)
(a) Vol. **84**, No. 1, p. K71—K74 (1984).

S. Kh. Milshtein, D. C. Joy, S. D. Ferris, and L. C. Kimerling
Defect Characterization Using SEM-CCM. Relative Contrast Measurements
(1.4 and 10.2; 14.3.3; 22.1.2)
(a) Vol. **84**, No. 2, p. 363—369 (1984).

P. Krispin and J. Maege
*V_P Related Defects in Diffused GaP:N Diodes
(13.4 and 14.3.3; 22.2.2)
(a) Vol. **84**, No. 2, p. 573—584 (1984).

M. Aslam, R. Singh, and P. Balk
*Nature of Electron and Hole Traps in MOS-Systems with Poly-Si Electrode
(14.3.3; 13.4; 22.1.2)
(a) Vol. **84**, No. 2, p. 659—668 (1984).

V. M. Ivastchenko, R. V. Konakova, and V. V. Mitin
*Peculiarities of Reverse Current–Voltage Characteristics of GaAs Lo–Hi–Lo Impatt Diodes
(14.3.3 and 14.3.4; 22.2.1)
(a) Vol. **84**, No. 2, p. 669—675 (1984).

F. Thuselt and K. Unger
High Carrier Densities and Gap Shrinkage in p–i–n-Type Structures
(13.5; 14.3.3; 22.2.2)
(a) Vol. **84**, No. 2, p. 677—691 (1984).

J. Šikula, P. Vašina, V. Musilová, Z. Chobola, and M. Rothbaufr
*1/*f* Noise in GaAs Schottky Diodes
(14.3.3; 22.2.1)
(a) Vol. **84**, No. 2, p. 693—696 (1984).

S. I. Radautsan, S. A. Rebrov, and A. E. Tsurkan
*Some Properties of ZnTe–InP Heterojunctions
(14.3.3; 20.3; 22.2.2; 22.4.3)
(a) Vol. **84**, No. 2, p. K169—K171 (1984).

G. I. Ibaev, S. I. Kulieva, and T. G. Magerramov
*Frequency Dependences of the Dynamic Current–Voltage Characteristic of the Se–CdSe Heterostructure Based on a Single-Crystal Selenium Film
(14.3.1 and 14.3.3; 22.1.3; 22.4.2)
(a) Vol. **84**, No. 2, p. K185—K188 (1984).

J. L. Lagzdons, G. E. Bajars, and A. R. Lūsis
*Modelling of the Solid State Electrochromic System $WO_3/HSbO_3 \cdot 2\ H_2O/Ni(OH)_2$
(14.3.3 and 14.3.4; 14.4; 22.8.1)
(a) Vol. **84**, No. 2, p. K197—K200 (1984).

K. Žďánský and N. T. Thuc Hien
*Analysis of DLTS Curves of Aggregated Deep Level Impurities
(13.4 and 14.3.3; 22.4.3)
(a) Vol. **85**, No. 1, p. 219—226 (1984).

J. Pułtorak
*Current–Voltage Characteristic of the p–n Junction with an Exponential Impurity Distribution at Its Base Contact
(14.3.3; 22)
(a) Vol. **85**, No. 1, p. K93—K96 (1984).

R. Pickenhain, K. Jacobs, W. Seifert, V. Chernyi, and S. Brehme
*Green and Yellow Light Emitting Diodes Produced from Vapour Phase Epitaxial GaP:N (I): Deep Levels in Diodes Produced by either Diffusion or Implantation of Zinc
(14.3.3 and 20.3; 13.4; 22.2.2)
(a) Vol. **85**, No. 2, p. 627—639 (1984).

P. SIRCAR and M. AUBIN
*Pulsed UV Laser Processed Gold and Aluminium Ohmic Contacts on n^+-GaAs
(14.3.3; 14.3.4; 22.2.1)
(a) Vol. **85**, No. 2, p. 649—654 (1984).

N. P. ESINA, N. V. ZOTOVA, S. A. KARANDASHEV, and G. M. FILARETOVA
*Tunneling Surface Recombination in MS Structures on p-InAs
(14.3.3 and 20.3; 16; 22.2.1)
(a) Vol. **85**, No. 2, p. 655—660 (1984).

S. KAČIULIS, K. KIBICKAS, and J. PARŠELIŪNAS
*Dependence of *I–U* Characteristics of GaAs n^+–n–n^+ Diodes on the Active Region Length
(14.3.3 and 14.3.4; 22.2.1)
(a) Vol. **85**, No. 2, p. K179—K182 (1984).

W. BRAUNE, M. SCHNÜRER, N. KUBICKI, and R. HERRMANN
**C–U* Measurements of Anodized InSb MOS Structures at 77 and 4.2 K
(13.3 and 14.3.3; 22.2.3)
(a) Vol. **86**, No. 1, p. 427—431 (1984).

W. TORBICZ, D. SOBCZYNSKA, A. OLSZYNA, G. FORTUNATO, and A. D'AMICO
*A Stable Pd–BN–Si_3N_4–SiO_2–Si FET for Hydrogen Detection
(14.3.3; 22.1.2)
(a) Vol. **86**, No. 1, p. 453—459 (1984).

D. RAY CHAUDHURI, J. B. ROY, and P. K. BASU
*Density of the 2D Electron Gas in Modulation Doped GaInAs/AlInAs Layers from a Charge Control Analysis
(14.3.3; 14.3.2; 22.2.4)
(a) Vol. **86**, No. 1, p. K79—K82 (1984).

S. A. GRUSHA, A. M. EVSTIGNEEV, K. A. ISMAILOV, R. V. KONAKOVA, A. N. KRASIKO, I. K. SINISHCHUK, YU. A. TKHORIK, and V. I. FAINBERG
*Electric Field Effect on Radiation-Induced Gettering of Defects in n^+–n–n^{++}-GaAs
(11 and 14.3.3; 22.2.1)
(a) Vol. **86**, No. 1, p. K83—K86 (1984).

S. B. SEVASTIANOV, N. N. GERASIMENKO, and N. V. VERSHININA
*Inhomogeneities in Irradiated SiO_2–Si Structures
(11 and 14.3.3; 14.3.4; 22.1.2)
(a) Vol. **86**, No. 2, p. 717—727 (1984).

H. TERLETZKI, A. BODEN, F. WULF, and W. R. FAHRNER
*Laterally Inhomogeneous Charge Build-Up in CMOS Inverters during Ionizing Irradiation
(11 and 14.3.3; 14.3.2; 22.1.2)
(a) Vol. **86**, No. 2, p. 789—794 (1984).

I. THURZO and E. PINČÍK
*Some Peculiar Features of Small-Signal Charge DLTS Response of GaAs MOS Capacitors
(13.3 and 14.3.3; 22.2.1)
(a) Vol. **86**, No. 2, p. 795—804 (1984).

G. Oelgart, B. Rheinländer, H. Haefner, and R. Pickenhain
*Degradation of Green Light Emitting Diodes LPE-GaP:N (I): Electron Beam Characterization and Deep-Level Spectroscopy
(14.3.3 and 20.3; 22.2.2)
(a) Vol. **86**, No. 2, p. 833—840 (1984).

W. Zagozdzon-Wosik, E. Mizera, and W. Kużmicz
*Emitter Efficiency of Si Bipolar Transistors — Structural and Electrical Investigations
(14.3.3; 22.1.2)
(a) Vol. **86**, No. 2, p. K125—K128 (1984).

14.3.4 High-Field Phenomena, Space-Charge Effects, Inhomogeneities, Injected Carriers (Electroluminescence See 20.3; Junctions See 14.3.3)

Review Article

D. A. Aronov, P. I. Knigin, Yu. S. Korolev, and V. V. Rubinov
*Exclusion Effect in Semiconductors with Non-Injecting Contacts
(14.3.3 and 14.3.4; 16; 22.1.1; 22.1.2; 22.2.2)
(a) Vol. **81**, No. 1, p. 11—45 (1985).

Original Papers and Short Notes

V. F. Petrenko and I. A. Ryzhkin
*Dielectric Properties of Ice in the Presence of Space Charge
(14.3.4 and 14.4; 22.6)
(b) Vol. **121**, No. 1, p. 421—427 (1984).

V. K. Malyutenko, Yu. M. Malozovskii, and G. I. Teslenko
*Negative Differential Photoconductivity in Magnetoconcentration Effect
(14.3.4 and 16; 22.2.3)
(a) Vol. **81**, No. 1, p. 391—398 (1984).

M. Siradjuddin, V. K. Raju, and P. J. Reddy
*Switching in Solution-Grown PVF Films
(14.3.4; 22.9)
(a) Vol. **81**, No. 1, p. K37—K41 (1984).

H. K. Phan, P. H. Binh, and L. H. Phu
A Simple Method for Investigating Charge Storage Effect in MIS Switching Diodes
(14.3.3; 14.3.4; 22.1.2)
(a) Vol. **81**, No. 1, p. K81—K84 (1984).

D. A. Aronov and B. R. Mamatkulov
*The Capacitance and Characteristic Relaxation Times at Carrier Exclusion in Compensated Semiconductors with Deep Traps and Non-Injecting Contacts
(14.3.4; 22.1.2)
(a) Vol. **81**, No. 1, p. K85—K89 (1984).

P. K. Basu and J. B. Roy
*Warm Electron Coefficient of Two-Dimensional Electron Gas in Silicon Inversion Layer at Low Temperatures
(14.3.2 and 14.3.4; 22.1.2)
(b) **121**, No. 2, p. 743—748 (1984).

G. O. Müller
*Basics of Electron-Impact-Excited Luminescence Devices
(14.3.4 and 20.3; 22.4.1)
(a) Vol. **81**, No. 2, p. 597—608 (1984).

R. MACH and G. O. MÜLLER
*Efficiency and Saturation in AC Thin Film El Structures
(14.3.4 and 20.3; 22.4.1)
(a) Vol. **81**, No. 2, p. 609—623 (1984).

P. J. DEAN
*Comparison of MOCVD-Grown with Conventional II–VI Materials Parameters for EL Thin Films
(14.3.4 and 20.3; 3; 13.4; 22.4.1; 22.4.2)
(a) Vol. **81**, No. 2, p. 625—646 (1984).

D. THEIS
*Selected Analytical Tools Yield a Better Insight into Electroluminescent Thin Films
(1.5 and 14.3.4; 20.3; 22.4.1)
(a) Vol. **81**, No. 2, p. 647—655 (1984).

N. A. VLASENKO, Z. L. DENISOVA, and V. S. KHOMCHENKO
*Electroluminescence of SiO_x–LnF_3 Thin Films
(14.3.4 and 20.3; 14.3.1; 22.6)
(a) Vol. **81**, No. 2, p. 657—660 (1984).

N. A. VLASENKO, YU. V. KOPYTKO, and V. S. PEKAR
*Concentration and Field Dependences of Electroluminescence Decay Kinetics in ZnS:Mn Thin Film Structures
(14.3.4 and 20.3; 22.4.1)
(a) Vol. **81**, No. 2, p. 661—667 (1984).

F. J. BRYANT and A. KRIER
*Direct Current Electroluminescence in Rare-Earth-Doped Zinc Sulphide
(14.3.4 and 20.3; 22.4.1)
(a) Vol. **81**, No. 2, p. 681—686 (1984).

W. E. HAGSTON
*Mechanism of Excitation of Rare-Earth Complexes in Electroluminescence Devices
(14.3.4 and 20.3; 13.4; 22.4)
(a) Vol. **81**, No. 2, p. 687—693 (1984).

R. MACH, G. O. MÜLLER, and G. SCHULZ
*On the Time Behaviour of Electron Impact Excitation in Thin Film Electroluminescent Devices
(14.3.4 and 20.3; 14.3.1; 22.4.1)
(a) Vol. **81**, No. 2, p. 723—732 (1984).

R. MACH, G. O. MÜLLER, G. SCHULZ, J. von KALBEN, and W. GERICKE
*On the Mechanism of Electron-Impact-Excited Luminescence in AC Thin Film Devices: ZnSe:Mn
(14.3.4 and 20.3; 22.4.2)
(a) Vol. **81**, No. 2, p. 733—738 (1984).

M. S. YUNUSOV, S. N. ABDURAKHMANOVA, M. A. ZAIKOVSKAYA, and B. L. OKSENGENDLER
*The Influence of Minority Carrier Injection on Subthreshold Defects in Silicon
(11 and 14.3.4; 22.1.2)
(a) Vol. **81**, No. 2, p. K145—K149 (1984).

R. F. Konopleva, V. A. Evseev, B. A. Borisov, I. V. Nazarkin, and V. A. Chekanov
Electrophysical Properties of Disordered Systems Based on Crystalline Germanium Compensated by Fast Neutron Irradiation
(2 and 14.3; 14.3.4; 22.1.1)
(b) Vol. **122**, No. 1, p. 349—359 (1984).

A. V. Nabok, B. A. Nesterenko, Yu. M. Shirshov, Yu. V. Goltvianskii, and A. P. Dubchak
*Profile of Trapped Charge in Silicon Nitride Films in MNOS Structures
(13.4 and 14.3.4; 14.4; 22)
(a) Vol. **82**, No. 1, p. 221—227 (1984).

L. S. Gasanov
*Effect of Light and Magnetic Field on Double Injection in Semiconductor Layers
(14.3.4; 22)
(a) Vol. **82**, No. 1, p. 307—317 (1984).

G. Arnold and G. Oelgart
Ionization Rates for Electrons and Holes in GaAsP
(14.3.3; 14.3.4; 22.2.4)
(a) Vol. **82**, No. 1, p. 319—325 (1984).

V. V. Baidakov, V. N. Ermakov, N. N. Grigorev, V. V. Kolomoets, and T. A. Kudykina
*Breakdown of Impurity States of As and Sb in Germanium at Uniaxial Compression ***P*** || [001] || ***E***
(13.4 and 14.3.4; 22.1.1)
(b) Vol. **122**, No. 2, p. K163—K167 (1984).

M. Hrobár, M. Grendel, and I. Thurzo
Current DLTS of MIS Structures Due to Dielectric Polarization of the Insulator
(14.3.3; 14.3.4)
(a) Vol. **82**, No. 2, p. 519—525 (1984).

I. Zólomy
Negative Differential Resistance in the *I–U* Characteristics of Surface Oxide Transistors
(14.3.3; 14.3.4; 22.1.2)
(a) Vol. **82**, No. 2, p. K209—K211 (1984).

D. D. Askerov, A. M. Agaev, T. D. Dzhafarov, and M. G. Ramazanzade
*Sm–GaSe Barrier Structures
(14.3.4; 16; 22.3)
(a) Vol. **83**, No. 1, p. 323—328 (1984).

G. S. Nadkarni, S. Radhakrishnan, and N. Sankarraman
*Space Charge Limited Conduction in Thin Film V_2O_5–P_2O_5–ZnO Glasses
(14.3.1 and 14.3.4; 22.6.1)
(a) Vol. **83**, No. 1, p. 335—340 (1984).

G. Kaden
*Zur Berücksichtigung der Oberflächenbandverbiegung im Silizium bei Photoinjektion in MOS-Strukturen
(Consideration of Surface Band Bending in Silicon in Case of Internal Photoemission in MOS Structures)
(14.3.3 and 14.3.4; 13.3; 22.1.2)
(a) Vol. **83**, No. 1, p. 375—386 (1984).

A. M. Evstigneev, L. F. Linnik, L. G. Linnik, E. A. Salkov, A. N. Krasiko, and O. V. Snitko
Observation of Anisotropic Internal Stresses by Electroreflectance and Transverse Photovoltaic Effect Measurements
(10.2; 14.3.4; 16; 20.1; 22.1.1)
(a) Vol. **83**, No. 1, p. K31—K33 (1984).

R. Brazis, P. Pipinys, and A. Rimeika
*Temperature Dependence of Currents in Thin Films of Silicon Oxide
(13.4 and 14.3.4; 22.6)
(a) Vol. **83**, No. 1, p. K69—K71 (1984).

S. A. Azimov, P. M. Karageorgy-Alkalaev, A. Yu. Leiderman, and V. S. Rubin
Computer Simulation of Generation–Recombination Currents in Amorphous Silicon p–n Diode Structures
(2 and 14.3; 14.3.4; 22.1.2)
(a) Vol. **83**, No. 1, p. K81—K86 (1984).

B. Giese and M. Mocker
*Carrier Heating in Lead Chalcogenides
(13.1 and 14.3.4; 22)
(b) Vol. **123**, No. 2, p. 755—762 (1984).

V. L. Bonch-Bruevich and Le Vu Ky
*Stochastic Self-Oscillations in a Two-Temperature Electron–Hole Plasma in a Semiconductor
(13 and 14.3.4)
(b) Vol. **124**, No. 1, p. 111—116 (1984).

H. Stramska
*The Voigt Effect under a Strong Electric Field in Semiconductors
(14.3.4 and 18; 22)
(b) Vol. **124**, No. 1, p. 431—437 (1984).

P. Středa
*Critical Current of the Quantum Hall Regime
(14.3.2 and 14.3.4; 22.2.1)
(b) Vol. **124**, No. 1, p. K97—K99 (1984).

D. A. Aronov, P. I. Knigin, and B. R. Mamatkulov
*Exclusion Effect Quenching in Compensated Semiconductors under the Action of Impurity Illumination Generating Majority Current Carriers
(14.3.3 and 14.3.4; 16; 22)
(a) Vol. **84**, No. 1, p. 301—310 (1984).

A. A. Simashkevich and S. D. Shutov
*Evidence of Schottky Barrier Formation at Contact of Metal with Chalcogenide Glassy Semiconductor
(14.3.4; 2; 16; 22)
(a) Vol. **84**, No. 1, p. 343—352 (1984).

V. V. Aristov, I. E. Bondarenko, N. N. Dryomova, V. V. Kazmiruk, and E. B. Yakimov
The EBIC Contrast of Dislocation Slip Planes in Silicon
(1.4 and 10.2; 14.3.4; 22.1.2)
(a) Vol. **84**, No. 1, p. K43—K45 (1984).

V. I. Sokolov, A. N. Mamedov, T. P. Surkova, G. A. Emelchenko, and L. G. Kolinova
Fine Structure of the Photoionization Band Edge in Ni-Doped ZnO
(13.5.1 and 20.1; 14.3.4; 22.4)
(b) Vol. **124**, No. 2, p. K155—K159 (1984).

M. S. Kamara, M. G. Lukashevich, and V. F. Stelmakh
*Electric Field Influence on n-GaAs Magnetoresistance in Impurity Conduction
(14.3.4; 22.2.1)
(a) Vol. **84**, No. 2, p. 613—619 (1984).

B. Olejníková
*Non-Exponential Capacitance Transient Due to Field-Enhanced Emission of Electrons from Deep Traps
(13.4 and 14.3.4; 22.2.1)
(a) Vol. **84**, No. 2, p. 621—630 (1984).

L. B. Kiss, M. I. Török, L. Nánai, and I. Hevesi
*The $1/f$ Noise of Point Contact between Substances of Anisotropic Conductivity
(14.3.4)
(a) Vol. **84**, No. 2, p. 639—643 (1984).

V. M. Ivastchenko, R. V. Konakova, and V. V. Mitin
*Peculiarities of Reverse Current–Voltage Characteristics of GaAs Lo–Hi–Lo Impatt Diodes
(14.3.3 and 14.3.4; 22.2.1)
(a) Vol. **84**, No. 2, p. 669—675 (1984).

J. L. Lagzdons, G. E. Bajārs, and A. R. Lūsis
*Modelling of the Solid State Electrochromic System $WO_3/HSbO_3 \cdot 2\,H_2O/Ni(OH)_2$
(14.3.3 and 14.3.4; 14.4; 22.8.1)
(a) Vol. **84**, No. 2, p. K197—K200 (1984).

J. Kočka and K. Čermák
Drift Mobility of a-Si:H
(2 and 14.3.1; 14.3.4; 22.1.2)
(b) Vol. **125**, No. 1, p. K89—K91 (1984).

B. P. Tyagi and K. Sen
*Grain Size and Temperature Dependence of the Electrical Behaviour of Polysilicon
(14.3.4; 13.4; 22.1.2)
(a) Vol. **85**, No. 2, p. 603—608 (1984).

S. Nešpůrek, O. Zmeškal, and F. Schauer
*An Ohmic Contact Test Using the Activation Energy of Steady-State Space-Charge-Limited Current
(14.3.4; 13.4; 22.1.2; 22.4.2)
(a) Vol. **85**, No. 2, p. 619—625 (1984).

P. Sircar and M. Aubin
Pulsed UV Laser Processed Gold and Aluminium Ohmic Contacts on n^+-GaAs
(14.3.3; 14.3.4; 22.2.1)
(a) Vol. **85**, No. 2, p. 649—654 (1984).

E. SANDOW, W. WESCH, and E. NEBAUER
*Ohmic AuGe Contacts on n-GaAs Produced by Nanosecond Laser Pulse Irradiation
(14.3.4; 9; 11; 22.2.1)
(a) Vol. **85**, No. 2, p. K169—K173 (1984).

S. KAČIULIS, K. KIBICKAS, and J. PARŠELIŪNAS
*Dependence of *I–U* Characteristics of GaAs n^+–n–n^+ Diodes on the Active Region Length
(14.3.3 and 14.3.4; 22.2.1)
(a) Vol. **85**, No. 2, p. K179—K182 (1984).

S. V. VOITIKOV and V. P. GRIBKOVSKII
*On the Theory of Crystal Excitation by Electric Discharges
(14.3.4)
(b) Vol. **126**, No. 1, p. 415—423 (1984).

M. I. TÖRÖK
*On the Poole-Frenkel Constant
(14.3.4)
(b) Vol. **126**, No. 1, p. K97—K99 (1984).

P. F. LUGAKOV and V. V. SHUSHA
*A Model of Nonequilibrium Charge-Carrier Recombination in Semiconductors Containing Nonuniformities
(14.3.4; 13.4; 22.1.2)
(a) Vol. **86**, No. 1, p. 309—312 (1984).

U. BRAATZ and D. ZAPPE
*Investigation of Space Charge Limited, Poole-Frenkel, and Isothermal Currents on HgI_2 Single Crystals
(14.3.4; 13.4; 22.5)
(a) Vol. **86**, No. 1, p. 407—411 (1984).

A. J. MUHAMMAD, D. V. MORGAN, and A. E. GUILE
*The Effect of Overlap of Ohmic Contacts on the Characteristics of a MIM Structure Having a Single Discrete Shallow Trap Level
(14.3.2 and 14.3.4)
(a) Vol. **86**, No. 1, p. 439—452 (1984).

TRAN CHOT
*Effect of Vacuum Annealing on Electrical Characteristics of Ni–Si Schottky Diodes
(14.3.4; 22.1.2)
(a) Vol. **86**, No. 1, p. K73—K78 (1984).

E.-H. BÖTTCHER and W. F. SCHMIDT
*Charge Carrier Generation in Solid Xenon by Two-Photon Absorption
(14.3.4 and 20.1; 23)
(b) Vol. **126**, No. 2, p. K165—K169 (1984).

A. Zoul
Influence of Deep Surface Levels on *C–U* Schottky Doping Profiles of n-GaAs VPE Layers
(14.3.2; 14.3.4; 22.2.1)
(a) Vol. **86**, No. 2, p. 673—678 (1984).

S. B. Sevastianov, N. N. Gerasimenko, and N. V. Vershinina
Inhomogeneities in Irradiated SiO_2–Si Structures
(11 and 14.3.3; 14.3.4; 22.1.2)
(a) Vol. **86**, No. 2, p. 717—727 (1984).

B. R. Nag, S. R. Ahmed, and M. Deb Roy
*Submicron Electron Transport in Silicon at 300 and 77 K
(14.3.4; 6; 13.1; 22.1.2)
(a) Vol. **86**, No. 2, p. 773—779 (1984).

14.4 Dielectrics

Original Papers and Short Notes

V. F. PETRENKO and I. A. RYZHKIN
*Dielectric Properties of Ice in the Presence of Space Charge
(14.3.4 and 14.4; 22.6)
(b) Vol. **121**, No. 1, p. 421—427 (1984).

I. CUCULESCU, A. ENACHE, C. FLOREA, and R. BENA
*Optical and Electrohydrodynamical Effects in a Compensated Cholesteric Mixture
(14.4 and 20.1; 22.9)
(a) Vol. **81**, No. 1, p. 191—196 (1984).

H. W. HELBERG
Dielectric Relaxation Behavior of Dislocations in Semiconductors
(10.2 and 14.3; 14.4; 22)
(a) Vol. **81**, No. 1, p. 381—390 (1984).

A. CHANDORKAR, V. T. KARULKAR, A. S. VENGURLEKAR, and K. V. RAMANATHAN
MOS Capacitance–Voltage Characteristics and Dielectric Properties of Ion Implanted Thermal Oxides on Silicon
(14.3.3; 11; 14.4; 22.1.2)
(a) Vol. **81**, No. 1, p. 407—414 (1984).

A. S. LITOVCHENKO and V. V. MAZYKIN
*Dehydroxylation Effect upon Electrical Conduction of Mica
(14.4; 19; 22.8.1)
(a) Vol. **81**, No. 1, p. K47—K50 (1984).

J. L. MIANE, M. E. ACHOUR, and F. CARMONA
*Microwave Propagation in Random Particulate Composites
(14.4; 22)
(a) Vol. **81**, No. 1, p. K71—K76 (1984).

T. S. VERMA, V. KUMAR, and J. SHANKER
Evaluation of Photoelastic Constants and Strain Derivates of the Static Dielectric Constant of Silver and Thallium Halides
(12.2; 14.4; 22.5)
(b) Vol. **121**, No. 2, p. 487—491 (1984).

H. ZSCHEILE, R. GRÜNDLER, U. DAHMS, J. FRÖHNER, and G. LEHMANN
π-Plasmon Dispersion of Trans-$(CH)_x$ at Small Momentum Transfer by Electron Energy Loss Spectroscopy
(13.5.2; 14.4; 22.9)
(b) Vol. **121**, No. 2, p. K161—K164 (1984).

A. KESSLER and P. KIZLER
Investigation of Z Center Conversions in $KCl:SrCl_2$ (II): Dielectric Relaxation Caused by Mutual Conversions of Z_1, Z_2, and F Centers
(10.2; 14.4; 22.5.2)
(b) Vol. **122**, No. 1, p. 105—112 (1984).

A. V. NABOK, B. A. NESTERENKO, YU. M. SHIRSHOV, YU. V. GOLTVIANSKII, and A. P. DUBCHAK
Profile of Trapped Charge in Silicon Nitride Films in MNOS Structures
(13.4 and 14.3.4; 14.4; 22)
(a) Vol. **82**, No. 1, p. 221—227 (1984).

G. MICOCCI and A. TEPORE
Thermally Stimulated Current in PbI_2 at Low Electric Fields
(13.4; 14.4; 22.5)
(a) Vol. **82**, No. 1, p. K79—K82 (1984).

Y. MOMOSE and M. NOGUCHI
*Electron Emission during Frictional Contact between Metals and Polymers
(14.4 and 17; 21; 22.9)
(a) Vol. **82**, No. 1, p. K83—K86 (1984).

A. KORNETA and A. MILEWSKI
*Study of Materials with Uniaxial Anisotropy Using a Dielectric Resonator Arrangement
(14.4; 22.6)
(a) Vol. **82**, No. 1, p. K107—K110 (1984).

E. N. S. MUCCILLO and R. MUCCILLO
Thermally Stimulated Depolarization Currents in γ-Irradiated (Mrad Range) $KBr:Sr^{++}$ Crystals
(10.2 and 11; 14.4; 22.5.2)
(b) Vol. **122**, No. 2, p. K127—K132 (1984).

L. N. KANTOROVICH
*Multipole Theory of the Polarization of Solids by Point Defects (II): The Point Charge Approximation
(13.4 and 14.4; 22.5.2)
(b) Vol. **123**, No. 1, p. 325—334 (1984).

V. V. KRASNOGOLOVETS and P. M. TOMCHUK
*On the Proton Conductivity Mechanism in Systems Containing Hydrogen-Bonded Chains
(14.4; 6; 22.8.1)
(b) Vol. **123**, No. 1, p. 365—376 (1984).

R. BLINC, J. DOLINŠEK, G. LAHAJNAR, I. ZUPANČIČ, L. A. SHUVALOV, and A. I. BARANOV
*Spin-Lattice Relaxation and Self-Diffusion Study of the Protonic Superionic Conductors $CsHSeO_4$ and C_5HSO_4
(9 and 14.4; 19; 22.8.1)
(b) Vol. **123**, No. 1, p. K83—K87 (1984).

J. MAIER
*Surface Induced Defects in the Space Charge Region and the Enhancement of Ionic Conductivity in Two-Phase Systems
(14.4; 22)
(b) Vol. **123**, No. 1, p. K89—K91 (1984).
Erratum
(b) Vol. **124**, No. 2, p. K187 (1984).

H. FRANKE
*Li^+ Ionic Conductivity in $LiNbO_3$
(14.4; 22.8.1)
(a) Vol. **83**, No. 1, p. K73—K76 (1984).

N. M. RAVINDRA, J. F. MORHANGE, P. A. THOMAS, and C. ANCE
*Polarizability Studies of Amorphous Silicon
(2 and 14.4; 22.1.2)
(b) Vol. **123**, No. 2, p. 611—618 (1984).

R. R. NIGMATULLIN
*To the Theoretical Explanation of the "Universal Response"
(14.4)
(b) Vol. **123**, No. 2, p. 739—745 (1984).

M. BÖHM and A. SCHARMANN
Entropy Change in Order–Disorder Transitions of Dipolar Relaxations
(10.2; 14.4; 22.8.1)
(a) Vol. **83**, No. 2, p. 537—541 (1984).

P. J. REUTERS, L. DOTHANH, and P. BALK
Determination of Carrier Type in MIS Structures with Pyrolytic AlN Insulator Films
(14.3.3; 14.4; 22; 22.1.2)
(a) Vol. **83**, No. 2, p. 719—727 (1984).

R. R. NIGMATULLIN
*On the Theory of Relaxation for Systems with a "Remnant" Memory
(14.4)
(b) Vol. **124**, No. 1, p. 389—393 (1984).

A. SUBRAHMANYAM and K. V. S. BADARINATH
*Dielectric Properties of RbCl Single Crystals
(14.4; 11; 22.5.2)
(a) Vol. **84**, No. 1, p. K93—K96 (1984).

S. MINAGAWA
*Mechanical, Electrical, and Thermal Fields Produced by Moving Dislocations and Disclinations in Thermo-Piezoelectric Continua
(10.2 and 14.4)
(b) Vol. **124**, No. 2, p. 565—571 (1984).

E. B. YAKIMOV, N. A. YARYKIN, and V. I. NIKITENKO
The Influence of Thermal Treatment on Thermally Stimulated Depolarization Spectrum of the Electret State in Plastically Deformed Silicon
(10.2 and 13.4; 14.4; 22.1.2)
(a) Vol. **84**, No. 2, p. 443—450 (1984).

A. L. BHATTACHARYYA and N. CH. MITRA
*Dielectric Relaxations in CeO_2
(14.4; 22.6)
(a) Vol. **84**, No. 2, p. 493—499 (1984).

D. SCHWARZ, I. K[illegible]ER, and H.-G. UNRUH
*Ionic Conductivity of β-CuBr and α-CuI in the Microwave Region
(14.4; 22.5)
(a) Vol. **84**, No. 2, p. K177—K180 (1984).

J. L. LAGZDONS, G. E. BAJĀRS, and A. R. LŪSIS
Modelling of the Solid State Electrochromic System $WO_3/HSbO_3 \cdot 2\,H_2O/Ni(OH)_2$
(14.[illegible] and 14.3.4; 14.4; 22.8.1)
(a) Vol. **84**, No. 2, p. K197—K200 (1984).

A. ALEGRÍA, J. M. BARANDIARÁN, and J. COLMENERO
*Dielectric Relaxation at the Glass Transition as a Free Volume Process (II): A Continuous Distribution of Relaxation Times
(2 and 14.4; 22.9)
(b) Vol. **125**, No. 1, p. 409—419 (1984).

H. BOOYENS and J. H. BASSON
Piezoelectrically Induced Charge Distributions around Dislocations in CdTe and HgCdTe
(10.2 and 13.4; 14.4; 22.4.3; 22.4.4)
(a) Vol. **85**, No. 1, p. 243—248 (1984).

P. STREITENBERGER
Effective Exchange and Correlation Interaction and Dielectric Screening in the Electron Gas. A Functional Approach to the Local-Field Correction
(13; 14.4; 21)
(b) Vol. **125**, No. 2, p. 681—692 (1984).

K. TENNAKONE and M. KAHANDA
*Conductivity Transition in Cesium Doped Iodine
(14.3 and 14.4; 22.5)
(b) Vol. **125**, No. 2, p. K181—K183 (1984).

K. IKEZAKI, T. ISHII, and T. MIURA
*Thermal Influence of Vacuum Deposition of Metallic Electrodes on TSC from Positively Corona-Charged Polyethylene Films
(14.4; 22.9)
(a) Vol. **85**, No. 2, p. 615—618 (1984).

H.-G. UNRUH
*Dielectric Dispersion and Dynamics of Nitrite Ions in $K_2Ba(NO_2)_4$
(1.2 and 14.4; 6; 22.8.1)
(b) Vol. **126**, No. 1, p. 115—120 (1984).

A. A. KAMINSKII, E. L. BELOKONEVA, B. V. MILL, YU. V. PISAREVSKII, S. E. SARKISOV, I. M. SILVESTROVA, A. V. BUTASHIN, and G. G. KHODZHABAGYAN
*Pure and Nd^{3+} Doped $Ca_3Ga_2Ge_4O_{14}$ and $Sr_3Ga_2Ge_4O_{14}$ Single Crystals, Their Structure, Optical, Spectral Luminescence, Electromechanical Properties, and Stimulated Emission
(14.4 and 20.2; 1.1; 12.2; 20.1; 20.3; 22.8.1)
(a) Vol. **86**, No. 1, p. 345—362 (1984).

A. P. TYUTNEV, V. S. SAENKO, A. V. VANNIKOV, and V. E. OSKIN
*Radiation Induced Dielectric Effect in Polymers
(11 and 14.4; 22.9)
(a) Vol. **86**, No. 1, p. 363—374 (1984).

E. HARTMANN, L. KOVÁCS, and J. PAITZ
*Electrical Conductivity of Gadolinium–Gallium Garnet (GGG) Crystals
(10.2 and 14.4; 22.8.2)
(a) Vol. **86**, No. 1, p. 401—405 (1984).

M. TOPIĆ, A. MOGUŠ-MILANKOVIĆ, and Z. KATOVIĆ
*Thermally Stimulated Depolarization Current Study in Novolac Phenol-Formaldehyde Resin
(14.4; 22.9)
(a) Vol. **86**, No. 1, p. 413—420 (1984).

A. A. EVTUSHENKO, V. F. PETRENKO, and I. A. RYZHKIN
*Electric Polarization of Ice at Nonuniform Elastic Strains
(10.2 and 14.4; 12.2; 22)
(a) Vol. **86**, No. 1, p. K31—K34 (1984).

H. BOUDRIOT, R. GRÜNDLER, K. DEUS, T. STÖCKERT, and H. A. SCHNEIDER
Electron Energy Loss Measurements on $CdSnP_2$: Plasmon Data, Dielectric Function, Band Structure Data
(13.1 and 13.5.2; 14.4; 22.8)
(b) Vol. **126**, No. 2, p. K149—K154 (1984).

H. SODOLSKI
*Anomalous Polarization Current in Dielectrics
(14.4)
(a) Vol. **86**, No. 2, p. 703—708 (1984).

S. IHARA, Y. WARITA, and K. SUZUKI
*Ionic Conductivity in $AgI_{1-x}Cl_x$
(10.2 and 14.4; 22.5.4)
(a) Vol. **86**, No. 2, p. 729—734 (1984).

T. K. VASILEVA, M. M. NADOLIISKY, and S. D. TOSHEV
*Study of PZT Ceramics Doped with Cr_2O_3
(14.4; 22.8.1)
(a) Vol. **86**, No. 2, p. K109—K111 (1984).

14.4.1 Ferroelectrics

Review Article

C. Boulesteix
A Survey of Domains and Domain Walls Generated by Crystallographic Phase Transitions Causing a Change of the Lattice
(1.2 and 1.4; 14.4.1; 22; 22.6; 22.8)
(a) Vol. **86**, No. 1, p. 11—42 (1984).

Original Papers and Short Notes

J. M. Wesselinowa
*Damping of the Transverse and Longitudinal Excitations of the Ising Model in a Transverse Field
(13 and 14.4.1; 22.8.1)
(b) Vol. **121**, No. 1, p. 317—327 (1984).

V. K. Yanovskii
*Phase Transitions and Properties of Doped $KNbO_3$ Crystals
(1.2 and 14.4.1; 14.3; 15; 20.1; 22.8.1)
(a) Vol. **81**, No. 1, p. 399—406 (1984).

J. M. Wesselinowa
*Relaxation Time for H-Bonded Ferroelectrics
(13 and 14.4.1)
(b) Vol. **121**, No. 2, p. K157—K160 (1984).

A. Fousková, M. Pavel, and B. Březina
Light Scattering by Defects in TGS
(10.2; 14.4.1; 22.9)
(a) Vol. **81**, No. 2, p. K117—K119 (1984).

H. Cerva and W. Graeff
Contrast Investigations of Surface Acoustic Waves by Stroboscopic Topography (I): Orientation Contrast
(1 and 7; 14.4.1; 22.8.1)
(a) Vol. **82**, No. 1, p. 35—45 (1984).

G. Spörl, D. D. Chat, and E. Hegenbarth
*Thermal Conductivity of Pseudo-One-Dimensional Ferroelectric CsH_2PO_4
(8 and 14.4.1; 22.8.1)
(a) Vol. **82**, No. 1, p. K27—K31 (1984).

F. Smutný and M. Połomska
*Effects of Electric Field on the II-III Phase Transition in Lithium Ammonium Sulphate and Its Deuterated Analogue
(1.2 and 14.4.1; 22.8.1)
(a) Vol. **82**, No. 1, p. K33—K36 (1984).

K. S. V. Nambi
*Pyroelectroluminescence Induced by Tourmaline
(14.4.1 and 20.3; 22.8.2)
(a) Vol. **82**, No. 1, p. K71—K73 (1984).

A. Ioanid, M. Popescu, N. Vlahovici, and I. Bunget
On the Phase Transition of KH_2PO_4 at $T \approx 110$ °C
(1.2; 14.4.1; 22.8.1)
(a) Vol. **82**, No. 2, p. K125—K128 (1984).

J. Minge and S. Waplak
*Successive Phase Transitions in $(NH_4)_3H(SO_4)_2$ and $(ND_4)_3D(SO_4)_2$ Crystals Studied by EPR of the VO^{2+} Ion and SeO_3^- Radical
(14.4.1; 4; 19; 22.8.1)
(b) Vol. **123**, No. 1, p. 27—36 (1984).

L. A. Shuvalov, S. A. Gridnev, V. I. Kudryash, and B. N. Prasolov
*Ferroelastic Domain Structure Dynamics and Internal Friction of $KH_3(SeO_3)_2$ Single Crystals
(14.4.1; 10.2; 12.2; 22.8.1)
(a) Vol. **83**, No. 1, p. 131—137 (1984).

I. Henning and E. Hegenbarth
*Specific Heat of $SrTiO_3$ in the Range of 0.5 to 6 K
(8 and 14.4.1; 22.8.1)
(a) Vol. **83**, No. 1, p. K23—K26 (1984).

N. S. Buinov and S. R. Syrtsov
*Ferroelectric Phase Transition in the Field of Nonresonance Electromagnetic Waves
(13.1 and 14.4.1; 22)
(b) Vol. **123**, No. 2, p. K165—K167 (1984).

L. A. Shebanov, L. V. Korzunova, and E. Zh. Freidenfeld
*A New Perovskite-Like Ferroelectric $Bi_2Pb_{1/3}Bi_{2/3}Ti_{4/3}W_{2/3}O_9$ with the Mixed Bismuth Layer Structure
(14.4.1; 22.8.1)
(a) Vol. **83**, No. 2, p. K127—K129 (1984).

V. L. Aksenov, A. Yu. Didyk, and N. M. Plakida
*Ultrasonic Attenuation in Quasi-One-Dimensional Ferroelectrics
(7 and 14.4.1; 22.8.1)
(b) Vol. **124**, No. 1, p. 45—54 (1984).

U. Geppert and A. Kühnel
*The Effect of Relaxing Defects on the Spin-Lattice Relaxation at Ferroelectric Phase Transitions
(14.4.1; 19)
(b) Vol. **124**, No. 1, p. 55—59 (1984).

J. Barnaś
*On the Possibility of Resonant Excitation of Spin Waves by an External Electric Field in Electronic-Type Linear Ferromagnetoelectrics
(14.4.1 and 18.2; 19)
(b) Vol. **124**, No. 1, p. K31—K34 (1984).

J. D. Stephenson
Double Laue Pattern Topography of Top-Seeded $BaTiO_3$ Single Crystals Using White Synchrotron X-Radiation
(1.4 and 1.6; 14.4.1; 22.8.1)
(a) Vol. **84**, No. 1, p. 19—30 (1984).

R. Sanctuary and P. Günter
*Electrooptic Properties of Rb_2ZnCl_4 in the Incommensurate and Ferroelectric Phases
(14.4.1 and 20.1; 22.8)
(a) Vol. **84**, No. 1, p. 103—111 (1984).

V. I. Zametin
Absorption Edge Anomalies in Polar Semiconductors and Dielectrics at Phase Transitions
(20.1; 1.2; 14.4.1; 22; 22.5.2; 22.8.1; 22.9)
(b) Vol. **124**, No. 2, p. 625—640 (1984).

A. T. Borges and L. A. Amarante Ribeiro
*Effects of the $B_{ij}S_i^xS_j^y$ Interaction on the IMTF's Dynamics
(14.4.1)
(b) Vol. **124**, No. 2, p. 781—791 (1984).

R. L. Davidovich, P. S. Gordienko, J. Grigas, T. A. Kaidalova, V. Urbonavičius, and L. A. Zemnukhova
*Phase Transitions in $KSbF_4$
(1.2 and 14.4.1; 8; 19; 22.8)
(a) Vol. **84**, No. 2, p. 387—392 (1984).

H. Beige and A. Kühnel
*Electromechanical Coefficients at Ferroelectric Phase Transitions. Rochelle Salt and $RbHSO_4$
(14.4.1; 22.8.1)
(a) Vol. **84**, No. 2, p. 433—437 (1984).

Yu. N. Moskvich, A. A. Sukhovskii, and O. V. Rozanov
*Deuteron Magnetic Resonance of Highly Deuterated Ammonium Hydrogen Selenate
(1.1 and 14.4.1; 19; 22.8.1)
(a) Vol. **84**, No. 2, p. K109—K113 (1984).

L. N. Kamysheva, A. S. Sidorkin, and S. D. Milovidova
*Dielectric Relaxation in Ferroelectric TGS
(14.4.1; 22.9)
(a) Vol. **84**, No. 2, p. K115—K120 (1984).

F. Engelke, D. Michel, and F. Pille
*NMR Study of Dynamical Processes in Crystals of the TSCC Family
(14.4.1; 19; 22.9)
(b) Vol. **125**, No. 2, p. 483—492 (1984).

B. Wyncke, F. Brehat, M. El Sherif, and G. V. Kozlov
Far-Infrared and Submillimeter Dielectric Response of Sodium Nitrite
(6 and 20.1; 1.2; 14.4.1; 22.8.1)
(b) Vol. **125**, No. 2, p. 493—498 (1984).

K. Žičkus, A. Audzijonis, J. Batarūnas, and A. Šileika
*The Fundamental Absorption Edge Tail of Ferroelectric SbSI
(14.4.1; 20.1; 22.8)
(b) Vol. **125**, No. 2, p. 645—651 (1984).

S. K. Plachkova and L. A. Tersiev
*Phase Transitions and Intervalley Jahn-Teller Effect in the System $(GeTe)_{1-x}(AgSbTe_2)_x$
(1.2 and 14.4.1; 13.1; 22.8)
(b) Vol. **125**, No. 2, p. 659—664 (1984).

B. Topič, R. Blinc, and L. A. Shuvalov
^{87}Rb NMR in Monoclinic RbD_2PO_4
(1.2; 14.4.1; 19; 22.8.1)
(a) Vol. **85**, No. 2, p. 409—415 (1984).

G. Sorge, J. Janich, L. A. Shuvalov, and N. M. Shchagina
*The Influence of an Electric DC Field on the New Intermediate Phase between the α- and β-Phases of $NaH_3(SeO_3)_2$ Crystals
(1.2 and 14.4.1; 12.2; 22.8)
(a) Vol. **86**, No. 1, p. 111—115 (1984).

V. V. Aristov, L. S. Kokhanchik, and Yu. I. Voronovskii
*Voltage Contrast of Ferroelectric Domains of Lithium Niobate in SEM
(14.4.1; 22.8.1)
(a) Vol. **86**, No. 1, p. 133—141 (1984).

V. I. Torgashev, V. Dvořák, and F. Smutný
*On Phase Transitions in $LiNH_4SO_4$
(1.2 and 14.4.1; 22.8.1)
(b) Vol. **126**, No. 2, p. 459—466 (1984).

15. Thermoelectric and Thermomagnetic Properties

Review Articles

B. M. Askerov, B. I. Kuliev, and S. R. Figarova
Electron Transport Phenomena in Conductive Films
(14.3.1; 6; 13.1; 15; 22)
(b) Vol. **121**, No. 1, p. 11—37 (1984).

L. B. Dubrovskaya and S. Z. Nazarova
Some Regularities in the Behaviour of the Magnetic, Electrokinetic, and Superconducting Properties of Transition Metal Monocarbide and Mononitride Solid Solutions
(14.3 and 18; 13.1; 14.2; 15; 22)
(a) Vol. **83**, No. 1, p. 11—38 (1984).

Original Papers and Short Notes

E. M. Raspopova, V. T. Kalinnikov, T. N. Lyubovskaya, A. P. Chernov, and T. N. Fesenko
Glass Formation and Some Properties of the $(Cd_2GeAs_4)_{1-x}(TlAs)_x$ System
(2; 14.3; 15; 22.8)
(a) Vol. **81**, No. 1, p. 139—143 (1984).

V. K. Yanovskii
Phase Transitions and Properties of Doped $KNbO_3$ Crystals
(1.2 and 14.4.1; 14.3; 15; 20.1; 22.8.1)
(a) Vol. **81**, No. 1, p. 399—406 (1984).

S. K. Patapis and K. Alexopoulos
*Thermoelectric Power of Hematite at the Morin Transition
(15; 22.6)
(a) Vol. **81**, No. 1, p. K57—K61 (1984).

M. I. Auslender, V. V. Karyagin, A. B. Zolotovitskii, V. P. Kalashnikov, and B. A. Gizhevskii
*Longitudinal Nernst-Ettinghausen Effect in Heavily Doped EuO
(15; 22.6)
(b) Vol. **121**, No. 2, p. 737—741 (1984).

M. Stordeur and H. T. Langhammer
Erratum to: Anisotropy of Transport Properties Due to Direction-Dependent Band Nonparabolicity
(13.1; 14.3; 15; 22)
(b) Vol. **121**, No. 2, p. 765 (1984).

F. Völklein and E. Kessler
A Method for the Measurement of Thermal Conductivity, Thermal Diffusivity, and Other Transport Coefficients of Thin Films
(14.1 and 14.3.1; 8; 15; 21.7)
(a) Vol. **81**, No. 2, p. 585—596 (1984).

A. A. GHANI AWAD
*Thermopower and Resistivity in Correlation to B-Site Magnetic Ordering in Cu–Ni and Cu–Cd Ferrospinels
(14.3 and 15; 22.8.2)
(a) Vol. **81**, No. 2, p. K155—K158 (1984).

C. BANSAL and K. SURENDRANATH
*Thermoelectric Power and Electrical Resistivity of $Fe_2Nb_3Se_{15}$
(14 and 15; 22.8)
(b) Vol. **122**, No. 1, p. 345—348 (1984).

S. K. PLACHKOVA
Estimation of Some Band-Edge Structure Parameters of the System $(GeTe)_{1-x}(AgSbTe_2)_x$
(13.1; 14.3; 15; 22)
(b) Vol. **122**, No. 1, p. K53—K58 (1984).

J.-P. DOUMERC, J. HEJTMANEK, J.-P. CHAMINADE, M. POUCHARD, and M. KRUSSANOVA
A Photoelectrochemical Study of $CuWO_4$ Single Crystals
(14.3; 13.4; 15; 16; 22.8.1)
(a) Vol. **82**, No. 1, p. 285—294 (1984).

D. COQUILLAT, A. PRADEL, G. BRUN, and J. C. TEDENAC
Growth and Physical Properties of $TlBiTe_2$ Crystals
(13.1 and 14.3; 8; 15; 20.1; 22.8)
(a) Vol. **82**, No. 1, p. 295—299 (1984).

P. KUIVALAINEN, H. ISOTALO, and H. STUBB
*Thermoelectric Power in Inhomogeneous Metallic Polymers
(14.3 and 15; 22.9)
(b) Vol. **122**, No. 2, p. 791—797 (1983).

L. G. KHVOSTANTSEV and V. A. SIDOROV
Phase Transitions in Antimony at Hydrostatic Pressure up to 9 GPa
(1.2; 14.1; 15; 21.7)
(a) Vol. **82**, No. 2, p. 389—398 (1984).

L. URAY
The Effect of Void Size on the Resistivity of Sintered Tungsten
(1.3 and 14.1; 15; 21)
(a) Vol. **82**, No. 2, p. 545—552 (1984).

H. SÜSSMANN, A. PRIEMUTH, and U. PRÖHL
*Doping Properties of Pb and Ge in Bi_2Te_3 and Sb_2Te_3
(14.3 and 15; 22.7)
(a) Vol. **82**, No. 2, p. 561—567 (1984).

R. RICHTER and F. GOEDSCHE
*On the Thermoelectric Power of Amorphous Ferromagnets
(2 and 15; 14.1; 18.2; 21)
(b) Vol. **123**, No. 1, p. 143—154 (1984).

S. K. Plachkova
*Thermoelectric Figure of Merit of the System $(GeTe)_{1-x}(AgSbTe_2)_x$
(14.3 and 15; 8; 22.8)
(a) Vol. **83**, No. 1, p. 349—355 (1984).

J. Rosenzweig, J. Zhang, and U. Birkholz
Annealing Effects on Electrical Conductivity and Thermoelectric Power of Sintered PbTe
(14.3; 15; 22)
(a) Vol. **83**, No. 1, p. 357—363 (1984).

G. J. Ashwell
Electrical, Magnetic, and Structural Characterisation of [1,2-Bis(1-methyl-4-pyridinium)ethylene]$^{2+}$-$(TCNQ)_4^{2-}(H_2O)$ and the Properties of Two Isostoichiometric TCNQ Salts
(14.3 and 18.1; 1.1; 15; 22.9)
(a) Vol. **83**, No. 2, p. 669—673 (1984).

V. B. Gohel, C. K. Acharya, and A. R. Jani
On the Determination of Pseudopotential Parameters and the Comprehensive Study of Metallic Properties
(13.1; 2; 14.1; 15; 21)
(b) Vol. **124**, No. 1, p. 137—145 (1984).

A. Conan, A. Bonnet, M. Zoaeter, and D. Ramoul
*Dependence of the Total Mobility in a One-Band Model. Application to n-Type $MoTe_2$
(14.3 and 15; 22)
(b) Vol. **124**, No. 1, p. 403—410 (1984).

M. Stordeur
Investigation of the Weak-Field Charge Transport in Semiconducting V_2–VI_3 Compounds with Trigonal Symmetry (1): Expanded Model for the Transport Coefficients
(14.3; 15; 22.7)
(b) Vol. **124**, No. 1, p. 439—446 (1984).

N. B. Mustafaev and M. G. Shakhtakhtinskii
*Thermopower of a Thin Bi Film in a Classical Strong Magnetic Field
(15; 13.1; 21.7)
(b) Vol. **124**, No. 1, p. K93—K96 (1984).

L. Uray and M. Menyhárd
The Segregation of Iron in Tungsten
(1.3; 1.4; 9; 14.1; 15; 21; 21.1.1)
(a) Vol. **84**, No. 1, p. 65—72 (1984).

A. A. Ghani, S. A. Mazen, and A. H. Ashour
*The Annealing Effect on the Conduction Mechanism of Copper Ferrite
(14.3 and 15; 22.8.2)
(a) Vol. **84**, No. 1, p. 337—341 (1984).

M. STRORDEUR and G. SIMON
Investigation of the Weak-Field Charge Transport in Semiconducting V_2–VI_3 Compounds with Trigonal Symmetry (II): Interpretation of the Weak-Field Charge Transport in Sb_2Te_3 Single Crystals
(14.3; 13.1; 15; 22.7)
(b) Vol. **124**, No. 2, p. 799—806 (1984).

B. G. MULIMANI and R. HARRIS
*A Note on the Electron–Phonon Enhancement of the Thermopower in Amorphous Alloys
(15; 2; 21)
(b) Vol. **124**, No. 2, p. K181—K185 (1984).

H. S. MAITI, S. RAJENDRAN, and V. SITAKARA RAO
Electrical Conductivity and Thermoelectric Power Studies of Iron(III)–Nickel(II) Mixed Oxides
(14.3; 15; 22.6.1)
(a) Vol. **84**, No. 2, p. 631—638 (1984).

R. SATYANARAYANA and S. RAMANA MURTHY
Electrical Conductivity of Barium–Strontium Ferrites
(14.3; 15; 22.8.2)
(a) Vol. **84**, No. 2, p. 655—658 (1984).

N. N. KOREN, V. I. LEVCHENKO, V. V. DIKAREVA, and A. V. KAZUSHCHIK
*Physical Properties of Germanium Telluride Crystals Grown from Gas Phase
(14.3 and 15; 10.2; 22)
(a) Vol. **84**, No. 2, p. K173—K176 (1984).

A. C. GUPTA and A. N. NIGAM
Effect of Cold Work on Hydrogenation of Nickel as Seen in the Thermoelectric Behaviour
(13.2; 12.1; 15; 21.1)
(b) Vol. **125**, No. 1, p. 207—209 (1984).

M. M. BABAEV, T. M. GASYMOV, and A. A. KATANOV
*Thermopower and Thermomagnetic Effects of Hot Carriers under Conditions of Mutual Electron–Phonon Drag
(15; 6; 13; 22)
(b) Vol. **125**, No. 1, p. 421—429 (1984).

P. STŘEDA
Universal Transport Properties of Inversion Layers in Quantizing Magnetic Field
(13.3; 8; 14.1; 15; 18; 21)
(b) Vol. **125**, No. 2, p. 849—860 (1984).

L. V. VASSILEV and ST. K. DIMITROVA
*Observation of Anderson-Like Transition in the System $Pb_{1-x}Ge_xTe$
(14.3 and 15; 1.2; 22.3.1)
(b) Vol. **125**, No. 2, p. K177—K180 (1984).

E. CHILLA and W. SCHILLER
*The Influence of Electron–Phonon Interaction on the Thermoelectric Power at High Temperatures
(14.1 and 15; 21)
(b) Vol. **126**, No. 1, p. K91—K96 (1984).

O. A. Golikova, N. E. Solovev, Ya. A. Ugai, and V. A. Feigelman
*Electrophysical Properties of α-Rhombohedral Boron
(14.3 and 15; 22.1)
(a) Vol. **86**, No. 1, p. K51—K54 (1984).

P. I. Baranskii, T. G. Grishchenko, V. V. Savyak, and Yu. V. Simonenko
*Temperature Dependence of the Thermoelectric Characteristics of n-Si Elastically Strained in the [100] Direction
(14.3 and 15; 22.1.2)
(a) Vol. **86**, No. 1, p. K63—K67 (1984).

O. P. Hansen and I. F. I. Mikhail
Galvanomagnetic Effects of Bismuth in Non-Quantising Magnetic Fields. Pseudo-Parabolic Model
(14.1; 15; 21.7)
(b) Vol. **126**, No. 2, p. 721—728 (1984).

E. Kaden and H.-L. Günter
*The Thermoelectric Power of p-Ge at Low Temperatures
(6 and 15; 14.3; 22.1.1)
(b) Vol. **126**, No. 2, p. 733—740 (1984).

L. Borisova
Density-of-States Effective Mass in Solid Solutions of the $(Ag_{x/2}Pb_{1-x}Sb_{x/2})Te$ Type
(13.1; 14.3; 15; 22)
(b) Vol. **126**, No. 2, p. K155—K158 (1984).

L. G. Khvostantsev, V. A. Sidorov, and O. B. Tsiok
*Gapless State and Anomalous Behavior of Thermopower of Black Phosphorus at Pressure up to 12 GPa
(13.1 and 15; 14.3; 22.1)
(a) Vol. **86**, No. 2, p. K99—K104 (1984).

P. G. Rustamov, F. M. Sadigov, and Ch. I. Abilov
*Electrophysical Properties of $CeBiTe_3$
(14.3 and 15; 22.8)
(a) Vol. **86**, No. 2, p. K113—K115 (1984).

16. Photoconductivity. Photovoltaic Effects

Review Article

D. A. ARONOV, P. I. KNIGIN, YU. S. KOROLEV, and V. V. RUBINOV
Exclusion Effect in Semiconductors with Non-Injecting Contacts
(14.3.3 and 14.3.4; 16; 22.1.1; 22.1.2; 22.2.2)
(a) Vol. **81**, No. 1, p. 11—45 (1984).

Original Papers and Short Notes

G. M. SHMELEV, G. I. TSURKAN, and NGUYEN QUOC ANH
*Photostimulated Planar Acoustomagnetoelectric Effect in Semiconductors
(16; 22)
(b) Vol. **121**, No. 1, p. K97—K102 (1984).

V. K. MALYUTENKO, YU. M. MALOZOVSKII, and G. I. TESLENKO
*Negative Differential Photoconductivity in Magnetoconcentration Effect
(14.3.4 and 16; 22.2.3)
(a) Vol. **81**, No. 1, p. 391—398 (1984).

M. P. MIKALKEVIČIUS and K. K. TUMKEVIČIUS
*A Mechanism of Current Percolation in a Heterogeneous Layer of PbTe
(14.3.1 and 16; 22)
(a) Vol. **81**, No. 1, p. K43—K45 (1984).

J. P. SCHAFFER, H. PARK, J. H. LIND, and P. L. JONES
*Hydrogenated Amorphous Boron: Transient and Steady State Photoconductivity
(2 and 16; 22.1)
(a) Vol. **81**, No. 1, p. K51—K55 (1984).

V. D. EGOROV, G. O. MÜLLER, H. H. WEBER, M. A. JACOBSON, and A. F. DITE
*Control of Carrier Concentration by Stimulated Emission in Highly Excited Direct Gap Semiconductors
(16. and 20.3; 22.4.1)
(b) Vol. **122**, No. 1, p. 183—192 (1984).

A. N. GEORGOBIANI, A. N. GRUZINTSEV, Z. P. ILYUKHINA, V. E. TEZLEVAN, and I. M. TIGINYANU
*Photoconductivity and Non-Equilibrium Carrier Recombination in $CdIn_2S_4$ Single Crystals
(13.4 and 16; 22.8)
(a) Vol. **82**, No. 1, p. 207—212 (1984).

M. NOWAK
*Oscillatory Dependence of the Interference Photomagnetoelectric Effect and Photoconductivity on Magnetic Field
(16; 20.1; 22.4.4)
(a) Vol. **82**, No. 1, p. 249—256 (1984).

J. P. Doumerc, J. Hejtmanek, J.-P Chaminade, M. Pouchard, and M. Krussanova
A Photoelectrochemical Study of $CuWO_4$ Single Crystals
(14.3; 13.4; 15; 16; 22.8.1)
(a) Vol. **82**, No. 1, p. 285—294 (1984).

A. I. Isaev, T. M. Guseinov, S. D. Mekhtieva, and D. Sh. Abdinov
*Negative Photoconductivity in Selenium Single Crystals
(16; 22.1.3)
(a) Vol. **82**, No. 1, p. K87—K90 (1984).

V. I. Berezkin and A. I. Grachev
*Properties of Al-Doped Bismuth Silicon Oxide
(16 and 20.1; 22.8)
(a) Vol. **82**, No. 1, p. K95—K99 (1984).

Ya. I. Verzimacha, A. V. Kovalshuk, M. V. Kurik, C. Hamann, and A. Mrwa
*Solar Cells of Schottky Type with Ni–PbPc Interface
(14.3.3 and 16; 22.9)
(a) Vol. **82**, No. 1, p. K111—K115 (1984).

W. Kautek, F. Decker, and H. Gerischer
Optical Anisotropy of Transition Metal Dichalcogenides. A Photoelectrochemical Determination
(20.1; 13.1; 16; 22)
(b) Vol. **122**, No. 2, p. 651—659 (1984).

V. I. Sokolov and V. V. Chernyaev
Detection of Manganese Photoionization in $Cd_{1-x}Mn_xTe$ Solid Solutions
(13.5.1; 16; 20.1; 22.4.4)
(b) Vol. **122**, No. 2, p. 703—708 (1984).

W. Knap, J. Kossut, and J. Mycielski
Photoelectromagnetic Effect and Photoconductivity in Quantizing Magnetic Fields
(13.1; 16; 22)
(b) Vol. **122**, No. 2, p. 761—773 (1984).

Yu. S. Bulyshev, I. M. Kashirskii, and V. V. Sinitskii
*Photoconductivity and Energy Level Structure of 2,4,7-Trinitro-9-Fluorenone Thin Films
(14.3.1 and 16; 13.1; 22.9)
(a) Vol. **82**, No. 2, p. 537—544 (1984).

C. Moţoc, R. Mitroi, and C. Roşu
*Photovoltaic Effects in Nematic and Cholesteric Mesophases Containing Nonmesogenic Compounds
(16; 2; 22.9)
(a) Vol. **82**, No. 2, p. 575—579 (1984).

K. Lischka, R. Durstberger, G. Lindemann, and H. Staudinger
Defect States in $Pb_{1-x}Sn_xTe$
(13.4; 16; 22.8)
(b) Vol. **123**, No. 1, p. 319—324 (1984).

V. A. Kovarskii, L. V. Chernysh, and A. V. Belousov
About the Low-Temperature Dependence of Activation for Capture Cross-Sections of Charge Carriers in Semiconductors with Persistent Photoconductivity
(13.4; 16; 22)
(b) Vol. **123**, No. 1, p. 345—351 (1984).

D. D. Askerov, A. M. Agaev, T. D. Dzhafarov, and M. G. Ramazanzade
Sm-GaSe Barrier Structures
(14.3.4; 16; 22.3)
(a) Vol. **83**, No. 1, p. 323—328 (1984).

P. Mialhe, A. Khoury, and J. P. Charles
*A Review of Techniques to Determine the Series Resistance of Solar Cells
(14.3.3 and 16; 22.1.2)
(a) Vol. **83**, No. 1, p. 403—409 (1984).

A. M. Evstigneev, L. F. Linnik, L. G. Linnik, E. A. Salkov, A. N. Krasiko, and O. V. Snitko
Observation of Anistropic Internal Stresses by Electroreflectance and Transverse Photovoltaic Effect Measurements
(10.2; 14.3.4; 16; 20.1; 22.1.1)
(a) Vol. **83**, No. 1, p. K31—K33 (1984).

J. Misiewicz, Nella Mirowska, and Z. Gumienny
*Fundamental Optical and Photoelectric Properties of Poly- and Mono-Crystals of Zn_3P_2
(16 and 20.1; 22.3)
(a) Vol. **83**, No. 1, p. K51—K56 (1984).

B. G. Tagiev, G. M. Niftiev, and S. A. Abushov
Photoelectrical and Optical Properties of GaSe: Yb Single Crystals
(13.4; 16; 20.3; 22.3)
(a) Vol. **83**, No. 1, p. K61—K64 (1984).

P. C. Dhanasekaran, P. R. Vaya, and B. S. V. Gopalam
*Spectral Sensitivity Calculations on n^+–p and n^+–p–p^+ Silicon Solar Cells
(14.3.3 and 16; 22.1.2)
(a) Vol. **83**, No. 2, p. 693—700 (1984).

D. T. Rasulov and R. Akhmedov
*Thin-Film Solar Cell on the Basis of Polycrystalline Cadmium Telluride
(14.3.3 and 16; 22.4.1; 22.4.3; 22.4.4)
(a) Vol. **83**, No. 2, p. K211—K213 (1984).

R. Piotrzkowski
*Pressure Study of Photoconductivity in Indium Antimonide (I): Low-Concentration n-Type at 77 K. Pressure-Dependent Capture Coefficient of a Hole-Trapping Center
(13.4 and 16; 12.2; 22.2.3)
(b) Vol. **124**, No. 1, p. 411—421 (1984).

O. Yu. Borkovskaya, N. L. Dmitruk, V. G. Litovchenko, and O. I. Maeva
*The Influence of Deep Levels on Photomemory Effect in Structures with a Potential Barrier
(16; 13.4; 22.2.1)
(a) Vol. 84, No. 1, p. 285—290 (1984).

M. Böhm, O. Erb, and A. Scharmann
*Electrical Conduction in $PbMoO_4$ Crystals
(16 and 20.1; 13.4; 22.8.1)
(a) Vol. 84, No. 1, p. 291—299 (1984).

D. A. Aronov, P. I. Knigin, and B. R. Mamatkulov
Exclusion Effect Quenching in Compensated Semiconductors under the Action of Impurity Illumination Generating Majority Current Carriers
(14.3.3 and 14.3.4; 16; 22)
(a) Vol. 84, No. 1, p. 301—310 (1984).

D. P. Joshi and R. S. Srivastava
*Effective Diffusion Length of Minority Carriers in Polycrystalline Silicon
(14.3.3 and 16; 1.4; 22.1.2)
(a) Vol. 84, No. 1, p. 311—317 (1984).

A. A. Simashkevich and S. D. Shutov
Evidence of Schottky Barrier Formation at Contact of Metal with Chalcogenide Glassy Semiconductor
(14.3.4; 2; 16; 22)
(a) Vol. 84, No. 1, p. 343—352 (1984).

S. El-Halawany, R. Bacewicz, J. Filipowicz, and R. Trykozko
*Photoconductivity of Flash Evaporated $CuInSe_2$ Thin Films
(16; 14.3.1; 20.1; 22.8)
(a) Vol. 85, No. 1, K89—K92 (1984).

T. K. Bandyopadhyay and S. R. Chaudhuri
Electronic Properties of Anodic n-CdS Films from Photoelectrochemical Measurements
(14.3.1; 16; 22.4.1)
(a) Vol. 84, No. 2, p. 651—654 (1984).

A. J. Szadkowski and A. Lubomirska-Wittlin
*Compensation of p-Type $Cd_{1-x}Mn_xTe$ by Diffusion of Ag
(14.3 and 16; 9; 13.4; 22.4.4)
(a) Vol. 84, No. 2, p. K181—K184 (1984).

C. Chiang
Photo-Illumination and Current Excitation of Switching in Amorphous Thin Films
(2 and 14.3.1; 16)
(a) Vol. 84, No. 2, p. K189—K191 (1984).

S. Huant, L. Dmowski, M. Baj, and L. C. Brunel
Intraconduction Band Magneto-Optical Study of InSb under Hydrostatic Pressure
(13.1 and 19; 16; 22.2.3)
(b) Vol. 125, No. 1, p. 215—219 (1984).

K. MOCHIZUKI and K. SUZUKI
*Effect of the Stoichiometry Control on the Photoelectrical Properties of ZnS_xSe_{1-x}
(16 and 20.3; 22.4.4)
(a) Vol. **85**, No. 1, p. 249—256 (1984).

D. S. MISRA, A. KUMAR, and S. C. AGARWAL
*Intensity and Temperature Dependence of Steady-State Photoconductivity Down to 8 K and DOS Distribution Obtained from These Measurements in a-Si:H
(16; 2; 22.1.2)
(a) Vol. **85**, No. 1, p. 297—303 (1984).

M. ADOLF, J. ROSENZWEIG, and U. BIRKHOLZ
*Bipolar Photoconductivity and Hall Effect of the Acoustoelectric Current in CdTe
(7 and 16; 14.3; 22.4.3)
(a) Vol. **85**, No. 1, p. K81—K84 (1984).

I. N. GRINCHESHEN, N. S. POPOVICH, and A. A. SHTANOV
*Relaxation Peculiarities in $TlSbSe_2$ Crystals
(14.3 and 16; 22.8)
(a) Vol. **85**, No. 1, p. K85—K88 (1984).

G. MICOCCI, A. RIZZO, and A. TEPORE
*Photoconductivity in Halogenated and Hydrogenated Amorphous Silicon Films
(16; 2; 22.1.2)
(a) Vol. **85**, No. 2, p. 609—614 (1984).

N. P. ESINA, N. V. ZOTOVA, S. A. KARANDASHEV, and G. M. FILARETOVA
Tunneling Surface Recombination in MS Structures on p-InAs
(14.3.3 and 20.3; 16; 22.2.1)
(a) Vol. **85**, No. 2, p. 655—660 (1984).

G. M. SHMELEV, NGUYEN HONG SHON, and G. I. TSURKAN
*Magnetic Oscillations of Acoustoelectrical Current of Photoelectrons in p-GaAs Type Semiconductors
(6 and 16; 7; 22.2.1)
(b) Vol. **126**, No. 1, p. K105—K108 (1984).

B. T. KOLOMIETS, V. M. LYUBIN, M. A. TAGUYRDZHANOV, and G. ZENTAI
*A Comparative Study of Photoconductivity in Vitreous As_2Se_3 and a-Si:H Thin Films
(16; 2; 14.3.1; 22; 22.1.2)
(a) Vol. **86**, No. 1, p. 383—387 (1984).

JUH TZENG LUE
Time-Resolved Reflectivity Measurements on RF Sputtered a-Si:H Alloy
(13.4 and 20.1; 2; 16; 22.1.2)
(b) Vol. **126**, No. 2, p. 713—720 (1984).

T. K. BANDYOPADHYAY and S. R. CHAUDHURI
*Effect of Annealing Ambients on the Performance of Thin Film n-CdSe Photoanodes
(16; 22.4.2)
(a) Vol. **86**, No. 2, p. 827—832 (1984).

17. Emission of Electrons and Ions

Original Papers and Short Notes

V. Cháb and I. Bartoš
*Band Structure and Lifetime Determinations for GeS from Angle-Resolved Photoemission
(13.1 and 17; 22)
(b) Vol. **121**, No. 1, p. 301—306 (1984).

H. J. Fitting, J. Boyde, and J. Reinhardt
*Monte-Carlo Approach of Electron Emission from SiO_2
(17; 22.6)
(a) Vol. **81**, No. 1, p. 323—332 (1984).

G. L. Gutsev, Yu. M. Shulga, and Yu. G. Borodko
The Discrete Variational X_α-Method Interpretation of the C KVV and Ti $L_{2,3}M_{2,3}V$ Auger Spectra of TiC
(13.1; 17; 21.1)
(b) Vol. **121**, No. 2, p. 595—601 (1984).

H. Neumann, D. W. Bullett, H. Sommer, R. D. Tomlinson, and W. John
Valence Bands of the Cu–III–VI_2 Chalcopyrites Studied by Photoemission Spectra, X-Ray Emission Spectra, and Electronic Structure Calculations
(13.1; 17; 20; 22.8)
(b) Vol. **121**, No. 2, p. 641—647 (1984).

M. V. Kovalchuk and E. Kh. Mukhamedzhanov
*The Use of a Gas-Flow Proportional Counter for the Energetical Analysis of Photoelectrons under X-Ray Diffraction Conditions
(1.6 and 17; 22.1.1)
(a) Vol. **81**, No. 2, p. 427—432 (1984).

Y. Momose and M. Noguchi
*Electron Emission during Frictional Contact between Metals and Polymers
(14.4 and 17; 21; 22.9)
(a) Vol. **82**, No. 1, p. K83—K86 (1984).

V. Atzrodt and H. Lange
Principal Component Analysis for Refractory Metal Silicide Investigations with Auger Electron Spectroscopy
(1.6; 17; 21; 22.1.2)
(a) Vol. **82**, No. 2, p. 373—378 (1984).

B. Sorkin and H. Käämbre
*V_K Centres in Thermo-Exoemission from CsBr:Tl
(10.2 and 17; 13.4; 22.5.2)
(a) Vol. **82**, No. 2, p. K149—K152 (1984).

M. Aoki, T. Nishikawa, and S. Okabe
*Pattern Variation of Evaprated Thin Films and Their Non-Pre-Excited TSEE
(1.5 and 17; 20.3; 22.5.2; 22.5.3)
(a) Vol. **83**, No. 1, p. 85—91 (1984).

L. Červinka, V. Cháb, and S. Koc
Erratum to: XPS Investigation of the Amorphous $TlSe_x$ System
(2 and 4; 17; 22.3)
(a) Vol. **83**, No. 1, p. 418 (1984).

A. M. Afanasev, R. M. Imamov, and E. Kh. Mukhamedzhanov
The Yield of Photoelctrons of Different Energies in the X-Ray Laue Diffraction
(1.6; 17; 22.1.2)
(a) Vol. **83**, No. 1, p. K5—K9 (1984).

B. Sorkin and H. Käämbre
A New Evidence for Hole-Induced Thermo-Exoemission
(10.2; 17; 20.3; 22.5.2)
(a) Vol. **83**, No. 1, p. K39—K42 (1984).

N. K. Bondareva, I. V. Krylova, and V. B. Golubev
*Thermostimulated Exoemission and Radial Recombination in Polyethylene Irradiated with an Electron Beam
(13.3 and 17; 11; 19; 22.9)
(a) Vol. **83**, No. 2, p. 589—595 (1984).

B. Sorkin and H. Käämbre
*Thermo-Exoemission of NaCl:Ag Crystals, Cleaved before and after X-Irradiation
(17; 11; 13.3; 13.4; 22.5.2)
(a) Vol. **83**, No. 2, p. K171—K174 (1984).

K. S. V. Nambi
*Comment on "The Role of Sulphur-Oxyradicals in Thermoluminescence and Exoelectron Emission of $CaSO_4$ Phosphors" by R. Huzimura and K. Atarashi
(17 and 20.3; 11; 13.4; 19; 22.8.1)
(a) Vol. **83**, No. 2, p. K175—K177 (1984).

V. Drchal and J. Kudrnovský
The Auger Spectra from Alloys. Effect of Partial Long-Range Order
(13.1; 17; 21)
(b) Vol. **124**, No. 1, p. 179—189 (1984).

A. M. Afanasev, R. M. Imamov, A. V. Maslov, and E. M. Pashaev
*Photoeffect in X-Ray Grazing Incidence Diffraction
(1 and 17; 22.1.2)
(a) Vol. **84**, No. 1, p. 73—78 (1984).

W. Frentrup, M. Griepentrog, H. Klose, G. Kreysch, and U. Müller-Jahreis
*The Influence of Alkali Atoms Implanted in Silicon on the Negative Secondary Ion Emission
(17; 13.3; 22.1.2)
(a) Vol. **84**, No. 1, p. 269—272 (1984).

M. OKUSAWA
*Splitting of Core-Level XPS Lines of Transition-Metal Ions in Compounds and Moment of the Lines
(13 and 17; 22.5)
(b) Vol. **124**, No. 2, p. 673—681 (1984).

K. HRICOVÍNI and J. KREMPASKÝ
On the Electronic Structure of Iron–Boron Alloys
(13.1; 2; 17; 21.1.1)
(b) Vol. **124**, No. 2, p. K147—K150 (1984).

T. CHASSÉ, U. BERG, and O. BRÜMMER
*Angular-Resolved X-Ray Photoelectron Spectra of Valence and Core Levels from GaS
(13.1 and 17; 22.3)
(b) Vol. **125**, No. 1, p. 267—274 (1984).

M. GABER and H.-J. FITTING
*Energy–Depth Relation of Electrons in Bulk Targets by Monte-Carlo Calculations
(17; 11; 21)
(a) Vol. **85**, No. 1, p. 195—198 (1984).

G. BERGER and V. ATZRODT
In Vitro Characterization of Bioactivity of Glassy or Glass–Crystalline Implant Materials Using Auger Electron Spectroscopy
(1.5; 2; 17; 22.6.1)
(a) Vol. **85**, No. 1, p. K9—K13 (1984).

K. UHLMANN and M. SCHMIDT
Influence of Work Function Change Due to Oxygen Chemisorption on the Secondary-Ion Emission Probability
(1.5 and 13.3; 17; 21.1.1)
(a) Vol. **85**, No. 1, p. K19—K22 (1984).

M. SCROCCO
*X-Ray Photoelectron Spectra of Ag, Ga, In, and β-Sn Metals. Multielectron Excitations
(13.1 and 17; 21; 21.6)
(b) Vol. **125**, No. 2, p. 713—720 (1984).

S. J. MALI and D. G. KANHERE
*Calculation of X-Ray Photoemission Shifts for Some Metallic Elements Using Slater's Transition State Theory
(13 and 17; 21)
(b) Vol. **125**, No. 1, p. K143—K145 (1984).

Y. FUKUDA, A. TOMITA, and N. TAKEUCHI
*Thermoluminescence and Thermally Stimulated Exoelectron Emission in Glass and Sintered CaB_4O_7 $:CuCl_2$
(17 and 20.3; 2; 22.6)
(a) Vol. **85**, No. 2, p. K141—K144 (1984).

A. R. Krasnaya, L. N. Oster, and V. Ya. Yaskolko
*Exoelectron Emission from Calcium Sulfate Excited by Slow Electrons
(17; 10.2; 22.8.1)
(a) Vol. **85**, No. 2, p. K145—K148 (1984).

G. Rudlof and H. Glaefeke
*Remarks about a New, Constructive Method for the Analysis of Thermally Stimulated Relaxation Processes (TSRP)
(13.4 and 17; 22)
(a) Vol. **85**, No. 2, p. K149—K153 (1984).

G. Thörner and G. Borstel
*Relativistic Theory of Photoemission from Solids (I): Key Aspects of the Model
(17; 13.3)
(b) Vol. **126**, No. 2, p. 617—628 (1984).

M. Rösler and W. Brauer
*On the Theory of Ion-Induced Electron Emission
(17; 13.5.2; 21)
(b) Vol. **126**, No. 2, p. 629—642 (1984).

18. Magnetic Properties

Review Article

L. B. Dubrovskaya and S. Z. Nazarova
*Some Regularities in the Behaviour of the Magnetic, Electrokinetic, and Superconducting Properties of Transition Metal Monocarbide and Mononitride Solid Solutions
(14.3 and 18; 13.1; 14.2; 15; 22)
(a) Vol. **83**, No. 1, p. 11—38 (1984).

Original Papers and Short Notes

H. J. Krokoszinski and K. Bärner
*Characterization of the Magnetic Phases of $MnAs_{1-x}P_x$ Crystals
(18; 21.1)
(b) Vol. **121**, No. 1, p. 157—160 (1984).

J. Pozo and R. Ferrer
*Specific Heat of Amorphous Magnets with Mixed Exchange Interactions
(8 and 18; 2)
(b) Vol. **121**, No. 1, p. K39—K41 (1984).

L. N. Blinov, L. A. Baydakov, Z. Cimpl, and F. Kosek
*A Note on the Magnetic Susceptibility of Tetrahedral Semiconductors and Ionic Crystals
(18; 22)
(b) Vol. **121**, No. 1, p. K43—K45 (1984).

A. V. Andreev, M. Zelený, L. Havela, and J. Hřebík
*Magnetic Properties of UGaNi
(18; 22.7)
(a) Vol. **81**, No. 1, p. 307—311 (1984).

T. Datta, S. D. LeVine, D. Thornberry, and E. R. Jones, Jr.
*Experimental Determination of the Edwards-Anderson Order Parameter in a High Temperature Spin Glass
(2 and 18; 21.1.1)
(b) Vol. **121**, No. 2, p. K125—K130 (1984).

G. Bhat, A. Mody, and A. Rangwala
*A Cluster Model of Spin Glass and Replica Symmetry Breaking
(18)
(b) Vol. **121**, No. 2, p. K135—K139 (1984).

B. Dusowski and G. Heber
*Magnetic Properties of a Random Ensemble of Two-Spin Clusters. Molecular-Field Approach
(13 and 18)
(b) Vol. **122**, No. 1, p. 133—140 (1984).

E. L. Ivchenko, V. P. Kochereshko, G. V. Mikhailov, and I. N. Uraltsev
*Resonance Magneto-Spatial Dispersion of Crystals
(18 and 20.1; 13.5.1; 22.4.2)
(b) Vol. **122**, No. 1, p. 221—230 (1984).

G. I. Tsurkan, G. M. Shmelev, and Nguyen Hong Shon
Photostimulated Uneven Magnetoresistance at Inelastic Scattering of Electrons in Semiconductors
(14.3; 18; 22)
(b) Vol. **122**, No. 1, p. 361—366 (1984).

A. V. Andreev, L. Havela, M. Zelený, and J. Hřebík
*Magnetic Properties of UGaCo
(18; 22.7)
(a) Vol. **82**, No. 1, p. 191—194 (1984).

D. Roy, Devendra Kumar, and D. Bahadur
*Crystallisation Behaviour and Magnetic Properties of $Al_2O_3 \cdot Fe_2O_3 \cdot MnO \cdot SiO_2$ Glass
(1.2 and 18; 22.8.2)
(a) Vol. **82**, No. 1, p. K11—K15 (1984).

E. Burzo and I. Lupsa
*On the Thermal Variation of Spontaneous Magnetization in Ternary NiCuD (D = Zn, Al, or Si) Alloys
(18; 18.2; 21.1)
(a) Vol. **82**, No. 1, p. K49—K52 (1984).

P. V. Nutsubidze, V. I. Chechernikov, L. G. Chachkhiani, Z. B. Chachkhiani, D. G. Barakadze, and V. K. Slovyanskikh
*Magnetic Properties of Mixed Chalcogenides of Rare-Earth Elements and Uranium
(18; 22.8)
(a) Vol. **82**, No. 1, p. K53—K56 (1984).

T. Tsuji, K. Naito, and K. Ishigure
*Effect of Particle Size on Mössbauer Parameters of α-Fe_2O_3
(5 and 18; 22.6)
(a) Vol. **82**, No. 1, p. K57—K61 (1984).

S. Guha and C. Basu
Stimulated Brillouin Scattering of Laser Radiation by an Electron-Acoustic Wave in Magnetised Semiconductor Plasmas
(13 and 20; 18)
(b) Vol. **122**, No. 2, p. 799—804 (1984).

Z. A. Matysina, D. R. Rizdvyanetskii, and S. Y. Zaginaichenko
*The Structural Diagrams of Magnetic Decompositive Alloys
(1.2 and 18; 21; 21.1)
(a) Vol. **82**, No. 2, p. K163—K168 (1984).

W. Nolting and M. Matlak
*Complete Analytical Solution for the Zero Bandwidth s–f Model
(13 and 18)
(b) Vol. **123**, No. 1, p. 155—168 (1984).

H. Fehske, E. Kolley, and W. Kolley
*Spin-Glass Behaviour in Disordered Hubbard Alloys
(18; 13; 21)
(b) Vol. **123**, No. 2, p. 533—540 (1984).

K. Bärner, C. Santandrea, U. Neitzel, and E. Gmelin
*Thermal and Magnetic Properties of $Cr_xMn_{1-x}As$ Compounds
(8 and 18; 21.1)
(b) Vol. **123**, No. 2, p. 541—552 (1984).

J. Šlechta
*On the Magnetic Part of C_V of Amorphous Materials
(8 and 18; 2)
(b) Vol. **123**, No. 2, p. K133—K137 (1984).

M. Mondal and K. P. Ghatak
*Influence of Magnetic Quantization on the Effective Electron Mass in Zero-Gap Semiconductors
(13.1 and 18; 22)
(b) Vol. **123**, No. 2, p. K143—K148 (1984).

J. Sznajd
*Effective Hamiltonian for the Two-Dimensional Arbitrary Spin Ising Model
(18)
(b) Vol. **124**, No. 1, p. 103—110 (1984).

V. P. Kochereshko, J. Röseler, I. N. Uraltsev, and K. Henneberger
Dipole Forbidden Transmission of Magnetopolaritons in CdSe
(13.5.1 and 20.1; 18; 22.4.2)
(b) Vol. **124**, No. 1, p. 213—219 (1984).

M. J. Stillman, P. W. M. Jacobs, K. Oyama Gannon, and D. J. Simkin
*Absorption and Magnetic Circular Dichroism Spectra of $CsI:Tl^+$
(18 and 20.1; 22.5.2)
(b) Vol. **124**, No. 1, p. 261—270 (1984).

M. Kibler, G. Grenet, J. C. Gâcon, and B. Jacquier
Energy Levels of Sm^{2+} in BaClF and SrClF Crystals under Magnetic Field (I): Theory
(13.4 and 20.3; 18; 22.5.4)
(b) Vol. **124**, No. 1, p. 325—334 (1984).

H. Stramska
*The Voigt Effect under a Strong Electric Field in Semiconductors
(14.3.4 and 18; 22)
(b) Vol. **124**, No. 1, p. 431—437 (1984).

R. B. Prabhu and M. N. Ghatikar
*Isomer Shift and Susceptibility of Mix-Valent $EuCu_2Si_2$
(13 and 18; 22.8)
(b) Vol. **124**, No. 1, p. K25—K29 (1984).

E. M. GOLDYS
The Donor Wave Functions in Gallium Phosphide and the Real Band Structure
(13.4 and 20.1; 18; 22.2.2)
(b) Vol. **124**, No. 1, p. K79—K82 (1984).

D. V. GITSU and F. M. MUNTYANU
Magnetoresistance of $Bi_{1-x}Sb_x$ Alloys ($0.08 \leqq x \leqq 0.15$) in Transversal Quantizing Magnetic Fields
(14.3; 18; 22.7)
(b) Vol. **124**, No. 1, p. K87—K91 (1984).

R. Z. SADYKHOV, L. M. VALIEV, and A. O. ISMAILOV
*Magnetic Properties of the $Fe_xCo_{1-x}Cr_2S_4$ ($0 < x < 1$) System
(18; 22.8)
(a) Vol. **84**, No. 1, p. K51—K54 (1984).

B. G. TAGIEV, G. D. SULTANOV, G. M. NIFTIEV, I. M. ASKEROV, and G. K. ASLANOV
*The Mössbauer Effect and Magnetic Susceptibilities in $Ga_2Se_3:^{57}Fe$ Single Crystals
(5 and 18; 22.3)
(a) Vol. **84**, No. 1, p. K59—K62 (1984).

G. MARTÍNEZ and R. RAMÍREZ
*Correlation Effects in the Rotationally Invariant Asymmetric Anderson Model
(13.4 and 18)
(b) Vol. **124**, No. 2, p. 573—586 (1984).

M. KIBLER, J. C. GÂCON, G. GRENET, B. JACQUIER, and L. C. BRUNEL
Energy Levels of Sm^{2+} in BaClF and SrClF Crystals under Magnetic Field (II): Results
(13.4 and 20.3; 18; 22.5.4)
(b) Vol. **124**, No. 2, p. 731—739 (1984).

H. LUCHIAN
*Magnetic Symmetry in the A15 Type of Structure
(14.2 and 18; 21)
(b) Vol. **125**, No. 2, p. K125—K128 (1984).

D. HAJDUKOVIĆ and I. M. SAVIĆ
*On the Location of the Transition Points of the Ising Model with Multiple Transitions
(18)
(b) Vol. **124**, No. 2, p. K129—K131 (1984).

L. JAHN and D. HINZ
*Rotational Hysteresis and Anisotropy Field of Mn–Al–C Magnets
(18; 21)
(a) Vol. **84**, No. 2, p. K133—K134 (1984).

N. M. L. KÖCHE, P. C. MORAIS, and K. SKEFF NETO
*The Usefulness of the Photoacoustic Cell for Magnetic Measurements
(7 and 18; 19; 22.8.2)
(a) Vol. **85**, No. 1, p. K65—K67 (1984).

M. I. Vladimir, V. A. Moskalenko, and V. V. Skafaru
*The Distribution of the Exchange Magnetic Fields in Spin Glasses. The Account of Local State Probabilities
(18)
(b) Vol. **125**, No. 2, p. 543—550 (1984).

P. Středa
Universal Transport Properties of Inversion Layers in Quantizing Magnetic Field
(13.3; 8; 14.1; 15; 18; 21)
(b) Vol. **125**, No. 2, p. 849—860 (1984).

D. Hajduković and I. M. Savić
*Ground-State Degeneracy of the One-Dimensional Potts and Ising Magnets with Competing Interactions
(13 and 18)
(b) Vol. **125**, No. 2, p. K113—K116 (1984).

O. F. Antonov and R. Kh. Sabirov
*Contribution of the Higher-Order Secular Spin–Spin Interaction to the First and the Second Magnetic Resonance Line Moments
(18 and 19)
(b) Vol. **125**, No. 2, p. K117—K120 (1984).

V. S. Rukhlov and G. B. Teitelbaum
*Effect of Spin–Spin Reservoir on the Magnetic Bistability
(13 and 18)
(b) Vol. **125**, No. 2, p. K121—K124 (1984).

A. Z. Maksymowicz, T. Stobiecki, E. Jarocki, and W. Karaś
*Charge Transfer and Concentration Dependence of Magnetization in Amorphous Fe–B Alloys
(13 and 18; 2; 21.1.1)
(b) Vol. **126**, No. 1, p. 191—195 (1984).

I. N. Karnaukhov and A. F. Zhuravlev
*Magnetic State of Crystals with Dislocations
(10.1 and 18; 19)
(b) Vol. **126**, No. 1, p. 207—212 (1984).

B. Lorenz and H. Fehske
*Critical Study of the Static Functional Integral Method in the Hubbard Model
(13 and 18; 8)
(b) Vol. **126**, No. 1, p. 235—245 (1984).

L. R. Evangelista and V. K. Saxena
*Random-Field Ising Model with a Zero Temperature Multicritical Point
(18)
(b) Vol. **126**, No. 1, p. K29—K31 (1984).

M. MONDAL and K. P. GHATAK
*Magnetic Susceptibility of Electrons in Ultrathin Films of Semiconductors in the Presence of a Quantizing Magnetic Field
(13 and 18; 22.2.1)
(b) Vol. **126**, No. 1, p. K41—K45 (1984).

E. ZSCHECH, W. BLAU, H. VEGA, K. KLEINSTÜCK, S. MAGER, M. A. KOZLOV, and M. A. SHEROMOV
EXAFS and X-Ray Diffractional Investigation of the Heusler-Type Alloys Co_2MnSi and $Fe_{2.4}Mn_{0.6}Al$. Determination of the Ordering Probabilities
(1.3; 18; 21.1; 21.1.1)
(a) Vol. **86**, No. 1, p. 117—124 (1984).

H. GRUBER, E. KRAUTZ, H. P. FRITZER, K. GATTERER, and A. POPITSCH
Electronic Property Changes of η-Mo_4O_{11} by Doping with Tungsten, Rhenium, and Vanadium
(14.1; 1.2; 13.1; 18; 20.1; 22.6.1)
(a) Vol. **86**, No. 2, p. 749—757 (1984).

18.1 Paramagnetic Properties

Original Papers and Short Notes

G. J. Ashwell
(2,2-DHPE Cl)$_x$(2,2-DPE)$_{1-x}$(TCNQ), a Pauli Paramagnetic Organic Conductor
(14.3; 1.1; 18.1; 22.9)
(a) Vol. **81**, No. 1, p. 361—366 (1984).

V. I. Pecherskaya, D. N. Bolshutkin, V. A. Desnenko, and V. Ya. Ilichev
*Electrical and Magnetic Properties of the Fe18Cr(30–40)Ni Alloys at Low Temperatures
(18.1 and 18.2; 14.1; 21.1; 21.1.1)
(a) Vol. **81**, No. 2, p. 507—512 (1984).

A. S. Litovchenko, A. V. Brodovoi, V. D. Tkachenko, and V. V. Mazykin
*Temperature Dependence of the Magnetic Susceptibility of Teniolite
(18.1; 22.8.1)
(a) Vol. **81**, No. 2, p. K151—K153 (1984).

P. S. Riseborough
De Haas-Van Alphen Effect in Nearly Magnetic Transition Metals
(13; 18.1; 21; 21.4)
(b) Vol. **122**, No. 1, p. 161—170 (1984).

Y. Watanabe
ESR and Magnetic Investigation of Lithium Intercalated VSe_2
(13.4; 10.2; 18.1; 19; 22.8)
(b) Vol. **122**, No. 1, p. 309—316 (1984).

E. Kolley and W. Kolley
*Critical Behaviour of a Random Itinerant-Electron Spin Model
(18.1 and 18.2; 21)
(b) Vol. **122**, No. 1, p. K27—K31 (1984).

Z. Onyszkiewicz
*Theory of the Dependence of the Phase Boundary of $FeBr_2$ Metamagnet on Hydrostatic Pressure
(18.1 and 18.4; 22.5)
(b) Vol. **122**, No. 2, p. 543—547 (1984).

G. C. Upreti and R. S. Saraswat
EPR Study of VO^{2+} in Some Paramagnetic Tutton Salt Single Crystals
(19; 18.1; 22.8.1)
(b) Vol. **122**, No. 2, p. K169—K173 (1984).

R. Niedzielski and R. Troć
*Magnetic Susceptibility of PrP, PrS, NdP, NdS, and $U_{1-x}Pr(Nd)_xP$ in the Paramagnetic Region
(18.1; 21)
(b) Vol. **123**, No. 1, p. K29—K34 (1984).

G. J. Ashwell and I. W. Nowell
Electrical, Magnetic, and Structural Characterisation of [1,2-Bis(1-ethyl-4-pyridinium)ethane]$^{2+}$ (7,7,8,8-Tetracyano-p-quinodimethanide)$_4^{2-}$
(1.1; 14.3; 18.1; 22.9)
(a) Vol. **83**, No. 1, p. 39—45 (1984).

E. K. Sadykov
The Theory of Double Gamma Resonance in Paramagnetics
(5 and 19; 18.1)
(b) Vol. **123**, No. 2, p. 703—709 (1984).

H. Kido, T. Hoshikawa, M. Shimada, and M. Koizumi
*Crystal Structure and Magnetic and Electrical Properties of Pseudo-Ternary $GdCu_{1-x}Zn_xSi$ Compounds
(14.1 and 18.1; 1.3; 18.2; 22.8)
(a) Vol. **83**, No. 2, p. 561—565 (1984).

C. Miyake, H. Takeuchi, K. Fuji, and S. Imoto
*Electron Spin Resonance and Magnetic Susceptibility of M–U–O Ternary Mixed Oxides (II): Anomalies of Electron Spin Resonance and Magnetic Susceptibility of $Na_{2n-5}UO_n$ ($n = 3$ and 4)
(18.1 and 19; 18.4; 22.6.1)
(a) Vol. **83**, No. 2, p. 567—572 (1984).

G. J. Ashwell
*Electrical, Magnetic, and Structural Characterisation of [1,2-Bis(1-methyl-4-pyridinium)ethylene]$^{2+}$(TCNQ)$_4^{2-}$(H_2O) and the Properties of Two Isostoichiometric TCNQ Salts
(14.3 and 18.1; 1.1; 15; 22.9)
(a) Vol. **83**, No. 2, p. 669—673 (1984).

C. Miyake, H. Takeuchi, H. Ohya-Nishiguchi, and S. Imoto
*Erratum to: Electron Spin Resonance and Magnetic Susceptibility of M–U–O Ternary Mixed Oxides (I): Anomalies of Electron Spin Resonance and Magnetic Susceptibility of $Li_{2n-5}UO_n$ ($n = 3, 4,$ and 6)
(18.1 and 19; 18.4; 22.6.1)
(a) Vol. **83**, No. 2, p. 729 (1984).

A. N. Bogdanov, A. V. Zhuravlev, and V. T. Telepa
*First-Order Phase Transition from Antiferromagnetic to Paramagnetic Phase in $(C_2H_5NH_3)_2CuCl_4$
(18.1 and 18.4; 1.2; 22.8)
(a) Vol. **83**, No. 2, p. K135—K140 (1984).

L. Kowalewski and R. Wojciechowski
*Mode–Mode Coupling Effects in the Singlet–Triplet Paramagnet in the Vicinity of the Curie Point
(18.1)
(b) Vol. **124**, No. 2, p. 599—608 (1984).

D. A. Garanin and V. S. Lutovinov
*Diagrammatic Approach to the NQR Linewidths in Magnetically Ordered Materials
(18.1 and 19)
(b) Vol. **124**, No. 2, p. K133—K137 (1984).

A. I. Gusev and A. A. Rempel
A Study of the Atomic Ordering in the Niobium Carbide Using the Magnetic Susceptibility Method
(1.3; 18.1; 22)
(a) Vol. **84**, No. 2, p. 527—534 (1984).

A. V. Salker, E. B. Mirza, V. V. Deshpande, and H. V. Keer
Thermodynamic Analysis of the Phase Transition in V_4O_7
(1.2; 8; 14.1; 14.3; 18.1; 22.6)
(a) Vol. **84**, No. 2, p. K105—K108 (1984).

V. E. Sidorov, V. S. Gushchin, and B. A. Baum
*Magnetic Structure of γ-Iron Containing Oxygen
(18.1; 21.1.1)
(a) Vol. **85**, No. 2, p. 497—501 (1984).

M. Takahashi, Y. Sasaki, and S. Ishio
*Magnetic Freezing in Solution-Quenched Dilute Fe–Cu Alloys
(18.1; 21.1; 21.1.1)
(a) Vol. **86**, No. 1, p. 289—294 (1984).

K. Iwahana
Magnetic Field Dependence of the Spin Polarization of the F Center
(10.2; 18.1; 22.5.2)
(b) Vol. **126**, No. 2, p. 693—702 (1984).

T. Pannaparayil, M. Oskooie-Tabrizi, C. Lo, and L. N. Mulay
Infrared Spectroscopic, Magnetic, and Mössbauer Studies of Aluminosilicate Cage Structures (Fe/H-Mordenites) for Syngas Conversion
(5; 18.1; 20.1; 22.8.1)
(a) Vol. **86**, No. 2, p. 663—671 (1984).

18.2 Ferromagnetic Properties

Review Articles

K. N. Shrivastava
*Origin of Superexchange Interactions in 3d Magnetic Solids (I)
(6 and 18.2; 13; 22.5)
(b) Vol. **125**, No. 1, p. 11—43 (1984).

K. N. Shrivastava
*Origin of Superexchange Interactions in 3d Magnetic Solids (II)
(6 and 18.2; 13; 22.5)
(b) Vol. **125**, No. 2, p. 441—460 (1984).

Original Papers and Short Notes

A. A. Kazakov
Two Types of Low-Temperature Phase Transitions in Uniaxial Ferromagnets with High Single-Ion Anisotropy
(13.5.4; 18.2)
(b) Vol. **121**, No. 1, p. 177—183 (1984).

A. I. Mitsek, D. I. Sirota, and N. P. Kolmakova
*Domain Wall Stabilization in Crystals with Complicated Magnetic Phase Diagram
(18.2 and 18.4)
(b) Vol. **121**, No. 1, p. 185—195 (1984).

R. Honmura, E. F. Sarmento, I. P. Fittipaldi, and T. Kaneyoshi
*Phase Diagram for the Ising Model with a Free Surface
(18.2; 13.3)
(b) Vol. **121**, No. 1, p. 197—202 (1984).

H. E. Borges and P. R. Silva
*Bidimensional Spin —1/2 Ising Model in a Random Field
(13 and 18.2)
(b) Vol. **121**, No. 1, p. K19—K24 (1984).

H. E. Borges and P. R. Silva
*Tridimensional Spin —1/2 Ising Model in a Random Field
(13 and 18.2)
(b) Vol. **121**, No. 1, p. K25—K28 (1984).

H. A. Brown
*Additional Temarks to the Schrödinger Spin-Exchange Ferromagnet
(18.2)
(b) Vol. **121**, No. 1, p. K29—K31 (1984).

A. MODY
*A Numerical Study of a Two-Dimensional Bond-Disordered Ising Ferromagnet
(18.2)
(b) Vol. **121**, No. 1, p. K33—K37 (1984).

H. J. BLYTHE, F. WALZ, and H. KRONMÜLLER
*Magnetic Investigation of the Annealing Behaviour of Some Dilute Iron–Nickel Alloys after Low-Temperature Electron Irradiation
(10.1 and 18.2; 11; 21.1; 21.1.1)
(a) Vol. **81**, No. 1, p. 227—238 (1984).

A. I. LUKANIN and M. V. MEDVEDEV
*Spin Waves in an Easy-Axis Ferromagnet with a Finite Concentration of Impurities with Competing Anisotropy
(18.2; 21.4)
(b) Vol. **121**, No. 2, p. 525—537 (1984).

A. A. KAZAKOV and S. M. KLESHEV
*Thermodynamical Behaviour of the Uniaxial Crystals of "Easy Axis" Type with Diverse Origins of Magnetic Anisotropy
(18.2)
(b) Vol. **121**, No. 2, p. 539—545 (1984).

A. I. LUKANIN and M. V. MEDVEDEV
*Magnetic States of a Binary Ferromagnetic Alloy with Competition of Easy-Axis and Easy-Plane Single-Ion Anisotropies
(18.2; 21.4)
(b) Vol. **121**, No. 2, p. 573—582 (1984).

R. G. PILLAY, S. H. DEVARE, F. PLEITER, and H. G. DEVARE
*Transferred Hyperfine Field at ^{111}Cd in Ordered RNi_5 Compounds (R = Rare-Earth)
(18.2; 22.7)
(b) Vol. **121**, No. 2, p. K141—K145 (1984).

J. M. WESSELINOWA
*Magnon Damping of the s–d Model above T_C
(18.2; 13.5.4; 22)
(b) Vol. **121**, No. 2, p. K147—K149 (1984).

P. R. SILVA, F. C. SÁ BARRETO, and H. E. BORGES
*Simulation of the Site Diluted Ising Model through an Infinite Random Field
(13 and 18.2)
(b) Vol. **121**, No. 2, p. K151—K155 (1984).

J. KUČERA, L. KOZÁK, and H. MEHRER
Magnetic Anomalies of Self-Diffusion and Co Heterodiffusion in α-Fe
(9; 10.1; 18.2; 21.1.1)
(a) Vol. **81**, No. 2, p. 497—505 (1984).

V. I. PECHERSKAYA, D. N. BOLSHUTKIN, V. A. DESNENKO, and V. YA. ILICHEV
*Electrical and Magnetic Properties of the Fe18Cr(30–40)Ni Alloys at Low Temperatures
(18.1 and 18.2; 14.1; 21.1; 21.1.1)
(a) Vol. **81**, No. 2, p. 507—512 (1984).

A. H. WAFIK, G. GRAISS, and A. RIZK
*Barkhausen Jumps in a Dual-Phase Steel Treated in the Intercritical Temperature Range
(18.2; 21.1.1)
(a) Vol. **81**, No. 2, p. 513—517 (1984).

S. MOTOGI and G. A. MAUGIN
*Effects of Magnetostriction on Vibrations of Bloch and Néel Walls
(18.2; 12.2; 22.1.2; 22.6)
(a) Vol. **81**, No. 2, p. 519—532 (1984).

K. SUMIYAMA and Y. NAKAMURA
High Concentration Fe–Cu and Fe–Ag Alloys Produced by Vapor Quenching
(1.3; 5; 18.2; 21.1; 21.1.1; 21.6)
(a) Vol. **81**, No. 2, p. K109—K112 (1984).
Erratum
(a) Vol. **83**, No. 2, p. K219 (1984).

M. J. ŠKRINJAR, D. V. KAPOR, and J. P. ŠETRAJČIĆ
*The Validity of the Low-Temperature Expansions for the Anisotropic Heisenberg Models
(18.2)
(b) Vol. **122**, No. 1, p. 125—131 (1984).

E. KOLLEY and W. KOLLEY
*Critical Behaviour of a Random Itinerant-Electron Spin Model
(18.1 and 18.2; 21)
(b) Vol. **122**, No. 1, p. K27—K31 (1984).

F. WALZ
*High-Sensitivity Two-Component Measuring Technique of Inductances as Applied to the Investigation of Magnetic After-Effects
(18.2; 18.3; 21.1.1)
(a) Vol. **82**, No. 1, p. 179—189 (1984).

E. BURZO and I. LUPSA
On the Thermal Variation of Spontaneous Magnetization in Ternary NiCuD (D = Zn, Al, or Si) Alloys
(18; 18.2; 21.1)
(a) Vol. **82**, No. 1, p. K49—K52 (1984).

V. D. SEDYKH and A. V. SEREBRYAKOV
*The Effect of Thermal Treatment on the Magnetic Anisotropy in Amorphous Fe–B Alloys
(2 and 18.2; 5; 21.1.1)
(a) Vol. **82**, No. 1, p. K63—K66 (1984).

R. S. Gekht
*Low-Temperature Properties of One-Dimensional Heisenberg Systems with Random Anisotropy
(13 and 18.2)
(b) Vol. **122**, No. 2, p. 509—517 (1984).

A. Z. Maksymowicz
*Short-Range Order and Magnetic Properties in Amorphous Heisenberg Ferromagnets
(2 and 18.2)
(b) Vol. **122**, No. 2, p. 519—526 (1984).

V. Brédimas, H. Gamari-Seale, and C. Papatriantafillou
*The Magnetic Behaviour of the Pseudobinary Compounds $Gd_2Au_xAl_{1-x}$
(18.2 and 18.4; 21.4)
(b) Vol. **122**, No. 2, p. 527—534 (1984).

D. L. Alov, S. I. Gubarev, and V. B. Timofeev
*Determination of the s–d Exchange Interaction Sign in $Cd_{1-x}Mn_xS$ Crystals
(13.4 and 18.2; 22.4.4)
(b) Vol. **122**, No. 2, p. K133—K136 (1984).

F. Kh. Abdullaev and A. A. Abdumalikov
*Dynamics of Magnetic Solitons in Planar Ferromagnets with Random Anisotropy
(13.5 and 18.2; 6; 13.5.1)
(b) Vol. **122**, No. 2, p. K141—K146 (1984).

U. Hartmann and H. H. Mende
*The Influence of Anisotropy-Controlled Néel Relaxation on Magnetostatic Properties of Ferrofluids
(18.2)
(a) Vol. **82**, No. 2, p. 481—489 (1984).

V. Hajko, A. Zentko, M. Timko, and V. Hajko, Jr.
*Isotopic Effect in Neutron-Irradiated $Fe_{40}Ni_{40}B_{20}$ Alloys
(11 and 18.2; 2; 21.1.1)
(a) Vol. **82**, No. 2, p. K159—K162 (1984).

T. Soumura, S. Saito, T. Kuraishi, and T. Maeda
*Magnetic Moment and Its Temperature Variation of Amorphous $(Fe_{0.5}Ni_{0.5})_{1-x}B_x$ Alloys
(2 and 18.2; 21.1; 21.1.1)
(a) Vol. **82**, No. 2, p. K169—K172 (1984).

R. Krishnan
*Magnetic Moment of Amorphous $Co_{1-x}Nb_x$ Alloys and Friedel's Model
(2 and 18.2; 21; 21.1)
(a) Vol. **82**, No. 2, p. K173—K175 (1984).

M. Sh. Erukhimov and S. G. Ovchinnikov
Atomic Operators in the Theory of Narrow-Band Magnetic Semiconductors
(13.5.4; 18.2; 22)
(b) Vol. **123**, No. 1, p. 105—114 (1984).

L. PŮST
Low-Temperature FMR and FMAR Measurements of Metal Single Crystals (II): Saturation Magnetization, Magnetocrystalline Anisotropy, and g-Factor Determination
(19; 18.2; 21)
(b) Vol. **123**, No. 1, p. 115—122 (1984).

R. RICHTER and F. GOEDSCHE
On the Thermoelectric Power of Amorphous Ferromagnets
(2 and 15; 14.1; 18.2; 21)
(b) Vol. **123**, No. 1, p. 143—154 (1984).

I. M. SAVIĆ and D. HAJDUKOVIĆ
*On the Location of the Critical Point of the q-State Potts Model on the B.C.C. and F.C.C. Lattice
(18.2)
(b) Vol. **123**, No. 1, p. K35—K38 (1984).

M. SEIDEL and G. NOCKE
*A Semi-Empirical Relation beween the Magnetic Susceptibility and Porosity of Sintered Iron
(18.2; 21.1.1)
(a) Vol. **83**, No. 1, p. K43—K46 (1984).

T. KANEYOSHI and I. TAMURA
*The Distribution of Magnetic Moments and the Temperature Dependence of Reduced Hyperfine Fields in Amorphous Ferromagnets
(2 and 18.2; 21.1.1)
(b) Vol. **123**, No. 2, p. 525—531 (1984).

R. PLUMIER
*Study at 0 K and Various Magnetic Fields of the Triangular Magnetic Configurations
(18.2 and 18.4)
(b) Vol. **123**, No. 2, p. 553—562 (1984).

KH. I. PUSHKAROV and M. T. PRIMATAROWA
*Solidary Clusters of Spin Deviations and Lattice Deformation in an Anharmonic Ferromagnetic Chain
(13.5.4 and 18.2)
(b) Vol. **123**, No. 2, p. 573—584 (1984).

J. M. WESSELINOWA
*Electron–Phonon Interaction in Ferromagnetic Semiconductors
(18.2; 6; 13; 22)
(b) Vol. **123**, No. 2, p. 585—593 (1984).

M. ŠOB, S. SZUSZKIEWICZ, and M. SZUSZKIEWICZ
Polarized Positron Annihilation and Enhancement Effects in Ferromagnetic Iron
(13; 18.2; 21.1.1)
(b) Vol. **123**, No. 2, p. 649—652 (1984).

J. M. WESSELINOWA
*Relaxation Time for Ferromagnetic Semiconductors
(13 and 18.2; 22)
(b) Vol. **123**, No. 2, p. K129—K132 (1984).

G. Chadda
*Planar Heisenberg Ferromagnet with Biquadratic Interactions
(18.2)
(b) Vol. **123**, No. 2, p. K139—K142 (1984).

H. Kido, T. Hoshikawa, M. Shimada, and M. Koizumi
Crystal Structure and Magnetic and Electrical Properties of Pseudo-Ternary $GdCu_{1-x}Zn_xSi$ Compounds
(14.1 and 18.1; 1.3; 18.2; 22.8)
(a) Vol. **83**, No. 2, p. 561—565 (1984).

L. Jahn and V. Christoph
*Rotational Hysteresis of 2–17–RE–Cobalt Magnets
(18.2; 21.1; 21.4)
(a) Vol. **83**, No. 2, p. K141—K144 (1984).

I. Ya. Korenblit and B. G. Tankhilevich
*Fluctuations in a Nonequilibrium System of Electrons and Magnons in Ferromagnetic Semiconductors
(13.5.4 and 18.2; 22)
(b) Vol. **124**, No. 1, p. 79—86 (1984).

J. Barnaś
*On the Possibility of Resonant Excitation of Spin Waves by an External Electric Field in Electronic-Type Linear Ferromagnetoelectrics
(14.4.1 and 18.2; 19)
(b) Vol. **124**, No. 1, p. K31—K34 (1984).

J. A. Plascak
*Spin-1/2 Anisotropic Heisenberg Model in a Transverse Field
(18.2)
(b) Vol. **124**, No. 1, p. K35—K38 (1984).

U. Lindner
Solitary Solutions of the Heisenberg Model
(13.5.4; 18.2)
(b) Vol. **124**, No. 1, p. K39—K44 (1984).

J. P. Šetrajčić, M. J. Skrinjar, and D. Mirjanić
On the Possibility of Bose-Einstein Condensation and the Superfluidity of Spin Waves
(13.5.1 and 13.5.4; 18.2)
(b) Vol. **124**, No. 1, p. K45—K48 (1984).

W. Winkler
*Mössbauer Studies of Small Magnetite Particles of a Magnetic Fluid
(18.2; 5; 22.6)
(a) Vol. **84**, No. 1, p. 193—198 (1984).

B. Chenevier, D. Fruchart, M. Bachmann, J. P. Senateur, P. Chaudouet et L. Lundgren
*Propriétés magnétiques de MnRhAs. Etude de la transition antiferro–ferromagnétique
(Magnetic Properties of MnRhAs. Study of the Antiferromagnetic–Ferromagnetic Transition)
(18.2 and 18.4; 1.2; 14.1; 21.1)
(a) Vol. **84**, No. 1, p. 199—206 (1984).

K. H. J. Buschow and D. B. de Mooij
*The Effect of H_2 Absorption on the Magnetic Properties of Gd–Pd Intermetallic Compounds
(18.2 and 18.4; 21; 21.4)
(a) Vol. **84**, No. 1, p. 207—213 (1984).

R. Zach, R. Duraj, and A. Szytuła
*Pressure Effect on Magnetic Transformations in the $Co_xNi_{1-x}MnGe$ System
(18.2 and 18.4; 1.2; 22.8)
(a) Vol. **84**, No. 1, p. 229—236 (1984).

T. Palewski
*The Magnetic Properties of $U_xTh_{1-x}Se$ Solid Solution
(18.2; 21.5)
(a) Vol. **84**, No. 1, p. K47—K50 (1984).

A. P. Ges, V. M. Derkachenko, and V. V. Fedotova
*Effect of Scandium Ions on Magnetic Properties of Dysprosium Orthoferrite
(18.2; 22.8.2)
(a) Vol. **84**, No. 1, p. K63—K65 (1984).

A. R. Ali, S. A. Mahmoud, and G. Said
*On the Role of Cu Atoms on the Behaviour of Domain Wall–Dislocation Interaction in Ni–Cu Alloy
(10.1 and 18.2; 21.1)
(a) Vol. **84**, No. 1, p. K67—K70 (1984).

D. Fraitová
*An Analytical Theory of FMR in Bulk Metals (Parallel Configuration) (III): Surface Impedance
(18.2 and 19)
(b) Vol. **124**, No. 2, p. 587—592 (1984).

S. L. Lock and B. S. Lee
*Two-Spin Cluster Approximation in the Blume-Capel Model
(18.2)
(b) Vol. **124**, No. 2, p. 593—598 (1984).

J. Kierul, J. Ledzion, and Z. Osetek
*Magnetic Susceptibility of Non-Uniform Media
(18.2; 21.1)
(b) Vol. **124**, No. 2, p. K121—K124 (1984).

H. E. Khodenkov, N. N. Kudelkin, and V. V. Randoshkin
*The Break-Down of the 360° Bloch Domain Wall in Bubble Magnetic Films
(18.2; 22.8.2)
(a) Vol. **84**, No. 2, p. K135—K138 (1984).

Yu. V. Zakharov and Yu. I. Mankov
Magnetoresistance of Ferromagnets with Domain Structure
(14.1; 18.2; 21)
(b) Vol. **125**, No. 1, p. 197—205 (1984).

D. V. Kapor, J. P. Šetrajčić, and M. J. Škrinjar
*Temperature Dependence of the Energy Gap of an Anisotropic Heisenberg Ferromagnetic Model
(18.2)
(b) Vol. **125**, No. 1, p. K45—K49 (1984).
Erratum
(b) Vol. **126**, No. 2, p. K175 (1984).

K. Oesterreicher and H. Oesterreicher
*Structure and Magnetic Properties of $Nd_2Fe_{14}BH_{2.7}$
(18.2; 21.1)
(a) Vol. **85**, No. 1, p. K61—K64 (1984).

G. I. Makovetskii and V. P. Dymont
*Crystallographic, Magnetic, and Electrical Properties of Nickel-Substituted Chromium Telluride
(14.1 and 18.2; 22.8)
(a) Vol. **85**, No. 1, p. K69—K71 (1984).

T. Iwashita and N. Uryû
*The Curie Temperature of the Ising Ferromagnet on the Honeycomb Lattice with Mixed Spins
(18.2)
(b) Vol. **125**, No. 2, p. 551—562 (1984).

V. N. Zabluda, A. V. Malakhovskii, and I. S. Edelman
*Magnetooptical Effects and Optical Absorption of $FeBO_3$ in the Region of ${}^6A_{1g} \rightarrow {}^4T_{1g}({}^4G), {}^4T_{2g}({}^4G)$ Transitions
(18.2 and 20.1; 13.4; 22.8.1)
(b) Vol. **125**, No. 2, p. 751—757 (1984).

S. L. Kakani and U. N. Upadhyaya
*Superconducting and Magnetic Order Parameters in the Coexistence Region of Ternary Superconductors
(14.2 and 18.2; 22.8)
(b) Vol. **125**, No. 2, p. 861—867 (1984).

F. Walz
*Transition-Induced Magnetic After-Effect in the Amorphous Alloy $Fe_{20}Ni_{56}B_{24}$ at the Critical Temperature
(18.2; 2; 10.1; 21.1; 21.1.1)
(a) Vol. **85**, No. 2, p. 503—509 (1984).

H. Taguchi, M. Shimada, and M. Koizumi
*Magnetic Properties in the System $(Sr_{1-x}Ca_x) CoO_{3-\delta}$
(18.2; 14.1; 22.6.1)
(a) Vol. **85**, No. 2, p. 523—527 (1984).

J. M. Wesselinowa
*Some Critical Properties of the s–d Model
(18.2; 22.8)
(b) Vol. **126**, No. 1, p. 229—233 (1984).

S. Trimper
*Anisotropic Interactions in $2 + \varepsilon$ Dimensions
(18.2; 1.2)
(b) Vol. **126**, No. 1, p. K23—K27 (1984).

M. B. Shah and M. S. C. Bose
*Magnetic NDT Technique to Evaluate Fatigue Damage
(10.1 and 18.2; 21)
(a) Vol. **86**, No. 1, p. 275—281 (1984).

A. V. Goltsev
*The Parisi-Sompolinsky Solution for an Infinite-Ranged Spin Glass Model with *m*-Component Spins
(18.2 and 18.4)
(b) Vol. **126**, No. 2, p. 537—545 (1984).

Zhen-ya Li and Chuan-zhang Yang
*Magnetic Properties of a Random Mixture of a One-Dimensional System with Anisotropic Heisenberg Exchange
(18.2; 18.4)
(b) Vol. **126**, No. 2, p. K125—K130 (1984).

M. S. C. Bose
*Effect of Saturated Magnetic Field on Fatigue Life of Carbon Steel
(12.1 and 18.2; 21.1.1)
(a) Vol. **86**, No. 2, p. 649—654 (1984).

H. M. van Noort, D. B. de Mooij, and K. H. J. Buschow
*Crystal Structure, Magnetic Properties, and ^{57}Fe Mössbauer Effect of PtFeSn
(1 and 18.2; 5; 22.8)
(a) Vol. **86**, No. 2, p. 655—662 (1984).

18.2.1 Ferromagnetic Films

Original Papers and Short Notes

W. MAASS, U. KREY, and H. HOFFMANN
*Magnetic Approach-to-Saturation in Thin Films
(18.2.1; 21.1.1; 21.4)
(b) Vol. **122**, No. 2, p. K137—K140 (1984).

H. RATAJCZAK and I. GOŚCIAŃSKA
*Planar Hall Effect in Thin Amorphous RE–TM Films with Perpendicular Anisotropy
(14.1 and 18.2.1; 21.4)
(a) Vol. **82**, No. 2, p. 491—496 (1984).

R. ŻUBEREK, F. STOBIECKI, and J. WOSIK
*Magnetoelastic Properties of $Fe_{1-x}B_x$ Amorphous Thin Films
(2 and 18.2.1; 21.1.1)
(a) Vol. **82**, No. 2, p. K177—K179 (1984).

Y. SHIMADA
*Amorphous Co–Metal Films Produced by Sputtering
(18.2.1; 2; 21.1)
(a) Vol. **83**, No. 1, p. 255—261 (1984).

Y. SHIMADA
*Initial Permeability of Amorphous Films with Multilayered Structure
(2 and 18.2.1; 21)
(a) Vol. **84**, No. 1, p. K55—K58 (1984).

T. LUCIŃSKI and J. BASZYŃSKI
Influence of Composition and Structural Transformation on Transport Properties in Amorphous $Fe_{1-x}Si_x$ Films
(14.1; 2; 18.2.1; 21.1.1)
(a) Vol. **84**, No. 2, p. 607—612 (1984).

J. BASZYŃSKI, L. SMARDZ, and F. STOBIECKI
*Magnetic Properties of Compositionally Modulated Ni/Cu Layers
(1.5 and 18.2.1; 19; 21.1)
(a) Vol. **84**, No. 2, p. K129—K131 (1984).

J. ČERMÁK, R. NOVÁK, P. NOVÁK, and M. NEVŘIVA
*Yttrium Iron Garnet Films Substituted by Gd and Mn
(18.2.1; 22.8.2)
(a) Vol. **85**, No. 1, p. 173—177 (1984).

L. PŮST and Z. FRAIT
*Method for Measurement of Magnetostriction Constants in (001) Thin Films Using FMR
(18.2.1; 19; 22.8.2)
(a) Vol. **85**, No. 1, p. 179—188 (1984).

J. J. Żebrowski and A. Sukiennicki
*On the Inhibition of Dynamic Conversion in Bubble Garnet Domains by a Thin Midplane Layer
(18.2.1; 22.8.2)
(a) Vol. **85**, No. 1, p. 189—193 (1984).

J. M. Alameda, M. C. Contreras, and H. Rubio
*Magnetic Properties of Amorphous Fe_xSi_{1-x} Films (I): Transverse Susceptibility, Structure Constant, and Local Anisotropy
(18.2.1; 2; 21.1.1)
(a) Vol. **85**, No. 2, p. 511—522 (1984).

18.3 Ferrimagnetic Properties

Original Papers and Short Notes

A. Kasten, D. Weschenfelder, and G. Hess
Spectroscopic Investigation of the Magnetic Properties of $Ho_2O_2SO_4$
(18.4; 13.4; 18.3; 22.8.1)
(b) Vol. **121**, No. 1, p. 145—155 (1984).

M. A. Sigal
*Magnetostatic Waves in a Coaxial Cylindrical Magnetic Domain
(18.3 and 19)
(b) Vol. **121**, No. 1, p. 161—171 (1984).

V. P. Gupta, A. K. Singh, K. Chandra, and S. K. Jaireth
Mössbauer Investigations on Pyrrhotite
(10.2; 5; 18.3; 18.4; 22.8)
(a) Vol. **81**, No. 1, p. 281—291 (1984).

M. A. Ivanov, V. Ya. Mitrofanov, and A. Ya. Fishman
*Magnetic Anisotropy of Disordered Magnetics with Jahn-Teller Ions
(18.3; 13.4; 22.8.2)
(b) Vol. **121**, No. 2, p. 547—559 (1984).

V. S. Kuleshov and V. A. Popov
*The Morin Point as Degenerate Critical Point
(18.3 and 18.4)
(b) Vol. **121**, No. 2, p. 561—572 (1984).

F. Walz
High-Sensitivity Two-Component Measuring Technique of Inductances as Applied to the Investigation of Magnetic After-Effects
(18.2; 18.3; 21.1.1)
(a) Vol. **82**, No. 1, p. 179—189 (1984).

A. Kasten and G. Hess
*Magnetic Properties of $Tb_2O_2SO_4$
(18.3 and 18.4; 22.8.1)
(b) Vol. **123**, No. 1, p. 123—134 (1984).

V. Crişan, I. Pop, V. Toşa, and N. Rusu
*Band Structure of $Gd_2Ni_2Al_{15}$ Intermetallic Compound
(13.1 and 18.3; 21.1; 21.4)
(b) Vol. **123**, No. 1, p. K53—K56 (1984).

A. Kasten and J. Hoffmann
*Sequence of Ferrimagnetic Phases in $Er_2O_2SO_4$
(18.3 and 18.4; 22.8.1)
(b) Vol. **123**, No. 2, p. 563—571 (1984).

J. Loos
*On the Fluctuations and Phase Transitions in Dicke-Like Models
(6 and 18.3)
(b) Vol. **123**, No. 2, p. 595—603 (1984).

G. Nicoară and V. Ghiordănescu
*The Presence of Ti^{3+} in LiTiZn Ferrites
(18.3 and 20.1; 22.8.2)
(a) Vol. **84**, No. 2, p. K139—K142 (1984).

V. I. Maltsev and V. G. Vologin
*Neutron Diffraction Studies of the Magnetic Ground State in Li–Zn Ferrite
(18.3; 22.8.2)
(a) Vol. **85**, No. 2, p. 529—534 (1984).

O. Kohmoto
Electrical Resistivity of Amorphous Co-Rich FeCo–SiB Alloys
(14.1; 2; 18.3; 21.1; 21.1.1)
(a) Vol. **85**, No. 2, p. K155—K157 (1984).

L. Merten, R. Kotzott, and A. Lehmayer
Dispersion Relations for Magnon-Polaritons in Uniaxial Ferrimagnetic Crystals
(13.5.4 and 20.1; 18.3)
(b) Vol. **126**, No. 1, p. 223—228 (1984).

A. A. Wafik and S. A. Mazen
*Barkhausen Jumps and Resistivity of Cu–Cd Ferrites
(18.3; 14.3; 22.8.2)
(a) Vol. **86**, No. 1, p. 271—274 (1984).

G. P. Kramar and Ya. I. Panova
*The Permeability Dispersion in Ferrites at Different Levels of Direct Current Magnetic Biasing
(18.3; 22.8.2)
(a) Vol. **86**, No. 1, p. 283—287 (1984).

18.4 Antiferromagnetic Properties

Original Papers and Short Notes

A. Kasten, D. Weschenfelder, and G. Hess
*Spectroscopic Investigation of the Magnetic Properties of $Ho_2O_2SO_4$
(18.4; 13.4; 18.3; 22.8.1)
(b) Vol. **121**, No. 1, p. 145—155 (1984).

A. I. Mitsek, D. I. Sirota, and N. P. Kolmakova
*Domain Wall Stabilization in Crystals with Complicated Magnetic Phase Diagram
(18.2 and 18.4)
(b) Vol. **121**, No. 1, p. 185—195 (1984).

V. P. Gupta, A. K. Singh, K. Chandra, and S. K. Jaireth
Mössbauer Investigations on Pyrrhotite
(10.2; 5; 18.3; 18.4; 22.8)
(a) Vol. **81**, No. 1, p. 281—291 (1984).

V. S. Kuleshov and V. A. Popov
*The Morin Point as Degenerate Critical Point
(18.3 and 18.4)
(b) Vol. **121**, No. 2, p. 561—572 (1984).

Z. Smetana, V. Šíma, and A. V. Andreev
*On the Magnetic Arrangement of the Intermetallic $ErCu_2$ Compound
(18.4; 21.4)
(b) Vol. **121**, No. 2, p. K131—K134 (1984).

G. A. Kraftmakher, V. V. Meriakri, and A. Ya. Chervonenkis
*Magnetic Field Sign Effect on Resonance Absorption in Domain Walls of Orthoferrites
(18.4; 22.8.2)
(a) Vol. **81**, No. 2, p. K159—K163 (1984).

S. N. Barilo, A. P. Ges, L. E. Soshnikov, and T. V. Smirnova
*Thermal Conductivity of Rare-Earth Orthoferrites at Reorientation Phase Transitions
(8 and 18.4; 7; 22.8.2)
(b) Vol. **122**, No. 2, p. 47—52 (1984).

T. Nakamura and K. Kawamura
*The Frustration of Antiferromagnetic Bonds in the Screw Dislocation
(10.2 and 18.4)
(b) Vol. **122**, No. 1, p. 141—150 (1984).

S. S. Aplesnin
*Thermodynamic Properties of Antiferromagnets with Random Interactions
(18.4)
(b) Vol. **122**, No. 1, p. K33—K36 (1984).

W. BAŻELA-WRÓBEL, A. SZYTUŁA, and J. LECIEJEWICZ
*Magnetic Properties of RhMnSi and CoTiSi
(18.4; 22.8)
(a) Vol. **82**, No. 1, p. 195—200 (1984).

V. BRÉDIMAS, H. GAMARI-SEALE, and C. PAPATRIANTAFILLOU
*The Magnetic Behaviour of the Pseudobinary Compounds $Gd_2Au_xAl_{1-x}$
(18.2 and 18.4; 21.4)
(b) Vol. **122**, No. 2, p. 527—534 (1984).

Z. ONYSZKIEWICZ
*Theory of the Dependence of the Phase Boundary of $FeBr_2$ Metamagnet on Hydrostatic Pressure
(18.1 and 18.4; 22.5)
(b) Vol. **122**, No. 2, p. 543—547 (1984).

A. F. BARABANOV, L. B. LITINSKII, and A. V. MIKHEENKOV
On the Atomic Phase of Crystalline Hydrogen
(1.2 and 13; 18.4; 23)
(b) Vol. **122**, No. 2, p. K109—K114 (1984).

M. NAWROCKI, R. PLANEL, F. MOLLOT, and M. J. KOZIELSKI
*Magnetization and Spin-Flip Raman Scattering in $Cd_{1-x}Mn_xS$
(18.4 and 20.1; 13.5; 22.8)
(b) Vol. **123**, No. 1, p. 99—103 (1984).

A. KASTEN and G. HESS
*Magnetic Properties of $Tb_2O_2SO_4$
(18.3 and 18,4; 22.8.1)
(b) Vol. **123**, No. 1, p. 123—134 (1984).

M. J. ŠKRINJAR, D. V. KAPOR, J. P. ŠETRAJČIĆ, and Z. IVIĆ
*New Approach to the Theory of Hybrid Excitations in Magnetic Dielectrics
(13 and 18.4; 20.1)
(b) Vol. **123**, No. 1, p. 135—141 (1984).

S. B. TARASENKO, V. T. TELEPA, and G. K. CHEPURNYKH
*The NMR Amplification Coefficient of Antiferromagnets at Superstrong Coupling of Magnetic and Elastic Waves
(18.4; 19)
(b) Vol. **123**, No. 1, p. K25—K28 (1984).

A. I. BELYAEVA, M. M. KOTLYARSKII, O. V. MILOSLAVSKAYA, YU. N. STELMAKHOV, and V. P. YUREV
*Investigations of the Crystal and Magnetic Structure Heterogeneities of $RbMnCl_3$
(18.4; 1.4; 22.5)
(a) Vol. **83**, No. 1, p. 245—253 (1984).

R. PLUMIER
*Study at 0 K and Various Magnetic Fields of the Triangular Magnetic Configurations
(18.2 and 18.4)
(b) Vol. **123**, No. 2, p. 553—562 (1984).

A. KASTEN and J. HOFFMANN
*Sequence of Ferrimagnetic Phases in $Er_2O_2SO_4$
(18.3 and 18.4; 22.8.1)
(b) Vol. **123**, No. 2, p. 563—571 (1984).

C. MIYAKE, H. TAKEUCHI, K. FUJI, and S. IMOTO
Electron Spin Resonance and Magnetic Susceptibility of M–U–O Ternary Mixed Oxides (II): Anomalies of Electron Spin Resonance and Magnetic Susceptibility of $Na_{2n-5}UO_n$ ($n = 3$ and 4)
(18.1 and 19; 18.4; 22.6.1)
(a) Vol. **83**, No. 2, p. 567—572 (1984).

C. MIYAKE, H. TAKEUCHI, H. OHYA-NISHIGUCHI, and S. IMOTO
Erratum to: Electron Spin Resonance and Magnetic Susceptibility of M–U–O Ternary Mixed Oxides (I): Anomalies of Electron Spin Resonance and Magnetic Susceptibility of $Li_{2n-5}UO_n$ ($n = 3, 4$, and 6)
(18.1 and 19; 18.4; 22.6.1)
(a) Vol. **83**, No. 2, p. 729 (1984).

A. N. BOGDANOV, A. V. ZHURAVLEV, and V. T. TELEPA
*First-Order Phase Transition from Antiferromagnetic to Paramagnetic Phase in $(C_2H_5NH_3)_2CuCl_4$
(18.1 and 18.4; 1.2; 22.8)
(a) Vol. **83**, No. 2, p. K135—K140 (1984).

A. VERVOITTE, J. C. CANIT, B. BRIAT, and U. KAMBLI
Optical and Magnetic Circular Dichroism Study of the 2D Antiferromagnet Rb_2MnCl_4
(13.5.1 and 13.5.4; 18.4; 20.1; 22.5)
(b) Vol. **124**, No. 1, p. 87—102 (1984).

B. CHENEVIER, D. FRUCHART, M. BACHMANN, J. P. SENATEUR, P. CHAUDOUET et L. LUNDGREN
*Propriétés magnétiques de MnRhAs. Etude de la transition antiferro–ferromagnétique
(Magnetic Properties of MnRhAs. Study of the Antiferromagnetic–Ferromagnetic Transition)
(18.2 and 18.4; 1.2; 14.1; 21.1)
(a) Vol. **84**, No. 1, p. 199—206 (1984).

K. H. J. BUSCHOW and D. B. DE MOOIJ
*The Effect of H_2 Absorption on the Magnetic Properties of Gd–Pd Intermetallic Compounds
(18.2 and 18.4; 21; 21.4)
(a) Vol. **84**, No. 1, p. 207—213 (1984).

V. N. DERKACHENKO, A. K. ZVEZDIN, A. M. KADOMTSEVA, N. M. KOVTUN, M. M. LUKINA, and A. A. MUKHIN
*Effect of Magnetic Vacancies on Magnetic Properties of Terbium Orthoferrites
(18.4; 22.8.2)
(a) Vol. **84**, No. 1, p. 215—222 (1984).

V. N. VASYUKOV, S. N. LUKIN, and V. T. TELEPA
*Distortion of the AFMR Line by the RF Field Power
(18.4 and 19; 22.5)
(a) Vol. **84**, No. 1, p. 223—227 (1984).

R. ZACH, R. DURAJ, and A. SZYTUŁA
*Pressure Effect on Magnetic Transformations in the $Co_xNi_{1-x}MnGe$ System
(18.2 and 18.4; 1.2; 22.8)
(a) Vol. 84, No. 1, p. 229—236 (1984).

U. LINDNER and S. SCHERF
*Two-Soliton Solution in a Modified Hubbard Chain
(13.5 and 18.4)
(b) Vol. 125, No. 1, p. 177—185 (1984).

D. HAJDUKOVIĆ and I. M. SAVIĆ
*Ground-State Degeneracy of the Ising Antiferromagnet on the Hypercubic Lattice
(18.4)
(b) Vol. 125, No. 1, p. K51—K53 (1984).

O. N. BORGES, L. A. AMARANTE RIBEIRO, and L. L. GONÇALVES
*The *XY* Model with Nearest and Next-Nearest Neighbour Interaction in a Transverse Field
(18.4)
(b) Vol. 125, No. 2, p. 563—572 (1984).

V. L. SAFONOV
*Interactions of Magnons in the Six-Sublattice Antiferromagnet
(13.5.4. and 18.4; 22.5)
(b) Vol. 126, No. 1, p. 197—206 (1984).

G. L. M. DASSEN and M. A. DE MOURA
*Soliton Solution for the Excited State in the Spin-Peierls $S = 1/2$ Antiferromagnetic Chain
(18.4)
(b) Vol. 126, No. 1, p. 213—221 (1984).

W. BÖHM, H. G. KAHLE, and W. WÜCHNER
Spectroscopic Study of the Crystal Field Splittings in $TbPO_4$
(1.2 and 13.4; 18.4; 20.1; 22.8.1)
(b) Vol. 126, No. 1, p. 381—392 (1984).

D. HAJDUKOVIĆ and M. MIJATOVIĆ
*On the Ground-State Properties of the Ising Antiferromagnet with General Spin on the Honeycomb Lattice
(18.4)
(b) Vol. 126, No. 1, p. K33—K35 (1984).

D. HAJDUKOVIĆ and I. M. SAVIĆ
*On the Critical Properties of the Potts Antiferromagnets of the Bond-Decorated B.C.C. and F.C.C. Lattice
(18.4)
(b) Vol. 126, No. 1, p. K37—K39 (1984).

D. GIGNOUX, J. C. GOMEZ-SAL, and J. RODRIGUEZ FERNANDEZ
*Magnetic Structure of the Er_2Pt Antiferromagnet
(18.4; 21.4)
(a) Vol. 86, No. 1, p. 295—300 (1984).

A. V. GOLTSEV
*The Parisi-Sompolinsky Solution for an Infinite Ranged Spin Glass Model with *m*-Component Spins
(18.2 and 18.4)
(b) Vol. **126**, No. 2, p. 537—545 (1984).

ZHEN-YA LI and CHUAN-ZHANG YANG
Magnetic Properties of a Random Mixture of a One-Dimensional System with Anisotropic Heisenberg Exchange
(18.2; 18.4)
(b) Vol. **126**, No. 2, p. K125—K130 (1984).

N. B. IVANOV
*Generalized Dispersion Relations for Surface Polaritons in Uniaxial Antiferromagnetic Crystals
(13.5.3 and 18.4)
(b) Vol. **126**, No. 2, p. K131—K134 (1984).

19. Magnetic Resonance

Original Papers and Short Notes

K. D. Becker
The Encounter Model in Impurity Diffusion. Correlation Effects in NMR
(9 and 10.2; 19; 22.5.1; 22.5.2)
(b) Vol. **121**, No. 1, p. 91—100 (1984).

M. A. Sigal
*Magnetostatic Waves in a Coaxial Cylindrical Magnetic Domain
(18.3 and 19)
(b) Vol. **121**, No. 1, p. 161—171 (1984).

D. Fraitová and L. Půst
*FMR Linewidth in Bulk Metals in Parallel Configuration
(19; 21.1; 21.1.1)
(b) Vol. **121**, No. 1, p. 173—176 (1984).

R. M. Bilbé, J. E. Nicholls, and J. J. Davies
*Electron Paramagnetic Resonance Investigation of Pb- and Ge-Doped CdTe
(13.4 and 19; 22.4.3)
(b) Vol. **121**, No. 1, p. 339—344 (1984).

D. V. Oparin, G. I. Pilipenko, F. F. Gavrilov, and V. I. Tyutyunnik
*Electron Paramagnetic Resonance of Mn^{2+} Ions in Lithium Hydride Single Crystals
(19; 22)
(b) Vol. **121**, No. 1, p. K75—K77 (1984).

S. Jurga
Molecular Motions and Polymorphic Phase Transitions in $(CH_3)_4NClO_4$ and $(CH_3)_3NHClO_4$ as Studied by NMR
(1.2; 19; 22.9)
(a) Vol. **81**, No. 1, p. 77—85 (1984).

C. Y. Chen, K. L. Sweeney, and L. E. Halliburton
Reduction and Radiation Effects in Lithium Tantalate
(10.2; 11; 19; 20.1; 22.8.1)
(a) Vol. **81**, No. 1, p. 253—257 (1984).

S. Simon and Al. Nicula
*EPR Study of Gd^{3+} and Cu^{2+} Ions from Lithium-Borate Vitroceramics
(1.4 and 19; 22.8.1)
(a) Vol. **81**, No. 1, p. K1—K5 (1984).

C. Jäger and K. Heide
*Temperature Dependence of the ^{23}Na NMR Central Lineshape in $Na_2Ca(CO_3)_2$
(6 and 19; 22.8.1)
(a) Vol. **81**, No. 1, p. K7—K10 (1984).

F. Momo, A. Sotgiu, M. Bettinelli, and A. Montenero
*Distribution of Magnetic Parameters in Some Copper Containing Glasses
(2 and 19)
(a) Vol. **81**, No. 1, p. K27—K30 (1984).

A. S. Litovchenko and V. V. Mazykin
Dehydroxylation Effect upon Electrical Conduction of Mica
(14.4; 19; 22.8.1)
(a) Vol. **81**, No. 1, p. K47—K50 (1984).

Sang Don Choi and Ok Hee Chung
*Theory of Cyclotron Resonance Lineshape in an Electron–Phonon System
(13.1 and 19; 22.2.3)
(b) Vol. **121**, No. 2, p. K181—K185 (1984).

V. K. Jain, V. P. Seth, and R. S. Bansal
*Electron Paramagnetic Resonance of Cr^{3+} in $Al_2(SO_4)_3 \cdot 18\ H_2O$ Single Crystals
(19; 13.4; 22.8.1)
(b) Vol. **121**, No. 2, p. K201—K204 (1984).

A. A. Alybakov, V. A. Gubanova, M. Zhanibekov, and N. Toichev
*Formation and Stability of Paramagnetic Centers in KBr:CrO_4^{2-}, Ca^{2+} Crystals
(19 and 20.1; 6; 13.4; 22.5.2)
(b) Vol. **121**, No. 2, p. K205—K208 (1984).

S. Orzeszko, W. Bała, K. Fabisiak, and F. Rozpłoch
DC Conductivity and ESR of Hydrogenated Amorphous Carbon Films
(2 and 14.3.1; 19; 22.1)
(a) Vol. **81**, No. 2, p. 579—584 (1984).

J. Kluge and J. Donecker
*Transfer Effects in ODMR Spectra of ZnSe:Mn
(19 and 20.3; 13.4; 22.4.2)
(a) Vol. **81**, No. 2, p. 675—679 (1984).

J. Kreissl and W. Gehlhoff
*EPR Investigations of ZnS:Mn and ZnSe:Mn. Single Crystals, Powders, Thin-Film Structures
(13.4 and 19; 22.4.1; 22.4.2)
(a) Vol. **81**, No. 2, p. 701—707 (1984).

C. Sanchez, M. Henry, R. Morineau, and M. C. Leroy
Small Polaron Mobility in α-$Li_xV_2O_5$
(13.4 and 13.5.3; 14.3; 19; 22.6.1)
(b) Vol. **122**, No. 1, p. 175—182 (1984).

M. Korkmaz, R. Bessege, and A. Gavaix
*EPR Studies of Cu^{2+} in Anhydrous Lead Tartrate
(13.4 and 19; 22.8.1)
(b) Vol. **122**, No. 1, p. 255—262 (1984).

W. Schrittenlacher, H. Nelkowski, and H. Pradella
*ESR and ODMR of Zinc–Vacancy-Associated Defects in ZnSe
(13.4 and 19; 22.4.2)
(b) Vol. **122**, No. 1, p. 285—294 (1984).

Y. Watanabe
ESR and Magnetic Investigation of Lithium Intercalated VSe_2
(13.4; 10.2; 18.1; 19; 22.8)
(b) Vol. **122**, No. 1, p. 309—316 (1984).

H. J. A. Koopmans, M. M. A. Perik, B. Nieuwenhuijse, and P. J. Gellings
*The Simulation and Interpretation of the EPR Powder Spectra of Gd^{3+}-Doped $LaAlO_3$
(13.4 and 19; 22.8.1)
(b) Vol. **122**, No. 1, p. 317—330 (1984).

V. Hinkel and D. Ploumbidis
Temperature Dependence of the Magnetic Part of the Hyperfine Field at ^{123}Sb in the Liquid Sb–Te System
(2 and 13; 19; 21.7; 22.1)
(b) Vol. **122**, No. 1, p. K45—K48 (1984).

R. Rünger and D. Ploumbidis
*High Temperature NMR in the V_3Au Compound
(13 and 19; 21.1)
(b) Vol. **122**, No. 1, p. K49—K52 (1984).

N. N. Kurkin and N. K. Solovarov
*The Variation of Light Transmission of Heavily Doped Ruby at Magnetic Resonance
(19; 20.1; 22.6)
(b) Vol. **122**, No. 1, p. K69—K72 (1984).

S. J. Lee and Chung-In Um
Impurity and Thickness Effects on the Cyclotron Resonance Linewidth of Electrons in Si Inversion Layers
(13.3; 14.3.2; 19; 22.1.2)
(b) Vol. **122**, No. 1, p. K73—K78 (1984).

M. B. Zapart, W. Zapart, and A. I. Zvyagin
*Phase Transitions in Ferroelastic $RbIn(MoO_4)_2$ Crystals by Electron Paramagnetic Resonance of Cr^{3+} Ions
(1.2 and 19; 13.4; 22.8.1)
(a) Vol. **82**, No. 1, p. 67—73 (1984).

L. Půst and Z. Frait
*Low-Temperature FMR and FMAR Measurements of Metal Single Crystals (I): General Consideration, Experimental Techniques
(19; 21)
(b) Vol. **122**, No. 2, p. 535—541 (1984).

A. K. Petrosyan, R. M. Khachatryan, and E. G. Sharoyan
Jahn-Teller Effect in EPR and Optical Absorption Spectra of Cu^{2+} in $LiNbO_3$ Single Crystals
(13.4; 19; 20.1; 22.8.1)
(b) Vol. **122**, No. 2, p. 725—734 (1984).

W. Schrittenlacher
Analysis of the Hyperfine Interactions of the Vacancy–Iodine Complex in ZnSe
(13.4; 19; 22.4.2)
(b) Vol. **122**, No. 2, p. 735—743 (1984).

E. A. Harris, J. H. Mellor, and S. Parke
*Electron Paramagnetic Resonance of Tetravalent Praseodymium in Zircon
(13.4 and 19; 22.8.1)
(b) Vol. **122**, No. 2, p. 757—760 (1984).

V. M. Polyanovskii
Stark Cyclotron-Phonon Oscillations of Sound Absorption in Semiconductors with Superlattice
(6 and 7; 19; 22)
(b) Vol. **122**, No. 2, p. K115—K118 (1984).

G. C. Upreti and R. S. Saraswat
*EPR Study of VO^{2+} in Some Paramagnetic Tutton Salt Single Crystals
(19; 18.1; 22.8.1)
(b) Vol. **122**, No. 2, p. K169—K173 (1987).

M. E. De Gouvêa and A. S. T. Pires
*EPR Shift in TMMC with the Applied Field Parallel to the Chain Axis
(19; 22.9)
(b) Vol. **122**, No. 2, p. K175—K177 (1984).

W. Pastusiak
*Erratum to: The Relativistic Rigid Lattice Contributions to the Spin–Lattice Coupling Coefficients of the S-State $4f^7$ Ions
(19; 22.5.3)
(b) Vol. **122**, No. 2, p. K195 (1984).

L. C. Otero-Diaz, J. Soria, and M. A. Alario-Franco
Extended Defects in Vanadium Doped Rutile. An EPR and TEM Study
(1.3 and 10.2; 19; 22.6.1)
(a) Vol. **82**, No. 2, p. 379—388 (1984).

P. I. Baranskii, V. M. Babich, N. P. Baran, A. A. Bugay, Yu. P. Dotsenko, and V. B. Kovalchuk
The Effect of Heat Treatment on Compensated CZ Silicon
(13.4; 14.3; 19; 22.1.2)
(a) Vol. **82**, No. 2, p. 533—536 (1984).

O. J. Żogał, Ch. Jäger, H. Döhler, and B. Schnabel
*Multipulse ^1H NMR Study of Proton Self-Diffusion in Scandium and Lutetium Dihydride
(9 and 19; 21.4)
(a) Vol. **82**, No. 2, p. K153—K157 (1984).

J. MINGE and S. WAPLAK
Successive Phase Transitions in $(NH_4)_3H(SO_4)_2$ and $(ND_4)_3D(SO_4)_2$ Crystals Studied by EPR of the VO^{2+} Ion and SeO_3^- Radical
(14.4.1; 4; 19; 22.8.1)
(b) Vol. **123**, No. 1, p. 27—36 (1984).

L. PŮST
*Low-Temperature FMR and FMAR Measurements of Metal Single Crystals (II): Saturation Magnetization, Magnetocrystalline Anisotropy, and *g*-Factor Determination
(19; 18.2; 21)
(b) Vol. **123**, No. 1, p. 115—122 (1984).

V. S. GRUNIN and I. B. PATRINA
*Impurities and Charge Compensation in Some Molybdenium Oxides
(13.4 and 19; 22.6; 22.8.1)
(b) Vol. **123**, No. 1, p. 353—363 (1984).

S. B. TARASENKO, V. T. TELEPA, and G. K. CHEPURNYKH
The NMR Amplification Coefficient of Antiferromagnets at Superstrong Coupling of Magnetic and Elastic Waves
(18.4; 19)
(b) Vol. **123**, No. 1, p. K25—K28 (1984).

R. BLINC, J. DOLINŠEK, G. LAHAJNAR, I. ZUPANČIČ, L. A. SHUVALOV, and A. I. BARANOV
Spin–Lattice Relaxation and Self-Diffusion Study of the Protonic Superionic Conductors $CsHSeO_4$ and $CsHSO_4$
(9 and 14.4; 19; 22.8.1)
(b) Vol. **123**, No. 1, p. K83—K87 (1984).

J. J. KLEPERIS, P. D. CIKMACH, and A. R. LUSIS
Colour Centres in Amorphous Tungsten Trioxide Thin Films
(13.4 and 20.1; 14.3.1; 19; 22.6)
(a) Vol. **83**, No. 1, p. 291—297 (1984).

E. K. SADYKOV
*The Theory of Double Gamma Resonance in Paramagnetics
(5 and 19; 18.1)
(b) Vol. **123**, No. 2, p. 703—709 (1984).

A. GOLZENÉ, B. MEYER, and C. SCHWAB
*Experimental Evidence for an Associated Defect Model for the Neutron Generated As_{Ga} Center in Gallium Arsenide
(11 and 19; 13.4; 22.2.1)
(b) Vol. **123**, No. 2, p. K125—K128 (1984).

C. MYAKE, H. TAKEUCHI, K. FUJI, and S. IMOTO
*Electron Spin Resonance and Magnetic Susceptibility of M–U–O Ternary Mixed Oxides (II): Anomalies of Electron Spin Resonance and Magnetic Susceptibility of $Na_{2n-5}UO_n$ (n = 3 and 4)
(18.1 and 19; 18.4; 22.6.1)
(a) Vol. **83**, No. 2, p. 567—572 (1984).

N. K. Bondareva, I. V. Krylova, and V. B. Golubev
Thermostimulated Exoemission and Radical Recombination in Polyethylene Irradiated with an Electron Beam
(13.3 and 17; 11; 19; 22.9)
(a) Vol. **83**, No. 2, p. 589—595 (1984).

C. Miyake, H. Takeuchi, H. Ohya-Nishiguchi, and S. Imoto
*Erratum to: Electron Spin Resonance and Magnetic Susceptibility of M–U–O Ternary Mixed Oxides (I): Anomalies of Electron Spin Resonance and Magnetic Susceptibility of $Li_{2n-5}UO_n$ ($n = 3, 4$, and 6)
(18.1 and 19; 18.4; 22.6.1)
(a) Vol. **83**, No. 2, p. 729 (1984).

K. S. V. Nambi
Comment on "The Role of Sulphur-Oxyradicals in Thermoluminescence and Exoelectron Emission of $CaSO_4$ Phosphors" by R. Huzimura and K. Atarashi
(17 and 20.3; 11; 13.4; 19; 22.8.1)
(a) Vol. **83**, No. 2, p. K175—K177 (1984).

U. Geppert and A. Kühnel
The Effect of Relaxing Defects on the Spin–Lattice Relaxation at Ferroelectric Phase Transitions
(14.4.1; 19)
(b) Vol. **124**, No. 1, p. 55—59 (1984).

M. Heming and G. Lehmann
*Correlation of Zero Field Splittings and Site Distortions. Interstitial Incorporation of Mn^{2+} in Urea-Ammoniumchloride $(H_2N)_2CO \cdot NH_4Cl$
(13.4 and 19; 22.8)
(b) Vol. **124**, No. 1, p. 315—323 (1984).

M. Budzynski, E. A. Goremychkin, O. I. Kochetov, A. Latuszynski, P. Mikolajczak, E. Mühle, A. I. Muminov, and M. Subotowicz
Electric Field Gradient at Gd in Gadolinium and Rare Earth Trifluoride Single Crystals
(13.4; 19; 22.5)
(b) Vol. **124**, No. 1, p. 355—362 (1984).

J. Barnaś
On the Possibility of Resonant Excitation of Spin Waves by an External Electric Field in Electronic-Type Linear Ferromagnetoelectrics
(14.4.1 and 18.2; 19)
(b) Vol. **124**, No. 1, p. K31—K34 (1984).

A. Watterich, E. Szontagh, and É. Kocsárdy
Paramagnetic Centres in Aluminium-Oxide Layers Prepared by Anodic Oxidation
(10.2; 19; 20.1; 22.6)
(a) Vol. **84**, No. 1, p. 165—169 (1984).

V. N. Vasyukov, S. N. Lukin, and V. T. Telepa
*Distortion of the AFMR Line by the RF Field Power
(18.4 and 19; 22.5)
(a) Vol. **84**, No. 1, p. 223—227 (1984).

D. Fraitová
*An Analytical Theory of FMR in Bulk Metals (Parallel Configuration) (III): Surface Impedance
(18.2 and 19)
(b) Vol. **124**, No. 2, p. 587—592 (1984).

M. Jirsa
*On the Angular Dependence of the Effective Spin-Pinning in Thin Insulator Film
(13.3 and 19)
(b) Vol. **124**, No. 2, p. 609—618 (1984).

V. G. Baryshevskii
Positronium Spin Rotation and Time Oscillation of the Normal to the Three-Photon Positronium Decay Plane in Magnetic Field
(13; 19)
(b) Vol. **124**, No. 2, p. 619—623 (1984).

E. Burzo, M. Chipară, D. Ungur, and I. Ardelean
*Electron Paramagnetic Resonance Study of $xFe_2O_3(1 - x) [B_2O_3 \cdot PbO]$ Glasses
(2 and 19; 22.8)
(b) Vol. **124**, No. 2, p. K117—K120 (1984).

D. A. Garanin and V. S. Lutovinov
*Diagrammatic Approach to the NQR Linewidths in Magnetically Ordered Materials
(18.1 and 19)
(b) Vol. **124**, No. 2, p. K133—K137 (1984).

R. Rai, S. K. Gupta, D. K. Suri, and S. Z. Ali
*Electron Spin Resonance Study of $CuGa_xIn_{1-x}Se_2$ Chalcopyrite Semiconductors
(19; 22.8)
(b) Vol. **124**, No. 2, p. K167—K170 (1984).

R. L. Davidovich, P. S. Gordienko, J. Grigas, T. A. Kaidalova, V. Urbonavičius, and L. A. Zemnukhova
Phase Transitions in $KSbF_4$
(1.2 and 14.4.1; 8; 19; 22.8)
(a) Vol. **84**, No. 2, p. 387—392 (1984).

M. Messaoudi, A. Kuhn, J. Toulouse, and J. Duran
An Isomer of the Organic Semiconductor Cs_2TCNQ_3. X-Ray, Optic, and EPR Analysis
(1.1; 19; 20.1; 22.9)
(a) Vol. **84**, No. 2, p. 411—416 (1984).

L. S. Grigoryan, M. V. Simonyan, and E. G. Sharoyan
Electrical, Magnetic, and Spectral Properties of the New Class of Charge Transfer Salts Based on Phthalocyanines and Iodine
(2 and 14.3.1; 19; 20.1; 22.9)
(a) Vol. **84**, No. 2, p. 597—606 (1984).

A. V. Parygin, G. V. Chikvaidze, G. K. Semin, B. V. Lokshin, and Ya. A. Eiduss
^{75}As-NQR Study of Yellow Crystalline Arsenic
(1.2; 19; 21.7)
(a) Vol. **84**, No. 2, p. K101—K103 (1984).

YU. N. MOSKVICH, A. A. SUKHOVSKII, and O. V. ROZANOV
Deuteron Magnetic Resonance of Highly Deuterated Ammonium Hydrogen Selenate
(1.1 and 14.4.1; 19; 22.8.1)
(a) Vol. 84, No. 2, p. K109—K113 (1984).

J. BASZYŃSKI, L. SMARDZ, and F. STOBIECKI
Magnetic Properties of Compositionally Modulated Ni/Cu Layers
(1.5 and 18.2.1; 19; 21.1)
(a) Vol. 84, No. 2, p. K129—K131 (1984).

M. JIRSA
*Angular Dependence of SSWR in Thin Metallic Films. Approximate Analytical Solution
(13.3 and 19; 21)
(b) Vol. 125, No. 1, p. 187—196 (1984).

S. HUANT, L. DMOWSKI, M. BAJ, and L. C. BRUNEL
*Intraconduction Band Magneto-Optical Study of InSb under Hydrostatic Pressure
(13.1 and 19; 16; 22.2.3)
(b) Vol. 125, No. 1, p. 215—219 (1984).

L. ABELLO, S. J. SCHWERDTFEGER, and C. F. SCHWERDTFEGER
*ESR Study of Cu^{2+} Diffused into Sodium β''-Alumina
(13.4 and 19; 22.6.1)
(b) Vol. 125, No. 1, p. 333—340 (1984).

A. DAVITULIANI, M. KIKNADZE, R. MIRIANASHVILI, and T. SANADZE
*Ligand Hyperfine Interaction of Er^{3+} in BaF_2 Single Crystals
(19; 22.5.3)
(b) Vol. 125, No. 1, p. K63—K65 (1984).

L. PŮST and Z. FRAIT
Method for Measurement of Magnetostriction Constants in (001) Thin Films Using FMR
(18.2.1; 19; 22.8.2)
(a) Vol. 85, No. 1, p. 179—188 (1984).

G. VÖLKEL, S. A. DZUBA, A. BARTL, W. BRUNNER, and J. FRÖHNER
*Time-Resolved EPR on Polyacetylene
(19; 22.9)
(a) Vol. 85, No. 1, p. 257—263 (1984).

S. V. VINTSENTS, V. F. KISELEV, and G. S. PLOTNIKOV
Energy Transfer between Excited Adsorbed Dye Molecules and Charged Defects in Insulator–Semiconductor Structures
(13.4; 19; 20.3; 22.1.1; 22.1.2)
(a) Vol. 85, No. 1, p. 273—281 (1984).

V. I. GATALSKAYA, E. M. GOLOLOBOV, and L. A. KUROCHKIN
Relaxation Time and Electron Scattering Cross Section in Irradiated p-Silicon at Temperatures of Liquid Helium from Cyclotron Resonance Data
(11; 19; 22.1.2)
(a) Vol. 85, No. 1, p. K51—K55 (1984).

N. M. L. Köche, P. C. Morais, and K. Skeff Neto
The Usefulness of the Photoacoustic Cell for Magnetic Measurements
(7 and 18; 19; 22.8.2)
(a) Vol. **85**, No. 1, p. K65—K67 (1984).

F. Engelke, D. Michel, and F. Pille
NMR Study of Dynamical Processes in Crystals of the TSCC Family
(14.4.1; 19; 22.9)
(b) Vol. **125**, No. 2, p. 483—492 (1984).

G. Heder and J. M. Spaeth
*ENDOR Investigation of H_i^0(Li) Centres in KCl
(10.2 and 19; 13.4; 22.5.2)
(b) Vol. **125**, No. 2, p. 523—531 (1984).

O. F. Antonov and R. Kh. Sabirov
*Contribution of the Higher-Order Secular Spin–Spin Interaction to the First and the Second Magnetic Resonance Line Moments
(18 and 19)
(b) Vol. **125**, No. 2, p. K117—K120 (1984)

B. Topič, R. Blinc, and L. A. Shuvalov
^{87}Rb NMR in Monoclinic RbD_2PO_4
(1.2; 14.4.1; 19; 22.8.1)
(a) Vol. **85**, No. 2, p. 409—415 (1984).

I. N. Karnaukhov and A. F. Zhuravlev
Magnetic State of Crystals with Dislocations
(10.1 and 18; 19)
(b) Vol. **126**, No. 1, p. 207—212 (1984).

S. K. J. Al-Ani and C. A. Hogarth
The Spin Density and Optical Properties of Amorphous Thin Arsenic Silicate Films
(2 and 13.1; 19; 20.1; 22.6)
(b) Vol. **126**, No. 1, p. 293—304 (1984).

M. Korkmaz, B. Aktas et M. Dupont
*Le spectre de RPE de l'ion Mn^{2+} dans le tartrate de cadmium ($CdC_4H_4O_6 \cdot 5\,H_2O$)
(EPR Spectra of the Mn^{2+} Ion in Cadmium Tartrate ($CdC_4H_4O_6 \cdot 5\,H_2O$))
(13.4 and 19; 22.8.1)
(b) Vol. **126**, No. 1, p. 323—327 (1984).

D. Backs and A. Hertrampf
*Optically Induced EPR of Conduction and Donor Band Electrons in Zinc Sulphide
(13.4 and 19; 22.4.1)
(b) Vol. **126**, No. 1, p. 343—349 (1984).

J. Lech, A. Ślęzak, and Bójko
*EPR Powder Spectra of Low-Symmetric Mn^{2+} Complexes in Aragonite Lattice
(13.4 and 19; 22.8)
(b) Vol. **126**, No. 1, p. 371—380 (1984).

W. Pastusiak
*Dynamical Relativistic Crystal Field Contributions to the Zero Field Splitting of S-State $4f^7$ Ions
(13.4 and 19; 22.5.3)
(b) Vol. **126**, No. 1, p. K77—K81 (1984).

G. Hüber and D. Ploumbidis
*Electrical Resistivity and NMR Investigations of the V–Co System at High Temperatures
(14.1 and 19; 21.1)
(b) Vol. **126**, No. 1, p. K101—K104 (1984).

P. C. Morais, A. L. Tronconi, and K. Skeff Neto
Photoacoustic Signal Behavior near the Ferromagnetic Resonance Field
(7; 19; 22.8.2)
(a) Vol. **86**, No. 1, p. 163—168 (1984).

Ch. Lushchik, J. Kolk, A. Lushchik, and N. Lushchik
Radiational Creation of Frenkel Defects in KCl–Tl
(10.2; 19; 20.1; 20.3; 22.5.2)
(a) Vol. **86**, No. 1, p. 219—227 (1984).

A. V. Dvurechenskii, B. P. Kashnikov, and V. V. Suprunchik
*New EPR Defects in Si⟨Al⟩
(13.4 and 19; 11; 22.1.2)
(a) Vol. **86**, No. 1, p. 313—318 (1984).

A. K. Khitrin, V. V. Laiko, and B. N. Provotorov
*The Influence of Molecular Motions on Multipulse NMR Spectra in Solids
(19)
(b) Vol. **126**, No. 2, p. 481—486 (1984).

M. Jirsa and V. Kamberský
*Angular Dependence of Spin-Wave Resonance in Thin Films with Asymmetrical Boundary Conditions. Analytical Treatment
(13.3 and 19)
(b) Vol. **126**, No. 2, p. 547—555 (1984).

A. V. Skripov and A. P. Stepanov
*Electronic Structure of V_3Si and V_3Ge. Comparison of NMR Data with Band-Structure Calculations
(13.1 and 19; 21.1)
(b) Vol. **126**, No. 2, p. 557—563 (1984).

M. S. Jahan and D. W. Cooke
*Thermal Evolution of Low-Temperature Manganese Centers in X-Irradiated CaF_2:Mn
(10.2 and 19; 11; 20.3; 22.5.3)
(b) Vol. **126**, No. 2, p. 687—692 (1984).

V. M. Babich, N. P. Baran, A. A. Bugai, Yu. P. Dotsenko, V. B. Kovalchuk, and V. A. Shershel
On the Properties of Thermodonors-II in Cz-Si Crystals of High Carbon Content
(13.4; 14.3; 19; 22.1.2)
(a) Vol. **86**, No. 2, p. 679—683 (1984).

20. Optical Properties

Review Article

G. Dräger and O. Brümmer
*Polarized X-Ray Emission Spectra of Single Crystals
(13.1 and 20; 21; 22)
(b) Vol. **124**, No. 1, p. 11—28 (1984).

Original Papers and Short Notes

V. Prabhawalkar and B. D. Padalia
*X-Ray Spectroscopy Study of RMn_2Si_2 Intermetallics
(20; 21; 21.4)
(b) Vol. **121**, No. 1, p. K65—K68 (1984).

M. S. Islam and C. Mande
X-Ray Spectroscopic Study of Re_3W and Re_3Mo Systems
(4; 20; 21)
(a) Vol. **81**, No. 1, p. 197—201 (1984).

L. Merten and W. Goltsche
*Stimulated Raman Effect on Plasmons and Longitudinal Plasmon-Phonons in Simple Cubic Crystals
(6 and 20; 13.5.2; 22)
(b) Vol. **121**, No. 2, p. 471—480 (1984).

H. Neumann, D. W. Bullett, H. Sommer, R. D. Tomlinson, and W. John
Valence Bands of the Cu–III–VI_2 Chalcopyrites Studied by Photoemission Spectra, X-Ray Emission Spectra, and Electronic Structure Calculations
(13.1; 17; 20; 22.8)
(b) Vol. **121**, No. 2, p. 641—647 (1984).

P. Rennert, H. Schelle, and U.-H. Gläser
Calculation of Polarized K and L X-Ray-Emission Spectra and of Partial Densities of States of the H.C.P. Metals Be, Mg, and Zn
(13.1; 20; 21.3)
(b) Vol. **121**, No. 2, p. 673—684 (1984).

T. Chattopadhyay and A. R. Chetal
*Plasmon and X-Ray Chemical Shifts in Some Copper Systems
(13.5.2 and 20; 22; 22.5)
(b) Vol. **122**, No. 1, p. 193—196 (1984).

B. Darshan, B. D. Padalia, and Om Prakash
*X-Ray Spectroscopy Study of $Y_{3-x}Gd_{/x}Fe_5O_{12}$
(20; 22.8.2)
(b) Vol. **122**, No. 1, p. K59—K63 (1984).

O. P. Aggarwal and N. N. Saxena
*K-Absorption Spectra of Copper Oxides
(13 and 20; 22.6)
(b) Vol. **122**, No. 2, p. 669—673 (1984).

S. Guha and C. Basu
*Stimulated Brillouin Scattering of Laser Radiation by an Electron-Acoustic Wave in Magnetised Semiconductor Plasmas
(13 and 20; 18)
(b) Vol. **122**, No. 2, p. 799—804 (1984).

K. B. Garg, J. Leiro, M. Heinonen, and M. Richardson
X-Ray Emission Study of Bonding in SmT_2 Compounds (T = Fe, Co, Ni)
(4 and 13; 20; 21.1; 21.1.1; 21.4)
(a) Vol. **83**, No. 1, p. 279—284 (1984).

S. B. Maslenkov, A. I. Kozlenkov, S. A. Filin, and A. I. Shulgin
*Soft X-Ray Emission Spectra from Some Ni–Al Alloys
(13.1 and 20; 21; 21.1)
(b) Vol. **123**, No. 2, p. 605—609 (1984).

K. Ławniczak-Jabłońska, E. Minni, J. Pełka, E. Suoninen, and J. Auleytner
XPS and XES Study of the Valence Band of Cr–Al B.C.C. Solid Solutions
(13.1; 20; 21.1)
(b) Vol. **123**, No. 2, p. 627—634 (1984).

W. Huhn and A. Stahl
Self-Consistent Field Theory Applied to the Semiconductor Band Edge
(13.1 and 13.5; 14.3; 20; 22)
(b) Vol. **124**, No. 1, p. 167—177 (1984).

Hoang Xuan Nguyen and R. Zimmermann
*Nonlinear Exciton Transmission in CdS. Non-Homogeneous and Non-Stationary Model Calculations
(13.5.1 and 20; 22.4.1)
(b) Vol. **124**, No. 1, p. 191—200 (1984).

T. A. Matveeva
4f States in Spectra of Rare-Earth Metals
(13.1; 20; 21.4)
(b) Vol. **124**, No. 1, p. K59—K62 (1984).

J. Muto and Y. Kobayashi
*Photobleaching Behaviours of Rhodamine 6G Diffused in Polymethyl Methacrylate and in the Copolymer of Methyl Methacrylate with Methacrylic Acid
(10.2 and 20; 22.9)
(a) Vol. **84**, No. 1, p. K29—K33 (1984).

S. N. Gupta and S. Shah
Evidence of Open 4f Shells in Rare Earth Compounds
(13; 20; 21.4)
(b) Vol. **125**, No. 1, p. 211—214 (1984).

NGUYEN VAN HIEU and HA VINH TAN
*Polariton Effect in Non-Linear Acousto-Optical Phenomena
(13.5 and 20; 7)
(b) Vol. **125**, No. 1, p. 245—250 (1984).

D. E. POPOV, E. I. ROZUM, S. R. UGLOV and S. A. VOROBIEV
*Anomalous Radiation of Fast Channeled Electrons in Optically Transparent Crystals
(11 and 20; 13; 22.1.3; 22.5.2)
(b) Vol. **125**, No. 1, p. 251—258 (1984).

J. A. LEIRO
*The First Born Approximation in Estimates of the Threshold Anomalies in the X-Ray Spectra of Simple Metals
(20; 13; 21)
(b) Vol. **125**, No. 1, p. 305—310 (1984).

L. WENDLER
*Coupled Surface Plasmon–Phonon Polaritons in Bilayer Systems
(6 and 20; 13.5.2; 21.6; 22.2.2)
(b) Vol. **125**, No. 1, p. K27—K31 (1984).

G. S. KUMAR, S. SUDHAKAR, R. ETHIRAJ, and V. G. KRISHNA MURTHY
*Strain-Optical Constant (p_{44}) of Mixed Crystals of KCl–KBr of Equimolar Concentration
(12.2 and 20; 22.5.4)
(a) Vol. **85**, No. 1, p. K27—K30 (1984).

A. D. POGREBNYAK, SH. R. MASTOV, M. F. KUZNETSOV, K. P. AREFIEV, S. V. RAKITIN, and S. A. VOROBIEV
*Observation of RF Pulses from Solids during Laser Irradiation
(11 and 20; 22.1.2; 22.2.1)
(a) Vol. **85**, No. 1, p. K31—K33 (1984).

A. V. SOLDATOV and A. N. GUSATINSKII
*Energy Delocalization of the Rare-Earth f-States in Some Compounds
(13 and 20; 22)
(b) Vol. **125**, No. 2, p. K129—K132 (1984).

S. GHOSH and S. KHAN
*Amplitude Modulation and Demodulation of Electromagnetic Waves in Magnetised Semiconducting Plasmas
(14 and 20; 22)
(a) Vol. **85**, No. 2, p. 535—541 (1984).

S. CHATTOPADHYAY and C. MANDE
XANES and EXAFS Study of the Compound CoAsSe
(1; 20; 21.1)
(b) Vol. **126**, No. 1, p. 53—57 (1984).

T. K. HATWAR, B. D. PADALIA, M. N. GHATIKAR, and D. R. CHOPRA
*Extended X-Ray Absorption Fine Structure in RAl_2Ga_2 Intermetallics
(20;21.4)
(b) Vol. **126**, No. 1, p. 279—283 (1984).

M. ASCHE and O. G. SARBEI
*The Role of Electron–Hole Interaction in the Cooling Process of Highly Excited Carriers
(13 and 20; 22.2.1)
(b) Vol. **126**, No. 2, p. 607—616 (1984).

H. J. ENGELMANN, K. BERNDT, and F. STORBECK
Quantitative Chemical Analysis of 3d Transition Metal Alloys by SXAPS
(1.3 and 13.3; 20; 21.1)
(a) Vol. **86**, No. 2, p. 499—506 (1984).

Y. Q. SHENG, R. Y. LI, and P. MUNZ
*The Study of Mode Characteristics of Metal Clad Dielectric Waveguides
(13.5.2 and 20; 21.6; 22.5.2)
(a) Vol. **86**, No. 2, p. 813—819 (1984).

20.1 Spectra, Optical Constants (X-Ray Spectra See 20)

Original Papers and Short Notes

J. OTTO and F. FISCHER
*FIR Studies of OCN^-, SCN^-, and $SeCN^-$ Centre Gap Modes in KI Single Crystals
(6 and 20.1; 22.5.2)
(b) Vol. **121**, No. 1, p. 81—90 (1984).

A. KESSLER and P. KIZLER
Investigation of Z Center Conversions in $KCl:SrCl_2$ (I): Optical Measurement of the Kinetics of the Thermal $Z_1 \rightarrow F$, Z_2 Decay
(10.2; 20.1; 22.5.2)
(b) Vol. **121**, No. 1, p. 111—116 (1984).

A. WATTERICH, I. FÖLDVÁRI, G. CORRADI, and H. OPYRCHAŁ
*Comment on the Identification of Z Centres in LiF:Mg and LiF:Mg, Ti Single Crystals
(10.2 and 20.1; 22.5.2)
(b) Vol. **121**, No. 1, p. 117—125 (1984).

J. PRAKASH
*Photo-Chemical Reaction Studies in Selenium-Doped Potassium Iodide
(10.2 and 20.1; 22.5.2)
(b) Vol. **121**, No. 1, p. 127—132 (1984).

L. PAJASOVÁ, P. PAJAS, O. A. MAKAROV, and V. M. ZAKHAROV
*Optical Properties of Amorphous GeS in the Spectral Region of 18 to 38 eV
(2 and 20.1; 13.1; 22)
(b) Vol. **121**, No. 1, p. 293—300 (1984).

V. N. SALOMATOV and I. A. PARFIANOVICH
*The F_H-Centre Energy Structure in KCl and KBr
(13.4 and 20.1; 10.2; 22.5.2)
(b) Vol. **121**, No. 1, p. 401—406 (1984).

V. A. HAISLER, V. M. ZALETIN, A. F. KRAVCHENKO, and G. Y. YASHIN
*Raman Scattering of First and Second Order in Red Mercury Iodide
(6 and 20.1; 22.5)
(b) Vol. **121**, No. 1, p. K13—K17 (1984).

A. K. RAY, S. B. TRICKEY, and A. B. KUNZ
*Optical Absorption of Solid Xenon at High Pressure
(20.1; 13.1; 23)
(b) Vol. **121**, No. 1, p. K47—K50 (1984).

K. ŽIČKUS and A. AUDZIJONIS
*The Exponential Optical Absorption Edge Tail of SbSeI and BiSeI
(20.1; 13.1; 22.8)
(b) Vol. **121**, No. 1, p. K51—K53 (1984).

W. Hörig, H. Neumann, and G. Kühn
*The Fundamental Absorption Edge of $LiInSe_2$
(20.1; 13.1; 22.8)
(b) Vol. **121**, No. 1, p. K55—K58 (1984).

V. N. Abramov, A. N. Ermoshkin, A. I. Kuznetsov, and V. V. Mürk
Breakdown of Excitonic Absorption and Luminescence Induced by a d-Type Conduction Band in Sc_2O_3
(13.5.1 and 20.3; 20.1; 22.6)
(b) Vol. **121**, No. 1, p. K59—K63 (1984).

S. M. Basha, K. Purandar, and Y. P. Reddy
*Electronic Absorption Spectrum of Fe^{3+} Doped in Calcium Tartrate Tetrahydrate
(13.4 and 20.1; 22.9)
(b) Vol. **121**, No. 1, p. K79—K81 (1984).

I. Cuculescu, A. Enache, C. Florea, and R. Bena
*Optical and Electrohydrodynamical Effects in a Compensated Cholesteric Mixture
(14.4 and 20.1; 22.9)
(a) Vol. **81**, No. 1, p. 191—196 (1984).

C. Y. Chen, K. L. Sweeney, and L. E. Halliburton
Reduction and Radiation Effects in Lithium Tantalate
(10.2; 11; 19; 20.1; 22.8.1)
(a) Vol. **81**, No. 1, p. 253—257 (1984).

V. K. Yanovskii
Phase Transitions and Properties of Doped $KNbO_3$ Crystals
(1.2 and 14.4.1; 14.3; 15; 20.1; 22.8.1)
(a) Vol. **81**, No. 1, p. 399—406 (1984).

A. Lehmann, L. Schumann, and K. Hübner
*Optical Phonons in Amorphous Silicon Oxides (II): Calculation of Phonon Spectra and Interpretation of the IR Transmission of SiO_x
(6 and 20.1; 2; 22.6)
(b) Vol. **121**, No. 2, p. 505—511 (1984).

J. Y. Bigot and B. Hönerlage
On the Dielectric Function of Three-Level Systems under High Excitation (I): Theoretical Approach
(13.5.1; 20.1; 22.5)
(b) Vol. **121**, No. 2, p. 649—659 (1984).

F. Henneberger and H. Rossmann
*Resonatorless Optical Bistability Based on Increasing Nonlinear Absorption
(13.5.1 and 20.1; 22.4.1)
(b) Vol. **121**, No. 2, p. 685—693 (1984).

C. Albers and Th. Moritz
Comments on the Applicability of Energy Gap versus Refractive Index Relations to Lead Chalcogenides
(13.1; 20.1; 22)
(b) Vol. **121**, No. 2, p. K175—K180 (1984).

J. PULS and F. HENNEBERGER
*Saturation of the Band-Edge Absorption of CdS_xSe_{1-x} Mixed Crystals at Room Temperature
(20.1; 22.4.4)
(b) Vol. **121**, No. 2, p. K187—K190 (1984).

F. HENNEBERGER, J. PULS, H. ROSSMANN, and I. N. URALTSEV
*On the Refractive Index at the Exciton Resonance of Highly Excited CdS
(13.5.1 and 20.1; 22.4.1)
(b) Vol. **121**, No. 2, p. K191—K194 (1984).

A. A. ALYBAKOV, V. A. GUBANOVA, M. ZHANIBEKOV, and N. TOICHEV
*Formation and Stability of Paramagnetic Centers in KBr:CrO_4^{2-}, Ca^{2+} Crystals
(19 and 20.1; 6; 13.4; 22.5.2)
(b) Vol. **121**, No. 2, p. K205—K208 (1984).

S. DEMIAN, F. EL. AKKAD, B. FARAG, H. MATHIEU, M. ROUZEYRE, and J. CHEVALLIER
Photoluminescence and Optical Properties of Lithium-Doped $Mg_xZn_{1-x}Te$ Alloys
(13.4 and 20.3; 20.1; 22.4.4)
(a) Vol. **81**, No. 2, p. 549—555 (1984).

D. N. POPOV, N. M. NANCHEVA, M. R. MINEV, and N. S. BAKALOVA
On the Nature of the X-Centers in KCl, Studied by Positron Angular Correlation
(10.2; 20.1; 22.5.2)
(a) Vol. **81**, No. 2, p. K141—K144 (1984).

F. CERDEIRA, V. LEMOS, F. E. A. MELO, and M. CARDONA
*Effective Charges and Their Pressure Dependence in α-$LiIO_3$
(6 and 20.1; 22.8.1)
(b) Vol. **122**, No. 1, p. 53—64 (1984).

P. KLEINERT
Coherent-Potential Approximation for the Lattice Vibrations of Mixed III–V Crystals
(6; 20.1; 22.2.4)
(b) Vol. **122**, No. 1, p. 81—89 (1984).

M. G. ABRAMISHVILI, V. I. ALTUKHOV, T. L. KALABEGISHVILI, and V. G. KVACHADZE
Thermal Conductivity and Optical and Paramagnetic Absorption Spectra of Reactor Irradiated MgO Crystals
(8 and 11; 20.1; 22.6)
(b) Vol. **122**, No. 1, p. 95—104 (1984).

U. SLIWCZUK, H. STOLZ, and W. VON DER OSTEN
Indirect-Forbidden Exciton Transitions in AgBr
(13.5.1; 6; 20.1; 22.5.1)
(b) Vol. **122**, No. 1, p. 203—209 (1984).

E. L. IVCHENKO, V. P. KOCHERESHKO, G. V. MIKHAILOV, and I. N. URALTSEV
*Resonance Magneto-Spatial Dispersion of Crystals
(18 and 20.1; 13.5.1; 22.4.2)
(b) Vol. **122**, No. 1, p. 221—230 (1984).

A. V. EGRANOV and A. I. NEPOMNYACHIKH
Magnesium Colour Centres in NaF
(10.2 and 13.4; 20.1; 20.3; 22.5.2)
(b) Vol. **122**, No. 1, p. 249—254 (1984).

S. SHIMANUKI and N. NISHIMAKI
A Correlation between the Spectral Location of the Absorption Bands Due to the Localized Centers and the Lattice Constant in Alkali Halide Crystals. Anionic Impurity Centers with s^2-Configuration and F Center which Have Cubic Symmetry
(10.2 and 13.4; 20.1; 22.5.2)
(b) Vol. **122**, No. 1, p. 269—283 (1984).

J. M. BARANOWSKI and PHUONG AN
*The Ni Donor and Acceptor Levels in CdSe
(13.4 and 20.1; 22.4.2)
(b) Vol. **122**, No. 1, p. 331—336 (1984).

FAT DUEN HO
The Improved Point-Ion Potentials for the F^+-Centers in $LiNbO_3$ and $LiTaO_3$
(10.2 and 13.4; 20.1; 22.8.1)
(b) Vol. **122**, No. 1, p. 337—344 (1984).

J. E. GRIFFITHS, J. C. PHILLIPS, G. P. ESPINOSA, J. P. REMEIKA, and P. M. BRIDENBAUGH
*Assignment of the Companion A_1^c Line in $Ge_x(S, Se)_{1-x}$ Glasses
(6 and 20.1; 2; 22)
(b) Vol. **122**, No. 1, p. K11—K15 (1984).

A. KRÓL, W. NAZAREWICZ, and M. J. KOZIELSKI
*Al_{Zn}–V_{Zn} Pairs in Wurtzite-Type ZnS
(10.2 and 20.1; 6; 22.4.1)
(b) Vol. **122**, No. 1, p. K23—K25 (1984).

N. N. KURKIN and N. K. SOLOVAROV
The Variation of Light Transmission of Heavily Doped Ruby at Magnetic Resonance
(19; 20.1; 22.6)
(b) Vol. **122**, No. 1, p. K69—K72 (1984).

R. B. PODE and B. T. DESHMUKH
Study of Annealing and Deformation Effect on TL of NaCl:KCl
(20.3; 20.1; 22.5.2)
(b) Vol. **122**, No. 1, p. K79—K81 (1984).

G. BALDACCHINI, G. P. GALLERANO, U. M. GRASSANO, M. MEUCCI, A. SCACCO, F. SOMMA, A. TIOURINE, and M. TONELLI
Optical Pumping and Reorientation of the F_A Centers in KCl:Li^+
(10.2 and 13.4; 20.1; 22.5.2)
(b) Vol. **122**, No. 1, p. K83—K86 (1984).

E. MACHO, J. LLOPIS, A. REMÓN, C. BALLESTEROS, and J. PIQUERAS
Optical Study of Defects in Quenched MgO Crystals
(10.2 and 20.3; 20.1; 22.6)
(a) Vol. **82**, No. 1, p. 111—117 (1984).

W. Hohenau
*The Influence of Silver and Copper Electrolysis on the Radiation-Induced Optical Absorption of Natural Quartz
(11 and 20.1; 22.6)
(a) Vol. **82**, No. 1, p. 119—124 (1984).

P. Süptitz and A. Fischer
Lateral Diffusion of Photodoped Silver in Amorphous $Ge_{25}Se_{75}$ Layers
(9; 20.1; 22)
(a) Vol. **82**, No. 1, p. 157—161 (1984).

K. Somaiah and V. Hari Babu
Glow Curves and the Emission of Flux Grown BaFCl:Gd Crystals
(13.4 and 20.3; 20.1; 22.5.4)
(a) Vol. **82**, No. 1, p. 201—206 (1984).

M. Suezawa and K. Sumino
Nature of Thermal Donors in Silicon Crystals
(13.4; 20.1; 22.1.2)
(a) Vol. **82**, No. 1, p. 235—242 (1984).

D. Nesheva
*Studies of Low-Energy Optical Absorption and Gap State Density of a-CdS by the Constant Photocurrent Method
(13.1 and 20.1; 2; 14.3.1; 22.4.1)
(a) Vol. **82**, No. 1, p. 243—248 (1984).

M. Nowak
Oscillatory Dependence of the Interference Photomagnetoelectric Effect and Photoconductivity on Magnetic Field
(16; 20.1; 22.4.4)
(a) Vol. **82**, No. 1, p. 249—256 (1984).

D. Coquillat, A. Pradel, G. Brun, and J. C. Tedenac
Growth and Physical Properties of $TlBiTe_2$ Crystals
(13.1 and 14.3; 8; 15; 20.1; 22.8)
(a) Vol. **82**, No. 1, p. 295—299 (1984).

V. I. Berezkin and A. I. Grachev
*Properties of Al-Doped Bismuth Silicon Oxide
(16 and 20.1; 22.8)
(a) Vol. **82**, No. 1, p. K95—K99 (1984).

Z. Černošek and E. Černošková
*Short-Range Order Model of Ge_xS_{1-x} Glasses and Utilization of This Model in Calculation of the Optical Gap and the IR 367 cm^{-1} Band Intensity
(2 and 20.1; 22)
(b) Vol. **122**, No. 2, p. 405—410 (1984).

P. P. Lottici
*On the Optic Vibrational Modes in Amorphous Arsenic
(6 and 20.1; 2; 22.1)
(b) Vol. **122**, No. 2, p. 431–434 (1984).

K. Aoki and M. Cardona
*Effect of Pressure on the Raman Scattering Efficiency of GaP
(6 and 20.1; 22.2.2)
(b) Vol. **122**, No. 2, p. 441—447 (1984).

C. Patel, T. J. Parker, H. Jamshidi, and W. F. Sherman
*Phonon Frequencies in GaAs
(6 and 20.1; 22.2.1)
(b) Vol. **122**, No. 2, p. 461—467 (1984).

T. Eshita, K. Tanimura, and N. Itoh
*Photo-Induced Transformation of Close Frenkel Pairs in Strontium Fluorite
(10.2 and 20.1; 20.3; 22.5.3)
(b) Vol. **122**, No. 2, p. 489—500 (1984).

R. M. Vlasova, V. A. Kapustin, V. N. Semkin, G. Sekretarczyk, L. S. Agroskin, G. V. Papayan, and L. P. Rautian
*Molecular Excitons in Quasi-One-Dimensional Conducting Crystals (Semiconductors and Metals)
(20.1; 13.5.1; 22.9)
(b) Vol. **122**, No. 2, p. 549—557 (1984).

V. I. Belitsky, A. V. Goltsev, I. G. Lang, and S. T. Pavlov
*Two-Phonon Resonance Raman Scattering and Quantum Interference Effects in Polar Semiconductors in High Magnetic Fields
(6 and 20.1; 22)
(b) Vol. **122**, No. 2, p. 581—590 (1984).

V. G. Tsirelson, O. V. Korolkova, I. S. Rez, and R. P. Ozerov
Extended Model of Bond Charges and Its Application in Calculation of Optical Properties of Crystals with Different Types of Chemical Bonds
(4; 20.1; 22; 22.5.2; 22.6; 22.8.1)
(b) Vol. **122**, No. 2, p. 599—612 (1984).

S. Logothetidis, S. Ves, and J. Spyridelis
*Ellipsometric Determination of the Optical Constants of $ZnIn_2S_4$
(20.1; 13.1; 22.8)
(b) Vol. **122**, No. 2, p. 613—622 (1984).

G. Valiukonis, F. M. Gashimzade, D. A. Guseinova, G. Krivaitė, M. M. Mamedov, and A. Šileika
*Electroreflectance and Thermoreflectance Spectra of SnSe Crystals
(13.1 and 20.1; 22)
(b) Vol. **122**, No. 2, p. 623—627 (1984).

W. Kautek, F. Decker, and H. Gerischer
*Optical Anisotropy of Transition Metal Dichalcogenides. A Photoelectrochemical Determination
(20.1; 13.1; 16; 22)
(b) Vol. **122**, No. 2, p. 651—659 (1984).

F. Lukeš, E. Schmidt, J. Humlíček, and P. Dub
*Optical Properties of GeS
(13.1 and 20.1; 22)
(b) Vol. **122**, No. 2, p. 675—686 (1984).

V. I. Sokolov and V. V. Chernyaev
Detection of Manganese Photoionization in $Cd_{1-x}Mn_xTe$ Solid Solutions
(13.5.1; 16; 20.1; 22.4.4)
(b) Vol. **122**, No. 2, p. 703—708 (1984).

R. K. Chaudhary, A. V. Krishna Rao, and K. V. Rao
*Effect of Magnetic Field on the F-Band Absorption in NaCl and KCl Single Crystals Irradiated with X-Rays under DC Fields
(10.2 and 20.1; 22.5.2)
(b) Vol. **122**, No. 2, p. 709—713 (1984).

A. Akilbekov and A. Elango
Low-Temperature Pair Associates of H Centres in KBr
(10.2 and 13.4; 20.1; 22.5.2)
(b) Vol. **122**, No. 2, p. 715—723 (1984).

A. K. Petrosyan, R. M. Khachatryan, and E. G. Sharoyan
Jahn-Teller Effect in EPR and Optical Absorption Spectra of Cu^{2+} in $LiNbO_3$ Single Crystals
(13.4; 19; 20.1; 22.8.1)
(b) Vol. **122**, No. 2, p. 725—734 (1984).

R. Bacewicz
*Excitonic Wavelength Modulated Reflectance of $CdGa_2Se_4$
(13.5.1 and 20.1; 22.8)
(b) Vol. **122**, No. 2, p. K155—K158 (1984).

T. Lideikis, V. Noreika, A. Petraitiene, I. Šimkiene, and K. Valacka
*Optical Absorption in Indium Doped Lead Tin Telluride
(20.1; 13.4; 22)
(b) Vol. **122**, No. 2, p. K179—K182 (1984).

Mukesh Jain, A. V. R. Warrier, and H. K. Sehgal
*Composition Dependence of the Optical Energy Gap in $Pb_{1-x}Hg_x$ Te Alloy Thin Films
(20.1; 13.1; 22.8)
(a) Vol. **82**, No. 2, p. K181—K184 (1984).

A. A. Kaminskii, A. G. Petrosyan, K. L. Ovanesyan, G. O. Shironyan, V. A. Fedorov, and S. E. Sarkisov
Concentrational 3 μm Stimulated Emission Tuning in the $(Gd_{1-x}Er_x)_3Al_5O_{12}$ Crystal System
(20.2 and 20.3; 20.1; 22.8)
(a) Vol. **82**, No. 2, p. K185—K190 (1984).

S. H. PAWAR, S. P. TAMHANKAR, and C. D. LOKHANDE
Deposition of Bi_2CdS_4 Films by the Spray Pyrolysis Technique
(14.3.1; 20.1; 22.8)
(a) Vol. **82**, No. 2, p. K195—K198 (1984).

G. JUNGK and A. RÖSELER
*Ellipsometric Investigation of Lattice Vibrations in Anisotropic Systems. α-Quartz
(6 and 20.1; 22.6)
(b) Vol. **123**, No. 1, p. 65—70 (1984).

R. RAUCH
Photochemical Conversion of M_A Centres and Identification of M_A^+ Centres in CaF_2:Na
(10.2; 20.1; 20.3; 22.5.3)
(b) Vol. **123**, No. 1, p. 93—98 (1984).

M. NAWROCKI, R. PLANEL, F. MOLLOT, and M. J. KOZIELSKI
*Magnetization and Spin-Flip Raman Scattering in $Cd_{1-x}Mn_xS$
(18.4 and 20.1; 13.5; 22.8)
(b) Vol. **123**, No. 1, p. 99—103 (1984).

M. J. ŠKRINJAR, D. V. KAPOR, J. P. ŠETRAJČIĆ, and Z. IVIĆ
New Approach to the Theory of Hybrid Excitations in Magnetic Dielectrics
(13 and 18.4; 20.1)
(b) Vol. **123**, No. 1, p. 135—141 (1984).

A. I. BOBRYSHEVA, V. V. BALTAGA, and M. V. GRODETSKII
Biexciton Two-Photon Absorption and Raman Scattering in CuBr
(13.5.1; 20.1; 22.5)
(b) Vol. **123**, No. 1, p. 169—181 (1984).

B. HÖNERLAGE and J. Y. BIGOT
On the Dielectric Function of Three-Level Systems under High Excitation (II): Exciton–Biexciton Transitions and Population Effects in CuCl
(13.5.1; 20.1; 22.5)
(b) Vol. **123**, No. 1, p. 201—208 (1984).

TH. HAYNE and G. WEISER
*Dispersion of Exciton Polaritons in a Monoclinic Cyanine Dye Crystal. Evaluation of the Reflectivity on Arbitrary Faces
(13.5 and 20.1; 22.9)
(b) Vol. **123**, No. 1, p. 271—280 (1984).

S. A. MAZEN, N. A. HAKEEM, and B. A. SABRAH
*The IR Absorption Spectra of $Cu_{1-x}Cd_xFe_2O_4$
(6 and 20.1; 22.8.2)
(b) Vol. **123**, No. 1, p. K1—K4 (1984).

M. PAGANNONE, B. FORNARI, G. MATTEI, and L. MATTIOLI
*Radiative (Brewster-Type) Surface Mode Dispersion in Sodium Chlorate
(6 and 20.1; 13.5; 22.8.1)
(b) Vol. **123**, No. 1, p. K13—K16 (1984).

A. Król and R. Zielińska
A New Defect Complex in CdTe: In, Li
(6 and 10.2; 20.1; 22.4.3)
(b) Vol. **123**, No. 1, p. K17—K20 (1984).

H. T. Langhammer, M. Stordeur, H. Sobotta, and V. Riede
*IR Transmission Investigations of Sb_2Te_3 Single Crystals
(20.1; 13.1; 22.7)
(b) Vol. **123**, No. 1, p. K47—K51 (1984).

M. Załużny
*On the Influence of the Depolarization Field on the Intersubband Absorption in Thin Semiconductor Films
(13.1 and 20.1; 22; 22.2.1)
(b) Vol. **123**, No. 1, p. K57—K61 (1984).

W. Henrion and H. Lange
*Optical Properties of $CrSi_2$
(20.1; 13.1; 21)
(b) Vol. **123**, No. 1, p. K69—K73 (1984).

S. V. J. Lakshman and A. S. Jacob
*Electronic Absorption Spectrum of Cu^{2+} in Sodium Zinc Sulphate Tetrahydrate Single Crystals
(13.4 and 20.1; 22.8.1)
(b) Vol. **123**, No. 1, p. K75—K77 (1984).

M. M. El-Oker, F. A. Sultan, S. A. Yousef, and S. A. El-Sahhar
*Surface Plasmon Modes and Energy Loss Function of Indium Thin Films on Mica Substrates
(13.5.2 and 20.1; 13.1; 21)
(a) Vol. **83**, No. 1, p. 263—267 (1984).

G. Debiais
*Caractérisation optique et électrique de couches mixtes de sulfure de mercure
(Optical and Electrical Characterization of Mercury Sulfide Mixed Films)
(14.3.1 and 20.1; 13.1; 22.4.4)
(a) Vol. **83**, No. 1, p. 269—278 (1984).

F. S. Nasredinov, A. A. Andreev, A. N. Kurmantaev, A. R. Regel, and P. P. Seregin
Tin Impurity Atoms in Vacuum Deposited Amorphous Silicon
(13.4 and 14.3.1; 2; 20.1; 22.1.2)
(a) Vol. **83**, No. 1, p. 285—289 (1984).

J. J. Kleperis, P. D. Cikmach, and A. R. Lusis
*Colour Centres in Amorphous Tungsten Trioxide Thin Films
(13.4 and 20.1; 14.3.1; 19; 22.6)
(a) Vol. **83**, No. 1, p. 291—297 (1984).

A. L. Shabalov and M. S. Feldman
*Optical Properties of Thin GeO_x Films
(6 and 20.1; 22.6)
(a) Vol. **83**, No. 1, p. K11—K14 (1984).

E. Culea, Al. Nicula, and I. Bratu
*An Infrared Study of $xV_2O_5 \cdot (1 - x) B_2O_3$ Glasses
(6 and 20.1; 2; 22.6.1)
(a) Vol. 83, No. 1, p. K15—K18 (1984).

A. M. Evstigneev, L. F. Linnik, L. G. Linnik, E. A. Salkov, A. N. Krasiko, and O. V. Snitko
Observation of Anisotropic Internal Stresses by Electroreflectance and Transverse Photovoltaic Effect Measurements
(10.2; 14.3.4; 16; 20.1; 22.1.1)
(a) Vol. 83, No. 1, p. K31—K33 (1984).

J. Teteris
*Photoinduced Optical Absorption in As–Se
(20.1; 2; 22)
(a) Vol. 83, No. 1, p. K47—K50 (1984).

J. Misiewicz, Nella Mirowska, and Z. Gumienny
*Fundamental Optical and Photoelectric Properties of Poly- and Mono-Crystals of Zn_3P_2
(16 and 20.1; 22.3)
(a) Vol. 83, No. 1, p. K51—K56 (1984).

S. Chaudhuri, S. K. Biswas, and A. Choudhury
*Effects of Heat Treatment on the Optical Absorption of Amorphous $In_{30}Se_{70}$ Films
(2 and 20.1; 22.3)
(a) Vol. 83, No. 1, p. K57—K60 (1984).

L. Wendler
*Phonon-Polaritons in Bilayer Systems. p-Polarized Guided Wave Phonon-Polaritons and Guided Wave Surface Phonon-Polaritons
(6 and 20.1; 13.5; 22.2.1; 22.2.2; 22.5.3)
(b) Vol. 123, No. 2, p. 469—477 (1984).

B. B. S. Negi and P. N. Ram
First-Order Raman Scattering in RbCl Containing F-Centres
(10.2; 6; 20.1; 22.5.2)
(b) Vol. 123, No. 2, p. 491—495 (1984).

S. P. Gwet, Y. Mathey, and C. Sourisseau
*The Infrared, Raman, Resonance Raman Spectra, and the Valence Force Field of the HfS_3 Layer-Type Compound
(6 and 20.1; 22)
(b) Vol. 123, No. 2, p. 503—517 (1984).

L. Bosi, P. Gagliardelli, and M. Nimis
New Results on the Decay Properties of Perturbed and Unperturbed M-Centers in $NaCl:CaCl_2$
(10.2; 20.1; 20.3; 22.5.2)
(b) Vol. 123, No. 2, p. 519—524 (1984).

S. K. J. Al-Ani, C. A. Hogarth, and D. N. Waters
*The Effect of Temperature on the Optical Absorption Edge of Amorphous Thin Films of Silicon Monoxide
(20.1; 2; 13; 22.6)
(b) Vol. **123**, No. 2, p. 653—658 (1984).

V. A. Kovarskii, E. P. Sinyavskii, and L. V. Chernysh
Optical Properties of Bielectron–Impurity Complexes
(13.4; 20.1; 22.5)
(b) Vol. **123**, No. 2, p. 671—677 (1984).

R. Borromei and E. Cavalli
*Low-Temperature Absorption Spectrum of Ni^{2+} Ion in Single Crystals of Ni_2NbBO_6
(13.4 and 20.1; 22.8.1)
(b) Vol. **123**, No. 2, p. 679—689 (1984).

N. I. Likholit, H.-E. Ponath, V. L. Strizhevskii, and Yu. N. Yashkir
*Transient Coherent Anti-Stokes Raman Spectroscopy of Surface Polaritons
(13.5 and 20.1; 6; 22)
(b) Vol. **123**, No. 2, p. 691—701 (1984).

W. Wojdowski
*Infrared Spectra of Thin Films of $Bi_{12}GeO_{20}$ and $Bi_{12}SiO_{20}$
(6 and 20.1; 22.8)
(b) Vol. **123**, No. 2, p. K101—K104 (1984).

Sh. M. Efendiev
*Raman Spectra in $Bi_{12}TiO_{20}$ Single Crystals
(6 and 20.1; 22.8)
(b) Vol. **123**, No. 2, p. K105—K108 (1984).

V. Riede, H. Neumann, H. Sobotta, W. Kissinger, M. S. Omar, and B. R. Pamplin
*Infrared Study of Lattice and Free Carrier Effects in p-Type $CuGe_2P_3$
(6 and 20.1; 13.1; 22.8)
(b) Vol. **123**, No. 1, p. K119—K123 (1984).

A. Anagnostopoulos, K. Kambas, and B. Ploss
*Optical Measurements and Temperature Dependence of the Energy Gap in $ZnIn_2S_4$ Layered Compound
(13.1 and 20.1; 22.8)
(b) Vol. **123**, No. 2, p. K155—K159 (1984).

P. Široký, E. Schmidt, F. Lukeš, and J. Humlíček
*Optical Properties of Ga-Substituted Magnetoplumbites
(20.1; 22.8.2)
(a) Vol. **83**, No. 2, p. 581—588 (1984).

M. M. El-Oker
*Detection of Surface Plasmons in Some Indium Films by Attenuated Total Reflection
(13.5.2 and 20.1; 21)
(a) Vol. **83**, No. 2, p. 613—616 (1984).

M. Vaněček, A. Abrahám, O. Štika, J. Stuchlík, and J. Kočka
*Gap States Density in a-Si:H Deduced from Subgap Optical Absorption Measurement on Schottky Solar Cells
(13.1 and 20.1; 2; 22.1.2)
(a) Vol. **83**, No. 2, p. 617—623 (1984).

M. T. Mohammad and A. K. Abass
*Optical Properties of Chemically Deposited Conducting Glass for SIS Solar Cells
(20.1; 22.6)
(a) Vol. **83**, No. 2, p. 681—683 (1984).

K. Ots and M. Seilenthal
On the Structural Model of CVD a-Si:O:H Films
(1.5 and 2; 20.1; 22.1.2; 22.6)
(a) Vol. **83**, No. 2, p. K109—K111 (1984).

N. Yu. Konstantinov, L. G. Karaseva, V. V. Gromov, and A. V. Yakovlev
*Interconfiguration Transitions of $4f^3$–$4f^2 5d^1$ of the Nd^{3+} Ion in YAG Single Crystals
(13.4 and 20.1; 22.8.2)
(a) Vol. **83**, No. 2, p. K153—K157 (1984).

A. A. Kaminskii, A. G. Petrosyan, and K. L. Ovanesyan
Stimulated Emission of Pr^{3+}, Nd^{3+}, and Er^{3+} Ions in Crystals with Complex Anions
(20.2 and 20.3; 20.1; 22.8.1)
(a) Vol. **83**, No. 2, p. K159—K163 (1984).

A. A. Kaminskii, V. A. Timofeeva, A. B. Bykov, and S. E. Sarkisov
Luminescence and Stimulated Emission in the $^4F_{3/2} \rightarrow {}^4I_{11/2}$ and $^4F_{3/2} \rightarrow {}^4I_{3/2}$ Channels of Nd^{3+} Ions in Orthorhombic $NaYGeO_4$ Crystals
(20.2 and 20.3; 20.1; 22.8.1)
(a) Vol. **83**, No. 2, p. K165—K169 (1984).

H. Reimer and F. Fischer
*SERS from Ag Colloids in Alkali Halide Crystals
(6 and 20.1; 10.2; 22.5.2)
(b) Vol. **124**, No. 1, p. 61—67 (1984).

A. Vervoitte, J. C. Canit, B. Briat, and U. Kambli
Optical and Magnetic Circular Dichroism Study of the 2D Antiferromagnet Rb_2MnCl_4
(13.5.1 and 13.5.4; 18.4; 20.1; 22.5)
(b) Vol. **124**, No. 1, p. 87—102 (1984).

P. K. Sen
*Nonlinear Absorption in III–V Semiconductors
(13.5.1 and 20.1; 22.2)
(b) Vol. **124**, No. 1, p. 117—125 (1984).

H. S. Brandi, G. Jalbert, and O. L. Malta
*Multiphoton Absorption in Solids. Non-Perturbative Approach in the Weak-Field Limit
(13.1 and 20.1; 22)
(b) Vol. **124**, No. 1, p. 147—157 (1984).

V. P. Kochereshko, J. Röseler, I. N. Uraltsev, and K. Henneberger
*Dipole Forbidden Transmission of Magnetopolaritons in CdSe
(13.5.1 and 20.1; 18; 22.4.2)
(b) Vol. **124**, No. 1, p. 213—219 (1984).

B. Hönerlage and J. Y. Bigot
*Dispersion and Absorption Anomalies in Three-Level Systems
(13.5.1 and 20.1; 22.5)
(b) Vol. **124**, No. 1, p. 221—227 (1984).

K. M. Paraskevopoulos, D. Siapkas, and N. A. Economou
*Infrared Reflectivity Plasma Spectra of $TlBiTe_2$ Single Crystals
(13.5.2 and 20.1; 22.8)
(b) Vol. **124**, No. 1, p. 229—234 (1984).

J. P. Šetrajčić, D. V. Kapor, and D. Mirjanić
On the Possibility of Bose-Einstein Condensation and the Superfluidity of Frenkel Excitons
(13.5.1; 20.1)
(b) Vol. **124**, No. 1, p. 235—239 (1984).

N. V. Baramidze, V. L. Bonch-Bruevich, and N. I. Kurdiani
*Optical Absorption in Neutron Irradiated GaAs
(13.4 and 20.1; 11; 22.2.1)
(b) Vol. **124**, No. 1, p. 241—247 (1984).

M. J. Stillman, P. W. M. Jacobs, K. Oyama Gannon, and D. J. Simkin
*Absorption and Magnetic Circular Dichroism Spectra of $CsI:Tl^+$
(18 and 20.1; 22.5.2)
(b) Vol. **124**, No. 1, p. 261—270 (1984).

C. M. Weinert and U. Scherz
*Calculation of the Jahn-Teller Coupling Forces for $ZnS:Cu^{2+}$ and $GaP:Ni^+$
(6 and 20.1; 13.4; 22.2.2; 22.4.1)
(b) Vol. **124**, No. 1, p. 287—296 (1984).

G. Jungk and A. Röseler
*LO–TO-Splitting and Many-Particle Effects
(6 and 20.1; 2; 22.6)
(b) Vol. **124**, No. 1, p. K1—K4 (1984).

W. Giriat and J. Stankiewicz
*Absorption Edge in $Mn_xZn_{1-x}Te$
(20.1; 13.1; 22.4.4)
(b) Vol. **124**, No. 1, p. K53—K57 (1984).

E. M. Goldys
*The Donor Wave Functions in Gallium Phosphide and the Real Band Structure
(13.4 and 20.1; 18; 22.2.2)
(b) Vol. **124**, No. 1, p. K79—K82 (1984).

R. Sanctuary and P. Günter
*Electrooptic Properties of Rb_2ZnCl_4 in the Incommensurate and Ferroelectric Phases
(14.4.1 and 20.1; 22.8)
(a) Vol. **84**, No. 1, p. 103—111 (1984).

D. Senulienė, G. Babonas, E. I. Leonov, I. Muminov, and V. M. Orlov
*Optical Properties of $Bi_{12}Si_{1-x}Ti_xO_{20}$ Single Crystals
(6 and 20.1; 22.8)
(a) Vol. **84**, No. 1, p. 113—117 (1984).

A. Watterich, E. Szontagh, and F. Kocsárdy
Paramagnetic Centres in Aluminium-Oxide Layers Prepared by Anodic Oxidation
(10.2; 19; 20.1; 22.6)
(a) Vol. **84**, No. 1, p. 165—169 (1984).

R. Pareja, R. González, and M. A. Pedrosa
Study of Thermochemically Reduced and Electron-Irradiated $LiNbO_3$ Single Crystals by Positron Annihilation and Optical Absorption Measurements
(10.2 and 11; 20.1; 22.8.1)
(a) Vol. **84**, No. 1, p. 179—183 (1984).

W. Ulrici
*Evaluation of Chromium Concentration in Semi-Insulating GaAs:Cr from Optical Absorption
(13.4 and 20.1; 22.2.1)
(a) Vol. **84**, No. 1, p. 243—250 (1984).

M. Böhm, O. Erb, and A. Scharmann
*Electrical Conduction in $PbMoO_4$ Crystals
(16 and 20.1; 13.4; 22.8.1)
(a) Vol. **84**, No. 1, p. 291—299 (1984).

E. Culea, Al. Nicula, and I. Bratu
*An Infrared Study of $V_2O_5 \cdot As_2O_3$ Glasses
(6 and 20.1; 2; 22.6.1)
(a) Vol. **84**, No. 1, p. K17—K21 (1984).

S. El-Halawany, R. Bacewicz, J. Filipowicz, and R. Trykozko
Photoconductivity of Flash Evaporated $CuInSe_2$ Thin Films
(16; 14.3.1; 20.1; 22.8)
(a) Vol. **84**, No. 1, p. K89—K92 (1984).

R. Sudharsanan, T. K. K. Srinivasan, and S. Radhakrishna
*Raman and Infrared Studies on Ag_2HgI_4 and Cu_2HgI_4 Superionic Compounds
(6 and 20.1; 22.5.4)
(b) Vol. **124**, No. 2, p. 481—487 (1984).

V. I. Zametin
*Absorption Edge Anomalies in Polar Semiconductors and Dielectrics at Phase Transitions
(20.1; 1.2; 14.4.1; 22; 22.5.2; 22.8.1; 22.9)
(b) Vol. **124**, No. 2, p. 625—640 (1984).

R. Borromei, L. Oleari, and P. Day
*The Absorption Spectrum of Manganate(VI) in $BaSO_4$. A Reinterpretation
(13.4 and 20.1; 22.8.1)
(b) Vol. **124**, No. 2, p. 707—718 (1984).

T. F. Veremeichik
*Similarity of Ground- and Excited-State Absorption Spectra of Cr^{3+} Ions in a Strong Crystal Field
(13.4 and 20.1; 22.6)
(b) Vol. **124**, No. 2, p. 719—729 (1984).

M. Matos and B. Maffeo
Molecular Cluster-INDO Calculations of Substitutional Defects in Alkali Halides
(13.4; 20.1; 22.5.2)
(b) Vol. **124**, No. 2, p. 757—765 (1984).

K. Kambas, C. Julien, M. Jouanne, A. Likforman, and M. Guittard
*Raman Spectra of α- and γ-In_2Se_3
(6 and 20.1; 22.3)
(b) Vol. **124**, No. 2, p. K105—K108 (1984).

M. Ishii, M. Saeki, and I. Kawada
*Raman Study of Non-Stoichiometric Titanium Sulfides
(6 and 20.1; 22)
(b) Vol. **124**, No. 2, p. K109—K112 (1984).

K. Čermák
*The Exponential Optical Absorption Edge Tail of Tl_3VS_4 and Tl_3TaS_4
(20.1; 13.1; 22.8)
(b) Vol. **124**, No. 2, p. K143—K145 (1984).

N. B. Mustafaev and M. G. Shakhtakhtinskii
*Optical Absorption in Size-Quantized Bi Films
(13.1 and 20.1; 21.7)
(b) Vol. **124**, No. 2, p. K151—K154 (1984).

V. I. Sokolov, A. N. Mamedov, T. P. Surkova, G. A. Emelchenko, and L. G. Kolinova
*Fine Structure of the Photoionization Band Edge in Ni-Doped ZnO
(13.5.1 and 20.1; 14.3.4; 22.4)
(b) Vol. **124**, No. 2, p. K155—K159 (1984).

M. Messaoudi, A. Kuhn, J. Toulouse, and J. Duran
An Isomer of the Organic Semiconductor Cs_2TCNQ_3. X-Ray, Optic, and EPR Analysis
(1.1; 19; 20.1; 22.9)
(a) Vol. **84**, No. 2, p. 411—416 (1984).

M. Pardavi-Horváth I. Földvári, I. Fellegvári, L. Gosztonyi, and J. Paitz
*Spectroscopic Properties of Ca^{2+}-Doped GGG
(20.1; 13.4; 22.8.2)
(a) Vol. **84**, No. 2, p. 547—553 (1984).

L. S. Grigoryan, M. V. Simonyan, and E. G. Sharoyan
Electrical, Magnetic, and Spectral Properties of the New Class of Charge Transfer Salts Based on Phthalocyanines and Iodine
(2 and 14.3.1; 19; 20.1; 22.9)
(a) Vol. **84**, No. 2, p. 597—606 (1984).

G. Nicoară and V. Ghiordănescu
*The Presence of Ti^{3+} in LiTiZn Ferrites
(18.3 and 20.1; 22.8.2)
(a) Vol. **84**, No. 2, p. K139—K142 (1984).

P. Lošťák, J. Horák, and L. Koudelka
*Some Physical Properties and Point Defects in $Bi_2Te_{3-x}S_x$ Mixed Crystals
(20.1; 13.4; 14.3; 22.7.1)
(a) Vol. **84**, No. 2, p. K143—K147 (1984).

L. N. Skuja, A. N. Trukhin, and A. E. Plaudis
Luminescence in Germanium-Doped Glassy SiO_2
(20.3; 2; 10.2; 13.4; 20.1; 22.6)
(a) Vol. **84**, No. 2, p. K153—K157 (1984).

Tran Kim Anh, Nguyen Dai Hung, Nquy Huu Chi, and Nguyen Manh Son
*Energy Transfer between Tb^{3+} and Eu^{3+} in Rare Earth Pentaphosphates
(20.1 and 20.3; 13.4; 22.8.1)
(a) Vol. **84**, No. 2, p. K159—K163 (1984).

A. Ribeiro Filho and D. R. Tilley
*Thick Films near Phase Transitions. Green Functions, Fluctuation Spectra, and Light Scattering Theory
(1.2 and 20.1; 1.5)
(b) Vol. **125**, No. 1, p. 45—54 (1984).

Sh. M. Efendiev, V. M. Nagiev, and V. M. Burlakov
*Anharmonicity Effects in IR-Reflection Spectra of Sheelites
(6 and 20.1; 22.8.1)
(b) Vol. **125**, No. 1, p. 75—79 (1984).

M. H. Ben Ghozlen, Y. Mlik, and A. Daoud
*Raman Study of Mono-, Di-, and Trimethyl Ammonium Hexachlorostannates
(6 and 20.1; 22.9)
(b) Vol. **125**, No. 1, p. 87—96 (1984).

W. Kullmann, J. Geurts, W. Richter, N. Lehner, H. Rauh, U. Steigenberger, G. Eichhorn, and R. Geick
Effect of Hydrostatic and Uniaxial Pressure on Structural Properties and Raman Active Lattice Vibrations in Bi_2Te_3
(4; 1.1; 6; 20.1; 22.7)
(b) Vol. **125**, No. 1, p. 131—138 (1984).

T. R. Gawron and J. Mycielski
Shallow Impurity States in Strong Magnetic Field
(13.4; 20.1; 22)
(b) Vol. **125**, No. 1, p. 341—348 (1984).

R. Borromei
*The Electronic Spectrum of the CrO_4^{3-} Ion in the $Ca_2SiO_4 \cdot CaCl_2$ Host Lattice
(13.4 and 20.1; 22.8)
(b) Vol. **125**, No. 1, p. 349—355 (1984).

L. E. Kholodenkov and A. G. Makhanek
*Two-Photon Luminescence Excitation Spectra of Eu^{3+} in Yttrium Aluminium Garnet
(20.1 and 20.3; 13.4; 22.8.2)
(b) Vol. **125**, No. 1, p. 365—374 (1984).

A. Parisini, P. P. Lottici, and C. Razzetti
*Comments on the Optic Vibrational Modes in Defect Chalcopyrites
(6 and 20.1; 22.8)
(b) Vol. **125**, No. 1, p. K7—K11 (1984).

G. Corradi, E. Rzepka, et K. Polgár
*Influence des impuretés cationiques (Cu, Cr, Ni, Er) sur les spectres Raman du niobate de lithium
(Influence of Cationic Impurities (Cu, Cr, Ni, Er) on the Raman Spectra of Lithium Nitrate)
(6 and 20.1; 22.8.1)
(b) Vol. **125**, No. 1, p. K13—K16 (1984).

H. Sobotta, H. Neumann, N. N. Syrbu, and V. Riede
*Infrared Study of the Lattice Vibrations in Tetragonal CdP_2
(6 and 20.1; 22.3)
(b) Vol. **125**, No. 1, p. K17—K20 (1984).

Yu. M. Annenkov, V. F. Stolarenko, and T. S. Frangulian
*Effect of Powerful Pulsed Electron Irradiation on Creation of Hole Colour Centres in KCl at 300 K
(10.2 and 20.1; 11; 22.5.2)
(b) Vol. **125**, No. 1, p. K37—K40 (1984).

W. Burckhardt, P. Bussemer, and U. Dietrich
*Structure and Molar Refraction and Its Wavelength Dependence at Different Alkali and Ammonium Halides
(4 and 20.1; 22.5; 22.5.2)
(a) Vol. **85**, No. 1, p. 97—103 (1984).

P. Gaworzewski, E. Hild, F.-G. Kirscht, and L. Vecsernyés
Infrared Spectroscopical and TEM Investigations of Oxygen Precipitation in Silicon Crystals with Medium and High Oxygen Concentrations
(1.4 and 10.2; 20.1; 22.1.2)
(a) Vol. **85**, No. 1, p. 133—147 (1984).

V. G. Artyushenko, V. V. Voitsekhovskii, L. S. Kornienko, I. S. Lisitskii, and A. O. Rybaltovskii
*Effect of Ionizing Radiation on Optical Properties of Thallium Halides in the 0.36 to 15 μm Range
(11 and 20.1; 10.2; 22.5)
(a) Vol. **85**, No. 1, p. 167—171 (1984).

E. H. Sin, C. K. Ong, and H. S. Tan
*Temperature Dependence of Interband Optical Absorption of Silicon at 1152, 1064, 750, and 694 nm
(20.1; 22.1.2)
(a) Vol. **85**, No. 1, p. 199—204 (1984).

K. Somaiah and V. Hari Babu
Two Types of F-Centres and Thermoluminescence in BaFCl Crystals
(10.2; 20.1; 20.3; 22.5.4)
(a) Vol. **85**, No. 1, p. 265—271 (1984).

V. M. Nagiev, Sh. M. Efendiev, and V. M. Burlakov
*Vibrational Spectra of Crystals with Sheelite Structure and the Solid Solutions on Their Basis
(6 and 20.1; 22.8.1)
(b) Vol. **125**, No. 2, p. 467—475 (1984).

B. Wyncke, F. Brehat, M. El Sherif, and G. V. Kozlov
*Far-Infrared and Submillimeter Dielectric Response of Sodium Nitrite
(6 and 20.1; 1.2; 14.4.1; 22.8.1)
(b) Vol. **125**, No. 2, p. 493—498 (1984).

G. Jalbert and H. S. Brandi
*Absorption of Circularly Polarized Light by Solids
(20.1; 20.2; 22.2.1; 22.4.1)
(b) Vol. **125**, No. 2, p. 605—611 (1984).

K. Žičkus, A. Audzijonis, J. Batarūnas, and A. Šileika
The Fundamental Absorption Edge Tail of Ferroelectric SbSI
(14.4.1; 20.1; 22.8)
(b) Vol. **125**, No. 2, p. 645—651 (1984).

D. Fröhlich, A. Nöthe, and K. Reimann
*Determination of Valence Band Parameters in ZnTe
(13.5.1 and 20.1; 13.1; 22.4.3)
(b) Vol. **125**, No. 2, p. 653—657 (1984).

M. P. Lisitsa, A. V. Stolyarenko, S. F. Terekhova, and N. A. Onishchenko
Refraction Index Variation in the Exciton Resonance of CdS and $TlInS_2$ Single Crystals under Laser Excitation
(13.5.1; 20.1; 22.3.1; 22.4.1)
(b) Vol. **125**, No. 2, p. 705—711 (1984).

V. N. Zabluda, A. V. Malakhovskii, and I. S. Edelman
*Magnetooptical Effects and Optical Absorption of $FeBO_3$ in the Region of $^6A_{1g} \rightarrow {}^4T_{1g}(^4G), {}^4T_{2g}(^4G)$ Transitions
(18.2 and 20.1; 13.4; 22.8.1)
(b) Vol. **125**, No. 2, p. 751—757 (1984).

P. Petelenz
Optical Detrapping in Molecular Crystals (II): Second Conduction Band
(13.4; 20.1; 22.9)
(b) Vol. **125**, No. 2, p. 765—771 (1984).

U. W. Pohl, H.-E. Gumlich, and W. Busse
Zero-Phonon Lines in the Excitation Band Centred at 23200 cm^{-1} Due to Mn^{2+} in Polymorphic ZnS
(13.4; 20.1; 22.4.1)
(b) Vol. **125**, No. 2, p. 773—777 (1984).

H. Werheit and R. Franz
States of Thermal Equilibrium in β-Rhombohedral Boron and Their Alteration by Optical Excitation
(1.2; 13.4; 20.1; 22.1)
(b) Vol. **125**, No. 2, p. 779—784 (1984).

D. Vignaud and J. L. Farvacque
*Optical Absorption Studies of Plastically Deformed InSb
(13.4 and 20.1; 10.2; 22.2.3)
(b) Vol. **125**, No. 2, p. 785—794 (1984).

V. N. Ermakov, D. V. Korbutyak, V. G. Litovchenko, O. Yu. Mikityuk, and V. V. Nitsovich
A Study of the Reflection Spectra of Surface Exciton–Impurity Complexes
(13.4 and 13.5.1; 13.3; 20.1; 22.4.1)
(b) Vol. **125**, No. 2, p. 815—822 (1984).

V. A. Haisler, V. M. Zaletin, and A. F. Kravchenko
*Raman Study of Anharmonic Effects in Red Mercury Iodide
(6 and 20.1; 22.5)
(b) Vol. **125**, No. 2, p. K103—K108 (1984).

I. J. Lalov
Temperature Vibronic Spectra of Molecular Crystals
(6 and 13.5.1; 20.1; 22.9)
(b) Vol. **125**, No. 2, p. K147—K150 (1984).

M. Suezawa and K. Sumino
On the Annihilation of Thermal Donors in Silicon Crystals
(10.2 and 13.4; 20.1; 22.1.2)
(a) Vol. **85**, No. 2, p. 469—472 (1984).

A. A. Kaminskii, S. E. Sarkisov, G. A. Denisenko, V. V. Ryabchenkov, V. A. Lomonov, Yu. E. Perlin, M. G. Blazha, D. Schultze, B. Hermoneit, and P. Reiche
*Growth, Spectral and Luminescence Study of Cubic $Bi_4Ge_3O_{12}:Pr^{3+}$ Crystals
(20.1 and 20.3; 13.4; 22.8)
(a) Vol. **85**, No. 2, p. 553—567 (1984).

V. V. Emtsev, Yu. N. Daluda, P. Gaworzewski, and K. Schmalz
Electrical Properties of Thermal Donors Formed in CZ-Si during Heat Treatment at 450 °C
(13.4; 14.3; 20.1; 22.1.2)
(a) Vol. **85**, No. 2, p. 575—584 (1984).

J. MÁLEK, L. TICHÝ, and J. KLIKORKA
*Exponential Absorption Edge in $Ge_{20}Sb_{15}S_{65}$ Glass Depending on Sample Preparation
(2 and 20.1; 22.8)
(a) Vol. **85**, No. 2, p. K123—K127 (1984).

A. BĂRBULESCU and LUCIA SINCAN
*Absorption Edge Shift in the PbO–SiO_2 System
(20.1; 2; 22.6.1)
(a) Vol. **85**, No. 2, p. K129—K131 (1984).

S. MOCHIZUKI
*Radiative Optical Modes of Vibration in NiO Thin Film
(20.1; 6; 13.5; 22.6)
(b) Vol. **126**, No. 1, p. 105—114 (1984).

D. DELLA SALA, C. GIOVANNELLA, and F. EVANGELISTI
Study of Bonding Configurations in Amorphous Ge_xSi_{1-x}:H Alloys
(2; 4; 20.1; 22.1.1; 22.1.2)
(b) Vol. **126**, No. 1, p. 125—131 (1984).

L. MERTEN, R. KOTZOTT, and A. LEHMEYER
*Dispersion Relations for Magnon-Polaritons in Uniaxial Ferrimagnetic Crystals
(13.5.4 and 20.1; 18.3)
(b) Vol. **126**, No. 1, p. 223—228 (1984).

N. M. KHAMIDULLINA
Calculation of Some Characteristics of an Intrinsic Random Field in Hydrogenated Amorphous Silicon
(2; 20.1; 22.1.2)
(b) Vol. **126**, No. 1, p. 253—260 (1984).

T. N. DIMOV, I. A. ILIEV, and H. LANGE
*Peculiarities of Polarization Properties of CdS Single Crystals around the Isotropic Point
(20.1; 22.4.1)
(b) Vol. **126**, No. 1, p. 261—271 (1984).

K. ČERMÁK
The Width of the Energy Gap of $Tl_3^{I}B^{V}C_4^{VI}$-Type Compounds
(4 and 13.1; 20.1; 22.8)
(b) Vol. **126**, No. 1, p. 273—277 (1984).

S. K. J. AL-ANI and C. A. HOGARTH
The Spin Density and Optical Properties of Amorphous Thin Arsenic Silicate Films
(2 and 13.1; 19; 20.1; 22.6)
(b) Vol. **126**, No. 1, p. 293—304 (1984).

D. FRÖHLICH, H.-J. WEBER, R. WILLE, and R. FRANCINI
*Temperature Dependence of the Exciton Lineshape of GeO_2
(13.5.1 and 20.1; 22.6)
(b) Vol. **126**, No. 1, p. 305—310 (1984).

D. Vignaud and J. L. Farvacque
*Free-Carrier Optical Absorption Induced by Dislocation Scattering Mechanisms in III–V Compounds
(13.1 and 20.1; 10.2; 22.2)
(b) Vol. **126**, No. 1, p. 351—359 (1984).

W. Böhm, H. G. Kahle, and W. Wüchner
Spectroscopic Study of the Crystal Field Splittings in $TbPO_4$
(1.2 and 13.4; 18.4; 20.1; 22.8.1)
(b) Vol. **126**, No. 1, p. 381—392 (1984).

V. V. Sobolev and A. I. Kozlov
*Free Exciton and Exciton–Impurity Complexes of Zinc Diphosphide and Zinc Diarsenide
(13.5.1 and 20.1; 22.3)
(b) Vol. **126**, No. 1, p. K59—K62 (1984).

D. Fasold, K. Hehl, and S. Jetschke
*Optical Reflection and Transmission through a Flat System with a Thick Fluctuating Layer
(20.1)
(a) Vol. **86**, No. 1, p. 125—131 (1984).

Ch. Lushchik, J. Kolk, A. Lushchik, and N. Lushchik
Radiational Creation of Frenkel Defects in KCl–Tl
(10.2; 19; 20.1; 20.3; 22.5.2)
(a) Vol. **86**, No. 1, p. 219—227 (1984).

I. Tale and J. Rosa
Fractional Glow Technique Spectroscopy of Traps in Heavily Doped AlN:O
(13.4; 20.1; 20.3; 22.2)
(a) Vol. **86**, No. 1, p. 319—326 (1984).

R. Capelletti and M. Manfredi
*The Role of Eu^{++} Aggregation on Optical Absorption, Emission, Luminescence Time Decay, and ITC Plots in NaCl:Eu^{++}
(20.1 and 20.3; 13.4; 22.5.2)
(a) Vol. **86**, No. 1, p. 333—343 (1984).

A. A. Kaminskii, E. L. Belokoneva, B. V. Mill, Yu. V. Pisarevskii, S. E. Sarkisov, I. M. Silvestrova, A. V. Butashin, and G. G. Khodzhabagyan
Pure and Nd^{3+} Doped $Ca_3Ga_2Ge_4O_{14}$ and $Sr_3Ga_2Ge_4O_{14}$ Single Crystals, Their Structure, Optical, Spectral Luminescence, Electromechanical Properties, and Stimulated Emission
(14.4 and 20.2; 1.1; 12.2; 20.1; 20.3; 22.8.1)
(a) Vol. **86**, No. 1, p. 345—362 (1984).

Dang Tran Quan
Electrical Properties and Optical Absorption of SnSe Evaporated Thin Films
(14.3.1; 20.1; 22)
(a) Vol. **86**, No. 1, p. 421—426 (1984).

J. W. TOMM, K. HERRMANN, C. BARTHEL, and U. BARTHEL
*On the Dispersion of the Refractive Index in Active Layers of Lead-Salt Injection Lasers
(20.1; 20.2; 22)
(a) Vol. **86**, No. 1, p. 433—438 (1984).

W. S. TSE and C. C. CHEN
*Lattice Vibrations of Crystalline Silicon and Germanium Tetrachlorides
(6 and 20.1; 22)
(a) Vol. **86**, No. 1, p. K23—K25 (1984).

V. M. YARTSEV
Charge Transfer and Electron–Molecular Vibration Coupling in Tetramerized Quasi-1D Semiconductors
(14.3; 6; 20.1; 22.9)
(b) Vol. **126**, No. 2, p. 501—510 (1984).

L. E. KHOLODENKOV, A. G. MAKHANEK, and A. A. KAMINSKII
*One- and Two-Photon Luminescence Excitation of Eu^{3+} in a LaF_3 Monocrystal
(20.1 and 20.3; 13.4; 22.5)
(b) Vol. **126**, No. 2, p. 659—667 (1984).

JUH TZENG LUE
*Time-Resolved Reflectivity Measurements on RF Sputtered a-Si:H Alloy
(13.4 and 20.1; 2; 16; 22.1.2)
(b) Vol. **126**, No. 2, p. 713—720 (1984).

M. P. DOS SANTOS, C. HIRLIMANN, and M. BALKANSKI
*Raman Detection of the Nitrogen Localized Mode in GaP:N
(6 and 20.1; 22.2.2)
(b) Vol. **126**, No. 2, p. K113—K116 (1984).

R. BALTRAMIEJŪNAS, I. RÜCKMANN, G. TAMULAITIS, and A. ŽUKAUSKAS
*Sample-Dependent Luminescence Spectra in Highly Photoexcited CdS and CdSe
(13.5.1 and 20.1; 22.4.1; 22.4.2)
(b) Vol. **126**, No. 2, p. K135—K138 (1984).

K. R. ALLAKHVERDIEV, S. S. BABAEV, N. A. BAKHYSHOV, T. G. MAMEDOV, E. YU. SALAEV, and I. K. EFENDIEVA
*The Influence of Hydrostatic Pressure on the Fundamental Absorption Edge of Crystals with TlSe Type Structure
(13.1 and 20.1; 12.2; 22; 22.8)
(b) Vol. **126**, No. 2, p. K139—K143 (1984).

H. NAKANISHI and T. IRIE
*Pressure Effect on the Absorption Edges of $CdIn_2S_4$ and Related Compounds
(13.1 and 20.1; 12.2; 22.8.2)
(b) Vol. **126**, No. 2, p. K145—K148 (1984).

E.-H. Böttcher and W. F. Schmidt
*Charge Carrier Generation in Solid Xenon by Two-Photon Absorption
(14.3.4 and 20.1; 23)
(b) Vol. **126**, No. 2, p. K165—K169 (1984).

D. Fasold, K. Hehl, and S. Jetschke
*Refractive Index and Absorption Measurement in a Thin Layer by Two Reflection Measurements
(20.1; 11; 22.8.1)
(a) Vol. **86**, No. 2, p. 485—490 (1984).

G. Berger and M. Giehler
*Characterization of in Vitro Corroded Surfaces of Bioactive Glasses (P_2O_5–CaO–SiO_2) with Infrared Reflection Spectroscopy
(2 and 20.1; 4; 6; 22.6.1)
(a) Vol. **86**, No. 2, p. 531—542 (1984).

T. Pannaparayil, M. Oskooie-Tabrizi, C. Lo, and L. N. Mulay
Infrared Spectroscopic, Magnetic, and Mössbauer Studies of Aluminosilicate Cage Structures (Fe/H-Mordenites) for Syngas Conversion
(5; 18.1; 20.1; 22.8.1)
(a) Vol. **86**, No. 2, p. 663—671 (1984).

D. Wruck, M. Seifert, and W. Ulrici
Microinhomogeneities of Donor and Acceptor Distributions in n-Type LEC-GaAs from Free-Carrier Infrared Absorption
(1.4 and 13.4; 20.1; 22.2.1)
(a) Vol. **86**, No. 2, p. 691—702 (1984).

H. Gruber, E. Krautz, H. P. Fritzer, K. Gatterer, and A. Popitsch
Electronic Property Changes of η-Mo_4O_{11} by Doping with Tungsten, Rhenium, and Vanadium
(14.1; 1.2; 13.1; 18; 20.1; 22.6.1)
(a) Vol. **86**, No. 2, p. 749—757 (1984).

M. E. Abu-Zeid, M. M. El-Oker, A. I. Abu-Shady, and M. S. Adili
*Photoacoustic and Ellipsometric Study of Black Nickel Coatings on Copper Substrates
(7 and 20.1; 21.1)
(a) Vol. **86**, No. 2, p. 805—811 (1984).

R. M. Bilbé, B. Lunn, J. E. Nicholls, and D. H. J. Totterdell
The Role of Copper as an Impurity in High Resistivity Cadmium Telluride and Its Influence on γ-Ray Spectrometer Performance
(13.4 and 14.3; 20.1; 22.4.3)
(a) Vol. **86**, No. 2, p. 821—826 (1984).

20.2 Lasers

Original Papers and Short Notes

A. A. Kaminskii, H. D. Kürsten, and D. Schultze
*Stimulated Emission of Ferroelectric $Pb_5Ge_3O_{11}:Nd^{3+}, K^+$
(20.2 and 20.3; 22.8.1)
(a) Vol. **81**, No. 1, p. K19—K21 (1984).

A. A. Kaminskii, A. G. Petrosyan, K. L. Ovanesyan, G. O. Shironyan, V. A. Fedorov, and S. E. Sarkisov
*Concentrational 3μm Stimulated Emission Tuning in the $(Gd_{1-x}Er_x)_3Al_5O_{12}$ Crystal System
(20.2 and 20.3; 20.1; 22.8)
(a) Vol. **82**, No. 2, p. K185—K190 (1984).

A. A. Kaminskii, L. P. Kozeeva, and A. A. Pavlyuk
*Stimulated Emission of Er^{3+} and Ho^{3+} Ions in $KLa(MoO_4)_2$ Crystals
(20.2 and 20.3; 13.4; 22.8)
(a) Vol. **83**, No. 1, p. K65—K67 (1984).

A. A. Kaminskii, A. G. Petrosyan, and K. L. Ovanesyan
*Stimulated Emission of Pr^{3+}, Nd^{3+}, and Er^{3+} Ions in Crystals with Complex Anions
(20.2 and 20.3; 20.1; 22.8.1)
(a) Vol. **83**, No. 2, p. K159—K163 (1984).

A. A. Kaminskii, V. A. Timofeeva, A. B. Bykov, and S. E. Sarkisov
*Luminescence and Stimulated Emission in the ${}^4F_{3/2} \rightarrow {}^4I_{11/2}$ and ${}^4F_{3/2} \rightarrow {}^4I_{13/2}$ Channels of Nd^{3+} Ions in Orthorhombic $NaYGeO_4$ Crystals
(20.2 and 20.3; 20.1; 22.8.1)
(a) Vol. **83**, No. 2, p. K165—K169 (1984).

T. Kurobori and N. Takeuchi
*Room Temperature, Stable Laser Action Using F_2 Centers in LiF:Mg, OH Crystals
(20.2; 10.2; 22.5.2)
(a) Vol. **83**, No. 2, p. K179—K182 (1984).

A. A. Kaminskii, Z. I. Zhmurova, V. A. Lomonov, and S. E. Sarkisov
*Two Stimulated Emission ${}^4F_{3/2} \rightarrow {}^4I_{11/2,\,13/2}$ Channels of Nd^{3+} Ions in Crystals of the CaF_2–ScF_3 System
(20.2 and 20.3; 13.4; 22.8)
(a) Vol. **84**, No. 1, p. K81—K84 (1984).

S. S. Dimov, L. I. Pavlov, K. V. Stamenov, and G. B. Altshuller
Dispersion of Laser-Induced Nonlinear Optical Susceptibility in CdS
(13.4; 20.2; 22.4.1)
(a) Vol. **84**, No. 2, p. 555—560 (1984).

G. Jalbert and H. S. Brandi
Absorption of Circularly Polarized Light by Solids
(20.1; 20.2; 22.2.1; 22.4.1)
(b) Vol. 125, No. 2, p. 605—611 (1984).

A. A. Kaminskii, E. L. Belokoneva, B. V. Mill, Yu. V. Pisarevskii, S. E. Sarkisov, I. M. Silvestrova, A. V. Butashin, and G. G. Khodzhabagyan
*Pure and Nd^{3+} Doped $Ca_3Ga_2Ge_4O_{14}$ and $Sr_3Ga_2Ge_4O_{14}$ Single Crystals, Their Structure, Optical, Spectral Luminescence, Electromechanical Properties, and Stimulated Emission
(14.4 and 20.2; 1.1; 12.2; 20.1; 20.3; 22.8.1)
(a) Vol. 86, No. 1, p. 345—362 (1984).

J. W. Tomm, K. Herrmann, C. Barthel, and U. Barthel
On the Dispersion of the Refractive Index in Active Layers of Lead-Salt Injection Lasers
(20.1; 20.2; 22)
(a) Vol. 86, No. 1, p. 433—438 (1984).

20.3 Luminescence (X-Ray Spectra See 20)

Original Papers and Short Notes

J. Manam and V. V. Ratnam
*V_3 and F_A Centers in KCl:Na Crystals. A Thermoluminescence Study
(13.4 and 20.3; 10.2; 22.5.2)
(b) Vol. **121**, No. 1, p. 393—399 (1984).

V. N. Abramov, A. N. Ermoshkin, A. I. Kuznetsov, and V. V. Mürk
*Breakdown of Excitonic Absorption and Luminescence Induced by a d-Type Conduction Band in Sc_2O_3
(13.5.1 and 20.3; 20.1; 22.6)
(b) Vol. **121**, No. 1, p. K59—K63 (1984).

V. D. Tkachev, A. V. Mudryi, and N. S. Minaev
Noble Gas Atoms as Chemical Impurities in Silicon
(10.2 and 13.4; 20.3; 22.1.2)
(a) Vol. **81**, No. 1, p. 313—321 (1984).

A. Abdel-Kader, F. J. Bryant, and J. H. C. Hogg
Tellurium Doping and Implantation of Zinc Sulphide
(13.4; 20.3; 22.4.1)
(a) Vol. **81**, No. 1, p. 333—342 (1984).

J. Oswald, H. Feichtinger, and R. Czaputa
A Useful Method to Improve Convergence in Least-Squares Fitting Procedures
(13.4; 14.3; 20.3; 22.8)
(a) Vol. **81**, No. 1, p. 343—351 (1984).

A. A. Kaminskii, H. D. Kürsten, and D. Schultze
*Stimulated Emission of Ferroelectric $Pb_5Ge_3O_{11}:Nd^{3+}, K^+$
(20.2 and 20.3; 22.8.1)
(a) Vol. **81**, No. 1, p. K19—K21 (1984).

A. U. Grinfelds, A. R. Silin, L. N. Skuya, and V. G. Plekhanov
*Low Temperature Recombination Luminescence in Amorphous and Crystalline SiO_2
(20.3; 2; 13.5.3; 22.6)
(a) Vol. **81**, No. 1, p. K23—K26 (1984).

B. Kaufmann
*Photoluminescence of a Deep Level in Undoped PbTe
(13.4 and 20.3; 22)
(a) Vol. **81**, No. 1, p. K31—K35 (1984).

B. G. Tagiev, G. M. Niftiev, and S. A. Abushov
*Photoluminescence of GaSe:Mn Single Crystals
(20.3; 13.4; 22.3)
(b) Vol. **121**, No. 2, p. K195—K199 (1984).

B. S. ACHARYA, M. L. MUKHERJEE, C. M. SUNTA, and S. P. KATHURIA
*Thermoluminescence Emission and Decay Kinetics of NaCl:$SnCl_2$ Single Crystals
(13.4 and 20.3; 22.5.2)
(a) Vol. **81**, No. 2, p. 533—540 (1984).

R. STEGMANN, G. OELGART, W. SAUER, A. SCHEIFLER, and M. TRAPP
*Investigation of the Recombination Mechanism in $GaAs_{0.11}P_{0.89}$:N
(13.5.1 and 20.3; 22.2.4)
(a) Vol. **81**, No. 2, p. 541—548 (1984).

S. DEMIAN, F. EL AKKAD, B. FARAG, H. MATHIEU, M. ROUZEYRE, and J. CHEVALLIER
*Photoluminescence and Optical Properties of Lithium-Doped $Mg_xZn_{1-x}Te$ Alloys
(13.4 and 20.3; 20.1; 22.4.4)
(a) Vol. **81**, No. 2, p. 549—555 (1984).

G. O. MÜLLER
*Basics of Electron-Impact-Excited Luminescence Devices
(14.3.4 and 20.3; 22.4.1)
(a) Vol. **81**, No. 2, p. 597—608 (1984).

R. MACH and G. O. MÜLLER
*Efficiency and Saturation in AC Thin Film El Structures
(14.3.4 and 20.3; 22.4.1)
(a) Vol. **81**, No. 2, p. 609—623 (1984).

P. J. DEAN
*Comparison of MOCVD-Grown with Conventional II-VI Materials Parameters for EL Thin Films
(14.3.4 and 20.3; 3; 13.4; 22.4.1; 22.4.2)
(a) Vol. **81**, No. 2, p. 625—646 (1984).

D. THEIS
Selected Analytical Tools Yield a Better Insight into Electroluminescent Thin Films
(1.5 and 14.3.4; 20.3; 22.4.1)
(a) Vol. **81**, No. 2, p. 647—655 (1984).

N. A. VLASENKO, Z. L. DENISOVA, and V. S. KHOMCHENKO
*Electroluminescence of SiO_x–LnF_3 Thin Films
(14.3.4 and 20.3; 14.3.1; 22.6)
(a) Vol. **81**, No. 2, p. 657—660 (1984).

N. A. VLASENKO, YU. V. KOPYTKO, and V. S. PEKAR
*Concentration and Field Dependences of Electroluminescence Decay Kinetics in ZnS:Mn Thin Film Structures
(14.3.4 and 20.3; 22.4.1)
(a) Vol. **81**, No. 2, p. 661—667 (1984).

A. S. POPOV
Degradation Model of Red GaP LEDs
(13.4 and 14.3.3; 20.3; 22.2.2)
(a) Vol. **81**, No. 2, p. 669—674 (1984).

J. KLUGE and J. DONECKER
*Transfer Effects in ODMR Spectra of ZnSe:Mn
(19 and 20.3; 13.4; 22.4.2)
(a) Vol. **81**, No. 2, p. 675—679 (1984).

F. J. BRYANT and A. KRIER
*Direct Current Electroluminescence in Rare-Earth-Doped Zinc Sulphide
(14.3.4 and 20.3; 22.4.1)
(a) Vol. **81**, No. 2, p. 681—686 (1984).

W. E. HAGSTON
*Mechanism of Excitation of Rare-Earth Complexes in Electroluminescence Devices
(14.3.4 and 20.3; 13.4; 22.4)
(a) Vol. **81**, No. 2, p. 687—693 (1984).

DANG DINH THONG, W. HEIMBRODT, D. HOMMEL, and O. GOEDE
*Optical Study of ZnS:Mn Thin Films with High Mn Concentrations
(20.3; 22.4.1)
(a) Vol. **81**, No. 2, p. 695—700 (1984).

V. S. PEKAR
*On the Polarization of Center Radiation in an Optically Thin Layer. The Resonator Effect on the Phototransition Probability
(20.3; 22)
(a) Vol. **81**, No. 2, p. 709—713 (1984).

W. GERICKE, J. von KALBEN, U. REINSPERGER, B. SELLE, L. REDLICH, and V. GUTAN
*Annealing of Luminescent ZnSe:Mn Thin Films by a Scanning CW Laser Beam
(20.3; 22.4.1)
(a) Vol. **81**, No. 2, p. 715—721 (1984).

R. MACH, G. O. MÜLLER, and G. SCHULZ
*On the Time Behaviour of Electron Impact Excitation in Thin Film Electroluminescent Devices
(14.3.4 and 20.3; 14.3.1; 22.4.1)
(a) Vol. **81**, No. 2, p. 723—732 (1984).

R. MACH, G. O. MÜLLER, G. SCHULZ, J. von KALBEN, and W. GERICKE
*On the Mechanism of Electron-Impact-Excited Luminescence in AC Thin Film Devices: ZnSe:Mn
(14.3.4 and 20.3; 22.4.2)
(a) Vol. **81**, No. 2, p. 733—738 (1984).

U. WOGGON, K. WANDEL, and H. WEINERT
*Minority Carrier Lifetime Measurements by Photoluminescence at Room Temperature in $GaAs_{1-x}P_x$:N, Te ($x \geqq 0,70$)
(20.3; 22.2.4)
(a) Vol. **81**, No. 2, p. 739—749 (1984).

A. E. AKSENOV, YU. M. FEDOROV, and A. A. LEKSIKOV
*Systems of Interacting Centers of Photoluminescence in $RbMnF_3$ and $KMnF_3$
(20.3; 13.4; 22.5)
(a) Vol. **81**, No. 2, p. K171—K174 (1984).

D. HOMMEL, H. HARTMANN, and J. M. LANGER
*Emission Patterns of Tb Doped ZnS Bulk Crystals
(20.3; 22.4.1)
(a) Vol. **81**, No. 2, p. K199—K203 (1984).

V. D. EGOROV, G. O. MÜLLER, H. H. WEBER, M. A. JACOBSON, and A. F. DITE
*Control of Carrier Concentration by Stimulated Emission in Highly Excited Direct Gap Semiconductors
(16 and 20.3; 22.4.1)
(b) Vol. **122**, No. 1, p. 183—192 (1984).

A. V. EGRANOV and A. I. NEPOMNYACHIKH
Magnesium Colour Centres in NaF
(10.2 and 13.4; 20.1; 20.3; 22.5.2)
(b) Vol. **122**, No. 1, p. 249—254 (1984).

Q. KIM and D. W. LANGER
*Effects of Excitation Intensity on Photoluminescence of Pure CdTe
(13.5.1 and 20.3; 22.4.3)
(b) Vol. **122**, No. 1, p. 263—268 (1984).

R. B. PODE and B. T. DESHMUKH
*Study of Annealing and Deformation Effect on TL of NaCl:KCl
(20.3; 20.1; 22.5.2)
(b) Vol. **122**, No. 1, p. K79—K81 (1984).

E. MACHO, J. LLOPIS, A. REMÓN, C. BALLESTEROS, and J. PIQUERAS
*Optical Study of Defects in Quenched MgO Crystals
(10.2 and 20.3; 20.1; 22.6)
(a) Vol. **82**, No. 1, p. 111—117 (1984).

K. SOMAIAH and V. HARI BABU
*Glow Curves and the Emission of Flux Grown BaFCl:Gd Crystals
(13.4 and 20.3; 20.1; 22.5.4)
(a) Vol. **82**, No. 1, p. 201—206 (1984).

H. L. OCZKOWSKI
*The Retrapping Process in a Thermoluminescence Model with an Intra-Pair Metastable State
(13.4 and 20.3; 22.4.2)
(a) Vol. **82**, No. 1, p. 213—219 (1984).

K. NARASIMHA REDDY, U. V. SUBBA RAO, and M. LAKSHMIPATHI RAO
*Pre- and Post-Irradiation Deformation Effects on Thermoluminescence Studies in Barium-Doped NaCl Crystals
(10.2 and 20.3; 22.5.2)
(a) Vol. **82**, No. 1, p. 229—233 (1984).

K. S. V. NAMBI
*Pyroelectroluminescence Induced by Tourmaline
(14.4.1 and 20.3; 22.8.2)
(a) Vol. **82**, No. 1, p. K71—K73 (1984).

S. G. Abdullaeva, N. T. Mamedov, F. A. Mustafaev, and E. Yu. Salaev
*Photoluminescence of Ternary Layered Crystals $TlGaS_2$
(20.3; 22.8)
(a) Vol. **82**, No. 1, p. K75—K77 (1984).

T. Eshita, K. Tanimura, and N. Itoh
Photo-Induced Transformation of Close Frenkel Pairs in Strontium Fluorite
(10.2 and 20.1; 20.3; 22.5.3)
(b) Vol. **122**, No. 2. p. 489—500 (1984).

E. Ya. Glushko
*Metallic Quenching of Exciton Luminescence in Anthracene
(13.5.1 and 20.3; 22.9)
(b) Vol. **122**, No. 2, p. 569—580 (1984).

R. Schwabe, J. L. Staehli, I. Streit, and F. Thuselt
*EHP Luminescence of GaP:N under Resonant Free-Exciton Excitation
(13.5.1 and 20.3; 22.2.2)
(b) Vol. **122**, No. 2, p. 629—639 (1984).

K. D. Glinchuk, A. V. Prokhorovich, and N. S. Zayats
*A New (Non-Copper-Induced) 1.35 eV Emission Band in n-Type GaAs. Origin and Characteristics
(13.4 and 20.3; 22.2.1)
(a) Vol. **82**, No. 2, p. 503—510 (1984).

E. Chimczak, W. S. Gordon, and M. Bertrandt-Żytkowiak
*Model of the Electroluminescence in Short-Pulse-Excited Thin-Film Structures Based on ZnS:Mn
(14.3.3 and 20.3; 13.4; 22.4.1)
(a) Vol. **82**, No. 2, p. 527—531 (1984).

A. A. Kaminskii, A. G. Petrosyan, K. L. Ovanesyan, G. O. Shironyan, V. A. Fedorov, and S. E. Sarkisov
*Concentrational 3μm Stimulated Emission Tuning in the $(Gd_{1-x}Er_x)_3Al_5O_{12}$ Crystal System
(20.2 and 20.3; 20.1; 22.8)
(a) Vol. **82**, No. 2, p. K185—K190 (1984).

C. Ortiz, C. N. Afonso, R. M. Shelby, A. C. Tropper, and G. C. Bjorklund
*Photochemical Hole Burning in Polycrystalline and Neutron Irradiated NaF
(10.2 and 20.3; 22.5.2)
(b) Vol. **123**, No. 1, p. 79—84 (1984).

R. Rauch
Photochemical Conversion of M_A Centres and Identification of M_A^+ Centres in CaF_2:Na
(10.2; 20.1; 20.3; 22.5.3)
(b) Vol. **123**, No. 1, p. 93—98 (1984).

M. Beiler, M. Mocker, and O. Ziep
Theory of Second-Order Auger Recombination in Strong Degenerate Small-Gap Semiconductors. Application to PbSSe
(13.1 and 14.3; 20.3; 22.8)
(b) Vol. **123**, No. 1, p. 247—255 (1984).

M. J. SPRINGIS and J. A. VALBIS
*Visible Luminescence of Colour Centres in Sapphire
(10.2 and 20.3; 22.6)
(b) Vol. **123**, No. 1, p. 335—343 (1984).

YU. P. GNATENKO, P. A. SKUBENKO, Z. D. KOVALYUK, and V. M. KAMINSKII
*An Isothermal Annealing Effect on the Luminescence Spectra of GaSe Single Crystals
(13.5.1 and 20.3; 22.3)
(b) Vol. **123**, No. 1, p. K63—K67 (1984).

E. RADZHABOV
*Impurity Centers in LiF: Li_2O
(13.4 and 20.3; 10.2; 22.5.2)
(b) Vol. **123**, No. 1, p. K79—K82 (1984).

M. AOKI, T. NISHIKAWA, and S. OKABE
Pattern Variation of Evaporated Thin Films and Their Non-Pre-Excited TSEE
(1.5 and 17; 20.3; 22.5.2; 22.5.3)
(a) Vol. **83**, No. 1, p. 85—91 (1984).

K. LÖHNERT and E. KUBALEK
*The Cathodoluminescence Contrast Formation of Localized Non-Radiative Defects in Semiconductors
(10.2 and 20.3; 13.4; 22.2)
(a) Vol. **83**, No. 1, p. 307—314 (1984).

A. KRIER and F. J. BRYANT
*The Hot Electron Energy Distribution Function in ZnS:Cu:Cl:Er Thin Films
(20.3; 13.4; 22.4.1)
(a) Vol. **83**, No. 1, p. 315—322 (1984).

B. P. CHANDRA, R. D. VERMA, B. MAJUMDAR, and V. P. BHATT
*Dependence of Mechanoluminescence on the Microhardness and Dislocation Density in Crystals
(12.2 and 20.3; 10.2; 22.5.2; 22.8.1)
(a) Vol. **83**, No. 1, p. K19—K22 (1984).

B. SORKIN and H. KÄÄMBRE
A New Evidence for Hole-Induced Thermo-Exoemission
(10.2; 17; 20.3; 22.5.2)
(a) Vol. **83**, No. 1, p. K39—K42 (1984).

B. G. TAGIEV, G. M. NIFTIEV, and S. A. ABUSHOV
Photoelectrical and Optical Properties of GaSe:Yb Single Crystals
(13.4; 16; 20.3; 22.3)
(a) Vol. **83**, No. 1, p. K61—K64 (1984).

A. A. KAMINSKII, L. P. KOZEEVA, and A. A. PAVLYUK
*Stimulated Emission of Er^{3+} and Ho^{3+} Ions in $KLa(MoO_4)_2$ Crystals
(20.2 and 20.3; 13.4; 22.8)
(a) Vol. **83**, No. 1, p. K65—K67 (1984).

L. Bosi, P. Gagliardelli, and M. Nimis
New Results on the Decay Properties of Perturbed and Unperturbed M-Centers in $NaCl:CaCl_2$
(10.2; 20.1; 20.3; 22.5.2)
(b) Vol. **123**, No. 2, p. 519—524 (1984).

M. Ferrari, M. Montagna, L. Gonzo, G. Viliani, O. Pilla, and G. Farace
Resonant Energy Transfer in Ruby under Magnetic Field
(13.4; 20.3; 22.6)
(b) Vol. **123**, No. 2, p. 659—662 (1984).

R. L. Calvert and R. J. Danby
*Thermoluminescence and Radiophotoluminescence from Eu- and Sm-Doped $CaSo_4$
(20.3; 11; 13.4; 22.8.1)
(a) Vol. **83**, No. 2, p. 597—604 (1984).

S. Bhushan, D. Diwan, and S. P. Kathuria
*Thermoluminescence of ZnO:Pr and ZnO:Gd under UV, β- and γ-Ray Irradiations
(20.3; 11; 13.4; 22.4)
(a) Vol. **83**, No. 2, p. 605—611 (1984).

K. D. Glinchuk, A. V. Prokhorovich, and N. S. Zayats
*Effect of 2.2 MeV Electron Irradiation and Annealing on the 1.37 eV Emission Band in Zinc-Doped p-Type GaAs
(13.4 and 20.3; 11; 22.2.1)
(a) Vol. **83**, No. 2, p. 625—630 (1984).

H.-D. Autenrieth, S. Kemmler-Sack und W. Wischert
*Lumineszenz und sensibilisierte Emission der dreiwertigen Seltenen Erden in $Ca_3La_2W_2O_{12}$
(Luminescence and Sensitized Emission of Trivalent Rare Earth Ions in $Ca_3La_2W_2O_{12}$)
(13.4 and 20.3; 22.8)
(a) Vol. **83**, No. 2, p. 631—636 (1984).

C. Ballesteros, J. Piqueras, J. Llopis, and R. González
*Cathodoluminescence from MgO Single Crystals Containing a High Concentration of Anion Vacancies
(20.3; 10.2; 13.4; 22.6)
(a) Vol. **83**, No. 2, p. 645—649 (1984).

R. K. Gartia and R. Singha
*Analysis of Thermoluminescence of Topaz
(20.3; 13.4; 22.8.1)
(a) Vol. **83**, No. 2, p. 651—655 (1984).

J. Benoit, P. Benalloul, A. Geoffroy, N. Balbo, C. Barthou, J. P. Denis, and B. Blanzat
*Study of Highly Concentrated ZnS:Mn ACTFEL Devices
(20.3; 22.4.1)
(a) Vol. **83**, No. 2, p. 709—717 (1984).

P. Madhusudana Rao and K. V. Reddy
*Photoluminescence Study of Ammonium Tetrabromozincate
(20.3; 22.8)
(a) Vol. **83**, No. 2, p. K145—K148 (1984).

A. A. Kaminskii, A. G. Petrosyan, and K. L. Ovanesyan
*Stimulated Emission of Pr^{3+}, Nd^{3+}, and Er^{3+} Ions in Crystals with Complex Anions
(20.2 and 20.3; 20.1; 22.8.1)
(a) Vol. **83**, No. 2, p. K159—K163 (1984).

A. A. Kaminskii, V. A. Timofeeva, A. B. Bykov, and S. E. Sarkisov
*Luminescence and Stimulated Emission in the $^4F_{3/2} \rightarrow {}^4I_{11/2}$ and $^4F_{3/2} \rightarrow {}^4I_{13/2}$ Channels of Nd^{3+} Ions in Orthorhombic $NaYGeO_4$ Crystals
(20.2 and 20.3; 20.1; 22.8.1)
(a) Vol. **83**, No. 2, p. K165—K169 (1984).

K. S. V. Nambi
*Comment on „The Role of Sulphur-Oxyradicals in Thermoluminescence and Exoelectron Emission of $CaSO_4$ Phosphors" by R. Huzimura and K. Atarashi
(17 and 20.3; 11; 13.4; 19; 22.8.1)
(a) Vol. **83**, No. 2, p. K175—K177 (1984).

I. Baltog and L. Mihuţ
*Luminescence Decay in NaI:Pb
(20.3; 13.4; 22.5.2)
(b) Vol. **124**, No. 1, p. 307—313 (1984).

M. Kibler, G. Grenet, J. C. Gâcon, and B. Jacquier
*Energy Levels of Sm^{2+} in BaClF and SrClF Crystals under Magnetic Field (I): Theory
(13.4 and 20.3; 18; 22.5.4)
(b) Vol. **124**, No. 1, p. 325—334 (1984).

R. Gaethke, P. Gürtler, R. Kink, E. Roick, and G. Zimmerer
*Vibrational Relaxation and Hot Luminescence of Ne_2^* Centers in Solid Neon
(13.5.1 and 20.3; 23)
(b) Vol. **124**, No. 1, p. 335—342 (1984).

O. Goede and Dang Dinh Thong
*Energy Transfer Processes in (Zn, Mn)S Mixed Crystals
(13.4 and 20.3; 22.4.1)
(b) Vol. **124**, No. 1, p. 343—353 (1984).

R. Jankowiak, B. Ries, and H. Bässler
*Spectral Diffusion and Triplet Exciton Localization in an Organic Glass
(13.5.1 and 20.3; 2; 22.9)
(b) Vol. **124**, No. 1, p. 363—371 (1984).

T. Ikari and S. Shigetomi
*Photoluminescence of n-Type InSe
(20.3; 13.5.1; 22.3)
(b) Vol. **124**, No. 1, p. K49—K51 (1984).

V. F. Zolin, V. M. Markushev, M. N. Popova, R. A. Aliev, G. D. Guseinov, and R. M. Sardarly
*Neodymium Luminescence in the Ternary Semiconducting Chalcogenides $TlGaS_2$ and $TlGaSe_2$
(13.4 and 20.3; 22.8)
(b) Vol. **124**, No. 1, p. K69—K73 (1984).

A. A. ALYBAKOV, V. A. GUBANOVA, G. S. DENISOV, and B. S. UMURZAKOV
*Influence of Oxygen Centers on the Thermoluminescence of NaF Crystals
(13.4 and 20.3; 10.2; 22.5.2)
(b) Vol. **124**, No. 1, p. K75—K78 (1984).

M. MORENO and M. T. BARRIUSO
An Insight into the 10*Dq* Value Found in $CaZnF_4:Mn^{2+}$
(10.2 and 13.4; 20.3; 22.5)
(b) Vol. **124**, No. 1, p. K83—K86 (1984).

M. A. S. SWEET, C. AUST, I. T. LIDDELL, and B. RAY
*Thermoluminescence Studies of CaS Single Crystals in the Temperature Range 8 to 275 K
(20.3; 22.4.1)
(a) Vol. **84**, No. 1, p. K75—K79 (1984).

A. A. KAMINSKII, Z. I. ZHMUROVA, V. A. LOMONOV, and S. E. SARKISOV
*Two Stimulated Emission $^4F_{3/2} \rightarrow {}^4I_{11/2,\,13/2}$ Channels of Nd^{3+} Ions in Crystals of the CaF_2–ScF_3 System
(20.2 and 20.3; 13.4; 22.8)
(a) Vol. **84**, No. 1, p. K81—K84 (1984).

M. KIBLER, J. C. GÂCON, G. GRENET, B. JACQUIER, and L. C. BRUNEL
*Energy Levels of Sm^{2+} in BaClF and SrClF Crystals under Magnetic Field (II): Results
(13.4 and 20.3; 18; 22.5.4)
(b) Vol. **124**, No. 2, p. 731—739 (1984).

L. O. SCHWAN, R. FLORIAN, D. SCHMID, E. BETZ, W. SCHROF, B. WALKER, and H. PORT
Dynamics of the Optical Excitation–Deexcitation Cycle of O_2^- Centers in KCl
(10.2 and 13.4; 20.3; 22.5.2)
(b) Vol. **124**, No. 2, p. 741—745 (1984).

H.-J. MÜLLER and H.-J. WÜNSCHE
Model Calculations for Bound Excitons and Bound Excitonic Molecules in $GaAs_{1-x}P_x$:N
(13.4 and 13.5.1; 20.3; 22.2.4)
(b) Vol. **124**, No. 2, p. 747—756 (1984).

L. F. GUDYMENKO and E. G. GULE
*The Nature of Non-Linear Radiation of $A^{II}B^{VI}$ Single Crystals
(13.5 and 20.3; 22.4.1)
(b) Vol. **124**, No. 2, p. K139—K142 (1984).

E. I. CHAIKINA, S. SH. EGEMBERDIEVA, and A. S. FILIPCHENKO
*Photoluminescence of Te-Doped Gallium Antimonide Crystals
(20.3; 22.2.3)
(a) Vol. **84**, No. 2, p. 541—545 (1984).

K. D. GLINCHUK, A. V. PROKHOROVICH, and N. S. ZAYATS
*Effect of 2.2 MeV Electron Irradiation and Annealing on the Non-Copper-Induced 1.35 eV Emission Band in n-Type GaAs
(13.4 and 20.3; 11; 22.2.1)
(a) Vol. **84**, No. 2, p. 567—572 (1984).

L. N. Skuja, A. N. Trukhin, and A. E. Plaudis
*Luminescence in Germanium-Doped Glassy SiO_2
(20.3; 2; 10.2; 13.4; 20.1; 22.6)
(a) Vol. **84**, No. 2, p. K153—K157 (1984).

Tran Kim Anh, Nguyen Dai Hung, Nquy Huu Chi, and Nguyen Manh Son
*Energy Transfer between Tb^{3+} and Eu^{3+} in Rare Earth Pentaphosphates
(20.1 and 20.3; 13.4; 22.8.1)
(a) Vol. **84**, No. 2, p. K159—K163 (1984).

S. I. Radautsan, S. A. Rebrov, and A. E. Tsurkan
Some Properties of ZnTe–InP Heterojunctions
(14.3.3; 20.3; 22.2.2; 22.4.3)
(a) Vol. **84**, No. 2, p. K169—K171 (1984).

V. Nagirnyi, T. Soovik, and S. Zazubovich
*Photostimulated Reorientation of $H_z^{[001]}(Pb^{2+})$ Centres in a $KCl:PbCl_2$ Crystal
(10.2 and 20.3; 22.5.2)
(b) Vol. **125**, No. 1, p. 139—145 (1984).

Th. Kleinefeld and W. von der Osten
*Localized and Free Biexcitons in Silver Bromide
(13.5.1 and 20.3; 22.5.1)
(b) Vol. **125**, No. 1, p. 293—303 (1984).

D. Thébault, J. Barrau, G. Armelles, N. Lauret, and J. P. Noguier
*The No-Phonon Luminescence at 0.793 eV from Si:Au. Uniaxial Stress and Zeeman Spectroscopy Experiments
(13.4 and 20.3; 22.1.2)
(b) Vol. **125**, No. 1, p. 357—364 (1984).

L. E. Kholodenkov and A. G. Makhanek
*Two-Photon Luminescence Excitation Spectra of Eu^{3+} in Yttrium Aluminium Garnet
(20.1 and 20.3; 13.4; 22.8.2)
(b) Vol. **125**, No. 1, p. 365—374 (1984).

E. Mugeński and R. Cywiński
*Low-Temperature Photoluminescence of the Aggregate Centres in Eu^{2+}-Doped KCl Crystals
(20.3; 10.2; 22.5.2)
(b) Vol. **125**, No. 1, p. 381—386 (1984).

P. A. M. Berdowski, J. van Herk, L. Jansen, and G. Blasse
*Luminescence and Energy Transfer Characteristics of TbOCl
(20.3; 22.8)
(b) Vol. **125**, No. 1, p. 387—391 (1984).

H. Englisch
Instabilities and δ-Distributions in Solid State Physics
(13; 20.3; 22)
(b) Vol. **125**, No. 1, p. 393—400 (1984.

V. A. Chepurnoi, S. N. Mysovskii, I. A. Parflianovich, V. N. Salomatov, and A. P. Shkadarevich
Influence of Oxygen Defects on the Position and Form of the F_2^+ Center Zero-Phonon Lines in LiF
(10.2 and 13.4; 20.3; 22.5.2)
(b) Vol. **125**, No. 1, p. K41—K44 (1984).

P. Lilley, M. R. Czerniak, and J. E. Nicholls
*Photoluminescence and Electrical Properties of Vapour Phase Epitaxial ZnSe Grown on GaAs
(14.3.1 and 20.3; 13.4; 22.4.2)
(a) Vol. **85**, No. 1, p. 235—242 (1984).

K. Mochizuki and K. Suzuki
*Effect of the Stoichiometry Control on the Photoelectrical Properties of ZnS_xSe_{1-x}
(16 and 20.3; 22.4.4)
(a) Vol. **85**, No. 1, p. 249—256 (1984).

K. Somaiah and V. Hari Babu
Two Types of F-Centres and Thermoluminescence in BaFCl Crystals
(10.2; 20.1; 20.3; 22.5.4)
(a) Vol. **85**, No. 1, p. 265—271 (1984).

S. V. Vintsents, V. F. Kiselev, and G. S. Plotnikov
Energy Transfer between Excited Adsorbed Dye Molecules and Charged Defects in Insulator–Semiconductor Structures
(13.4; 19; 20.3; 22.1.1; 22.1.2)
(a) Vol. **85**, No. 1, p. 273—281 (1984).

V. A. Bespalov, V. F. Kiselev, G. S. Plotnikov, and A. M. Saletskii
*Fluorescence Decay of Adsorbed Dye Molecules on Solid Surfaces
(13.3 and 20.3; 22.1.1; 22.6)
(a) Vol. **85**, No. 1, p. K73—K76 (1984).

F. R. Kessler, U. Barkow, and K. Jürgens
*Cathodoluminescence of InGaAsP
(20.3; 22.2.4)
(a) Vol. **85**, No. 1, p. K77—K79 (1984).

M. S. Brodin, V. M. Bandura, and M. G. Matsko
*Excitonic Surface Polaritons in Luminescence from ZnTe Crystals
(13.5.1 and 20.3; 13.3; 22.4.3)
(b) Vol. **125**, No. 2, p. 613—618 (1984).

Y. Unuma, Y. Abe, Y. Masumoto, and S. Shionoya
*Dynamical Behaviour of the High Density Electron–Hole System in CdSe
(13.5.1 and 20.3; 22.4.2)
(b) Vol. **125**, No. 2, p. 735—742 (1984).

G. Ambrazevičius, G. Babonas, and Yu. V. Rud
*Excitation Spectra of Photoluminescence Bands in $Cd_{1-x}Mn_xTe$
(13.4 and 20.3; 22.4.4)
(b) Vol. **125**, No. 2, p. 759—764 (1984).

G. Milchberg, K. Saminadayar, E. Molva, and H. R. Zelsmann
*Optical Determination of the Valence-Band Parameters in CdTe
(13.4 and 20.3; 13.1; 22.4.3)
(b) Vol. **125**, No. 2, p. 795—803 (1984).

S. Sapru and K. L. Tan
*Eu^{2+} Ions Doped in Alkali Halides Act as Donors on Irradiation
(20.3; 11; 13.4; 22.5.2)
(b) Vol. **125**, No. 2, p. K161—K163 (1984).

M. J. Springis and J. A. Valbis
*Blue Luminescence of Color Centres in Sapphire
(20.3; 22.6)
(b) Vol. **125**, No. 2, p. K165—K169 (1984).

H. Hess
*Luminescent Properties of Zinc Silicate Doped with Gallium
(20.3; 22.8.1)
(a) Vol. **85**, No. 2, p. 543—551 (1984).

A. A. Kaminskii, S. E. Sarkisov, G. A. Denisenko, V. V. Ryabchenkov, V. A. Lomonov, Yu. E. Perlin, M. G. Blazha, D. Schultze, B. Hermoneit, and P. Reiche
*Growth, Spectral and Luminescence Study of Cubic $Bi_4Ge_3O_{12}:Pr^{3+}$ Crystals
(20.1 and 20.3; 13.4; 22.8)
(a) Vol. **85**, No. 2, p. 553—567 (1984).

R. Pickenhain, K. Jacobs, W. Seifert, V. Chernyi, and S. Brehme
*Green and Yellow Light Emitting Diodes Produced from Vapour Phase Epitaxial GaP:N (I): Deep Levels in Diodes Produced by either Diffusion or Implantation of Zinc
(14.3.3 and 20.3; 13.4; 22.2.2)
(a) Vol. **85**, No. 2, p. 627—639 (1984).

W. Hergert and L. Pasemann
*Theoretical Study of the Information Depth of the Cathodoluminescence Signal in Semiconductor Materials
(20.3; 10.2; 14.3; 22)
(a) Vol. **85**, No. 2, p. 641—648 (1984).

N. P. Esina, N. V. Zotova, S. A. Karandashev, and G. M. Filaretova
*Tunneling Surface Recombination in MS Structures on p-InAs
(14.3.3 and 20.3; 16; 22.2.1)
(a) Vol. **85**, No. 2, p. 655—660 (1984).

Y. Fukuda, A. Tomita, and N. Takeuchi
*Thermoluminescence and Thermally Stimulated Exoelectron Emission in Glass and Sintered CaB_4O_7: $CuCl_2$
(17 and 20.3; 2; 22.6)
(a) Vol. **85**, No. 2, p. K141—K144 (1984).

R. Chen, V. K. Mathur, J. F. Rhodes, M. D. Brown, S. W. McKeever, and R. K. Bull
*Thermoluminescence Governed by Simultaneous Thermal Stimulation of Electrons and Holes
(13.4 and 20.3; 22)
(b) Vol. **126**, No. 1, p. 361—369 (1984).

M. Suhrke and B. Esser
*Time-Resolved Luminescence of Charge Carriers in Localized States in Amorphous Semiconductors
(20.3; 2; 13.4; 22)
(b) Vol. **126**, No. 1, p. 393—406 (1984).

Zh. Egemberdiev, V. Nagirnyi, T. Soovik, and S. Zazubovich
*Decay Kinetics and Polarization of the A_T Emission of Pb^{2+} Centres of Different Structure in KBr: $PbBr_2$
(13.4 and 20.3; 22.5.4)
(b) Vol. **126**, No. 1, p. 407—414 (1984).

Zh. Egemberdiev, V. Nagirnyi, T. Soovik, and S. Zazubovich
*Slow Decay Component of the A_1 Emission in KBr:Pb Crystals
(13.4 and 20.3; 22.5.2)
(b) Vol. **126**, No. 1, p. K63—K66 (1984).

Ch. Lushchik, J. Kolk, A. Lushchik, and N. Lushchik
Radiational Creation of Frenkel Defects in KCl–Tl
(10.2; 19; 20.1; 20.3; 22.5.2)
(a) Vol. **86**, No. 1, p. 219—227 (1984).

A. Kozanecki and H. Rzewuski
*Distant DAP Bands in Electron-Irradiated and Ion-Implanted CdSe
(11 and 20.3; 10.2; 22.4.2)
(a) Vol. **86**, No. 1, p. 263—270 (1984).

I. Tale and J. Rosa
Fractional Glow Technique Spectroscopy of Traps in Heavily Doped AlN:O
(13.4; 20.1; 20.3; 22.2)
(a) Vol. **86**, No. 1, p. 319—326 (1984).

S. C. Sabharwal, S. P. Kathuria, and B. Ghosh
*Kinetics of Thermoluminescence in γ-Irradiated NaI(Tl)
(20.3; 11; 13.4; 22.5.2)
(a) Vol. **86**, No. 1, p. 327—331 (1984).

R. Capelletti and M. Manfredi
*The Role of Eu^{++} Aggregation on Optical Absorption, Emission, Luminescence Time Decay, and ITC Plots in NaCl:Eu^{++}
(20.1 and 20.3; 13.4; 22.5.2)
(a) Vol. **86**, No. 1, p. 333—343 (1984).

A. A. Kaminskii, E. L. Belokoneva, B. V. Mill, Yu.V. Pisarevskii, S. E. Sarkisov, I. M. Silvestrova, A. V. Butashin, and G. G. Khodzhabagyan
Pure and Nd^{3+} Doped $Ca_3Ga_2Ge_4O_{14}$ and $Sr_3Ga_2Ge_4O_{14}$ Single Crystals, Their Structure, Optical, Spectral Luminescence, Electromechanical Properties, and Stimulated Emission
(14.4 and 20.2; 1.1; 12.2; 20.1; 20.3; 22.8.1)
(a) Vol. **86**, No. 1, p. 345—362 (1984).

T. Ngoc, V. T. Bich, T. K. Anh, N. N. Dinh, and P. Long
*Kinetics of Luminescence Quenching in Neodymium-Doped Fluorophosphate Glasses
(20.3; 22.8.1)
(a) Vol. **86**, No. 1, p. K45—K50 (1984).

V. Nagirnyi, T. Soovik, and S. Zazubovich
*Decay Kinetics and Polarization of the A_x Emission of KI: In Crystal
(13.4 and 20.3; 22.5.2)
(b) Vol. **126**, No. 2, p. 653—658 (1984).

L. E. Kholodenkov, A. G. Makhanek, and A. A. Kaminskii
*One- and Two-Photon Luminescence Excitation of Eu^{3+} in a LaF_3 Monocrystal
(20.1 and 20.3; 13.4; 22.5)
(b) Vol. **126**, No. 2, p. 659—667 (1984).

M. S. Jahan and D. W. Cooke
Thermal Evolution of Low-Temperature Manganese Centers in X-Irradiated CaF_2:Mn
(10.2 and 19; 11; 20.3; 22.5.3)
(b) Vol. **126**, No. 2, p. 687—692 (1984).

O. Goede, W. Heimbrodt, and Dang Dinh Thong
*Non-Exponential ZnS: Mn Luminescence Decay Due to Energy Transfer
(20.3; 22.4.1)
(b) Vol. **126**, No. 2, p. K159—K163 (1984).

D. Wolf and S. Kemmler-Sack
*IR Emission and Sensitized Luminescence of the Trivalent Rare Earth Ions in 18R-$Ba_6B_2W_3O_{18}$ (B = Y, Er, Yb)
(13.4 and 20.3; 22.8)
(a) Vol. **86**, No. 2, p. 685—690 (1984).

G. Oelgart, B. Rheinländer, H. Haefner, and R. Pickenhain
*Degradation of Green Light Emitting Diodes LPE-GaP: N (I): Electron Beam Characterization and Deep-Level Spectroscopy
(14.3.3 and 20.3; 22.2.2)
(a) Vol. **86**, No. 2, p. 833—840 (1984).

Sh. M. Efendiev, N. G. Darvishov, and V. T. Gabrielyan
*Photoluminescence in $PbMo_xW_{1-x}O_4$ Mixed Crystals
(20.3; 22.8.1)
(a) Vol. **86**, No. 2, p. K105—K108 (1984).

21. Metals. Alloys

Review Articles

G. Dräger and O. Brümmer
Polarized X-Ray Emission Spectra of Single Crystals
(13.1 and 20; 21; 22)
(b) Vol. **124**, No. 1, p. 11—28 (1984).

V. A. Pavlov, V. I. Shalaev, and V. M. Alyabiev
Effect of Dislocation Structure on Creep and Fracture of Metals and Alloys (I): Normal Deformation Conditions
(10.1; 9; 21)
(a) Vol. **85**, No. 1, p. 11—27 (1984).

V. A. Pavlov, V. I. Shalaev, and V. M. Alyabiev
Effect of Dislocation Structure on Creep and Fracture of Metals and Alloys (II): Irradiation
(10.1; 11; 21)
(a) Vol. **85**, No. 2, p. 315—333 (1984).

Original Papers and Short Notes

K. C. Gupta and R. P. S. Rathore
Lattice Equilibrium and Phonons in Molybdenum and Tungsten
(4 and 6; 12.1; 21)
(b) Vol. **121**, No. 1, p. 73—79 (1984).

W. John, V. V. Nemoshkalenko, V. N. Antonov, and Vl. N. Antonov
Electron–Phonon Interaction in Transition Metals. Results of Relativistic APW Band Structure Calculations
(13.1; 21)
(b) Vol. **121**, No. 1, p. 233—239 (1984).

E. V. Chulkov, I. Yu. Sklyadneva, and V. E. Panin
The First-Principles Calculation of the Norm-Conserving Non-Local Singular Atomic Pseudopotentials
(13; 21; 22)
(b) Vol. **121**, No. 1, p. 265—274 (1984).

H. Stachowiak
Screening of a Proton Moving through the Electron Gas
(11 and 13.4; 21)
(b) Vol. **121**, No. 1, p. 307—315 (1984).

V. V. Gudkov and V. D. Fil
Ultrasonic Wave Field Transport by Conduction Electrons in Indium
(7 and 14.1; 13.2; 21)
(b) Vol. **121**, No. 1, p. 433—440 (1984).

V. Prabhawalkar and B. D. Padalia
X-Ray Spectroscopy Study of RMn_2Si_2 Intermetallics
(20; 21; 21.4)
(b) Vol. **121**, No. 1, p. K65—K68 (1984).

H. Stachowiak
On the Positron Schrödinger Equation in a Metal
(13; 21)
(b) Vol. **121**, No. 1, p. K69—K73 (1984).

V. Christoph and H. Wegener
Conductivity of Strongly Disordered Metallic Alloys
(14.1; 21)
(b) Vol. **121**, No. 1, p. K103—K107 (1984).

S. Simozar and J. A. Alonso
On the Factors Controlling Glass Forming Ability of Metallic Alloys Formed by Fast Liquid Quenching
(1.3 and 2; 8; 21; 21.1)
(a) Vol. **81**, No. 1, p. 55—61 (1984).

C.-D. Lien, M-A. Nicolet, and S. S. Lau
Low Temperature Formation of $NiSi_2$ from Evaporated Silicon
(1.5 and 3; 21)
(a) Vol. **81**, No. 1, p. 123—128 (1984).

Cz. Marucha and J. Rafałowicz
Deviation from Matthiessen's Rule for Aluminium Thermal Conductivity
(8; 21)
(a) Vol. **81**, No. 1, p. 185—189 (1984).

M. S. Islam and C. Mande
X-Ray Spectroscopic Study of Re_3W and Re_3Mo Systems
(4; 20; 21)
(a) Vol. **81**, No. 1, p. 197—201 (1984).

D. Segers, P. van Mourik, M. H. van Wijngaarden, and B. M. Rao
Precipitation of Silicon in a Solid Quenched Aluminium–Silicon (1.3 at%) Alloy Studied by Positron Annihilation
(10.1; 21)
(a) Vol. **81**, No. 1, p. 209—216 (1984).

S. V. Tsivinskii
The Theory of Strain Hardening in Single and Polycrystals
(10.1; 1.4; 21; 21.1; 21.1.1; 22.1.1; 22.1.2)
(a) Vol. **81**, No. 1, p. 299—306 (1984).

V. M. Chernov and M. M. Savin
Erratum to: Elastic Interaction of Edge and Screw Dislocations with Point Defects in Cubic Crystals
(10.1; 21)
(a) Vol. **81**, No. 1, p. 415 (1984).

V. Z. Bengus, E. D. Tabachnikova, and V. I. Startsev
Mechanical Behaviour of Some Metallic Glasses at 4.2 to 300 K
(2 and 12.1; 10.1; 21)
(a) Vol. **81**, No. 1, p. K11—K13 (1984).

N. V. Ershov, A. L. Ageev, A. V. Serikov, Yu. A. Babanov, and V. V. Vasin
Regular Method in X-Ray Scattering Analysis of Amorphous and Liquid Metals
(2; 21; 21.1)
(b) Vol. **121**, No. 2, p. 451—460 (1984).

A. Baldan
Microstructural Properties of the β-NiAl Phase in the Ternary Ni-Based Ni–Al–Ta Superalloys
(1.4; 21; 21.1)
(a) Vol. **81**, No. 2, p. 451—462 (1984).

K. Iwasaki
High-Temperature Internal Friction Peaks of Pure Aluminium Single-, Bi-, and Polycrystals Measured with an Inverted Flexure Pendulum
(10.1 and 12.1; 21)
(a) Vol. **81**, No. 2, p. 485—496 (1984).

E. G. Vyatkin, A. M. Taratin, and S. A. Vorobiev
Computer Simulation of Glancing Scattering of a Fast Proton Beam by a Crystal Surface
(11; 21)
(b) Vol. **122**, No. 1, p. 29—33 (1984).

P. S. Riseborough
De Haas-Van Alphen Effect in Nearly Magnetic Transition Metals
(13; 18.1; 21; 21.4)
(b) Vol. **122**, No. 1, p. 161—170 (1984).

W. John
Rigid Muffin-Tin Approximation for the Electron–Phonon Matrix Element. Nonlocal and Nonspherical Potentials
(13.1; 21)
(b) Vol. **122**, No. 1, p. 197—201 (1984).

E. Kolley and W. Kolley
Critical Behaviour of a Random Itinerant-Electron Spin Model
(18.1 and 18.2; 21)
(b) Vol. **122**, No. 1, p. K27—K31 (1984).

P. Seidel and H.-G. Meyer
Critical Josephson Current of Tunnel Junctions with Proximity Electrodes Containing Magnetic Impurities
(14.2; 21)
(b) Vol. **122**, No. 1, p. K105—K107 (1984).

G. K. Schmidt
Method for Obtaining the Spectrum of Plastic Strains of Mesoscopic Volume Elements in Fatigued Materials Investigated by X-Rays (I): Plastic and Elastic Strains in Fatigued Material
(10.1 and 12.1; 1.4; 21)
(a) Vol. **82**, No. 1, p. 75—84 (1984).

R. Tognato
Some Considerations on the Interfacial Free Energy between Discolation Core and Bulk Material at High Temperatures
(1.6; 10.1; 21)
(a) Vol. **82**, No. 1, p. K17—K19 (1984).

O. Simmich
Influence of Non-Linear Terms in the Flux Equation on the Particle Growth
(1.4 and 9; 3; 21)
(a) Vol. **82**, No. 1, p. K21—K25 (1984).

Y. Momose and M. Noguchi
Electron Emission during Frictional Contact between Metals and Polymers
(14.4 and 17; 21; 22.9)
(a) Vol. **82**, No. 1, p. K83—K86 (1984).

L. Půst and Z. Frait
Low-Temperature FMR and FMAR Measurements of Metal Single Crystals (I): General Consideration, Experimental Techniques
(19; 21)
(b) Vol. **122**, No. 2, p. 535—541 (1984).

H. C. Saxena and R. P. S. Rathore
Lattice Vibrations in Hafnium
(6; 21)
(b) Vol. **122**, No. 2, p. K119—K124 (1984).

G. N. Gaidukov, A. A. Podrezov, and J. P. Hirth
A Volterra-Type Dislocation Model of a Low-Angle [001] Twist Boundary in an F.C.C. Crystal
(1.3; 10.1; 21)
(a) Vol. **82**, No. 2, p. 355—363 (1984).

V. Atzrodt and H. Lange
Principal Component Analysis for Refractory Metal Silicide Investigations with Auger Electron Spectroscopy
(1.6; 17; 21; 22.1.2)
(a) Vol. **82**, No. 2, p. 373—378 (1984).

G. K. Schmidt
Method for Obtaining the Spectrum of Plastic Strains of Mesoscopic Volume Elements in Fatigued Materials Investigated by X-Rays (II): Distributions of Elastic Strains and X-Ray Reflex Profiles
(10.1 and 12.1; 1.4; 21)
(a) Vol. **82**, No. 2, p. 413—420 (1984).

TH. WIRTH, V. ATZRODT, and H. LANGE
Dynamic Alterations of the Surface Composition during Sputtering of Silicides
(1.6 and 11; 21)
(a) Vol. **82**, No. 2, p. 459—466 (1984).

M. JAŁOCHOWSKI
Electron Tunnelling in Pb/Ag Ultrathin Layered Structures
(14.3.1; 14.3.3; 21; 21.6)
(a) Vol. **82**, No. 2, p. 497—502 (1984).

L. URAY
The Effect of Void Size on the Resistivity of Sintered Tungsten
(1.3 and 14.1; 15; 21)
(a) Vol. **82**, No. 2, p. 545—552 (1984).

Z. A. MATYSINA, D. R. RIZDVYANETSKII, and S. Y. ZAGINAICHENKO
The Structural Diagrams of Magnetic Decompositive Alloys
(1.2 and 18; 21; 21.1)
(a) Vol. **82**, No. 2, p. K163—K168 (1984).

R. KRISHNAN
Magnetic Moment of Amorphous $Co_{1-x}Nb_x$ Alloys and Friedel's Model
(2 and 18.2; 21; 21.1)
(a) Vol. **82**, No. 2, p. K173—K175 (1984).

S. K. ROY and D. L. BHATTACHARYA
Debye-Waller Factor for Small Concentration of Mössbauer Impurities in Metallic Solids
(6; 5; 10.1; 21)
(b) Vol. **123**, No. 1, p. 45—51 (1984).

U. POTZEL, J. VÖLKL, H. WIPF, and A. MAGERL
Tracer Diffusion and Chemical Diffusion of Hydrogen in TaH_x
(9; 21)
(b) Vol. **123**, No. 1, p. 85—92 (1984).

L. PŮST
Low-Temperature FMR and FMAR Measurements of Metal Single Crystals (II): Saturation Magnetization, Magnetocrystalline Anistropy, and *g*-Factor Determination
(19; 18.2; 21)
(b) Vol. **123**, No. 1, p. 115—122 (1984).

R. RICHTER and F. GOEDSCHE
On the Thermoelectric Power of Amorphous Ferromagnets
(2 and 15; 14.1; 18.2; 21)
(b) Vol. **123**, No. 1, p. 143—154 (1984).

E. M. KOGAN, YU. N. GORNOSTIRIEV, and V. V. USTINOV
Conduction Electron Scattering by a Short-Range Ordered Metal Surface
(13.3; 21)
(b) Vol. **123**, No. 1, p. 289—293 (1984).

D. Gerlich
Fourth-Order Elastic Constants of Solids
(12.1 and 12.2; 21; 21.6; 22.5.2)
(b) Vol. **123**, No. 1, p. K5—K7 (1984).

R. Niedzielski and R. Troć
Magnetic Susceptibility of PrP, PrS, NdP, NdS, and $U_{1-x}Pr(Nd)_xP$ in the Paramagnetic Region
(18.1; 21)
(b) Vol. **123**, No. 1, p. K29—K34 (1984).

W. Henrion and H. Lange
Optical Properties of $CrSi_2$
(20.1; 13.1; 21)
(b) Vol. **123**, No. 1, p. K69—K73 (1984).

A. Baldan
The Microstructural Properties of (β + (Ni, Al)$_7$Hf$_2$) Eutectics
(1.3; 10.1; 21; 21.1)
(a) Vol. **83**, No. 1, p. 47—57 (1984).

K. S. Vahvaselkä
An X-Ray Investigation of Structural Properties and Electrical Resistivities of Liquid Al, Liquid Cu, and Liquid Al–Cu Alloys
(1.3 and 2; 14.1; 21; 21.1)
(a) Vol. **83**, No. 1, p. 103—112 (1984).

D. O. Northwood and I. O. Smith
Internal and Friction Stresses for Al–Mg Class I High Temperature Creep Alloys
(10.1 and 12.1; 21)
(a) Vol. **83**, No. 1, p. 227—234 (1984).

P. Delobelle, D. Varchon, and C. Oytana
Short-Range Reordering of Heavy Interstitials in Ta, Nb, and Fe during Relaxation and Static Strain Aging
(10.1; 12.1; 21; 21.1.1)
(a) Vol. **83**, No. 1, p. 235—243 (1984).

M. M. El-Oker, F. A. Sultan, S. A. Yousef, and S. A. El-Sahhar
Surface Plasmon Modes and Energy Loss Function of Indium Thin Films on Mica Substrates
(13.5.2 and 20.1; 13.1; 21)
(a) Vol. **83**, No. 1, p. 263—267 (1984).

T. Soma, T. Itoh, and H.-Matsuo Kagaya
Lattice Dynamics of Al
(6; 8; 21)
(b) Vol. **123**, No. 2, p. 463—467 (1984).

H. Fehske, E. Kolley, and W. Kolley
Spin-Glass Behaviour in Disordered Hubbard Alloys
(18; 13; 21)
(b) Vol. **123**, No. 2, p. 533—540 (1984).

S. B. Maslenkov, A. I. Kozlenkov, S. A. Filin, and A. I. Shulgin
Soft X-Ray Emission Spectra from Some Ni–Al Alloys
(13.1 and 20; 21; 21.1)
(b) Vol. **123**, No. 2, p. 605—609 (1984).

H.-G. Bartel, G. Scholz, R. Neumann, and W. Schmidt
The Electronic Structure of Octahedral Platinum and Iridium Clusters (Pt_6, Ir_6). A Comparison between Xα-SW and EHT Calculations
(13.1; 21)
(b) Vol. **123**, No. 2, p. 641—648 (1984).

G. Knuyt
Electrical Resistivity Calculations for Amorphous Cu_xAl_{1-x}
(2 and 14.1; 21; 21.1)
(b) Vol. **123**, No. 2, p. 763—771 (1984).

Th. Karakostas and P. Delavignette
Annealing Structures of Grain Boundaries in Zirconium
(1.4; 21)
(a) Vol. **83**, No. 2, p. 485—497 (1984).

A. Baldan
An Investigation Mechanism of Thermally Activated Deformation in Certain Ni–Al–Ta and Ni–Al–Hf Superalloys
(10.1; 1.3; 21; 21.1)
(a) Vol. **83**, No. 2, p. 507—511 (1984).

M. M. El-Oker
Detection of Surface Plasmons in Some Indium Films by Attenuated Total Reflection
(13.5.2 and 20.1; 21)
(a) Vol. **83**, No. 2, p. 613—616 (1984).

K. Mukherjee, M. Chandrasekaran, and M. Kato
Electrical Resistivity of an Equiatomic Solid and Liquid Au–Cd Alloy
(14.1; 2; 8; 21; 21.6)
(a) Vol. **83**, No. 2, p. 657—667 (1984).

Z. Bojarski, H. Morawiec, and T. J. Panek
Positron Annihilation Study of Martensitic Transformation in Cu–Zn–Al Alloy
(1.2 and 10.1; 21; 21.1)
(a) Vol. **83**, No. 2, p. K97—K101 (1984).

V. Valvoda
X-Ray Debye Temperatures of HfC and α-W_2C
(6 and 8; 21)
(a) Vol. **83**, No. 2, p. K123—K125 (1984).

V. B. Gohel, C. K. Acharya, and A. R. Jani
On the Determination of Pseudopotential Parameters and the Comprehensive Study of Metallic Properties
(13.1; 2; 14.1; 15; 21)
(b) Vol. **124**, No. 1, p. 137—145 (1984).

A. Rubaszek and H. Stachowiak
Application of the Quadratic Response Theory to the Screening of a Positron in an Electron Gas of Metallic Density
(13; 13.2; 21)
(b) Vol. **124**, No. 1, p. 159—166 (1984).

V. Drchal and J. Kudrnovský
The Auger Spectra from Alloys. Effect of Partial Long-Range Order
(13.1; 17; 21)
(b) Vol. **124**, No. 1, p. 179—189 (1984).

U. H. Gläser
On the Calculation of the Residual Resistivity of Precipitates
(10.1 and 14.1; 21; 21.1)
(b) Vol. **124**, No. 1, p. 423—429 (1984).

P. Mrozek, M. Menyhárd, J. Wernisch, and A. Jabłoński
Surface Composition of the Ordered Al50–Co50 Alloy
(1.3 and 1.6; 21; 21.1)
(a) Vol. **84**, No. 1, p. 39—48 (1984).

L. Uray and M. Menyhárd
The Segregation of Iron in Tungsten
(1.3; 1.4; 9; 14.1; 15; 21; 21.1.1)
(a) Vol. **84**, No. 1, p. 65—72 (1984).

R. Khanna
Separation of Thermal Component in Diffuse Scattering from Disordered Binary Alloys
(1.3; 6; 21)
(a) Vol. **84**, No. 1, p. 95—101 (1984).

T. S. Kê, P. Cui, and C. M. Su
Internal Friction in High-Purity Aluminium Single Crystals
(10.1 and 12.1; 21)
(a) Vol. **84**, No. 1, p. 157—164 (1984).

K. H. J. Buschow and D. B. de Mooij
The Effect of H_2 Absorption on the Magnetic Properties of Gd–Pd Intermetallic Compounds
(18.2 and 18.4; 21; 21.4)
(a) Vol. **84**, No. 1, p. 207—213 (1984).

Y. Shimada
Initial Permeability of Amorphous Films with Multilayered Structure
(2 and 18.2.1; 21)
(a) Vol. **84**, No. 1, p. K55—K58 (1984).

E. Canessa, D. F. Mariani, and J. Vignolo
Structure of Non-Simple Liquid Metals
(2; 21; 21.7)
(b) Vol. **124**, No. 2, p. 465—472 (1984).

P. Chatterjee
Superconducting Properties of the Hexagonal-Close-Packed Alloy System TcZr
(14.2; 21)
(b) Vol. **124**, No. 2, p. 773—779 (1984).

H. Luchian
Magnetic Symmetry in the A15 Type of Structure
(14.2 and 18; 21)
(b) Vol. **124**, No. 2, p. K125—K128 (1984).

B. G. Mulimani and R. Harris
A Note on the Electron–Phonon Enhancement of the Thermopower in Amorphous Alloys
(15; 2; 21)
(b) Vol. **124**, No. 2, p. K181—K185 (1984).

G. Cocco, S. Enzo, C. Antonione, L. Battezzati, and G. Riontino
Structure Features in the Palladium–Beryllium Amorphous System
(2; 21; 21.3)
(a) Vol. **84**, No. 2, p. 371—377 (1984).

F. Reynaud
Application of the Grand Canonical Ensemble in Classical Statistical Physics to the Study of Order–Disorder Transitions in Binary Substitutional Alloys
(1.2 and 1.3; 21)
(a) Vol. **84**, No. 2, p. 393—400 (1984).

V. I. Kononenko, A. L. Sukhman, S. L. Gruverman, and V. V. Torokin
Density and Surface Tension of Liquid Rare Earth Metals, Scandium and Yttrium
(2 and 12.1; 1.2; 21; 21.4)
(a) Vol. **84**, No. 2, p. 423—432 (1984).

T. S. Kê, L. D. Zhang, P. Cui, Q. Huang, and B. S. Zhang
Internal-Friction Peak in Polygonized High-Purity Aluminium and In-Situ Observation by Transmission Electron Microscope
(10.1 and 12.1; 21)
(a) Vol. **84**, No. 2, p. 465—474 (1984).

D. O. Northwood, L. Moerner, and I. O. Smith
The Breakdown in Creep by Viscous Glide in Al–Mg Solid Solution Alloys at High Stress Levels
(10.1; 21)
(a) Vol. **84**, No. 2, p. 509—515 (1984).

O. L. de Lange and C. J. Young
Interaction between Dislocation Arrays
(10.1; 21)
(a) Vol. **84**, No. 2, p. 517—526 (1984).

L. Jahn and D. Hinz
Rotational Hysteresis and Anisotropy Field of Mn–Al–C Magnets
(18; 21)
(a) Vol. **84**, No. 2, p. K133—K134 (1984).

A. I. ROMANÈNKO, F. S. RAKHMENKULOV, I. N. KUROPYATNIK, V. E. FEDOROV, and A. V. MISHCHENKO
Electrical Resistance of the Mo-Chain Compound Mo_2S_3 in Equilibrium and Non-Equilibrium States
(14.1; 21)
(a) Vol. **84**, No. 2, p. K165—K167 (1984).

T. SOMA, T. ITOH, and H.-MATSUO KAGAYA
Pressure Effect on Phonon Frequencies and Elastic Stiffness Constants of Al
(6; 12.1; 21)
(b) Vol. **125**, No. 1, p. 107—112 (1984).

M. JIRSA
Angular Dependence of SSWR in Thin Metallic Films. Approximate Analytical Solution
(13.3 and 19; 21)
(b) Vol. **125**, No. 1, p. 187—196 (1984).

YU. V. ZAKHAROV and YU. I. MANKOV
Magnetoresistance of Ferromagnets with Domain Structure
(14.1; 18.2; 21)
(b) Vol. **125**, No. 1, p. 197—205 (1984).

J. A. LEIRO
The First Born Approximation in Estimates of the Threshold Anomalies in the X-Ray Spectra of Simple Metals
(20; 13; 21)
(b) Vol. **125**, No. 1, p. 305—310 (1984).

F. C. LOVEY, G. VAN TENDELOO, and S. AMELINCKX
The Nature of Some Planar Defects in 2H Martensite of Cu–Al Alloys as Determined by HREM
(1.4; 10.1; 21; 21.1)
(a) Vol. **85**, No. 1, p. 29—37 (1984).

N. KURAMOCHI, H. MIZUBAYASHI, and S. OKUDA
Dislocation Pinning at Low Temperatures in Nb
(10.1 and 12.1; 21)
(a) Vol. **85**, No. 1, p. 121—131 (1984).

M. GABER and H.-J. FITTING
Energy–Depth Relation of Electrons in Bulk Targets by Monte-Carlo Calculations
(17; 11; 21)
(a) Vol. **85**, No. 1, p. 195—198 (1984).

P. MICHEL, B. WEBER, and G. GÖTZ
Formation of Pt Silicides by a Millisecond Laser Pulse
(1.6 and 11; 21)
(a) Vol. **85**, No. 1, p. K1—K4 (1984).

R. TOGNATO
Influence of Dislocation Core Density on Overheating Absence
(10.1; 21)
(a) Vol. **85**, No. 1, p. K47—K50 (1984).

B. A. Vaid and K. C. Sharma
Ultrasonic Attenuation in Liquid Metals
(2 and 7; 21)
(b) Vol. **125**, No. 2, p. 505—512 (1984).

P. Streitenberger
Effective Exchange and Correlation Interaction and Dielectric Screening in the Electron Gas. A Functional Approach to the Local-Field Correction
(13; 14.4; 21)
(b) Vol. **125**, No. 2, p. 681—692 (1984).

M. Scrocco
X-Ray Photoelectron Spectra of Ag, Ga, In, and β-Sn Metals. Multielectron Excitations
(13.1 and 17; 21; 21.6)
(b) Vol. **125**, No. 2, p. 713—720 (1984).

G. Schubert and J. Richter
The Electrical Resistivity of Clustering Alloys
(1.3 and 14.1; 21)
(b) Vol. **125**, No. 2, p. 839—847 (1984).

P. Středa
Universal Transport Properties of Inversion Layers in Quantizing Magnetic Field
(13.3; 8; 14.1; 15; 18; 21)
(b) Vol. **125**, No. 2, p. 849—860 (1984).

S. J. Mali and D. G. Kanhere
Calculation of X-Ray Photoemission Shifts for Some Metallic Elements Using Slater's Transition State Theory
(13 and 17; 21)
(b) Vol. **125**, No. 2, p. K143—K145 (1984).

I. M. Chapnik
Some Regularities in the Occurrence of Superconductivity
(14.2; 21)
(b) Vol. **125**, No. 2, p. K189—K191 (1984).

J. M. López and J. A. Alonso
The Atomic Size-Mismatch Contribution to the Enthalpy of Formation of Concentrated Substitutional Metallic Solid Solutions
(1.3 and 8; 21; 21.1)
(a) Vol. **85**, No. 2, p. 423—428 (1984).

B. Rauschenbach, G. Blasek, and R. Dietsch
Modification of Concentration Profiles in Iron and Aluminium by High-Fluence Implantation of Nitrogen and Boron Ions
(1.3 and 11; 21; 21.1.1)
(a) Vol. **85**, No. 2, p. 473—480 (1984).

I. Baker and E. M. Schulson
On Intrinsic Stacking Faults in Polycrystalline Ni_3Al
(10.1; 1.4; 21; 21.1)
(a) Vol. **85**, No. 2, p. 481—490 (1984).

A. Pawełek
A Simple Model of Small Yield Point Due to Strain Rate Change Based on the Dynamics of the Source Generated Dislocation Group
(10.1 and 12.1; 21)
(a) Vol. **85**, No. 2, p. K117—K121 (1984).

K. Helming and S. Matthies
On the Interpretation of Orientation Distributions and Qualitative Ghost Corrections for Hexagonal-Orthorhombic Textures
(1.3 and 1.4; 21)
(b) Vol. **126**, No. 1, p. 43—52 (1984).

T. Soma, T. Itoh, and H.-Matsuo Kagaya
Pressure Dependence of the Melting Point of Al
(6; 8; 21)
(b) Vol. **126**, No. 1, p. 71—75 (1984).

A. M. Monti
Divacancy in H.C.P. Metals
(10.1; 21)
(b) Vol. **126**, No. 1, p. 163—175 (1984).

D. Korn
Electrical Resistivity of Amorphous SnMn
(2 and 14.1; 21)
(b) Vol. **126**, No. 1, p. 425—432 (1984).

E. Chilla and W. Schiller
The Influence of Electron–Phonon Interaction on the Thermoelectric Power at High Temperatures
(14.1 and 15; 21)
(b) Vol. **126**, No. 1, p. K91—K96 (1984).

P. Malétras and P. Duval
Comparison of Temper-Hardened Cu–Al Alloys before and after Thinning-Down
(1.3; 21; 21.1)
(a) Vol. **86**, No. 1, p. 85—92 (1984).

H. Yamaguchi, T. Tsuchida, I. Hashimoto, Y. Ohtani, and S. Ohtaki
Stage II Recovery of Defects in Cold-Worked Aluminium
(10.1; 14.1; 21)
(a) Vol. **86**, No. 1, p. 237—244 (1984).

M. B. Shah and M. S. C. Bose
Magnetic NDT Technique to Evaluate Fatigue Damage
(10.1 and 18.2; 21)
(a) Vol. **86**, No. 1, p. 275—281 (1984).

V. E. Arkhipov, V. I. Voronin, B. N. Goshchitskii, A. E. Karkin, V. N. Kozhanov, A. V. Mirmelshtein, and E. P. Romanov
The Structure, Superconducting Properties, and Electrical Resisitivity of V_2Zr and V_2Hf Irradiated with Fast Neutrons
(14.1 and 14.2; 1.2; 21; 21.1)
(a) Vol. **86**, No. 1, p. K59—K62 (1984).

S. Palchaudhuri, R. N. Joarder, and R. V. Gopala Rao
Partial Structure Factors and Self-Diffusion in Liquid Copper–Lead Alloys
(2 and 9; 21; 21.1)
(b) Vol. **126**, No. 2, p. 443—452 (1984).

S. Mahajan and S. Prakash
Heat of Solution of Hydrogen in Aluminum
(8; 10.1; 21)
(b) Vol. **126**, No. 2, p. 467—480 (1984).

M. Rösler and W. Brauer
On the Theory of Ion-Induced Electron Emission
(17; 13.5.2; 21)
(b) Vol. **126**, No. 2, p. 629—642 (1984).

K. Masuda-Jindo
Tight-Binding Electronic Theory for Lattice Defects in Transition Metals. Correlation Effects
(13.4; 21; 21.1.1)
(b) Vol. **126**, No. 2, p. 643—652 (1984).

U. H. Gläser and H. Löffler
Effective Medium Approach for the Residual Resistivity of Alloys Containing Guinier-Preston Zones
(14.1; 13; 21; 21.1)
(b) Vol. **126**, No. 2, p. 741—751 (1984).

M. F. Zhu, F. C. T. So, E. T-S. Pan, and M-A. Nicolet
Investigation of Amorphous $W_{60}Zr_{40}$ Film as a Diffusion Barrier in Metallization Schemes
(1.5; 2; 21)
(a) Vol. **86**, No. 2, p. 471—476 (1984).

P. Siebert, M. Werner, and H. Löffler
Heterogeneous Nucleation of Precipitates in an Al–12 at% Zn Alloy at Dislocation Loops
(1.3; 10.1; 21; 21.1)
(a) Vol. **86**, No. 2, p. 543—551 (1984).

F. C. Lovey, G. van Tendeloo, J. van Landuyt, L. Delaey, and S. Amelinckx
On the Nature of Various Stacking Defects in 18R Martensite in Cu–Al Alloys. A Study by High Resolution Electron Microscopy
(1.3; 10.1; 21; 21.1)
(a) Vol. **86**, No. 2, p. 553—564 (1984).

T. S. Kê, P. Cui, S. C. Yan, and Q. Huang
Internal Friction Peaks of Super-High-Purity Aluminium at Medium Temperatures
(10.1 and 12.1; 21)
(a) Vol. **86**, No. 2, p. 593—601 (1984).

K. Benhaddane and P. Beauchamp
A Computation of (110) Intrinsic Stacking Fault Energies in B2 Ordered Alloys
(1.4; 10.1; 21; 21.1.1)
(a) Vol. **86**, No. 2, p. 613—621 (1984).

K. Iwasaki
High-Temperature Internal Friction Peaks of Dilute Aluminium Alloys
(10.1 and 12.1; 21)
(a) Vol. **86**, No. 2, p. 637—648 (1984).

Gyanesh Chandra and O. P. Katyal
Applicability of the Mayadas-Shatzkes Model on Electrical Resistivity of Polycrystalline Tin–Lead Alloy Films
(14.1; 1.3; 21)
(a) Vol. **86**, No. 2, p. 759—764 (1984).

Gyanesh Chandra and O. P. Katyal
The Study of Electrical Properties of Polycrystalline Tin Films
(14.1; 21)
(a) Vol. **86**, No. 2, p. 765—772 (1984).

21.1 Transition Metals (Group Sc to Zn) and Their Alloys

Original Papers and Short Notes

H. J. KROKOSZINSKI and K. BÄRNER
Characterization of the Magnetic Phases of $MnAs_{1-x}P_x$ Crystals
(18; 21.1)
(b) Vol. **121**, No. 1, p. 157—160 (1984).

D. FRAITOVÁ and L. PŮST
FMR Linewidth in Bulk Metals in Parallel Configuration
(19; 21.1; 21.1.1)
(b) Vol. **121**, No. 1, p. 173—176 (1984).

S. B. SHRIVASTAVA and V. K. OJHA
Band Structure Calculation of the Momentum Density in F.C.C. Manganese
(13.1; 21.1)
(b) Vol. **121**, No. 1, p. 225—231 (1984).

S. SIMOZAR and J. A. ALONSO
On the Factors Controlling Glass Forming Ability of Metallic Alloys Formed by Fast Liquid Quenching
(1.3 and 2; 8; 21; 21.1)
(a) Vol. **81**, No. 1, p. 55—61 (1984).

N. TANAKA and K. OHSHIMA
High Resolution Electron Microscopic Observations of Disordered Cu–Pd Alloys
(1.3; 21.1)
(a) Vol. **81**, No. 1, p. 129—138 (1984).

S. BANERJEE and K. URBAN
Kinetics of Order–Disorder Transformation in Alloys under Electron Irradiation
(1.2; 1.3; 11; 21.1)
(a) Vol. **81**, No. 1, p. 145—162 (1984).

V. VALVODA and P. ČAPKOVÁ
X-Ray Diffraction Study of Charge Distribution in Titanium Carbide
(4; 21.1)
(a) Vol. **81**, No. 1, p. 203—208 (1984).

H. J. BLYTHE, F. WALZ, and H. KRONMÜLLER
Magnetic Investigation of the Annealing Behaviour of Some Dilute Iron–Nickel Alloys after Low-Temperature Electron Irradiation
(10.1 and 18.2; 11; 21.1; 21.1.1)
(a) Vol. **81**, No. 1, p. 227—238 (1984).

P. WEISSBRODT, F. STORBECK, and W. HAUFFE
Sputtering Behaviour of Vanadium Silicide Single Crystals under 10 keV Kr^+ Ion Bombardment
(1.6 and 11; 21.1)
(a) Vol. **81**, No. 1, p. 259—265 (1984).

S. V. Tsivinskii
The Theory of Strain Hardening in Single and Polycrystals
(10.1; 1.4; 21; 21.1; 21.1.1; 22.1.1; 22.1.2)
(a) Vol. **81**, No. 1, p. 299—306 (1984).

G. L. Gutsev, Yu. M. Shulga, and Yu. G. Borodko
The Discrete Variational X_α-Method Interpretation of the C KVV and Ti $L_{2,3}M_{2,3}V$ Auger Spectra of TiC
(13.1; 17; 21.1)
(b) Vol. **121**, No. 2, p. 595—601 (1984).

Z. Pawłowska
Positron Annihilation in Zinc. Theory and Experiment
(13.1; 21.1)
(b) Vol. **121**, No. 2, p. 695—703 (1984).

A. Baldan
Microstructural Properties of the β-NiAl Phase in the Ternary Ni-Based Ni–Al–Ta Superalloys
(1.4; 21; 21.1)
(a) Vol. **81**, No. 2, p. 451—462 (1984).

P. M. Hazzledine, H. P. Karnthaler, and A. Korner
The Application of the Wulff Construction to Dislocation Problems
(10.1; 12.1; 21.1; 21.6)
(a) Vol. **81**, No. 2, p. 473—484 (1984).

V. I. Pecherskaya, D. N. Bolshutkin, V. A. Desnenko, and V. Ya. Ilichev
Electrical and Magnetic Properties of the Fe18Cr(30–40)Ni Alloys at Low Temperatures
(18.1 and 18.2; 14.1; 21.1; 21.1.1)
(a) Vol. **81**, No. 2, p. 507—512 (1984).

K. Sumiyama and Y. Nakamura
High Concentration Fe–Cu and Fe–Ag Alloys Produced by Vapor Quenching
(1.3; 5; 18.2; 21.1; 21.1.1; 21.6)
(a) Vol. **81**, No. 2, p. K109—K112 (1984).
Erratum
(a) Vol. **83**, No. 2, p. K219 (1984).

P. Čapková and V. Valvoda
The Surface Roughness Effect in Texture Measurements by the Schulz X-Ray Reflection Method
(1.6; 21.1)
(a) Vol. **81**, No. 2, p. K113—K115 (1984).

R. Rünger and D. Ploumbidis
High Temperature NMR in the V_3Au Compound
(13 and 19; 21.1)
(b) Vol. **122**, No. 1, p. K49—K52 (1984).

L. J. BALK, D. G. DAVIES, and N. KULTSCHER
The Dependence of Scanning Electron Acoustic Microscopy (SEAM) Imaging on Chopping and Detection Frequency for Metal Samples
(1 and 7; 21.1)
(a) Vol. 82, No. 1, p. 23—33 (1984).

G. VAN TENDELOO, S. AMELINCKX, A. F. MARSHALL, and R. G. WALMSLEY
A Complex High-Temperature Phase in Cu_9Zr_2
(1.4; 21.1)
(a) Vol. 82, No. 1, p. 55—61 (1984).

N. K. GOBRAN, M. M. DANIAL, and R. KAMEL
Kinetics of the Amorphous–Crystalline Transition in the Metallic Glass $Fe_{40}Ni_{40}B_{20}$
(1.2; 2; 21.1; 21.1.1)
(a) Vol. 82, No. 1, p. 63—66 (1984).

A. NOHARA, M. IZUMI, H. SAKA, and T. IMURA
Plastic Deformation Behavior of β-CuZn Single Crystals at Low and High Temperatures
(10.1; 21.1)
(a) Vol. 82, No. 1, p. 163—170 (1984).

E. BURZO and I. LUPSA
On the Thermal Variation of Spontaneous Magnetization in Ternary NiCuD (D = Zn, Al, or Si) Alloys
(18; 18.2; 21.1)
(a) Vol. 82, No. 1, p. K49—K52 (1984).

S. CHAKRABORTY, A. MANNA, and A. K. GHOSH
Charge Transfer and Energy of Formation in Binary Alloys: CsCl Intermetallic Compounds
(1.3 and 4; 21.1; 21.1.1; 21.6)
(b) Vol. 122, No. 2, p. 469—480 (1984).

J. MOLENDA
Electronic Processes in Electrode Materials of A_xMX_2-Type
(13.1; 1.3; 21.1; 21.2)
(b) Vol. 122, No. 2, p. 591—598 (1984).

A. D. KOROTAEV, YU. I. CHUMLYAKOV, V. F. ESIPENKO, and L. S. BUSHNEV
Superelasticity Effects in Alloy Single Crystals of Cu–15% Al–2% Co with Non-Coherent Particles Due to Twinning
(12.1; 21.1)
(a) Vol. 82, No. 2, p. 405—412 (1984).

Z. A. MATYSINA, D. R. RIZDVYANETSKII, and S. Y. ZAGINAICHENKO
The Structural Diagrams of Magnetic Decompositive Alloys
(1.2 and 18; 21; 21.1)
(a) Vol. 82, No. 2, p. K163—K168 (1984).

T. Soumura, S. Saito, T. Kuraishi, and T. Maeda
Magnetic Moment and Its Temperature Variation of Amorphous $(Fe_{0.5}Ni_{0.5})_{1-x}B_x$ Alloys
(2 and 18.2; 21.1; 21.1.1)
(a) Vol. **82**, No. 2, p. K169—K172 (1984).

R. Krishnan
Magnetic Moment of Amorphous $Co_{1-x}Nb_x$ Alloys and Friedel's Model
(2 and 18.2; 21; 21.1)
(a) Vol. **82**, No. 2, p. K173—K175 (1984).

V. Crişan, I. Pop, V. Toşa, and N. Rusu
Band Structure of $Gd_2Ni_2Al_{15}$ Intermetallic Compound
(13.1 and 18.3; 21.1; 21.4)
(b) Vol. **123**, No. 1, p. K53—K56 (1984).

A. Baldan
The Microstructural Properties of (β + $(Ni, Al)_7Hf_2$) Eutectics
(1.3; 10.1; 21; 21.1)
(a) Vol. **83**, No. 1, p. 47—57 (1984).

H. Q. Ye, Z. Zhang, and K. H. Kuo
A High Resolution Electron Microscopic Study of Ordered Structures in $Ni_3(Ti, V)$ Alloys
(1.3 and 1.4; 21.1)
(a) Vol. **83**, No. 1, p. 77—84 (1984).

K. S. Vahvaselkä
An X-Ray Investigation of Structural Properties and Electrical Resistivities of Liquid Al, Liquid Cu, and Liquid Al–Cu Alloys
(1.3 and 2; 14.1; 21; 21.1)
(a) Vol. **83**, No. 1, p. 103—112 (1984).

M. Jergel and P. Mrafko
The Temperature Dependence of the Prepeak in the Diffraction Pattern of the Amorphous $Ti_{61}Cu_{16}Ni_{23}$ and $Ti_{62.5}Cu_{12}Ni_{23}Si_{2.5}$ Alloys
(2; 1.4; 21.1)
(a) Vol. **83**, No. 1, p. 113—121 (1984).

K. Maeda, S. Kobayashi, S. Takeuchi, T. Chiba, and K. Osamura
Absence of Dilatational Defects in Deformed $Cu_{57}Zr_{43}$ Amorphous Alloy
(2 and 10.1; 21.1)
(a) Vol. **83**, No. 1, p. 219—226 (1984).

Y. Shimada
Amorphous Co–Metal Films Produced by Sputtering
(18.2.1; 2; 21.1)
(a) Vol. **83**, No. 1, p. 255—261 (1984).

K. B. Garg, J. Leiro, M. Heinonen, and M. Richardson
X-Ray Emission Study of Bonding in SmT_2 Compounds (T ≡ Fe, Co, Ni)
(4 and 13; 20; 21.1; 21.1.1; 21.4)
(a) Vol. **83**, No. 1, p. 279—284 (1984).

K. Bärner, C. Santandrea, U. Neitzel, and E. Gmelin
Thermal and Magnetic Properties of Cr_xMn_{1-x} As Compounds
(8 and 18; 21.1)
(b) Vol. **123**, No. 2, p. 541—552 (1984).

S. B. Maslenkov, A. I. Kozlenkov, S. A. Filin, and A. I. Shulgin
Soft X-Ray Emission Spectra from Some Ni–Al Alloys
(13.1 and 20; 21; 21.1)
(b) Vol. **123**, No. 2, p. 605—609 (1984).

K. Ławniczak-Jabłońska, E. Minni, J. Pełka, E. Suoninen, and J. Auleytner
XPS and XES Study of the Valence Band of Cr–Al B.C.C. Solid Solutions
(13.1; 20; 21.1)
(b) Vol. **123**, No. 2, p. 627—634 (1984).

G. Knuyt
Electrical Resistivity Calculations for Amorphous Cu_xAl_{1-x}
(2 and 14.1; 21; 21.1)
(b) Vol. **123**, No. 2, p. 763—771 (1984).

I. M. Chapnik
Superconductivity and the Electronic Structure
(14.2; 21.1)
(b) Vol. **123**, No. 2, p. K183—K187 (1984).

R. Pérez, S. S. Sheinin, and K. Z. Botros
Effects of Systematic Reflections on Diffraction Contrast in H.C.P. Cobalt
(1.1; 21.1)
(a) Vol. **83**, No. 2, p. 445—453 (1984).

A. Baldan
An Investigation of the Mechanism of Thermally Activated Deformation in Certain Ni–Al–Ta and Ni–Al–Hf Superalloys
(10.1; 1.3; 21; 21.1)
(a) Vol. **83**, No. 2, p. 507—511 (1984).

M. Doi, Y. Itoh, D.-Y. Chang, and T. Imura
SIMS Studies on the Impurity Diffusion in Amorphous Alloys
(9; 2; 21.1; 21.1.1)
(a) Vol. **83**, No. 2, p. 529—536 (1984).

Z. Bojarski, H. Morawiec, and T. J. Panek
Positron Annihilation Study of Martensitic Transformation in Cu–Zn–Al Alloy
(1.2 and 10.1; 21; 21.1)
(a) Vol. **83**, No. 2, p. K97—K101 (1984).

L. Jahn and V. Christoph
Rotational Hysteresis of 2–17–RE–Cobalt Magnets
(18.2; 21.1; 21.4)
(a) Vol. **83**, No. 2, p. K141—K144 (1984).

G. L. F. Fraga, J. V. Kunzler, F. Ogiba, and D. E. Brandão
Electrical Resistivity of the Ni_2MnIn Heusler Alloy
(14.1; 21.1)
(a) Vol. **83**, No. 2, p. K187—K190 (1984).

T. Soumura, S. Saito, and T. Maeda
Electrical Resistivity and Structural Stability for Amorphous $(Fe_{0.5}Ni_{0.5})_{1-x}B_x$ Alloys
(2 and 14.1; 21.1; 21.1.1)
(a) Vol. **83**, No. 2, p. K195—K198 (1984).

W. Kienle and W. Frank
Calculation of the Intensity of the Small-Angle X-Ray Scattering by Small Clusters of Intrinsic Point Defects
(10.1; 21.1)
(b) Vol. **124**, No. 1, p. 69—77 (1984).

J. Lopez, J. C. Le Bossé, and R. Baudoing
Critical Analysis of the Excitation Map Concept for Studying Threshold Effects in VLEED
(13.3; 21.1)
(b) Vol. **124**, No. 1, p. 381—388 (1984).

U. H. Gläser
On the Calculation of the Residual Resistivity of Precipitates
(10.1 and 14.1; 21; 21.1)
(b) Vol. **124**, No. 1, p. 423—429 (1984).

H. Solbrig
Comment on "Critical Analysis of the Excitation Map Concept for Studying Threshold Effects in VLEED" by J. Lopez, J. C. Le Bossé, and R. Baudoing
(13.3; 21.1)
(b) Vol. **124**, No. 1, p. K63—K67 (1984).

P. Mrozek, M. Menyhárd, J. Wernisch, and A. Jabłoński
Surface Composition of the Ordered Al50–Co50 Alloy
(1.3 and 1.6; 21; 21.1)
(a) Vol. **84**, No. 1, p. 39—48 (1984).

C. A. Hewett, S. S. Lau, R. Fastow, and J. W. Mayer
Pulsed Ion Beam Annealing of Ti–Au Alloys
(1.3 and 11; 1.5; 2; 21.1; 21.6)
(a) Vol. **84**, No. 1, p. 49—54 (1984).

G. van Tendeloo and S. Amelinckx
A New Long-Period Derivative of the Au_5Mn_2 Structure
(1.1 and 1.4; 21.1; 21.6)
(a) Vol. **84**, No. 1, p. 185—192 (1984).

B. Chenevier, D. Fruchart, M. Bacmann, J. P. Senateur, P. Chaudouet, and L. Lundgren
Magnetic Properties of MnRhAs. Study of the Antiferromagnetic–Ferromagnetic Transition
(18.2 and 18.4; 1.2; 14.1; 21.1)
(a) Vol. **84**, No. 1, p. 199—206 (1984).

R. Basu
A Mechanism for Drag on Dislocations in Zinc
(10.1; 21.1)
(a) Vol. **84**, No. 1, p. K35—K38 (1984).

A. R. Ali, S. A. Mahmoud, and G. Said
On the Role of Cu Atoms on the Behaviour of Domain Wall–Dislocation Interaction in Ni–Cu Alloy
(10.1 and 18.2; 21.1)
(a) Vol. **84**, No. 1, p. K67—K70 (1984).

B. Vasvári
A Simple Muffin-Tin Model for Liquid Metals
(13; 2; 21.1)
(b) Vol. **124**, No. 2, p. 683—692 (1984).

J. Kierul, J. Ledzion, and Z. Osetek
Magnetic Susceptibility of Non-Uniform Media
(18.2; 21.1)
(b) Vol. **124**, No. 2, p. K121—K124 (1984).

R. Zürcher, V. Gröger, and F. Stangler
Discontinuous Flow in Cu–Be2 between 4.2 and 12.5 K
(10.1; 21.1)
(a) Vol. **84**, No. 2, p. 475—480 (1984).

A. Nohara and T. Imura
Yield-Point Phenomenon in a β-CuZn Single Crystal at High Temperatures
(10.1; 12.1; 21.1)
(a) Vol. **84**, No. 2, p. 501—507 (1984).

M. Z. Butt, P. Feltham, and S. M. Raza
Stress Relaxation in Nickel between 15 and 300 K
(10.1; 21.1)
(a) Vol. **84**, No. 2, p. K125—K127 (1984).

J. Baszyński, L. Smardz, and F. Stobiecki
Magnetic Properties of Compositionally Modulated Ni/Cu Layers
(1.5 and 18.2.1; 19; 21.1)
(a) Vol. **84**, No. 2, p. K129—K131 (1984).

U. Walzer
Elastic and Bonding Properties of Transition Metals
(4; 12.1; 13.1; 21.1)
(b) Vol. **125**, No. 1, p. 55—64 (1984).
Erratum
(b) Vol. **126**, No. 2, p. 752 (1984).

S. Chakraborty, A. Manna, and A. K. Ghosh
Charge Transfer and Energy of Formation in Ordered Binary Alloys
(4 and 13.1; 21.1; 21.6)
(b) Vol. **125**, No. 1, p. 97—100 (1984).

A. PASTUREL, C. COLINET, C. ALLIBERT, P. HICTER, A. PERCHERON-GUEGAN, and J. C. ACHARD
A Theoretical and Experimental Study of the Enthalpies of Formation of $LaNi_5$-Type Compounds
(4 and 13.1; 21.1; 21.4)
(b) Vol. **125**, No. 1, p. 101—106 (1984).

A. C. GUPTA and A. N. NIGAM
Effect of Cold Work on Hydrogenation of Nickel as Seen in the Thermoelectric Behaviour
(13.2; 12.1; 15; 21.1)
(b) Vol. **125**, No. 1, p. 207—209 (1984).

F. C. LOVEY, G. VAN TENDELOO, and S. AMELINCKX
The Nature of Some Planar Defects in 2H Martensite of Cu–Al Alloys as Determined by HREM
(1.4; 10.1; 21; 21.1)
(a) Vol. **85**, No. 1, p. 29—37 (1984).

V. G. CHUDINOV and V. I. PROTASOV
Kinetics of Vacancy Mechanism of the Solid–Liquid Transition in F.C.C. Metals
(1.2; 10.1; 21.1)
(a) Vol. **85**, No. 1, p. 105—112 (1984).

A. I. GUSEV
Structural Vacancies in Nonstoichiometric Compounds at High Pressure. Thermodynamic Model
(10.1; 8; 12.1; 21.1)
(a) Vol. **85**, No. 1, p. 159—166 (1984).

K. OESTERREICHER and H. OESTERREICHER
Structure and Magnetic Properties of $Nd_2Fe_{14}BH_{2.7}$
(18.2; 21.1)
(a) Vol. **85**, No. 1, p. K61—K64 (1984).

J. V. YAKHMI, I. K. GOPALAKRISHNAN, and A. K. GROVER
Electrical Resistivity Studies on the Heusler Alloys $Co_2T_{1-x}Al_{1+x}$ (T = Ti or Zr)
(14.1; 21.1)
(a) Vol. **85**, No. 1, p. K89—K92 (1984).

M. A. ANDREEVA and R. N. KUZMIN
Mössbauer Conversion Electron Spectra at Total Reflection
(5; 1.6; 13.3; 21.1)
(b) Vol. **125**, No. 2, p. 461—465 (1984).

K. K. CHOPRA and S. BOUARAB
Off-Symmetry Phonon Frequencies of Copper
(6; 21.1)
(b) Vol. **125**, No. 2, p. 499—504 (1984).

CH. OPITZ, H. MÜLLER, and H. MIESSNER
Bonding Mechanism for Interstitial Hydrogen in Nickel
(4; 10.1; 13.4; 21.1)
(b) Vol. **125**, No. 2, p. K93—K97 (1984).

J. M. López and J. A. Alonso
The Atomic Size-Mismatch Contribution to the Enthalpy of Formation of Concentrated Substitutional Metallic Solid Solutions
(1.3 and 8; 21; 21.1)
(a) Vol. **85**, No. 2, p. 423—428 (1984).

V. G. Chudinov, N. V. Moseev, B. N. Goshchitskii, and V. I. Protasov
Possible Amorphisation and Phase Separation of Intermetallic Mo_3Si Compound under Radiation. Molecular Dynamics Study
(11; 2; 21.1)
(a) Vol. **85**, No. 2, p. 435—440 (1984).

I. Baker and E. M. Schulson
On Intrinsic Stacking Faults in Polycrystalline Ni_3Al
(10.1; 1.4; 21; 21.1)
(a) Vol. **85**, No. 2, p. 481—490 (1984).

F. Walz
Transition-Induced Magnetic After-Effect in the Amorphous Alloy $Fe_{20}Ni_{56}B_{24}$ at the Critical Temperature
(18.2; 2; 10.1; 21.1; 21.1.1)
(a) Vol. **85**, No. 2, p. 503—509 (1984).

R. A. Dunlap, K. Dini, and P. Hargraves
Hydrogen Absorption and Crystallisation in Amorphous $Co_{90}Zr_{10}$
(1.3; 2; 21.1)
(a) Vol. **85**, No. 2, p. K101—K103 (1984).

O. Kohmoto
Electrical Resistivity of Amorphous Co-Rich FeCo–SiB Alloys
(14.1; 2; 18.3; 21.1; 21.1.1)
(a) Vol. **85**, No. 2, p. K155—K157 (1984).

S. Chattopadhyay and C. Mande
XANES and EXAFS Study of the Compound CoAsSe
(1; 20; 21.1)
(b) Vol. **126**, No. 1, p. 53—57 (1984).

G. Hüber and D. Ploumbidis
Electrical Resistivity and NMR Investigations of the V–Co System at High Temperatures
(14.1 and 19; 21.1)
(b) Vol. **126**, No. 1, p. K101—K104 (1984).

M. Takeda and H. Hashimoto
A High Resolution Electron Microscopy Study of Domain Structure in CuAuI
(1.4; 21.1; 21.6)
(a) Vol. **86**, No. 1, p. 67—73 (1984).

P. Malètras and P. Duval
Comparison of Temper-Hardened Cu–Al Alloys before and after Thinning-Down
(1.3; 21; 21.1)
(a) Vol. **86**, No. 1, p. 85—92 (1984).

E. Zschech, W. Blau, H. Vega, K. Kleinstück, S. Mager, M. A. Kozlov, and M. A. Sheromov
EXAFS and X-Ray Diffractional Investigation of the Heusler-Type Alloys Co_2MnSi and $Fe_{2.4}Mn_{0.6}Al$. Determination of the Ordering Probabilities
(1.3; 18; 21.1; 21.1.1)
(a) Vol. **86**, No. 1, p. 117—124 (1984).

S. V. Sokolskii, F. F. Lavrentev, and O. P. Salita
Jump-Like Deformation and Stability of the Hardened and Structural States of Zinc Single Crystals with Forest Dislocations in the Temperature Range 293 to 4.2 K
(10.1; 21.1)
(a) Vol. **86**, No. 1, p. 177—183 (1984).

H. M. Eissa and M. M. El-Oker
The Interaction of Straight Dislocations with Dislocation Loops in an Anisotropic Material (Zinc)
(10.1; 21.1)
(a) Vol. **86**, No. 1, p. 207—218 (1984).

M. Takahashi, Y. Sasaki, and S. Ishio
Magnetic Freezing in Solution-Quenched Dilute Fe–Cu Alloys
(18.1; 21.1; 21.1.1)
(a) Vol. **86**, No. 1, p. 289—294 (1984).

V. E. Arkhipov, V. I. Voronin, B. N. Goshchitskii, A. E. Karkin, V. N. Kozhanov, A. V. Mirmelshtein, and E. P. Romanov
The Structure, Superconducting Properties, and Electrical Resistivity of V_2Zr and V_2Hf Irradiated with Fast Neutrons
(14.1 and 14.2; 1.2; 21; 21.1)
(a) Vol. **86**, No. 1, p. K59—K62 (1984).

S. Palchaudhuri, R. N. Joarder, and R. V. Gopala Rao
Partial Structure Factors and Self-Diffusion in Liquid Copper–Lead Alloys
(2 and 9; 21; 21.1)
(b) Vol. **126**, No. 2, p. 443—452 (1984).

A. O. O. Mekhrabov and M. Doyama
Electronic Theory of Atomic Short-Range Order for Ternary Alloys Using the Pseudopotential Approximation and Its Comparison with Experiments
(1.3 and 13; 21.1; 21.1.1)
(b) Vol. **126**, No. 2, p. 453—458 (1984).

A. I. Melker and S. N. Romanov
Athernal Vacancy Cluster Formation in F.C.C. Metals by High Energy PKA
(11; 10.1; 21.1; 21.6)
(b) Vol. **126**, No. 2, p. 517—525 (1984).

A. V. Skripov and A. P. Stepanov
Electronic Structure of V_3Si and V_3Ge. Comparison of NMR Data with Band-Structure Calculations
(13.1 and 19; 21.1)
(b) Vol. **126**, No. 2, p. 557—563 (1984).

M. A. Alam and M. Tomak
Electrical Resistivity of Liquid Noble Metal Alloys
(2 and 14.1; 21.1; 21.6)
(b) Vol. **126**, No. 2, p. 729—732 (1984).

U. H. Gläser and H. Löffler
Effective Medium Approach for the Residual Resistivity of Alloys Containing Guinier-Preston Zones
(14.1; 13; 21; 21.1)
(b) Vol. **126**, No. 2, p. 741—751 (1984).

H. J. Engelmann, K. Berndt, and F. Storbeck
Quantitative Chemical Analysis of 3d Transition Metal Alloys by SXAPS
(1.3 and 13.3; 20; 21.1)
(a) Vol. **86**, No. 2, p. 499—506 (1984).

D. M. Vanderwalker
The Nucleation of Ni Silicides on Dislocations and Twins in Silicon
(1.6 and 10.2; 1.2; 21.1; 22.1.2)
(a) Vol. **86**, No. 2, p. 507—516 (1984).

P. Siebert, M. Werner, and H. Löffler
Heterogeneous Nucleation of Precipitates in an Al–12 at% Zn Alloy at Dislocation Loops
(1.3; 10.1; 21; 21.1)
(a) Vol. **86**, No. 2, p. 543—551 (1984).

F. C. Lovey, G. van Tendeloo, J. van Landuyt, L. Delaey, and S. Amelinckx
On the Nature of Various Stacking Defects in 18R Martensite in Cu–Al Alloys. A Study by High Resolution Electron Microscopy
(1.3; 10.1; 21; 21.1)
(a) Vol. **86**, No. 2, p. 553—564 (1984).

R. Döhl, M.-P. Macht, and V. Naundorf
Measurement of the Diffusion Coefficient of Cobalt in Copper
(9; 10.1; 21.1)
(a) Vol. **86**, No. 2, p. 603—612 (1984).

M. E. Abu-Zeid, M. M. El-Oker, A. I. Abu-Shady, and M. S. Adili
Photoacoustic and Ellipsometric Study of Black Nickel Coatings on Copper Substrates
(7 and 20.1; 21.1)
(a) Vol. **86**, No. 2, p. 805—811 (1984).

21.1.1 Fe and Fe Alloys

Original Papers and Short Notes

D. Fraitová and L. Půst
FMR Linewidth in Bulk Metals in Parallel Configuration
(19; 21.1; 21.1.1)
(b) Vol. **121**, No. 1, p. 173—176 (1984).

H. J. Blythe, F. Walz, and H. Kronmüller
Magnetic Investigation of the Annealing Behaviour of Some Dilute Iron–Nickel Alloys after Low-Temperature Electron Irradiation
(10.1 and 18.2; 11; 21.1; 21.1.1)
(a) Vol. **81**, No. 1, p. 227—238 (1984).

S. V. Tsivinskii
The Theory of Strain Hardening in Single and Polycrystals
(10.1; 1.4; 21; 21.1; 21.1.1; 22.1.1; 22.1.2)
(a) Vol. **81**, No. 1, p. 299—306 (1984).

N. V. Ershov, A. L. Ageev, A. V. Serikov, Yu. A. Babanov, and V. V. Vasin
Regular Method in X-Ray Scattering Analysis of Amorphous and Liquid Metals
(2; 21; 21.1.1)
(b) Vol. **121**, No. 2, p. 451—460 (1984).

T. Datta, S. D. LeVine, D. Thornberry, and E. R. Jones, Jr.
Experimental Determination of the Edwards-Anderson Order Parameter in a High Temperature Spin Glass
(2 and 18; 21.1.1)
(b) Vol. **121**, No. 2, p. K125—K130 (1984).

E. Zedler and G. Lehmann
Influence of Annealing Treatments on the Density of Amorphous $Fe_{40}Ni_{40}P_{14}B_6$ and $Fe_5Co_{70}Si_{15}B_{10}$ Alloys
(2; 21.1.1)
(a) Vol. **81**, No. 2, p. 445—449 (1984).

J. Kučera, L. Kozák, and H. Mehrer
Magnetic Anomalies of Self-Diffusion and Co Heterodiffusion in α-Fe
(9; 10.1; 18.2; 21.1.1)
(a) Vol. **81**, No. 2, p. 497—505 (1984).

V. I. Pecherskaya, D. N. Bolshutkin, V. A. Desnenko, and V. Ya. Ilichev
Electrical and Magnetic Properties of the Fe18Cr(30–40)Ni Alloys at Low Temperatures
(18.1 and 18.2; 14.1; 21.1; 21.1.1)
(a) Vol. **81**, No. 2, p. 507—512 (1984).

A. H. Wafik, G. Graiss, and A. Rizk
Barkhausen Jumps in a Dual-Phase Steel Treated in the Intercritical Temperature Range
(18.2; 21.1.1)
(a) Vol. **81**, No. 2, p. 513—517 (1984).

K. Sumiyama and Y. Nakamura
High Concentration Fe–Cu and Fe–Ag Alloys Produced by Vapor Quenching
(1.3; 5; 18.2; 21.1; 21.1.1; 21.6)
(a) Vol. **81**, No. 2, p. K109—K112 (1984).
Erratum
(a) Vol. **83**, No. 2, p. K219 (1984).

N. K. Gobran, M. M. Danial, and R. Kamel
Kinetics of the Amorphous–Crystalline Transition in the Metallic Glass $Fe_{40}Ni_{40}B_{20}$
(1.2; 2; 21.1; 21.1.1)
(a) Vol. **82**, No. 1, p. 63—66 (1984).

F. Walz
High-Sensitivity Two-Component Measuring Technique of Inductances as Applied to the Investigation of Magnetic After-Effects
(18.2; 18.3; 21.1.1)
(a) Vol. **82**, No. 1, p. 179—189 (1984).

V. D. Sedykh and A. V. Serebryakov
The Effect of Thermal Treatment on the Magnetic Anisotropy in Amorphous Fe–B Alloys
(2 and 18.2; 5; 21.1.1)
(a) Vol. **82**, No. 1, p. K63—K66 (1984).

S. Chakraborty, A. Manna, and A. K. Ghosh
Charge Transfer and Energy of Formation in Binary Alloys: CsCl Intermetallic Compounds
(1.3 and 4; 21.1; 21.1.1; 21.6)
(b) Vol. **122**, No. 2, p. 469—480 (1984).

W. Maass, U. Krey, and H. Hoffmann
Magnetic Approach-to-Saturation in Thin Films
(18.2.1; 21.1.1; 21.4)
(b) Vol. **122**, No. 2, p. K137—K140 (1984).

V. Hajko, A. Zentko, M. Timko, and V. Hajko, Jr.
Isotopic Effect in Neutron-Irradiated $Fe_{40}Ni_{40}B_{20}$ Alloys
(11 and 18.2; 2; 21.1.1)
(a) Vol. **82**, No. 2, p. K159—K162 (1984).

T. Soumura, S. Saito, T. Kuraishi, and T. Maeda
Magnetic Moment and Its Temperature Variation of Amorphous $(Fe_{0.5}Ni_{0.5})_{1-x}B_x$ Alloys
(2 and 18.2; 21.1; 21.1.1)
(a) Vol. **82**, No. 2, p. K169—K172 (1984).

R. Żuberek, F. Stobiecki, and J. Wosik
Magnetoelastic Properties of $Fe_{1-x}B_x$ Amorphous Thin Films
(2 and 18.2.1; 21.1.1)
(a) Vol. **82**, No. 2, p. K177—K179 (1984).

P. Delobelle, D. Varchon, and C. Oytana
Short-Range Reordering of Heavy Interstitials in Ta, Nb, and Fe during Relaxation and Static Strain Aging
(10.1; 12.1; 21; 21.1.1)
(a) Vol. **83**, No. 1, p. 235—243 (1984).

K. B. Garg, J. Leiro, M. Heinonen, and M. Richardson
X-Ray Emission Study of Bonding in SmT_2 Compounds (T ≡ Fe, Co, Ni)
(4 and 13; 20; 21.1; 21.1.1; 21.4)
(a) Vol. **83**, No. 1, p. 279—284 (1984).

M. Seidel and G. Nocke
A Semi-Empirical Relation between the Magnetic Susceptibility and Porosity of Sintered Iron
(18.2; 21.1.1)
(a) Vol. **83**, No. 1, p. K43—K46 (1984).

T. Kaneyoshi and I. Tamura
The Distribution of Magnetic Moments and the Temperature Dependence of Reduced Hyperfine Fields in Amorphous Ferromagnets
(2 and 18.2; 21.1.1)
(b) Vol. **123**, No. 2, p. 525—531 (1984).

M. Šob, S. Szuskiewicz, and M. Szuskiewicz
Polarized Positron Annihilation and Enhancement Effects in Ferromagnetic Iron
(13; 18.2; 21.1.1)
(b) Vol. **123**, No. 2, p. 649—652 (1984).

M. Doi, Y. Itoh, D.-Y. Chang, and T. Imura
SIMS Studies on the Impurity Diffusion of Boron in Amorphous Alloys
(9; 2; 21.1; 21.1.1)
(a) Vol. **83**, No. 2, p. 529—536 (1984).

C. Hidalgo, N. de Diego, and M. A. Ochando
A Positron Study of Iron Alloys
(1.4 and 10.1; 21.1.1)
(a) Vol. **83**, No. 2, p. K93—K96 (1984).

T. Soumura, S. Saito, and T. Maeda
Electrical Resistivity and Structural Stability for Amorphous $(Fe_{0.5}Ni_{0.5})_{1-x}B_x$ Alloys
(2 and 14.1; 21.1; 21.1.1)
(a) Vol. **83**, No. 2, p. K195—K198 (1984).

L. Cser, I. Gladkih, and C. Hargitai
Impurity Charge Screening in 3d Alloys
(13.4; 5; 21.1.1)
(b) Vol. **124**, No. 1, p. 271—278 (1984).

L. Uray and M. Menyhárd
The Segregation of Iron in Tungsten
(1.3; 1.4; 9; 14.1; 15; 21; 21.1.1)
(a) Vol. **84**, No. 1, p. 65—72 (1984).

K. Hricovíni and J. Krempaský
On the Electronic Structure of Iron–Boron Alloys
(13.1; 2; 17; 21.1.1)
(b) Vol. **124**, No. 2, p. K147—K150 (1984).

T. Luciński and Baszyński
Influence of Composition and Structural Transformation on Transport Properties in Amorphous $Fe_{1-x}Si_x$ Films
(14.1; 2; 18.2.1; 21.1.1)
(a) Vol. **84**, No. 2, p. 607—612 (1984).

H. M. Ledbetter
Monocrystal–Polycrystal Elastic Constants of a Stainless Steel
(12.1; 21.1.1)
(a) Vol. **85**, No. 1, p. 89—96 (1984).

K. Uhlmann and M. Schmidt
Influence of Work Function Change Due to Oxygen Chemisorption on the Secondary-Ion Emission Probability
(1.5 and 13.3; 17; 21.1.1)
(a) Vol. **85**, No. 1, p. K19—K22 (1984).

I. G. Ratishvili
Annihilation of Frenkel Pairs in B.C.C. Metals (α-Iron) (I): The Case of the Unstressed Lattice (General Consideration)
(10.1; 21.1.1)
(b) Vol. **125**, No. 2, p. 533—542 (1984).

A. Paja and T. Stobiecki
Electrical Resistivity of Amorphous $Fe_{1-x}B_x$ Alloys
(14.1; 2; 21.1.1)
(b) Vol. **125**, No. 2, p. K171—K176 (1984).

B. Rauschenbach, G. Blasek, and R. Dietsch
Modification of Concentration Profiles in Iron and Aluminium by High-Fluence Implantation of Nitrogen and Boron Ions
(1.3 and 11; 21; 21.1.1)
(a) Vol. **85**, No. 2, p. 473—480 (1984).

V. E. Sidorov, V. S. Gushchin, and B. A. Baum
Magnetic Structure of γ-Iron Containing Oxygen
(18.1; 21.1.1)
(a) Vol. **85**, No. 2, p. 497—501 (1984).

F. Walz
Transition-Induced Magnetic After-Effect in the Amorphous Alloy $Fe_{20}Ni_{56}B_{24}$ at the Critical Temperature
(18.2; 2; 10.1; 21.1; 21.1.1)
(a) Vol. **85**, No. 2, p. 503—509 (1984).

J. M. ALAMEDA, M. C. CONTRERAS, and H. RUBIO
Magnetic Properties of Amorphous Fe_xSi_{1-x} Films (I): Transverse Susceptibility, Structure Constant, and Local Anisotropy
(18.2.1; 2; 21.1.1)
(a) Vol. **85**, No. 2, p. 511—522 (1984).

O. KOHMOTO
Electrical Resistivity of Amorphous Co-Rich FeCo–SiB Alloys
(14.1; 2; 18.3; 21.1; 21.1.1)
(a) Vol. **85**, No. 2, p. K155—K157 (1984).

A. Z. MAKSYMOWICZ, T. STOBIECKI, E. JAROCKI, and W. KARAŚ
Charge Transfer and Concentration Dependence of Magnetization in Amorphous Fe–B Alloys
(13 and 18; 2; 21.1.1)
(b) Vol. **126**, No. 1, p. 191—195 (1984).

E. ZSCHECH, W. BLAU, H. VEGA, K. KLEINSTÜCK, S. MAGER, M. A. KOZLOV, and M. A. SHEROMOV
EXAFS and X-Ray Diffractional Investigation of the Heusler-Type Alloys Co_2MnSi and $Fe_{2.4}Mn_{0.6}Al$. Determination of the Ordering Probabilities
(1.3; 18; 21.1; 21.1.1)
(a) Vol. **86**, No. 1, p. 117—124 (1984).

V. V. KIRSANOV, S. B. KISLITSIN, and E. M. KISLITSINA
Interstitial Migration in a Stress Gradient
(9 and 10.1; 21.1.1)
(a) Vol. **86**, No. 1, p. 199—206 (1984).

M. TAKAHASHI, Y. SASAKI, and S. ISHIO
Magnetic Freezing in Solution-Quenched Dilute Fe–Cu Alloys
(18.1; 21.1; 21.1.1)
(a) Vol. **86**, No. 1, p. 289—294 (1984).

A. O. O. MEKHRABOV and M. DOYAMA
Electronic Theory of Atomic Short-Range Order for Ternary Alloys Using the Pseudopotential Approximation and Its Comparison with Experiments
(1.3 and 13; 21.1; 21.1.1)
(b) Vol. **126**, No. 2, p. 453—458 (1984).

K. MASUDA-JINDO
Tight-Binding Electronic Theory for Lattice Defects in Transition Metals. Correlation Effects
(13.4; 21; 21.1.1)
(b) Vol. **126**, No. 2, p. 643—652 (1984).

C. I. NISTOR, V. SPÂNU, S. MANDACHE, G. FILOTI, and V. FLORESCU
Hyperfine Field Distribution in Metglasses
(5; 2; 21.1.1)
(a) Vol. **86**, No. 2, p. 525—529 (1984).

K. Benhaddane and P. Beauchamp
A Computation of (110) Intrinsic Stacking Fault Energies in B2 Ordered Alloys
(1.4; 10.1; 21; 21.1.1)
(a) Vol. **86**, No. 2, p. 613—621 (1984).

M. S. C. Bose
Effect of Saturated Magnetic Field on Fatigue Life of Carbon Steel
(12.1 and 18.2; 21.1.1)
(a) Vol. **86**, No. 2, p. 649—654 (1984).

21.2 Alkali Metals

Original Papers and Short Notes

XIAN-WU ZOU
Narrowing and Splitting of the 2S Band in Lithium with Increasing Atomic Separation
(13.1; 21.2)
(b) Vol. **121**, No. 2, p. K171—K174 (1984).

B. PRASAD
Thermal Expansion and Mode Grüneisen Parameters in Lithium
(8; 21.2)
(b) Vol. **122**, No. 1, p. 91—93 (1984).

G. ANGONOA, R. DOVESI, C. PISANI, and C. ROETTI
A New Technique for the Evaluation of Densities of States in ab initio Calculations of Periodic Systems
(13.1; 13.2; 21.2; 21.3)
(b) Vol. **122**, No. 1, p. 211—220 (1984).

J. MOLENDA
Electronic Processes in Electrode Materials of A_xMX_2-Type
(13.1; 1.3; 21.1; 21.2)
(b) Vol. **122**, No. 2, p. 591—598 (1984).

J. A. ALONSO, D. J. GONZÁLEZ, and M. P. IÑIGUEZ
*Role of the Excess Volume of Formation on Alloying
(4; 1.3; 21.2; 21.3)
(b) Vol. **123**, No. 2, p. 485—489 (1984).

B. PRASAD
First-Principles Thermal Expansion in Rb and Cs
(6 and 8; 21.2)
(b) Vol. **124**, No. 2, p. 497—501 (1984).

T. SOMA, H. OHSUGI, and H.-MATSUO KAGAYA
Phonon Dispersion Curves of Cs–K, Cs–Rb, and Rb–K Alloys
(6; 8; 21.2)
(b) Vol. **124**, No. 2, p. 525—529 (1984).

N. P. KOVALENKO and L. M. KUZMINA
The Influence of Many-Body Interactions on the Speed of Sound in Liquid Metals
(2 and 12.1; 7; 21.2)
(b) Vol. **124**, No. 2, p. 537—544 (1984).

DO YI-JING, CHEN LI-RONG, and YEN TZU-TUNG
The Relation between the Lindemann Coefficient and the Temperature for the Alkali Metals
(8; 21.2)
(b) Vol. **124**, No. 2, p. K113—K116 (1984).

B. R. Gelchinskii and R. M. Ryskulbekova
On Correlation between Structural Features of Liquid Metals and the Effective Ion–Ion Potential
(2 and 4; 21.2; 21.7)
(b) Vol. **125**, No. 1, p. 81—86 (1984).

R. N. Joarder, S. Palchaudhuri, and R. V. Gopala Rao
Entropy and Heat Capacity of Liquid Metals
(2 and 8; 21.2)
(b) Vol. **125**, No. 1, p. K33—K36 (1984).

T. Soma and H.-Matsuo Kagaya
Compression Effect on Mode Grüneisen Parameters of Alkali Metals
(6 and 8; 12.1; 21.2)
(b) Vol. **126**, No. 1, p. K5—K7 (1984).

T. Soma and H.-Matsuo Kagaya
Mode Grüneisen Parameter for Local and Band Modes of Rb–K, Cs–K, and Cs–Rb Alloy Systems
(6; 8; 12.1; 21.2)
(b) Vol. **126**, No. 2, p. K109—K112 (1984).

21.3 Alkali-Earth Metals

Original Papers and Short Notes

E. J. Savino and A. M. Monti
Calculation of Lattice Defect Saddle-Point Configurations
(10.1; 9; 21.3)
(b) Vol. **121**, No. 2, p. 513—524 (1984).

P. Rennert, H. Schelle, and U.-H. Gläser
Calculation of Polarized K and L X-Ray-Emission Spectra and of Partial Densities of States of the H.C.P. Metals Be, Mg, and Zn
(13.1; 20; 21.3)
(b) Vol. **121**, No. 2, p. 673—684 (1984).

G. Angonoa, R. Dovesi, C. Pisani, and C. Roetti
A New Technique for the Evaluation of Densities of States in ab initio Calculations of Periodic Systems
(13.1; 13.2; 21.2; 21.3)
(b) Vol. **122**, No. 1, p. 211—220 (1984).

B. Prasad
Phonon Frequencies and Binding Energy of F.C.C. Calcium
(6; 4; 21.3)
(b) Vol. **123**, No. 1, p. K9—K11 (1984).

J. A. Alonso, D. J. González, and M. P. Iñiguez
Role of the Excess Volume of Formation on Alloying
(4; 1.3; 21.2; 21.3)
(b) Vol. **123**, No. 2, p. 485—489 (1984).

G. Cocco, S. Enzo, C. Antonione, L. Battezzati, and G. Riontino
Structure Features in the Palladium–Beryllium Amorphous System
(2; 21; 21.3)
(a) Vol. **84**, No. 2, p. 371—377 (1984).

D. O. Northwood and I. O. Smith
Steady-State Creep and Strain Transients for Stress-Dip Tests in Polycrystalline Magnesium at 300 °C
(10.1 and 12.1; 21.3)
(a) Vol. **85**, No. 1, p. 149—158 (1984).

G. Fantozzi, C. Esnouf, J. Lauzier, and C. Minier
Internal Friction at Low Temperature in High Purity Magnesium
(10.1 and 12.1; 21.3)
(a) Vol. **85**, No. 2, p. 463—468 (1984).

S. P. Singh and R. P. S. Rathore
Phonons in Face-Centred Cubic Calcium and Strontium
(6; 12.1; 21.3)
(b) Vol. **126**, No. 1, p. 87—90 (1984).

H. C. González
Hardening of Single Crystals of Magnesium by Low Neutron Doses at 77 K
(10.1 and 11; 21.3)
(a) Vol. **86**, No. 1, p. 169—175 (1984).

21.4 Rare-Earth Metals

Review Article

R. Ramji Rao and A. Ramanand
Thermal and Elastic Properties of Rare-Earth Metals
(8 and 12.1; 6; 21.4)
(b) Vol. **122**, No. 1, p. 11—27 (1984).

Original Papers and Short Notes

O. V. Farberovich, G. P. Nizhnikova, S. V. Vlasov, and E. P. Domashevskaya
Local-Density-Functional Approximation to the Energy Band Structure of TmS Using the Self-Consistent Relativistic Linearized-Augmented-Plane-Wave Method
(13.1; 21.4)
(b) Vol. **121**, No. 1, p. 241—253 (1984).

V. Prabhawalkar and B. D. Padalia
X-Ray Spectroscopy Study of RMn_2Si_2 Intermetallics
(20; 21; 21.4)
(b) Vol. **121**, No. 1, p. K65—K68 (1984).

A. I. Lukanin and M. V. Medvedev
Spin Waves in an Easy-Axis Ferromagnet with a Finite Concentration of Impurities with Competing Anisotropy
(18.2; 21.4)
(b) Vol. **121**, No. 2, p. 525—537 (1984).

A. I. Lukanin and M. V. Medvedev
Magnetic States of a Binary Ferromagnetic Alloy with Competition of Easy-Axis and Easy-Plane Single-Ion Anisotropies
(18.2; 21.4)
(b) Vol. **121**, No. 2, p. 573—582 (1984).

E. A. Goremychkin, E. Mühle, P. G. Ivanitskii, V. T. Krotenko, M. V. Pasechnik, V. V. Slisenko, A. A. Vasilkevich, B. Lippold, O. D. Chistyakov, and E. M. Savitskii
Crystal Electric Field Splitting in $TbNi_5$ and $ErNi_5$ Studied by Inelastic Neutron Scattering
(13.4; 21.4)
(b) Vol. **121**, No. 2, p. 623—631 (1984).

Z. Smetana, V. Šíma, and A. V. Andreev
On the Magnetic Arrangement of the Intermetallic $ErCu_2$ Compound
(18.4; 21.4)
(b) Vol. **121**, No. 2, p. K131—K134 (1984).

P. S. Riseborough
De Haas-Van Alphen Effect in Nearly Magnetic Transition Metals
(13; 18.1; 21; 21.4)
(b) Vol. **122**, No. 1, p. 161—170 (1984).

M. A. SAVCHENKO and A. V. STEFANOVICH
Fluctuation Theory of Magnetic Superconductors
(14.2; 21.4)
(b) Vol. **122**, No. 1, p. 367—376 (1984).

B. LIPPOLD and H. MÜLLER
Kondo-Like Behaviour of the Electrical Resistivity of the Compound $Pr_{0.25}La_{0.75}Al_3$
(14; 21.4)
(b) Vol. **122**, No. 1, p. 377—382 (1984).

V. BRÉDIMAS, H. GAMARI-SEALE, and C. PAPATRIANTAFILLOU
The Magnetic Behaviour of the Pseudobinary Compounds $Gd_2Au_xAl_{1-x}$
(18.2 and 18.4; 21.4)
(b) Vol. **122**, No. 2, p. 527—534 (1984).

W. MAASS, U. KREY, and H. HOFFMANN
Magnetic Approach-to-Saturation in Thin Films
(18.2.1; 21.1.1; 21.4)
(b) Vol. **122**, No. 2, p. K137—K140 (1984).

H. RATAJCZAK and I. GOŚCIAŃSKA
Planar Hall Effect in Thin Amorphous RE–TM Films with Perpendicular Anisotropy
(14.1 and 18.2.1; 21.4)
(a) Vol. **82**, No. 2, p. 491—496 (1984).

O. J. ŻOGAL, CH. JÄGER, H. DÖHLER, and B. SCHNABEL
Multipulse ^1H NMR Study of Proton Self-Diffusion in Scandium and Lutetium Dihydride
(9 and 19; 21.4)
(a) Vol. **82**, No. 2, p. K153—K157 (1984).

J. NAUDTS, A. WERNER, and H. D. HOCHHEIMER
Pressure–Volume Relation in Intermediate Valence Compounds
(13.1; 21.4)
(b) Vol. **123**, No. 1, p. 209—217 (1984).

V. CRIŞAN, I. POP, V. TOŞA, and N. RUSU
Band Structure of $Gd_2Ni_2Al_{15}$ Intermetallic Compound
(13.1 and 18.3; 21.1; 21.4)
(b) Vol. **123**, No. 1, p. K53—K56 (1984).

K. B. GARG, J. LEIRO, M. HEINONEN, and M. RICHARDSON
X-Ray Emission Study of Bonding in SmT_2 Compounds (T ≡ Fe, Co, Ni)
(4 and 13; 20; 21.1; 21.1.1; 21.4)
(a) Vol. **83**, No. 1, p. 279—284 (1984).

E. BORCHI and S. DE GENNARO
Superconducting Properties of Rare-Earth Amorphous Alloys
(14.2; 21.4)
(b) Vol. **123**, No. 2, p. 723—731 (1984).

I. O. Bashkin, M. E. Kost, and E. G. Ponyatovskii
Phase Transitions in Non-Stoichiometric Cerium Hydrides
(1.2; 21.4)
(a) Vol. **83**, No. 2, p. 461—465 (1984).

A. M. Gómez and J. Serna
Positron Lifetime in Polycrystalline Gadolinium
(10.1; 21.4)
(a) Vol. **83**, No. 2, p. K131—K133 (1984).

L. Jahn and V. Christoph
Rotational Hysteresis of 2–17–RE–Cobalt Magnets
(18.2; 21.1; 21.4)
(a) Vol. **83**, No. 2, p. K141—K144 (1984).

T. A. Matveeva
4f States in Spectra of Rare-Earth Metals
(13.1; 20; 21.4)
(b) Vol. **124**, No. 1, p. K59—K62 (1984).

K. H. J. Buschow and D. B. de Mooij
The Effect of H_2 Absorption on the Magnetic Properties of Gd–Pd Intermetallic Compounds
(18.2 and 18.4; 21; 21.4)
(a) Vol. **84**, No. 1, p. 207—213 (1984).

V. I. Kononenko, A. L. Sukhman, S. L. Gruverman, and V. V. Torokin
Density and Surface Tension of Liquid Rare Earth Metals, Scandium and Yttrium
(2 and 12.1; 1.2; 21; 21.4)
(a) Vol. **84**, No. 2, p. 423—432 (1984).

A. Pasturel, C. Colinet, C. Allibert, P. Richter, A. Percheron-Guegan, and J. C. Achard
A Theoretical and Experimental Study of the Enthalpies of Formation of $LaNi_5$-Type Compounds
(4 and 13.1; 21.1; 21.4)
(b) Vol. **125**, No. 1, p. 101—106 (1984).

S. N. Gupta and S. Shah
Evidence of Open 4f Shells in Rare Earth Compounds
(13; 20; 21.4)
(b) Vol. **125**, No. 1, p. 211—214 (1984).

T. K. Hatwar, B. D. Padalia, M. N. Ghatikar, and D. R. Chopra
Extended X-Ray Absorption Fine Structure in RAl_2Ga_2 Intermetallics
(20; 21.4)
(b) Vol. **126**, No. 1, p. 279—283 (1984).

D. Gignoux, J. C. Gomez-Sal, and J. Rodriguez Fernandez
Magnetic Structure of the Er_2Pt Antiferromagnet
(18.4; 21.4)
(a) Vol. **86**, No. 1, p. 295—300 (1984).

21.5 Actinides

Short Note

T. Palewski
The Magnetic Properties of $U_xTh_{1-x}Se$ Solid Solution
(18.2; 21.5)
(a) Vol. **84**, No. 1, p. K47—K50 (1984).

21.6 Noble Metals

Original Papers and Short Notes

J. Kilmer, C. M. van Vliet, G. Bosman, A. van der Ziel, and P. H. Handel
Evidence of Electromagnetic Quantum $1/f$ Noise Found in Gold Thin Films
(14.1; 21.6)
(b) Vol. **121**, No. 1, p. 429—432 (1984).

G. Seifert, H.-G. Fritsche, P. Ziesche, and V. Heera
On the Electronic Structure of Palladium–Hydrogen and Platinum–Hydrogen Systems
(13; 21.6)
(b) Vol. **121**, No. 2, p. 705—715 (1984).

P. M. Hazzledine, H. P. Karnthaler, and A. Korner
The Application of the Wulff Construction to Dislocation Problems
(10.1; 12.1; 21.1; 21.6)
(a) Vol. **81**, No. 2, p. 473—484 (1984).

K. Sumiyama and Y. Nakamura
High Concentration Fe–Cu and Fe–Ag Alloys Produced by Vapor Quenching
(1.3; 5; 18.2; 21.1; 21.1.1; 21.6)
(a) Vol. **81**, No. 2, p. K109—K112 (1984).
Erratum
(a) Vol. **83**, No. 2, p. K219 (1984).

J. J. Kleperis, S. J. Takeris, A. R. Lusis, and J. P. Stradins
The Investigation of Hydrogen Diffusion in Palladium by the Chemichromic Effect
(9; 21.6)
(a) Vol. **81**, No. 2, p. K121—K125 (1984).

V. E. Antonov, T. E. Antonova, I. T. Belash, V. Yu. Malyshev, and V. I. Rashupkin
Superconductivity and Atomic Ordering of Pd–Cu–H Solid Solutions
(14.2; 1.2; 21.6)
(a) Vol. **81**, No. 2, p. K185—K190 (1984).

S. Chakraborty, A. Manna, and A. K. Ghosh
Charge Transfer and Energy of Formation in Binary Alloys: CsCl Intermetallic Compounds
(1.3 and 4; 21.1; 21.1.1; 21.6)
(b) Vol. **122**, No. 2, p. 469—480 (1984).

G. Knuyt and A. Keupers
Calculation of the Resistivity of Liquid and Amorphous $Au_{1-x}Si_x$ Using Ziman Theory
(2 and 14.1; 21.6)
(b) Vol. **122**, No. 2, p. 781—790 (1984).

M. Jałochowski
Electron Tunnelling in Pb/Ag Ultrathin Layered Structures
(14.3.1; 14.3.3; 21; 21.6)
(a) Vol. **82**, No. 2, p. 497—502 (1984).

D. Gerlich
Fourth-Order Elastic Constants of Solids
(12.1 and 12.2; 21; 21.6; 22.5.2)
(b) Vol. **123**, No. 1, p. K5—K7 (1984).

V. V. Nemoshkalenko, A. E. Krasovskii, V. N. Antonov, Vl. N. Antonov, U. Fleck, H. Wonn, and P. Ziesche
Addendum to: The Relativistic Linear Muffin-Tin Orbital Method. Application to Au
(13.1; 21.6)
(b) Vol. **123**, No. 2, p. 787 (1984).

K. Mukherjee, M. Chandrasekaran, and M. Kato
Electrical Resistivity of an Equiatomic Solid and Liquid Au–Cd Alloy
(14.1; 2; 8; 21; 21.6)
(a) Vol. **83**, No. 2, p. 657—667 (1984).

C. A. Hewett, S. S. Lau, R. Fastow, and J. W. Mayer
Pulsed Ion Beam Annealing of Ti–Au Alloys
(1.3 and 11; 1.5; 2; 21.1; 21.6)
(a) Vol. **84**, No. 1, p. 49—54 (1984).

G. van Tendeloo and S. Amelinckx
A New Long-Period Derivative of the Au_5Mn_2 Structure
(1.1 and 1.4; 21.1; 21.6)
(a) Vol. **84**, No. 1, p. 185—192 (1984).

H. Berndt and H.-J. Hunger
Experimental Determination of the M-Shell Ionization Cross Section
(13; 21.6)
(a) Vol. **84**, No. 2, p. K149—K152 (1984).

S. Chakraborty, A. Manna, and A. K. Ghosh
Charge Transfer and Energy of Formation in Ordered Binary Alloys
(4 and 13.1; 21.1; 21.6)
(b) Vol. **125**, No. 1, p. 97—100 (1984).

L. Wendler
Coupled Surface Plasmon–Phonon Polaritons in Bilayer Systems
(6 and 20; 13.5.2; 21.6; 22.2.2)
(b) Vol. **125**, No. 1, p. K27—K31 (1984).

M. Scrocco
X-Ray Photoelectron Spectra of Ag, Ga, In, and β-Sn Metals. Multielectron Excitations
(13.1 and 17; 21; 21.6)
(b) Vol. **125**, No. 2, p. 713—720 (1984).

Y. Matsuoka, S. Terakado, and S. Tanigawa
Positron Age–Momentum Correlation in Ultrafine Particles of Silver and Gold
(10.1; 13.3; 21.6)
(b) Vol. **125**, No. 2, p. 823—829 (1984).

M. TAKEDA and H. HASHIMOTO
A High Resolution Electron Microscopy Study of Domain Structure in CuAuI
(1.4; 21.1; 21.6)
(a) Vol. **86**, No. 1, p. 67—73 (1984).

A. I. MELKER and S. N. ROMANOV
Athermal Vacancy Cluster Formation in F.C.C. Metals by High Energy PKA
(11; 10.1; 21.1; 21.6)
(b) Vol. **126**, No. 2, p. 517—525 (1984).

M. A. ALAM and M. TOMAK
Electrical Resistivity of Liquid Noble Metal Alloys
(2 and 14.1; 21.1; 21.6)
(b) Vol. **126**, No. 2, p. 729—732 (1984).

Y. QU. SHENG, R. Y. LI, and P. MUNZ
The Study of Mode Characteristics of Metal Clad Dielectric Waveguides
(13.5.2 and 20; 21.6; 22.5.2)
(a) Vol. **86**, No. 2, p. 813—819 (1984).

21.7 Semi-Metals
(Semiconducting Alloys See 22.7)

Original Papers and Short Notes

F. Völklein and E. Kessler
A Method for the Measurement of Thermal Conductivity, Thermal Diffusivity, and Other Transport Coefficients of Thin Films
(14.1 and 14.3.1; 8; 15; 21.7)
(a) Vol. **81**, No. 2, p. 585—596 (1984).

V. Hinkel and D. Ploumbidis
Temperature Dependence of the Magnetic Part of the Hyperfine Field at ^{123}Sb in the Liquid Sb–Te System
(2 and 13; 19; 21.7; 22.1)
(b) Vol. **122**, No. 1, p. K45—K48 (1984).

L. G. Khvostantsev and V. A. Sidorov
Phase Transitions in Antimony at Hydrostatic Pressure up to 9 GPa
(1.2; 14.1; 15; 21.7)
(a) Vol. **82**, No. 2, p. 389—398 (1984).

J. Dominec, K. Misek, W. Kraak, and R. Herrmann
Ultrasound Attenuation and Quantum Oscillations in Bi–Sb Alloys
(7 and 13; 21.7)
(b) Vol. **123**, No. 2, p. 635—640 (1984).

V. K. Arora and H. N. Spector
Quantum Freeze-Out of Carriers in Semimetallic and Semiconducting Thin Wires
(14.1 and 14.3; 21.7)
(b) Vol. **123**, No. 2, p. 747—753 (1984).

N. B. Mustafaev and M. G. Shakhtakhtinskii
Thermopower of a Thin Bi Film in a Classical Strong Magnetic Field
(15; 13.1; 21.7)
(b) Vol. **124**, No. 1, p. K93—K96 (1984).

E. Canessa, D. F. Mariani, and J. Vignolo
Structure of Non-Simple Liquid Metals
(2; 21; 21.7)
(b) Vol. **124**, No. 2, p. 465—472 (1984).

N. B. Mustafaev and M. G. Shakhtakhtinskii
Optical Absorption in Size-Quantized Bi Films
(13.1 and 20.1; 21.7)
(b) Vol. **124**, No. 2, p. K151—K154 (1984).

A. V. Parygin, G. V. Chikvaidze, G. K. Semin, B. V. Lokshin, and Ya. A. Eiduss
^{75}As-NQR Study of Yellow Crystalline Arsenic
(1.2; 19; 21.7)
(a) Vol. **84**, No. 2, p. K101—K103 (1984).

B. R. Gelchinskii and R. M. Ryskulbekova
On Correlation between Structural Features of Liquid Metals and the Effective Ion–Ion Potential
(2 and 4; 21.2; 21.7)
(b) Vol. **125**, No. 1, p. 81—86 (1984).

H. Berger, Yu. A. Kulyupin, S. A. Nepijko, and N. N. Sedov
Quantitative Interference Electron Microscopy for Spherical Objects
(1.5 and 13.3; 21.7)
(a) Vol. **85**, No. 1, p. 45—50 (1984).

P. Enders
Electronic Structure of the Group-V Semimetals: Tight-Binding Parameter Estimation from Chemical Trends
(13.1; 21.7)
(b) Vol. **125**, No. 2, p. K133—K135 (1984).
Erratum
(b) Vol. **126**, No. 2, p. K177 (1984).

E. Canessa, D. F. Mariani, and J. Vignolo
Liquid Structure Factor of Sb: Application of the Canessa-Mariani-Vignolo Model
(2; 21.7)
(b) Vol. **126**, No. 1, p. K1—K4 (1984).

O. P. Hansen and I. F. I. Mikhail
Galvanomagnetic Effects of Bismuth in Non-Quantising Magnetic Fields. Pseudo-Parabolic Model
(14.1; 15; 21.7)
(b) Vol. **126**, No. 2, p. 721—728 (1984).

22. Semiconductors and Ionic Crystals

Review Articles

B. M. ASKEROV, B. I. KULIEV, and S. R. FIGAROVA
Electron Transport Phenomena in Conductive Films
(14.3.1; 6; 13.1; 15; 22)
(b) Vol. **121**, No. 1, p. 11—37 (1984).

L. B. DUBROVSKAYA and S. Z. NAZAROVA
Some Regularities in the Behaviour of the Magnetic, Electrokinetic, and Superconducting Properties of Transition Metal Monocarbide and Mononitride Solid Solutions
(14.3 and 18; 13.1; 14.2; 15; 22)
(a) Vol. **83**, No. 1, p. 11—38 (1984).

G. DRÄGER and O. BRÜMMER
Polarized X-Ray Emission Spectra of Single Crystals
(13.1 and 20; 21; 22)
(b) Vol. **124**, No. 1, p. 11—28 (1984).

C. BOULESTEIX
A Survey of Domains and Domain Walls Generated by Crystallographic Phase Transitions Causing a Change of the Lattice
(1.2 and 1.4; 14.4.1; 22; 22.6; 22.8)
(a) Vol. **86**, No. 1, p. 11—42 (1984).

Original Papers and Short Notes

P. ENDERS
Structure and Bonding in Cubic IV—VI Crystals (II): Instability of the TO (Γ) Mode: Unified Universal-Parameter Treatment of the Group-V Semimetals and Cubic IV—VI Semiconductors
(1.2; 4; 13; 22)
(b) Vol. **121**, No. 1, p. 39—46 (1984).

E. V. CHUŁKOV, I. YU. SKLYADNEVA, and V. E. PANIN
The First-Principles Calculation of the Norm-Conserving Non-Local Singular Atomic Pseudopotentials
(13; 21; 22)
(b) Vol. **121**, No. 1, p. 265—274 (1984).

L. PAJASOVÁ, P. PAJAS, O. A. MAKAROV, and V. M. ZAKHAROV
Optical Properties of Amorphous GeS in the Spectral Region of 18 to 38eV
(2 and 20.1; 13.1; 22)
(b) Vol. **121**, No. 1, p. 293—300 (1984).

V. CHÁB and I. BARTOŠ
Band Structure and Lifetime Determinations for GeS from Angle-Resolved Photoemission
(13.1 and 17; 22)
(b) Vol. **121**, No. 1, p. 301—306 (1984).

J. Monecke
Dielectric Function and Conductivity of p-Type Semiconductors
(13.1 and 14.3; 22)
(b) Vol. **121**, No. 1, p. 329—338 (1984).

H.-J. Unger
Localization of Electronic States in Disordered Tight-Binding Systems (II): Mobility Edges
(2 and 13.4; 22)
(b) Vol. **121**, No. 1, p. 371—383 (1984).

Y. Ishihara, Y. Ohno, and I. Nakada
Anisotropic Electrical Properties of GeSe
(13.4 and 14.3; 22)
(b) Vol. **121**, No. 1, p. 407—412 (1984).

H. Böttger and D. Wegener
On the Percolation Approach to the Hopping Conduction in Strong Electrical Fields
(2 and 14.3; 6; 22)
(b) Vol. **121**, No. 1, p. 413—419 (1984).

V. Čápek
Initial Condition Term and Method of Artificial Temperature for Generalized Master Equations
(6 and 13; 22)
(b) Vol. **121**, No. 1, p. K7—K12 (1984).

L. N. Blinov, L. A. Baydakov, Z. Cimpl, and F. Kosek
A Note on the Magnetic Susceptibility of Tetrahedral Semiconductors and Ionic Crystals
(18; 22)
(b) Vol. **121**, No. 1, p. K43—K45 (1984).

D. V. Oparin, G. I. Pilipenko, F. F. Gavrilov, and V. I. Tyutyunnik
Electron Paramagnetic Resonance of Mn^{2+} Ions in Lithium Hydride Single Crystals
(19; 22)
(b) Vol. **121**, No. 1, p. K75—K77 (1984).

G. M. Shmelev, G. I. Tsurkan, and Nguyen Quoc Anh
Photostimulated Planar Acoustomagnetoelectric Effect in Semiconductors
(16; 22)
(b) Vol. **121**, No. 1, p. K97—K102 (1984).

N. Kaiser
Nucleation of Antimony on Amorphous Carbon
(1.5 and 3; 2; 22)
(a) Vol. **81**, No. 1, p. 99—103 (1984).

S. K. Zhilinskii, M. E. Itkis, and F. Ya. Nad
Frequency Dependence of Conductivity in TaS_3 at Helium Temperatures
(14; 22)
(a) Vol. **81**, No. 1, p. 367—375 (1984).

H. W. Helberg
Dielectric Relaxation Behavior of Dislocations in Semiconductors
(10.2 and 14.3; 14.4; 22)
(a) Vol. **81**, No. 1, p. 381—390 (1984).

B. Kaufmann
Photoluminescence of a Deep Level in Undoped PbTe
(13.4 and 20.3; 22)
(a) Vol. **81**, No. 1, p. K31—K35 (1984).

M. P. Mikalkevičius and K. K. Tumkevičius
A Mechanism of Current Percolation in a Heterogeneous Layer of PbTe
(14.3.1 and 16; 22)
(a) Vol. **81**, No. 1, p. K43—K45 (1984).

J. L. Miane, M. E. Achour, and F. Carmona
Microwave Propagation in Random Particulate Composites
(14.4; 22)
(a) Vol. **81**, No. 1, p. K71—K76 (1984).

P. Enders
Structure and Bonding in Cubic IV–VI Crystals (III): On the Cubic to Rhombohedral Phase Transition
(1.2; 4; 6; 22)
(b) Vol. **121**, No. 2, p. 461—470 (1984).

L. Merten and W. Goltsche
Stimulated Raman Effect on Plasmons and Longitudinal Plasmon-Phonons in Simple Cubic Crystals
(6 and 20; 13.5.2; 22)
(b) Vol. **121**, No. 2, p. 471—480 (1984).

V. I. Ivashchenko, A. A. Lisenko, and E. A. Zhurakovskii
Electronic Structure of YC_x
(13.1; 22)
(b) Vol. **121**, No. 2, p. 583—588 (1984).

W. Richert and W. Ebeling
Thermodynamic Functions of the Electron Fluid for a Wide Density–Temperature Range
(13; 13.5; 22)
(b) Vol. **121**, No. 2, p. 633—639 (1984).

E. Gerlach
Scattering by Ions near Interfaces
(14.3.1; 22)
(b) Vol. **121**, No. 2, p. 757—764 (1984).

M. Stordeur and H. T. Langhammer
Erratum to: Anisotropy of Transport Properties Due to Direction-Dependent Band Nonparabolicity
(13.1; 14.3; 15; 22)
(b) Vol. **121**, No. 2, p. 765 (1984).

J. M. Wesselinowa
Magnon Damping of the s–d Model above T_C
(18.2; 13.5.4; 22)
(b) Vol. **121**, No. 2, p. K147—K149 (1984).

S. W. Gu and J. Zhang
Note on Strong Coupling Polarons in a Semi-Infinite Ionic Crystal
(13.5.3; 22)
(b) Vol. **121**, No. 2, p. K165—K170 (1984).

C. Albers and Th. Moritz
Comments on the Applicability of Energy Gap versus Refractive Index Relations to Lead Chalcogenides
(13.1; 20.1; 22)
(b) Vol. **121**, No. 2, p. K175—K180 (1984).

G. M. Shmelev, Nguyen Quoz Anh, G. I. Tsurkan, and S. Y. Mensah
"Currentless" Amplification of Hypersound in a Planar Configuration by Inelastic Scattering of Electrons
(7 and 14.3; 22)
(b) Vol. **121**, No. 2, p. K209—K213 (1984).

V. S. Pekar
On the Polarization of Center Radiation in an Optically Thin Layer. The Resonator Effect on the Phototransition Probability
(20.3; 22)
(a) Vol. **81**, No. 2, p. 709—713 (1984).

D. J. Easton and J. T. Edmond
Glass–Crystal Transitions in Materials in the System As_2S_3–As_2Te_3
(2 and 8; 22)
(a) Vol. **81**, No. 2, p. K103—K107 (1984).

T. Chattopadhyay and A. R. Chetal
Plasmon and X-Ray Chemical Shifts in Some Copper Systems
(13.5.2 and 20; 22; 22.5)
(b) Vol. **122**, No. 1, p. 193—196 (1984).

Yu. B. Gaididei and A. A. Vakhnenko
Nonequilibrium Kinetics of Exciton Wave Packets in Crystals
(13.5.1; 6; 22)
(b) Vol. **122**, No. 1, p. 239—247 (1984).

G. I. Tsurkan, G. M. Shmelev, and Nguyen Hong Shon
Photostimulated Uneven Magnetorestistance at Inelastic Scattering of Electrons in Semiconductors
(14.3; 18; 22)
(b) Vol. **122**, No. 1, p. 361—366 (1984).

J. E. GRIFFITHS, J. C. PHILLIPS, G. P. ESPINOSA, J. P. REMEIKA, and P. M. BRIDENBAUGH
Assignment of the Companion A_1^c Line in $Ge_x(S, Se)_{1-x}$ Glasses
(6 and 20.1; 2; 22)
(b) Vol. **122**, No. 1, p. K11—K15 (1984).

S. W. GU and S. L. BAN
The Ground State Effective Hamiltonian of Strong Coupling One-Dimensional Optical Exciton
(13.5.3; 22)
(b) Vol. **122**, No. 1, p. K41—K44 (1984).

S. K. PLACHKOVA
Estimation of Some Band-Edge Structure Parameters of the System $(GeTe)_{1-x}(AgSbTe_2)_x$
(13.1; 14.3; 15; 22)
(b) Vol. **122**, No. 1, p. K53—K58 (1984).

M. MONDAL and K. P. GHATAK
On the Modification of the Einstein Relation for a Semiconductor Superlattice
(14.3; 22)
(b) Vol. **122**, No. 1, p. K93—K97 (1984).

J. KELBER, CH. SCHNITTLER, and G. PAASCH
The Completeness Relation and the Local Density Approximation for Layer Structures in Microelectronics
(13.3; 22)
(b) Vol. **122**, No. 1, p. K99—K103 (1984).

P. SÜPTITZ and A. FISCHER
Lateral Diffusion of Photodoped Silver in Amorphous $Ge_{25}Se_{75}$ Layers
(9; 20.1; 22)
(a) Vol. **82**, No. 1, p. 157—161 (1984).

A. V. NABOK, B. A. NESTERENKO, YU. M. SHIRSHOV, YU. V. GOLTVIANSKII, and A. P. DUBCHAK
Profile of Trapped Charge in Silicon Nitride Films in MNOS Structures
(13.4; 14.3.4; 14.4; 22)
(a) Vol. **82**, No. 1, p. 221—227 (1984).

S. S. SIMEONOV and M. D. IVANOVICH
Recombination Current in Abrupt Semiconductor p–n Junctions
(14.3.3; 22)
(a) Vol. **82**, No. 1, p. 275—284 (1984).

L. S. GASANOV
Effect of Light and Magnetic Field on Double Injection in Semiconductor Layers
(14.3.4; 22)
(a) Vol. **82**, No. 1, p. 307—317 (1984).

Z. ČERNOŠEK and E. ČERNOŠKOVÁ
Short-Range Order Model of Ge_xS_{1-x} Glasses and Utilization of This Model in Calculation of the Optical Gap and the IR 367 cm^{-1} Band Intensity
(2 and 20.1; 22)
(b) Vol. **122**, No. 2, p. 405—410 (1984).

R. A. Evarestov and V. P. Smirnov
Application of the Band Representations of Space Groups in the Theory of Electronic States of Crystalline Solids (II): Band Representations for Some Space Groups
(13.1; 22)
(b) Vol. 122, No. 2, p. 559—567 (1984).

V. I. Belitsky, A. V. Goltsev, I. G. Lang, and S. T. Pavlov
Two-Phonon Resonance Raman Scattering and Quantum Interference Effects in Polar Semiconductors in High Magnetic Fields
(6 and 20.1; 22)
(b) Vol. 122, No. 2, p. 581—590 (1984).

V. G. Tsirelson, O. V. Korolkova, I. S. Rez, and R. P. Ozerov
Extended Model of Bond Charges and Its Application in Calculation of Optical Properties of Crystals with Different Types of Chemical Bonds
(4; 20.1; 22; 22.5.2; 22.6; 22.8.1)
(b) Vol. 122, No. 2, p. 599—612 (1984).

G. Valiukonis, F. M. Gashimzade, D. A. Guseinova, G. Krivaitė, M. M. Mamedov, and A. Šileika
Electroreflectance and Thermoreflectance Spectra of SnSe Crystals
(13.1 and 20.1; 22)
(b) Vol. 122, No. 2, p. 623—627 (1984).

B. Giese
Electron Scattering Interaction and Elementary Excitations in Lead Chalcogenides
(13.1 and 13.5; 6; 22)
(b) Vol. 122, No. 2, p. 641—650 (1984).

W. Kautek, F. Decker, and H. Gerischer
Optical Anisotropy of Transition Metal Dichalcogenides. A Photoelectrochemical Determination
(20.1; 13.1; 16; 22)
(b) Vol. 122, No. 2, p. 651—659 (1984).

F. Lukeš, E. Schmidt, J. Humlíček, and P. Dub
Optical Properties of GeS
(13.1 and 20.1; 22)
(b) Vol. 122, No. 2, p. 675—686 (1984).

A. Mandelis
Self-Consistent Semi-Classical Dynamic Theory of Non-Radiative Capture and Emission Statistics in Defect Semiconductors
(13.4; 22)
(b) Vol. 122, No. 2, p. 687—701 (1984).

W. Lorenz and C. Engler
Quantum-Theoretical Model of Adiabatic Rate Processes in Condensed Systems or at Interfaces
(13.3; 4; 6; 8; 22)
(b) Vol. 122, No. 2, p. 745—755 (1984).

W. KNAP, J. KOSSUT, and J. MYCIELSKI
Photoelectromagnetic Effect and Photoconductivity in Quantizing Magnetic Fields
(13.1; 16; 22)
(b) Vol. **122**, No. 2, p. 761—773 (1984).

V. M. POLYANOVSKII
Stark Cyclotron-Phonon Oscillations of Sound Absorption in Semiconductors with Superlattice
(6 and 7; 19; 22)
(b) Vol. **122**, No. 2, p. K115—K118 (1984).

M. HANKE, D. HENNIG, and R. SCHUCHARDT
Cluster-Size Dependence of the Electronic Density of States within the Bethe-Lattice Approach
(13.1; 22)
(b) Vol. **122**, No. 2, p. K151—K154 (1984).

I. BARTOŠ
On Inhomogeneous Damping of Electrons in Crystals
(13.1; 22)
(b) Vol. **122**, No. 2, p. K159—K162 (1984).

T. LIDEIKIS, V. NOREIKA, A. PETRAITIENE, I. ŠIMKIENE, and K. VALACKA
Optical Absorption in Indium Doped Lead Tin Telluride
(20.1; 13.4; 22)
(b) Vol. **122**, No. 2, p. K179—K182 (1984).

M. SH. ERUKHIMOV and S. G. OVCHINNIKOV
Atomic Operators in the Theory of Narrow-Band Magnetic Semiconductors
(13.5.4; 18.2; 22)
(b) Vol. **123**, No. 1, p. 105—114 (1984).

S. YU. POTAPENKO and A. M. SATANIN
Electron States and Conductivity of Size Quantized Systems with Rough Boundaries
(13.1; 13.3; 14.3.1; 22)
(b) Vol. **123**, No. 1, p. 191—199 (1984).

V. A. KOVARSKII, L. V. CHERNYSH, and A. V. BELOUSOV
About the Low-Temperature Dependence of Activation for Capture Cross-Sections of Charge Carriers in Semiconductors with Persistent Photoconductivity
(13.4; 16; 22)
(b) Vol. **123**, No. 1, p. 345—351 (1984).

M. ZAŁUŻNY
On the Influence of the Depolarization Field on the Intersubband Absorption in Thin Semiconductor Films
(13.1 and 20.1; 22; 22.2.1)
(b) Vol. **123**, No. 1, p. K57—K61 (1984).

J. Maier
Surface Induced Defects in the Space Charge Region and the Enhancement of Ionic Conductivity in Two-Phase Systems
(14.4; 22)
(b) Vol. **123**, No. 1, p. K89—K91 (1984).
Erratum
(b) Vol. **124**, No. 2, p. K187 (1984).

D. G. Chambaere and E. De Grave
A Study of the Non-Stoichiometrical Halogen and Water Content of β-FeOOH
(1; 5; 22)
(a) Vol. **83**, No. 1, p. 93—102 (1984).

K. R. Zhdanov, A. V. Mishenko, F. S. Rakhmenkulov, and V. E. Fedorov
Structural Anisotropy and Heat Capacity of NbS_3
(6 and 8; 22)
(a) Vol. **83**, No. 1, p. 147—152 (1984).

B. V. Kobrin and V. P. Shantarovich
Study of Ge_2S_3 Glass Undoped and Doped with Metals by Positron Annihilation Technique
(2 and 10.2; 22)
(a) Vol. **83**, No. 1, p. 159—164 (1984).

J. Rosenzweig, J. Zhang, and U. Birkholz
Annealing Effects on Eletrical Conductivity and Thermoelectric Power of Sintered PbTe
(14.3; 15; 22)
(a) Vol. **83**, No. 1, p. 357—363 (1984).

J. Teteris
Photoinduced Optical Absorption in As–Se
(20.1; 2; 22)
(a) Vol. **83**, No. 1, p. K47—K50 (1984).

D. W. Field
Electron Transfer and Thermal Vibration Parameters in Titanium Nitride: An X-Ray Diffraction Study
(4; 6; 22)
(b) Vol. **123**, No. 2, p. 479—483 (1984).

S. P. Gwet, Y. Mathey, and C. Sourisseau
The Infrared, Raman, Resonance Raman Spectra, and the Valence Force Field of the HfS_3 Layer-Type Compound
(6 and 20.1; 22)
(b) Vol. **123**, No. 2, p. 503—517 (1984).

J. M. Wesselinowa
Electron–Phonon Interaction in Ferromagnetic Semiconductors
(18.2; 6; 13; 22)
(b) Vol. **123**, No. 2, p. 585—593 (1984).

A. E. SMITH
A Calculation of Absorption Potentials for Low Energy Electrons in MgO, 2H-MoS_2, and 2H-$NbSe_2$
(1.6 and 13; 22; 22.6)
(b) Vol. **123**, No. 2, p. 619—626 (1984).

N. I. LIKHOLIT, H.-E. PONATH, V. L. STRIZHEVSKII, and YU. N. YASHKIR
Transient Coherent Anti-Stokes Raman Spectroscopy of Surface Polaritons
(13.5 and 20.1; 6; 22)
(b) Vol. **123**, No. 2, p. 691—701 (1984).

B. GIESE and M. MOCKER
Carrier Heating in Lead Chalcogenides
(13.1 and 14.3.4; 22)
(b) Vol. **123**, No. 2, p. 755—762 (1984).

J. M. WESSELINOWA
Relaxation Time for Ferromagnetic Semiconductors
(13 and 18.2; 22)
(b) Vol. **123**, No. 2, p. K129—K132 (1984).

M. MONDAL and K. P. GHATAK
Influence of Magnetic Quantization on the Effective Electron Mass in Zero-Gap Semiconductors
(13.1 and 18; 22)
(b) Vol. **123**, No. 2, p. K143—K148 (1984).

N. S. BUINOV and S. R. SYRTSOV
Ferroelectric Phase Transition in the Field of Nonresonance Electromagnetic Waves
(13.1 and 14.4.1; 22)
(b) Vol. **123**, No. 2, p. K165—K167 (1984).

H.-U. TSCHIRNER and K. UHLIG
Some Remarks on Activation Energies Observed in the Liquid Semiconductor Ga_2Te_3
(2 and 14.3; 13.4; 22)
(b) Vol. **123**, No. 2, p. K177—K181 (1984).

P. J. REUTERS, L. DOTHANH, and P. BALK
Determination of Carrier Type in MIS Structures with Pyrolytic AlN Insulator Films
(14.3.3; 14.4; 22; 22.1.2)
(a) Vol. **83**, No. 2, p. 719—727 (1984).

M. RICK, J. ROSENZWEIG, and U. BIRKHOLZ
Anisotropy of Electrical Conductivity in Si_2Te_3
(14.3; 22)
(a) Vol. **83**, No. 2, p. K183—K186 (1984).

E. A. AKOPYAN, A. SH. MEKHTIEV, and A. L. POIMANOV
Photon Storage Mode in a Forward-Biased p–n Junction
(14.3.3; 22)
(a) Vol. **83**, No. 2, p. K215—K217 (1984).

I. Ya. Korenblit and B. G. Tankhilevich
Fluctuations in a Nonequilibrium System of Electrons and Magnons in Ferromagnetic Semiconductors
(13.5.4 and 18.2; 22)
(b) Vol. **124**, No. 1, p. 79—86 (1984).

H. S. Brandi, G. Jalbert, and O. L. Malta
Multiphoton Absorption in Solids. Non-Perturbative Approach in the Weak-Field Limit
(13.1 and 20.1; 22)
(b) Vol. **124**, No. 1, p. 147—157 (1984).

W. Huhn and A. Stahl
Self-Consistent Field Theory Applied to the Semiconductor Band Edge
(13.1 and 13.5; 14.3; 20; 22)
(b) Vol. **124**, No. 1, p. 167—177 (1984).

A. Conan, A. Bonnet, M. Zoaeter, and D. Ramoul
Dependence of the Total Mobility in a One-Band Model. Application to n-Type $MoTe_2$
(14.3 and 15; 22)
(b) Vol. **124**, No. 1, p. 403—410 (1984).

H. Stramska
The Voigt Effect under a Strong Electric Field in Semiconductors
(14.3.4 and 18; 22)
(b) Vol. **124**, No. 1, p. 431—437 (1984).

P. Enders
On the Dielectric Theory: Does the Static Ionic Charge Scale with $f_i^{1/2}$?
(4; 22)
(b) Vol. **124**, No. 1, p. K21—K23 (1984).

D. A. Aronov, P. I. Knigin, and B. R. Mamatkulov
Exclusion Effect Quenching in Compensated Semiconductors under the Action of Impurity Illumination Generating Majority Current Carriers
(14.3.3 and 14.3.4; 16; 22)
(a) Vol. **84**, No. 1, p. 301—310 (1984).

A. A. Simashkevich and S. D. Shutov
Evidence of Schottky Barrier Formation at Contact of Metal with Chalcogenide Glassy Semiconductor
(14.3.4; 2; 16; 22)
(a) Vol. **84**, No. 1, p. 343—352 (1984).

C. H. Ling, C. Y. Kwok, and K. Prasad
Observation of an Anomalously High Etch Rate in Plasma-Enhanced CVD Silicon Nitride Films
(1.5; 22)
(a) Vol. **84**, No. 1, p. K1—K4 (1984).

V. I. Zametin
Absorption Edge Anomalies in Polar Semiconductors and Dielectrics at Phase Transitions
(20.1; 1.2; 14.4.1; 22; 22.5.2; 22.8.1; 22.9)
(b) Vol. **124**, No. 2, p. 625—640 (1984).

A. E. Smith and D. F. Lynch
Low-Energy Electron Reflection Intersities from 2H-MoS_2 as Derived by Scattering Matrix Calculations
(1.6 and 13; 22)
(b) Vol. **124**, No. 2, p. 649—655 (1984).

M. Ishii, M. Saeki, and I. Kawada
Raman Study of Non-Stoichiometric Titanium Sulfides
(6 and 20.1; 22)
(b) Vol. **124**, No. 2, p. K109—K112 (1984).

A. I. Gusev and A. A. Rempel
A Study of the Atomic Ordering in the Niobium Carbide Using the Magnetic Susceptibility Method
(1.3; 18.1; 22)
(a) Vol. **84**, No. 2, p. 527—534 (1984).

P. Gille and M. Schenk
Structural Incorporation and Electrical Efficiency of TI in PbTe
(10.2; 14.3; 22)
(a) Vol. **84**, No. 2, p. K121—K124 (1984).

N. N. Koren, V. I. Levchenko, V. V. Dikareva, and A. V. Kazushchik
Physical Properties of Germanium Telluride Crystals Grown from Gas Phase
(14.3 and 15; 10.2; 22)
(a) Vol. **84**, No. 2, p. K173—K176 (1984).

E. E. Matyas and T. E. Borisenko
Electrical Properties of PbTe–SnS Solid Solutions
(14.3; 22)
(a) Vol. **84**, No. 2, p. K193—K196 (1984).

R. P. Gairola
Low-Temperature Lattice Thermal Conductivity of Nonmetallic Solids with Isotopic Impurities
(6 and 8; 10.2; 22; 22.1.1)
(b) Vol. **125**, No. 1, p. 65—74 (1984).

N. Agnihotri, R. K. Singh, and S. P. Sanyal
Interionic Forces and Elastic Properties of Na_2S
(4; 6; 12.2; 22; 22.5.3)
(b) Vol. **125**, No. 1, p. 113—119 (1984).

T. R. Gawron and J. Mycielski
Shallow Impurity States in Strong Magnetic Field
(13.4; 20.1; 22)
(b) Vol. **125**, No. 1, p. 341—348 (1984).

B. Lorenz
Charge Ordering in the Anti-Th_3P_4 Structure
(4; 1.2; 22)
(b) Vol. **125**, No. 1, p. 375—379 (1984).

H. Englisch
Instabilities and δ-Distributions in Solid State Physics
(13; 20.3; 22)
(b) Vol. **125**, No. 1, p. 393—400 (1984).

M. M. Babaev, T. M. Gasymov, and A. A. Katanov
Thermopower and Thermomagnetic Effects of Hot Carriers under Conditions of Mutual Electron–Phonon Drag
(15; 6; 13; 22)
(b) Vol. **125**, No. 1, p. 421—429 (1984).

V. P. Gupta and O. P. Singh
Phillips Ionicity and Structural Classification of Layered Transition Metal Dichalcogenides
(1 and 4; 22)
(b) Vol. **125**, No. 1, p. K1—K5 (1984).

S. A. Alterovitz, G. H. Bu-Abbud, J. A. Wollam, and D. C. Liu
Structural and Chemical Analysis of a Silicon Nitride Film on GaAs by Null Ellipsometry
(1.3 and 1.5; 22)
(a) Vol. **85**, No. 1, p. 69—76 (1984).

T. Jung and L. Roth
DC Plasma Nitridation of Thin Aluminium Films
(1.5; 11; 22)
(a) Vol. **85**, No. 1, p. K15—K18 (1984).

J. Pułtorak
Current–Voltage Characteristic of the p–n Junction with an Exponential Impurity Distribution at Its Base Contact
(14.3.3; 22)
(a) Vol. **85**, No. 1, p. K93—K96 (1984).

G. Manzke, V. May, and K. Henneberger
Excited States and Collective Excitations in a Dense Gas of Excitons
(13.5.1; 22)
(b) Vol. **125**, No. 2, p. 693—704 (1984).

A. V. Soldatov and A. N. Gusatinskii
Energy Delocalization of the Rare-Earth f-States in Some Compounds
(13 and 20; 22)
(b) Vol. **125**, No. 2, p. K129—K132 (1984).

S. Ghosh and S. Khan
Amplitude Modulation and Demodulation of Electromagnetic Waves in Magnetised Semiconducting Plasmas
(14 and 20; 22)
(a) Vol. **85**, No. 2, p. 535—541 (1984).

W. HERGERT and L. PASEMANN
Theoretical Study of the Information Depth of the Cathodoluminescence Signal in Semiconductor Materials
(20.3; 10.2; 14.3; 22)
(a) Vol. **85**, No. 2, p. 641—648 (1984).

D. BALTRŪNAS
A Study of the Possibility of Restoration of Composition of SnTe Oxidized Epitaxial Layers by Means of NGR Spectroscopy
(3 and 5; 22)
(a) Vol. **85**, No. 2, p. K97—K99 (1984).

G. RUDLOF and H. GLAEFEKE
Remarks about a New, Constructive Method for the Analysis of Thermally Stimulated Relaxation Processes (TSRP)
(13.4 and 17; 22)
(a) Vol. **85**, No. 2, p. K149—K153 (1984).

A. R. VASCONCELLOS and R. LUZZI
On Laser-Induced Lattice Heating in a Polar Semiconductor
(8 and 11; 6; 22; 22.2.1)
(b) Vol. **126**, No. 1, p. 63—70 (1984).

R. CHEN, V. K. MATHUR, J. F. RHODES, M. D. BROWN, S. W. MCKEEVER, and R. K. BULL
Thermoluminescence Governed by Simultaneous Thermal Stimulation of Electrons and Holes
(13.4 and 20.3; 22)
(b) Vol. **126**, No. 1, p. 361—369 (1984).

M. SUHRKE and B. ESSER
Time-Resolved Luminescence of Charge Carriers in Localized States in Amorphous Semiconductors
(20.3; 2; 13.4; 22)
(b) Vol. **126**, No. 1, p. 393—406 (1984).

M. MONDAL and K. P. GHATAK
On the Effective Mass in Tetragonal Semiconductors in the Presence of an Arbitrarily Oriented Quantizing Magnetic Field
(13.1; 22)
(b) Vol. **126**, No. 1, p. K47—K51 (1984).

B. T. KOLOMIETS, V. M. LYUBIN, M. A. TAGUYRDZHANOV, and G. ZENTAI
A Comparative Study of Photoconductivity in Vitreous As_2Se_3 and a-Si:H Thin Films
(16; 2; 14.3.1; 22; 22.1.2)
(a) Vol. **86**, No. 1, p. 383—387 (1984).

DANG TRAN QUAN
Electrical Properties and Optical Absorption of SnSe Evaporated Thin Films
(14.3.1; 20.1; 22)
(a) Vol. **86**, No. 1, p. 421—426 (1984).

J. W. Tomm, K. Herrmann, C. Barthel, and U. Barthel
On the Dispersion of the Refractive Index in Active Layers of Lead-Salt Injection Lasers
(20.1; 20.2; 22)
(a) Vol. **86**, No. 1, p. 433—438 (1984).

A. A. Rempel, S. Z. Nazarova, and A. I. Gusev
Effect of Atomic Ordering on the Heat Capacity of Non-Stoichiometric Niobium Carbide
(1.2 and 8; 6; 22)
(a) Vol. **86**, No. 1, p. K11—K14 (1984).

W. S. Tse and C. C. Chen
Lattice Vibrations of Crystalline Silicon and Germanium Tetrachlorides
(6 and 20.1; 22)
(a) Vol. **86**, No. 1, p. K23—K25 (1984).

A. A. Evtushenko, V. F. Petrenko, and I. A. Ryzhkin
Electric Polarization of Ice at Nonuniform Elastic Strains
(10.2 and 14.4; 12.2; 22)
(a) Vol. **86**, No. 1, p. K31—K34 (1984).

R. Tognato
Overheating in Semiconductors with a Diamond-Like Structure
(10.2; 22)
(a) Vol. **86**, No. 1, p. K35—K37 (1984).

F. Bechstedt, R. Enderlein, and O. Heinrich
The Effect of Many-Electron Correlation on Photothresholds of Semiconductors and Valence Band Discontinuities at Heterojunctions
(13.1; 22)
(b) Vol. **126**, No. 2, p. 575—585 (1984).

K. R. Allakhverdiev, S. S. Babaev, N. A. Bakhyshov, T. G. Mamedov, E. Yu. Salaev, and I. K. Efendieva
The Influence of Hydrostatic Pressure on the Fundamental Absorption Edge of Crystals with TlSe Type Structure
(13.1 and 20.1; 12.2; 22; 22.8)
(b) Vol. **126**, No. 2, p. K139—K143 (1984).

L. Borisova
Density-of-States Effective Mass in Solid Solutions of the $(Ag_{x/2}Pb_{1-x}Sb_{x/2})Te$ Type
(13.1; 14.3; 15; 22)
(b) Vol. **126**, No. 2, p. K155—K158 (1984).

I. A. Ryzhkin
Propagation of Elastic Waves in Ice
(7; 22)
(a) Vol. **86**, No. 2, p. K87—K90 (1984).

22.1 Elements

Review Article

A. Blacha, H. Presting, and M. Cardona
Deformation Potentials of $k = 0$ States of Tetrahedral Semiconductors
(6 and 13.1; 22.1; 22.2; 22.4)
(b) Vol. **126**, No. 1, p. 11—36 (1984).

Original Papers and Short Notes

J. P. Schaffer, H. Park, J. H. Lind, and P. L. Jones
Hydrogenated Amorphous Boron: Transient and Steady State Photoconductivity
(2 and 16; 22.1)
(a) Vol. **81**, No. 1, p. K51—K55 (1984).

S. Orzeszko, W. Bała, K. Fabisiak, and F. Rozpłoch
DC Conductivity and ESR of Hydrogenated Amorphous Carbon Films
(2 and 14.3.1; 19; 22.1)
(a) Vol. **81**, No. 2, p. 579—584 (1984).

V. Hinkel and D. Ploumbidis
Temperature Dependence of the Magnetic Part of the Hyperfine Field at ^{123}Sb in the Liquid Sb–Te System
(2 and 13; 19; 21.7; 22.1)
(b) Vol. **122**, No. 1, p. K45—K48 (1984).

P. P. Lottici
On the Optic Vibrational Modes in Amorphous Arsenic
(6 and 20.1; 2; 22.1)
(b) Vol. **122**, No. 2, p. 431—434 (1984).

S. N. Sahu, J. T. Borenstein, V. A. Singh, and J. W. Corbett
Semi-Empirical Tight Binding Calculations for the Energy Bands of the Diamond and Zincblende Type Semiconductors
(13.1; 22.1; 22.2)
(b) Vol. **122**, No. 2, p. 661—667 (1984).

V. Frei and P. Deus
A Frequency Sum Rule and Spectral Moments in Lattice Dynamics
(6; 22.1; 22.1.1; 22.2.2; 22.5.2; 22.8)
(b) Vol. **125**, No. 1, p. 121—130 (1984).

A. V. Mudryi, A. L. Pushkarchuk, V. D. Tkachev, and A. G. Ulyashin
Noble Gas Atoms as Interstitials in Silicon and Diamond
(10.2 and 13.4; 22.1; 22.1.2)
(b) Vol. **125**, No. 1, p. K75—K78 (1984).

H. Werheit and R. Franz
States of Thermal Equilibrium in β-Rhombohedral Boron and Their Alteration by Optical Excitation
(1.2; 13.4; 20.1; 22.1)
(b) Vol. **125**, No. 2, p. 779—784 (1984).

G. Bambakidis and W. T. McKeowin
C_5H_9 Cluster as a Model for the (111) 1 × 1 Diamond Surface
(13.3; 22.1)
(b) Vol. **126**, No. 1, p. K67—K70 (1984).

O. A. Golikova, N. E. Solovev, Ya. A. Ugai, and V. A. Feigelman
Electrophysical Properties of α-Rhombohedral Boron
(14.3 and 15; 22.1)
(a) Vol. **86**, No. 1, p. K51—K54 (1984).

B. Holland, H. S. Greenside, and M. Schlüter
Cohesive Properties of Ge, Si, and Diamond Calculated with Minimum Basis Sets
(4 and 13; 12.2; 22.1; 22.1.1; 22.1.2)
(b) Vol. **126**, No. 2, p. 511—515 (1984).

V. E. Borisenko, V. V. Gribkovskii, V. A. Labunov, and S. G. Yudin
Pulsed Heating of Semiconductors
(8; 11; 22.1; 22.2)
(a) Vol. **86**, No. 2, p. 573—583 (1984).

L. G. Khvostantsev, V. A. Sidorov, and O. B. Tsiok
Gapless State and Anomalous Behaviour of Thermopower of Black Phosphorus at Pressure up to 12 GPa
(13.1 and 15; 14.3; 22.1)
(a) Vol. **86**, No. 2, p. K99—K104 (1984).

22.1.1 Germanium

Review Articles

D. A. Aronov, P. I. Knigin, Yu. S. Korolev, and V. V. Rubinov
Exclusion Effect in Semiconductors with Non-Injecting Contacts
(14.3.3 and 14.3.4; 16; 22.1.1; 22.1.2; 22.2.2)
(a) Vol. **81**, No. 1, p. 11—45 (1984).

J. Pułtorak
The l–h Junction as an Element of the Semiconductor Device
(14.3.3; 22.1.1)
(a) Vol. **82**, No. 1, p. 11—22 (1984).

Original Papers and Short Notes

S. V. Tsivinskii
The Theory of Strain Hardening in Single and Polycrystals
(10.1; 1.4; 21; 21.1; 21.1.1; 22.1.1; 22.1.2)
(a) Vol. **81**, No. 1, p. 299—306 (1984).

M. Rösler, R. Zimmermann, and W. Richert
The Electron–Hole Plasma at Finite Temperatures. Calculation of Exchange-Correlation Energies and Padé Approximants
(13.5; 22.1.1; 22.2.1)
(b) Vol. **121**, No. 2, p. 609—621 (1984).

M. V. Kovalchuk and E. Kh. Mukhamedzhanov
The Use of a Gas-Flow Proportional Counter for the Energetical Analysis of Photoelectrons under X-Ray Diffraction Conditions
(1.6 and 17; 22.1.1)
(a) Vol. **81**, No. 2, p. 427—432 (1984).

V. E. Gusakov and V. Yu. Yavid
The Role of Nonelastic Nuclear Processes in the Formation of Defect Clusters in Semiconductors Irradiated with High-Energy Protons
(11; 22.1.1)
(a) Vol. **81**, No. 2, p. 469—472 (1984).

R. F. Konopleva, V. A. Evseev, B. A. Borisov, I. V. Nazarkin, and V. A. Chekanov
Electrophysical Properties of Disordered-Systems Based on Crystalline Germanium Compensated by Fast Neutron Irradiation
(2 and 14.3; 14.3.4; 22.1.1)
(b) Vol. **122**, No. 1, p. 349—359 (1984).

Y. Kamiura, F. Hashimoto, S. Yoneyama, and T. Nobusada
Reduction of Copper Contamination in Germanium on Heat Treatments above 700 °C
(10.2; 13.4; 22.1.1)
(a) Vol. **82**, No. 1, p. 135—141 (1984).

O. BOSTANJOGLO and E. ENDRUSCHAT
Time-Resolved TEM of Laser-Annealing
(1.2 and 1.4; 2; 11; 22.1.1)
(a) Vol. **82**, No. 1, p. K1—K3 (1984).

K. S. JONES and S. J. PEARTON
Grain Boundaries in Germanium: Effects of Exposure to Plasmas
(10.2 and 14.3; 22.1.1)
(a) Vol. **82**, No. 1, p. K101—K105 (1984).

B. A. VAID and K. C. SHARMA
Structural Disorder Model of Amorphous Semiconductors: Ge and Si
(2; 22.1.1; 22.1.2)
(b) Vol. **122**, No. 2, p. 423—429 (1984).

T. SOMA, Y. KITANI, and H.-MATSUO KAGAYA
Local and Band Mode Frequencies of $Si_{1-x}Ge_x$ Solid Solution
(6; 22.1.1; 22.1.2)
(b) Vol. **122**, No. 2, p. K125—K126 (1984).

V. V. BAIDAKOV, V. N. ERMAKOV, N. N. GRIGOREV, V. V. KOLOMOETS, and T. A. KUDYKINA
Breakdown of Impurity States of As and Sb in Germanium at Uniaxial Compression $\boldsymbol{P} \parallel [001] \parallel \boldsymbol{E}$
(13.4 and 14.3.4; 22.1.1)
(b) Vol. **122**, No. 2, p. K163—K167 (1984).

F. THUSELT
Low-Density Approach to RPA Self-Energies
(13.5.2; 22.1.1; 22.1.2; 22.2.1; 22.2.2)
(b) Vol. **123**, No. 1, p. 257—269 (1984).

S. R. BAKHCHIEVA, N. P. KEKELIDZE, and M. G. KEKUA
Thermal Conductivity of Germanium Doped with Silicon, Tin, and Aluminium
(8; 22.1.1)
(a) Vol. **83**, No. 1, p. 139—145 (1984).

A. M. EVSTIGNEEV, L. F. LINNIK, L. G. LINNIK, E. A. SALKOV, A. N. KRASIKO, and O. V. SNITKO
Observation of Anisotropic Internal Stresses by Electroreflectance and Transverse Photovoltaic Effect Measurements
(10.2; 14.3.4; 16; 20.1; 22.1.1)
(a) Vol. **83**, No. 1, p. K31—K33 (1984).

H. BARTSCH, D. HOEHL, and G. KÄSTNER
Radiation-Induced Rod-Like Defects in Silicon and Germanium
(1.4 and 11; 22.1.1; 22.1.2)
(a) Vol. **83**, No. 2, p. 543—551 (1984).

S. N. KOZLOV and N. L. LEVSHIN
Isotopic Effect in the Recombination of Nonequilibrium Charge Carriers on the Germanium Surface
(13.3; 22.1.1)
(a) Vol. **83**, No. 2, p. K149—K152 (1984).

N. Pashov, G. Kästner, M. Kalitzova, S. Simov, M. Mikhailov, and K. Kolev
Electron Microscope and Electron Diffraction Studies of Amorphization Phenomena in Ge Implanted with Te Ions
(1.4 and 11; 2; 22.1.1)
(a) Vol. **84**, No. 1, p. K23—K27 (1984).

R. P. Gairola
Low-Temperature Lattice Thermal Conductivity of Nonmetallic Solids with Isotopic Impurities
(6 and 8; 10.2; 22; 22.1.1)
(b) Vol. **125**, No. 1, p. 65—74 (1984).

V. Frei and P. Deus
A Frequency Sum Rule and Spectral Moments in Lattice Dynamics
(6; 22.1; 22.1.1; 22.2.2; 22.5.2; 22.8)
(b) Vol. **125**, No. 1, p. 121—130 (1984).

Y. Kamiura and F. Hashimoto
Origin of a Shallow Acceptor in Quenched Germanium
(10.2 and 13.4; 14.3; 22.1.1)
(a) Vol. **85**, No. 1, p. 227—234 (1984).

S. V. Vintsents, V. F. Kiselev, and G. S. Plotnikov
Energy Transfer between Excited Adsorbed Dye Molecules and Charged Defects in Insulator–Semiconductor Structures
(13.4; 19; 20.3; 22.1.1; 22.1.2)
(a) Vol. **85**, No. 1, p. 273—281 (1984).

V. A. Bespalov, V. F. Kiselev, G. S. Plotnikov, and A. M. Saletskii
Fluorescence Decay of Adsorbed Dye Molecules on Solid Surfaces
(13.3 and 20.3; 22.1.1; 22.6)
(a) Vol. **85**, No. 1, p. K73—K76 (1984).

T. Soma, Y. Kitani, and H.-Matsuo Kagaya
Mode Grüneisen Parameter and Pressure-Induced Phase Transition of Si–Ge Solid Solution
(1.2 and 6; 8; 22.1.1; 22.1.2)
(b) Vol. **125**, No. 1, p. K99—K101 (1984).

P. A. Besirganyan, R. Ts. Gabrielyan, V. G. Kohn, and A. H. Toneyan
Six-Beam X-Ray Diffraction in Ge Single Crystals
(1.1; 22.1.1)
(a) Vol. **85**, No. 2, p. 349—358 (1984).

D. Della Sala, C. Giovannella, and F. Evangelisti
Study of Bonding Configurations in Amorphous Ge_xSi_{1-x}:H Alloys
(2; 4; 20.1; 22.1.1; 22.1.2)
(b) Vol. **126**, No. 1, p. 125—131 (1984).

R. A. Vardanian and G. G. Kirakosian
Electron Scattering by a Charged Dislocation
(10.2 and 14.3; 22.1.1)
(b) Vol. **126**, No. 1, p. K83—K86 (1984).

A. M. KONIN and A. P. SAŠČIUK
Thermogradient Magnetoconcentration Effect
(14.3; 22.1.1; 22.2.3)
(a) Vol. **86**, No. 1, p. K55—K58 (1984).

B. HOLLAND, H. S. GREENSIDE, and M. SCHLÜTER
Cohesive Properties of Ge, Si, and Diamond Calculated with Minimum Basis Sets
(4 and 13; 12.2; 22.1; 22.1.1; 22.1.2)
(b) Vol. **126**, No. 2, p. 511—515 (1984).

E. KADEN and H.-L. GÜNTER
The Thermoelectric Power of p-Ge at Low Temperatures
(6 and 15; 14.3; 22.1.1)
(b) Vol. **126**, No. 2, p. 733—740 (1984).

Y. KAMIURA and F. HASHIMOTO
Annealing Behaviour of Shallow Acceptors in Quenched Germanium
(10.2; 13.4; 14.3; 22.1.1)
(a) Vol. **86**, No. 2, p. K95—K97 (1984).

22.1.2 Silicon

Review Article

D. A. Aronov, P. I. Knigin, Yu. S. Korolev, and V. V. Rubinov
Exclusion Effect in Semiconductors with Non-Injecting Contacts
(14.3.3 and 14.3.4; 16; 22.1.1; 22.1.2; 22.2.2)
(a) Vol. **81**, No. 1, p. 11—45 (1984).

Original Papers and Short Notes

M. Cahay and E. Kartheuser
Polarizability of Donors in Polar Semiconductors: Effect of a Weak Magnetic Field
(13.4; 22.1.2; 22.2.2; 22.4.2)
(b) Vol. **121**, No. 1, p. 345—356 (1984).

S. Žurauskas and A. Dargys
Field Ionization of Phosphorus Atoms in Silicon
(13.4; 14.3.3; 22.1.2)
(b) Vol. **121**, No. 1, p. 385—391 (1984).

P. A. Aleksandrov, A. M. Afanasiev, M. K. Melkonyan, and S. A. Stepanov
X-Ray Diffraction under Specular Reflection Conditions on Crystals with an Amorphous Surface Film
(1.6 and 2; 10.2; 22.1.2)
(a) Vol. **81**, No. 1, p. 47—53 (1984).

V. Yu. Balandin, A. V. Dvurechenskii, and L. N. Aleksandrov
Dynamics of Self-Sustaining Crystallization of Amorphous Silicon Layers
(2 and 3; 1.5; 22.1.2)
(a Vol. **81**, No. 1, p. 63—68 (1984).

A. D. Pogrebnyak, E. Yu. Boyarko, Yu. Yu. Kryuchkov, and S. A. Vorobev
Positron Annihilation and Profiles of Radiation Damages in GaAs and Si Crystals Irradiated by Supercurrent Proton or Electron Beams
(11; 14.3; 22.1.2; 22.2.1)
(a) Vol. **81**, No. 1, p. 217—225 (1984).

W. Bergholz, S. Damgaard, J. W. Petersen, and G. Weyer
Reactions of Cobalt in Silicon with Electron-Irradiation-Induced Defects
(10.2; 5; 22.1.2)
(a) Vol. **81**, No. 1, p. 239—246 (1984).

E. Wieser, H. Syhre, F. G. Rüdenauer, and W. Steiger
Influence of Flash-Lamp Annealing on the Diffusion Behaviour of Ion Implanted Boron Profiles
(9; 11; 22.1.2)
(a) Vol. **81**, No. 1, p. 247—252 (1984).

M. Pasemann, W. Bergholz, and W. Schröter
Mössbauer Spectroscopical and Electron Microscopical Investigations of the Behaviour of Cobalt in Silicon
(1.4 and 10.2; 5; 22.1.2)
(a) Vol. **81**, No. 1, p. 273—280 (1984).

S. V. Tsivinskii
The Theory of Strain Hardening in Single and Polycrystals
(10.1; 1.4; 21; 21.1; 21.1.1; 22.1.1; 22.1.2)
(a) Vol. **81**, No. 1, p. 299—306 (1984).

V. D. Tkachev, A. V. Mudryi, and N. S. Minaev
Noble Gas Atoms as Chemical Impurities in Silicon
(10.2 and 13.4; 20.3; 22.1.2)
(a) Vol. **81**, No. 1, p. 313—321 (1984).

A. Chandorkar, V. T. Karulkar, A. S. Vengurlekar, and K. V. Ramanathan
MOS Capacitance–Voltage Characteristics and Dielectric Properties of Ion Implanted Thermal Oxides on Silicon
(14.3.3; 11; 14.4; 22.1.2)
(a) Vol. **81**, No. 1, p. 407—414 (1984).

A. F. Khokhlov, V. A. Panteleev, E. V. Dobrokhotov, G. A. Maksimov, and V. A. Sidorov
Phosphorus Diffusion in Amorphous Silicon
(2 and 9; 22.1.2)
(a) Vol. **81**, No. 1, p. K15—K18 (1984).

H. K. Phan, P. H. Binh, and L. H. Phu
A Simple Method for Investigating Charge Storage Effect in MIS Switching Diodes
(14.3.3; 14.3.4; 22.1.2)
(a) Vol. **81**, No. 1, p. K81—K84 (1984).

D. A. Aronov and B. R. Mamatkulov
The Capacitance and Characteristic Relaxation Times at Carrier Exclusion in Compensated Semiconductors with Deep Traps and Non-Injecting Contacts
(14.3.4; 22.1.2)
(a) Vol. **81**, No. 1, p. K85—K89 (1984).

P. K. Basu and J. B. Roy
Warm Electron Coefficient of Two-Dimensional Electron Gas in Silicon Inversion Layer at Low Temperatures
(14.3.2 and 14.3.4; 22.1.2)
(b) Vol. **121**, No. 2, p. 743—748 (1984).

L. Wendler and E. Jäger
Surface Polaritons in a SiO_2/Si Bilayer System
(6 and 13.5; 22.1.2; 22.6)
(b) Vol. **121**, No. 2, p. K119—K124 (1984).

A. A. Sitnikova, L. M. Sorokin, I. E. Talanin, E. G. Sheikhet, and E. S. Falkevich
Electron-Microscopic Study of Microdefects in Silicon Single Crystals Grown at High Speed
(1.4; 10.2; 22.1.2)
(a) Vol. **81**, No. 2, p. 433—438 (1984).

S. Motogi and G. A. Maugin
Effect of Magnetostriction on Vibrations of Bloch and Néel Walls
(18.2; 12.2; 22.1.2; 22.6)
(a) Vol. **81**, No. 2, p. 519—532 (1984).

Chu Hao, J. Zimmermann, M. Charef, R. Fauquembergue, and E. Constant
A Semi-Classical Model for Simulating Inversion Carrier Transport in Si MOS Devices
(14.3.2 and 14.3.3; 22.1.2)
(a) Vol. **81**, No. 2, p. 569—577 (1984).

T. Gessner, M. Pasemann, and B. Schmidt
Erratum to: Calculation of Neutron-Induced Defect Clusters in Silicon and Comparison with TEM Investigations
(11; 1.4; 22.1.2)
(a) Vol. **81**, No. 2, p. 751 (1984).

H. J. Goldsmid, H. Hora, and G. L. Paul
Anomalous Heat Conduction of Ion-Implanted Amorphous Layers in Silicon Crystals Using a Laser-Probe Technique
(2 and 11; 8; 22.1.2)
(a) Vol. **81**, No. 2, p. K127—K130 (1984).

M. Kittler, K.-W. Schröder, E. Bugiel, and C. Becker
Quantitative EBIC Investigations on Bulk Stacking Faults in Silicon
(1.4 and 10.2; 14.3; 22.1.2)
(a) Vol. **81**, No. 2, p. K131—K135 (1984).

M. S. Yunusov, S. N. Abdurakhmanova, M. A. Zaikovskaya, and B. L. Oksengendler
The Influence of Minority Carrier Injection on Subthreshold Defects in Silicon
(11 and 14.3.4; 22.1.2)
(a) Vol. **81**, No. 2, p. K145—K149 (1984).

K. Schmalz, P. Gaworzewski, and F.-G. Kirscht
Deep Levels in Czochralski p-Si Due to Heat Treatment at 600 to 900 °C
(10.2 and 13.4; 14.3; 22.1.2)
(a) Vol. **81**, No. 2, p. K165—K169 (1984).

A. P. Pathak
Position Dependence of Channeling Stopping Power
(11 and 13; 22.1.2)
(b) Vol. **122**, No. 1, p. 171—174 (1984).

S. J. Lee and Chung-In Um
Impurity and Thickness Effects on the Cyclotron Resonance Linewidth of Electrons in Si Inversion Layers
(13.3; 14.3.2; 19; 22.1.2)
(b) Vol. **122**, No. 1, p. K73—K78 (1984).

D. M. Vanderwalker
The Influence of Cu on Dislocation Motion in Si
(10.2; 22.1.2)
(a) Vol. **82**, No. 1, p. 93—99 (1984).

H. Kerkow, G. Kreysch, and B. Lukasch
The Influence of Foreign Atoms on the Epitaxial Annealing of Ion-Implanted Silicon
(11; 3; 22.1.2)
(a) Vol. **82**, No. 1, p. 125—133 (1984).

C. Wagner and A. El-Sadek
The Annealing Behavior of Arsenic-Implanted Silicon between 600 and 800 °C Characterized by Carrier Density Profiles
(11 and 14.3; 22.1.2)
(a) Vol. **82**, No. 1, p. 149—155 (1984).

E. Wieser and D. Pannin
Electrical Activation and Damage Annealing of Boron-Implanted Silicon by Flash-Lamp Irradiation
(11; 14.3; 22.1.2)
(a) Vol. **82**, No. 1, p. 171—177 (1984).

M. Suezawa and K. Sumino
Nature of Thermal Donors in Silicon Crystals
(13.4; 20.1; 22.1.2)
(a) Vol. **82**, No. 1, p. 235—242 (1984).

W. Scharff, J.-W. Erben, A. Wolf, M. Heber, C. Hamann, C. Weissmantel, M. Voelskov, J. Matthäi, R. Kögler, R. Klabes, and E. Wieser
Growth of Single-Crystalline Regions on Amorphous Insulating Substrates by Zone-Melting Recrystallization
(2 and 3; 11; 22.1.2)
(a) Vol. **82**, No. 1, p. K5—K9 (1984).

B. A. Vaid and K. C. Sharma
Structural Disorder Model of Amorphous Semiconductors: Ge and Si
(2; 22.1.1; 22.1.2)
(b) Vol. **122**, No. 2, p. 423—429 (1984).

T. Soma, Y. Kitani, and H.-Matsuo Kagaya
Local and Band Mode Frequencies of $Si_{1-x}Ge_x$ Solid Solution
(6; 22.1.1; 22.1.2)
(b) Vol. **122**, No. 2, p. K125—K126 (1984).

I. G. Kaverina, V. V. Korobtsov, V. G. Zavodinskii, and A. V. Zotov
Solid Phase Epitaxial Growth Anisotropy of Vacuum-Deposited Amorphous Silicon
(1.5 and 3; 2; 22.1.2)
(a) Vol. **82**, No. 2, p. 345—353 (1984).

N. M. Olekhnovich and A. L. Karpei
Dynamical Effects of X-Ray Scattering in Laue Geometry for Si Crystals with Structure Defects
(10.2; 22.1.2)
(a) Vol. **82**, No. 2, p. 365—371 (1984).

V. Atzrodt and H. Lange
Principal Component Analysis for Refractory Metal Silicide Investigations with Auger Electron Spectroscopy
(1.6; 17; 21; 22.1.2)
(a) Vol. **82**, No. 2, p. 373—378 (1984).

D. J. H. Cockayne, P. Pirouz, Z. Liu, G. R. Anstis, and P. Karnthaler
A Quantitative Analysis of Image Contrast from Extrinsic Stacking Faults
(1.4 and 10.2; 22.1.2)
(a) Vol. **82**, No. 2, p. 425—439 (1984).

V. I. Kuznetsov, P. F. Lugakov, and A. V. Tsikunov
Interaction of Radiation Defects with the Surface of Silicon
(11 and 14.3.2; 13.4; 22.1.2)
(a) Vol. **82**, No. 2, p. 467—473 (1984).

G. A. Kachurin, V. A. Mayer, S. I. Romanov, M. Voelskow, R. Klabes, and E. Wieser
Low-Energy Implantation of Arsenic in Silicon
(11; 14.3; 22.1.2)
(a) Vol. **82**, No. 2, p. 475—480 (1984).

L. A. Kazakevich, P. F. Lugakov, and I. M. Filippov
The Effect of Dislocations on the Charge-Carrier Recombination Processes in Irradiated Silicon
(10.2 and 13.4; 11; 14.3; 22.1.2)
(a) Vol. **82**, No. 2, p. 511—518 (1984).

P. I. Baranskii, V. M. Babich, N. P. Baran, A. A. Bugay, Yu. P. Dotsenko, and V. B. Kovalchuk
The Effect of Heat Treatment on Compensated CZ Silicon
(13.4; 14.3; 19; 22.1.2)
(a) Vol. **82**, No. 2, p. 533—536 (1984).

R. Klabes, J. Matthäi, M. Voelskow, E. Wieser, J.-W. Erben, W. Scharff, and C. Weissmantel
Epitaxial Regrowth of Amorphous or Polycrystalline Silicon Layers on Silicon Single Crystals and Bridging Epitaxy by Flash Lamp Irradiation
(1.5 and 3; 2; 11; 22.1.2)
(a) Vol. **82**, No. 2, p. K121—K123 (1984).

F. Eichhorn, J. Kulda, and P. Mikula
Small Angle Neutron Scattering on a Reactor Beam Irradiated Silicon Single Crystal
(11; 22.1.2)
(a) Vol. **82**, No. 2, p. K141—K143 (1984).

Dao Khac An
An Approximate Method for Determining the Segregation Coefficient of the Boron Drive Diffusion in Oxidizing Ambients
(9; 22.1.2; 22.6)
(a) Vol. **82**, No. 2, p. K145—K148 (1984).

I. ZÓLOMY
Negative Differential Resistance in the $I-U$ Characteristics of Surface Oxide Transistors
(14.3.3; 14.3.4; 22.1.2)
(a) Vol. **82**, No. 2, p. K209—K211 (1984).

F. THUSELT
Low-Density Approach to RPA Self-Energies
(13.5.2; 22.1.1; 22.1.2; 22.2.1; 22.2.2)
(b) Vol. **123**, No. 1, p. 257—269 (1984).

M. WAGNER and G. GÖTZ
Crystallization of High-Dose Antimony-Implanted Silicon by Millisecond Pulse Laser Irradiation
(1.2 and 3; 22.1.2)
(a) Vol. **83**, No. 1, p. 69—76 (1984).

V. RYBKA, V. HNATOWICZ, J. KVÍTEK, J. VACÍK, and B. SCHMIDT
Determination of the Range Profiles of Boron Implanted into Si and SiO_2
(11; 22.1.2; 22.6)
(a) Vol. **83**, No. 1, p. 165—171 (1984).

V. L. INDENBOM, V. M. KAGANER, W. MÖHLING, and E. V. SUVOROV
Plane-Wave X-Ray Images of Dislocations Parallel to the Diffraction Vector
(1.4 and 10.2; 22.1.2)
(a) Vol. **83**, No. 1, p. 195—205 (1984).

K. YASUTAKE, M. UMENO, and H. KAWABE
Oxygen Precipitation and Microdefects in Czochralski-Grown Silicon Crystals
(10.2; 22.1.2)
(a) Vol. **83**, No. 1, p. 207—217 (1984).

F. S. NASREDINOV, A. A. ANDREEV, A. N. KURMANTAEV, A. R. REGEL, and P. P. SEREGIN
Tin Impurity Atoms in Vacuum Deposited Amorphous Silicon
(13.4 and 14.3.1; 2; 20.1; 22.1.2)
(a) Vol. **83**, No. 1, p. 285—289 (1984).

I. I. KOLKOVSKII, P. F. LUGAKOV, and V. V. SHUSHA
Charge-Carrier Recombination in Silicon Irradiated with γ-Rays of Different Energies
(11 and 13.4; 22.1.2)
(a) Vol. **83**, No. 1, p. 299—306 (1984).

G. KADEN
Consideration of Surface Band Bending in Silicon in Case of Internal Photoemission in MOS Structures
(14.3.3 and 14.3.4; 13.3; 22.1.2)
(a) Vol. **83**, No. 1, p. 375—386 (1984).

P. MIALHE, A. KHOURY, and J. P. CHARLES
A Review of Techniques to Determine the Series Resistance of Solar Cells
(14.3.3 and 16; 22.1.2)
(a) Vol. **83**, No. 1, p. 403—409 (1984).

S. Kashchieva, P. Danesh, and A. Dyakov
RF Annealing of the Implantation-Induced Defects in Silicon Using Hydrogen Plasma
(11 and 13.4; 14.3.3; 22.1.2)
(a) Vol. **83**, No. 1, p. 411—417 (1984).

M. Wagner, H.-D. Geiler, G. Andrä, and G. Götz
Control of Explosive Liquid-Phase Crystallization of Ion-Implanted Silicon Layers by Double Pulse Laser Irradiation
(3 and 11; 22.1.2)
(a) Vol. **83**, No. 1, p. K1—K3 (1984).

A. M. Afanasev, R. M. Imamov, and E. Kh. Mukhamedzhanov
The Yield of Photoelectrons of Different Energies in the X-Ray Laue Diffraction
(1.6; 17; 22.1.2)
(a) Vol. **83**, No. 1, p. K5—K9 (1984).

S. A. Azimov, P. M. Karageorgy-Alkalaev, A. Yu. Leiderman, and V. S. Rubin
Computer Simulation of Generation–Recombination Currents in Amorphous Silicon p–n Diode Structures
(2 and 14.3; 14.3.4; 22.1.2)
(a) Vol. **83**, No. 1, p. K81—K86 (1984).

N. M. Ravindra, J. F. Morhange, P. A. Thomas, and C. Ance
Polarizability Studies of Amorphous Silicon
(2 and 14.4; 22.1.2)
(b) Vol. **123**, No. 2, p. 611—618 (1984).

A. Turos, H. Frey, O. Meyer, W. Müller, and J. M. Pirrung
Ion Beam Analysis of Amorphous Silicon Films Produced by Magnetron Sputtering
(2 and 14.3.1; 3; 22.1.2)
(a) Vol. **83**, No. 2, p. 437—443 (1984).

Yu. A. Aleksandrov, B. Chalupa, J. Kulda, T. A. Machekhina, R. Michalec, P. Mikula, L. N. Sedláková, J. Vávra, and M. Vrána
Backscattering Diffraction of a Pulsed Neutron Beam on an Elastically Bent Single Crystal
(1.1; 22.1.2)
(a) Vol. **83**, No. 2, p. 455—460 (1984).

P. F. Lugakov and V. V. Lukyanitsa
Efficiency of Radiation Defect Formation in Silicon at Various Intensities of Electron and γ-Ray Irradiation
(11; 13.4; 22.1.2)
(a) Vol. **83**, No. 2, p. 521—528 (1984).

H. Bartsch, D. Hoehl, and G. Kästner
Radiation-Induced Rod-Like Defects in Silicon and Germanium
(1.4 and 11; 22.1.1; 22.1.2)
(a) Vol. **83**, No. 2, p. 543—551 (1984).

V. N. SOLOVEV
Atomic Particle Delocalisation Effect in Disordered Media
(2 and 9; 22.1.2)
(a) Vol. **83**, No. 2, p. 553—559 (1984).

V. G. GRYGORIEV, F. G. NESHOV, A. A. PUZANOV, and A. R. URMANOV
Effects of Electronic Structure on Ion Stopping Cross-Section in Solids
(11; 22.1.2)
(a) Vol. **83**, No. 2, p. 573—580 (1984).

M. VANĚČEK, A. ABRAHÁM, O. ŠTIKA, J. STUCHLÍK, and J. KOČKA
Gap States Density in a-Si:H Deduced from Subgap Optical Absorption Measurement on Schottky Solar Cells
(13.1 and 20.1; 2; 22.1.2)
(a) Vol. **83**, No. 2, p. 617—623 (1984).

H. LEMKE
Properties of Substitutional Manganese in Silicon
(13.4; 14.3; 22.1.2)
(a) Vol. **83**, No. 2, p. 637—643 (1984).

P. C. DHANASEKARAN, P. R. VAYA, and B. S. V. GOPALAM
Spectral Sensitivity Calculations on n^+–p and n^+–p–p^+ Silicon Solar Cells
(14.3.3 and 16; 22.1.2)
(a) Vol. **83**, No. 2, p. 693—700 (1984).

P. J. REUTERS, L. DOTHANH, and P. BALK
Determination of Carrier Type in MIS Structures with Pyrolytic AlN Insulator Films
(14.3.3; 14.4; 22; 22.1.2)
(a) Vol. **83**, No. 2, p. 719—727 (1984).

K. OTS and M. SEILENTHAL
On the Structural Model of CVD a-Si:O:H Films
(1.5 and 2; 20.1; 22.1.2; 22.6)
(a) Vol. **83**, No. 2, p. K109—K111 (1984).

A. M. AFANASEV, R. M. IMAMOV, A. V. MASLOV, and E. M. PASHAEV
Photoeffect in X-Ray Grazing Incidence Diffraction
(1 and 17; 22.1.2)
(a) Vol. **84**, No. 1, p. 73—78 (1984).

G. KOWALSKI
X-Ray Diffraction Contrast and Ray-Paths in the Case of a Dislocation in a Particular Position
(10.2; 22.1.2)
(a) Vol. **84**, No. 1, p. 119—132 (1984).

L. PASEMANN
Theoretical Study of the Temperature Dependence of EBIC Contrast from Individual, Surface-Parallel Dislocations in a Schottky Diode
(1.4 and 10.2; 14.3.3; 22.1.2)
(a) Vol. **84**, No. 1, p. 133—142 (1984).

E. BUGIEL, M. KITTLER, A. BORCHARDT, and H. RICHTER
On the Gettering Efficiency of Crystal Defects in Silicon
(10.2; 14.3.3; 22.1.2)
(a) Vol. **84**, No. 1, p. 143—147 (1984).

V. V. KVEDER, R. LABUSCH, and YU. A. OSSIPYAN
The Exodiffusion of Hydrogen in Dislocated Crystalline Silicon
(10.2 and 13.4; 22.1.2)
(a) Vol. **84**, No. 1, p. 149—156 (1984).

A. V. DVURECHENSKII, R. GRÖTZSCHEL, N. M. IGONINA, B. A. KOVAL, and N. I. LEBEDEVA
Diffusion of Impurities in Undercooled Melt of Pulse Heated Ion-Implanted Silicon
(9; 11; 22.1.2)
(a) Vol. **84**, No. 1, p. 171—177 (1984).

V. V. BORIMSKII, K. D. GLINCHUK, N. M. LITOVCHENKO, and Z. A. SALNIK
New Oxygen-Induced Recombination Centres in 600 to 800 °C Heat-Treated Silicon
(13.4; 22.1.2)
(a) Vol. **84**, No. 1, p. 237—241 (1984).

V. M. BABICH, N. P. BARAN, V. L. BORBLIK, YU. P. DOTSENKO, V. B. KOVALCHUK, and V. A. SHERSHEL
Specific Features of Piezogalvanomagnetic Effects in Oxygen-Containing Silicon Crystals in the Prescence of Thermal Donors-I and -II
(13.4; 14.3; 22.1.2)
(a) Vol. **84**, No. 1, p. 263—268 (1984).

W. FRENTRUP, M. GRIEPENTROG, H. KLOSE, G. KREYSCH, and U. MÜLLER-JAHREIS
The Influence of Alkali Atoms Implanted in Silicon on the Negative Secondary Ion Emission
(17; 13.3; 22.1.2)
(a) Vol. **84**, No. 1, p. 269—272 (1984).

D. P. JOSHI and R. S. SRIVASTAVA
Effective Diffusion Length of Minority Carriers in Polycrystalline Silicon
(14.3.3 and 16; 1.4; 22.1.2)
(a) Vol. **84**, No. 1, p. 311—317 (1984).

V. V. ARISTOV, I. E. BONDARENKO, N. N. DRYOMOVA, V. V. KAZMIRUK, and E. B. YAKIMOV
The EBIC Contrast of Dislocation Slip Planes in Silicon
(1.4 and 10.2; 14.3.4; 22.1.2)
(a) Vol. **84**, No. 1, p. K43—K45 (1984).

K. WINER and F. WOOTEN
Structure and Vibrational Spectra of a Model of a-Si:H with Periodic Boundary Conditions
(2 and 6; 22.1.2)
(b) Vol. **124**, No. 2, p. 473—480 (1984).

A. M. TARATIN and S. A. VOROBEV
Multiple Scattering of Relativistic Electrons under Axial Channeling in a Crystal. Model of the Lattice of the Atomic Rows
(11; 13; 22.1.2)
(b) Vol. **124**, No. 2, p. 641—648 (1984).

A. Stern and K. Hübner
Electron Density of States of an $Si–SiO_2$ Interface with a Gradual Chemical Transition
(13.3; 22.1.2; 22.6)
(b) Vol. **124**, No. 2, p. K171—K175 (1984).

S. Kh. Milshtein, D. C. Joy, S. D. Ferris, and L. C. Kimerling
Defect Characterization Using SEM–CCM. Relative Contrast Measurements
(1.4 and 10.2; 14.3.3; 22.1.2)
(a) Vol. **84**, No. 2, p. 363—369 (1984).

E. B. Yakimov, N. A. Yarykin, and V. I. Nikitenko
The Influence of Thermal Treatment on Thermally Stimulated Depolarization Spectrum of the Electret State in Plastically Deformed Silicon
(10.2 and 13.4; 14.4; 22.1.2)
(a) Vol. **84**, No. 2, p. 443—450 (1984).

P. F. Lugakov and V. V. Lukyanitsa
Peculiarities of Radiation Defect Formation and Annealing in n-Si Due to Their Interaction with Each Other and Defect Clusters
(10.2 and 11; 14.3; 22.1.2)
(a) Vol. **84**, No. 2, p. 457—464 (1984).

P. Grosbras, J. L. Demenet, H. Garem, and J. C. Desoyer
Static Equilibrium between Partial Dislocations in Silicon Deformed by Uniaxial Stress
(10.2; 22.1.2)
(a) Vol. **84**, No. 2, p. 481—492 (1984).

S. Aleksandrova, A. Szekeres, and L. Kassamakova
RF Plasma Induced Lateral Nonuniformities in $Si–SiO_2$ Structures
(13.3 and 14.3.2; 22.1.2)
(a) Vol. **84**, No. 2, p. 561—565 (1984).

M. Aslam, R. Singh, and P. Balk
Nature of Electron and Hole Traps in MOS Systems with Poly-Si Electrode
(14.3.3; 13.4; 22.1.2)
(a) Vol. **84**, No. 2, p. 659—668 (1984).

M. A. Chubisov and A. F. Akkerman
Simulation of Z_1-Oscillations in Low-Energy Heavy Ion Ranges in Solids
(11; 22.1.2)
(b) Vol. **125**, No. 1, p. 169—175 (1984).

D. Thébault, J. Barrau, G. Armelles, N. Lauret, and J. P. Noguier
The No-Phonon Luminescence at 0.793 eV from Si:Au. Uniaxial Stress and Zeeman Spectroscopy Experiments
(13.4 and 20.3; 22.1.2)
(b) Vol. **125**, No. 1, p. 357—364 (1984).

A. V. Mudryi, A. L. Pushkarchuk, V. D. Tkachev, and A. G. Ulyashin
Noble Gas Atoms as Interstitials in Silicon and Diamond
(10.2 and 13.4; 22.1; 22.1.2)
(b) Vol. **125**, No. 1, p. K75—K78 (1984).

J. Kočka and K. Čermák
Drift Mobility of a-Si:H
(2 and 14.3.1; 14.3.4; 22.1.2)
(b) Vol. **125**, No. 1, p. K89—K91 (1984).

D. M. Vanderwalker
A Dark-Field TEM Method for Crystal Structure Determination of Cu-Rich Phases on Twins in Silicon
(1.4; 22.1.2)
(a) Vol. **85**, No. 1, p. 61—67 (1984).

Z. Janáček
Coherence Properties of Radiation Diffracted by an Elastically Bent Crystal
(1.1; 22.1.2)
(a) Vol. **85**, No. 1, p. 77—82 (1984).

P. Gaworzewski, E. Hild, F.-G. Kirscht, and L. Vecsernyés
Infrared Spectroscopical and TEM Investigations of Oxygen Precipitation in Silicon Crystals with Medium and High Oxygen Concentrations
(1.4 and 10.2; 20.1; 22.1.2)
(a) Vol. **85**, No. 1, p. 133—147 (1984).

E. H. Sin, C. K. Ong, and H. S. Tan
Temperature Dependence of Interband Optical Absorption of Silicon at 1152, 1064, 750, and 694 nm
(20.1; 22.1.2)
(a) Vol. **85**, No. 1, p. 199—204 (1984).

S. V. Vintsents, V. F. Kiselev, and G. S. Plotnikov
Energy Transfer between Excited Adsorbed Dye Molecules and Charged Defects in Insulator–Semiconductor Structures
(13.4; 19; 20.3; 22.1.1; 22.1.2)
(a) Vol. **85**, No. 1, p. 273—281 (1984).

D. S. Misra, A. Kumar, and S. C. Agarwal
Intensity and Temperature Dependence of Steady-State Photoconductivity Down to 8 K and DOS Distribution Obtained from These Measurements in a-Si:H
(16; 2; 22.1.2)
(a) Vol. **85**, No. 1, p. 297—303 (1984).

A. D. Pogrebnyak, Sh. R. Mastov, M. F. Kuznetsov, K. P. Arefiev, S. V. Rakitin, and S. A. Vorobiev
Observation of RF Pulses from Solids during Laser Irradiation
(11 and 20; 22.1.2; 22.2.1)
(a) Vol. **85**, No. 1, p. K31—K33 (1984).

D. Hildebrandt, H. Strusny, R. Groetzschel, E. Kotal, and F. Paszti
Damage and Trapping Behaviour of Crystalline Silicon at Low Energy Deuterium Implantation
(10.2 and 11; 22.1.2)
(a) Vol. **85**, No. 1, p. K35—K38 (1984).

A. V. Dvurechenskii, B. P. Kashnikov, G. P. Pokhil, V. P. Popov, A. V. Tulinov, and A. A. Turinge
Defect Structure Study with Planar Channeling in Pulse-Annealed Ion-Implanted Silicon
(10.2 and 11; 22.1.2)
(a) Vol. **85**, No. 1, p. K39—K42 (1984).

V. D. Tkachev, G. Hölzer, and A. R. Chelyadinskii
Damage Profiles in Ion Implanted Silicon
(10.2 and 11; 22.1.2)
(a) Vol. **85**, No. 1, p. K43—K46 (1984).

V. I. Gatalskaya, E. M. Gololobov, and L. A. Kurochkin
Relaxation Time and Electron Scattering Cross Section in Irradiated p-Silicon at Temperatures of Liquid Helium from Cyclotron Resonance Data
(11; 19; 22.1.2)
(a) Vol. **85**, No. 1, p. K51—K55 (1984).

T. Soma, Y. Kitani, and H.-Matsuo Kagaya
Mode Grüneisen Parameter and Pressure-Induced Phase Transition of Si–Ge Solid Solution
(1.2 and 6; 8; 22.1.1; 22.1.2)
(b) Vol. **125**, No. 2, p. K99—K101 (1984).

P. F. Lugakov and T. A. Lukashevich
Formation and Parameters of Boron–Divacancy Complexes in Irradiated p-Si
(11; 13.4; 22.1.2)
(a) Vol. **85**, No. 2, p. 441—444 (1984).

M. Suezawa and K. Sumino
On the Annihilation of Thermal Donors in Silicon Crystals
(10.2 and 13.4; 20.1; 22.1.2)
(a) Vol. **85**, No. 2, p. 469—472 (1984).

V. V. Emtsev, Yu. N. Daluda, P. Gaworzewski, and K. Schmalz
Electrical Properties of Thermal Donors Formed in CZ-Si during Heat Treatment at 450 °C
(13.4; 14.3; 20.1; 22.1.2)
(a) Vol. **85**, No. 2, p. 575—584 (1984).

V. I. Gubskaya, P. V. Kuchinskii, and V. M. Lomako
The Effect of the Vacancy Charge State on the Radiation Defect Formation in Silicon
(11 and 13.4; 22.1.2)
(a) Vol. **85**, No. 2, p. 585—590 (1984).

B. P. Tyagi and K. Sen
Grain Size and Temperature Dependence of the Electrical Behaviour of Polysilicon
(14.3.4; 13.4; 22.1.2)
(a) Vol. **85**, No. 2, p. 603—608 (1984).

G. Micocci, A. Rizzo, and A. Tepore
Photoconductivity in Halogenated and Hydrogenated Amorphous Silicon Films
(16; 2; 22.1.2)
(a) Vol. **85**, No. 2, p. 609—614 (1984).

S. Nešpůrek, O. Zmeškal, and F. Schauer
An Ohmic Contact Test Using the Activation Energy of Steady-State Space-Charge-Limited Current
(14.3.4; 13.4; 22.1.2; 22.4.2)
(a) Vol. **85**, No. 2, p. 619—625 (1984).

J. L. Lindström, B. G. Svensson, J. W. Corbett, and G. S. Oehrlein
On the Complex of the Oxygen Interstitial and the Silicon Interstitial in Silicon
(10.2; 22.1.2)
(a) Vol. **85**, No. 2, p. K109—K111 (1984).

H. Lemke
Properties of Energy Levels of Cobalt in Silicon
(13.4; 14.3; 22.1.2)
(a) Vol. **85**, No. 2, p. K133—K136 (1984).

H. Lemke
Properties of Energy Levels of Rhodium and Iridium in Silicon
(13.4; 14.3; 22.1.2)
(a) Vol. **85**, No. 2, p. K137—K140 (1984).

A. P. Mayer and R. K. Wehner
Calculation of Grüneisen Constants in Si
(6; 22.1.2)
(b) Vol. **126**, No. 1, p. 91—103 (1984).

J. Estel, J. Kalus, and L. Pintschovius
Measurement of Phonon Energies in Silicon under Uniaxial Stress
(6; 22.1.2)
(b) Vol. **126**, No. 1, p. 121—124 (1984).

D. Della Sala, C. Giovannella, and F. Evangelisti
Study of Bonding Configurations in Amorphous Ge_xSi_{1-x}:H Alloys
(2; 4; 20.1; 22.1.1; 22.1.2)
(b) Vol. **126**, No. 1, p. 125—131 (1984).

K. Wesser, S. N. Sahu, J. P. Karins, and J. W. Corbett
Comparison of an Extended Interstitial with a Split Interstitial in Silicon
(10.2 and 13.4; 22.1.2)
(b) Vol. **126**, No. 1, p. 157—161 (1984).

N. M. Khamidullina
Calculation of Some Characteristics of an Intrinsic Random Field in Hydrogenated Amorphous Silicon
(2; 20.1; 22.1.2)
(b) Vol. **126**, No. 1, p. 253—260 (1984).

A. SH. MAKHMUDOV, Z. M. KHAKIMOV, A. A. LEVIN, M. S. YUNUSOV, and P. K. KHABIBULLAEV
Green's Function Method in the Equivalent Orbital Basis. Point Defects and Complexes of Two Point Defects in Silicon
(13.1 and 13.4; 22.1.2)
(b) Vol. **126**, No. 1, p. 311—321 (1984).

P. A. ALEKSANDROV, A. M. AFANASIEV, and S. A. STEPANOV
Bragg-Laue Diffraction in Inclined Geometry
(1.1; 1.6; 22.1.2)
(a) Vol. **86**, No. 1, p. 143—154 (1984).

V. V. KALININ and N. N. GERASIMENKO
On the Effect of Doping Impurities on the Formation of Rod-Like Defects in Silicon
(10.2 and 11; 22.1.2)
(a) Vol. **86**, No. 1, p. 185—190 (1984).

A. D. POGREBNYAK
Positron and Positronium States in Semiconductors Irradiated by Supercurrent Beams of Charged Particles
(11; 22.1.2; 22.2.1)
(a) Vol. **86**, No. 1, p. 191—198 (1984).

H. BENDER
Investigation of the Oxygen-Related Lattice Defects in Czochralski Silicon by Means of Electron Microscopy Techniques
(10.2; 1.4; 22.1.2)
(a) Vol. **86**, No. 1, p. 245—261 (1984).

P. F. LUGAKOV and V. V. SHUSHA
A Model of Nonequilibrium Charge-Carrier Recombination in Semiconductors Containing Non-uniformities
(14.3.4; 13.4; 22.1.2)
(a) Vol. **86**, No. 1, p. 309—312 (1984).

A. V. DVURECHENSKII, B. P. KASHENIKOV, and V. V. SUPRUNCHIK
New EPR Defects in Si⟨Al⟩
(13.4 and 19; 11; 22.1.2)
(a) Vol. **86**, No. 1, p. 313—318 (1984).

I. P. ZVYAGIN
Planar Conduction in Inhomogeneous Films. Application to Amorphous Silicon
(2 and 14.3.2; 22.1.2)
(a) Vol. **86**, No. 1, p. 375—381 (1984).

B. T. KOLOMIETS, V. M. LYUBIN, M. A. TAGUYRDZHANOV, and G. ZENTAI
A Comparative Study of Photoconductivity in Vitreous As_2Se_3 and a-Si:H Thin Films
(16; 2; 14.3.1; 22; 22.1.2)
(a) Vol. **86**, No. 1, p. 383—387 (1984).

W. TORBICZ, D. SOBCZYNSKA, A. OLSZYNA, G. FORTUNATO, and A. D'AMICO
A Stable Pd–BN–Si_3N_4–SiO_2–Si FET for Hydrogen Detection
(14.3.3; 22.1.2)
(a) Vol. **86**, No. 1, p. 453—459 (1984).

H. LEMKE
On the Possible Existence of "Repulsive" Centres with Large Capture Cross Section in Silicon
(13.4; 14.3; 22.1.2)
(a) Vol. **86**, No. 1, p. K39—K43 (1984).

P. I. BARANSKII, T. G. GRISHCHENKO, V. V. SAVYAK, and YU. V. SIMONENKO
Temperature Dependence of the Thermoelectric Characteristics of n-Si Elastically Strained in the [100] Direction
(14.3 and 15; 22.1.2)
(a) Vol. **86**, No. 1, p. K63—K67 (1984).

TRAN CHOT
Effect of Vacuum Annealing on Electrical Characteristics of Ni–Si Schottky Diodes
(14.3.4; 22.1.2)
(a) Vol. **86**, No. 1, p. K73—K78 (1984).

B. HOLLAND, H. S. GREENSIDE, and M. SCHLÜTER
Cohesive Properties of Ge, Si, and Diamond Calculated with Minimum Basis Sets
(4 and 13; 12.2; 22.1; 22.1.1; 22.1.2)
(b) Vol. **126**, No. 2, p. 511—515 (1984).

V. V. KAPLIN, S. B. NURMAGAMBETOV, V. I. GRIDNEV, E. I. ROZUM, S. PAK, S. R. UGLOV, and S. A. VOROBIEV
Angular Distributions of Fast Electrons at Planar Channeling in Silicon Crystals
(11; 13; 22.1.2)
(b) Vol. **126**, No. 2, p. 565—574 (1984).

U. PIETSCH
Simulation of Dynamical Bond Charge Transfer Properties in Elemental Semiconductors by Means of a Simple Quantum-Mechanical Model
(4; 13; 22.1.2)
(b) Vol. **126**, No. 2, p. 595—605 (1984).

C. TUNCAY and M. TOMAK
Ideal Vacancy States in Si. A Study of the Empirical Tight-Binding Method
(13.4; 22.1.2)
(b) Vol. **126**, No. 2, p. 677—685 (1984).

JUH TZENG LUE
Time-Resolved Reflectivity Measurements on RF Sputtered a-Si:H Alloy
(13.4 and 20.1; 2; 16; 22.1.2)
(b) Vol. **126**, No. 2, p. 713—720 (1984).

D. M. VANDERWALKER
The Nucleation of Ni Silicides on Dislocations and Twins in Silicon
(1.6 and 10.2; 1.2; 21.1; 22.1.2)
(a) Vol. **86**, No. 2, p. 507—516 (1984).

V. V. BORIMSKII, K. D. GLINCHUK, N. M. LITOVCHENKO, and Z. A. SALNIK
Influence of Carbon and Preannealing on the Formation of Oxygen-Induced Recombination Centres in Heat-Treated Silicon
(13.4; 22.1.2)
(a) Vol. **86**, No. 2, p. 623—627 (1984).

V. M. BABICH, N. P. BARAN, A. A. BUGAI, YU. P. DOTSENKO, V. B. KOVALCHUK, and V. A. SHERSHEL
On the Properties of Thermodonors-II in Cz-Si Crystals of High Carbon Content
(13.4; 14.3; 19; 22.1.2)
(a) Vol. **86**, No. 2, p. 679—683 (1984).

S. B. SEVASTIANOV, N. N. GERASIMENKO, and N. V. VERSHININA
Inhomogeneities in Irradiated SiO_2–Si Structures
(11 and 14.3.3; 14.3.4; 22.1.2)
(a) Vol. **86**, No. 2, p. 717—727 (1984).

B. R. NAG, S. R. AHMED, and M. DEB ROY
Submicron Electron Transport in Silicon at 300 and 77 K
(14.3.4; 6; 13.1; 22.1.2)
(a) Vol. **86**, No. 2, p. 773—779 (1984).

M. VOELSKOW, J. MATTHÄI, and R. KLABES
Electrical Properties of Ion Implanted and Short Time Annealed Polycrystalline Silicon
(11 and 14.3.1; 22.1.2)
(a) Vol. **86**, No. 2, p. 781—788 (1984).

H. TERLETZKI, A. BODEN, F. WULF, and W. R. FAHRNER
Laterally Inhomogeneous Charge Build-Up in CMOS Inverters during Ionizing Irradiation
(11 and 14.3.3; 14.3.2; 22.1.2)
(a) Vol. **86**, No. 2, p. 789—794 (1984).

V. M. BABICH, YU. P. DOTSENKO, and V. B. KOVALCHUK
Some Characteristic Properties of Thermal Donor Formation in Oxygen-Containing Silicon at 450 °C
(10.2; 22.1.2)
(a) Vol. **86**, No. 2, p. K91—K94 (1984).

W. ZAGOZDZON-WOSIK, E. MIZERA, and W. KUŻMICZ
Emitter Efficiency of Si Bipolar Transistors — Structural and Electrical Investigations
(14.3.3; 22.1.2)
(a) Vol. **86**, No. 2, p. K125—K128 (1984).

22.1.3 Group VI Elements

Original Papers and Short Notes

A. I. ISAEV, T. M. GUSEINOV, S. D. MEKHTIEVA, and D. SH. ABDINOV
Negative Photoconductivity in Selenium Single Crystals
(16; 22.1.3)
(a) Vol. **82**, No. 1, p. K87—K90 (1984).

R. M. MEHRA, HEMANT KUMAR, SURINDER KOUL, and P. C. MATHUR
Thickness Dependence of DC Conductivity of Amorphous Se and Binary Amorphous Se–Te, Se–Ge, and Se–Sb Films
(14.3.1; 2; 22.1.3)
(a) Vol. **83**, No. 1, p. 341—347 (1984).

G. FLEURY, A. HAMOU, C. LHERMITTE, and C. VIGER
Bismuth Influence on a-Se Film Crystallization
(2 and 14.3.1; 22.1.3)
(a) Vol. **83**, No. 2, p. K103—K108 (1984).

G. I. IBAEV, S. I. KULIEVA, and T. G. MAGERRAMOV
Frequency Dependences of the Dynamic Current–Voltage Characteristic of the Se–CdSe Heterostructure Based on a Single-Crystal Selenium Film
(14.3.1 and 14.3.3; 22.1.3; 22.4.2)
(a) Vol. **84**, No. 2, p. K185—K188 (1984).

D. E. POPOV, E. I. ROZUM, S. R. UGLOV, and S. A. VOROBIEV
Anomalous Radiation of Fast Channeled Electrons in Optically Transparent Crystals
(11 and 20; 13; 22.1.3; 22.5.2)
(b) Vol. **125**, No. 1, p. 251—258 (1984).

P. MICHON, H. ATMANI, and C. VAUTIER
Crystallization Study of Evaporated a-Se Films by Means of DTA Technique
(1.5 and 3; 2; 22.1.3)
(a) Vol. **85**, No. 2, p. 399—407 (1984).

G. PARTHASARATHY and E. S. R. GOPAL
Unusual Effect of Tellurium Doping on the Electrical Properties of Bulk Amorphous Selenium
(2 and 14.3; 22.1.3)
(a) Vol. **85**, No. 2, p. K165—K167 (1984).

P. KOŠTIAL
The Influence of Heating on the Attenuation of Surface Waves in Electrographic Se Layer
(1.5 and 7; 2; 22.1.3)
(a) Vol. **86**, No. 1, p. K7—K10 (1984).

22.2 III–V Compounds

Review Article

A. Blacha, H. Presting, and M. Cardona
Deformation Potentials of $k = 0$ States of Tetrahedral Semiconductors
(6 and 13.1; 22.1; 22.2; 22.4)
(b) Vol. **126**, No. 1, p. 11—36 (1984).

Original Papers and Short Notes

J. Walachová and D. Nohavica
Erratum to: Determination of p–n Junctions in InP/GaInAsP Laser Heterostructures Using a Profilometer
(14.3.3; 22.2)
(a) Vol. **81**, No. 1, p. K91 (1984).

S. N. Sahu, J. T. Borenstein, V. A. Singh, and J. W. Corbett
Semi-Empirical Tight Binding Calculations for the Energy Bands of the Diamond and Zincblende Type Semiconductors
(13.1; 22.1; 22.2)
(b) Vol. **122**, No. 2, p. 661—667 (1984).

K. Löhnert and E. Kubalek
The Cathodoluminescence Contrast Formation of Localized Non-Radiative Defects in Semiconductors
(10.2 and 20.3; 13.4; 22.2)
(a) Vol. **83**, No. 1, p. 307—314 (1984).

H.-Matsuo Kagaya and T. Soma
Lattice Dynamics of Tetrahedral Compounds with the Local Heine-Abarenkov Model Potential
(6; 22.2; 22.4)
(b) Vol. **124**, No. 1, p. 37—44 (1984).

P. K. Sen
Nonlinear Absorption in III–V Semiconductors
(13.5.1 and 20.1; 22.2)
(b) Vol. **124**, No. 1, p. 117—125 (1984).

C. Prasad and J. D. Dubey
Electronic Structure and Properties of Cubic Boron Nitride
(13.1; 22.2)
(b) Vol. **125**, No. 2, p. 629—638 (1984).

I. S. Smirnova
Dislocations in $A^{II}B^{VI}$ and $A^{III}B^{V}$ Piezoelectric Crystals
(10.2 and 14.3; 22.2; 22.4)
(b) Vol. **126**, No. 1, p. 177—189 (1984).

F. DUJARDIN, B. STÉBÉ, and G. MUNSCHY
Neutral Bound Excitons in a Low Magnetic Field
(13.4 and 13.5.1; 22.2; 22.4)
(b) Vol. **126**, No. 1, p. 329—334 (1984).

D. VIGNAUD and J. L. FARVACQUE
Free-Carrier Optical Absorption Induced by Dislocation Scattering Mechanisms in III–V Compounds
(13.1 and 20.1; 10.2; 22.2)
(b) Vol. **126**, No. 1, p. 351—359 (1984).

H.-MATSUO KAGAYA and T. SOMA
Dynamical Elastic Constants of Covalent Compounds
(12.2; 6; 22.2; 22.4)
(b) Vol. **126**, No. 1, p. K9—K14 (1984).

I. TALE and J. ROSA
Fractional Glow Technique Spectroscopy of Traps in Heavily Doped AlN:O
(13.4; 20.1; 20.3; 22.2)
(a) Vol. **86**, No. 1, p. 319—326 (1984).

V. E. BORISENKO, V. V. GRIBKOVSKII, V. A. LABUNOV, and S. G. YUDIN
Pulsed Heating of Semiconductors
(8; 11; 22.1; 22.2)
(a) Vol. **86**, No. 2, p. 573—583 (1984).

22.2.1 Arsenides

Original Papers and Short Notes

T. Soma and H.-M. Kagaya
NaCl-Type Lattice of GaAs and InSb under Pressure
(1.2; 22.2.1; 22.2.3)
(b) Vol. **121**, No. 1, p. K1—K5 (1984).

A. D. Pogrebnyak, E. Yu. Boyarko, Yu. Yu. Kryuchkov, and S. A. Vorobev
Positron Annihilation and Profiles of Radiation Damages in GaAs and Si Crystals Irradiated by Supercurrent Proton or Electron Beams
(11; 14.3; 22.1.2; 22.2.1)
(a) Vol. **81**, No. 1, p. 217—225 (1984).

M. Rösler, R. Zimmermann, and W. Richert
The Electron–Hole Plasma at Finite Temperatures. Calculation of Exchange-Correlation Energies and Padé Approximants
(13.5; 22.1.1; 22.2.1)
(b) Vol. **121**, No. 2, p. 609—621 (1984).

C. Patel, T. J. Parker, H. Jamshidi, and W. F. Sherman
Phonon Frequencies in GaAs
(6 and 20.1; 22.2.1)
(b) Vol. **122**, No. 2, p. 461—467 (1984).

K. Bhattacharyya and P. K. Basu
Variational Calculation of Ground Subband Energy of Two-Dimensional Electron Gas in $p–n^+–p$ Structure
(13.1 and 14.3.3; 22.2.1)
(b) Vol. **122**, No. 2, p. K191—K194 (1984).

V. I. Osinskii, F. M. Katsapov, and E. A. Tyavlovskaya
Structural Perfection of Selective GaAs Regions in Si Substrate Windows
(1.4 and 1.6; 22.2.1)
(a) Vol. **82**, No. 2, p. 399—403 (1984).

K. D. Glinchuk, A. V. Prokhorovich, and N. S. Zayats
A New (Non-Copper-Induced) 1.35 eV Emission Band in n-Type GaAs. Origin and Characteristics
(13.4 and 20.3; 22.2.1)
(a) Vol. **82**, No. 2, p. 503—510 (1984).

V. N. Brudnyi, M. A. Krivov, A. I. Potapov, and Yu. V. Rud
Electrical Properties of Electron- and Proton-Irradiated GaAs and $ZnGeAs_2$ Solid Solutions
(11 and 14.3; 22.2.1; 22.8)
(a) Vol. **82**, No. 2, p. K191—K194 (1984).

F. Thuselt
Low-Density Approach to RPA Self-Energies
(13.5.2; 22.1.1; 22.1.2; 22.2.1; 22.2.2)
(b) Vol. **123**, No. 1, p. 257—269 (1984).

Yu. V. Vorobev, E. I. Tolpygo, and M. K. Sheinkman
The Auger Process with Energy Transfer to the Bound Charge Carriers in Multi-Valley Semiconductors
(13.4 and 14.3; 22.2.1)
(b) Vol. **123**, No. 1, p. 295—308 (1984).

M. Załużny
On the Influence of the Depolarization Field on the Intersubband Absorption in Thin Semiconductor Films
(13.1 and 20.1; 22; 22.2.1)
(b) Vol. **123**, No. 1, p. K57—K61 (1984).

E. Ejimanya and D. V. Morgan
Damage Induced Degradation of Metal Contacts on Gallium Arsenide
(14.3.3; 11; 22.2.1)
(a) Vol. **83**, No. 1, p. K87—K92 (1984).

L. Wendler
Phonon-Polaritons in Bilayer Systems. p-Polarized Guided Wave Phonon-Polaritons and Guided Wave Surface Phonon-Polaritons
(6 and 20.1; 13.5; 22.2.1; 22.2.2; 22.5.3)
(b) Vol. **123**, No. 2, p. 469—477 (1984).

A. Golzené, B. Meyer, and C. Schwab
Experimental Evidence for an Associated Defect Model for the Neutron Generated As_{Ga} Center in Gallium Arsenide
(11 and 19; 13.4; 22.2.1)
(b) Vol. **123**, No. 2, p. K125—K128 (1984).

K. D. Glinchuk, A. V. Prokhorovich, and N. S. Zayats
Effect of 2.2 MeV Electron Irradiation and Annealing on the 1.37 eV Emission Band in Zinc-Doped p-Type GaAs
(13.4 and 20.3; 11; 22.2.1)
(a) Vol. **83**, No. 2, p. 625—630 (1984).

N. G. Filonov, N. K. Maksimova, A. P. Vyatkin, I. D. Romanova, and A. M. Mysik
Transport Mechanism in GaAs Schottky Diodes. Deep Centres Effects
(14.3.3; 13.4; 22.2.1)
(a) Vol. **83**, No. 2, p. 701—708 (1984).

H. Stolz and R. Zimmermann
Screening in a System on Many Particles Composed of Two Charged Fermions. Applications to Excitons
(13.5; 13.5.1; 22.2.1)
(b) Vol. **124**, No. 1, p. 201—212 (1984).

N. V. Baramidze, V. L. Bonch-Bruevich, and N. I. Kurdiani
Optical Absorption in Neutron Irradiated GaAs
(13.4 and 20.1; 11; 22.2.1)
(b) Vol. **124**, No. 1, p. 241—247 (1984).

T. Soma, Y. Takahashi, and H.-Matsuo Kagaya
Equation of State and Pressure Derivative of the Bulk Modulus for III–V Covalent Crystals
(1.2 and 4; 12.2; 22.2.1; 22.2.3)
(b) Vol. **124**, No. 1, p. K11—K14 (1984).

P. Středa
Critical Current of the Quantum Hall Regime
(14.3.2 and 14.3.4; 22.2.1)
(b) Vol. **124**, No. 1, p. K97—K99 (1984).

S. K. Maksimov, M. Ziegler, I. I. Khodos, I. I. Snighiryova, and M. Sh. Shikhsaidov
High Resolution Electron Microscopic Investigations of Dislocations in Deformed GaAs Single Crystals Doped with Te
(10.2; 1.4; 22.2.1)
(a) Vol. **84**, No. 1, p. 79—86 (1984).

W. Ulrici
Evaluation of Chromium Concentration in Semi-Insulating GaAs:Cr from Optical Absorption
(13.4 and 20.1; 22.2.1)
(a) Vol. **84**, No. 1, p. 243—250 (1984).

O. Yu. Borkovskaya, N. L. Dmitruk, V. G. Litovchenko, and O. I. Maeva
The Influence of Deep Levels on Photomemory Effect in Structures with a Potential Barrier
(16; 13.4; 22.2.1)
(a) Vol. **84**, No. 1, p. 285—290 (1984).

Yu. B. Bolkhovityanov, Yu. D. Vaulin, B. Z. Olshanetskii, S. I. Stenin, and S. A. Teys
Auger Analysis of a GaAs (111) A Surface after Its Contact with the In–Ga–As–P Saturated Liquid
(1.6 and 2; 22.2.1; 22.2.4)
(a) Vol. **84**, No. 1, p. K13—K16 (1984).

E. Nebauer and M. Trapp
Ge and Au Profiles in GaAs Produced from AuGe Contacts and Studied by SIMS
(9; 14.3.2; 22.2.1)
(a) Vol. **84**, No. 1, p. K39—K42 (1984).

P. Kleinert
Cluster-Bethe-Lattice Treatment of the Intersitial Boron Centre in Irradiated Gallium Arsenide
(10.2; 11; 22.2.1)
(b) Vol. **124**, No. 2, p. 559—564 (1984).

K. D. Glinchuk, A. V. Prokhorovich, and N. S. Zayats
Effect of 2.2 MeV Electron Irradiation and Annealing on the Non-Copper-Induced 1.35 eV Emission Band in n-Type GaAs
(13.4 and 20.3; 11; 22.2.1)
(a) Vol. **84**, No. 2, p. 567—572 (1984).

M. S. Kamara, M. G. Lukashevich, and V. F. Stelmakh
Electric Field Influence on n-GaAs Magnetoresistance in Impurity Conduction
(14.3.4; 22.2.1)
(a) Vol. **84**, No. 2, p. 613—619 (1984).

B. Olejníková
Non-Exponential Capacitance Transient Due to Field-Enhanced Emission of Electrons from Deep Traps
(13.4 and 14.3.4; 22.2.1)
(a) Vol. **84**, No. 2, p. 621—630 (1984).

V. M. Ivastchenko, R. V. Konakova, and V. V. Mitin
Peculiarities of Reverse Current–Voltage Characteristics of GaAs Lo–Hi–Lo Impatt Diodes
(14.3.3 and 14.3.4; 22.2.1)
(a) Vol. **84**, No. 2, p. 669—675 (1984).

J. Šikula, P. Vašina, V. Musilová, Z. Chobola, and M. Rothbauer
$1/f$ Noise in GaAs Schottky Diodes
(14.3.3; 22.2.1)
(a) Vol. **84**, No. 2, p. 693—696 (1984).

M. de Dios Leyva, R. P. Alvarez, and J. L. Gondar
Fundamental Zero Energy Gap States in Superlattices
(13.1; 22.2.1; 22.2.3)
(b) Vol. **125**, No. 1, p. 221—228 (1984).

G. A. Adegboyega
Electrical Measurements on Silver-Diffused GaAs
(13.4 and 14.3; 22.2.1)
(a) Vol. **85**, No. 1, p. 215—218 (1984).

A. D. Pogrebnyak, Sh. R. Mastov, M. F. Kuznetsov, K. P. Arefiev, S. V. Rakitin, and S. A. Vorobiev
Observation of RF Pulses from Solids during Laser Irradiation
(11 and 20; 22.1.2; 22.2.1)
(a) Vol. **85**, No. 1, p. K31—K33 (1984).

V. M. Lomako and P. Ya. Starostin
Temperature Dependence of the Charge-Carrier Removal Rate in n-GaAs Containing Defect Clusters
(11 and 13.4; 22.2.1)
(a) Vol. **85**, No. 1, p. K57—K60 (1984).

G. Jalbert and H. S. Brandi
Absorption of Circularly Polarized Light by Solids
(20.1; 20.2; 22.2.1; 22.4.1)
(b) Vol. **125**, No. 2, p. 605—611 (1984).

P. Sircar and M. Aubin
Pulsed UV Laser Processed Gold and Aluminium Ohmic Contacts on n^+-GaAs
(14.3.3; 14.3.4; 22.2.1)
(a) Vol. **85**, No. 2, p. 649—654 (1984).

N. P. Esina, N. V. Zotova, S. A. Karandashev, and G. M. Filaretova
Tunneling Surface Recombination in MS Structures on p-InAs
(14.3.3 and 20.3; 16; 22.2.1)
(a) Vol. **85**, No. 2, p. 655—660 (1984).

E. SANDOW, W. WESCH, and E. NEBAUER
Ohmic AuGe Contacts on n-GaAs Produced by Nanosecond Laser Pulse Irradiation
(14.3.4; 9; 11; 22.2.1)
(a) Vol. **85**, No. 2. p. K169—K173 (1984).

S. KAČIULIS, K. KIBICKAS, and J. PARŠELIŪNAS
Dependence of *I–U* Characteristics of GaAs n^+–n–n^+ Diodes on the Active Region Length
(14.3.3 and 14.3.4; 22.2.1)
(a) Vol. **85**, No. 2, p. K179—K182 (1984).

A. R. VASCONCELLOS and R. LUZZI
On Laser-Induced Lattice Heating in a Polar Semiconductor
(8 and 11; 6; 22; 22.2.1)
(b) Vol. **126**, No. 1, p. 63—70 (1984).

M. MONDAL and K. P. GHATAK
Magnetic Susceptibility of Electrons in Ultrathin Films of Semiconductors in the Presence of a Quantizing Magnetic Field
(13 and 18; 22.2.1)
(b) Vol. **126**, No. 1, p. K41—K45 (1984).

G. M. SHMELEV, NGUYEN HONG SHON, and G. I. TSURKAN
Magnetic Oscillations of Acoustoelectrical Current of Photoelectrons in p-GaAs Type Semiconductors
(6 and 16; 7; 22.2.1)
(b) Vol. **126**, No. 1, p. K105—K108 (1984).

A. D. POGREBNYAK
Positron and Positronium States in Semiconductors Irradiated by Supercurrent Beams of Charged Particles
(11; 22.1.2; 22.2.1)
(a) Vol. **86**, No. 1, p. 191—198 (1984).

H.-R. HÖCHE and J. SCHREIBER
Anisotropic Deformation Behaviour of GaAs
(10.2; 22.2.1)
(a) Vol. **86**, No. 1, p. 229—236 (1984).

A. M. AFANASIEV, S. M. AFANASIEV, P. A. ALEKSANDROV, R. M. IMAMOV, and E. M. PASHAEV
Grazing Bragg-Laue Diffraction for Studying the Crystal Structure of Thin Films
(1.6 and 10.2; 22.2.1; 22.2.2)
(a) Vol. **86**, No. 1, p. K1—K5 (1984).

S. A. GRUSHA, A. M. EVSTIGNEEV, K. A. ISMAILOV, R. V. KONAKOVA, A. N. KRASIKO, I. K. SINISHCHUK, YU. A. TKHORIK, and V. I. FAINBERG
Electric Field Effect on Radiation-Induced Gettering of Defects in n^+–n–n^{++}-GaAs
(11 and 14.3.3; 22.2.1)
(a) Vol. **86**, No. 1, p. K83—K86 (1984).

M. Asche and O. G. Sarbei
The Role of Electron–Hole Interaction in the Cooling Process of Highly Excited Carriers
(13 and 20; 22.2.1)
(b) Vol. **126**, No. 2, p. 607—616 (1984).

D. Shaw
Alternative Mechanisms for the Diffusion of Sn and Zn in GaAs
(9; 10.2; 22.2.1)
(a) Vol. **86**, No. 2, p. 629—635 (1984).

A. Zoul
Influence of Deep Surface Levels on *C–U* Schottky Doping Profiles of n-GaAs VPE Layers
(14.3.2; 14.3.4; 22.2.1)
(a) Vol. **86**, No. 2, p. 673—678 (1984).

D. Wruck, M. Seifert, and W. Ulrici
Microinhomogeneities of Donor and Acceptor Distributions in n-Type LEC-GaAs from Free-Carrier Infrared Absorption
(1.4 and 13.4; 20.1; 22.2.1)
(a) Vol. **86**, No. 2, p. 691—702 (1984).

I. Thurzo and E. Pinčík
Some Peculiar Features of Small-Signal Charge DLTS Response of GaAs MOS Capacitors
(13.3 and 14.3.3; 22.2.1)
(a) Vol. **86**, No. 2, p. 795—804 (1984).

V. Rybka, F. Deml, and P. Hampl
Dual Implantation of Si and P into GaAs
(11 and 14.3; 22.2.1)
(a) Vol. **86**, No. 2, p. K117—K119 (1984).

22.2.2 Phosphides

Review Article

D. A. ARONOV, P. I. KNIGIN, YU. S. KOROLEV, and V. V. RUBINOV
Exclusion Effect in Semiconductors with Non-Injecting Contacts
(14.3.3 and 14.3.4; 16; 22.1.1; 22.1.2; 22.2.2)
(a) Vol. **81**, No. 1, p. 11—45 (1984).

Original Papers and Short Notes

M. CAHAY and E. KARTHEUSER
Polarizability of Donors in Polar Semiconductors: Effect of a Weak Magnetic Field
(13.4; 22.1.2; 22.2.2; 22.4.2)
(b) Vol. **121**, No. 1, p. 345—356 (1984).

K. ŽĎÁNSKÝ and N. T. THUC HIEN
Analysis of DLTS Curves Corresponding to Non-Exponential Transients
(13.4 and 14.3.3; 22.2.2)
(a) Vol. **81**, No. 1, p. 353—360 (1984).

H.-M. KAGAYA and T. SOMA
Lattice Dynamics of InP and ZnTe
(6; 22.2.2; 22.4.3)
(b) Vol. **121**, No. 2, p. K113—K118 (1984).

A. S. POPOV
Degradation Model of Red GaP LEDs
(13.4 and 14.3.3; 20.3; 22.2.2)
(a) Vol. **81**, No. 2, p. 669—674 (1984).

K. AOKI and M. CARDONA
Effect of Pressure on the Raman Scattering Efficiency of GaP
(6 and 20.1; 22.2.2)
(b) Vol. **122**, No. 2, p. 441—447 (1984).

R. SCHWABE, J. L. STAEHLI, I. STREIT, and F. THUSELT
EHP Luminescence of GaP:N under Resonant Free-Exciton Excitation
(13.5.1 and 20.3; 22.2.2)
(b) Vol. **122**, No. 2, p. 629—639 (1984).

F. THUSELT
Low-Density Approach to RPA Self-Energies
13.5.2; 22.1.1; 22.1.2; 22.2.1; 22.2.2)
(b) Vol. **123**, No. 1, p. 257—269 (1984).

L. WENDLER
Phonon-Polaritons in Bilayer Systems. p-Polarized Guided Wave Phonon-Polaritons and Guided Wave Surface Phonon-Polaritons
(6 and 20.1; 13.5; 22.2.1; 22.2.2; 22.5.3)
(b) Vol. **123**, No. 2, p. 469—477 (1984).

T. Stöckert, R. Gründler, H. Boudriot, and H. A. Schneider
Volume Plasmon Dispersion of Polycrystalline GaP and InP
(13.5.2; 22.2.2)
(b) Vol. **123**, No. 2, p. K161—K164 (1984).

C. M. Weinert and U. Scherz
Calculation of the Jahn-Teller Coupling Forces for $ZnS:Cu^{2+}$ and $GaP:Ni^{+}$
(6 and 20.1; 13.4; 22.2.2; 22.4.1)
(b) Vol. **124**, No. 1, p. 287—296 (1984).

E. M. Goldys
The Donor Wave Functions in Gallium Phosphide and the Real Band Structure
(13.4 and 20.1; 18; 22.2.2)
(b) Vol. **124**, No. 1, p. K79—K82 (1984).

W. Siegel and G. Kühnel
Influence of the Temperature Dependence of the Thermal Impurity Activation Energy on the Analysis of Hall Measurements on GaP
(13.4 and 14.3; 22.2.2)
(a) Vol. **84**, No. 1, p. 251—262 (1984).

A. Tanaka, T. Sukegawa, and T. Sugiura
Phosphorus Pressure Effects on GaP Crystal Quality in the Solution Growth
(3; 10.2; 22.2.2)
(a) Vol. **84**, No. 1, p. K9—K12 (1984).

P. Krispin and J. Maege
V_P Related Defects in Diffused GaP:N Diodes
(13.4 and 14.3.3; 22.2.2)
(a) Vol. **84**, No. 2, p. 573—584 (1984).

F. Thuselt and K. Unger
High Carrier Densities and Gap Shrinkage in p–i–n-Type Structures
(13.5; 14.3.3; 22.2.2)
(a) Vol. **84**, No. 2, p. 677—691 (1984).

S. I. Radautsan, S. A. Rebrov, and A. E. Tsurkan
Some Properties of ZnTe–InP Heterojunctions
(14.3.3; 20.3; 22.2.2; 22.4.3)
(a) Vol. **84**, No. 2, p. K169—K171 (1984).

V. Frei and P. Deus
A Frequency Sum Rule and Spectral Moments in Lattice Dynamics
(6; 22.1; 22.1.1; 22.2.2; 22.5.2; 22.8)
(b) Vol. **125**, No. 1, p. 121—130 (1984).

I. Gorczyca
Calculation of the Effect of Hydrostatic Pressure on the Band Structure of GaP and GaSb
(13.1; 12.2; 22.2.2; 22.2.3)
(b) Vol. **125**, No. 1, p. 229—235 (1984).

L. Wendler
Coupled Surface Plasmon–Phonon Polaritons in Bilayer Systems
(6 and 20; 13.5.2; 21.6; 22.2.2)
(b) Vol. **125**, No. 1, p. K27—K31 (1984).

F. Schneider and S. Unterricker
Radiation Damage and Its Annealing Behaviour in InP after Recoil Implantation of ^{118}Sb and ^{111}In Observed by TDPAC
(10.2 and 11; 5; 22.2.2)
(a) Vol. **85**, No. 2, p. 455—461 (1984).

R. Pickenhain, K. Jacobs, W. Seifert, V. Chernyi, and S. Brehme
Green and Yellow Light Emitting Diodes Produced from Vapour Phase Epitaxial GaP:N (I): Deep Levels in Diodes Produced by either Diffusion or Implantation of Zinc
(14.3.3 and 20.3; 13.4; 22.2.2)
(a) Vol. **85**, No. 2, p. 627—639 (1984).

S. A. Azimov, D. A. Aronov, V. M. Rubinov, and M. Tuichiev
On the Mechanism of Current Passage in Gallium Phosphide Films Excited by Light or a Beam of Accelerated Electrons
(14.3.1; 11; 22.2.2)
(a) Vol. **85**, No. 2, p. K159—K164 (1984).

A. M. Afanasiev, S. M. Afanasiev, P. A. Aleksandrov, R. M. Imamov, and E. M. Pashaev
Grazing Bragg-Laue Diffraction for Studying the Crystal Structure of Thin Films
(1.6 and 10.2; 22.2.1; 22.2.2)
(a) Vol. **86**, No. 1, p. K1—K5 (1984).

S. Jalaj, I. Singh, and S. K. Joshi
Tight-Binding Calculations for the Angle-Resolved Local Density of States on InP(110) Surfaces
(13.3; 22.2.2)
(b) Vol. **126**, No. 2, p. 669—676 (1984).

M. P. Dos Santos, C. Hirlimann, and M. Balkanski
Raman Detection of the Nitrogen Localized Mode in GaP:N
(6 and 20.1; 22.2.2)
(b) Vol. **126**, No. 2, p. K113—K116 (1984).

G. Oelgart, B. Rheinländer, H. Haefner, and R. Pickenhain
Degradation of Green Light Emitting Diodes LPE-GaP:N (I): Electron Beam Characterization and Deep-Level Spectroscopy
(14.3.3 and 20.3; 22.2.2)
(a) Vol. **86**, No. 2, p. 833—840 (1984).

22.2.3 Antimonides

Original Papers and Short Notes

T. Soma and H.-M. Kagaya
NaCl-Type Lattice of GaAs and InSb under Pressure
(1.2; 22.2.1; 22.2.3)
(b) Vol. **121**, No. 1, p. K1—K5 (1984).

V. K. Malyutenko, Yu. M. Malozovskii, and G. I. Teslenko
Negative Differential Photoconductivity in Magnetoconcentration Effect
(14.3.4 and 16; 22.2.3)
(a) Vol. **81**, No. 1, p. 391—398 (1984).

S. Ghosh and R. B. Saxena
Parametric Conversion of an Electromagnetic Wave into an Acoustic Wave in Magnetised Cubic Piezoelectric Semiconducting Plasma
(13.5.2; 7; 22.2.3)
(b) Vol. **121**, No. 2, p. 661—672 (1984).

Sang Don Choi and Ok Hee Chung
Theory of Cyclotron Resonance Lineshape in an Electron–Phonon System
(13.1 and 19; 22.2.3)
(b) Vol. **121**, No. 2, p. K181—K185 (1984).

T. Jung, W. Braune, M. Schnürer, and J. Schulze
Film and Interface Analysis of InSb MOS Structures
(1.6 and 14.3.2; 22.2.3)
(a) Vol. **81**, No. 2, p. 463—468 (1984).

S. R. Rao, M. Nagabhooshanam, and V. Hari Babu
Electrical Properties of Annealed InSb Thin Films
(14.3.1; 2; 22.2.3)
(a) Vol. **81**, No. 2, p. K195—K197 (1984).

U. Schaller, W. Kraak, and R. Herrmann
Metal–Insulator-Transition in n-InSb under High Hydrostatic Pressure
(14.1 and 14.3; 12.2; 22.2.3)
(b) Vol. **122**, No. 1, p. K87—K91 (1984).

S. Guha, C. Basu, and M. Bose
Parametric Excitation of Electron-Acoustic Waves in Magnetised Semiconductor Plasmas
(7 and 14.3; 22.2.3)
(a) Vol. **83**, No. 2, p. 675—680 (1984).

R. Herrmann, G. Nachtwei, and W. Kraak
On the Barrier Height in Grain Boundaries of InSb Bicrystals
(14.3; 13.1; 22.2.3)
(a) Vol. **83**, No. 2, p. K207—K210 (1984).

S. Ghosh and S. Dixit
Parametric Decay of a High-Power Helicon Wave into a Transverse Acoustic and Another Helicon Waves in Longitudinally Magnetised Cubic Piezoelectric Semiconducting Plasmas
(13.5.2; 7; 22.2.3)
(b) Vol. **124**, No. 1, p. 395—402 (1984).

R. Piotrzkowski
Pressure Study of Photoconductivity in Indium Antimonide (I): Low-Concentration n-Type at 77 K. Pressure-Dependent Capture Coefficient of a Hole-Trapping Center
(13.4 and 16; 12.2; 22.2.3)
(b) Vol. **124**, No. 1, p. 411—421 (1984).

T. Soma, Y. Takahashi, and H.-Matsuo Kagaya
Equation of State and Pressure Derivative of the Bulk Modulus for III–V Covalent Crystals
(1.2 and 4; 12.2; 22.2.1; 22.2.3)
(b) Vol. **124**, No. 1, p. K11—K14 (1984).

M. S. Pathak, S. Khan, and S. Ghosh
Longitudinal Phonon–Plasmon Interactions in Semiconducting Plasmas
(6 and 13.5.2; 7; 22.2.3)
(b) Vol. **124**, No. 2, p. 489—496 (1984).

E. I. Chaikina, S. Sh. Egemberdieva, and A. S. Filipchenko
Photoluminescence of Te-Doped Gallium Antimonide Crystals
(20.3; 22.2.3)
(a) Vol. **84**, No. 2, p. 541—545 (1984).

S. Huant, L. Dmowski, M. Baj, and L. C. Brunel
Intraconduction Band Magneto-Optical Study of InSb under Hydrostatic Pressure
(13.1 and 19; 16; 22.2.3)
(b) Vol. **125**, No. 1, p. 215—219 (1984).

M. De Dios Leyva, R. P. Alvarez, and J. L. Gondar
Fundamental Zero Energy Gap States in Superlattices
(13.1; 22.2.1; 22.2.3)
(b) Vol. **125**, No. 1, p. 221—228 (1984).

I. Gorczyca
Calculation of the Effect of Hydrostatic Pressure on the Band Structure of GaP and GaSb
(13.1; 12.2; 22.2.2; 22.2.3)
(b) Vol. **125**, No. 1, p. 229—235 (1984).

R. Herrmann, W. Kraak, and M. Gliński
Quantized Hall Effect in the n-Inversion Layer in InSb Grain Boundaries
(13.3 and 14.3; 22.2.3)
(b) Vol. **125**, No. 1, p. K85—K88 (1984).

D. Vignaud and J. L. Farvacque
Optical Absorption Studies of Plastically Deformed InSb
(13.4 and 20.1; 10.2; 22.2.3)
(b) Vol. **125**, No. 2, p. 785—794 (1984).

W. KRAAK, U. SCHALLER, and R. HERRMANN
Sensitive Pressure Gauge on the Basis of Pure n-InSb
(12.2 and 14.3; 22.2.3)
(a) Vol. **85**, No. 2, p. K183—K187 (1984).

D. I. ALADASHVILI, L. KONCZEWICZ, and S. POROWSKI
Studies of the Deep Levels in p-Type InSb under Pressure
(13.4; 14.3; 22.2.3)
(a) Vol. **86**, No. 1, p. 301—308 (1984).

W. BRAUNE, M. SCHNÜRER, N. KUBICKI, and R. HERRMANN
C–U Measurements of Anodized InSb MOS Structures at 77 and 4.2 K
(13.3 and 14.3.3; 22.2.3)
(a) Vol. **86**, No. 1, p. 427—431 (1984).

A. M. KONIN and A. P. SAŠČIUK
Thermogradient Magnetoconcentration Effect
(14.3; 22.1.1; 22.2.3)
(a) Vol. **86**, No. 1, p. K55—K58 (1984).

22.2.4 Mixed Crystals

Original Papers and Short Notes

R. STEGMANN, G. OELGART, W. SAUER, A. SCHEIFLER, and M. TRAPP
Investigation of the Recombination Mechanism in $GaAs_{0.11}P_{0.89}$:N
(13.5.1 and 20.3; 22.2.4)
(a) Vol. **81**, No. 2, p. 541—548 (1984).

U. WOGGON, K. WANDEL, and H. WEINERT
Minority Carrier Lifetime Measurements by Photoluminescence at Room Temperature in $GaAs_{1-x}P_x$:N, Te ($x \geqq 0.70$)
(20.3; 22.2.4)
(a) Vol. **81**, No. 2, p. 739—749 (1984).

P. KLEINERT
Coherent-Potential Approximation for the Lattice Vibrations of Mixed III–V Crystals
(6; 20.1; 22.2.4)
(b) Vol. **122**, No. 1, p. 81—89 (1984).

G. ARNOLD and G. OELGART
Ionization Rates for Electrons and Holes in GaAsP
(14.3.3; 14.3.4; 22.2.4)
(a) Vol. **82**, No. 1, p. 319—325 (1984).

U. PIETSCH, J. BAK-MISIUK, and V. GOTTSCHALCH
The Linear Thermal Expansion Coefficient of a $Ga_xIn_{1-x}As_yP_{1-y}$ Layer on InP:Sn Substrate
(8; 1.1; 22.2.4)
(a) Vol. **82**, No. 2, p. K137—K140 (1984).

U. WOGGON, K. WANDEL, and H. WEINERT
Temperature Dependence of Minority Carrier Lifetime in VPE–GaP:N,Te
(14.3.3; 22.2.4)
(a) Vol. **82**, No. 2, p. K205—K208 (1984).

G. OELGART and G. ARNOLD
Internal Electric Field-Enhanced Impurity Diffusion in $GaAs_{1-x}P_x$
(13.4 and 14.3.3; 9; 22.2.4)
(a) Vol. **83**, No. 1, p. 393—402 (1984).

S. K. MAKSIMOV
Correlation between Thermodynamic Instability of $A^{III}B^{V}$ Heterocompositions and Degradation of Devices Based on These Materials
(10.2 and 14.3.2; 9; 22.2.4)
(a) Vol. **83**, No. 2, p. 685—692 (1984).

YU. B. BOLKHOVITYANOV, YU. D. VAULIN, B. Z. OLSHANETSKII, S. I. STENIN, and S. A. TEYS
Auger Analysis of a GaAs (111) A Surface after Its Contact with the In–Ga–As–P Saturated Liquid
(1.6 and 2; 22.2.1; 22.2.4)
(a) Vol. **84**, No. 1, p. K13—K16 (1984).

H.-J. MÜLLER and H.-J. WÜNSCHE
Model Calculations for Bound Excitons and Bound Excitonic Molecules in $GaAs_{1-x}P_x$:N
(13.4 and 13.5.1; 20.3; 22.2.4)
(b) Vol. **124**, No. 2, p. 747—756 (1984).

G. OELGART and U. WERNER
Kilovolt Electron Energy Loss Distribution in GaAsP
(11; 13; 22.2.4)
(a) Vol. **85**, No. 1, p. 205—213 (1984).

F. R. KESSLER, U. BARKOW, and K. JÜRGENS
Cathodoluminescence of InGaAsP
(20.3; 22.2.4)
(a) Vol. **85**, No. 1, p. K77—K79 (1984).

D. RAY CHAUDHURI, J. B. ROY, and P. K. BASU
Density of the 2D Electron Gas in Modulation Doped GaInAs/AlInAs Layers from a Charge Control Analysis
(14.3.3; 14.3.2; 22.2.4)
(a) Vol. **86**, No. 1, p. K79—K82 (1984).

22.3 IV–IV, II–V, and III–VI Compounds

Original Papers and Short Notes

M. Gatulle and M. Fischer
Elastic Constants of the Layered Compounds GaS, GaSe, InSe, and Their Pressure Dependence (II): Theoretical Part
(12.2; 22.3)
(b) Vol. **121**, No. 1, p. 59—71 (1984).

J. Cisowski, J. C. Portal, E. K. Arushanov, J. M. Broto, S. Huant, and L. C. Brunel
Shubnikov-De Haas Effect and Cyclotron Resonance in p-CdSb
(13.1; 22.3)
(b) Vol. **121**, No. 1, p. 289—292 (1984).

T. Soma, Y. Kitani, and H.-Matsuo Kagaya
Lattice Dynamics and Thermal Properties of Si–Ge Solution
(6; 8; 22.3)
(b) Vol. **121**, No. 2, p. 501—504 (1984).

B. G. Tagiev, G. M. Niftiev, and S. A. Abushov
Photoluminescence of GaSe:Mn Single Crystals
(20.3; 13.4; 22.3)
(b) Vol. **121**, No. 2, p. K195—K199 (1984).

A. K. Sarychev and A. P. Vinogradov
Erratum to: Effective Medium Theory for the Magnetoconductivity Tensor of Disordered Materials
(14.3; 22.3)
(b) Vol. **121**, No. 2, p. K215 (1984).

Z. Januškevičius, V. Koval, A. Kudin, A. Sodeika, V. Tartachnik, and I. Tychina
Radiation Defects in Single Crystals of Zinc Diphosphide
(11 and 14.3; 22.3)
(a) Vol. **81**, No. 2, p. K191—K193 (1984).

A. N. Pilyankevich and V. F. Britun
Motion of Partial Dislocations in Silicon Carbide
(1.4 and 10.2; 22.3)
(a) Vol. **82**, No. 2, p. 449—457 (1984).

J. Leotin, M. Goiran, and S. Askenazy
Cyclotron Resonance of Holes in Cadmium Antimonide
(13.1; 22.3)
(b) Vol. **123**, No. 1, p. K43—K46 (1984).

Yu. P. Gnatenko, P. A. Skubenko, Z. D. Kovalyuk, and V. M. Kaminskii
An Isothermal Annealing Effect on the Luminescence Spectra of GaSe Single Crystals
(13.5.1 and 20.3; 22.3)
(b) Vol. **123**, No. 1, p. K63—K67 (1984).

D. D. Askerov, A. M. Agaev, T. D. Dzhafarov, and M. G. Ramazanzade
Sm–GaSe Barrier Structures
(14.3.4; 16; 22.3)
(a) Vol. **83**, No. 1, p. 323—328 (1984).

L. Červinka, V. Cháb, and S. Koc
Erratum to: XPS Investigation of the Amorphous $TlSe_x$ System
(2 and 4; 17; 22.3)
(a) Vol. **83**, No. 1, p. 418 (1984).

J. Misiewicz, Nella Mirowska, and Z. Gumienny
Fundamental Optical and Photoelectric Properties of Poly- and Mono-Crystals of Zn_3P_2
(16 and 20.1; 22.3)
(a) Vol. **83**, No. 1, p. K51—K56 (1984).

S. Chaudhuri, S. K. Biswas, and A. Choudhury
Effects of Heat Treatment on the Optical Absorption of Amorphous $In_{30}Se_{70}$ Films
(2 and 20.1; 22.3)
(a) Vol. **83**, No. 1, p. K57—K60 (1984).

B. G. Tagiev, G. M. Niftiev, and S. A. Abushov
Photoelectrical and Optical Properties of GaSe:Yb Single Crystals
(13.4; 16; 20.3; 22.3)
(a) Vol. **83**, No. 1, p. K61—K64 (1984).

T. Ikari and S. Shigetomi
Photoluminescence of n-Type InSe
(20.3; 13.5.1; 22.3)
(b) Vol. **124**, No. 1, p. K49—K51 (1984).

B. G. Tagiev, G. D. Sultanov, G. M. Niftiev, I. M. Askerov, and G. K. Aslanov
The Mössbauer Effect and Magnetic Susceptibilities in Ga_2Se_3:^{57}Fe Single Crystals
(5 and 18; 22.3)
(a) Vol. **84**, No. 1, p. K59—K62 (1984).

B. A. Lukiyanets, K. D. Tovstyuk, V. V. Boichuk, and K. K. Danilevich-Tovstyuk
The Jahn-Teller Effect in Layer Crystals
(6 and 13; 22.3)
(b) Vol. **124**, No. 2, p. 693—698 (1984).

K. Kambas, C. Julien, M. Jouanne, A. Likforman, and M. Guittard
Raman Spectra of α- and γ-In_2Se_3
(6 and 20.1; 22.3)
(b) Vol. **124**, No. 2, p. K105—K108 (1984).

M. T. Sebastian and P. Krishna
X-Ray Diffraction from a SiC Crystal Undergoing the 3C–6H Solid State Transformation by Non-Random Microtwinning
(1.2; 10.2; 22.3)
(a) Vol. **84**, No. 2, p. 401—409 (1984).

T. Chassé, U. Berg, and O. Brümmer
Angular-Resolved X-Ray Photoelectron Spectra of Valence and Core Levels from GaS
(13.1 and 17; 22.3)
(b) Vol. **125**, No. 1, p. 267—274 (1984).

H. Sobotta, H. Neumann, N. N. Syrbu, and V. Riede
Infrared Study of the Lattice Vibrations in Tetragonal CdP_2
(6 and 20.1; 22.3)
(b) Vol. **125**, No. 1, p. K17—K20 (1984).

V. V. Sobolev, and A. I. Kozlov
Free Exciton and Exciton–Impurity Complexes of Zinc Diphosphide and Zinc Diarsenide
(13.5.1 and 20.1; 22.3)
(b) Vol. **126**, No. 1, p. K59—K62 (1984).

S. Shigetomi, T. Ikari, Y. Koga, and S. Shigetomi
Annealing Behaviour of Electrical Properties of n-InSe Single Crystals
(14.3; 13.4; 22.3)
(a) Vol. **86**, No. 1, p. K69—K72 (1984).

A. U. Sheleg and V. V. Zaretskii
X-Ray Study of the Commensurate–Incommensurate Phase Transitions in α-ZnP_2
(1.2; 22.3)
(a) Vol. **86**, No. 2, p. 517—523 (1984).

22.3.1 Mixed Crystals

Original Papers and Short Notes

G. M. Niftiev, B. G. Tagiev, A. O. Khalilov, and S. A. Abushov
Lasting Relaxations and Residual Conductivity in $InSe_{1-x}S_x$ Single Crystals
(14.3; 22.3.1)
(a) Vol. **81**, No. 2, p. K175—K178 (1984).

M. P. Lisitsa, A. V. Stolyarenko, S. F. Terekhova, and N. A. Onishchenko
Refraction Index Variation in the Exciton Resonance of CdS and $TlInS_2$ Single Crystals under Laser Excitation
(13.5.1; 20.1; 22.3.1; 22.4.1)
(b) Vol. **125**, No. 2, p. 705—711 (1984).

L. V. Vassilev and St. K. Dimitrova
Observation of Anderson-Like Transition in the System $Pb_{1-x}Ge_xTe$
(14.3 and 15; 1.2; 22.3.1)
(b) Vol. **125**, No. 2, p. K177—K180 (1984).

22.4 II–VI Compounds

Review Article

A. Blacha, H. Presting, and M. Cardona
Deformation Potentials of $k = 0$ States of Tetrahedral Semiconductors
(6 and 13.1; 22.1; 22.2; 22.4)
(b) Vol. **126**, No. 1, p. 11—36 (1984).

Original Papers and Short Notes

W. E. Hagston
Mechanism of Excitation of Rare-Earth Complexes in Electroluminescence Devices
(14.3.4 and 20.3; 13.4; 22.4)
(a) Vol. **81**, No. 2, p. 687—693 (1984).

S. Bhushan, D. Diwan, and S. P. Kathuria
Thermoluminescence of ZnO:Pr and ZnO:Gd under UV, β- and γ-Ray Irradiations
(20.3; 11; 13.4; 22.4)
(a) Vol. **83**, No. 2, p. 605—611 (1984).

H.-Matsuo Kagaya and T. Soma
Lattice Dynamics of Tetrahedral Compounds with the Local Heine-Abarenkov Model Potential
(6; 22.2; 22.4)
(b) Vol. **124**, No. 1, p. 37—44 (1984).

V. I. Sokolov, A. N. Mamedov, T. P. Surkova, G. A. Emelchenko, and L. G. Kolinova
Fine Structure of the Photoionization Band Edge in Ni-Doped ZnO
(13.5.1 and 20.1; 14.3.4; 22.4)
(b) Vol. **124**, No. 2, p. K155—K159 (1984).

I. S. Smirnova
Dislocations in $A^{II}B^{VI}$ and $A^{III}B^{V}$ Piezoelectric Crystals
(10.2 and 14.3; 22.2; 22.4)
(b) Vol. **126**, No. 1, p. 177—189 (1984).

F. Dujardin, B. Stébé, and G. Munschy
Neutral Bound Excitons in a Low Magnetic Field
(13.4 and 13.5.1; 22.2; 22.4)
(b) Vol. **126**, No. 1, p. 329—334 (1984).

H.-Matsuo Kagaya and T. Soma
Dynamical Elastic Constants of Covalent Compounds
(12.2; 6; 22.2; 22.4)
(b) Vol. **126**, No. 1, p. K9—K14 (1984).

22.4.1 Sulphides

Original Papers and Short Notes

A. ABDEL-KADER, F. J. BRYANT, and J. H. C. HOGG
Tellurium Doping and Implantation of Zinc Sulphide
(13.4; 20.3; 22.4.1)
(a) Vol. **81**, No. 1, p. 333—342 (1984).

F. HENNEBERGER and H. ROSSMANN
Resonatorless Optical Bistability Based on Increasing Nonlinear Absorption
(13.5.1 and 20.1; 22.4.1)
(b) Vol. **121**, No. 2, p. 685—693 (1984).

F. HENNEBERGER, J. PULS, H. ROSSMANN, and I. N. URALTSEV
On the Refractive Index at the Exciton Resonance of Highly Excited CdS
(13.5.1 and 20.1; 22.4.1)
(b) Vol. **121**, No. 2, p. K191—K194 (1984).

G. O. MÜLLER
Basics of Electron-Impact-Excited Luminescence Devices
(14.3.4 and 20.3; 22.4.1)
(a) Vol. **81**, No. 2, p. 597—608 (1984).

R. MACH and G. O. MÜLLER
Efficiency and Saturation in AC Thin Film El Structures
(14.3.4 and 20.3; 22.4.1)
(a) Vol. **81**, No. 2, p. 609—623 (1984).

P. J. DEAN
Comparison of MOCVD-Grown with Conventional II–VI Materials Parameters for EL Thin Films
(14.3.4 and 20.3; 3; 13.4; 22.4.1; 22.4.2)
(a) Vol. **81**, No. 2, p. 625—646 (1984).

D. THEIS
Selected Analytical Tools Yield a Better Insight into Electroluminescent Thin Films
(1.5 and 14.3.4; 20.3; 22.4.1)
(a) Vol. **81**, No. 2, p. 647—655 (1984).

N. A. VLASENKO, YU. V. KOPYTKO, and V. S. PEKAR
Concentration and Field Dependences of Electroluminescence Decay Kinetics in ZnS:Mn Thin Film Structures
(14.3.4 and 20.3; 22.4.1)
(a) Vol. **81**, No. 2, p. 661—667 (1984).

F. J. BRYANT and A. KRIER
Direct Current Electroluminescence in Rare-Earth-Doped Zinc Sulphide
(14.3.4 and 20.3; 22.4.1)
(a) Vol. **81**, No. 2, p. 681—686 (1984).

DANG DINH THONG, W. HEIMBRODT, D. HOMMEL, and O. GOEDE
Optical Study of ZnS:Mn Thin Films with High Mn Concentrations
(20.3; 22.4.1)
(a) Vol. **81**, No. 2, p. 695—700 (1984).

J. KREISSL and W. GEHLHOFF
EPR Investigations of ZnS:Mn and ZnSe:Mn. Single Crystals, Powders, Thin-Film Structures
(13.4 and 19; 22.4.1; 22.4.2)
(a) Vol. **81**, No. 2, p. 701—707 (1984).

W. GERICKE, J. VON KALBEN, U. REINSPERGER, B. SELLE, L. REDLICH, and V. GUTAN
Annealing of Luminescent ZnSe:Mn Thin Films by a Scanning CW Laser Beam
(20.3; 22.4.1)
(a) Vol. **82**, No. 2, p. 715—721 (1984).

R. MACH, G. O. MÜLLER, and G. SCHULZ
On the Time Behaviour of Electron Impact Excitation in Thin Film Electroluminescent Devices
(14.3.4 and 20.3; 14.3.1; 22.4.1)
(a) Vol. **81**, No. 2, p. 723—732 (1984).

D. HOMMEL, H. HARTMANN, and J. M. LANGER
Emission Patterns of Tb Doped ZnS Bulk Crystals
(20.3; 22.4.1)
(a) Vol. **81**, No. 2, p. K199—K203 (1984).

V. D. EGOROV, G. O. MÜLLER, H. H. WEBER, M. A. JACOBSON, and A. F. DITE
Control of Carrier Concentration by Stimulated Emission in Highly Excited Direct Gap Semiconductors
(16 and 20.3; 22.4.1)
(b) Vol. **122**, No. 1, p. 183—192 (1984).

A. KRÓL, W. NAZAREWICZ, and M. J. KOZIELSKI
Al_{Zn}–V_{Zn} Pairs in Wurtzite-Type ZnS
(10.2 and 20.1; 6; 22.4.1)
(b) Vol. **122**, No. 1, p. K23—K25 (1984).

D. NESHEVA
Studies of Low-Energy Optical Absorption and Gap State Density of a-CdS by the Constant Photocurrent Method
(13.1 and 20.1; 2; 14.3.1; 22.4.1)
(a) Vol. **82**, No. 1, p. 243—248 (1984).

E. CHIMCZAK, W. S. GORDON, and M. BERTRANDT-ŻYTKOWIAK
Model of the Electroluminescence in Short-Pulse-Excited Thin Film Structures Based on ZnS:Mn
(14.3.3 and 20.3; 13.4; 22.4.1)
(a) Vol. **82**, No. 2, p. 527—531 (1984).

A. KRIER and F. J. BRYANT
The Hot Electron Energy Distribution Function in ZnS:Cu:Cl:Er Thin Films
(20.3; 13.4; 22.4.1)
(a) Vol. **83**, No. 1, p. 315—322 (1984).

J. Benoit, P. Benalloul, A. Geoffroy, N. Balbo, C. Barthou, J. P. Denis, and B. Blanzat
Study of Highly Concentrated ZnS:Mn ACTFEL Devices
(20.3; 22.4.1)
(a) Vol. **83**, No. 2, p. 709—717 (1984).

C. Butti, A. Raymond, and G. Massé
Conductivity Anisotropy in α-HgS
(14.3; 22.4.1)
(a) Vol. **83**, No. 2, p. K199—K202 (1984).

D. T. Rasulov and R. Akhmedov
Thin-Film Solar Cell on the Basis of Polycrystalline Cadmium Telluride
(14.3.3 and 16; 22.4.1; 22.4.3; 22.4.4)
(a) Vol. **83**, No. 2, p. K211—K213 (1984).

Hoang Xuan Nguyen and R. Zimmermann
Nonlinear Exciton Transmission in CdS. Non-Homogeneous and Non-Stationary Model Calculations
(13.5.1 and 20; 22.4.1)
(b) Vol. **124**, No. 1, p. 191—200 (1984).

C. M. Weinert and U. Scherz
Calculation of the Jahn-Teller Coupling Forces for $ZnS:Cu^{2+}$ and $GaP:Ni^{+}$
(6 and 20.1; 13.4; 22.2.2; 22.4.1)
(b) Vol. **124**, No. 1, p. 287—296 (1984).

O. Goede and Dang Dinh Thong
Energy Transfer Processes in (Zn, Mn)S Mixed Crystals
(13.4 and 20.3; 22.4.1)
(b) Vol. **124**, No. 1, p. 343—353 (1984).

M. A. S. Sweet, C. Aust, I. T. Liddell, and B. Ray
Thermoluminescence Studies of CaS Single Crystals in the Temperature Range 8 to 275 K
(20.3; 22.4.1)
(a) Vol. **84**, No. 1, p. K75—K79 (1984).

L. F. Gudymenko and E. G. Gule
The Nature of Non-Linear Radiation of $A^{II}B^{VI}$ Single Crystals
(13.5 and 20.3; 22.4.1)
(b) Vol. **124**, No. 2, p. K139—K142 (1984).

A. K. Sharma and D. P. Rewari
Stimulated Brillouin Scattering of an Electromagnetic Wave in CdS. Thermal Nonlinearity
(7; 22.4.1)
(a) Vol. **84**, No. 2, p. 439—442 (1984).

S. S. Dimov, L. I. Pavlov, K. V. Stamenov, and G. B. Altshuller
Dispersion of Laser-Induced Nonlinear Optical Susceptibility in CdS
(13.4; 20.2; 22.4.1)
(a) Vol. **84**, No. 2, p. 555—560 (1984).

T. K. Bandyopadhyay and S. R. Chaudhuri
Electronic Properties of Anodic n-CdS Films from Photoelectrochemical Measurements
(14.3.1; 16; 22.4.1)
(a) Vol. **84**, No. 2, p. 651—654 (1984).

G. Jalbert and H. S. Brandi
Absorption of Circularly Polarized Light by Solids
(20.1; 20.2; 22.2.1; 22.4.1)
(b) Vol. **125**, No. 2, p. 605—611 (1984).

M. P. Lisitsa, A. V. Stolyarenko, S. F. Terekhova, and N. A. Onishchenko
Refraction Index Variation in the Exciton Resonance of CdS and $TlInS_2$ Single Crystals under Laser Excitation
(13.5.1; 20.1; 22.3.1; 22.4.1)
(b) Vol. **125**, No. 2, p. 705—711 (1984).

U. W. Pohl, H.-E. Gumlich, and W. Busse
Zero-Phonon Lines in the Excitation Band Centred at 23200 cm^{-1} Due to Mn^{2+} in Polymorphic ZnS
(13.4; 20.1; 22.4.1)
(b) Vol. **125**, No. 2, p. 773—777 (1984).

V. N. Ermakov, D. V. Korbutyak, V. G. Litovchenko, O. Yu. Mikityuk, and V. V. Nitsovich
A Study of the Reflection Spectra of Surface Exciton–Impurity Complexes
(13.4 and 13.5.1; 13.3; 20.1; 22.4.1)
(b) Vol. **125**, No. 2, p. 815—822 (1984).

T. N. Dimov, I. A. Iliev, and H. Lange
Peculiarities of Polarization Properties of CdS Single Crystals around the Isotropic Point
(20.1; 22.4.1)
(b) Vol. **126**, No. 1, p. 261—271 (1984).

D. Backs and A. Hertrampf
Optically Induced EPR of Conduction and Donor Band Electrons in Zinc Sulphide
(13.4 and 19; 22.4.1)
(b) Vol. **126**, No. 1, p. 343—349 (1984).

R. Baltramiejūnas, I. Rückmann, G. Tamulaitis, and A. Žukauskas
Sample-Dependent Luminescence Spectra in Highly Photoexcited CdS and CdSe
(13.5.1 and 20.1; 22.4.1; 22.4.2)
(b) Vol. **126**, No. 2, p. K135—K138 (1984).

O. Goede, W. Heimbrodt, and Dang Dinh Thong
Non-Exponential ZnS:Mn Luminescence Decay Due to Energy Transfer
(20.3; 22.4.1)
(b) Vol. **126**, No. 2, p. K159—K163 (1984).

22.4.2 Selenides

Original Papers and Short Notes

M. CAHAY and E. KARTHEUSER
Polarizability of Donors in Polar Semiconductors: Effect of a Weak Magnetic Field
(13.4; 22.1.2; 22.2.2; 22.4.2)
(b) Vol. **121**, No. 1, p. 345—356 (1984).

P. J. DEAN
Comparison of MOCVD-Grown with Conventional II–VI Materials Parameters for EL Thin Films
(14.3.4 and 20.3; 3; 13.4; 22.4.1; 22.4.2)
(a) Vol. **81**, No. 2, p. 625—646 (1984).

J. KLUGE and J. DONECKER
Transfer Effects in ODMR Spectra of ZnSe:Mn
(19 and 20.3; 13.4; 22.4.2)
(a) Vol. **81**, No. 2, p. 675—679 (1984).

J. KREISSL and W. GEHLHOFF
EPR Investigations of ZnS:Mn and ZnSe:Mn. Single Crystals, Powders, Thin-Film Structures
(13.4 and 19; 22.4.1; 22.4.2)
(a) Vol. **81**, No. 2, p. 701—707 (1984).

R. MACH, G. O. MÜLLER, G. SCHULZ, J. VON KALBEN, and W. GERICKE
On the Mechanism of Electron-Impact-Excited Luminescence in AC Thin Film Devices: ZnSe:Mn
(14.3.4 and 20.3; 22.4.2)
(a) Vol. **81**, No. 2, p. 733—738 (1984).

E. L. IVCHENKO, V. P. KOCHERESHKO, G. V. MIKHAILOV, and I. N. URALTSEV
Resonance Magneto-Spatial Dispersion of Crystals
(18 and 20.1; 13.5.1; 22.4.2)
(b) Vol. **122**, No. 1, p. 221—230 (1984).

W. SCHRITTENLACHER, H. NELKOWSKI, and H. PRADELLA
ESR and ODMR of Zinc–Vacancy-Associated Defects in ZnSe
(13.4 and 19; 22.4.2)
(b) Vol. **122**, No. 1, p. 285—294 (1984).

J. M. BARANOWSKI and PHUONG AN
The Ni Donor and Acceptor Levels in CdSe
(13.4 and 20.1; 22.4.2)
(b) Vol. **122**, No. 1, p. 331—336 (1984).

H. L. OCZKOWSKI
The Retrapping Process in a Thermoluminescence Model with an Intra-Pair Metastable State
(13.4 and 20.3; 22.4.2)
(a) Vol. **82**, No. 1, p. 213—219 (1984).

W. Schrittenlacher
Analysis of the Hyperfine Interactions of the Vacancy–Iodine Complex in ZnSe
(13.4; 19; 22.4.2)
(b) Vol. **122**, No. 2, p. 735—743 (1984).

V. P. Kochereshko, J. Röseler, I. N. Uraltsev, and K. Henneberger
Dipole Forbidden Transmission of Magnetopolaritons in CdSe
(13.5.1 and 20.1; 18; 22.4.2)
(b) Vol. **124**, No. 1, p. 213—219 (1984).

G. I. Ibaev, S. I. Kulieva, and T. G. Magerramov
Frequency Dependences of the Dynamic Current–Voltage Characteristic of the Se–CdSe Heterostructure Based on a Single-Crystal Selenium Film
(14.3.1 and 14.3.3; 22.1.3; 22.4.2)
(a) Vol. **84**, No. 2, p. K185—K188 (1984).

E. Mizera, M. Sundberg, and P. Werner
Defect Structure of ZnSe Crystals Investigated by Electron Microscopy
(1.4; 10.2; 22.4.2)
(a) Vol. **85**, No. 1, p. 83—88 (1984).

P. Lilley, M. R. Czerniak, and J. E. Nicholls
Photoluminescence and Electrical Properties of Vapour Phase Epitaxial ZnSe Grown on GaAs
(14.3.1 and 20.3; 13.4; 22.4.2)
(a) Vol. **85**, No. 1, p. 235—242 (1984).

Y. Unuma, Y. Abe, Y. Masumoto, and S. Shionoya
Dynamical Behaviour of the High Density Electron–Hole System in CdSe
(13.5.1 and 20.3; 22.4.2)
(b) Vol. **125**, No. 2, p. 735—742 (1984).

S. W. Biernacki
Born-von Karman Cell Model for Deep Impurities
(13.4; 22.4.2)
(b) Vol. **125**, No. 2, p. 805—814 (1984).

V. A. Kasiyan, D. D. Nedeoglo, A. V. Simashkevich, and I. N. Timchenko
Hopping Conductivity and Electron Localization in Heavily Doped ZnSe
(13.4 and 14.3; 22.4.2)
(b) Vol. **125**, No. 2, p. 869—878 (1984).

S. Nešpůrek, O. Zmeškal, and F. Schauer
An Ohmic Contact Test Using the Activation Energy of Steady-State Space-Charge-Limited Current
(14.3.4; 13.4; 22.1.2; 22.4.2)
(a) Vol. **85**, No. 2, p. 619—625 (1984).

A. Kozanecki and H. Rzewuski
Distant DAP Bands in Electron-Irradiated and Ion-Implanted CdSe
(11 and 20.3; 10.2; 22.4.2)
(a) Vol. **86**, No. 1, p. 263—270 (1984).

R. BALTRAMIEJŪNAS, I. RÜCKMANN, G. TAMULAITIS, and A. ŽUKAUSKAS
Sample-Dependent Luminescence Spectra in Highly Photoexcited CdS and CdSe
(13.5.1 and 20.1; 22.4.1; 22.4.2)
(b) Vol. **126**, No. 2, p. K135—K138 (1984).

T. K. BANDYOPADHYAY and S. R. CHAUDHURI
Effect of Annealing Ambients on the Performance of Thin Film n-CdSe Photoanodes
(16; 22.4.2)
(a) Vol. **86**, No. 2, p. 827—832 (1984).

22.4.3 Tellurides

Original Papers and Short Notes

R. M. Bilbé, J. E. Nicholls, and J. J. Davies
Electron Paramagnetic Resonance Investigation of Pb- and Ge-Doped CdTe
(13.4 and 19; 22.4.3)
(b) Vol. **121**, No. 1, p. 339—344 (1984).

H.-M. Kagaya and T. Soma
Lattice Dynamics of InP and ZnTe
(6; 22.2.2; 22.4.3)
(b) Vol. **121**, No. 2, p. K113—K118 (1984).

Q. Kim and D. W. Langer
Effects of Excitation Intensity on Photoluminescence of Pure CdTe
(13.5.1 and 20.3; 22.4.3)
(b) Vol. **122**, No. 1, p. 263—268 (1984).

R. Jašinskaitė, A. Sakalas, A. Martinaitis, and P. Höschl
Native Acceptor Defects in Chlorine Doped Single Crystals of CdTe
(14.3; 13.4; 22.4.3)
(a) Vol. **82**, No. 2, p. K199—K204 (1984).

A. Król and R. Zielińska
A New Defect Complex in CdTe:In, Li
(6 and 10.2; 20.1; 22.4.3)
(b) Vol. **123**, No. 1, p. K17—K20 (1984).

D. T. Rasulov and R. Akhmedov
Thin-Film Solar Cell on the Basis of Polycrystalline Cadmium Telluride
(14.3.3 and 16; 22.4.1; 22.4.3; 22.4.4)
(a) Vol. **83**, No. 2, p. K211—K213 (1984).

S. I. Radautsan, S. A. Rebrov, and A. E. Tsurkan
Some Properties of ZnTe–InP Heterojunctions
(14.3.3; 20.3; 22.2.2; 22.4.3)
(a) Vol. **84**, No. 2, p. K169—K171 (1984).

K. Žďánský and N. T. Thuc Hien
Analysis of DLTS Curves of Aggregated Deep Level Impurities
(13.4 and 14.3.3; 22.4.3)
(a) Vol. **85**, No. 1, p. 219—226 (1984).

H. Booyens and J. H. Basson
Piezoelectrically Induced Charge Distributions around Dislocations in CdTe and HgCdTe
(10.2 and 13.4; 14.4; 22.4.3; 22.4.4)
(a) Vol. **85**, No. 1, p. 243—248 (1984).

M. ADOLF, J. ROSENZWEIG, and U. BIRKHOLZ
Bipolar Photoconductivity and Hall Effect of the Acoustoelectric Current in CdTe
(7 and 16; 14.3; 22.4.3)
(a) Vol. **85**, No. 1, p. K81—K84 (1984).

V. SURIYANARAYANAN, Y. MADHAVAN, and K. RAMACHANDRAN
Isotope Diffusion in Cadmium Telluride
(9; 22.4.3)
(b) Vol. **125**, No. 2, p. 513—516 (1984).

M. S. BRODIN, V. M. BANDURA, and M. G. MATSKO
Excitonic Surface Polaritons in Luminescence from ZnTe Crystals
(13.5.1 and 20.3; 13.3; 22.4.3)
(b) Vol. **125**, No. 2, p. 613—618 (1984).

D. FRÖHLICH, A. NÖTHE, and K. REIMANN
Determination of Valence Band Parameters in ZnTe
(13.5.1 and 20.1; 13.1; 22.4.3)
(b) Vol. **125**, No. 2, p. 653—657 (1984).

G. MILCHBERG, K. SAMINADAYAR, E. MOLVA, and H. R. ZELSMANN
Optical Determination of the Valence-Band Parameters in CdTe
(13.4 and 20.3; 13.1; 22.4.3)
(b) Vol. **125**, No. 2, p. 795—803 (1984).

R. M. BILBÉ, B. LUNN, J. E. NICHOLLS, and D. H. J. TOTTERDELL
The Role of Copper as an Impurity in High Resistivity Cadmium Telluride and Its Influence on γ-Ray Spectrometer Performance
(13.4 and 14.3; 20.1; 22.4.3)
(a) Vol. **86**, No. 2, p. 821—826 (1984).

22.4.4 Mixed Crystals

Original Papers and Short Notes

J. PULS and F. HENNEBERGER
Saturation of the Band-Edge Absorption of CdS_xSe_{1-x} Mixed Crystals at Room Temperature
(20.1; 22.4.4)
(b) Vol. **121**, No. 2, p. K187—K190 (1984).

S. DEMIAN, F. EL AKKAD, B. FARAG, H. MATHIEU, M. ROUZEYRE, and J. CHEVALLIER
Photoluminescence and Optical Properties of Lithium-Doped $Mg_xZn_{1-x}Te$ Alloys
(13.4 and 20.3; 20.1; 22.4.4)
(a) Vol. **81**, No. 2, p. 549—555 (1984).

M. NOWAK
Oscillatory Dependence of the Interference Photomagnetoelectric Effect and Photoconductivity on Magnetic Field
(16; 20.1; 22.4.4)
(a) Vol. **82**, No. 1, p. 249—256 (1984).

V. I. SOKOLOV and V. V. CHERNYAEV
Detection of Manganese Photoionization in $Cd_{1-x}Mn_xTe$ Solid Solutions
(13.5.1; 16; 20.1; 22.4.4)
(b) Vol. **122**, No. 2, p. 703—708 (1984).

D. L. ALOV, S. I. GUBAREV, and V. B. TIMOFEEV
Determination of the s–d Exchange Interaction Sign in $Cd_{1-x}Mn_xS$ Crystals
(13.4 and 18.2; 22.4.4)
(b) Vol. **122**, No. 2, p. K133—K136 (1984).

P. ŚWIDERSKI and A. TWARDOWSKI
The Excited States of Free Exciton in $Zn_{0.972}Mn_{0.028}Te$
(13.5.1; 22.4.4)
(b) Vol. **122**, No. 2, p. K147—K150 (1984).

G. DEBIAIS
Optical and Electrical Characterization of Mercury Sulfide Mixed Films
(14.3.1 and 20.1; 13.1; 22.4.4)
(a) Vol. **83**, No. 1, p. 269—278 (1984).

D. T. RASULOV and R. AKHMEDOV
Thin-Film Solar Cell on the Basis of Polycrystalline Cadmium Telluride
(14.3.3 and 16; 22.4.1; 22.4.3; 22.4.4)
(a) Vol. **83**, No. 2, p. K211—K213 (1984).

W. GIRIAT and J. STANKIEWICZ
Absorption Edge in $Mn_xZn_{1-x}Te$
(20.1; 13.1; 22.4.4)
(b) Vol. **124**, No. 1, p. K53—K57 (1984).

T. M. Razykov and B. Kh. Kadirov
Energy Band Diagrams of Cu_2S–$Zn_xCd_{1-x}S$ ($0 \leqq x \leqq 1$) Heterojunctions
(13.1 and 14.3.3; 22.4.4)
(a) Vol. **84**, No. 1, p. K71—K74 (1984).

A. J. Szadkowski and A. Lubomirska-Wittlin
Compensation of p-Type $Cd_{1-x}Mn_xTe$ by Diffusion of Ag
(14.3 and 16; 9; 13.4; 22.4.4)
(a) Vol. **84**, No. 2, p. K181—K184 (1984).

H. Booyens and J. H. Basson
Piezoelectrically Induced Charge Distributions around Dislocations in CdTe and HgCdTe
(10.2 and 13.4; 14.4; 22.4.3; 22.4.4)
(a) Vol. **85**, No. 1, p. 243—248 (1984).

K. Mochizuki and K. Suzuki
Effect of the Stoichiometry Control on the Photoelectrical Properties of ZnS_xSe_{1-x}
(16 and 20.3; 22.4.4)
(a) Vol. **85**, No. 1, p. 249—256 (1984).

G. Ambrazevičius, G. Babonas, and Yu. V. Rud
Excitation Spectra of Photoluminescence Bands in $Cd_{1-x}Mn_xTe$
(13.4 and 20.3; 22.4.4)
(b) Vol. **125**, No. 2, p. 759—764 (1984).

C. Fau, J. F. Dame, M. De Carvalho, J. Calas, M. Averous, and B. A. Lombos
Three-Band Model Applied to Narrow-Gap HgCdTe
(14.3; 13.1; 22.4.4)
(b) Vol. **125**, No. 2, p. 831—838 (1984).

H. Booyens and J. H. Basson
On the Optical Choice of Substrate for $Hg_{0.8}Cd_{0.2}Te$ Epitaxial Layers
(1.5 and 10.2; 3; 22.4.4)
(a) Vol. **85**, No. 2, p. 449—454 (1984).

22.5 Halides

Review Articles

K. N. Shrivastava
Origin of Superexchange Interactions in 3d Magnetic Solids (I)
(6 and 18.2; 13; 22.5)
(b) Vol. **125**, No. 1, p. 11—43 (1984).

K. N. Shrivastava
Origin of Superexchange Interactions in 3d Megnetic Solids (II)
(6 and 18.2; 13; 22.5)
(b) Vol. **125**, No. 2, p. 441—460 (1984).

Original Papers and Short Notes

V. A. Haisler, V. M. Zaletin, A. F. Kravchenko, and G. Y. Yashin
Raman Scattering of First and Second Order in Red Mercury Iodide
(6 and 20.1; 22.5)
(b) Vol. **121**, No. 1, p. K13—K17 (1984).

G. Gevers, A. S. Barrière, J. Grannec, L. Lozano, and B. Blanchard
Analysis of FeF_3 and NiF_2 Thin Films by Using RBS, (α, X) Reactions, and SIMS Measurements
(1.3; 14.3.1; 22.5)
(a) Vol. **81**, No. 1, p. 105—122 (1984).

T. S. Verma, V. Kumar, and J. Shanker
Evaluation of Photoelastic Constants and Strain Derivatives of the Static Dielectric Constant of Silver and Thallium Halides
(12.2; 14.4; 22.5)
(b) Vol. **121**, No. 2, p. 487—491 (1984).

J. Y. Bigot and B. Hönerlage
On the Dielectric Function of Three-Level Systems under High Excitation (I): Theoretical Approach
(13.5.1; 20.1; 22.5)
(b) Vol. **121**, No. 2, p. 649—659 (1984).

A. E. Aksenov, Yu. M. Fedorov, and A. A. Leksikov
Systems of Interacting Centers of Photoluminescence in $RbMnF_3$ and $KMnF_3$
(20.3; 13.4; 22.5)
(a) Vol. **81**, No. 2, p. K171—K174 (1984).

T. Chattopadhyay and A. R. Chetal
Plasmon and X-Ray Chemical Shifts in Some Copper Systems
(13.5.2 and 20; 22; 22.5)
(b) Vol. **122**, No. 1, p. 193—196 (1984).

G. Micocci and A. Tepore
Thermally Stimulated Current in PbI_2 at Low Electric Fields
(13.4; 14.4; 22.5)
(a) Vol. **82**, No. 1, p. K79—K82 (1984).

Z. Onyszkiewicz
Theory of the Dependence of the Phase Boundary of $FeBr_2$ Metamagnet on Hydrostatic Pressure
(18.1 and 18.4; 22.5)
(b) Vol. **122**, No. 2, p. 543—547 (1984).

A. I. Bobrysheva, V. V. Baltaga, and M. V. Grodetskii
Biexciton Two-Photon Absorption and Raman Scattering in CuBr
(13.5.1; 20.1; 22.5)
(b) Vol. **123**, No. 1, p. 169—181 (1984).

B. Hönerlage and J. Y. Bigot
On the Dielectric Function of Three-Level Systems under High Excitation (II): Exciton–Biexciton Transitions and Population Effects in CuCl
(13.5.1; 20.1; 22.5)
(b) Vol. **123**, No. 1, p. 201—208 (1984).

N. E. Christensen
First-Principles Calculation of Deformation Potentials of Copper Halides
(13.1; 22.5)
(b) Vol. **123**, No. 1, p. 281—288 (1984).

A. S. Barrière and G. Gevers
Impurity Conduction along the Least Resistance Paths of Granular Insulating Thin Films
(14.3.1; 13.4; 22.5)
(b) Vol. **123**, No. 1, p. 377—391 (1984).

A. I. Belyaeva, M. M. Kotlyarskii, O. V. Miloslavskaya, Yu. N. Stelmakhov, and V. P. Yurev
Investigations of the Crystal and Magnetic Structure Heterogeneities of $RbMnCl_3$
(18.4; 1.4; 22.5)
(a) Vol. **83**, No. 1, p. 245—253 (1984).

Chia-Nan Chang
The Lattice Dynamics of Solid HCl and HBr in Their Low-Temperature Phase
(6; 4; 22.5)
(b) Vol. **123**, No. 2, p. 435—444 (1984).

S. Ghosh, S. Chatterjee, and A. N. Basu
Phase Transition and Soft Mode Behaviour in Superionic Copper Halides
(1.2 and 6; 22.5)
(b) Vol. **123**, No. 2, p. 445—451 (1984).

V. A. Kovarskii, E. P. Sinyavskii, and L. V. Chernysh
Optical Properties of Bielectron–Impurity Complexes
(13.4; 20.1; 22.5)
(b) Vol. **123**, No. 2, p. 671—677 (1984).

R. P. Goyal and B. R. K. Gupta
Evaluation of Ionic Compressibilities and Static Polarizabilities of Cations and Anions of Ammonium Halides
(4; 22.5)
(b) Vol. **123**, No. 2, p. K109—K114 (1984).

R. K. Singh and P. Khare
Analysis of Anharmonic Properties of Some Partially Covalent Crystals
(4 and 12.2; 22.5; 22.5.2)
(b) Vol. **124**, No. 1, p. 29—36 (1984).

A. Vervoitte, J. C. Canit, B. Briat, and U. Kambli
Optical and Magnetic Circular Dichroism Study of the 2D Antiferromagnet Rb_2MnCl_4
(13.5.1 and 13.5.4; 18.4; 20.1; 22.5)
(b) Vol. **124**, No. 1, p. 87—102 (1984).

B. Höhnerlage and J. Y. Bigot
Dispersion and Absorption Anomalies in Three-Level Systems
(13.5.1 and 20.1; 22.5)
(b) Vol. **124**, No. 1, p. 221—227 (1984).

M. Budzynski, E. A. Goremychkin, O. I. Kochetov, A. Latuszynski, P. Mikolajczak, E. Mühle, A. I. Muminov, and M. Subotowicz
Electric Field Gradient at Gd in Gadolinium and Rare Earth Trifluoride Single Crystals
(13.4; 19; 22.5)
(b) Vol. **124**, No. 1, p. 355—362 (1984).

M. Moreno and M. T. Barriuso
An Insight into the $10Dq$ Value Found in $CaZnF_4:Mn^{2+}$
(10.2 and 13.4; 20.3; 22.5)
(b) Vol. **124**, No. 1, p. K83—K86 (1984).

V. N. Vasyukov, S. N. Lukin, and V. T. Telepa
Distortion of the AFMR Line by the RF Field Power
(18.4 and 19; 22.5)
(a) Vol. **84**, No. 1, p. 223—227 (1984).

A. S. Barrière and G. Gevers
Spatial Distribution of Localized Electronic States Responsible for the Polarization of Granular Insulating Thin Films
(13.4 and 14.3.1; 22.5)
(a) Vol. **84**, No. 1, p. 273—284 (1984).

F. Prokert and K. S. Aleksandrov
Neutron Scattering Studies on Phase Transition and Phonon Dispersion in $Cs_2NaBiCl_6$
(1.2 and 6; 22.5)
(b) Vol. **124**, No. 2, p. 503—513 (1984).

M. Okusawa
Splitting of Core-Level XPS Lines of Transition-Metal Ions in Compounds and Moment of the Lines
(13 and 17; 22.5)
(b) Vol. **124**, No. 2, p. 673—681 (1984).

D. Schwarz, H. Roemer, and H.-G. Unruh
Ionic Conductivity of β-CuBr and α-CuI in the Microwave Region
(14.4; 22.5)
(a) Vol. **84**, No. 2, p. K177—K180 (1984).

N. E. Christensen
Deformation Potentials of the Valence Band Edge in CuBr
(13.1; 22.5)
(b) Vol. **125**, No. 1, p. K59—K61 (1984).

R. V. Sharma
Partial Higher-Order Structure Factors in the Long-Wavelength Limit for Molten Cuprous Chloride
(2; 22.5)
(a) Vol. **85**, No. 1, p. 39—43 (1984).

W. Burckhardt, P. Bussemer, and U. Dietrich
Structure and Molar Refraction and Its Wavelenght Dependence at Different Alkali and Ammonium Halides
(4 and 20.1; 22.5; 22.5.2)
(a) Vol. **85**, No. 1, p. 97—103 (1984).

V. G. Artyushenko, V. V. Voitsekhovskii, L. S. Kornienko, I. S. Lisitskii, and A. O. Rybaltovskii
Effect of Ionizing Radiation on Optical Properties of Thallium Halides in the 0.36 to 15 μm Range
(11 and 20.1; 10.2; 22.5)
(a) Vol. **85**, No. 1, p. 167—171 (1984).

V. A. Haisler, V. M. Zaletin, and A. F. ravchenko
Raman Study of Anharmonic Effects in Red Mercury Iodide
(6 and 20.1; 22.5)
(b) Vol. **125**, No. 2, p. K103—K108 (1984).

K. Tennakone and M. Kahanda
Conductivity Transition in Cesium Doped Iodine
(14.3 and 14.4; 22.5)
(b) Vol. **125**, No. 2, p. K181—K183 (1984).

V. L. Safonov
Interactions of Magnons in the Six-Sublattice Antiferromagnet
(13.5.4 and 18.4; 22.5)
(b) Vol. **126**, No. 1, p. 197—206 (1984).

B. A. Orłowski and P. Plenkiewicz
Electronic Band Structure of CdF_2. Photoemission Experiment and Pseudopotential Calculations
(13.1; 22.5)
(b) Vol. **126**, No. 1, p. 285—292 (1984).

U. Braatz and D. Zappe
Investigation of Space Charge Limited, Poole-Frenkel, and Isothermal Currents on HgI_2 Single Crystals
(14.3.4; 13.4; 22.5)
(a) Vol. **86**, No. 1, p. 407—411 (1984).

L. E. Kholodenkov, A. G. Makhanek, and A. A. Kaminskii
One- and Two-Photon Luminescence Excitation of Eu^{3+} in a LaF_3 Monocrystal
(20.1 and 20.3; 13.4; 22.5)
(b) Vol. **126**, No. 2, p. 659—667 (1984).

22.5.1 Silver Halides

Original Papers

K. D. Becker
The Encounter Model in Impurity Diffusion. Correlation Effects in NMR
(9 and 10.2; 19; 22.5.1; 22.5.2)
(b) Vol. **121**, No. 1, p. 91—100 (1984).

U. Sliwczuk, H. Stolz, and W. von der Osten
Indirect-Forbidden Exciton Transitions in AgBr
(13.5.1; 6; 20.1; 22.5.1)
(b) Vol. **122**, No. 1, p. 203—209 (1984).

Th. Kleinefeld and W. von der Osten
Localized and Free Biexitons in Silver Bromide
(13.5.1 and 20.3; 22.5.1)
(b) Vol. **125**, No. 1, p. 293—303 (1984).

22.5.2 Alkali Halides

Review Article

J. SHANKER and G. G. AGRAWAL
Van der Waals Potentials in Ionic Crystals
(4; 6; 12.2; 22.5.2)
(b) Vol. **123**, No. 1, p. 11—26 (1984).

Original Papers and Short Notes

J. OTTO and F. FISCHER
FIR Studies of OCN^-, SCN^- and $SeCN^-$ Centre Gap Modes in KI Single Crystals
(6 and 20.1; 22.5.2)
(b) Vol. **121**, No. 1, p. 81—90 (1984).

K. D. BECKER
The Encounter Model in Impurity Diffusion. Correlation Effects in NMR
(9 and 10.2; 19; 22.5.1; 22.5.2)
(b) Vol. **121**, No. 1, p. 91—100 (1984).

A. KESSLER and P. KIZLER
Investigation of Z Center Conversions in KCl:$SrCl_2$ (I): Optical Measurements of the Kinetics of the Thermal $Z_1 \rightarrow F$, Z_2 Decay
(10.2; 20.1; 22.5.2)
(b) Vol. **121**, No. 1, p. 111—116 (1984).

A. WATTERICH, I. FÖLDVÁRI, G. CORRADI, and H. OPYRCHAL
Comment on the Identification of Z Centres in LiF:Mg and LiF:Mg, Ti Single Crystals
(10.2 and 20.1; 22.5.2)
(b) Vol. **121**, No. 1, p. 117—125 (1984).

J. PRAKASH
Photo-Chemical Reaction Studies in Selenium-Doped Potassium Iodide
(10.2 and 20.1; 22.5.2)
(b) Vol. **121**, No. 1, p. 127—132 (1984).

J. MANAM and V. V. RATNAM
V_3 and F_A Centers in KCl:Na Crystals. A Thermoluminescence Study
(13.4 and 20.3; 10.2; 22.5.2)
(b) Vol. **121**, No. 1, p. 393—399 (1984).

V. N. SALOMATOV and I. A. PARFIANOVICH
The F_H-Centre Energy Structure in KCl and KBr
(13.4 and 20.1; 10.2; 22.5.2)
(b) Vol. **121**, No. 1, p. 401—406 (1984).

R. P. SINGH, J. P. S. RANA, and G. B. VARSHNEY
The Contribution of Zero-Point Vibration toward the Static and Thermodynamical Properties of Potassium Halides
(4; 12.2; 22.5.2)
(b) Vol. **121**, No. 2, p. 493—500 (1984).

A. A. ALYBAKOV, V. A. GUBANOVA, M. ZHANIBEKOV, and N. TOICHEV
Formation and Stability of Paramagnetic Centers in KBr: CrO_4^{2-}, Ca^{2+} Crystals
(19 and 20.1; 6; 13.4; 22.5.2)
(b) Vol. **121**, No. 2, p. K205—K208 (1984).

B. S. ACHARYA, M. L. MUKHERJEE, C. M. SUNTA, and S. P. KATHURIA
Thermoluminescence Emission and Decay Kinetics of NaCl: $SnCl_2$ Single Crystals
(13.4 and 20.3; 22.5.2)
(a) Vol. **81**, No. 2, p. 533—540 (1984).

D. N. POPOV, N. M. NANCHEVA, M. R. MINEV, and N. S. BAKALOVA
On the Nature of the X-Centers in KCl, Studied by Positron Angular Correlation
(10.2; 20.1; 22.5.2)
(a) Vol. **81**, No. 2, p. K141—K144 (1984).

R. K. VARSHNEY and J. SHANKER
Temperature Dependence of Third-Order Elastic Constants of NaCl-Structure Alkali Halide Crystals
(12.2; 22.5.2)
(b) Vol. **122**, No. 1, p. 65—72 (1984).

A. KESSLER and P. KIZLER
Investigation of Z Center Conversions in KCl: $SrCl_2$ (II): Dielectric Relaxation Caused by Mutual Conversions of Z_1, Z_2 and F Centers
(10.2; 14.4; 22.5.2)
(b) Vol. **122**, No. 1, p. 105—112 (1984).

S. BANDYOPADHYAY, C. NEOGY, and S. K. DEB
Activation Energy of Substitutional Defects in Alkali Chlorides
(10.2; 22.5.2)
(b) Vol. **122**, No. 1, p. 113—123 (1984).

A. V. EGRANOV and A. I. NEPOMNYACHIKH
Magnesium Colour Centres in NaF
(10.2 and 13.4; 20.1; 20.3; 22.5.2)
(b) Vol. **122**, No. 1, p. 249—254 (1984).

S. SHIMANUKI and N. NISHIMAKI
A Correlation between the Spectral Location of the Absorption Bands Due to the Localized Centers and the Lattice Constant in Alkali Halide Crystals. Anionic Impurity Centers with s^2-Configuration and F Center which Have Cubic Symmetry
(10.2 and 13.4; 20.1; 22.5.2)
(b) Vol. **122**, No. 1, p. 269—283 (1984).

R. B. Pode and B. T. Deshmukh
Study of Annealing and Deformation Effect on TL of NaCl:KCl
(20.3; 20.1; 22.5.2)
(b) Vol. **122**, No. 1, p. K79—K81 (1984).

G. Baldacchini, G. P. Gallerano, U. M. Grassano, M. Meucci, A. Scacco, F. Somma, A. Tiourine, and M. Tonelli
Optical Pumping and Reorientation of the F_A Centers in KCl:Li^+
(10.2 and 13.4; 20.1; 22.5.2)
(b) Vol. **122**, No. 1, p. K83—K86 (1984).

J. A. Gorri, A. Paz, and F. Guiu
On the Stress Dependence of the Dislocation Velocity in Alkali Halide Crystals
(10.2; 22.5.2)
(a) Vol. **82**, No. 1, p. 85—91 (1984).

K. Narasimha Reddy, U. V. Subba Rao, and M. Lakshmipathi Rao
Pre- and Post-Irradiation Deformation Effects on Thermoluminescence Studies in Barium-Doped NaCl Crystals
(10.2 and 20.3; 22.5.2)
(a) Vol. **82**, No. 1, p. 229—233 (1984).

G. A. Bielig
Room Temperature Effects of γ-Rays on the Dislocation Damping Coefficient in LiF Crystals
(10.2 and 11; 22.5.2)
(a) Vol. **82**, No. 1, p. K43—K47 (1984).

J. Shanker, J. Narain, and A. K. Rajauria
Theory of Compression in Ionic Crystals
(4; 22.5.2)
(b) Vol. **122**, No. 2, p. 435—440 (1984).

S. Dzhumanov and P. K. Khabibullaev
The Coexistence of Unstable, Metastable, and Separated Frenkel Pair Defects in Solids (I): Basic Characteristics
(10.2; 13.4; 22.5.2)
(b) Vol. **122**, No. 2, p. 481—488 (1984).

A. Tóth, T. Keszthelyi, P. Kálmán, and J. Sárközi
Diffusion Model for Charge Transport by Moving Dislocations in Simple Ionic Crystals
(10.2; 22.5.2)
(b) Vol. **122**, No. 2, p. 501—508 (1984).

V. G. Tsirelson, O. V. Korolkova, I. S. Rez, and R. P. Ozerov
Extended Model of Bond Charges and Its Application in Calculation of Optical Properties of Crystals with Different Types of Chemical Bonds
(4; 20.1; 22; 22.5.2; 22.6; 22.8.1)
(b) Vol. **122**, No. 2, p. 599—612 (1984).

R. K. Chaudhary, A. V. Krishna Rao, and K. V. Rao
Effect of Magnetic Field on the F-Band Absorption in NaCl and KCl Single Crystals Irradiated with X-Rays under DC Fields
(10.2 and 20.1; 22.5.2)
(b) Vol. **122**, No. 2, p. 709—713 (1984).

A. Akilbekov and A. Elango
Low-Temperature Pair Associates of H Centres in KBr
(10.2 and 13.4; 20.1; 22.5.2)
(b) Vol. **122**, No. 2, p. 715—723 (1984).

E. N. S. Muccillo and R. Muccillo
Thermally Stimulated Depolarization Currents in γ-Irradiated (Mrad Range) KBr:Sr^{++} Crystals
(10.2 and 11; 14.4; 22.5.2)
(b) Vol. **122**, No. 2, p. K127—K132 (1984).

V. N. Salomatov, N. N. Kristoffel, and I. A. Parfianovich
On the Nature of Metastable Color Centers in KBr
(13.4; 10.2; 22.5.2)
(b) Vol. **122**, No. 2, p. K183—K185 (1984).

B. I. Smirnov, T. N. Snezhkova, and O. L. Volkova
Fracture in LiF Bicrystals at Plastic Deformation
(10.2; 22.5.2)
(a) Vol. **82**, No. 2, p. 441—447 (1984).

B. Sorkin and H. Käämbre
V_K Centres in Thermo-Exoemission from CsBr:Tl
(10.2 and 17; 13.4; 22.5.2)
(a) Vol. **82**, No. 2, p. K149—K152 (1984).

A. J. Kaur, W. N. Bhende, P. S. Bakhshi, and J. Shanker
Analysis of the Short-Range Repulsive Potential in Alkali Halides
(4; 6; 8; 22.5.2)
(b) Vol. **123**, No. 1, p. 71—78 (1984).

C. Ortiz, C. N. Afonso, R. M. Shelby, A. C. Tropper, and G. C. Bjorklund
Photochemical Hole Burning in Polycrystalline and Neutron Irradiated NaF
(10.2 and 20.3; 22.5.2)
(b) Vol. **123**, No. 1, p. 79—84 (1984).

L. N. Kantorovich
Multipole Theory of the Polarization of Solids by Points Defects (II): The Point Charge Approximation
(13.4 and 14.4; 22.5.2)
(b) Vol. **123**, No. 1, p. 325—334 (1984).

D. Gerlich
Fourth-Order Elastic Constants of Solids
(12.1 and 12.2; 21; 21.6; 22.5.2)
(b) Vol. **123**, No. 1, p. K5—K7 (1984).

E. Radzhabov
Impurity Centers in $LiF:Li_2O$
(13.4 and 20.3; 10.2; 22.5.2)
(b) Vol. **123**, No. 1, p. K79—K82 (1984).

M. Aoki, T. Nishikawa, and S. Okabe
Pattern Variation of Evaporated Thin Films and Their Non-Pre-Excited TSEE
(1.5 and 17; 20.3; 22.5.2; 22.5.3)
(a) Vol. **83**, No. 1, p. 85—91 (1984).

B. P. Chandra, R. D. Verma, B. Majumdar, and V. P. Bhatt
Dependence of Mechanoluminescence on the Microhardness and Dislocation Density in Crystals
(12.2 and 20.3; 10.2; 22.5.2; 22.8.1)
(a) Vol. **83**, No. 1, p. K19—K22 (1984).

B. Sorkin and H. Käämbre
A New Evidence for Hole-Induced Thermo-Exoemission
(10.2; 17; 20.3; 22.5.2)
(a) Vol. **83**, No. 1, p. K39—K42 (1984).

B. B. S. Negi and P. N. Ram
First-Order Raman Scattering in RbCl Containing F-Centres
(10.2; 6; 20.1; 22.5.2)
(b) Vol. **123**, No. 2, p. 491—495 (1984).

L. Bosi, P. Gagliardelli, and M. Nimis
New Results on the Decay Properties of Perturbed and Unperturbed M-Centers in $NaCl:CaCl_2$
(10.2; 20.1; 20.3; 22.5.2)
(b) Vol. **123**, No. 2, p. 519—524 (1984).

R. V. Gopala Rao and Sobhen Kumar Pal
Molten Rubidium Bromide — A Rigorous Study of Its Partial Structure and Allied Function
(2 and 4; 22.5.2)
(a) Vol. **83**, No. 2, p. 429—436 (1984).

B. Sorkin and H. Käämbre
Thermo-Exoemission of NaCl:Ag Crystals, Cleaved before and after X-Irradiation
(17; 11; 13.3; 13.4; 22.5.2)
(a) Vol. **83**, No. 2, p. K171—K174 (1984).

T. Kurobori and N. Takeuchi
Room Temperature, Stable Laser Action Using F_2 Centers in LiF:Mg, OH Crystals
(20.2; 10.2; 22.5.2)
(a) Vol. **83**, No. 2, p. K179—K182 (1984).

R. K. Singh and P. Khare
Analysis of Anharmonic Properties of Some Partially Covalent Crystals
(4 and 12.2; 22.5; 22.5.2)
(b) Vol. **124**, No. 1, p. 29—36 (1984).

H. Reimer and F. Fischer
SERS from Ag Colloids in Alkali Halide Crystals
(6 and 20.1; 10.2; 22.5.2)
(b) Vol. **124**, No. 1, p. 61—67 (1984).

M. J. Stillman, P. W. M. Jacobs, K. Oyama Gannon, and D. J. Simkin
Absorption and Magnetic Circular Dichroism Spectra of $CsI:Tl^+$
(18 and 20.1; 22.5.2)
(b) Vol. **124**, No. 1, p. 261—270 (1984).

P. Moretti, A. Ranfagni, and A. Agresti
Ground and Excited States of Tl^+ Impurity in KCl Crystal by the Use of the Green's Function Method
(13.4; 22.5.2)
(b) Vol. **124**, No. 1, p. 297—306 (1984).

I. Baltog and L. Mihuţ
Luminescence Decay in NaI:Pb
(20.3; 13.4; 22.5.2)
(b) Vol. **124**, No. 1, p. 307—313 (1984).

A. A. Alybakov, V. A. Gubanova, G. S. Denisov, and B. S. Umurzakov
Influence of Oxygen Centers on the Thermoluminescence of NaF Crystals
(13.4 and 20.3; 10.2; 22.5.2)
(b) Vol. **124**, No. 1, p. K75—K78 (1984).

A. Subrahmanyam and K. V. S. Badarinath
Dielectric Properties of RbCl Single Crystals
(14.4; 11; 22.5.2)
(a) Vol. **84**, No. 1, p. K93—K96 (1984).

N. Dass and M. Kumari
On the Isothermal Anderson-Grüneisen Parameter in Solids
(8; 22.5.2)
(b) Vol. **124**, No. 2, p. 531—536 (1984).

S. Bandyopadhyay and S. K. Deb
Solution Enthalpies of Divalent Defects in Alkali Fluorides and Chlorides
(10.2; 22.5.2)
(b) Vol. **124**, No. 2, p. 545—557 (1984).

V. I. Zametin
Absorption Edge Anomalies in Polar Semiconductors and Dielectrics at Phase Transitions
(20.1; 1.2; 14.4.1; 22; 22.5.2; 22.8.1; 22.9)
(b) Vol. **124**, No. 2, p. 625—640 (1984).

L. O. Schwan, R. Florian, D. Schmid, E. Betz, W. Schrof, B. Walker, and H. Port
Dynamics of the Optical Excitation–Deexcitation Cycle of O_2^- Centers in KCl
(10.2 and 13.4; 20.3; 22.5.2)
(b) Vol. **124**, No. 2, p. 741—745 (1984).

M. MATOS and B. MAFFEO
Molecular Cluster-INDO Calculations of Substitutional Defects in Alkali Halides
(13.4; 20.1; 22.5.2)
(b) Vol. **124**, No. 2, p. 757—765 (1984).

V. FREI and P. DEUS
A Frequency Sum Rule and Spectral Moments in Lattice Dynamics
(6; 22.1; 22.1.1; 22.2.2; 22.5.2; 22.8)
(b) Vol. **125**, No. 1, p. 121—130 (1984).

V. NAGIRNYI, T. SOOVIK, and S. ZAZUBOVICH
Photostimulated Reorientiom of $H_z^{[001]}(Pb^{2+})$ Centres in a $KCl:PbCl_2$ Crystal
(10.2 and 20.3; 22.5.2)
(b) Vol. **125**, No. 1, p. 139—145 (1984).

G. R. AUGST and YU. I. VLADISLAVSKII
A Theoretical Study of Simple Complexes of Point Defects in Alkali Halides by the Lattice Static Method
(10.2; 22.5.2)
(b) Vol. **125**, No. 1, p. 155—160 (1984).

D. E. POPOV, E. I. ROZUM, S. R. UGLOV, and S. A. VOROBIEV
Anomalous Radiation of Fast Channeled Electrons in Optically Transparent Crystals
(11 and 20; 13; 22.1.3; 22.5.2)
(b) Vol. **125**, No. 1, p. 251—258 (1984).

E. MUGEŃSKI and R. CYWIŃSKI
Low-Temperature Photoluminescence of the Aggregate Centres in Eu^{2+}-Doped KCl Crystals
(20.3; 10.2; 22.5.2)
(b) Vol. **125**, No. 1, p. 381—386 (1984).

G. I. BERSUKER and V. Z. POLINGER
Multimode Pseudo Jahn-Teller Effect for Off-Centre Impurities in Crystals
(6 and 13.4; 10.2; 22.5.2)
(b) Vol. **125**, No. 1, p. 401—408 (1984).

J. SHANKER, W. N. BHENDE, and P. S. BAKHSHI
Analysis of the Equation of State for LiF Crystal
(4; 12.2; 22.5.2)
(b) Vol. **125**, No. 1, p. K21—K26 (1984).

YU. M. ANNENKOV, V. F. STOLARENKO, and T. S. FRANGULIAN
Effect of Powerful Pulsed Electron Irradiation on Creation of Hole Colour Centres in KCl at 300 K
(10.2 and 20.1; 11; 22.5.2)
(b) Vol. **125**, No. 1, p. K37—K40 (1984).

V. A. CHEPURNOI, S. N. MYSOVSKII, I. A. PARFIANOVICH, V. N. SALOMATOV, and A. P. SHKADAREVICH
Influence of Oxygen Defects on the Position and Form of the F_2^+ Center Zero-Phonon Lines in LiF
(10.2 and 13.4; 20.3; 22.5.2)
(b) Vol. **125**, No. 1, p. K41—K44 (1984).

W. BURCKHARDT, P. BUSSEMER, and U. DIETRICH
Structure and Molar Refraction and Its Wavelength Dependence at Different Alkali and Ammonium Halides
(4 and 20.1; 22.5; 22.5.2)
(a) Vol. **85**, No. 1, p. 97—103 (1984).

T. SATOW, O. UEMURA, K. HOSHINO, and T. WATANABE
The Neutron Diffraction of Molten TlCl, TlBr, and TlI
(1 and 2; 22.5.2)
(a) Vol. **85**, No. 1, p. K5—K7 (1984).

G. HEDER and J. M. SPAETH
ENDOR Investigation of H_i^0(Li) Centres in KCl
(10.2 and 19; 13.4; 22.5.2)
(b) Vol. **125**, No. 2, p. 523—531 (1984).

O. V. BOEV and K. P. AREFIEV
Delocalized Positronium in Alkali Halides
(13; 22.5.2)
(b) Vol. **125**, No. 2, p. 619—627 (1984).

S. SAPRU and K. L. TAN
Eu^{2+} Ions Doped in Alkali Halides Act as Donors on Irradiation
(20.3; 11; 13.4; 22.5.2)
(b) Vol. **125**, No. 2, p. K161—K163 (1984).

A. A. EL-SHARKAWY, I. H. RASHED, M. S. ZAGHLOUL, and M. H. GHONIEM
Study of Thermal Properties of Polycrystalline KCl, KBr, and KI in the Temperature Range 300 to 600 K
(8; 22.5.2)
(a) Vol. **85**, No. 2, p. 429—434 (1984).

A. K. ARORA
Isochronal Annealing Study of Aggregation in Doped Alkali Halide Crystals
(10.2; 22.5.2)
(a) Vol. **85**, No. 2, p. 491—495 (1984).

B. A. VAID, K. C. SHARMA, and V. K. SYAL
A Disorder Model of Melting (II): Alkali Halides
(2; 8; 22.5.2)
(b) Vol. **126**, No. 1, p. 59—62 (1984).

R. K. VARSHNEY and J. SHANKER
Temperature Dependence of Third-Order Elastic Constants of CsCl, CsBr, and CsI Crystals
(12.2; 22.5.2)
(b) Vol. **126**, No. 1, p. 77—82 (1984).

ZH. EGEMBERDIEV, V. NAGIRNYI, T. SOOVIK, and S. ZAZUBOVICH
Slow Decay Component of the A_T Emission in KBr:Pb Crystals
(13.4 and 20.3; 22.5.2)
(b) Vol. **126**, No. 1, p. K63—K66 (1984).

R. V. G. Rao and S. Kumar Pal
A New Method of Evaluation of the Structure of Molten Sodium Chloride
(2; 4; 22.5.2)
(a) Vol. **86**, No. 1, p. 43—54 (1984).

Ch. Lushchik, J. Kolk, A. Lushchik, and N. Lushchik
Radiational Creation of Frenkel Defects in KCl–Tl
(10.2; 19; 20.1; 20.3; 22.5.2)
(a) Vol. **86**, No. 1, p. 219—227 (1984).

S. C. Sabharwal, S. P. Kathuria, and B. Ghosh
Kinetics of Thermoluminescence in γ-Irradiated NaI(Tl)
(20.3; 11; 13.4; 22.5.2)
(a) Vol. **86**, No. 1, p. 327—331 (1984).

R. Capelletti and M. Manfredi
The Role of Eu^{++} Aggregation on Optical Absorption, Emission, Luminescence Time Decay, and ITC Plots in NaCl:Eu^{++}
(20.1 and 20.3; 13.4; 22.5.2)
(a) Vol. **86**, No. 1, p. 333—343 (1984).

R. P. Singh, J. P. S. Rana, A. Daud, and G. B. Varshney
The Role of the Zero-Point Vibration towards Crystal Properties of Rubidium Halide Crystals
(4; 12.2; 22.5.2)
(b) Vol. **126**, No. 2, p. 495—499 (1984).

V. Nagirnyi, T. Soovik, and S. Zazubovich
Decay Kinetics and Polarization of the A_x Emission of KI:In Crystal
(13.4 and 20.3; 22.5.2)
(b) Vol. **126**, No. 2, p. 653—658 (1984).

K. Iwahana
Magnetic Field Dependence of the Spin Polarization of the F Center
(10.2; 18.1; 22.5.2)
(b) Vol. **126**, No. 2, p. 693—702 (1984).

S. Bandyopadhyay and S. K. Deb
Solution Enthalpies of Trivalent Defects in Alkali Chlorides
(10.2; 22.5.2)
(b) Vol. **126**, No. 2, p. K117—K121 (1984).

Y. Q. Sheng, R. Y. Li, and P. Munz
The Study of Mode Characteristics of Metal Clad Dielectric Waveguides
(13.5.2 and 20; 21.6; 22.5.2)
(a) Vol. **86**, No. 2, p. 813—819 (1984).

22.5.3 Alkali-Earth Halides

Original Papers and Short Notes

T. Eshita, K. Tanimura, and N. Itoh
Photo-Induced Transformation of Close Frenkel Pairs in Strontium Fluorite
(10.2 and 20.1; 20.3; 22.5.3)
(b) Vol. **122**, No. 2, p. 489—500 (1984).

W. Pastusiak
Erratum to: The Relativistic Rigid Lattice Contributions to the Spin–Lattice Coupling Coefficients of the S-State $4f^7$ Ions
(19; 22.5.3)
(b) Vol. **122**, No. 2, p. K195 (1984).

R. Rauch
Photochemical Conversion of M_A Centres and Identification of M_A^+ Centres in CaF_2 : Na
(10.2; 20.1; 20.3; 22.5.3)
(b) Vol. **123**, No. 1, p. 93—98 (1984).

M. Aoki, T. Nishikawa, and S. Okabe
Pattern Variation of Evaporated Thin Films and Their Non-Pre-Excited TSEE
(1.5 and 17; 20.3; 22.5.2; 22.5.3)
(a) Vol. **83**, No. 1, p. 85—91 (1984).

L. Wendler
Phonon-Polaritons in Bilayer Systems. p-Polarized Guided Wave Phonon-Polaritons and Guided Wave Surface Phonon-Polaritons
(6 and 20.1; 13.5; 22.2.1; 22.2.2; 22.5.3)
(b) Vol. **123**, No. 2, p. 469—477 (1984).

N. Agnihotri, R. K. Singh, and S. P. Sanyal
Interionic Forces and Elastic Properties of Na_2S
(4; 6; 12.2; 22; 22.5.3)
(b) Vol. **125**, No. 1, p. 113—119 (1984).

A. Davituliani, M. Kiknadze, R. Mirianashvili, and T. Sanadze
Ligand Hyperfine Interaction of Er^{3+} in BaF_2 Single Crystals
(19; 22.5.3)
(b) Vol. **125**, No. 1, p. K63—K65 (1984).

P. Varotsos, K. Alexopoulos, and M. Lazaridou
On the Proportionality of the Migration Volume and the Migration Enthalpy in Fluorides
(10.2; 9; 22.5.3)
(b) Vol. **125**, No. 2, p. K109—K112 (1984).

W. Pastusiak
Dynamical Relativistic Crystal Field Contributions to the Zero Field Splitting of S-State $4f^7$ Ions
(13.4 and 19; 22.5.3)
(b) Vol. **126**, No. 1, p. K77—K81 (1984).

M. S. JAHAN and D. W. COOKE
Thermal Evolution of Low-Temperature Manganese Centers in X-Irradiated CaF_2:Mn
(10.2 and 19; 11; 20.3; 22.5.3)
(b) Vol. **126**, No. 2, p. 687—692 (1984).

M. LAZARIDOU
Connection of the Defect Volumes to the Bulk Properties in CaF_2
(10.2; 22.5.3)
(b) Vol. **126**, No. 2, p. K123—K124 (1984).

22.5.4 Mixed Crystals

Original Papers and Short Notes

M. Hidaka, S. Yamashita, and Y. Okamoto
Study of Structural Phase Transitions of $KCaF_3$
(1.2; 6; 8; 22.5.4)
(a) Vol. **81**, No. 1, p. 177—183 (1984).

K. Somaiah and V. Hari Babu
Glow Curves and the Emission of Flux Grown BaFCl:Gd Crystals
(13.4 and 20.3; 20.1; 22.5.4)
(a) Vol. **82**, No. 1, p. 201—206 (1984).

S. N. Maiti and D. Roy
Lattice Dilatation in High-Concentration Binary Solid Solutions
(4; 12.2; 22.5.4)
(b) Vol. **122**, No. 2, p. 449—460 (1984).

R. K. Singh, Neeta Agnihotri, S. V. Suryanarayana, and S. P. Sanyal
Study of Cohesion, Harmonic and Anharmonic Elastic Properties of CaF_2–SrF_2 Mixed Crystals
(4; 12.2; 22.5.4)
(b) Vol. **123**, No. 2, p. 453—462 (1984).

M. Kibler, G. Grenet, J. C. Gâcon, and B. Jacquier
Energy Levels of Sm^{2+} in BaClF and SrClF Crystals under Magnetic Field (I): Theory
(13.4 and 20.3; 18; 22.5.4)
(b) Vol. **124**, No. 1, p. 325—334 (1984).

R. Sudharsanan, T. K. K. Srinivasan, and S. Radhakrishna
Raman and Infrared Studies on Ag_2HgI_4 and Cu_2HgI_4 Superionic Compounds
(6 and 20.1; 22.5.4)
(b) Vol. **124**, No. 2, p. 481—487 (1984).

M. Kibler, J. C. Gâcon, G. Grenet, B. Jacquier, and L. C. Brunel
Energy Levels of Sm^{2+} in BaClF and SrClF Crystals under Magnetic Field (II): Results
(13.4 and 20.3; 18; 22.5.4)
(b) Vol. **124**, No. 2, p. 731—739 (1984).

K. Somaiah and V. Hari Babu
Two Types of F-Centres and Thermoluminescence in BaFCl Crystals
(10.2; 20.1; 20.3; 22.5.4)
(a) Vol. **85**, No. 1, p. 265—271 (1984).

G. S. Kumar, S. Sudhakar, R. Ethiraj, and V. G. Krishna Murthy
Strain-Optical Constant (p_{44}) of Mixed Crystals of KCl–KBr of Equimolar Concentration
(12.2 and 20; 22.5.4)
(a) Vol. **85**, No. 1, p. K27—K30 (1984).

ZH. EGEMBERDIEV, V. NAGIRNYI, T. SOOVIK, and S. ZAZUBOVICH
Decay Kinetics and Polarization of the A_T Emission of Pb^{2+} Centres of Different Structure in KBr: $PbBr_2$
(13.4 and 20.3; 22.5.4)
(b) Vol. **126**, No. 1, p. 407—414 (1984).

M. HIDAKA, H. FUJII, B. J. GARRARD, and B. M. WANKLYN
Structural Phase Transitions of $RbVF_4$
(1.2; 22.5.4)
(a) Vol. **86**, No. 1, p. 75—83 (1984).

S. IHARA, Y. WARITA, and K. SUZUKI
Ionic Conductivity in $AgI_{1-x}Cl_x$
(10.2 and 14.4; 22.5.4)
(a) Vol. **86**, No. 2, p. 729—734 (1984).

22.6 Simple Oxides (CdO, ZnO See 22.4)

Review Article

C. Boulesteix
A Survey of Domains and Domain Walls Generated by Crystallographic Phase Transitions Causing a Change of the Lattice
(1.2 and 1.4; 14.4.1; 22; 22.6; 22.8)
(a) Vol. **86**, No. 1, p. 11—42 (1984).

Original Papers and Short Notes

D. Ihle
Polaronic Effects on the Verwey Transition in Fe_3O_4
(13.5.3; 22.6)
(b) Vol. **121**, No. 1, p. 217—224 (1984).

V. F. Petrenko and I. A. Ryzhkin
Dielectric Properties of Ice in the Presence of Space Charge
(14.3.4 and 14.4; 22.6)
(b) Vol. **121**, No. 1, p. 421—427 (1984).

V. N. Abramov, A. N. Ermoshkin, A. I. Kuznetsov, and V. V. Mürk
Breakdown of Excitonic Absorption and Luminescence Induced by a d-Type Conduction Band in Sc_2O_3
(13.5.1 and 20.3; 20.1; 22.6)
(b) Vol. **121**, No. 1, p. K59—K63 (1984).

G. Nihoul and C. Cesari
High Resolution Images for Large Cell Crystals Exhibiting Some Limits of the Projected Potential Approximation
(1.1 and 1.4; 22.6)
(a) Vol. **81**, No. 1, p. 87—98 (1984).

V. B. Kokshenev, K. Balcerek, T. Tyc, and A. Jeżowski
Discussion of the Heat Flux Rectification in the Solid–Solid System in the Acoustic-Mismatch Theory Framework
(8; 22.6)
(a) Vol. **81**, No. 1, p. 171—176 (1984).

H.-J. Fitting, J. Boyde, and J. Reinhardt
Monte-Carlo Approach of Electron Emission from SiO_2
(17; 22.6)
(a) Vol. **81**, No. 1, p. 323—332 (1984).

A. U. Grinfelds, A. R. Silin, L. N. Skuya, and V. G. Plekhanov
Low Temperature Recombination Luminescence in Amorphous and Crystalline SiO_2
(20.3; 2; 13.5.3; 22.6)
(a) Vol. **81**, No. 1, p. K23—K26 (1984).

S. K. Patapis and K. Alexopoulos
Thermoelectric Power of Hematite at the Morin Transition
(15; 22.6)
(a) Vol. **81**, No. 1, p. K57—K61 (1984).

I. Hevesi, L. B. Kiss, M. I. Török, and L. Nánai
Electron Concentration in V_2O_5 Single Crystals as Determined by $1/f$ Noise Measurements
(14.3; 22.6)
(a) Vol. **81**, No. 1, p. K67—K70 (1984).

A. Lehmann, L. Schumann, and K. Hübner
Optical Phonons in Amorphous Silicon Oxides (II): Calculation of Phonon Spectra and Interpretation of the IR Transmission of SiO_x
(6 and 20.1; 2; 22.6)
(b) Vol. **121**, No. 2, p. 505—511 (1984).

S. Wysocki
Kinetics of the Decay of Surface γ-Conductivity of Magnesium Oxide
(13.4; 14.3.2; 22.6)
(b) Vol. **121**, No. 2, p. 731—736 (1984).

M. I. Auslender, V. V. Karyagin, A. B. Zolotovitskii, V. P. Kalashnikov, and B. A. Gizhevskii
Longitudinal Nernst-Ettingshausen Effect in Heavily Doped EuO
(15; 22.6)
(b) Vol. **121**, No. 2, p. 737—741 (1984).

L. Wendler and E. Jäger
Surface Polaritons in a SiO_2/Si Bilayer System
(6 and 13.5; 22.1.2; 22.6)
(b) Vol. **121**, No. 2, p. K119—K124 (1984).

S. Motogi and G. A. Maugin
Effects of Magnetostriction on Vibrations of Bloch and Néel Walls
(18.2; 12.2; 22.1.2; 22.6)
(a) Vol. **81**, No. 2, p. 519—532 (1984).

N. A. Vlasenko, Z. L. Denisova, and V. S. Khomchenko
Electroluminescence of SiO_x–LnF_3 Thin Films
(14.3.4 and 20.3; 14.3.1; 22.6)
(a) Vol. **81**, No. 2, p. 657—660 (1984).

M. G. Abramishvili, V. I. Altukhov, T. L. Kalabegishvili, and V. G. Kvachadze
Thermal Conductivity and Optical and Paramagnetic Absorption Spectra of Reactor Irradiated MgO Crystals
(8 and 11; 20.1; 22.6)
(b) Vol. **122**, No. 1, p. 95—104 (1984).

M. Nylén
Electronic Structure of an Unreconstructed $(10\bar{1}0)$ α-Quartz Surface
(13.1 and 13.3; 22.6)
(b) Vol. **122**, No. 1, p. 301—308 (1984).

K. Feldmann and L. Fuentes
On the Possibility of Determining the Odd Part of the Orientation Distribution Function of Texturized Samples for Quartz-Like Structures Using Anomalous Scattering
(1.3 and 1.4; 22.6; 22.8.1)
(b) Vol. **122**, No. 1, p. K5—K9 (1984).

H. Szymczak
Acoustical Relaxation Due to the Bound Small Polaron in Cubic Crystals
(7 and 13.5.3; 22.6)
(b) Vol. **122**, No. 1, p. K37—K40 (1984).

N. N. Kurkin and N. K. Solovarov
The Variation of Light Transmission of Heavily Doped Ruby at Magnetic Resonance
(19; 20.1; 22.6)
(b) Vol. **122**, No. 1, p. K69—K72 (1984).

R. D. Baëta
Steady-State Creep in Quartz
(10.2; 22.6)
(a) Vol. **82**, No. 1, p. 101—109 (1984).

E. Macho, J. Llopis, A. Remón, C. Ballesteros, and J. Piqueras
Optical Study of Defects in Quenched MgO Crystals
(10.2 and 20.3; 20.1; 22.6)
(a) Vol. **82**, No. 1, p. 111—117 (1984).

W. Hohenau
The Influence of Silver and Copper Electrolysis on the Radiation-Induced Optical Absorption of Natural Quartz
(11 and 20.1; 22.6)
(a) Vol. **82**, No. 1, p. 119—124 (1984).

G. Brauer, F. Kerbe, Zs. Kajcsos, and A. Ashry
Positron Annihilation Studies on BeO Powders
(10.2; 22.6)
(a) Vol. **82**, No. 1, p. 143—148 (1984).

N. Keddar, D. L. Kirk, and S. G. Johnson
The Surface Degradation of the β''-Al_2O_3 Solid Electrolyte Utilised in Na–S Battery Cells
(1.6 and 14.3.2; 22.6)
(a) Vol. **82**, No. 1, p. 327—334 (1984).

T. Tsuji, K. Naito, and K. Ishigure
Effect of Particle Size on Mössbauer Parameters of α-Fe_2O_3
(5 and 18; 22.6)
(a) Vol. **82**, No. 1, p. K57—K61 (1984).

T. Wiktorczyk and C. Wesołowska
Application of TSEE and TSD Methods to Investigations on Trap Levels in Yb_2O_3 Thin Films Deposited by Electron Beam Evaporation
(13.4; 22.6)
(a) Vol. **82**, No. 1, p. K67—K70 (1984).

A. Korneta and A. Milewski
Study of Materials with Uniaxial Anisotropy Using a Dielectric Resonator Arrangement
(14.4; 22.6)
(a) Vol. **82**, No. 1, p. K107—K110 (1984).

V. G. Tsirelson, O. V. Korolkova, I. S. Rez, and R. P. Ozerov
Extended Model of Bond Charges and Its Application in Calculation of Optical Properties of Crystals with Different Types of Chemical Bonds
(4; 20.1; 22; 22.5.2; 22.6; 22.8.1)
(b) Vol. **122**, No. 2, p. 599—612 (1984).

O. P. Aggarwal and N. N. Saxena
K-Absorption Spectra of Copper Oxides
(13 and 20; 22.6)
(b) Vol. **122**, No. 2, p. 669—673 (1984).

Young Hwan Choi, Youn Soo Lim, Chul Koo Kim, and Kyun Nahm
Electrical Conductivity of Magnetite below the Verwey Transition Temperature
(14.3; 22.6)
(b) Vol. **122**, No. 2, p. K187—K190 (1984).

Dao Khac An
An Approximate Method for Determining the Segregation Coefficient of the Boron Drive Diffusion in Oxidizing Ambients
(9; 22.1.2; 22.6)
(a) Vol. **82**, No. 2, p. K145—K148 (1984).

G. Jungk and A. Röseler
Ellipsometric Investigation of Lattice Vibrations in Anisotropic Systems. α-Quartz
(6 and 20.1; 22.6)
(b) Vol. **123**, No. 1, p. 65—70 (1984).

F. López-Aguilar and J. Costa-Quintana
A First Principles Pseudopotential Model for the Strong Intrasite Interaction Applied to the $4f^{13}$ Configuration (Yb_2O_3)
(13.1; 22.6)
(b) Vol. **123**, No. 1, p. 219—228 (1984).

M. J. Springis and J. A. Valbis
Visible Luminescence of Colour Centres in Sapphire
(10.2 and 20.3; 22.6)
(b) Vol. **123**, No. 1, p. 335—343 (1984).

V. S. Grunin and I. B. Patrina
Impurities and Charge Compensation in Some Molybdenium Oxides
(13.4 and 19; 22.6; 22.8.1)
(b) Vol. **123**, No. 1, p. 353—363 (1984).

F. Appel, M. Bartsch, U. Messerschmidt, E. M. Nadgornyi, and S. N. Valkovskii
Dislocation Motion and Plasticity in MgO Single Crystals
(10.2; 22.6)
(a) Vol. **83**, No. 1, p. 179—194 (1984).

V. Rybka, V. Hnatowicz, J. Kvítek, J. Vacík, and B. Schmidt
Determination of the Range Profiles of Boron Implanted into Si and SiO_2
(11; 22.1.2; 22.6)
(a) Vol. **83**, No. 1, p. 165—171 (1984).

J. J. Kleperis, P. D. Cikmach, and A. R. Lusis
Colour Centres in Amorphous Tungsten Trioxide Thin Films
(13.4 and 20.1; 14.3.1; 19; 22.6)
(a) Vol. **83**, No. 1, p. 291—297 (1984).

A. L. Shabalov and M. S. Feldman
Optical Properties of Thin GeO_x Films
(6 and 20.1; 22.6)
(a) Vol. **83**, No. 1, p. K11—K14 (1984).

R. Brazis, P. Pipinys, and A. Rimeika
Temperature Dependence of Currents in Thin Films of Silicon Oxide
(13.4 and 14.3.4; 22.6)
(a) Vol. **83**, No. 1, p. K69—K71 (1984).

S. K. Agrawal, J. Narain, and J. Shanker
Investigation of the Equation of State and the Grüneisen Parameters for Transition Metal Oxides
(4; 6; 12.2; 22.6)
(b) Vol. **123**, No. 2, p. 497—501 (1984).

A. E. Smith
A Calculation of Absorption Potentials for Low Energy Electrons in MgO, 2H-MoS_2, and 2H-$NbSe_2$
(1.6 and 13; 22; 22.6)
(b) Vol. **123**, No. 2, p. 619—626 (1984).

S. K. J. Al-Ani, C. A. Hogarth, and D. N. Waters
The Effect of Temperature on the Optical Absorption Edge of Amorphous Thin Films of Silicon Monoxide
(20.1; 2; 13; 22.6)
(b) Vol. **123**, No. 2, p. 653—658 (1984).

M. Ferrari, M. Montagna, L. Gonzo, G. Viliani, O. Pilla, and G. Farace
Resonant Energy Transfer in Ruby under Magnetic Field
(13.4; 20.3; 22.6)
(b) Vol. **123**, No. 2, p. 659—662 (1984).

T. Grafe and R. Strehlow
Influence of Si–Si Bonds on the Local Electronic Density of States of a-SiO_2
(13.1 and 13.4; 22.6)
(b) Vol. **123**, No. 2, p. 663—669 (1984).

B. Müller, P. Grau, and G. Kluge
The Hardness of Chromium- and Titanium-Doped Sapphire Crystals
(12.2; 22.6)
(a) Vol. **83**, No. 2, p. 499—506 (1984).

C. Ballesteros, J. Piqueras, J. Llopis, and R. González
Cathodoluminescence from MgO Single Crystals Containing a High Concentration of Anion Vacancies
(20.3; 10.2; 13.4; 22.6)
(a) Vol. **83**, No. 2, p. 645—649 (1984).

M. T. Mohammad and A. K. Abass
Optical Properties of Chemically Deposited Conducting Glass for SIS Solar Cells
(20.1; 22.6)
(a) Vol. **83**, No. 2, p. 681—683 (1984).

K. Ots and M. Seilenthal
On the Structural Model of CVD a-Si:O:H Films
(1.5 and 2; 20.1; 22.1.2; 22.6)
(a) Vol. **83**, No. 2, p. K109—K111 (1984).

G. Jungk and A. Röseler
LO–TO–Splitting and Many-Particle Effects
(6 and 20.1; 2; 22.6)
(b) Vol. **124**, No. 1, p. K1—K4 (1984).

S. A. Fysh and P. E. Clark
On Mössbauer Analysis of Mineral Mixtures Having Environmentally Broadened Spectral Lines. Application to Bauxite
(1.3; 5; 22.6)
(a) Vol. **84**, No. 1, p. 31—38 (1984).

A. Watterich, E. Szontagh, and É. Kocsárdy
Paramagnetic Centres in Aluminium-Oxide Layers Prepared by Anodic Oxidation
(10.2; 19; 20.1; 22.6)
(a) Vol. **84**, No. 1, p. 165—169 (1984).

W. Winkler
Mössbauer Studies of Small Magnetite Particles of a Magnetic Fluid
(18.2; 5; 22.6)
(a) Vol. **84**, No. 1, p. 193—198 (1984).

A. L. Pushkarchuk, P. P. Mardilovich, A. I. Trokhimets, G. M. Zhidomirov, and S. G. Gagarin
The Energy Band Structure of the Corundum (0001) Surface with Different Extent of Hydroxylation
(13.1 and 13.3; 22.6)
(b) Vol. **124**, No. 2, p. 699—705 (1984).

T. F. Veremeichik
Similarity of Ground- and Excited-State Absorption Spectra of Cr^{3+} Ions in a Strong Crystal Field
(13.4 and 20.1; 22.6)
(b) Vol. **124**, No. 2, p. 719—729 (1984).

A. Stern and K. Hübner
Electron Density of States of an Si–SiO_2 Interface with a Gradual Chemical Transition
(13.3; 22.1.2; 22.6)
(b) Vol. **124**, No. 2, p. K171—K175 (1984).

W. Brodkorb, J. Salm, and E. Steinbeiss
Discussion of Possible Surface Processes during the Reactive SiO_x Film Deposition in a Planar Magnetron Sputtering Equipment
(1.5 and 2; 22.6)
(a) Vol. **84**, No. 2, p. 379—386 (1984).

G. Brauer, F. Kerbe, Zs. Kajcsos, and A. Ashry
Positron Annihilation Studies on Al_2O_3 Powders
(10.2; 13.3; 22.6)
(a) Vol. **84**, No. 2, p. 451—455 (1984).

A. L. Bhattacharyya and N. Ch. Mitra
Dielectric Relaxations in CeO_2
(14.4; 22.6)
(a) Vol. **84**, No. 2, p. 493—499 (1984).

A. V. Salker, E. B. Mirza, V. V. Deshpande, and H. V. Keer
Thermodynamic Analysis of the Phase Transition in V_4O_7
(1.2; 8; 14.1; 14.3; 18.1; 22.6)
(a) Vol. **84**, No. 2, p. K105—K108 (1984).

L. N. Skuja, A. N. Trukhin, and A. E. Plaudis
Luminescence in Germanium-Doped Glassy SiO_2
(20.3; 2; 10.2; 13.4; 20.1; 22.6)
(a) Vol. **84**, No. 2, p. K153—K157 (1984).

I. M. Antonova
On the Effective Potential for the Anion in MgO and CaO Crystals
(4 and 13; 22.6)
(b) Vol. **125**, No. 1, p. K55—K58 (1984).

V. Bethkenhagen and K. Hübner
Bethe-Lattice Approach to the Electronic Structure of Intrinsic Defects in v-SiO_2
(10.2 and 13.4; 22.6)
(b) Vol. **125**, No. 1, p. K79—K83 (1984).

V. A. Bespalov, V. F. Kiselev, G. S. Plotnikov, and A. M. Saletskii
Fluorescence Decay of Adsorbed Dye Molecules on Solid Surfaces
(13.3 and 20.3; 22.1.1; 22.6)
(a) Vol. **85**, No. 1, p. K73—K76 (1984).

M. J. Springis and J. A. Valbis
Blue Luminescence of Color Centres in Sapphire
(20.3; 22.6)
(b) Vol. **125**, No. 2, p. K165—K169 (1984).

B. Lorenz
Correct Low Temperature Expansion for the DC Conductivity of Magnetite
(14.3; 22.6)
(b) Vol. **125**, No. 2, p. K193—K195 (1984).

K. N. Jog, R. K. Singh, and S. P. Sanyal
The Relative Stability and Phase Transition Studies of MgO and CaO Crystals
(1.2 and 12.2; 22.6)
(a) Vol. **85**, No. 2, p. 417—421 (1984).

H. Krause, M. Schell, and H.-Chr. Golinsky
Hopping Conduction in SiO_2 Films Thermally Grown on Phosphorus-Doped Polycrystalline Silicon
(13.4 and 14.3.1; 22.6)
(a) Vol. **85**, No. 2, p. 569—573 (1984).

Y. Fukuda, A. Tomita, and N. Takeuchi
Thermoluminescence and Thermally Stimulated Exoelectron Emission in Glass and Sintered $CaB_4O_7:CuCl_2$
(17 and 20.3; 2; 22.6)
(a) Vol. **85**, No. 2, p. K141—K144 (1984).

S. Mochizuki
Radiative Optical Modes of Vibration in NiO Thin Film
(20.1; 6; 13.5; 22.6)
(b) Vol. **126**, No. 1, p. 105—114 (1984).

S. K. J. Al-Ani and C. A. Hogarth
The Spin Density and Optical Properties of Amorphous Thin Arsenic Silicate Films
(2 and 13.1; 19; 20.1; 22.6)
(b) Vol. **126**, No. 1, p. 293—304 (1984).

D. Fröhlich, H.-J. Weber, R. Wille, and R. Francini
Temperature Dependence of the Exciton Lineshape of GeO_2
(13.5.1 and 20.1; 22.6)
(b) Vol. **126**, No. 1, p. 305—310 (1984).

V. Bethkenhagen and H. Kübner
Cluster-Bethe-Lattice Approach to the Electronic Structure of a Twofold Coordinated Silicon Atom as a New Intrinsic Defect in v-SiO_2
(13.4; 10.2; 22.6)
(b) Vol. **126**, No. 1, p. K71—K76 (1984).

N. V. Rumak, V. V. Khatko, and V. N. Plotnikov
Structure and Properties of Silicon Dioxide Thermal Films (I): SiO_2 Films of up to 50 nm Thickness
(1.5 and 2; 3; 4; 22.6)
(a) Vol. **86**, No. 1, p. 93—100 (1984).

M. G. Blanchin and L. A. Bursill
Extended versus Small Defect Equilibria in Non-Stoichiometric Rutile (I): Dissolution and Re-Precipitation of Extended Defects Observed by In-Situ Transmission Electron Microscopy
(10.2; 1.3; 22.6)
(a) Vol. **86**, No. 1, p. 101—109 (1984).

N. V. RUMAK and V. V. KHATKO
Structure and Properties of Silicon Dioxide Thermal Films (II): 110 nm Thick SiO_2 Films
(1.5 and 2; 3; 22.6)
(a) Vol. **86**, No. 2, p. 477—484 (1984).

M. G. BLANCHIN and L. A. BURSILL
Extended Versus Small Defect Equilibria in Non-Stoichiometric Rutile (II): Structural Mechanisms for Dissolution and Reprecipitation of Extended Defects
(10.2; 1.3; 22.6)
(a) Vol. **86**, No. 2, p. 491—498 (1984).

22.6.1 Mixed Crystals

Original Papers and Short Notes

C. SANCHEZ, M. HENRY, R. MORINEAU, and M. C. LEROY
Small Polaron Mobility in α-$Li_xV_2O_5$
(13.4 and 13.5.3; 14.3; 19; 22.6.1)
(b) Vol. **122**, No. 1, p. 175—182 (1984).

L. C. OTERO-DIAZ, J. SORIA, and M. A. ALARIO-FRANCO
Extended Defects in Vanadium Doped Rutile. An EPR and TEM Study
(1.3 and 10.2; 19; 22.6.1)
(a) Vol. **82**, No. 2, p. 379—388 (1984).

I. N. GROZNOV, S. P. IONOV, I. B. KEVDINA, V. S. LUBIMOV, E. F. MAKAROV, O. V. MARCHENKO, N. N. MIKHAILOV, V. A. ONISHCHUK, and V. P. SHANTAROVICH
Mössbauer and Positron Studies of Superconducting Oxides $BaPb_{1-x}Bi_xO_3$
(14.2; 5; 22.6.1)
(b) Vol. **123**, No. 1, p. 183—190 (1984).

H. REUTHER
Contribution to the Thermotransport in Silicate Glasses
(2 and 9; 22.6.1)
(a) Vol. **83**, No. 1, p. 173—178 (1984).

G. S. NADKARNI, S. RADHAKRISHNAN, and N. SANKARRAMAN
Space Charge Limited Conduction in Thin Film V_2O_5–P_2O_5–ZnO Glasses
(14.3.1 and 14.3.4; 22.6.1)
(a) Vol. **83**, No. 1, p. 335—340 (1984).

E. CULEA, AL. NICULA, and I. BRATU
An Infrared Study of $xV_2O_5 \cdot (1-x)\,B_2O_3$ Glasses
(6 and 20.1; 2; 22.6.1)
(a) Vol. **83**, No. 1, p. K15—K18 (1984).

C. MIYAKE, H. TAKEUCHI, K. FUJI, and S. IMOTO
Electron Spin Resonance and Magnetic Susceptibility of M–U–O Ternary Mixed Oxides (II): Anomalies of Electron Spin Resonance and Magnetic Susceptibility of $Na_{2n-5}UO_n$ ($n = 3$ and 4)
(18.1 and 19; 18.4; 22.6.1)
(a) Vol. **83**, No. 2, p. 567—572 (1984).

C. MIYAKE, H. TAKEUCHI, H. OHYA-NISHIGUCHI, and S. IMOTO
Erratum to: Electron Spin Resonance and Magnetic Susceptibility of M–U–O Ternary Mixed Oxides (I): Anomalies of Electron Spin Resonance and Magnetic Susceptibility of $Li_{2n-5}UO_n$ ($n = 3$, 4, and 6)
(18.1 and 19; 18.4; 22.6.1)
(a) Vol. **83**, No. 2, p. 729 (1984).

E. CULEA, AL. NICULA, and I. BRATU
An Infrared Study of $V_2O_5 \cdot As_2O_3$ Glasses
(6 and 20.1; 2; 22.6.1)
(a) Vol. **84**, No. 1, p. K17—K21 (1984).

H. S. Maiti, S. Rajendran, and V. Sitakara Rao
Electrical Conductivity and Thermoelectric Power Studies of Iron(III)–Nickel(II) Mixed Oxides
(14.3; 15; 22.6.1)
(a) Vol. **84**, No. 2, p. 631—638 (1984).

L. Abello, S. J. Schwerdtfeger, and C. F. Schwerdtfeger
ESR Study of Cu^{2+} Diffused into Sodium β''-Alumina
(13.4 and 19; 22.6.1)
(b) Vol. **125**, No. 1, p. 333—340 (1984).

G. Berger and V. Atzrodt
In Vitro Characterization of Bioactivity of Glassy or Glass–Crystalline Implant Materials Using Auger Electron Spectroscopy
(1.5; 2; 17; 22.6.1)
(a) Vol. **85**, No. 1, p. K9—K13 (1984).

A. Dominguez-Rodriguez, J. Cabrera-Caño, R. Marquez, and J. Castaing
High-Temperature Creep of Equimolecular NiO–CoO Solid Solution
(10.2 and 12.2; 22.6.1)
(a) Vol. **85**, No. 2, p. 445—448 (1984).

H. Taguchi, M. Shimada, and M. Koizumi
Magnetic Properties in the System $(Sr_{1-x}Ca_x)\ CoO_{3-\delta}$
(18.2; 14.1; 22.6.1)
(a) Vol. **85**, No. 2, p. 523—527 (1984).

A. Bărbulescu and Lucia Sincan
Absorption Edge Shift in the PbO–SiO_2 System
(20.1: 2: 22.6.1)
(a) Vol. **85**, No. 2, p. K129—K131 (1984).

N. Kameswari, K. V. Ramanujachari, A. Meenakshi Sundaram, and C. S. Swamy
Synthesis and Structural Characteristics of $La_2Cu_{1-x}Co_xO_4$ ($x = 0.2$, 0.4, and 0.5)
(1.1 and 3; 22.6.1)
(a) Vol. **86**, No. 1, p. K15—K17 (1984).

G. Berger and M. Giehler
Characterization of in Vitro Corroded Surfaces of Bioactive Glasses (P_2O_5–CaO–SiO_2) with Infrared Reflection Spectroscopy
(2 and 20.1; 4; 6; 22.6.1)
(a) Vol. **86**, No. 2, p. 531—542 (1984).

H. Gruber, E. Krautz, H. P. Fritzer, K. Gatterer, and A. Popitsch
Electronic Property Changes of η-Mo_4O_{11} by Doping with Tungsten, Rhenium, and Vanadium
(14.1; 1.2; 13.1; 18; 20.1; 22.6.1)
(a) Vol. **86**, No. 2, p. 749—757 (1984).

22.7 Semiconducting Intermetallic Compounds

Original Papers and Short Notes

A. V. ANDREEV, M. ZELENÝ, L. HAVELA, and J. HŘEBÍK
Magnetic Properties of UGaNi
(18; 22.7)
(a) Vol. **81**, No. 1, p. 307—311 (1984).

R. G. PILLAY, S. H. DEVARE, F. PLEITER, and H. G. DEVARE
Transferred Hyperfine Field at ^{111}Cd in Ordered RNi_5 Compounds (R = Rare-Earth)
(18.2; 22.7)
(b) Vol. **121**, No. 2, p. K141—K145 (1984).

A. V. ANDREEV, L. HAVELA, M. ZELENÝ, and J. HŘEBÍK
Magnetic Properties of UGaCo
(18; 22.7)
(a) Vol. **82**, No. 1, p. 191—194 (1984).

H. SÜSSMANN, A. PRIEMUTH, and U. PRÖHL
Doping Properties of Pb and Ge in Bi_2Te_3 and Sb_2Te_3
(14.3 and 15; 22.7)
(a) Vol. **82**, No. 2, p. 561—567 (1984).

H. T. LANGHAMMER, M. STORDEUR, H. SOBOTTA, and V. RIEDE
IR Transmission Investigations of Sb_2Te_3 Single Crystals
(20.1; 13.1; 22.7)
(b) Vol. **123**, No. 1, p. K47—K51 (1984).

S. SANTHANAM and A. K. CHAUDHURI
Electrical Conduction in SnTe Thin Films
(14.3.1; 22.7)
(a) Vol. **83**, No. 1, p. K77—K80 (1984).

M. STORDEUR
Investigation of the Weak-Field Charge Transport in Semiconducting V_2–VI_3 Compounds with Trigonal Symmetry (I): Expanded Model for the Transport Coefficients
(14.3; 15; 22.7)
(b) Vol. **124**, No. 1, p. 439—446 (1984).

D. V. GITSU and F. M. MUNTYANU
Magnetoresistance of $Bi_{1-x}Sb_x$ Alloys ($0.08 \leqq x \leqq 0.15$) in Transversal Quantizing Magnetic Fields
14.3; 18; 22.7)
(b) Vol. **124**, No. 1, p. K87—K91 (1984).

W. HOYER, A. MÜLLER, W. MATZ, and M. WOBST
Neutron Scattering Investigation on Molten Ga–Te Alloys
(2; 1.3; 22.7)
(a) Vol. **84**, No. 1, p. 11—18 (1984).

M. Stordeur and G. Simon

Investigation of the Weak-Field Charge Transport in Semiconducting V_2–VI_3 Compounds with Trigonal Symmetry (II): Interpretation of the Weak-Field Charge Transport in Sb_2Te_3 Single Crystals
(14.3; 13.1; 15; 22.7)
(b) Vol. **124**, No. 2, p. 799—806 (1984).

A. Mueller, W. Hoyer, E. Thomas, and M. Wobst

X-Ray Scattering Investigation on Molten Ga–Te
(1.3 and 2; 22.7)
(a) Vol. **84**, No. 2, p. K97—K100 (1984).

W. Kullman, J. Geurts, W. Richter, N. Lehner, H. Rauh, U. Steigenberger, G. Eichhorn, and R. Geick

Effect of Hydrostatic and Uniaxial Pressure on Structural Properties and Raman Active Lattice Vibrations in Bi_2Te_3
(4; 1.1; 6; 20.1; 22.7)
(b) Vol. **125**, No. 1, p. 131—138 (1984).

J. Dominec

Semiconductor–Semimetal Transition in Bi–Sb Alloys Observed in Ultrasound Attenuation
(7 and 14.3; 14.1; 22.7)
(b) Vol. **126**, No. 1, p. K87—K90 (1984).

Shu Zhen and G. J. Davies

Calculation of Potential Energy Parameters for Some Intermetallic Compounds
(1.3 and 4; 22.7)
(a) Vol. **86**, No. 1, p. 155—161 (1984).

22.7.1 Mixed Crystals

Short Note

P. Lošťák, J. Horák, and L. Koudelka
Some Physical Properties and Point Defects in $Bi_2Te_{3-x}S_x$ Mixed Crystals
(20.1; 13.4; 14.3; 22.7.1)
(a) Vol. **84**, No. 2, p. K143—K147 (1984).

22.8 Tertiary and Higher Compounds

Review Article

C. Boulesteix
A Survey of Domains and Domain Walls Generated by Crystallographic Phase Transitions Causing a Change of the Lattice
(1.2 and 1.4; 14.4.1; 22; 22.6; 22.8)
(a) Vol. **86**, No. 1, p. 11—42 (1984).

Original Papers and Short Notes

L. V. Vassilev
Influence of the Phase Transition in $Pb_{1-x}Ge_xTe$ on the Electronic Dispersion Relation
(13.1; 1.2; 22.8)
(b) Vol. **121**, No. 1, p. 203—207 (1984).

K. Žičkus and A. Audzijonis
The Exponential Optical Absorption Edge Tail of SbSeI and BiSeI
(20.1; 13.1; 22.8)
(b) Vol. **121**, No. 1, p. K51—K53 (1984).

W. Hörig, H. Neumann, and G. Kühn
The Fundamental Absorption Edge of $LiInSe_2$
(20.1; 13.1; 22.8)
(b) Vol. **121**, No. 1, p. K55—K58 (1984).

K. K. P. Srivastava and S. N. Mishra
Spin Relaxation Consideration in Mössbauer Spectra of $FeSiF_6 \cdot 6\,H_2O$
(5; 22.8)
(b) Vol. **121**, No. 1, p. K83—K86 (1984).

E. M. Raspopova, V. T. Kalinnikov, T. N. Lyubovskaya, A. P. Chernov, and T. N. Fesenko
Glass Formation and Some Properties of the $(Cd_2GeAs_4)_{1-x}(TlAs)_x$ System
(2; 14.3; 15; 22.8)
(a) Vol. **81**, No. 1, p. 139—143 (1984).

V. P. Gupta, A. K. Singh, K. Chandra, and S. K. Jaireth
Mössbauer Investigations on Pyrrhotite
(10.2; 5; 18.3; 18.4; 22.8)
(a) Vol. **81**, No. 1, p. 281—291 (1984).

J. Oswald, H. Feichtinger, and R. Czaputa
A Useful Method to Improve Convergence in Least-Squares Fitting Procedures
(13.4; 14.3; 20.3; 22.8)
(a) Vol. **81**, No. 1, p. 343—351 (1984).

C. Rincón and S. M. Wasim
Influence of Intrinsic Defects on the Electrical Properties of $CuInSe_2$
(13.4 and 14.3; 22.8)
(a) Vol. **81**, No. 1, p. K77—K80 (1984).

I. P. Makarova, S. V. Misyul, L. A. Muradyan, A. F. Bovina, V. I. Simonov, and K. S. Aleksandrov
Anharmonic Thermal Atomic Vibrations in the Cubic Phase of $Cs_2NaNdCl_6$ Single Crystals
(1.2 and 6; 22.8)
(b) Vol. **121**, No. 2, p. 481—486 (1984).

H. Neumann, D. W. Bullett, H. Sommer, R. D. Tomlinson, and W. John
Valence Bands of the Cu–III–VI_2 Chalcopyrites Studied by Photoemission Spectra, X-Ray Emission Spectra, and Electronic Structure Calculations
(13.1; 17; 20; 22.8)
(b) Vol. **121**, No. 2, p. 641—647 (1984).

Y. Watanabe
ESR and Magnetic Investigation of Lithium Intercalated VSe_2
(13.4; 10.2; 18.1; 19; 22.8)
(b) Vol. **122**, No. 1, p. 309—316 (1984).

C. Bansal and K. Surendranath
Thermoelectric Power and Electrical Resistivity of $Fe_2Nb_3Se_{15}$
(14 and 15; 22.8)
(b) Vol. **122**, No. 1, p. 345—348 (1984).

W. Bażela-Wróbel, A. Szytuła, and J. Leciejewicz
Magnetic Properties of RhMnSi and CoTiSi
(18.4; 22.8)
(a) Vol. **82**, No. 1, p. 195—200 (1984).

A. N. Georgobiani, A. N. Gruzintsev, Z. P. Ilyukhina, V. E. Tezlevan, and I. M. Tiginyanu
Photoconductivity and Non-Equilibrium Carrier Recombination in $CdIn_2S_4$ Single Crystals
(13.4 and 16; 22.8)
(a) Vol. **82**, No. 1, p. 207—212 (1984).

D. Coquillat, A. Pradel, G. Brun, and J. C. Tedenac
Growth and Physical Properties of $TlBiTe_2$ Crystals
(13.1 and 14.3; 8; 15; 20.1; 22.8)
(a) Vol. **82**, No. 1, p. 295—299 (1984).

P. V. Nutsubidze, V. I. Chechernikov, L. G. Chachkhiani, Z. B. Chachkhiani, D. G. Barakadze, and V. K. Slovyanskikh
Magnetic Properties of Mixed Chalcogenides of Rare-Earth Elements and Uranium
(18; 22.8)
(a) Vol. **82**, No. 1, p. K53—K56 (1984).

S. G. Abdullaeva, N. T. Mamedov, F. A. Mustafaev, and E. Yu. Salaev
Photoluminescence of Ternary Layered Crystals $TlGaS_2$
(20.3; 22.8)
(a) Vol. **82**, No. 1, p. K75—K77 (1984).

V. I. Berezkin and A. I. Grachev
Properties of Al-Doped Bismuth Silicon Oxide
(16 and 20.1; 22.8)
(a) Vol. **82**, No. 1, p. K95—K99 (1984).

S. Logothetidis, S. Ves, and J. Spyridelis
Ellipsometric Determination of the Optical Constants of $ZnIn_2S_4$
(20.1; 13.1; 22.8)
(b) Vol. **122**, No. 2, p. 613—622 (1984).

R. Bacewicz
Excitonic Wavelength Modulated Reflectance of $CdGa_2Se_4$
(13.5.1 and 20.1; 22.8)
(b) Vol. **122**, No. 2, p. K155—K158 (1984).

S. M. Wasim and A. Noguera
Transport Properties of n-Type $CuInSe_2$
(13.4 and 14.3; 8; 22.8)
(a) Vol. **82**, No. 2, p. 553—559 (1984).

H. Broda
Commensurate–Commensurate Phase Transitions in $(NH_4)_2CoCl_4$
(1.2; 22.8)
(a) Vol. **82**, No. 2, p. K117—K120 (1984).

P. Orlandi and A. Rigamonti
On the Phase Transition and Crystal Field Effects in $FeSiF_6 \cdot 6\,H_2O$
(1.2 and 13.4; 5; 22.8)
(a) Vol. **82**, No. 2, p. K129—K132 (1984).

B. Lambert-Andron, P. Chaudouet, R. Madar, and R. Fruchart
Determination of the Crystallographic Structure of the $Ni_{16}Mn_6As_7$ G-Type Phase
(1.1; 22.8)
(a) Vol. **82**, No. 2, p. K133—K136 (1984).

Mukesh Jain, A. V. R. Warrier, and H. K. Sehgal
Composition Dependence of the Optical Energy Gap in $Pb_{1-x}Hg_xTe$ Alloy Thin Films
(20.1; 13.1; 22.8)
(a) Vol. **82**, No. 2, p. K181—K184 (1984).

A. A. Kaminskii, A. G. Petrosyan, K. L. Ovanesyan, G. O. Shironyan, V. A. Fedorov, and S. E. Sarkisov
Concentrational 3 μm Stimulated Emission Tuning in the $(Gd_{1-x}Er_x)_3Al_5O_{12}$ Crystal System
(20.2 and 20.3; 20.1; 22.8)
(a) Vol. **82**, No. 2, p. K185—K190 (1984).

V. N. Brudnyi, M. A. Krivov, A. I. Potapov, and Yu. V. Rud
Electrical Properties of Electron- and Proton-Irradiated GaAs and $ZnGeAs_2$ Solid Solutions
(11 and 14.3; 22.2.1; 22.8)
(a) Vol. **82**, No. 2, p. K191—K194 (1984).

S. H. Pawar, S. P. Tamhankar, and C. D. Lokhande
Deposition of Bi_2CdS_4 Films by the Spray Pyrolysis Technique
(14.3.1; 20.1; 22.8)
(a) Vol. **82**, No. 2, p. K195—K198 (1984).

M. Nawrocki, R. Planel, F. Mollot, and M. J. Kozielski
Magnetization and Spin-Flip Raman Scattering in $Cd_{1-x}Mn_xS$
(18.4 and 20.1; 13.5; 22.8)
(b) Vol. **123**, No. 1, p. 99—103 (1984).

M. Beiler, M. Mocker, and O. Ziep
Theory of Second-Order Auger Recombination in Strong Degenerate Small Gap Semiconductors. Application to PbSSe
(13.1 and 14.3; 20.3; 22.8)
(b) Vol. **123**, No. 1, p. 247—255 (1984).

K. Lischka, R. Durstberger, G. Lindemann, and H. Staudinger
Defect States in $Pb_{1-x}Sn_xTe$
(13.4; 16; 22.8)
(b) Vol. **123**, No. 1, p. 319—324 (1984).

S. K. Plachkova
Thermoelectric Figure of Merit of the System $(GeTe)_{1-x}(AgSbTe_2)_x$
(14.3 and 15; 8; 22.8)
(a) Vol. **83**, No. 1, p. 349—355 (1984).

V. N. Brudnyi, V. A. Novikov, A. D. Pogrebnyak, and Yu. P. Surov
Positron Annihilation in Electron-Irradiated p-$ZnGeP_2$ Compound
(10.2 and 11; 14.3; 22.8)
(a) Vol. **83**, No. 1, p. K35—K38 (1984).

A. A. Kaminskii, L. P. Kozeeva, and A. A. Pavlyuk
Stimulated Emission of Er^{3+} and Ho^{3+} Ions in $KLa(MoO_4)_2$ Crystals
(20.2 and 20.3; 13.4; 22.8)
(a) Vol. **83**, No. 1, p. K65—K67 (1984).

W. Wojdowski
Infrared Spectra of Thin Films of $Bi_{12}GeO_{20}$ and $Bi_{12}SiO_{20}$
(6 and 20.1; 22.8)
(b) Vol. **123**, No. 2, p. K101—K104 (1984).

Sh. M. Efendiev
Raman Spectra in $Bi_{12}TiO_{20}$ Single Crystals
(6 and 20.1; 22.8)
(b) Vol. **123**, No. 2, p. K105—108 (1984).

V. Riede, H. Neumann, H. Sobotta, W. Kissinger, M. S. Omar, and B. R. Pamplin
Infrared Study of Lattice and Free Carrier Effects in p-Type $CuGe_2P_3$
(6 and 20.1; 13.1; 22.8)
(b) Vol. **123**, No. 2, p. K119—K123 (1984).

A. Anagnostopoulos, K. Kambas, and B. Ploss
Optical Measurements and Temperature Dependence of the Energy Gap in $ZnIn_2S_4$ Layered Compound
(13.1 and 20.1; 22.8)
(b) Vol. **123**, No. 2, p. K155—K159 (1984).

H. Kido, T. Hoshikawa, M. Shimada, and M. Koizumi
Crystal Structure and Magnetic and Electrical Properties of Pseudo-Ternary $GdCu_{1-x}Zn_xSi$ Compounds
(14.1 and 18.1; 1.3; 18.2; 22.8)
(a) Vol. **83**, No. 2, p. 561—565 (1984).

H.-D. Autenrieth, S. Kemmler-Sack und W. Wischert
Luminescence and Sensitized Emission of Trivalent Rare Earth Ions in $Ca_3La_2W_2O_{12}$
(13.4 and 20.3; 22.8)
(a) Vol. **83**, No. 2, p. 631—636 (1984).

A. N. Bogdanov, A. V. Zhuravlev, and V. T. Telepa
First-Order Phase Transition from Antiferromagnetic to Paramagnetic Phase in $(C_2H_5NH_3)_2CuCl_4$
(18.1 and 18.4; 1.2; 22.8)
(a) Vol. **83**, No. 2, p. K135—K140 (1984).

P. Madhusudana Rao and K. V. Reddy
Photoluminescence Study of Ammonium Tetrabromozincate
(20.3; 22.8)
(a) Vol. **83**, No. 2, p. K145—K148 (1984).

K. M. Paraskevopoulos, D. Siapkas, and N. A. Economou
Infrared Reflectivity Plasma Spectra of $TlBiTe_2$ Single Crystals
(13.5.2 and 20.1; 22.8)
(b) Vol. **124**, No. 1, p. 229—234 (1984).

M. Heming and G. Lehmann
Correlation of Zero Field Splittings and Site Distortions. Interstitial Incorporation of Mn^{2+} in Urea-Ammoniumchloride $(H_2N)_2CO \cdot NH_4Cl$
(13.4 and 19; 22.8)
(b) Vol. **124**, No. 1, p. 315—323 (1984).

R. B. Prabhu and M. N. Ghatikar
Isomer Shift and Susceptibility of Mix-Valent $EuCu_2Si_2$
(13 and 18; 22.8)
(b) Vol. **124**, No. 1, p. K25—K29 (1984).

V. F. Zolin, V. M. Markushev, M. N. Popova, R. A. Aliev, G. D. Guseinov, and R. M. Sardarly
Neodymium Luminescence in the Ternary Semiconducting Chalcogenides $TlGaS_2$ and $TlGaSe_2$
(13.4 and 20.3; 22.8)
(b) Vol. **124**, No. 1, p. K69—K73 (1984).

H. Nakanishi, S. Endo, and T. Irie
Piezoresistance in $CdIn_2S_4$
(14.3; 13.1; 22.8)
(b) Vol. **124**, No. 1, p. K101—K104 (1984).

H. NEUMANN, P. DEUS, R. D. TOMLINSON, G. KÜHN, and B. HINTZE
Thermal Expansion of $CuInTe_2$ from 30 to 300 K
(6 and 8; 22.8)
(a) Vol. **84**, No. 1, p. 87—93 (1984).

R. SANCTUARY and P. GÜNTER
Electrooptic Properties of Rb_2ZnCl_4 in the Incommensurate and Ferroelectric Phases
(14.4.1 and 20.1; 22.8)
(a) Vol. **84**, No. 1, p. 103—111 (1984).

D. SENULIENĖ, G. BABONAS, E. I. LEONOV, I. MUMINOV, and V. M. ORLOV
Optical Properties of $Bi_{12}Si_{1-x}Ti_xO_{20}$ Single Crystals
(6 and 20.1; 22.8)
(a) Vol. **84**, No. 1, p. 113—117 (1984).

R. ZACH, R. DURAJ, and A. SZYTUŁA
Pressure Effect on Magnetic Transformations in the $Co_xNi_{1-x}MnGe$ System
(18.2 and 18.4; 1.2; 22.8)
(a) Vol. **84**, No. 1, p. 229—236 (1984).

R. Z. SADYKHOV, L. M. VALIEV, and A. O. ISMAILOV
Magnetic Properties of the $Fe_xCo_{1-x}Cr_2S_4$ $(0 < x < 1)$ System
(18; 22.8)
(a) Vol. **84**, No. 1, p. K51—K54 (1984).

A. A. KAMINSKII, Z. I. ZHMUROVA, V. A. LOMONOV, and S. E. SARKISOV
Two Stimulated Emission $^4F_{3/2} \rightarrow {}^4I_{11/2,13/2}$ Channels of Nd^{3+} Ions in Crystals of the CaF_2–ScF_3 System
(20.2 and 20.3; 13.4; 22.8)
(a) Vol. **84**, No. 1, p. K81—K84 (1984).

S. EL-HALAWANY, R. BACEWICZ, J. FILIPOWICZ, and R. TRYKOZKO
Photoconductivity of Flash Evaporated $CuInSe_2$ Thin Films
(16; 14.3.1; 20.1; 22.8)
(a) Vol. **84**, No. 1, p. K89—K92 (1984).

E. BURZO, M. CHIPARĂ, D. UNGUR, and I. ARDELEAN
Electron Paramagnetic Resonance Study of $xFe_2O_3(1 - x)$ $[B_2O_3 \cdot PbO]$ Glasses
(2 and 19; 22.8)
(b) Vol. **124**, No. 2, p. K117—K120 (1984).

K. ČERMÁK
The Exponential Optical Absorption Edge Tail of Tl_3VS_4 and Tl_3TaS_4
(20.1; 13.1; 22.8)
(b) Vol. **124**, No. 2, p. K143—K145 (1984).

R. RAI, S. K. GUPTA, D. K. SURI, and S. Z. ALI
Electron Spin Resonance Study of $CuGa_xIn_{1-x}Se_2$ Chalcopyrite Semiconductors
(19; 22.8)
(b) Vol. **124**, No. 2, p. K167—K170 (1984).

R. L. DAVIDOVICH, P. S. GORDIENKO, J. GRIGAS, T. A. KAIDALOVA, V. URBONAVIČIUS, and L. A. ZEMNUKHOVA
Phase Transitions in $KSbF_4$
(1.2 and 14.4.1; 8; 19; 22.8)
(a) Vol. **84**, No. 2, p. 387—392 (1984).

V. FREI and P. DEUS
A Frequency Sum Rule and Spectral Moments in Lattice Dynamics
(6; 22.1; 22.1.1; 22.2.2; 22.5.2; 22.8)
(b) Vol. **125**, No. 1, p. 121—130 (1984).

R. BORROMEI
The Electronic Spectrum of the CrO_4^{3-} Ion in the $Ca_2SiO_4 \cdot CaCl_2$ Host Lattice
(13.4 and 20.1; 22.8)
(b) Vol. **125**, No. 1, p. 349—355 (1984).

P. A. M. BERDOWSKI, J. VAN HERK, L. JANSEN, and G. BLASSE
Luminescence and Energy Transfer Characteristics of TbOCl
(20.3; 22.8)
(b) Vol. **125**, No. 1, p. 387—391 (1984).

A. PARISINI, P. P. LOTTICI, and C. RAZZETTI
Comments on the Optic Vibrational Modes in Defect Chalcopyrites
(6 and 20.1; 22.8)
(b) Vol. **125**, No. 1, p. K7—K11 (1984).

G. I. MAKOVETSKII and V. P. DYMONT
Crystallographie, Magnetic, and Electrical Properties of Nickel-Substituted Chromium Telluride
(14.1 and 18.2; 22.8)
(a) Vol. **85**, No. 1, p. K69—K71 (1984).

I. N. GRINCHESHEN, N. S. POPOVICH, and A. A. SHTANOV
Relaxation Peculiarities in $TlSbSe_2$ Crystals
(14.3 and 16; 22.8)
(a) Vol. **85**, No. 1, p. K85—K88 (1984).

K. ŽIČKUS, A. AUDZIJONIS, J. BATARŪNAS, and A. ŠILEIKA
The Fundamental Absorption Edge Tail of Ferroelectric SbSI
(14.4.1; 20.1; 22.8)
(b) Vol. **125**, No. 2, p. 645—651 (1984).

S. K. PLACHKOVA and L. A. TERSIEV
Phase Transitions and Intervalley Jahn-Teller Effect in the System $(GeTe)_{1-x}(AgSbTe_2)_x$
(1.2 and 14.4.1; 13.1; 22.8)
(b) Vol. **125**, No. 2, p. 659—664 (1984).

S. L. KAKANI and U. N. UPADHYAYA
Superconducting and Magentic Order Parameters in the Coexistence Region of Ternary Superconductors
(14.2 and 18.2; 22.8)
(b) Vol. **125**, No. 2, p. 861—867 (1984).

S. K. Ghatak and M. N. Ghatikar
On the Valence State of Rare-Earth Ion (R) in $RuCu_2Si_2$ and RCu_2Ge_2
(13.1; 22.8)
(b) Vol. **125**, No. 2, p. K125—K127 (1984).

P. L. Loyzance and M. Couzi
Group-to-Subgroup Relations in Space Groups. Application to the Structural Phase Transitions in the Perovskite-Type Layer Compounds $RbAlF_4$ and $RbVF_4$
(1.2; 22.8)
(a) Vol. **85**, No. 2, p. 359—374 (1984).

A. A. Kaminskii, S. E. Sarkisov, G. A. Denisenko, V. V. Ryabchenkov, V. A. Lomonov, Yu. E. Perlin, M. G. Blazha, D. Schultze, B. Hermoneit, and P. Reiche
Growth, Spectral and Luminescence Study of Cubic $Bi_4Ge_3O_{12}:Pr^{3+}$ Crystals
(20.1 and 20.3; 13.4; 22.8)
(a) Vol. **85**, No. 2, p. 553—567 (1984).

J. Málek, L. Tichý, and J. Klikorka
Exponential Absorption Edge in $Ge_{20}Sb_{15}S_{65}$ Glass Depending on Sample Preparation
(2 and 20.1; 22.8)
(a) Vol. **85**, No. 2, p. K123—K127 (1984).

J. M. Wessilinowa
Some Critical Properties of the s–d Model
(18.2; 22.8)
(b) Vol. **126**, No. 1, p. 229—233 (1984).

K. Čermák
The Width of the Energy Gap of $Tl_3^IB^VC_4^{VI}$-Type Compounds
(4 and 13.1; 20.1; 22.8)
(b) Vol. **126**, No. 1, p. 273—277 (1984).

J. Lech, A. Ślęzak, and I. Bójko
EPR Powder Spectra of Low-Symmetric Mn^{2+} Complexes in Aragonite Lattice
(13.4 and 19; 22.8)
(b) Vol. **126**, No. 1, p. 371—380 (1984).

G. Sorge, J. Janich, L. A. Shuvalov, and N. M. Shchagina
The Influence of an Electric DC Field on the New Intermediate Phase between the α- and β-Phases of $NaH_3(SeO_3)_2$ Crystals
(1.2 and 14.4.1; 12.2; 22.8)
(a) Vol. **86**, No. 1, p. 111—115 (1984).

M. A. Aldzhanov, D. A. Guseinov, and R. K. Veliev
Low-Temperature Heat Capacity of $CdGa_2S_4$
(8; 22.8)
(a) Vol. **86**, No. 1, p. K19—K21 (1984).

A. M. El-Dib and H. F. Hassan
Temperature Dependence of the Electronic Energy Bands of $NaNO_2$ Crystal from 20 to 245 °C
(13.1; 22.8)
(b) Vol. **126**, No. 2, p. 587—593 (1984).

K. R. Allakhverdiev, S. S. Babaev, N. A. Bakhyshov, T. G. Mamedov, E. Yu. Salaev, and I. K. Efendieva
The Influence of Hydrostatic Pressure on the Fundamental Absorption Edge of Crystals with TlSe Type Structure
(13.1 and 20.1; 12.2; 22; 22.8)
(b) Vol. **126**, No. 2, p. K139—K143 (1984).

H. Boudriot, R. Gründler, K. Deus, T. Stöckert, and H. A. Schneider
Electron Energy Loss Measurements on $CdSnP_2$: Plasmon Data, Dielectric Function, Band Structure Data
(13.1 and 13.5.2; 14.4; 22.8)
(b) Vol. **126**, No. 2, p. K149—K154 (1984).

H. M. van Noort, D. B. de Mooij, and K. H. J. Buschow
Crystal Structure, Magnetic Properties, and ^{57}Fe Mössbauer Effect of PtFeSn
(1 and 18.2; 5; 22.8)
(a) Vol. **86**, No. 2, p. 655—662 (1984).

D. Wolf and S. Kemmler-Sack
IR Emission and Sensitized Luminescence of the Trivalent Rare Earth Ions in 18R-$Ba_6B_2W_3O_{18}$ (B = Y, Er, Yb)
(13.4 and 20.3; 22.8)
(a) Vol. **86**, No. 2, p. 685—690 (1984).

P. G. Rustamov, F. M. Sadigov, and Ch. I. Abilov
Electrophysical Properties of $CeBiTe_3$
(14.3 and 15; 22.8)
(a) Vol. **86**, No. 2, p. K113—K115 (1984).

K. Tennakone and P. de S. Kaviratne
Activation of Li^+ Ion Transport in Lithium Ferricyanide by Interstitial Water
(14; 22.8)
(a) Vol. **86**, No. 2, p. K121—K123 (1984).

22.8.1 Salts of Oxy-Acids (Carbonates, Phosphates, Silicates Including Mica etc.)

Original Papers and Short Notes

A. Kasten, D. Weschenfelder, and G. Hess
Spectroscopic Investigation of the Magnetic Properties of $Ho_2O_2SO_4$
(18.4; 13.4; 18.3; 22.8.1)
(b) Vol. **121**, No. 1, p. 145—155 (1984).

J. M. Wesselinowa
Damping of the Transverse and Longitudinal Excitations of the Ising Model in a Transverse Field
(13 and 14.4.1; 22.8.1)
(b) Vol. **121**, No. 1, p. 317—327 (1984).

C. Y. Chen, K. L. Sweeney, and L. E. Halliburton
Reduction and Radiation Effects in Lithium Tantalate
(10.2; 11; 19; 20.1; 22.8.1)
(a) Vol. **81**, No. 1, p. 253—257 (1984).

V. K. Yanovskii
Phase Transitions and Properties of Doped $KNbO_3$ Crystals
(1.2 and 14.4.1; 14.3; 15; 20.1; 22.8.1)
(a) Vol. **81**, No. 1, p. 399—406 (1984).

S. Simon and Al. Nicula
EPR Study of Gd^{3+} and Cu^{2+} Ions from Lithium-Borate Vitroceramics
(1.4 and 19; 22.8.1)
(a) Vol. **81**, No. 1, p. K1—K5 (1984).

C. Jäger and K. Heide
Temperature Dependence of the ^{23}Na NMR Central Lineshape in $Na_2Ca(CO_3)_2$
(6 and 19; 22.8.1)
(a) Vol. **81**, No. 1, p. K7—K10 (1984).

A. A. Kaminskii, H. D. Kürsten, and D. Schultze
Stimulated Emission of Ferroelectric $Pb_5Ge_3O_{11}:Nd^{3+}, K^+$
(20.2 and 20.3; 22.8.1)
(a) Vol. **81**, No. 1, p. K19—K21 (1984).

A. S. Litovchenko and V. V. Mazykin
Dehydroxylation Effect upon Electrical Conduction of Mica
(14.4; 19; 22.8.1)
(a) Vol. **81**, No. 1, p. K47—K50 (1984).

V. K. Jain, V. P. Seth, and R. S. Bansal
Electron Paramagnetic Resonance of Cr^{3+} in $Al_2(SO_4)_3 \cdot 18\ H_2O$ Single Crystals
(19; 13.4; 22.8.1)
(b) Vol. **121**, No. 2, p. K201—K204 (1984).

K. S. Raju and E. Pippel
High Voltage Electron Microscopic (HVEM) Observations of Dislocations in Monocrystalline Gypsum
(1.4; 10.2; 22.8.1)
(a) Vol. **81**, No. 2, p. K93—K95 (1984).

Q. S. Litovchenko, A. V. Brodovoi, V. D. Tkachenko, and V. V. Mazykin
Temperature Dependence of the Magnetic Susceptibility of Teniolite
(18.1; 22.8.1)
(a) Vol. **81**, No. 2, p. K151—K153 (1984).

F. Cerdeira, V. Lemos, F. E. A. Melo, and M. Cardona
Effective Charges and Their Pressure Dependence in α-$LiIO_3$
(6 and 20.1; 22.8.1)
(b) Vol. **122**, No. 1, p. 53—64 (1984).

B. Daudin, W. Mutscheller, and M. Wagner
Heat Conductivity of the Cooperative Vibronic System $TmVO_4$
(6 and 8; 22.8.1)
(b) Vol. **122**, No. 1, p. 73—79 (1984).

M. Korkmaz, R. Bessege, and A. Gavaix
EPR Studies of Cu^{2+} in Anhydrous Lead Tartrate
(13.4 and 19; 22.8.1)
(b) Vol. **122**, No. 1, p. 255—262 (1984).

H. J. A. Koopmans, M. M. A. Perik, B. Nieuwenhuijse, and P. J. Gellings
The Simulation and Interpretation of the EPR Powder Spectra of Gd^{3+}-Doped $LaAlO_3$
(13.4 and 19; 22.8.1)
(b) Vol. **122**, No. 1, p. 317—330 (1984).

Fat Duen Ho
The Improved Point-Ion Potentials for the F^+-Centers in $LiNbO_3$ and $LiTaO_3$
(10.2 and 13.4; 20.1; 22.8.1)
(b) Vol. **122**, No. 1, p. 337—344 (1984).

K. Feldmann and L. Fuentes
On the Possibility of Determining the Odd Part of the Orientation Distribution Function of Texturized Samples for Quartz-Like Structures Using Anomalous Scattering
(1.3 and 1.4; 22.6; 22.8.1)
(b) Vol. **122**, No. 1, p. K5—K9 (1984).

H. Cerva and W. Graeff
Contrast Investigations of Surface Acoustic Waves by Stroboscopic Topography (I): Orientation Contrast
(1 and 7; 14.4.1; 22.8.1)
(a) Vol. **82**, No. 1, p. 35—45 (1984).

M. B. Zapart, W. Zapart, and A. I. Zvyagin
Phase Transitions in Ferroelastic $RbIn(MoO_4)_2$ Crystals by Electron Paramagnetic Resonance of Cr^{3+} Ions
(1.2 and 19; 13.4; 22.8.1)
(a) Vol. **82**, No. 1, p. 67—73 (1984).

J.-P. Doumerc, J. Hejtmanek, J.-P. Chaminade, M. Pouchard, and M. Krussanova
A Photoelectrochemical Study of $CuWO_4$ Single Crystals
(14.3; 13.4; 15; 16; 22.8.1)
(a) Vol. **82**, No. 1, p. 285—294 (1984).

G. Spörl, D. D. Chat, and E. Hegenbarth
Thermal Conductivity of Pseudo-One-Dimensional Ferroelectric CsH_2PO_4
(8 and 14.4.1; 22.8.1)
(a) Vol. **82**, No. 1, p. K27—K31 (1984).

F. Smutný and M. Połomska
Effects of Electric Field on the II–III Phase Transition in Lithium Ammonium Sulphate and Its Deuterated Analogue
(1.2 and 14.4.1; 22.8.1)
(a) Vol. **82**, No. 1, p. K33—K36 (1984).

B. Žekš, B. B. Lavrenčič, and R. Blinc
Microscopic Model for the Ferroelastic Transition in $KLiSO_4$
(1.2; 22.8.1)
(b) Vol. **122**, No. 2, p. 399—404 (1984).

V. G. Tsirelson, O. V. Korolkova, I. S. Rez, and R. P. Ozerov
Extended Model of Bond Charges and Its Application in Calculation of Optical Properties of Crystals with Different Types of Chemical Bonds
(4; 20.1; 22; 22.5.2; 22.6; 22.8.1)
(b) Vol. **122**, No. 2, p. 599—612 (1984).

A. K. Petrosyan, R. M. Khachatryan, and E. G. Sharoyan
Jahn-Teller Effect in EPR and Optical Absorption Spectra of Cu^{2+} in $LiNbO_3$ Single Crystals
(13.4; 19; 20.1; 22.8.1)
(b) Vol. **122**, No. 2, p. 725—734 (1984).

E. A. Harris, J. H. Mellor, and S. Parke
Electron Paramagnetic Resonance of Tetravalent Praseodymium in Zircon
(13.4 and 19; 22.8.1)
(b) Vol. **122**, No. 2, p. 757—760 (1984).

G. C. Upreti and R. S. Saraswat
EPR Study of VO^{2+} in Some Paramagnetic Tutton Salt Single Crystals
(19; 18.1; 22.8.1)
(b) Vol. **122**, No. 2, p. K169—K173 (1984).

A. Ioanid, M. Popescu, N. Vlahovici, and I. Bunget
On the Phase Transition of KH_2PO_4 at $T \approx 110$ °C
(1.2; 14.4.1; 22.8.1)
(a) Vol. **82**, No. 2, p. K125—K128 (1984).

J. Minge and S. Waplak
Successive Phase Transitions in $(NH_4)_3H(SO_4)_2$ and $(ND_4)_3D(SO_4)_2$ Crystals Studied by EPR of the VO^{2+} Ion and SeO_3^- Radical
(14.4.1; 4; 19; 22.8.1)
(b) Vol. **123**, No. 1, p. 27—36 (1984).

A. Kasten and G. Hess
Magnetic Properties of $Tb_2O_2SO_4$
(18.3 and 18.4; 22.8.1)
(b) Vol. **123**, No. 1, p. 123—134 (1984).

V. S. Grunin and I. B. Patrina
Impurities and Charge Compensation in Some Molybdenium Oxides
(13.4 and 19; 22.6; 22.8.1)
(b) Vol. **123**, No. 1, p. 353—363 (1984).

V. V. Krasnogolovets and P. M. Tomchuk
On the Proton Conductivity Mechanism in Systems Containing Hydrogen-Bonded Chains
(14.4; 6; 22.8.1)
(b) Vol. **123**, No. 1, p. 365—376 (1984).

M. Pagannone, B. Fornari, G. Mattei, and L. Mattioli
Radiative (Brewster-Type) Surface Mode Dispersion in Sodium Chlorate
(6 and 20.1; 13.5; 22.8.1)
(b) Vol. **123**, No. 1, p. K13—K16 (1984).

S. V. J. Lakshman and A. S. Jacob
Electronic Absorption Spectrum of Cu^{2+} in Sodium Zinc Sulphate Tetrahydrate Single Crystals
(13.4 and 20.1; 22.8.1)
(b) Vol. **123**, No. 1, p. K75—K77 (1984).

R. Blinc, J. Dolinšek, G. Lahajnar, I. Zupančič, L. A. Shuvalov, and A. I. Baranov
Spin–Lattice Relaxation and Self-Diffusion Study of the Protonic Superionic Conductors $CsHSeO_4$ and $CsHSO_4$
(9 and 14.4; 19; 22.8.1)
(b) Vol. **123**, No. 1, p. K83—K87 (1984).

L. A. Shuvalov, S. A. Gridnev, V. I. Kudryash, and B. N. Prasolov
Ferroelastic Domain Structure Dynamics and Internal Friction of $KH_3(SeO_3)_2$ Single Crystals
(14.4.1; 10.2; 12.2; 22.8.1)
(a) Vol. **83**, No. 1, p. 131—137 (1984).

A. Ya. Neiman, A. A. Afanasiev, L. M. Feodorova, V. T. Gabrielian, and S. M. Karagezian
Deviation from Stoichiometry and Electron Transfer in $PbMoO_4$
(10.2 and 14.3; 22.8.1)
(a) Vol. **83**, No. 1, p. 153—158 (1984).

B. P. Chandra, R. D. Verma, B. Majumdar, and V. P. Bhatt
Dependence of Mechanoluminescence on the Microhardness and Dislocation Density in Crystals
(12.2 and 20.3; 10.2; 22.5.2; 22.8.1)
(a) Vol. **83**, No. 1, p. K19—K22 (1984).

I. Henning and E. Hegenbarth
Specific Heat of $SrTiO_3$ in the Range of 0.5 to 6 K
(8 and 14.4.1; 22.8.1)
(a) Vol. **83**, No. 1, p. K23—K26 (1984).

H. Franke
Li$^+$ Ionic Conductivity in $LiNbO_3$
(14.4; 22.8.1)
(a) Vol. **83**, No. 1, p. K73—K76 (1984).

A. Kasten and J. Hoffmann
Sequence of Ferrimagnetic Phases in $Er_2O_2SO_4$
(18.3 and 18.4; 22.8.1)
(b) Vol. **123**, No. 2, p. 563—571 (1984).

R. Borromei and E. Cavalli
Low-Temperature Absorption Spectrum of Ni^{2+} Ion in Single Crystals of Ni_2NbBO_6
(13.4 and 20.1; 22.8.1)
(b) Vol. **123**, No. 2, p. 679—689 (1984).

S. Tsunekawa, H. F. J. Watanabe, and H. Takei
Linear Thermal Expansion of $SrTiO_3$
(1.2; 12.2; 22.8.1)
(a) Vol. **83**, No. 2, p. 467—472 (1984).
Erratum
(a) Vol. **86**, No. 2, p. 841 (1984).

M. Böhm and A. Scharmann
Entropy Change in Order–Disorder Transitions of Dipolar Relaxations
(10.2; 14.4; 22.8.1)
(a) Vol. **83**, No. 2, p. 537—541 (1984).

R. L. Calvert and R. J. Danby
Thermoluminescence and Radiophotoluminescence from Eu- and Sm-Doped $CaSO_4$
(20.3; 11; 13.4; 22.8.1)
(a) Vol. **83**, No. 2, p. 597—604 (1984).

R. K. Gartia and R. Singha
Analysis of Thermoluminescence of Topaz
(20.3; 13.4; 22.8.1)
(a) Vol. **83**, No. 2, p. 651—655 (1984).

A. M. Balagurov, B. N. Savenko, M. Dlouhá, S. Vratislav, and Z. Jirák
A Neutron Diffraction Study of Crystal and Domain Structure in $LiKSO_4$
(1.2; 22.8.1)
(a) Vol. **83**, No. 2, p. K117—K121 (1984).

L. A. Shebanov, L. V. Korzunova, and E. Zh. Freidenfeld
A New Perovskite-Like Ferroelectric $Bi_2Pb_{1/3}Bi_{2/3}Ti_{4/3}W_{2/3}O_9$ with the Mixed Bismuth Layer Structure
(14.4.1; 22.8.1)
(a) Vol. **83**, No. 2, p. K127—K129 (1984).

A. A. Kaminskii, A. G. Petrosyan, and K. L. Ovanesyan
Stimulated Emission of Pr^{3+}, Nd^{3+}, and Er^{3+} Ions in Crystals with Complex Anions
(20.2 and 20.3; 20.1; 22.8.1)
(a) Vol. **83**, No. 2, p. K159—K163 (1984).

A. A. Kaminskii, V. A. Timofeeva, A. B. Bykov, and S. E. Sarkisov
Luminescence and Stimulated Emission in the $^4F_{3/2} \rightarrow {}^4I_{11/2}$ and $^4F_{3/2} \rightarrow {}^4I_{13/2}$ Channels of Nd^{3+} Ions in Orthorhombic $NaYGeO_4$ Crystals
(20.2 and 20.3; 20.1; 22.8.1)
(a) Vol. **83**, No. 2, p. K165—K169 (1984).

K. S. V. Nambi
Comment on "The Role of Sulphur-Oxyradicals in Thermoluminescence and Exoelectron Emission of $CaSO_4$ Phosphors" by R. Huzimura and K. Atarashi
(17 and 20.3; 11; 13.4; 19; 22.8.1)
(a) Vol. **83**, No. 2, p. K175—K177 (1984).

V. L. Aksenov, A. Yu. Didyk, and N. M. Plakida
Ultrasonic Attenuation in Quasi-One-Dimensional Ferroelectrics
(7 and 14.4.1; 22.8.1)
(b) Vol. **124**, No. 1, p. 45—54 (1984).

J. D. Stephenson
Double Laue Pattern Topography of Top-Seeded $BaTiO_3$ Single Crystals Using White Synchrotron X-Radiation
(1.4 and 1.6; 14.4.1; 22.8.1)
(a) Vol. **84**, No. 1, p. 19—30 (1984).

M. Shiojiri, S. Sekimoto, T. Maeda, Y. Ikeda, and K. Iwauchi
Crystal Structure of Fe_2TiO_5
(1.1; 22.8.1)
(a) Vol. **84**, No. 1, p. 55—64 (1984).

R. Pareja, R. González, and M. A. Pedrosa
Study of Thermochemically Reduced and Electron-Irradiated $LiNbO_3$ Single Crystals by Positron Annihilation and Optical Absorption Measurements
(10.2 and 11; 20.1; 22.8.1)
(a) Vol. **84**, No. 1, p. 179—183 (1984).

M. Böhm, O. Erb, and A. Scharmann
Electrical Conduction in $PbMoO_4$ Crystals
(16 and 20.1; 13.4; 22.8.1)
(a) Vol. **84**, No. 1, p. 291—299 (1984).

V. I. Zametin
Absorption Edge Anomalies in Polar Semiconductors and Dielectrics at Phase Transitions
(20.1; 1.2; 14.4.1; 22; 22.5.2; 22.8.1; 22.9)
(b) Vol. **124**, No. 2, p. 625—640 (1984).

R. Borromei, L. Oleari, and P. Day
The Absorption Spectrum of Manganate(VI) in $BaSO_4$. A Reinterpretation
(13.4 and 20.1; 22.8.1)
(b) Vol. **124**, No. 2, p. 707—718 (1984).

R. DOERFLER, D. L. NAGY, H.-D. PFANNES, A. PUTZKA, G. RITTER, and N. ZEMAN
Non-Equilibrium Population of Fe^{3+} Electronic States after the Electron Capture (I): ^{57}Co in $LiNbO_3$ and $LiTaO_3$
(13.4; 5; 22.8.1)
(b) Vol. **124**, No. 2, p. 767—772 (1984).

H. BEIGE and A. KÜHNEL
Electromechanical Coefficients at Ferroelectric Phase Transitions. Rochelle Salt and $RbHSO_4$
(14.4.1; 22.8.1)
(a) Vol. **84**, No. 2, p. 433—437 (1984).

YU. N. MOSKVICH, A. A. SUKHOVSKII, and O. V. ROZANOV
Deuteron Magnetic Resonance of Highly Deuterated Ammonium Hydrogen Selenate
(1.1 and 14.4.1; 19; 22.8.1)
(a) Vol. **84**, No. 2, p. K109—K113 (1984).

TRAN KIM ANH, NGUYEN DAI HUNG, NQUY HUU CHI, and NGUYEN MANH SON
Energy Transfer between Tb^{3+} and Eu^{3+} in Rare Earth Pentaphosphates
(20.1 and 20.3; 13.4; 22.8.1)
(a) Vol. **84**, No. 2, p. K159—K163 (1984).

J. L. LAGZDONS, G. E. BAJĀRS, and A. R. LŪSIS
Modelling of the Solid State Electrochromic System $WO_3/HSbO_3 \cdot 2\,H_2O/Ni(OH)_2$
(14.3.3 and 14.3.4; 14.4; 22.8.1)
(a) Vol. **84**, No. 2, p. K197—K200 (1984).

SH. M. EFENDIEV, V. M. NAGIEV, and V. M. BURLAKOV
Anharmonicity Effects in IR-Reflection Spectra of Sheelites
(6 and 20.1; 22.8.1)
(b) Vol. **125**, No. 1, p. 75—79 (1984).

G. CORRADI, E. RZEPKA, and K. POLGÁR
Influence of Cationic Impurities (Cu, Cr, Ni, Er) on the Raman Spectra of Lithium Nitrate
(6 and 20.1; 22.8.1)
(b) Vol. **125**, No. 1, p. K13—K16 (1984).

V. M. NAGIEV, SH. M. EFENDIEV, and V. M. BURLAKOV
Vibrational Spectra of Crystals with Sheelite Structure and the Solid Solutions on Their Basis
(6 and 20.1; 22.8.1)
(b) Vol. **125**, No. 2, p. 467—475 (1984).

B. WYNCKE, F. BREHAT, M. EL SHERIF, and G. V. KOZLOV
Far-Infrared and Submillimeter Dielectric Response of Sodium Nitrite
(6 and 20.1; 1.2; 14.4.1; 22.8.1)
(b) Vol. **125**, No. 2, p. 493—498 (1984).

V. N. ZABLUDA, A. V. MALAKHOVSKII, and I. S. EDELMAN
Magnetooptical Effects and Optical Absorption of $FeBO_3$ in the Region of $^6A_{1g} \rightarrow {}^4T_{1g}(^4G)$, $^4T_{2g}(^4G)$ Transitions
(18.2 and 20.1; 13.4; 22.8.1)
(b) Vol. **125**, No. 2, p. 751—757 (1984).

B. TOPIČ, R. BLINC, and L. A. SHUVALOV
^{87}Rb NMR in Monoclinic RbD_2PO_4
(1.2; 14.4.1; 19; 22.8.1)
(a) Vol. **85**, No. 2, p. 409—415 (1984).

H. HESS
Luminescent Properties of Zinc Silicate Doped with Gallium
(20.3; 22.8.1)
(a) Vol. **85**, No. 2, p. 543—551 (1984).

A. R. KRASNAYA, L. N. OSTER, and V. YA. YASKOLKO
Exoelectron Emission from Calcium Sulfate Excited by Slow Electrons
(17; 10.2; 22.8.1)
(a) Vol. **85**, No. 2, p. K145—K148 (1984).

H.-G. UNRUH
Dielectric Dispersion and Dynamics of Nitrite Ions in $K_2Ba(NO_2)_4$
(1.2 and 14.4; 6; 22.8.1)
(b) Vol. **126**, No. 1, p. 115—120 (1984).

M. KORKMAZ, B. AKTAS, and M. DUPONT
EPR Spectra of the Mn^{2+} Ion in Cadmium Tartrate ($CdC_4H_4O_6 \cdot 5\,H_2O$)
(13.4 and 19; 22.8.1)
(b) Vol. **126**, No. 1, p. 323—327 (1984).

W. BÖHM, H. G. KAHLE, and W. WÜCHNER
Spectroscopic Study of the Crystal Field Splittings in $TbPO_4$
(1.2 and 13.4; 18.4; 20.1; 22.8.1)
(b) Vol. **126**, No. 1, p. 381—392 (1984).

V. V. ARISTOV, L. S. KOKHANCHIK, and YU. I. VORONOVSKII
Voltage Contrast of Ferroelectric Domains of Lithium Niobate in SEM
(14.4.1; 22.8.1)
(a) Vol. **86**, No. 1, p. 133—141 (1984).

A. A. KAMINSKII, E. L. BELOKONEVA, B. V. MILL, YU. V. PISAREVSKII, S. E. SARKISOV, I. M. SILVESTROVA, A. V. BUTASHIN, and G. G. KHODZHABAGYAN
Pure and Nd^{3+} Doped $Ca_3Ga_2Ge_4O_{14}$ and $Sr_3Ga_2Ge_4O_{14}$ Single Crystals, Their Structure, Optical, Spectral Luminescence, Electromechanical Properties, and Stimulated Emission
(14.4 and 20.2; 1.1; 12.2; 20.1; 20.3; 22.8.1)
(a) Vol. **86**, No. 1, p. 345—362 (1984).

T. NGOC, V. T. BICH, T. K. ANH, N. N. DINH, and P. LONG
Kinetics of Luminescence Quenching in Neodymium-Doped Fluorophosphate Glasses
(20.3; 22.8.1)
(a) Vol. **86**, No. 1, p. K45—K50 (1984).

V. I. TORGASHEV, V. DVOŘÁK, and F. SMUTNÝ
On Phase Transitions in $LiNH_4SO_4$
(1.2 and 14.4.1; 22.8.1)
(b) Vol. **126**, No. 2, p. 459—466 (1984).

D. Fasold, K. Hehl, and S. Jetschke
Refractive Index and Absorption Measurement in a Thin Layer by Two Reflection Measurements
(20.1; 11; 22.8.1)
(a) Vol. **86**, No. 2, p. 485—490 (1984).

T. Pannaparayil, M. Oskooie-Tabrizi, C. Lo, and L. N. Mulay
Infrared Spectroscopic, Magnetic, and Mössbauer Studies of Aluminosilicate Cage Structures (Fe/H-Mordenites) for Syngas Conversion
(5; 18.1; 20.1; 22.8.1)
(a) Vol. **86**, No. 2, p. 663—671 (1984).

Sh. M. Efendiev, N. G. Darvishov, and V. T. Gabrielyan
Photoluminescence in $PbMo_xW_{1-x}O_4$ Mixed Crystals
(20.3; 22.8.1)
(a) Vol. **86**, No. 2, p. K105—K108 (1984).

T. K. Vasileva, M. M. Nadoliisky, and S. D. Toshev
Study of PZT Ceramics Doped with Cr_2O_3
(14.4; 22.8.1)
(a) Vol. **86**, No. 2, p. K109—K111 (1984).

22.8.2 Spinels, Garnets, Ferrites

Original Papers and Short Notes

M. A. Ivanov, V. Ya. Mitrofanov, and A. Ya. Fishman
Magnetic Anisotropy of Disordered Magnetics with Jahn-Teller Ions
(18.3; 13.4; 22.8.2)
(b) Vol. **121**, No. 2, p. 547—559 (1984).

S. Ramana Murthy and T. Seshagiri Rao
Internal Friction in Ni–Zn Ferrites
(10.2 and 12.2; 22.8.2)
(a) Vol. **81**, No. 2, p. K137—K140 (1984).

A. A. Ghani Awad
Thermopower and Resistivity in Correlation to B-Site Magnetic Ordering in Cu–Ni and Cu–Cd Ferrospinels
(14.3 and 15; 22.8.2)
(a) Vol. **81**, No. 2, p. K155—K158 (1984).

G. A. Kraftmakher, V. V. Meriakri, and A. Ya. Chervonenkis
Magnetic Field Sign Effect on Resonance Absorption in Domain Walls of Orthoferrites
(18.4; 22.8.2)
(a) Vol. **81**, No. 2, p. K159—K163 (1984).

S. N. Barilo, A. P. Ges, L. E. Soshnikov, and T. V. Smirnova
Thermal Conductivity of Rare-Earth Orthoferrites at Reorientation Phase Transitions
(8 and 18.4; 7; 22.8.2)
(b) Vol. **122**, No. 1, p. 47—52 (1984).

B. Darshan, B. D. Padalia, and Om Prakash
X-Ray Spectroscopy Study of $Y_{3-x}Gd_xFe_5O_{12}$
(20; 22.8.2)
(b) Vol. **122**, No. 1, p. K59—K63 (1984).

D. Roy, Devendra Kumar, and D. Bahadur
Crystallisation Behaviour and Magnetic Properties of $Al_2O_3 \cdot Fe_2O_3 \cdot MnO \cdot SiO_2$ Glass
(1.2 and 18; 22.8.2)
(a) Vol. **82**, No. 1, p. K11—K15 (1984).

P. V. Reddy and T. S. Rao
Elastic Behaviour of Mixed Manganese–Magnesium Ferrites at Room Temperature
(12.2; 22.8.2)
(a) Vol. **82**, No. 1, p. K37—K41 (1984).

K. S. V. Nambi
Pyroelectroluminescence Induced by Tourmaline
(14.4.1 and 20.3; 22.8.2)
(a) Vol. **82**, No. 1, p. K71—K73 (1984).

M. Krzesińska and T. Szuta-Buchacz
Hypersonic Wave Attenuation in Yttrium Aluminum Garnet and Gadolinium Gallium Garnet
(7; 6; 12.2; 22.8.2)
(a) Vol. **82**, No. 2, p. 421—424 (1984).

E. M. Zub, E. V. Kuchis, and S. V. Sukhvalo
Electrical and Galvanomagnetic Properties of Films in the $Cu_xFe_{1-x}Cr_2S_4$ System
(14.3.1; 1.5; 22.8.2)
(a) Vol. **82**, No. 2, p. 569—574 (1984).

S. A. Mazen, N. A. Hakeem, and B. A. Sabrah
The IR Absorption Spectra of $Cu_{1-x}Cd_xFe_2O_4$
(6 and 20.1; 22.8.2)
(b) Vol. **123**, No. 1, p. K1—K4 (1984).

M. Hervieu, N. Nguyen, V. Caignaert, and B. Raveau
Calcium Mangano-Ferrites $(CaMO_3)_mCaMO_2$ □, M = Fe, Mn, with an Oxygen-Defect Perovskite Structure Related to Brownmillerite. Electron Microscopy Investigation
(1.1 and 1.4; 22.8.2)
(a) Vol. **83**, No. 2, p. 473—483 (1984).

P. Široký, E. Schmidt, F. Lukeš, and J. Humlíček
Optical Properties of Ga-Substituted Magnetoplumbites
(20.1; 22.8.2)
(a) Vol. **83**, No. 2, p. 581—588 (1984).

N. Yu. Konstantinov, L. G. Karaseva, V. V. Gromov, and A. V. Yakovlev
Interconfiguration Transitions of $4f^3$–$4f^25d^1$ of the Nd^{3+} Ion in YAG Single Crystals
(13.4 and 20.1; 22.8.2)
(a) Vol. **83**, No. 2, p. K153—K157 (1984).

V. N. Derkachenko, A. K. Zvezdin, A. M. Kadomtseva, N. M. Kovtun, M. M. Lukina, and A. A. Mukhin
Effect of Magnetic Vacancies on Magnetic Properties of Terbium Orthoferrites
(18.4; 22.8.2)
(a) Vol. **84**, No. 1, p. 215—222 (1984).

A. A. Ghani, S. A. Mazen, and A. H. Ashour
The Annealing Effect on the Conduction Mechanism of Copper Ferrite
(14.3 and 15; 22.8.2)
(a) Vol. **84**, No. 1, p. 337—341 (1984).

A. P. Ges, V. M. Derkachenko, and V. V. Fedotova
Effect of Scandium Ions on Magnetic Properties of Dysprosium Orthoferrite
(18.2; 22.8.2)
(a) Vol. **84**, No. 1, p. K63—K65 (1984).

Pran Kishan, Chandra Prakash, J. S. Baijal, and K. K. Laroia
Mössbauer Studies on Hyperfine Interactions in Titanium Substituted Lithium Ferrites
(5; 22.8.2)
(a) Vol. **84**, No. 2, p. 535—540 (1984).

M. Pardavi-Horváth, I. Földvári, I. Fellegvári, L. Gosztonyi, and J. Paitz
Spectroscopic Properties of Ca^{2+}-Doped GGG
(20.1; 13.4; 22.8.2)
(a) Vol. **84**, No. 2, p. 547—553 (1984).

I. I. Zyatkov, V. P. Miroshkin, and Ya. I. Panova
High-Frequency Conductivity of Manganese–Zinc Ferrites
(14.3; 22.8.2)
(a) Vol. **84**, No. 2, p. 645—650 (1984).

R. Satyanarayana and S. Ramana Murthy
Electrical Conductivity of Barium–Strontium Ferrites
(14.3; 15; 22.8.2)
(a) Vol. **84**, No. 2, p. 655—658 (1984).

H. E. Khodenkov, N. N. Kudelkin, and V. V. Randoshkin
The Break-Down of the 360° Bloch Domain Wall in Bubble Magnetic Films
(18.2; 22.8.2)
(a) Vol. **84**, No. 2, p. K135—K138 (1984).

G. Nicoară and V. Ghiordănescu
The Presence of Ti^{3+} in LiTiZn Ferrites
(18.3 and 20.1; 22.8.2)
(a) Vol. **84**, No. 2, p. K139—K142 (1984).

L. E. Kholodenkov and A. G. Makhanek
Two-Photon Luminescence Excitation Spectra of Eu^{3+} in Yttrium Aluminium Garnet
(20.1 and 20.3; 13.4; 22.8.2)
(b) Vol. **125**, No. 1, p. 365—374 (1984).

J. Čermák, R. Novák, P. Novák, and M. Nevřiva
Yttrium Iron Garnet Films Substituted by Gd and Mn
(18.2.1; 22.8.2)
(a) Vol. **85**, No. 1, p. 173—177 (1984).

L. Půst and Z. Frait
Method for Measurement of Magnetostriction Constants in (001) Thin Films Using FMR
(18.2.1; 19; 22.8.2)
(a) Vol. **85**, No. 1, p. 179—188 (1984).

J. J. Żebrowski and A. Sukiennicki
On the Inhibition of Dynamic Conversion in Bubble Garnet Domains by a Thin Midplane Layer
(18.2.1; 22.8.2)
(a) Vol. **85**, No. 1, p. 189—193 (1984).

N. M. L. Köche, P. C. Morais, and K. Skeff Neto
The Usefulness of the Photoacoustic Cell for Magnetic Measurements
(7 and 18; 19; 22.8.2)
(a) Vol. **85**, No. 1, p. K65—K67 (1984).

V. I. Maltsev and V. G. Vologin
Neutron Diffraction Studies of the Magnetic Ground State in Li–Zn Ferrite
(18.3; 22.8.2)
(a) Vol. **85**, No. 2, p. 529—534 (1984).

P. C. Morais, A. L. Tronconi, and K. Skeff Neto
Photoacoustic Signal Behavior near the Ferromagnetic Resonance Field
(7; 19; 22.8.2)
(a) Vol. **86**, No. 1, p. 163—168 (1984).

A. A. Wafik and S. A. Mazen
Barkhausen Jumps and Resistivity of Cu–Cd Ferrites
(18.3; 14.3; 22.8.2)
(a) Vol. **86**, No. 1, p. 271—274 (1984).

G. P. Kramar and Ya. I. Panova
The Permeability Dispersion in Ferrites at Different Levels of Direct Current Magnetic Biasing
(18.3; 22.8.2)
(a) Vol. **86**, No. 1, p. 283—287 (1984).

E. Hartmann, L. Kovács, and J. Paitz
Electrical Conductivity of Gadolinium–Gallium Garnet (GGG) Crystals
(10.2 and 14.4; 22.8.2)
(a) Vol. **86**, No. 1, p. 401—405 (1984).

H. Nakanishi and T. Irie
Pressure Effect on the Absorption Edges of $CdIn_2S_4$ and Related Compounds
(13.1 and 20.1; 12.2; 22.8.2)
(b) Vol. **126**, No. 2, p. K145—K148 (1984).

22.9 Organic Semiconductors

Original Papers and Short Notes

M. C. Böhm
Symmetry Breaking in Organometallic Polymers Due to Interchain Interactions
(13.1; 4; 22.9)
(b) Vol. **121**, No. 1, p. 255—264 (1984).

S. M. Basha, K. Purandar, and Y. P. Reddy
Electronic Absorption Spectrum of Fe^{3+} Doped in Calcium Tartrate Tetrahydrate
(13.4 and 20.1; 22.9)
(b) Vol. **121**, No. 1, p. K79—K81 (1984).

S. Jurga
Molecular Motions and Polymorphic Phase Transitions in $(CH_3)_4$ $NClO_4$ and $(CH_3)_3NHClO_4$ as Studied by NMR
(1.2; 19; 22.9)
(a) Vol. **81**, No. 1, p. 77—85 (1984).

I. Cuculescu, A. Enache, C. Florea, and R. Bena
Optical and Electrohydrodynamical Effects in a Compensated Cholesteric Mixture
(14.4 and 20.1; 22.9)
(a) Vol. **81**, No. 1, p. 191—196 (1984).

P. Kindl
Positron Lifetime Studies on γ-Irradiated PTFE of Different Initial Crystallinity
(11; 10.2; 22.9)
(a) Vol. **81**, No. 1, p. 293—298 (1984).

G. J. Ashwell
$(2,2\text{-DHPE Cl})_x(2,2\text{-DPE})_{1-x}(\text{TCNQ})$, a Pauli Paramagnetic Organic Conductor
(14.3; 1.1; 18.1; 22.9)
(a) Vol. **81**, No. 1, p. 361—366 (1984).

G. F. Myachina, T. G. Ermakova, and V. A. Lopyrev
Compensation Effect in the Electrical Conduction in Some Conjugated Polymers
(14.3; 22.9)
(a) Vol. **81**, No. 1, p. 377—380 (1984).

M. Siradjuddin, V. K. Raju, and P. J. Reddy
Switching in Solution-Grown PVF Films
(14.3.4; 22.9)
(a) Vol. **81**, No. 1, p. K37—K41 (1984).

H. von Berlepsch
Trapping Model for the Surface Potential Kinetics in HDPE
(14.3.1 and 14.3.2; 22.9)
(a) Vol. **81**, No. 1, p. K63—K65 (1984).

H. Zscheile, R. Gründler, U. Dahms, J. Fröhner, and G. Lehmann
π-Plasmon Dispersion of Trans-CH)$_x$ at Small Momentum Transfer by Electron Energy Loss Spectroscopy
(13.5.2; 14.4; 22.9)
(b) Vol. **121**, No. 2, p. K161—K164 (1984).

G. J. Ashwell and C. M. Care
Optical Phonon Assisted Hopping in (DPPE) (TCNQ)$_5$(H$_2$O)
(6 and 14; 22.9)
(a) Vol. **81**, No. 2, p. 557—561 (1984).

G. J. Ashwell and I. W. Nowell
4-Cyano-1,3,6-triazacycl[3.3.3]azine — A Homomolecular Organic Semiconductor
(1.1; 14.3; 22.9)
(a) Vol. **81**, No. 2, p. K97—K101 (1984).

A. Fousková, M. Pavel, and B. Březina
Light Scattering by Defects in TGS
(10.2; 14.4.1; 22.9)
(a) Vol. **81**, No. 2, p. K117—K119 (1984).

B. Karvaly, B. Mallik, G. Kemeny, and S. Trivedi
Anomalous Effects of Adsorption on the Electric Conductivity of Some Organic Semiconductor Powders under Pressure
(14.3; 22.9)
(a) Vol. **81**, No. 2, p. K179—K184 (1984).

M. Jolivet, M. Cadene, J. M. Abadie, and M. Rolland
Morphology, Dopant Concentration, and Electrical Properties of H_2SO_4- and $BiCl_3$-Doped Polyacetylene Films
(10.2 and 14.3.1; 22.9)
(a) Vol. **82**, No. 1, p. 257—263 (1984).

G. J. Ashwell, J. G. Allen, E. P. Goodings, and I. W. Nowell
Crystal Structure and Electrical Conductivity of [1,1'-bis(p-fluorophenyl)-4,4'-bipyridinium]$^{2+}$ (TCNQ)$_2^{2-}$ and Electrical Characterisation of 51 TCNQ Simple Salts
(1.1 and 14.3; 22.9)
(a) Vol. **82**, No. 1, p. 301—306 (1984).

Y. Momose and M. Noguchi
Electron Emission during Frictional Contact between Metals and Polymers
(14.4 and 17; 21; 22.9)
(a) Vol. **82**, No. 1, p. K83—K86 (1984).

Ya. I. Verzimacha, A. V. Kovalshuk, M. V. Kurik, C. Hamann, and A. Mrwa
Solar Cells of Schottky Type with Ni–PbPc Interface
(14.3.3 and 16; 22.9)
(a) Vol. **82**, No. 1, p. K111–K115 (1984).

R. M. VLASOVA, V. A. KAPUSTIN, V. N. SEMKIN, G. SEKRETARCZYK, L. S. AGROSKIN, G. V. PAPAYAN, and L. P. RAUTIAN
Molecular Excitons in Quasi-One-Dimensional Conducting Crystals (Semiconductors and Metals)
(20.1; 13.5.1; 22.9)
(b) Vol. **122**, No. 2, p. 549—557 (1984).

E. YA. GLUSHKO
Metallic Quenching of Exciton Luminescence in Anthracene
(13.5.1 and 20.3; 22.9)
(b) Vol. **122**, No. 2, p. 569—580 (1984).

P. KUIVALAINEN, H. ISOTALO, and H. STUBB
Thermoelectric Power in Inhomogeneous Metallic Polymers
(14.3 and 15; 22.9)
(b) Vol. **122**, No. 2, p. 791—797 (1984).

M. E. DE GOUVÊA and A. S. T. PIRES
EPR Shift in TMMC with the Applied Field Parallel to the Chain Axis
(19; 22.9)
(b) Vol. **122**, No. 2, p. K175—K177 (1984).

YU. S. BULYSHEV, I. M. KASHIRSKII, and V. V. SINITSKII
Photoconductivity and Energy Level Structure of 2,4,7-Trinitro-9-Fluorenone Thin Films
(14.3.1 and 16; 13.1; 22.9)
(a) Vol. **82**, No. 2, p. 537—544 (1984).

C. MOŢOC, R. MITROI, and C. ROŞU
Photovoltaic Effects in Nematic and Cholesteric Mesophases Containing Nonmesogenic Compounds
(16; 2; 22.9)
(a) Vol. **82**, No. 2, p. 575—579 (1984).

TH. HAYNE and G. WEISER
Dispersion of Exciton Polaritons in a Monoclinic Cyanine Dye Crystal. Evaluation of the Reflectivity on Arbitrary Faces
(13.5 and 20.1; 22.9)
(b) Vol. **123**, No. 1, p. 271—280 (1984).

B. ČABRIĆ and LJ. M. RISTOVSKI
Resonant Energy Transfer between Localized Excitations in a Linear Polymer Chain
(13.4 and 13.5; 22.9)
(b) Vol. **123**, No. 1, p. 309—317 (1984).

G. J. ASHWELL and I. W. NOWELL
Electrical, Magnetic, and Structural Characterisation of [1,2-Bis(1-ethyl-4-pyridinium)ethane]$^{2+}$ (7,7,8,8-Tetracyano-p-quinodimethanide)$_4^{2-}$
(1.1; 14.3; 18.1; 22.9)
(a) Vol. **83**, No. 1, p. 39—45 (1984).

A. A. El-Sharkawy, A. S. Hilal, and S. R. Atalla
Modified Plane, Radial, and Angström Temperature-Wave Techniques for the Measurement of Thermal Properties of Non-Conducting Materials
(8; 22.9)
(a) Vol. **83**, No. 1, p. 123—129 (1984).

T. Słupkowski
Electrical Conductivity of Mixtures of Conducting and Insulating Particles
(14.3; 22.9)
(a) Vol. **83**, No. 1, p. 329—333 (1984).

A. P. Tyutnev, V. S. Saenko, A. I. Karpechin, G. S. Mingaleev, V. I. Arkhipov, A. I. Rudenko, and A. V. Vannikov
Radiation-Induced Conductivity in Polymers under Continuous Irradiation
(11 and 14.3; 22.9)
(a) Vol. **83**, No. 1, p. 365—373 (1984).

A. V. Belushkin, I. Natkaniec, J. Wąsicki, and T. Zaleski
The Phonon Density of States near the Structural Phase Transition of Solid Biphenyl
(1.2 and 6; 22.9)
(b) Vol. **123**, No. 2, p. K115—K118 (1984).

N. K. Bondareva, I. V. Krylova, and V. B. Golubev
Thermostimulated Exoemission and Radical Recombination in Polyethylene Irradiated with an Electron Beam
(13.3 and 17; 11; 19; 22.9)
(a) Vol. **83**, No. 2, p. 589—595 (1984).

G. J. Ashwell
Electrical, Magnetic, and Structural Characterisation of [1,2-Bis(1-methyl-4-pyridinium)ethylene]$^{2+}$-$(TCNQ)_4^{2-}(H_2O)$ and the Properties of Two Isostoichiometric TCNQ Salts
(14.3 and 18.1; 1.1; 15; 22.9)
(a) Vol. **83**, No. 2, p. 669—673 (1984).

Z. Gałdecki, P. Grochulski, Z. Wawrzak, and J. Stępień-Damm
Some Properties of Cyclotrimethylene–Trinitramine Single Crystals
(1.1; 22.9)
(a) Vol. **83**, No. 2, p. K113—K115 (1984).

R. Jankowiak, B. Ries, and H. Bässler
Spectral Diffusion and Triplet Exciton Localization in an Organic Glass
(13.5.1 and 20.3; 2; 22.9)
(b) Vol. **124**, No. 1, p. 363—371 (1984).

W. Possart and A. Röder
Measurement of Electrical Potential Distribution in a Polymer near the Contact to a Metal by Means of Scanning Electron Microscopy
(14.3.2; 22.9)
(a) Vol. **84**, No. 1, p. 319—325 (1984).

G. S. Mingaleev, A. P. Tyutnev, V. I. Arkhipov, A. I. Rudenko, A. V. Vannikov, and V. S. Saenko
Computer Simulation and Analytical Solutions in the Study of the Transient Radiation-Induced Conductivity in Polymers
(14.3; 11; 22.9)
(a) Vol. **84**, No. 1, p. 327—336 (1984).

P. Grochulski and Z. Wawrzak
Thermal Expansion of Sodium–Cadmium Formate. Powder Data
(1.1 and 8; 22.9)
(a) Vol. **84**, No. 1, p. K5—K6 (1984).

J. Muto and Y. Kobayashi
Photobleaching Behaviours of Rhodamine 6G Diffused in Polymethyl Methacrylate and in the Copolymer of Methyl Methacrylate with Methacrylic Acid
(10.2 and 20; 22.9)
(a) Vol. **84**, No. 1, p. K29—K33 (1984).

V. I. Zametin
Absorption Edge Anomalies in Polar Semiconductors and Dielectrics at Phase Transitions
(20.1; 1.2; 14.4.1; 22; 22.5.2; 22.8.1; 22.9)
(b) Vol. **124**, No. 2, p. 625—640 (1984).

P. Reineker and B. J. Schmid
The Mixing of Frenkel and Charge-Transfer Excited States (I): General Treatment of Singlet and Triplet Excitations and Analytic Solutions of Simple Models
(13.5.1; 22.9)
(b) Vol. **124**, No. 2, p. 657—671 (1984).

M. Messaoudi, A. Kuhn, J. Toulouse, and J. Duran
An Isomer of the Organic Semiconductor Cs_2TCNQ_3. X-Ray, Optic, and EPR Analysis
(1.1; 19; 20.1; 22.9)
(a) Vol. **84**, No. 2, p. 411—416 (1984).

A. I. Melker and A. V. Ivanov
Computer Simulation of the Fracture of Two Coupled Anharmonic Chains with a Strong Interaction
(12.2; 22.9)
(a) Vol. **84**, No. 2, p. 417—422 (1984).

A. P. Tyutnev, P. I. Dubenskov, V. S. Saenko, and A. V. Vannikov
Radiation-Induced Conductivity of Poly-(N-Vinylcarbazole) and Poly-(N-Epoxypropylcarbazole)
(11 and 14.3; 22.9)
(a) Vol. **84**, No. 2, p. 585—596 (1984).

L. S. Grigoryan, M. V. Simonyan, and E. G. Sharoyan
Electrical, Magnetic, and Spectral Properties of the New Class of Charge Transfer Salts Based on Phthalocyanines and Iodine
(2 and 14.3.1; 19; 20.1; 22.9)
(a) Vol. **84**, No. 2, p. 597—606 (1984).

L. N. Kamysheva, A. S. Sidorkin, and S. D. Milovidova
Dielectric Relaxation in Ferroelectric TGS
(14.4.1; 22.9)
(a) Vol. **84**, No. 2, p. K115—K120 (1984).

M. H. Ben Ghozlen, Y. Mlik, and A. Daoud
Raman Study of Mono-, Di- and Trimethyl Ammonium Hexachlorostannates
(6 and 20.1; 22.9)
(b) Vol. **125**, No. 1, p. 87—96 (1984).

M. C. Böhm
A Crystal Orbital (CO) Model for Low-Dimensional Segregated Donor and Acceptor Stacks
(13.1 and 13.4; 22.9)
(b) Vol. **125**, No. 1, p. 311—322 (1984).

A. Alegría, J. M. Barandiarán, and J. Colmenero
Dielectric Relaxation at the Glass Transition as a Free Volume Process (II): A Continuous Distribution of Relaxation Times
(2 and 14.4; 22.9)
(b) Vol. **125**, No. 1, p. 409—419 (1984).

G. Völkel, S. A. Dzuba, A. Bartl, W. Brunner, and J. Fröhner
Time-Resolved EPR on Polyacetylene
(19; 22.9)
(a) Vol. **85**, No. 1, p. 257—263 (1984).

I. Diaconu, D. D. Eley, and M. R. Willis
AC Conductivity in Single Crystals and Compactions of TCNQ Complex Salts
(14.3; 22.9)
(a) Vol. **85**, No. 1, p. 283—289 (1984).

M. Messaoudi and J. Duran
Conductivity in One-Dimensional Disordered Cs_2TCNQ_3 Salts
(14.3; 22.9)
(a) Vol. **85**, No. 1, p. 291—296 (1984).

F. Engelke, D. Michel, and F. Pille
NMR Study of Dynamical Processes in Crystals of the TSCC Family
(14.4.1; 19; 22.9)
(b) Vol. **125**, No. 2, p. 483—492 (1984).

V. Andreev and M. Pope
Generation of Hot Free Electrons Due to the Interaction between Charge-Transfer Excitons and Thermalized Free Electrons in Molecular Crystals
(13 and 13.5.1; 22.9)
(b) Vol. **125**, No. 2, p. 573—581 (1984).

B. J. Schmid and P. Reineker
The Mixing of Frenkel and Charge-Transfer Excited States (II): Numerical Treatment of Non-Symmetric Dimers
(13.5.1; 22.9)
(b) Vol. **125**, No. 2, p. 593—604 (1984).

P. PETELENZ
Optical Detrapping in Molecular Crystals (II): Second Conduction Band
(13.4; 20.1; 22.9)
(b) Vol. **125**, No. 2, p. 765—771 (1984).

V. ČÁPEK and V. SZÖCS
Is the Sink Model of Exciton Trapping in Molecular Condensates Satisfactory?
(13.5.1; 22.9)
(b) Vol. **125**, No. 2, p. K137—K142 (1984).

I. J. LALOV
Temperature Vibronic Spectra of Molecular Crystals
(6 and 13.5.1; 20.1; 22.9)
(b) Vol. **125**, No. 2, p. K147—K150 (1984).

M. J. ŠKRINJAR, D. V. KAPOR, Ž. M. ŠKRIBIĆ, AND J. P. ŠETRAJČIĆ
On the Problem of Exciton Coupling in Molecular Crystals
(6 and 13.5.1; 22.9)
(b) Vol. **125**, No. 2, p. K151—K154 (1984).

A. P. TYUTNEV, V. S. SAENKO, A. F. DUNAEV, E. D. POZHIDAEV, and A. V. VANNIKOV
Temperature Dependence of Transient Radiation-Induced Conductivity in Polymers
(14.3; 11; 22.9)
(a) Vol. **85**, No. 2, p. 591—602 (1984).

K. IKEZAKI, T. ISHII, and T. MIURA
Thermal Influence of Vacuum Deposition of Metallic Electrodes on TSC from Positively Corona-Charged Polyethylene Films
(14.4; 22.9)
(a) Vol. **85**, No. 2, p. 615—618 (1984).

A. SASAKI and M. IWATA
Anisotropy of Microhardness in Anthracene Single Crystals
(12.2; 22.9)
(a) Vol. **85**, No. 2, p. K105—K107 (1984).

A. P. TYUTNEV, V. S. SAENKO, A. V. VANNIKOV, and V. E. OSKIN
Radiation Induced Dielectric Effect in Polymers
(11 and 14.4; 22.9)
(a) Vol. **86**, No. 1, p. 363—374 (1984).

M. TOPIĆ, A. MOGUŠ-MILANKOVIĆ, and Z. KATOVIĆ
Thermally Stimulated Depolarization Current Study in Novolac Phenol-Formaldehyde Resin
(14.4; 22.9)
(a) Vol. **86**, No. 1, p. 413—420 (1984).

V. M. YARTSEV
Charge Transfer and Electron–Molecular Vibration Coupling in Tetramerized Quasi-1D Semiconductors
(14.3; 6; 20.1; 22.9)
(b) Vol. **126**, No. 2, p. 501—510 (1984).

A. P. Tyutnev, V. S. Saenko, A. V. Vannikov, and E. D. Pozhidaev
Concerning the Radiation-Induced Surface Conductivity in Polymers
(14.3.2; 11; 22.9)
(a) Vol. **86**, No. 2, p. 709—716 (1984).

Th. Frauenheim, C. Hamann, and M. Müller
Electric Field-Induced Disorder–Order Transition in Organic Polycrystalline Films of Quasi-One-dimensional Lead-Phthalocyanine
(14.3.1; 1.2; 22.9)
(a) Vol. **86**, No. 2, p. 735—747 (1984).

23. Solidified Gases

Original Papers and Short Notes

A. K. Ray, S. B. Trickey, and A. B. Kunz
Optical Absorption of Solid Xenon at High Pressure
(20.1; 13.1; 23)
(b) Vol. **121**, No. 1, p. K47—K50 (1984).

A. F. Barabanov, L. B. Litinskii, and A. V. Mikheenkov
On the Atomic Phase of Crystalline Hydrogen
(1.2 and 13; 18.4; 23)
(b) Vol. **122**, No. 2, p. K109—K114 (1984).

N. P. Tishchenko
Kinetics of Impurity Particles in Low-Temperature Matrices
(9; 10.2; 23)
(a) Vol. **83**, No. 2, p. 513—519 (1984).

A. M. Ratner and E. I. Tarasova
Local Optical Centers in Rare Gas Crystals
(13.4; 13.5.1; 23)
(b) Vol. **124**, No. 1, p. 249—260 (1984).

R. Gaethke, P. Gürtler, R. Kink, E. Roick, and G. Zimmerer
Vibrational Relaxation and Hot Luminescence of Ne_2^* Centers in Solid Neon
(13.5.1 and 20.3; 23)
(b) Vol. **124**, No. 1, p. 335—342 (1984).

J. Narain, G. G. Agrawal, and J. Shanker
Compressibility of Rare-Gas Solids unter the Effect of High Pressure
(4 and 12.2; 23)
(b) Vol. **126**, No. 1, p. 83—86 (1984).

R. K. Singh and D. K. Neb
Mean Square Displacements of Rare Gas Solids
(4 and 6; 8; 23)
(b) Vol. **126**, No. 1, p. K15—K18 (1984).

E.-H. Böttcher and W. F. Schmidt
Charge Carrier Generation in Solid Xenon by Two-Photon Absorption
(14.3.4 and 20.1; 23)
(b) Vol. **126**, No. 2, p. K165—K169 (1984).

S. N. Vaidya
Theory of Melting (I): Energy Balance Principle
(1.2 and 6; 8; 23)
(a) Vol. **86**, No. 2, p. 565—571 (1984).

Author Index

J. M. Abadie	(a) **82**, 257 (1984)	10.2 (p. 109) and 14.3.1 (p. 234); 22.9
A. K. Abass	(a) **83**, 681 (1984)	20.1 (p. 331); 22.6
A. Abdel-Kader	(a) **81**, 333 (1984)	13.4 (p. 173); 20.3; 22.4.1
D. Sh. Abdinov	(a) **82**, K87 (1984)	16 (p. 270); 22.1.3
F. Kh. Abdullaev	(b) **122**, K141 (1984)	13.5 (p. 193) and 18.2 (p. 290); 6; 13.5.1
S. G. Abdullaeva	(a) **82**, K75 (1984)	20.3 (p. 349); 22.8
A. A. Abdumalikov	(b) **122**, K141 (1984)	13.5 (p. 196) and 18.2 (p. 290); 6; 13.5.1
S. N. Abdurakhmanova	(a) **81**, K145 (1984)	11 (p. 123) and 14.3.4 (p. 249); 22.1.2
Y. Abe	(b) **125**, 735 (1984)	13.5.1 (p. 203) and 20.3 (p. 355); 22.4.2
L. Abello	(b) **125**, 333 (1984)	13.4 (p. 187) and 19 (p. 312); 22.6.1
Ch. I. Abilov	(a) **86**, K113 (1984)	14.3 (p. 232) and 15 (p. 268); 22.8
A. Abrahám	(a) **83**, 617 (1984)	13.1 (p. 160) and 20.1 (p. 331); 2; 22.1.2
M. G. Abramishvili	(b) **122**, 95 (1984)	8 (p. 85) and 11 (p. 124); 20.1; 22.6
V. N. Abramov	(b) **121**, K59 (1984)	13.5.1 (p. 199) and 20.3 (p. 345); 20.1; 22.6
A. I. Abu-Shady	(a) **86**, 805 (1984)	7 (p. 83) and 20.1 (p. 342); 21.1
S. A. Abushov	(b) **121**, K195 (1984)	20.3 (p. 345); 13.4; 22.3
	(a) **81**, K175 (1984)	14.3 (p. 222); 22.3.1
	(a) **83**, K61 (1984)	13.4 (p. 181); 16; 20.3; 22.3
M. E. Abu-Zeid	(a) **86**, 805 (1984)	7 (p. 83) and 20.1 (p. 342); 21.1
J. C. Achard	(b) **125**, 101 (1984)	4 (p. 61) and 13.1 (p. 162); 21.1; 21.4
B. S. Acharya	(a) **81**, 533 (1984)	13.4 (p. 174) and 20.3 (p. 346); 22.5.2
C. K. Acharya	(b) **124**, 137 (1984)	13.1 (p. 160); 2; 14.1; 15; 21
M. E. Achour	(a) **81**, K71 (1984)	14.4 (p. 255); 22
G. A. Adegboyega	(a) **85**, 215 (1984)	13.4 (p. 188) and 14.3 (p. 228); 22.2.1
M. S. Adili	(a) **86**, 805 (1984)	7 (p. 83) and 20.1 (p. 342); 21.1
M. Adolf	(a) **85**, K81 (1984)	7 (p. 83) and 16 (p. 273); 14.3; 22.4.3
A. M. Afanasev	(a) **83**, K5 (1984)	1.6 (p. 37); 17; 22.1.2
	(a) **84**, 73 (1984)	1 (p. 9) and 17 (p. 275); 22.1.2
A. A. Afanasiev	(a) **83**, 153 (1984)	10.2 (p. 111) and 14.3 (p. 225); 22.8.1
A. M. Afanasiev	(a) **81**, 47 (1984)	1.6 (p. 36) and 2 (p. 39); 10.2; 22.1.2
	(a) **86**, 143 (1984)	1.1 (p. 13); 1.6; 22.1.2
	(a) **86**, K1 (1984)	1.6 (p. 38) and 10.2 (p. 120); 22.2.1; 22.2.2
S. M. Afanasiev	(a) **86**, K1 (1984)	1.6 (p. 38) and 10.2 (p. 120); 22.2.1; 22.2.2
C. N. Afonso	(b) **123**, 79 (1984)	10.2 (p. 110) and 20.3 (p. 349); 22.5.2
A. M. Agaev	(a) **83**, 323 (1984)	14.3.4 (p. 250); 16; 22.3
S. C. Agarwal	(a) **85**, 297 (1984)	16 (p. 273); 2; 22.1.2
A. L. Ageev	(b) **121**, 451 (1984)	2 (p. 40); 21; 21.1.1
O. P. Aggarwal	(b) **122**, 669 (1984)	13 (p. 146) and 20 (p. 317); 22.6
Neeta Agnihotri	(b) **123**, 453 (1984)	4 (p. 60); 12.2; 22.5.4
N. Agnihotri	(b) **125**, 113 (1984)	4 (p. 61); 6; 12.2; 22; 22.5.3
G. G. Agrawal	(b) **123**, 11 (1984)	4 (p. 58); 6; 12.2; 22.5.2
	(b) **126**, 83 (1984)	4 (p. 62) and 12.2 (p. 142); 23
S. K. Agrawal	(b) **123**, 497 (1984)	4 (p. 60); 6; 12.2; 22.6

A. Agresti	(b) **124**,	297 (1984)	13.4 (p. 183); 22.5.2
L. S. Agroskin	(b) **122**,	549 (1984)	20.1 (p. 325); 13.5.1; 22.9
S. R. Ahmed	(a) **86**,	773 (1984)	14.3.4 (p. 254); 6; 13.1; 22.1.2
Nguyen Ai Viet	see under letter "N"		
R. Akhmedov	(a) **83**,	K211 (1984)	14.3.3 (p. 244) and 16 (p. 271); 22.4.1; 22.4.3; 22.4.4
A. Akilbekov	(b) **122**,	715 (1984)	10.2 (p. 109) and 13.4 (p. 177); 20.1; 22.5.2
F. El Akkad	see under letter "E"		
A. F. Akkerman	(b) **125**,	169 (1984)	11 (p. 129); 22.1.2
E. A. Akopyan	(a) **83**,	K215 (1984)	14.3.3 (p. 244); 22
A. E. Aksenov	(a) **81**,	K171 (1984)	20.3 (p. 347); 13.4; 22.5
V. L. Aksenov	(b) **124**,	45 (1984)	7 (p. 82) and 14.4.1 (p. 261); 22.8.1
B. Aktaş	(b) **126**,	323 (1984)	13.4 (p. 192) and 19 (p. 313); 22.8.1
D. I. Aladashvili	(a) **86**,	301 (1984)	13.4 (p. 193); 14.3; 22.2.3
M. A. Alam	(b) **126**,	729 (1984)	2 (p. 54) and 14.1 (p. 215); 21.1; 21.6
J. M. Alameda	(a) **85**,	511 (1984)	18.2.1 (p. 297); 2; 21.1.1
S. K. J. Al-Ani	(b) **123**,	653 (1984)	20.1 (p. 330); 2; 13; 22.6
	(b) **126**,	293 (1984)	2 (p. 53) and 13.1 (p. 164); 19; 20.1; 22.6
M. A. Alario-Franco	(a) **82**,	379 (1984)	1.3 (p. 23) and 10.2 (p. 110); 19; 22.6.1
C. Albers	(b) **121**,	K175 (1984)	13.1 (p. 155); 20.1; 22
E. L. Albuquerque	(b) **124**,	457 (1984)	1.6 (p. 37); 1.2
M. A. Aldzhanov	(a) **86**,	K19 (1984)	8 (p. 89); 22.8
A. Alegría	(b) **125**,	409 (1984)	2 (p. 50) and 14.4 (p. 258); 22.9
K. S. Aleksandrov	(b) **121**,	481 (1984)	1.2 (p. 15) and 6 (p. 68); 22.8
	(b) **124**,	503 (1984)	1.2 (p. 17) and 6 (p. 75); 22.5
L. N. Aleksandrov	(a) **81**,	63 (1984)	2 (p. 39) and 3 (p. 55); 1.5; 22.1.2
P. A. Aleksandrov	(a) **81**,	47 (1984)	1.6 (p. 36) and 2 (p. 39); 10.2; 22.1.2
	(a) **86**,	143 (1984)	1.1 (p. 13); 1.6; 22.1.2
	(a) **86**,	K1 (1984)	1.6 (p. 38) and 10.2 (p. 120); 22.2.1; 22.2.2
Yu. A. Aleksandrov	(a) **83**,	455 (1984)	1.1 (p. 12); 22.1.2
N. E. Alekseevskii	(a) **86**,	389 (1984)	14.2 (p. 218)
S. Alexandrova	(a) **84**,	561 (1984)	13.3 (p. 169) and 14.3.2 (p. 239); 22.1.2
K. Alexopoulos	(a) **81**,	K57 (1984)	15 (p. 264); 22.6
	(b) **125**,	K109 (1984)	10.2 (p. 117); 9; 22.5.3
A. R. Ali	(a) **84**,	K67 (1984)	10.1 (p. 100) and 18.2 (p. 293); 21.1
S. Z. Ali	(b) **124**,	K167 (1984)	19 (p. 311); 22.8
R. A. Aliev	(b) **124**,	K69 (1984)	13.4 (p. 184) and 20.3 (p. 352); 22.8
K. R. Allakhverdiev	(b) **126**,	K139 (1984)	13.1 (p. 165) and 20.1 (p. 341); 12.2; 22; 22.8
J. G. Allen	(a) **82**,	301 (1984)	1.1 (p. 11) and 14.3 (p. 223); 22.9
C. Allibert	(b) **125**,	101 (1984)	4 (p. 61) and 13.1 (p. 162); 21.1; 21.4
N. S. Almeida	(b) **124**,	457 (1984)	1.6 (p. 37); 1.2
J. A. Alonso	(a) **81**,	55 (1984)	1.3 (p. 22) and 2 (p. 39); 8; 21; 21.1
	(b) **123**,	485 (1984)	4 (p. 60); 1.3; 21.2; 21.3
	(a) **85**,	423 (1984)	1.3 (p. 25) and 8 (p. 89); 21; 21.1
D. L. Alov	(b) **122**,	K133 (1984)	13.4 (p. 178) and 18.2 (p. 290); 22.4.4
S. A. Alterovitz	(a) **85**,	69 (1984)	1.3 (p. 25) and 1.5 (p. 34); 22
G. B. Altshuller	(a) **84**,	555 (1984)	13.4 (p. 186); 20.2; 22.4.1
V. I. Altukhov	(b) **122**,	95 (1984)	8 (p. 85) and 11 (p. 124); 20.1; 22.6
R. P. Alvarez	(b) **125**,	221 (1984)	13.1 (p. 162); 22.2.1; 22.2.3
A. A. Alybakov	(b) **121**,	K205 (1984)	19 (p. 306) and 20.1 (p. 322); 6; 13.4; 22.5.2
	(b) **124**,	K75 (1984)	13.4 (p. 184) and 20.3 (p. 353); 10.2; 22.5.2
V. M. Alyabiev	(a) **85**,	11 (1984)	10.1 (p. 97); 9; 21
	(a) **85**,	315 (1984)	10.1 (p. 97); 11; 21
L. A. Amarante Ribeiro	(b) **124**,	781 (1984)	14.4.1 (p. 262)
	(b) **125**,	563 (1984)	18.4 (p. 303)
G. Ambrazevičius	(b) **125**,	759 (1984)	13.4 (p. 189) and 20.3 (p. 355); 22.4.4
S. Amelinckx	(a) **82**,	55 (1984)	1.4 (p. 28); 21.1
	(a) **84**,	185 (1984)	1.1 (p. 12) and 1.4 (p. 30); 21.1; 21.6

	(a) **85**, 29 (1984)	1.4 (p. 31); 10.1; 21; 21.1
	(a) **86**, 553 (1984)	1.3 (p. 26); 10.1; 21; 21.1
A. D' Amico	see under letter "D"	
Dao Khac An	see under letter "D"	
Phuong An	see under letter "P"	
A. Anagnostopoulos	(b) **123**, K155 (1984)	13.1 (p. 160) and 20.1 (p. 330); 22.8
C. Ance	(b) **123**, 611 (1984)	2 (p. 46) and 14.4 (p. 257); 22.1.2
G. Andrä	(a) **83**, K1 (1984)	3 (p. 56) and 11 (p. 126); 22.1.2
A. A. Andreev	(a) **83**, 285 (1984)	13.4 (p. 180) and 14.3.1 (p. 235); 2; 20.1; 22.1.2
A. V. Andreev	(b) **121**, K131 (1984)	18.4 (p. 300); 21.4
	(a) **81**, 307 (1984)	18 (p. 278); 22.7
	(a) **82**, 191 (1984)	18 (p. 279); 22.7
V. Andreev	(b) **125**, 573 (1984)	13 (p. 150) and 13.5.1 (p. 202); 22.9
M. A. Andreeva	(b) **125**, 461 (1984)	5 (p. 66); 1.6; 13.3; 21.1
G. Angonoa	(b) **122**, 211 (1984)	13.1 (p. 156); 13.2; 21.2; 21.3
Nguyen Quoc Anh	see under letter "N"	
T. K. Anh	(a) **86**, K45 (1984)	20.3 (p. 358); 22.8.1
Tran Kim Anh	see under letter "T"	
Yu. M. Annenkov	(b) **125**, K37 (1984)	10.2 (p. 116) and 20 (p. 336); 11; 22.5.2
G. R. Anstis	(a) **82**, 425 (1984)	1.4 (p. 29) and 10.2 (p. 110); 22.1.2
C. Antonione	(a) **84**, 371 (1984)	2 (p. 49); 21; 21.3
O. F. Antonov	(b) **125**, K117 (1984)	18 (p. 282) and 19 (p. 313)
V. E. Antonov	(a) **81**, K185 (1984)	14.2 (p. 216); 1.2; 21.6
V. N. Antonov	(b) **121**, 233 (1984)	13.1 (p. 153); 21
	(b) **123**, 787 (1984)	13.1 (p. 160); 21.6
Vl. N. Antonov	(b) **121**, 233 (1984)	13.1 (p. 153); 21
	(b) **123**, 787 (1984)	13.1 (p. 160); 21.6
V. V. Antonov-Romanovskii	(b) **121**, 133 (1984)	10 (p. 95)
	(b) **125**, 517 (1984)	10 (p. 96)
I. M. Antonova	(b) **125**, K55 (1984)	4 (p. 62) and 13 (p. 150); 22.6
T. E. Antonova	(a) **81**, K185 (1984)	14.2 (p. 216); 1.2; 21.6
K. Aoki	(b) **122**, 441 (1984)	6 (p. 70) and 20.1 (p. 325); 22.2.2
M. Aoki	(a) **83**, 85 (1984)	1.5 (p. 33) and 17 (p. 275); 20.3; 22.5.2; 22.5.3
S. S. Aplesnin	(b) **122**, K33 (1984)	18.4 (p. 300)
F. Appel	(a) **83**, 179 (1984)	10.2 (p. 111); 22.6
A. R. Arakelyan	(b) **125**, K67 (1984)	5 (p. 66)
I. Ardelean	(b) **124**, K117 (1984)	2 (p. 49) and 19 (p. 311); 22.8
K. P. Arefiev	(b) **125**, 619 (1984)	13 (p. 150); 22.5.2
	(a) **85**, K31 (1984)	11 (p. 130) and 20 (p. 318); 22.1.2; 22.2.1
V. V. Aristov	(a) **84**, K43 (1984)	1.4 (p. 31) and 10.2 (p. 114); 14.3.4; 22.1.2
	(a) **86**, 133 (1984)	14.4.1 (p. 263); 22.8.1
V. E. Arkhipov	(a) **86**, K59 (1984)	14.1 (p. 215) and 14.2 (p. 219); 1.2; 21; 21.1
V. I. Arkhipov	(a) **83**, 365 (1984)	11 (p. 126) and 14.3 (p. 225); 22.9
	(a) **84**, 327 (1984)	14.3 (p. 227); 11; 22.9
G. Armelles	(b) **125**, 357 (1984)	13.4 (p. 188) and 20.3 (p. 354); 22.1.2
G. Arnold	(a) **82**, 319 (1984)	14.3.3 (p. 242); 14.3.4; 22.2.4
	(a) **83**, 393 (1984)	13.4 (p. 180) and 14.3.3 (p. 243); 9; 22.2.4
D. A. Aronov	(a) **81**, 11 (1984)	14.3.3 (p. 241) and 14.3.4 (p. 248); 16; 22.1.1; 22.1.2; 22.2.2
	(a) **81**, K85 (1984)	14.3.4 (p. 248); 22.1.2
	(a) **84**, 301 (1984)	14.3.3 (p. 244) and 14.3.4 (p. 251); 16; 22
	(a) **85**, K159 (1984)	14.3.1 (p. 236); 11; 22.2.2
A. K. Arora	(a) **85**, 491 (1984)	10.2 (p. 118); 22.5.2
V. K. Arora	(b) **123**, 747 (1984)	14.1 (p. 212) and 14.3 (p. 225); 21.7
	(b) **126**, K171 (1984)	14.3.2 (p. 239)
V. G. Artyushenko	(a) **85**, 167 (1984)	11 (p. 129) and 20.1 (p. 337); 10.2; 22.5

E. K. ARUSHANOV	(b) **121**, 289 (1984)	13.1 (p. 154); 22.3
M. ASCHE	(b) **126**, 607 (1984)	13 (p. 152) and 20 (p. 319); 22.2.1
A. H. ASHOUR	(a) **84**, 337 (1984)	14.3 (p. 227) and 15 (p. 266); 22.8.2
A. ASHRY	(a) **82**, 143 (1984)	10.2 (p. 108); 22.6
	(a) **84**, 451 (1984)	10.2 (p. 115); 13.3; 22.6
G. J. ASHWELL	(a) **81**, 361 (1984)	14.3 (p. 220); 1.1; 18.1; 22.9
	(a) **81**, 557 (1984)	6 (p. 69) and 14 (p. 210); 22.9
	(a) **81**, K97 (1984)	1.1 (p. 11); 14.3; 22.9
	(a) **82**, 301 (1984)	1.1 (p. 11) and 14.3 (p. 223); 22.9
	(a) **83**, 39 (1984)	1.1 (p. 11); 14.3; 18.1; 22.9
	(a) **83**, 669 (1984)	14.3 (p. 226) and 18.1 (p. 285); 1.1; 15; 22.9
S. ASKENAZY	(b) **123**, K43 (1984)	13.1 (p. 158); 22.3
B. M. ASKEROV	(b) **121**, 11 (1984)	14.3.1 (p. 233); 6; 13.1; 15; 22
D. D. ASKEROV	(a) **83**, 323 (1984)	14.3.4 (p. 250); 16; 22.3
I. M. ASKEROV	(a) **84**, K59 (1984)	5 (p. 65) and 18 (p. 281); 22.3
M. ASLAM	(a) **84**, 659 (1984)	14.3.3 (p. 244); 13.4; 22.1.2
G. K. ASLANOV	(a) **84**, K59 (1984)	5 (p. 65) and 18 (p. 281); 22.3
H. A. ASLANYAN	(b) **123**, K97 (1984)	7 (p. 82)
S. R. ATALLA	(a) **83**, 123 (1984)	8 (p. 86); 22.9
H. ATMANI	(a) **85**, 399 (1984)	1.5 (p. 35) and 3 (p. 56); 2; 22.1.3
V. ATZRODT	(a) **82**, 373 (1984)	1.6 (p. 37); 17; 21; 22.1.2
	(a) **82**, 459 (1984)	1.6 (p. 37) and 11 (p. 125); 21
	(a) **85**, K9 (1984)	1.5 (p. 34); 2; 17; 22.6.1
M. AUBIN	(a) **85**, 649 (1984)	14.3.3 (p. 246); 14.3.4; 22.2.1
A. AUDZIJONIS	(b) **121**, K51 (1984)	20.1 (p. 320); 13.1; 22.8
	(b) **125**, 645 (1984)	14.4.1 (p. 263); 20.1; 22.8
G. R. AUGST	(b) **125**, 155 (1984)	10.2 (p. 116); 22.5.2
J. AULEYTNER	(b) **123**, 627 (1984)	13.1 (p. 159); 20; 21.1
M. I. AUSLENDER	(b) **121**, 737 (1984)	15 (p. 264); 22.6
C. AUST	(a) **84**, K75 (1984)	20.3 (p. 353); 22.4.1
H.-D. AUTENRIETH	(a) **83**, 631 (1984)	13.4 (p. 182) and 20.3 (p. 351); 22.8
M. AVEROUS	(b) **125**, 831 (1984)	14.3 (p. 229); 13.1; 22.4.4
A. A. GHANI AWAD	(a) **81**, K155 (1984)	14.3 (p. 222) and 15 (p. 265); 22.8.2
S. A. AZIMOV	(a) **83**, K81 (1984)	2 (p. 45) and 14.3 (p. 225); 14.3.4; 22.1.2
	(a) **85**, K159 (1984)	14.3.1 (p. 236); 11; 22.2.2
M. M. BABAEV	(b) **125**, 421 (1984)	15 (p. 267); 6; 13; 22
S. S. BABAEV	(b) **126**, K139 (1984)	13.1 (p. 165) and 20.1 (p. 341); 12.2; 22; 22.8
YU. A. BABANOV	(b) **121**, 451 (1984)	2 (p. 40); 21; 21.1.1
V. M. BABICH	(a) **82**, 533 (1984)	13.4 (p. 179); 14.3; 19; 22.1.2
	(a) **84**, 263 (1984)	13.4 (p. 185); 14.3; 22.1.2
	(a) **86**, 679 (1984)	13.4 (p. 195); 14.3; 19; 22.1.2
	(a) **86**, K91 (1984)	10.2 (p. 121); 22.1.2
G. BABONAS	(a) **84**, 113 (1984)	6 (p. 74) and 20.1 (p. 333); 22.8
	(b) **125**, 759 (1984)	13.4 (p. 189) and 20.3 (p. 355); 22.4.4
V. HARI BABU	(a) **81**, K195 (1984)	14.3.1 (p. 234); 2; 22.2.3
	(a) **82**, 201 (1984)	13.4 (p. 176) and 20.3 (p. 348); 20.1; 22.5.4
	(a) **85**, 265 (1984)	10.2 (p. 117); 20.1; 20.3; 22.5.4
R. BACEWICZ	(b) **122**, K155 (1984)	13.5.1 (p. 200) and 20.1 (p. 326); 22.8
	(a) **84**, K89 (1984)	16 (p. 272); 14.3.1; 20.1; 22.8
D. BACKS	(b) **126**, 343 (1984)	13.4 (p. 192) and 19 (p. 313); 22.4.1
M. BACMANN	(a) **84**, 199 (1984)	18.2 (p. 292) and 18.4 (p. 302); 1.2; 14.1; 21.1
K. V. S. BADARINATH	(a) **84**, K93 (1984)	14.4 (p. 257); 11; 22.5.2
R. D. BAËTA	(a) **82**, 101 (1984)	10.2 (p. 108); 22.6
D. BAHADUR	(a) **82**, K11 (1984)	1.2 (p. 15) and 18 (p. 279); 22.8.2
V. V. BAIDAKOV	(b) **122**, K163 (1984)	13.4 (p. 178) and 14.3.4 (p. 250); 22.1.1

J. S. BAIJAL	(a) **84**, 535 (1984)	5 (p. 66); 22.8.2
M. BAJ	(b) **125**, 215 (1984)	13.1 (p. 162) and 19 (p. 312); 16; 22.2.3
G. E. BAJĀRS	(a) **84**, K197 (1984)	14.3.3 (p. 245) and 14.3.4 (p. 252); 14.4; 22.8.1
N. S. BAKALOVA	(a) **81**, K141 (1984)	10.2 (p. 107); 20.1; 22.5.2
I. BAKER	(a) **85**, 481 (1984)	10.1 (p. 102); 1.4; 21; 21.1
S. R. BAKHCHIEVA	(a) **83**, 139 (1984)	8 (p. 86); 22.1.1
P. S. BAKHSHI	(b) **123**, 71 (1984)	4 (p. 59); 6; 8; 22.5.2
	(b) **125**, K21 (1984)	4 (p. 61); 12.2; 22.5.2
N. A. BAKHYSHOV	(b) **126**, K139 (1984)	13.1 (p. 165) and 20.1 (p. 341); 12.2; 22; 22.8
J. BAK-MISIUK	(a) **82**, K137 (1984)	8 (p. 86); 1.1; 22.2.4
W. BAŁA	(a) **81**, 579 (1984)	2 (p. 41) and 14.3.1 (p. 233); 19; 22.1
A. M. BALAGUROV	(a) **83**, K117 (1984)	1.2 (p. 17); 22.8.1
V. YU. BALANDIN	(a) **81**, 63 (1984)	2 (p. 39) and 3 (p. 55); 1.5; 22.1.2
N. BALBO	(a) **83**, 709 (1984)	20.3 (p. 351); 22.4.1
K. BALCEREK	(a) **81**, 171 (1984)	8 (p. 84); 22.6
G. BALDACCHINI	(b) **122**, K83 (1984)	10.2 (p. 108) and 13.4 (p. 176); 20.1; 22.5.2
A. BALDAN	(a) **81**, 451 (1984)	1.4 (p. 27); 21; 21.1
	(a) **83**, 47 (1984)	1.3 (p. 23); 10.1; 21; 21.1
	(a) **83**, 507 (1984)	10.1 (p. 99); 1.3; 21; 21.1
L. J. BALK	(a) **82**, 23 (1984)	1 (p. 9) and 7 (p. 81); 21.1
P. BALK	(a) **83**, 719 (1984)	14.3.3 (p. 243); 14.4; 22; 22.1.2
	(a) **84**, 659 (1984)	14.3.3 (p. 244); 13.4; 22.1.2
M. BALKANSKI	(b) **126**, K113 (1984)	6 (p. 80) and 20.1 (p. 341); 22.2.2
C. BALLESTEROS	(a) **82**, 111 (1984)	10.2 (p. 108) and 20.3 (p. 348); 20.1; 22.6
	(a) **83**, 645 (1984)	20.3 (p. 351); 10.2; 13.4; 22.6
V. V. BALTAGA	(b) **123**, 169 (1984)	13.5.1 (p. 200); 20.1; 22.5
I. BALTOG	(b) **124**, 307 (1984)	20.3 (p. 352); 13.4; 22.5.2
R. BALTRAMIEJŪNAS	(b) **126**, K135 (1984)	13.5.1 (p. 204) and 20.1 (p. 341); 22.4.1; 22.4.2
D. BALTRŪNAS	(a) **85**, K97 (1984)	3 (p. 56) and 5 (p. 66); 22
G. BAMBAKIDIS	(b) **126**, K67 (1984)	13.3 (p. 170); 22.1
S. L. BAN	(b) **122**, K41 (1984)	13.5.3 (p. 207); 22
V. M. BANDURA	(b) **125**, 613 (1984)	13.5.1 (p. 203) and 20.3 (p. 355); 13.3; 22.4.3
S. BANDYOPADHYAY	(b) **122**, 113 (1984)	10.2 (p. 107); 22.5.2
	(b) **124**, 545 (1984)	10.2 (p. 114); 22.5.2
	(b) **126**, K117 (1984)	10.2 (p. 120); 22.5.2
T. K. BANDYOPADHYAY	(a) **84**, 651 (1984)	14.3.1 (p. 236); 16; 22.4.1
	(a) **86**, 827 (1984)	16 (p. 273); 22.4.2
S. BANERJEE	(a) **81**, 145 (1984)	1.2 (p. 14); 1.3; 11; 21.1
C. BANSAL	(b) **122**, 345 (1984)	14 (p. 210) and 15 (p. 265); 22.8
R. S. BANSAL	(b) **121**, K201 (1984)	19 (p. 306); 13.4; 22.8.1
A. F. BARABANOV	(b) **122**, K109 (1984)	1.2 (p. 16) and 13 (p. 146); 18.4; 23
D. G. BARAKADZE	(a) **82**, K53 (1984)	18 (p. 279); 22.8
N. V. BARAMIDZE	(b) **124**, 241 (1984)	13.4 (p. 183) and 20.1 (p. 332); 11; 22.2.1
N. P. BARAN	(a) **82**, 533 (1984)	13.4 (p. 179); 14.3; 19; 22.1.2
	(a) **84**, 263 (1984)	13.4 (p. 185); 14.3; 22.1.2
	(a) **86**, 679 (1984)	13.4 (p. 195); 14.3; 19; 22.1.2
J. M. BARANDIARÁN	(b) **125**, 409 (1984)	2 (p. 50) and 14.4 (p. 258); 22.9
A. I. BARANOV	(b) **123**, K83 (1984)	9 (p. 92) and 14.4 (p. 256); 19; 22.8.1
J. M. BARANOWSKI	(b) **122**, 331 (1984)	13.4 (p. 176) and 20.1 (p. 323); 22.4.2
P. I. BARANSKII	(a) **82**, 533 (1984)	13.4 (p. 179); 14.3; 19; 22.1.2
	(a) **86**, K63 (1984)	14.3 (p. 231) and 15 (p. 268); 22.1.2
A. BĂRBULESCU	(a) **85**, K129 (1984)	20.1 (p. 339); 2; 22.6.1
S. N. BARILO	(b) **122**, 47 (1984)	8 (p. 85) and 18.4 (p. 300); 7; 22.8.2
U. BARKOW	(a) **85**, K77 (1984)	20.3 (p. 355); 22.2.4
J. BARNAŚ	(b) **124**, K31 (1984)	14.4.1 (p. 261) and 18.2 (p. 292); 19
K. BÄRNER	(b) **121**, 157 (1984)	18 (p. 278); 21.1
	(b) **123**, 541 (1984)	8 (p. 87) and 18 (p. 280); 21.1

J. Barrau	(b) **125**, 357 (1984)	13.4 (p. 188) and 20.3 (p. 354); 22.1.2
F. C. Sá Barreto	see under letter "S"	
A. S. Barrière	(a) **81**, 105 (1984)	1.3 (p. 22); 14.3.1; 22.5
	(b) **123**, 377 (1984)	14.3.1 (p. 235); 13.4; 22.5
	(a) **84**, 273 (1984)	13.4 (p. 185) and 14.3.1 (p. 235); 22.5
M. T. Barriuso	(b) **124**, K83 (1984)	10.2 (p. 113) and 13.4 (p. 184); 20.3; 22.5
H.-G. Bartel	(b) **123**, 641 (1984)	13.1 (p. 159); 21
C. Barthel	(a) **86**, 433 (1984)	20.1 (p. 341); 20.2; 22
U. Barthel	(a) **86**, 433 (1984)	20.1 (p. 341); 20.2; 22
C. Barthou	(a) **83**, 709 (1984)	20.3 (p. 351); 22.4.1
A. Bartl	(a) **85**, 257 (1984)	19 (p. 312); 22.9
I. Bartoš	(b) **121**, 301 (1984)	13.1 (p. 154) and 17 (p. 274); 22
	(b) **122**, K159 (1984)	13.1 (p. 157); 22
H. Bartsch	(a) **83**, 543 (1984)	1.4 (p. 30) and 11 (p. 127); 22.1.1; 22.1.2
M. Bartsch	(a) **83**, 179 (1984)	10.2 (p. 111); 22.6
V. G. Baryshevskii	(b) **124**, 619 (1984)	13 (p. 148); 19
S. M. Basha	(b) **121**, K79 (1984)	13.4 (p. 173) and 20.1 (p. 321); 22.9
I. O. Bashkin	(a) **83**, 461 (1984)	1.2 (p. 17); 21.4
R. Bass	(b) **122**, K1 (1984)	1.6 (p. 36)
H. Bässler	(b) **124**, 363 (1984)	13.5.1 (p. 202) and 20.3 (p. 352); 2; 22.9
J. H. Basson	(a) **85**, 243 (1984)	10.2 (p. 117) and 13.4 (p. 189); 14.4; 22.4.3; 22.4.4
	(a) **85**, 449 (1984)	1.5 (p. 35) and 10.2 (p. 118); 3; 22.4.4
A. N. Basu	(b) **123**, 445 (1984)	1.2 (p. 16) and 6 (p. 72); 22.5
C. Basu	(b) **122**, 799 (1984)	13 (p. 146) and 20 (p. 317); 18
	(a) **83**, 675 (1984)	7 (p. 82) and 14.3 (p. 226); 22.2.3
P. K. Basu	(b) **121**, 743 (1984)	14.3.2 (p. 238) and 14.3.4 (p. 248); 22.1.2
	(b) **122**, K191 (1984)	13.1 (p. 157) and 14.3.3 (p. 242); 22.2.1
	(a) **86**, K79 (1984)	14.3.3 (p. 246); 14.3.2; 22.2.4
R. Basu	(a) **84**, K35 (1984)	10.1 (p. 100); 21.1
J. Baszyński	(a) **84**, 607 (1984)	14.1 (p. 213); 2; 18.2.1; 21.1.1
	(a) **84**, K129 (1984)	1.5 (p. 34) and 18.2.1 (p. 296); 19; 21.1
J. Batarūnas	(b) **125**, 645 (1984)	14.4.1 (p. 263); 20.1; 22.8
L. Battezzati	(a) **84**, 371 (1984)	2 (p. 49); 21; 21.3
R. Baudoing	(b) **124**, 381 (1984)	13.3 (p. 168); 21.1
B. A. Baum	(a) **85**, 497 (1984)	18.1 (p. 286); 21.1.1
L. A. Baydakov	(b) **121**, K43 (1984)	18 (p. 278); 22
C. Bazan	(a) **86**, 389 (1984)	14.2 (p. 218)
W. Bażela-Wróbel	(a) **82**, 195 (1984)	18.4 (p. 301); 22.8
V. A. Bazylev	(b) **125**, 583 (1984)	11 (p. 130); 13
P. Beauchamp	(a) **86**, 613 (1984)	1.4 (p. 32); 10.1; 21; 21.1.1
F. Bechstedt	(b) **126**, 575 (1984)	13.1 (p. 164); 22
C. Becker	(a) **81**, K131 (1984)	1.4 (p. 28) and 10.2 (p. 106); 14.3; 22.1.2
K. D. Becker	(b) **121**, 91 (1984)	9 (p. 91) and 10.2 (p. 105); 19; 22.5.1; 22.5.2
H. Beige	(a) **84**, 433 (1984)	14.4.1 (p. 262); 22.8.1
M. Beiler	(b) **123**, 247 (1984)	13.1 (p. 158) and 14.3 (p. 224); 20.3; 22.8
I. T. Belash	(a) **81**, K185 (1984)	14.2 (p. 216); 1.2; 21.6
V. I. Belitsky	(b) **122**, 581 (1984)	6 (p. 70) and 20.1 (p. 325); 22
E. L. Belokoneva	(a) **86**, 345 (1984)	14.4 (p. 258) and 20.2 (p. 344); 1.1; 12.2; 20.1; 20.3; 22.8.1
A. V. Belousov	(b) **123**, 345 (1984)	13.4 (p. 179); 16; 22
A. V. Belushkin	(b) **123**, K115 (1984)	1.2 (p. 16) and 6 (p. 73); 22.9
A. I. Belyaeva	(a) **83**, 245 (1984)	18.4 (p. 301); 1.4; 22.5
M. H. Ben Ghozlen	see under letter "G"	
R. Bena	(a) **81**, 191 (1984)	14.4 (p. 255) and 20.1 (p. 321); 22.9
P. Benalloul	(a) **83**, 709 (1984)	20.3 (p. 351); 22.4.1
H. Bender	(a) **86**, 245 (1984)	10.2 (p. 119); 1.4; 22.1.2
V. Z. Bengus	(a) **81**, K11 (1984)	2 (p. 40) and 12.1 (p. 135); 10.1; 21

K. BENHADDANE	(a)	**86**,	613 (1984)	1.4 (p. 32); 10.1; 21; 21.1.1
J. BENOIT	(a)	**83**,	709 (1984)	20.3 (p. 351); 22.4.1
P. A. M. BERDOWSKI	(b)	**125**,	387 (1984)	20.3 (p. 354); 22.8
V. I. BEREZKIN	(a)	**82**,	K95 (1984)	16 (p. 270) and 20.1 (p. 324); 22.8
U. BERG	(b)	**125**,	267 (1984)	13.1 (p. 162) and 17 (p. 276); 22.3
G. BERGER	(a)	**85**,	K9 (1984)	1.5 (p. 34); 2; 17; 22.6.1
	(a)	**86**,	531 (1984)	2 (p. 54) and 20.1 (p. 342); 4; 6; 22.6.1
H. BERGER	(a)	**85**,	45 (1984)	1.5 (p. 34) and 13.3 (p. 169); 21.7
W. BERGHOLZ	(a)	**81**,	239 (1984)	10.2 (p. 105); 5; 22.1.2
	(a)	**81**,	273 (1984)	1.4 (p. 27) and 10.2 (p. 106); 5; 22.1.2
H. VON BERLEPSCH	(a)	**81**,	K63 (1984)	14.3.1 (p. 233) and 14.3.2 (p. 238); 22.9
H. BERNDT	(a)	**84**,	K149 (1984)	13 (p. 149); 21.6
K. BERNDT	(a)	**86**,	499 (1984)	1.3 (p. 26) and 13.3 (p. 171); 20; 21.1
G. I. BERSUKER	(b)	**125**,	401 (1984)	6 (p. 76) and 13.4 (p. 188); 10.2; 22.5.2
K.-H. BERTHEL	(a)	**85**,	K175 (1984)	14.2 (p. 218)
M. BERTRANDT-ŻYTKOWIAK	(a)	**82**,	527 (1984)	14.3.3 (p. 242) and 20.3 (p. 349); 13.4; 22.4.1
P. A. BESIRGANYAN	(a)	**85**,	349 (1984)	1.1 (p. 13); 22.1.1
V. A. BESPALOV	(a)	**85**,	K73 (1984)	13.3 (p. 169) and 20.3 (p. 355); 22.1.1; 22.6
R. BESSEGE	(b)	**122**,	255 (1984)	13.4 (p. 175) and 19 (p. 306); 22.8.1
V. BETHKENHAGEN	(b)	**125**,	K79 (1984)	10.2 (p. 116) and 13.4 (p. 188); 22.6
	(b)	**126**,	K71 (1984)	13.4 (p. 193); 10.2; 22.6
M. BETTINELLI	(a)	**81**,	K27 (1984)	2 (p. 40) and 19 (p. 306)
E. BETZ	(b)	**124**,	741 (1984)	10.2 (p. 114) and 13.4 (p. 186); 20.3; 22.5.2
G. BHAT	(b)	**121**,	K135 (1984)	18 (p. 278)
V. P. BHATT	(a)	**83**,	K19 (1984)	12.2 (p. 140) and 20.3 (p. 350); 10.2; 22.5.2; 22.8.1
D. L. BHATTACHARYA	(b)	**123**,	45 (1984)	6 (p. 71); 5; 10.1; 21
A. L. BHATTACHARYYA	(a)	**84**,	493 (1984)	14.4 (p. 257); 22.6
K. BHATTACHARYYA	(b)	**122**,	K191 (1984)	13.1 (p. 157) and 14.3.3 (p. 242); 22.2.1
W. N. BHENDE	(b)	**123**,	71 (1984)	4 (p. 59); 6; 8; 22.5.2
	(b)	**125**,	K21 (1984)	4 (p. 61); 12.2; 22.5.2
S. BHUSHAN	(a)	**83**,	605 (1984)	20.3 (p. 351); 11; 13.4; 22.4
V. T. BICH	(a)	**86**,	K45 (1984)	20.3 (p. 358); 22.8.1
G. A. BIELIG	(a)	**82**,	K43 (1984)	10.2 (p. 109) and 11 (p. 125); 22.5.2
S. W. BIERNACKI	(b)	**125**,	805 (1984)	13.4 (p. 190); 22.4.2
J. Y. BIGOT	(b)	**121**,	649 (1984)	13.5.1 (p. 199); 20.1; 22.5
	(b)	**123**,	201 (1984)	13.5.1 (p. 200); 20.1; 22.5
	(b)	**124**,	221 (1984)	13.5.1 (p. 201) and 20.1 (p. 332); 22.5
R. M. BILBÉ	(b)	**121**,	339 (1984)	13.4 (p. 172) and 19 (p. 305); 22.4.3
	(a)	**86**,	821 (1984)	13.4 (p. 195) and 14.3 (p. 232); 20.1; 22.4.3
P. H. BINH	(a)	**81**,	K81 (1984)	14.3.3 (p. 241); 14.3.4; 22.1.2
U. BIRKHOLZ	(a)	**83**,	357 (1984)	14.3 (p. 225); 15; 22
	(a)	**83**,	K183 (1984)	14.3 (p. 226); 22
	(a)	**85**,	K81 (1984)	7 (p. 83) and 16 (p. 273); 14.3; 22.4.3
S. K. BISWAS	(a)	**83**,	K57 (1984)	2 (p. 45) and 20.1 (p. 329); 22.3
G. C. BJORKLUND	(b)	**123**,	79 (1984)	10.2 (p. 110) and 20.3 (p. 349); 22.5.2
A. BLACHA	(b)	**126**,	11 (1984)	6 (p. 67) and 13.1 (p. 153); 22.1; 22.2; 22.4
B. BLANCHARD	(a)	**81**,	105 (1984)	1.3 (p. 22); 14.3.1; 22.5
M. G. BLANCHIN	(a)	**86**,	101 (1984)	10.2 (p. 119); 1.3; 22.6
	(a)	**86**,	491 (1984)	10.2 (p. 120); 1.3; 22.6
B. BLANZAT	(a)	**83**,	709 (1984)	20.3 (p. 351); 22.4.1
G. BLASEK	(a)	**85**,	473 (1984)	1.3 (p. 25) and 11 (p. 131); 21; 21.1.1
G. BLASSE	(b)	**125**,	387 (1984)	20.3 (p. 354); 22.8
W. BLAU	(a)	**86**,	117 (1984)	1.3 (p. 26); 18; 21.1; 21.1.1
M. G. BLAZHA	(a)	**85**,	553 (1984)	20.1 (p. 338) and 20.3 (p. 356); 13.4; 22.8
R. BLINC	(b)	**122**,	399 (1984)	1.2 (p. 15); 22.8.1

	(b) **123**,	K83 (1984)	9 (p. 92) and 14.4 (p. 256); 19; 22.8.1
	(a) **85**,	409 (1984)	1.2 (p. 19); 14.4.1, 19; 22.8.1
L. N. Blinov	(b) **121**,	K43 (1984)	18 (p. 278); 22
M. D. Blokh	(b) **126**,	487 (1984)	6 (p. 79)
H. J. Blythe	(a) **81**	227 (1984)	10.1 (p. 97) and 18.2 (p. 288); 11; 21.1; 21.1.1
A. I. Bobrysheva	(b) **123**,	169 (1984)	13.5.1 (p. 200); 20.1; 22.5
A. Boden	(a) **86**,	789 (1984)	11 (p. 133) and 14.3.3 (p. 246); 14.3.2; 22.1.2
O. V. Boev	(b) **125**,	619 (1984)	13 (p. 150); 22.5.2
A. N. Bogdanov	(a) **83**,	K135 (1984)	18.1 (p. 285) and 18.4 (p. 302); 1.2; 22.8
M. Böhm	(a) **83**,	537 (1984)	10.2 (p. 112); 14.4; 22.8.1
	(a) **84**,	291 (1984)	16 (p. 272) and 20.1 (p. 333); 13.4; 22.8.1
M. C. Böhm	(b) **121**,	255 (1984)	13.1 (p. 154); 4; 22.9
	(b) **125**,	311 (1984)	13.1 (p. 163) and 13.4 (p. 187); 22.9
W. Böhm	(b) **126**,	381 (1984)	1.2 (p. 19) and 13.4 (p. 193); 18.4; 20.1; 22.8.1
V. V. Boichuk	(b) **124**,	693 (1984)	6 (p. 75) and 13 (p. 149); 22.3
Z. Bojarski	(a) **83**,	K97 (1984)	1.2 (p. 17) and 10.1 (p. 99); 21; 21.1
I. Bójko	(b) **126**,	371 (1984)	13.4 (p. 192) and 19 (p. 314); 22.8
Yu. B. Bolkhovityanov	(a) **84**,	K13 (1984)	1.6 (p. 37) and 2 (p. 48); 22.2.1; 22.2.4
D. N. Bolshutkin	(a) **81**,	507 (1984)	18.1 (p. 284) and 18.2 (p. 289); 14.1; 21.1; 21.1.1
V. L. Bonch-Bruevich	(b) **124**,	111 (1984)	13 (p. 148) and 14.3.4 (p. 251)
	(b) **124**,	241 (1984)	13.4 (p. 183) and 20.1 (p. 332); 11; 22.2.1
I. E. Bondarenko	(a) **84**,	K43 (1984)	1.4 (p. 31) and 10.2 (p. 114); 14.3.4; 22.1.2
N. K. Bondareva	(a) **83**,	589 (1984)	13.3 (p. 168) and 17 (p. 275); 11; 19; 22.9
A. Bonnet	(b) **124**,	403 (1984)	14.3 (p. 226) and 15 (p. 266); 22
H. Booyens	(a) **85**,	243 (1984)	10.2 (p. 117) and 13.4 (p. 189); 14.4; 22.4.3; 22.4.4
	(a) **85**,	449 (1984)	1.5 (p. 35) and 10.2 (p. 118); 3; 22.4.4
V. L. Borblik	(a) **84**,	263 (1984)	13.4 (p. 185); 14.3; 22.1.2
A. Borchardt	(a) **84**,	143 (1984)	10.2 (p. 113); 14.3.3; 22.1.2
E. Borchi	(b) **123**,	723 (1984)	14.2 (p. 217); 21.4
J. T. Borenstein	(b) **122**,	661 (1984)	13.1 (p. 157); 22.1; 22.2
N. I. Borgardt	(a) **86**,	55 (1984)	1.4 (p. 32) and 10 (p. 96)
A. T. Borges	(b) **124**,	781 (1984)	14.4.1 (p. 262)
H. E. Borges	(b) **121**,	K19 (1984)	13 (p. 144) and 18.2 (p. 287)
	(b) **121**,	K25 (1984)	13 (p. 144) and 18.2 (p. 287)
	(b) **121**,	K151 (1984)	13 (p. 145) and 18.2 (p. 288)
O. N. Borges	(b) **125**,	563 (1984)	18.4 (p. 303)
V. V. Borimskii	(a) **84**,	237 (1984)	13.4 (p. 184); 22.1.2
	(a) **86**,	623 (1984)	13.4 (p. 195); 22.1.2
T. E. Borisenko	(a) **84**,	K193 (1984)	14.3 (p. 228); 22
V. E. Borisenko	(a) **86**,	573 (1984)	8 (p. 90); 11; 22.1; 22.2
B. A. Borisov	(b) **122**,	349 (1984)	2 (p. 41) and 14.3 (p. 222); 14.3.4; 22.1.1
L. Borisova	(b) **126**,	K155 (1984)	13.1 (p. 165); 14.3; 15; 22
O. Yu. Borkovskaya	(a) **84**,	285 (1984)	16 (p. 272); 13.4; 22.2.1
Yu. G. Borodko	(b) **121**,	595 (1984)	13.1 (p. 155); 17; 21.1
R. Borromei	(b) **123**,	679 (1984)	13.4 (p. 181) and 20.1 (p. 330); 22.8.1
	(b) **124**,	707 (1984)	13.4 (p. 185) and 20.1 (p. 334); 22.8.1
	(b) **125**,	349 (1984)	13.4 (p. 188) and 20.1 (p. 336); 22.8
G. Borstel	(b) **126**,	617 (1984)	17 (p. 277); 13.3
M. Bose	(a) **83**,	675 (1984)	7 (p. 82) and 14.3 (p. 226); 22.2.3
M. S. C. Bose	(a) **86**,	275 (1984)	10.1 (p. 103) and 18.2 (p. 295); 21
	(a) **86**,	649 (1984)	12.1 (p. 138) and 18.2 (p. 295); 21.1.1
L. Bosi	(b) **123**,	519 (1984)	10.2 (p. 112); 20.1; 20.3; 22.5.2
G. Bosman	(b) **121**,	429 (1984)	14.1 (p. 211); 21.6
J. C. Le Bossé	(b) **124**,	381 (1984)	13.3 (p. 168); 21.1
O. Bostanjoglo	(a) **82**,	K1 (1984)	1.2 (p. 15) and 1.4 (p. 28); 2; 11; 22.1.1
K. Z. Botros	(a) **83**,	445 (1984)	1.1 (p. 11); 21.1
E.-H. Böttcher	(b) **126**,	K165 (1984)	14.3.4 (p. 253) and 20.1 (p. 342); 23

H. Böttger	(b) **121**, 413 (1984)	2 (p. 39) and 14.3 (p. 220); 6; 22
S. Bouarab	(b) **125**, 499 (1984)	6 (p. 77); 21.1
H. Boudriot	(b) **123**, K161 (1984)	13.5.2 (p. 205); 22.2.2
	(b) **126**, K149 (1984)	13.1 (p. 165) and 13.5.2 (p. 206); 14.4; 22.8
C. Boulesteix	(a) **86**, 11 (1984)	1.2 (p. 14) and 1.4 (p. 27); 14.4.1; 22; 22.6; 22.8
A. F. Bovina	(b) **121**, 481 (1984)	1.2 (p. 15) and 6 (p. 68); 22.8
E. Yu. Boyarko	(a) **81**, 217 (1984)	11 (p. 122); 14.3; 22.1.2; 22.2.1
J. Boyde	(a) **81**, 323 (1984)	17 (p. 274); 22.6
U. Braatz	(a) **86**, 407 (1984)	14.3.4 (p. 253); 13.4; 22.5
D. E. Brandão	(a) **83**, K187 (1984)	14.1 (p. 212); 21.1
H. S. Brandi	(b) **124**, 147 (1984)	13.1 (p. 160) and 20.1 (p. 331); 22
	(b) **125**, 605 (1984)	20.1 (p. 337); 20.2; 22.2.1; 22.4.1
I. Bratu	(a) **83**, K15 (1984)	6 (p. 72) and 20.1 (p. 329); 2; 22.6.1
	(a) **84**, K17 (1984)	6 (p. 75) and 20.1 (p. 333); 2; 22.6.1
G. Brauer	(a) **82**, 143 (1984)	10.2 (p. 108); 22.6
	(a) **84**, 451 (1984)	10.2 (p. 115); 13.3; 22.6
W. Brauer	(b) **126**, 629 (1984)	17 (p. 277); 13.5.2; 21
W. Braune	(a) **81**, 463 (1984)	1.6 (p. 36) and 14.3.2 (p. 238); 22.2.3
	(a) **86**, 427 (1984)	13.3 (p. 170) and 14.3.3 (p. 246); 22.2.3
R. Brazis	(a) **83**, K69 (1984)	13.4 (p. 181) and 14.3.4 (p. 251); 22.6
V. Brédimas	(b) **122**, 527 (1984)	18.2 (p. 290) and 18.4 (p. 301); 21.4
F. Brehat	(b) **125**, 493 (1984)	6 (p. 77) and 20.1 (p. 337); 1.2; 14.4.1; 22.8.1
S. Brehme	(a) **85**, 627 (1984)	14.3.3 (p. 245) and 20.3 (p. 356); 13.4; 22.2.2
B. Březina	(a) **81**, K117 (1984)	10.2 (p. 106); 14.4.1; 22.9
B. Briat	(b) **124**, 87 (1984)	13.5.1 (p. 201) and 13.5.4 (p. 208); 18.4; 20.1; 22.5
P. M. Bridenbaugh	(b) **122**, K11 (1984)	6 (p. 70) and 20.1 (p. 323); 2; 22
V. F. Britun	(a) **82**, 449 (1984)	1.4 (p. 29) and 10.2 (p. 110); 22.3
H. Broda	(a) **82**, K117 (1984)	1.2 (p. 16); 22.8
M. S. Brodin	(b) **125**, 613 (1984)	13.5.1 (p. 203) and 20.3 (p. 355); 13.3; 22.4.3
W. Brodkorb	(a) **84**, 379 (1984)	1.5 (p. 34) and 2 (p. 49); 22.6
A. V. Brodovoi	(a) **81**, K151 (1984)	18.1 (p. 284); 22.8.1
J. M. Broto	(b) **121**, 289 (1984)	13.1 (p. 154); 22.3
H. A. Brown	(b) **121**, K29 (1984)	18.2 (p. 287)
M. D. Brown	(b) **126**, 361 (1984)	13.4 (p. 192) and 20.3 (p. 357); 22
V. N. Brudnyi	(a) **82**, K191 (1984)	11 (p. 125) and 14.3 (p. 224); 22.2.1; 22.8
	(a) **83**, K35 (1984)	10.2 (p. 112) and 11 (p. 126); 14.3; 22.8
O. Brümmer	(b) **124**, 11 (1984)	13.1 (p. 153) and 20 (p. 316); 21; 22
	(b) **125**, 267 (1984)	13.1 (p. 162) and 17 (p. 276); 22.3
G. Brun	(a) **82**, 295 (1984)	13.1 (p. 156) and 14.3 (p. 223); 8; 15; 20.1; 22.8
L. C. Brunel	(b) **121**, 289 (1984)	13.1 (p. 154); 22.3
	(b) **124**, 731 (1984)	13.4 (p. 186) and 20.3 (p. 353); 18; 22.5.4
	(b) **125**, 215 (1984)	13.1 (p. 162) and 19 (p. 312); 16; 22.2.3
W. Brunner	(a) **85**, 257 (1984)	19 (p. 312); 22.9
U. Brüstel	(b) **123**, 229 (1984)	13.1 (p. 158)
F. J. Bryant	(a) **81**, 333 (1984)	13.4 (p. 173); 20.3; 22.4.1
	(a) **81**, 681 (1984)	14.3.4 (p. 249) and 20.3 (p. 347); 22.4.1
	(a) **83**, 315 (1984)	20.3 (p. 350); 13.4; 22.4.1
G. H. Bu-Abbud	(a) **85**, 69 (1984)	1.3 (p. 25) and 1.5 (p. 34); 22
M. Budzynski	(b) **124**, 355 (1984)	13.4 (p. 184); 19; 22.5
A. A. Bugay	(a) **82**, 533 (1984)	13.4 (p. 179); 14.3; 19; 22.1.2
	(a) **86**, 679 (1984)	13.4 (p. 195); 14.3; 19; 22.1.2
E. Bugiel	(a) **81**, K131 (1984)	1.4 (p. 28) and 10.2 (p. 106); 14.3; 22.1.2
	(a) **84**, 143 (1984)	10.2 (p. 113); 14.3.3; 22.1.2
N. S. Buinov	(b) **123**, K165 (1984)	13.1 (p. 160) and 14.4.1 (p. 261); 22
R. K. Bull	(b) **126**, 361 (1984)	13.4 (p. 192) and 20.3 (p. 357); 22
D. W. Bullett	(b) **121**, 641 (1984)	13.1 (p. 155); 17; 20; 22.8
Yu. S. Bulyshev	(a) **82**, 537 (1984)	14.3.1 (p. 234) and 16 (p. 270); 13.1; 22.9

I. Bunget	(a) **82**, K125 (1984)	1.2 (p. 16); 14.4.1; 22.8.1
W. Burckhardt	(a) **85**, 97 (1984)	4 (p. 62) and 20.1 (p. 336); 22.5; 22.5.2
V. M. Burlakov	(b) **125**, 75 (1984)	6 (p. 76) and 20.1 (p. 335); 22.8.1
	(b) **125**, 467 (1984)	6 (p. 77) and 20.1 (p. 337); 22.8.1
L. A. Bursill	(a) **86**, 101 (1984)	10.2 (p. 119); 1.3; 22.6
	(a) **86**, 491 (1984)	10.2 (p. 120); 1.3; 22.6
E. Burzo	(a) **82**, K49 (1984)	18 (p. 279); 18.2; 21.1
	(b) **124**, K117 (1984)	2 (p. 49) and 19 (p. 311); 22.8
K. H. J. Buschow	(a) **84**, 207 (1984)	18.2 (p. 293) and 18.4 (p. 302); 21; 21.4
	(a) **86**, 655 (1984)	1 (p. 10) and 18.2 (p. 295); 5; 22.8
L. S. Bushnev	(a) **82**, 405 (1984)	12.1 (p. 135); 21.1
W. Busse	(b) **125**, 773 (1984)	13.4 (p. 190); 20.1; 22.4.1
P. Bussemer	(a) **85**, 97 (1984)	4 (p. 62) and 20.1 (p. 336); 22.5; 22.5.2
A. V. Butashin	(a) **86**, 345 (1984)	14.4 (p. 258) and 20.2 (p. 344); 1.1; 12.2; 20.1; 20.3; 22.8.1
D. Butler	(a) **83**, K203 (1984)	14.2 (p. 218)
M. Z. Butt	(a) **84**, K125 (1984)	10.1 (p. 100); 21.1
C. Butti	(a) **83**, K199 (1984)	14.3 (p. 226); 22.4.1
A. B. Bykov	(a) **83**, K165 (1984)	20.2 (p. 343) and 20.3 (p. 352); 20.1; 22.8.1
J. Cabrera-Caño	(a) **85**, 445 (1984)	10.2 (p. 117) and 12.2 (p. 142); 22.6.1
B. Čabrić	(b) **123**, 309 (1984)	13.4 (p. 179) and 13.5 (p. 196); 22.9
M. Cadene	(a) **82**, 257 (1984)	10.2 (p. 109) and 14.3.1 (p. 234); 22.9
M. Cahay	(b) **121**, 345 (1984)	13.4 (p. 172); 22.1.2; 22.2.2; 22.4.2
V. Caignaert	(a) **83**, 473 (1984)	1.1 (p. 12) and 1.4 (p. 30); 22.8.2
J. Calas	(b) **125**, 831 (1984)	14.3 (p. 229); 13.1; 22.4.4
R. L. Calvert	(a) **83**, 597 (1984)	20.3 (p. 351); 11; 13.4; 22.8.1
Hoang Ngoc Cam	see under letter "H"	
S. G. Canagaratna	(b) **126**, 141 (1984)	10 (p. 96)
E. Canessa	(b) **124**, 465 (1984)	2 (p. 48); 21; 21.7
	(b) **126**, K1 (1984)	2 (p. 53); 21.7
J. C. Canit	(b) **124**, 87 (1984)	13.5.1 (p. 201) and 13.5.4 (p. 208); 18.4; 20.1; 22.5
V. Čápek	(b) **121**, K7 (1984)	6 (p. 68) and 13 (p. 144); 22
	(b) **125**, 477 (1984)	6 (p. 77); 13.4
	(b) **125**, 639 (1984)	13.5 (p. 198)
	(b) **125**, K137 (1984)	13.5.1 (p. 203); 22.9
R. Capelletti	(a) **86**, 333 (1984)	20.1 (p. 340) and 20.3 (p. 357); 13.4; 22.5.2
P. Čapková	(a) **81**, 203 (1984)	4 (p. 58); 21.1
	(a) **81**, K113 (1984)	1.6 (p. 36); 21.1
M. Cardona	(b) **122**, 53 (1984)	6 (p. 69) and 20.1 (p. 322); 22.8.1
	(b) **122**, 441 (1984)	6 (p. 70) and 20.1 (p. 325); 22.2.2
	(b) **126**, 11 (1984)	6 (p. 67) and 13.1 (p. 153); 22.1; 22.2; 22.4
C. M. Care	(a) **81**, 557 (1984)	6 (p. 69) and 14 (p. 210); 22.9
F. Carmona	(a) **81**, K71 (1984)	14.4 (p. 255); 22
G. Carter	(a) **81**, 267 (1984)	11 (p. 123)
M. De Carvalho	(b) **125**, 831 (1984)	14.3 (p. 229); 13.1; 22.4.4
J. Castaing	(a) **85**, 445 (1984)	10.2 (p. 117) and 12.2 (p. 142); 22.6.1
E. Cavalli	(b) **123**, 679 (1984)	13.4 (p. 181) and 20.1 (p. 330); 22.8.1
M. Celasco	(b) **122**, 411 (1984)	2 (p. 42)
	(b) **123**, 403 (1984)	2 (p. 45)
F. Cerdeira	(b) **122**, 53 (1984)	6 (p. 69) and 20.1 (p. 322); 22.8.1
J. Čermák	(a) **85**, 173 (1984)	18.2.1 (p. 296); 22.8.2
K. Čermák	(b) **124**, K143 (1984)	20.1 (p. 334); 13.1; 22.8
	(b) **125**, K89 (1984)	2 (p. 50) and 14.3.1 (p. 236); 14.3.4; 22.1.2
	(b) **126**, 273 (1984)	4 (p. 62) and 13.1 (p. 164); 20.1; 22.8
Z. Černošek	(b) **122**, 405 (1984)	2 (p. 42) and 20.1 (p. 324); 22

E. ČERNOŠKOVÁ	(b) **122**,	405 (1984)	2 (p. 42) and 20.1 (p. 324); 22
H. CERVA	(a) **82**,	35 (1984)	1 (p. 9) and 7 (p. 81); 14.4.1; 22.8.1
L. ČERVINKA	(a) **83**,	418 (1984)	2 (p. 45) and 4 (p. 59); 17; 22.3
C. CESARI	(a) **81**,	87 (1984)	1.1 (p. 11) and 1.4 (p. 27); 22.6
V. CHÁB	(b) **121**,	301 (1984)	13.1 (p. 154) and 17 (p. 274); 22
	(a) **83**,	418 (1984)	2 (p. 45) and 4 (p. 59); 17; 22.3
L. G. CHACHKHIANI	(a) **82**,	K53 (1984)	18 (p. 279); 22.8
Z. B. CHACHKHIANI	(a) **82**,	K53 (1984)	18 (p. 279); 22.8
G. CHADDHA	(b) **123**,	K139 (1984)	18.2 (p. 292)
E. I. CHAIKINA	(a) **84**,	541 (1984)	20.3 (p. 353); 22.2.3
S. CHAKRABORTY	(b) **122**,	469 (1984)	1.3 (p. 22) and 4 (p. 59); 21.1; 21.1.1; 21.6
	(b) **125**,	97 (1984)	4 (p. 61) and 13.1 (p. 162); 21.1; 21.6
B. CHALUPA	(a) **83**,	455 (1984)	1.1 (p. 12); 22.1.2
D. G. CHAMBAERE	(a) **83**,	93 (1984)	1 (p. 9); 5; 22
J.-P. CHAMINADE	(a) **82**,	285 (1984)	14.3 (p. 223); 13.4; 15; 16; 22.8.1
A. CHANDORKAR	(a) **81**,	407 (1984)	14.3.3 (p. 241); 11; 14.4; 22.1.2
B. P. CHANDRA	(a) **83**,	K19 (1984)	12.2 (p. 140) and 20.3 (p. 350); 10.2; 22.5.2; 22.8.1
GYANESH CHANDRA	(a) **86**,	759 (1984)	14.1 (p. 215); 1.3; 21
	(a) **86**,	765 (1984)	14.1 (p. 215); 21
K. CHANDRA	(a) **81**,	281 (1984)	10.2 (p. 106); 5; 18.3; 18.4; 22.8
CHANDRA PRAKASH	see under letter "P"		
M. CHANDRASEKARAN	(a) **83**,	657 (1984)	14.1 (p. 212); 2; 8; 21; 21.6
CHIA-NAN CHANG	(b) **123**,	435 (1984)	6 (p. 72); 4; 22.5
D.-Y. CHANG	(a) **83**,	529 (1984)	9 (p. 93); 2; 21.1; 21.1.1
I. M. CHAPNIK	(b) **123**,	K183 (1984)	14.2 (p. 217); 21.1
	(b) **125**,	K189 (1984)	14.2 (p. 218); 21
M. CHARBONNEAU	(b) **122**,	151 (1984)	13.5.3 (p. 207)
M. CHAREF	(a) **81**,	569 (1984)	14.3.2 (p. 238) and 14.3.3 (p. 241); 22.1.2
J. P. CHARLES	(a) **83**,	403 (1984)	14.3.3 (p. 243) and 16 (p. 271); 22.1.2
G. J. CHASHOO	(b) **123**,	733 (1984)	14.2 (p. 217)
T. CHASSÉ	(b) **125**,	267 (1984)	13.1 (p. 162) and 17 (p. 276); 22.3
D. D. CHAT	(a) **82**,	K27 (1984)	8 (p. 85) and 14.4.1 (p. 260); 22.8.1
P. CHATTERJEE	(b) **124**,	773 (1984)	14.2 (p. 218); 21
S. CHATTERJEE	(b) **123**,	445 (1984)	1.2 (p. 16) and 6 (p. 72); 22.5
S. CHATTOPADHYAY	(b) **126**,	53 (1984)	1 (p. 10); 20; 21.1
T. CHATTOPADHYAY	(b) **122**,	193 (1984)	13.5.2 (p. 205) and 20 (p. 316); 22; 22.5
R. K. CHAUDHARY	(b) **122**,	709 (1984)	10.2 (p. 109) and 20.1 (p. 326); 22.5.2
D. RAY CHAUDHURI	(a) **86**,	K79 (1984)	14.3.3 (p. 246); 14.3.2; 22.2.4
A. K. CHAUDHURI	(a) **83**,	K77 (1984)	14.3.1 (p. 235); 22.7
S. CHAUDHURI	(a) **83**,	K57 (1984)	2 (p. 45) and 20.1 (p. 329); 22.3
S. R. CHAUDHURI	(a) **84**,	651 (1984)	14.3.1 (p. 236); 16; 22.4.1
	(a) **86**,	827 (1984)	16 (p. 273); 22.4.2
P. CHAUDOUET	(a) **82**,	K133 (1984)	11 (p. 11); 22.8
	(a) **84**,	199 (1984)	18.2 (p. 292) and 18.4 (p. 302); 1.2; 14.1; 21.1
V. I. CHECHERNIKOV	(a) **82**,	K53 (1984)	18 (p. 279); 22.8
V. A. CHEKANOV	(b) **122**,	349 (1984)	2 (p. 41) and 14.3 (p. 222); 14.3.4; 22.1.1
A. R. CHELYADINSKII	(a) **85**,	K43 (1984)	10.2 (p. 117) and 11 (p. 130); 22.1.2
C. C. CHEN	(a) **86**,	K23 (1984)	6 (p. 79) and 20.1 (p. 341); 22
C. Y. CHEN	(a) **81**,	253 (1984)	10.2 (p. 105); 11; 19; 20.1; 22.8.1
R. CHEN	(b) **126**,	361 (1984)	13.4 (p. 192) and 20.3 (p. 357); 22
CHEN LI-RONG	see under letter "L"		
B. CHENEVIER	(a) **84**,	199 (1984)	18.2 (p. 292) and 18.4 (p. 302); 1.2; 14.1; 21.1
V. A. CHEPURNOI	(b) **125**,	K41 (1984)	10.2 (p. 116) and 13.4 (p. 188); 20.3; 22.5.2
G. K. CHEPURNYKH	(b) **123**,	K25 (1984)	18.4 (p. 301); 19
A. P. CHERNOV	(a) **81**,	139 (1984)	2 (p. 39); 14.3; 15; 22.8
V. M. CHERNOV	(a) **81**,	415 (1984)	10.1 (p. 97); 21
V. V. CHERNYAEV	(b) **122**,	703 (1984)	13.5.1 (p. 200); 16; 20.1; 22.4.4

V. Chernyi	(a) **85**, 627 (1984)	14.3.3 (p. 245) and 20.3 (p. 356); 13.4; 22.2.2
L. V. Chernysh	(b) **123**, 345 (1984)	13.4 (p. 179); 16; 22
	(b) **123**, 671 (1984)	13.4 (p. 181); 20.1; 22.5
A. Ya. Chervonenkis	(a) **81**, K159 (1984)	18.4 (p. 300); 22.8.2
A. R. Chetal	(b) **122**, 193 (1984)	13.5.2 (p. 205) and 20 (p. 316); 22; 22.5
J. Chevallier	(a) **81**, 549 (1984)	13.4 (p. 174) and 20.3 (p. 346); 20.1; 22.4.4
Nquy Huu Chi	see under letter "N"	
C. Chiang	(a) **81**, 563 (1984)	2 (p. 41) and 14.3.1 (p. 233)
	(a) **84**, K189 (1984)	2 (p. 50) and 14.3.1 (p. 236); 16
T. Chiba	(a) **83**, 219 (1984)	2 (p. 44) and 10.1 (p. 99); 21.1
G. V. Chikvaidze	(a) **84**, K101 (1984)	1.2 (p. 18); 19; 21.7
E. Chilla	(b) **126**, K91 (1984)	14.1 (p. 214) and 15 (p. 267); 21
E. Chimczak	(a) **82**, 527 (1984)	14.3.3 (p. 242) and 20.3 (p. 349); 13.4; 22.4.1
M. Chipară	(b) **124**, K117 (1984)	2 (p. 49) and 19 (p. 311); 22.8
O. D. Chistyakov	(b) **121**, 623 (1984)	13.4 (p. 174); 21.4
Z. Chobola	(a) **84**, 693 (1984)	14.3.3 (p. 245); 22.2.1
Sang Don Choi	(b) **121**, K181 (1984)	13.1 (p. 155) and 19 (p. 306); 22.2.3
Young Hwan Choi	see under letter "Y"	
D. R. Chopra	(b) **126**, 279 (1984)	20 (p. 318); 21.4
K. K. Chopra	(b) **125**, 499 (1984)	6 (p. 77); 21.1
Tran Chot	see under letter "T"	
A. Choudhury	(a) **83**, K57 (1984)	2 (p. 45) and 20.1 (p. 329); 22.3
N. E. Christensen	(b) **123**, 281 (1984)	13.1 (p. 158); 22.5
	(b) **125**, K59 (1984)	13.1 (p. 163); 22.5
V. Christoph	(b) **121**, K103 (1984)	14.1 (p. 211); 21
	(a) **83**, K141 (1984)	18.2 (p. 292); 21.1; 21.4
Chu Hao	see under letter "H"	
Chuan-zhang Yang	(b) **126**, K125 (1984)	18.2 (p. 295); 18.4
M. A. Chubisov	(b) **125**, 169 (1984)	11 (p. 129); 22.1.2
V. G. Chudinov	(a) **85**, 105 (1984)	1.2 (p. 18); 10.1; 21.1
	(a) **85**, 435 (1984)	11 (p. 130); 2; 21.1
Chul Koo Kim	(b) **122**, K187 (1984)	14.3 (p. 223); 22.6
E. V. Chulkov	(b) **121**, 265 (1984)	13 (p. 144); 21; 22
Yu. I. Chumlyakov	(a) **82**, 405 (1984)	12.1 (p. 135); 21.1
Chung-In Um	(b) **122**, K73 (1984)	13.3 (p. 167); 14.3.2; 19; 22.1.2
Ok Hee Chung	(b) **121**, K181 (1984)	13.1 (p. 155) and 19 (p. 306); 22.2.3
P. D. Cikmach	(a) **83**, 291 (1984)	13.4 (p. 180) and 20.1 (p. 328); 14.3.1; 19; 22.6
Z. Cimpl	(b) **121**, K43 (1984)	18 (p. 278); 22
J. Cisowski	(b) **121**, 289 (1984)	13.1 (p. 154); 22.3
P. E. Clark	(a) **84**, 31 (1984)	1.3 (p. 24); 5; 22.6
G. Cocco	(a) **84**, 371 (1984)	2 (p. 49); 21; 21.3
D. J. H. Cockayne	(a) **82**, 425 (1984)	1.4 (p. 29) and 10.2 (p. 110); 22.1.2
C. Colinet	(b) **125**, 101 (1984)	4 (p. 61) and 13.1 (p. 162); 21.1; 21.4
J. Colmenero	(b) **125**, 409 (1984)	2 (p. 50) and 14.4 (p. 258); 22.9
A. Conan	(b) **124**, 403 (1984)	14.3 (p. 226) and 15 (p. 266); 22
E. Constant	(a) **81**, 569 (1984)	14.3.2 (p. 238) and 14.3.3 (p. 241); 22.1.2
M. C. Contreras	(a) **85**, 511 (1984)	18.2.1 (p. 297); 2; 21.1.1
D. W. Cooke	(b) **126**, 687 (1984)	10.2 (p. 120) and 19 (p. 314); 11; 20.3; 22.5.3
D. Coquillat	(a) **82**, 295 (1984)	13.1 (p. 156) and 14.3 (p. 223); 8; 15; 20.1; 22.8
J. W. Corbett	(b) **122**, 661 (1984)	13.1 (p. 157); 22.1; 22.2
	(a) **85**, K109 (1984)	10.2 (p. 118); 22.1.2
	(b) **126**, 157 (1984)	10.2 (p. 118) and 13.4 (p. 192); 22.1.2
G. Corradi	(b) **121**, 117 (1984)	10.2 (p. 105) and 20.1 (p. 320); 22.5.2
	(b) **125**, K13 (1984)	6 (p. 77) and 20.1 (p. 336); 22.8.1
J. Costa-Quintana	(b) **123**, 219 (1984)	13.1 (p. 158); 22.6
M. Couzi	(a) **85**, 359 (1984)	1.2 (p. 19); 22.8
V. Crişan	(b) **123**, K53 (1984)	13.1 (p. 159) and 18.3 (p. 298); 21.1; 21.4

F. Crowne	(b) **122**, K65 (1984)	13.3 (p. 167)
L. Cser	(b) **124**, 271 (1984)	13.4 (p. 183); 5; 21.1.1
I. Cuculescu	(a) **81**, 191 (1984)	14.4 (p. 255) and 20.1 (p. 321); 22.9
P. Cui	(a) **84**, 157 (1984)	10.1 (p. 100) and 12.1 (p. 136); 21
	(a) **84**, 465 (1984)	10.1 (p. 100) and 12.1 (p. 136); 21
	(a) **86**, 593 (1984)	10.1 (p. 103) and 12.1 (p. 138); 21
E. Culea	(a) **83**, K15 (1984)	6 (p. 72) and 20.1 (p. 329); 2; 22.6.1
	(a) **84**, K17 (1984)	6 (p. 75) and 20.1 (p. 333); 2; 22.6.1
R. Cywiński	(b) **125**, 381 (1984)	20.3 (p. 354); 10.2; 22.5.2
R. Czaputa	(a) **81**, 343 (1984)	13.4 (p. 173); 14.3; 20.3; 22.8
M. R. Czerniak	(a) **85**, 235 (1984)	14.3.1 (p. 236) and 20.3 (p. 355); 13.4; 22.4.2
U. Dahms	(b) **121**, K161 (1984)	13.5.2 (p. 205); 14.4; 22.9
Nguyen Dai Hung	see under letter "N"	
Yu. N. Daluda	(a) **85**, 575 (1984)	13.4 (p. 191 ; 14.3; 20.1; 22.1.2
J. F. Dame	(b) **125**, 831 (1984)	14.3 (p. 229); 13.1; 22.4.4
S. Damgaard	(a) **81**, 239 (1984)	10.2 (p. 105); 5; 22.1.2
A. D'Amico	(a) **86**, 453 (1984)	14.3.3 (p. 246); 22.1.2
R. J. Danby	(a) **83**, 597 (1984)	20.3 (p. 351); 11; 13.4; 22.8.1
P. Danesh	(a) **83**, 411 (1984)	11 (p. 126) and 13.4 (p. 181); 14.3.3; 22.1.2
Dang Dinh Thong	(a) **81**, 695 (1984)	20.3 (p. 347); 22.4.1
	(b) **124**, 343 (1984)	13.4 (p. 184) and 20.3 (p. 352); 22.4.1
	(b) **126**, K159 (1984)	20.3 (p. 358); 22.4.1
Dang Tran Quan	(a) **86**, 421 (1984)	14.3.1 (p. 237); 20.1; 22
M. M. Danial	(a) **82**, 63 (1984)	1.2 (p. 15); 2; 21.1; 21.1.1
K. K. Danilevich-Tovstyuk	(b) **124**, 693 (1984)	6 (p. 75) and 13 (p. 149); 22.3
V. I. Danilyuk	(a) **81**, 163 (1984)	3 (p. 55); 1.6
Dao Khac An	(a) **82**, K145 (1984)	9 (p. 92); 22.1.2; 22.6
A. Daoud	(b) **125**, 87 (1984)	6 (p. 76) and 20.1 (p. 335); 22.9
A. Dargys	(b) **121**, 385 (1984)	13.4 (p. 172); 14.3.3; 22.1.2
	(b) **125**, K71 (1984)	13.4 (p. 188)
B. Darshan	(b) **122**, K59 (1984)	20 (p. 316); 22.8.2
N. G. Darvishov	(a) **86**, K105 (1984)	20.3 (p. 358); 22.8.1
N. Dass	(b) **124**, 531 (1984)	8 (p. 88); 22.5.2
G. L. M. Dassen	(b) **126**, 213 (1984)	18.4 (p. 303)
T. Datta	(b) **121** K125 (1984)	2 (p. 40) and 18 (p. 278); 21.1.1
A. Daud	(b) **126**, 495 (1984)	4 (p. 63); 12.2; 22.5.2
B. Daudin	(b) **122**, 73 (1984)	6 (p. 69) and 8 (p. 85); 22.8.1
M. David	(a) **85**, 375 (1984)	1 (p. 10) and 10 (p. 96)
R. L. Davidovich	(a) **84**, 387 (1984)	1.2 (p. 18) and 14.4.1 (p. 262); 8; 19; 22.8
D. G. Davies	(a) **82**, 23 (1984)	1 (p. 9) and 7 (p. 81); 21.1
G. J. Davies	(a) **86**, 155 (1984)	1.3 (p. 26) and 4 (p. 62); 22.7
J. J. Davies	(b) **121**, 339 (1984)	13.4 (p. 172) and 19 (p. 305); 22.4.3
A. Davituliani	(b) **125**, K63 (1984)	19 (p. 312); 22.5.3
P. Day	(b) **124**, 707 (1984)	13.4 (p. 185) and 20.1 (p. 334); 22.8.1
P. J. Dean	(a) **81**, 625 (1984)	14.3.4 (p. 249) and 20.3 (p. 346); 3; 13.4; 22.4.1; 22.4.2
S. K. Deb	(b) **122**, 113 (1984)	10.2 (p. 107); 22.5.2
	(b) **124**, 545 (1984)	10.2 (p. 114); 22.5.2
	(b) **126**, K117 (1984)	10.2 (p. 120); 22.5.2
M. Deb Roy	(a) **86**, 773 (1984)	14.3.4 (p. 254); 6; 13.1; 22.1.2
G. Debiais	(a) **83**, 269 (1984)	14.3.1 (p. 235) and 20.1 (p. 328); 13.1; 22.4.4
F. Decker	(b) **122**, 651 (1984)	20.1 (p. 326); 13.1; 16; 22
L. Delaey	(a) **86**, 553 (1984)	1.3 (p. 26); 10.1; 21; 21.1
P. Delavignette	(a) **83**, 485 (1984)	1.4 (p. 30); 21
D. Della Sala	(b) **126**, 125 (1984)	2 (p. 52); 4; 20.1; 22.1.1; 22.1.2

P. Delobelle	(a) **83**, 235 (1984)	10.1 (p. 99); 12.1; 21; 21.1.1
J. L. Demenet	(a) **84**, 481 (1984)	10.2 (p. 115); 22.1.2
S. Demian	(a) **81**, 549 (1984)	13.4 (p. 174) and 20.3 (p. 346); 20.1; 22.4.4
F. Deml	(a) **86**, K117 (1984)	11 (p. 133) and 14.3 (p. 232); 22.2.1
A. V. Demura	(b) **125**, 583 (1984)	11 (p. 130); 13
J. P. Denis	(a) **83**, 709 (1984)	20.3 (p. 351); 22.4.1
G. A. Denisenko	(a) **85**, 553 (1984)	20.1 (p. 338) and 20.3 (p. 356); 13.4; 22.8
G. S. Denisov	(b) **124**, K75 (1984)	13.4 (p. 184) and 20.3 (p. 353); 10.2; 22.5.2
Z. L. Denisova	(a) **81**, 657 (1984)	14.3.4 (p. 249) and 20.3 (p. 346); 14.3.1; 22.6
V. M. Derkachenko	(a) **84**, K63 (1984)	18.2 (p. 293); 22.8.2
V. N. Derkachenko	(a) **84**, 215 (1984)	18.4 (p. 302); 22.8.2
B. T. Deshmukh	(b) **122**, K79 (1984)	20.3 (p. 348); 20.1; 22.5.2
V. V. Deshpande	(a) **84**, K105 (1984)	1.2 (p. 18); 8; 14.1; 14.3; 18.1; 22.6
V. A. Desnenko	(a) **81**, 507 (1984)	18.1 (p. 284) and 18.2 (p. 289); 14.1; 21.1; 21.1.1
J. C. Desoyer	(a) **84**, 481 (1984)	10.2 (p. 115); 22.1.2
K. Deus	(b) **126**, K149 (1984)	13.1 (p. 165) and 13.5.2 (p. 206); 14.4; 22.8
P. Deus	(a) **84**, 87 (1984)	6 (p. 74) and 8 (p. 87); 22.8
	(b) **125**, 121 (1984)	6 (p. 76); 22.1; 22.1.1; 22.2.2; 22.5.2; 22.8
H. G. Devare	(b) **121**, K141 (1984)	18.2 (p. 288); 22.7
S. H. Devare	(b) **121**, K141 (1984)	18.2 (p. 288); 22.7
Devendra Kumar	see under letter "K"	
P. C. Dhanasekaran	(a) **83**, 693 (1984)	14.3.3 (p. 243) and 16 (p. 271); 22.1.2
G. Dharmadurai	(a) **83**, K27 (1984)	6 (p. 72) and 8 (p. 86); 1.6
I. Diaconu	(a) **85**, 283 (1984)	14.3 (p. 229); 22.9
A. M. El-Dib	see under letter "E"	
A. Yu. Didyk	(b) **124**, 45 (1984)	7 (p. 82) and 14.4.1 (p. 261); 22.8.1
N. de Diego	(a) **83**, K93 (1984)	1.4 (p. 30) and 10.1 (p. 99); 21.1.1
U. Dietrich	(a) **85**, 97 (1984)	4 (p. 62) and 20.1 (p. 336); 22.5; 22.5.2
R. Dietsch	(a) **85**, 473 (1984)	1.3 (p. 25) and 11 (p. 131); 21; 21.1.1
V. V. Dikareva	(a) **84**, K173 (1984)	14.3 (p. 228) and 15 (p. 267); 10.2; 22
St. K. Dimitrova	(b) **125**, K177 (1984)	14.3 (p. 229) and 15 (p. 267); 1.2; 22.3.1
R. Dimmich	(a) **84**, K85 (1984)	14.1 (p. 213)
S. S. Dimov	(a) **84**, 555 (1984)	13.4 (p. 186); 20.2; 22.4.1
T. N. Dimov	(b) **126**, 261 (1984)	20.1 (p. 339); 22.4.1
N. N. Dinh	(a) **86**, K45 (1984)	20.3 (p. 358); 22.8.1
Dang Dinh Thong	see under letters "Da"	
K. Dini	(a) **85**, K101 (1984)	1.3 (p. 25); 2; 21.1
M. De Dios Leyva	(b) **125**, 221 (1984)	13.1 (p. 162); 22.2.1; 22.2.3
A. F. Dite	(b) **122**, 183 (1984)	16 (p. 269) and 20.3 (p. 348); 22.4.1
D. Diwan	(a) **83**, 605 (1984)	20.3 (p. 351); 11; 13.4; 22.4
S. Dixit	(b) **124**, 395 (1984)	13.5.2 (p. 206); 7; 22.2.3
M. Dlouhá	(a) **83**, K117 (1984)	1.2 (p. 17); 22.8.1
N. L. Dmitruk	(a) **84**, 285 (1984)	16 (p. 272); 13.4; 22.2.1
L. Dmowski	(b) **125**, 215 (1984)	13.1 (p. 162) and 19 (p. 312); 16; 22.2.3
E. V. Dobrokhotov	(a) **81**, K15 (1984)	2 (p. 40) and 9 (p. 91); 22.1.2
R. Doerfler	(b) **124**, 767 (1984)	13.4 (p. 186); 5; 22.8.1
R. Döhl	(a) **86**, 603 (1984)	9 (p. 94); 10.1; 21.1
H. Döhler	(a) **82**, K153 (1984)	9 (p. 92) and 19 (p. 308); 21.4
M. Doi	(a) **83**, 529 (1984)	9 (p. 93); 2; 21.1; 21.1.1
J. Dolinšek	(b) **123**, K83 (1984)	9 (p. 92) and 14.4 (p. 256); 19; 22.8.1
E. P. Domashevskaya	(b) **121**, 241 (1984)	13.1 (p. 153); 21.4
J. Dominec	(b) **123**, 635 (1984)	7 (p. 82) and 13 (p. 147); 21.7
	(b) **126**, K87 (1984)	7 (p. 83) and 14.3 (p. 230); 14.1; 22.7
A. Dominguez-Rodriguez	(a) **85**, 445 (1984)	10.2 (p. 117) and 12.2 (p. 142); 22.6.1
Sang Don Choi	see under letter "C"	
J. Donecker	(a) **81**, 675 (1984)	19 (p. 306) and 20.3 (p. 347); 13.4; 22.4.2
E. Donth	(b) **124**, 515 (1984)	2 (p. 48); 12.2

M. P. Dos Santos	(b) **126**, K113 (1984)	6 (p. 80) and 20.1 (p. 341); 22.2.2
L. Do Thanh	(a) **83**, 719 (1984)	14.3.3 (p. 243); 14.4; 22; 22.1.2
Yu. P. Dotsenko	(a) **82**, 533 (1984)	13.4 (p. 179); 14.3; 19; 22.1.2
	(a) **84**, 263 (1984)	13.4 (p. 185); 14.3; 22.1.2
	(a) **86**, 679 (1984)	13.4 (p. 195); 14.3; 19; 22.1.2
	(a) **86**, K91 (1984)	10.2 (p. 121); 22.1.2
J.-P. Doumerc	(a) **82**, 285 (1984)	14.3 (p. 223); 13.4; 15; 16; 22.8.1
R. Dovesi	(b) **122**, 211 (1984)	13.1 (p. 156); 13.2; 21.2; 21.3
M. Doyama	(b) **126**, 453 (1984)	1.3 (p. 26) and 13 (p. 151); 21.1; 21.1.1
Do Yi-Jing	see under letter "Y"	
G. Dräger	(b) **124**, 11 (1984)	13.1 (p. 153) and 20 (p. 316); 21; 22
V. Drchal	(b) **124**, 179 (1984)	13.1 (p. 161); 17; 21
N. N. Dryomova	(a) **84**, K43 (1984)	1.4 (p. 31) and 10.2 (p. 114); 14.3.4; 22.1.2
P. Dub	(b) **122**, 675 (1984)	13.1 (p. 157) and 20.1 (p. 326); 22
A. P. Dubchak	(a) **82**, 221 (1984)	13.4 (p. 177) and 14.3.4 (p. 250); 14.4; 22
P. I. Dubenskov	(a) **84**, 585 (1984)	11 (p. 129) and 14.3 (p. 227); 22.9
J. D. Dubey	(b) **125**, 629 (1984)	13.1 (p. 163); 22.2
L. B. Dubrovskaya	(a) **83**, 11 (1984)	14.3 (p. 220) and 18 (p. 278); 13.1; 14.2; 15; 22
Fat Duen Ho	see under letter "F"	
F. Dujardin	(b) **126**, 329 (1984)	13.4 (p. 192) and 13.5.1 (p. 204); 22.2; 22.4
A. F. Dunanev	(a) **85**, 591 (1984)	14.3 (p. 230); 11; 22.9
R. A. Dunlap	(a) **85**, K101 (1984)	1.3 (p. 25); 2; 21.1
M. Dupont	(b) **126**, 323 (1984)	13.4 (p. 192) and 19 (p. 313); 22.8.1
R. Duraj	(a) **84**, 229 (1984)	18.2 (p. 293) and 18.4 (p. 303); 1.2; 22.8
J. Duran	(a) **84**, 411 (1984)	1.1 (p. 12); 19; 20.1; 22.9
	(a) **85**, 291 (1984)	14.3 (p. 229); 22.9
R. Durstberger	(b) **123**, 319 (1984)	13.4 (p. 179); 16; 22.8
B. Dusowski	(b) **122**, 133 (1984)	13 (p. 145) and 18 (p. 278)
P. Duval	(a) **86**, 85 (1984)	1.3 (p. 25); 21; 21.1
V. Dvořák	(b) **126**, 459 (1984)	1.2 (p. 20) and 14.4.1 (p. 263); 22.8.1
A. V. Dvurechenskii	(a) **81**, 63 (1984)	2 (p. 39) and 3 (p. 55); 1.5; 22.1.2
	(a) **84**, 171 (1984)	9 (p. 93); 11; 22.1.2
	(a) **85**, K39 (1984)	10.2 (p. 117) and 11 (p. 130); 22.1.2
	(a) **86**, 313 (1984)	13.4 (p. 193) and 19 (p. 314); 11; 22.1.2
A. Dyakov	(a) **83**, 411 (1984)	11 (p. 126) and 13.4 (p. 181); 14.3.3; 22.1.2
V. P. Dymont	(a) **85**, K69 (1984)	14.1 (p. 213) and 18.2 (p. 294); 22.8
A. A. Dyshekov	(b) **126**, 37 (1984)	1.5 (p. 35); 12
T. D. Dzhafarov	(a) **83**, 323 (1984)	14.3.4 (p. 250); 16; 22.3
S. Dzhumanov	(b) **122**, 481 (1984)	10.2 (p. 109); 13.4; 22.5.2
S. A. Dzuba	(a) **85**, 257 (1984)	19 (p. 312); 22.9
A. Ya. Dzyublik	(b) **123**, 53 (1984)	5 (p. 65) and 6 (p. 71)
D. J. Easton	(a) **81**, K103 (1984)	2 (p. 41) and 8 (p. 85); 22
W. Ebeling	(b) **121**, 633 (1984)	13 (p. 145); 13.5; 22
R. C. Ecob	(a) **86**, 585 (1984)	10 (p. 96)
N. A. Economou	(b) **124**, 229 (1984)	13.5.2 (p. 205) and 20.1 (p. 332); 22.8
I. S. Edelman	(b) **125**, 751 (1984)	18.2 (p. 294) and 20.1 (p. 337); 13.4; 22.8.1
J. T. Edmond	(a) **81**, K103 (1884)	2 (p. 41) and 8 (p. 85); 22
Sh. M. Efendiev	(b) **123**, K105 (1984)	6 (p. 73) and 20.1 (p. 330); 22.8
	(b) **125**, 75 (1984)	6 (p. 76) and 20.1 (p. 335); 22.8.1
	(b) **125**, 467 (1984)	6 (p. 77) and 20.1 (p. 337); 22.8.1
	(a) **86**, K105 (1984)	20.3 (p. 358); 22.8.1
I. K. Efenl'ieva	(b) **126**, K139 (1984)	13.1 (p. 165) and 20.1 (p. 341); 12.2; 22; 22.8
Zh Egemberdiev	(b) **126**, 407 (1984)	13.4 (p. 193) and 20.3 (p. 357); 22.5.4
	(b) **126**, K63 (1984)	13.4 (p. 193) and 20.3 (p. 357); 22.5.2
S. Sh. Egemberdieva	(a) **84**, 541 (1984)	20.3 (p. 353); 22.2.3

V. D. Egorov	(b) **122**, 183 (1984)	16 (p. 269) and 20.3 (p. 348); 22.4.1
A. V. Egranov	(b) **122**, 249 (1984)	10.2 (p. 107) and 13.4 (p. 175); 20.1; 20.3; 22.5.2
F. Eichhorn	(a) **82**, K141 (1984)	11 (p. 125); 22.1.2
G. Eichhorn	(b) **125**, 131 (1984)	4 (p. 61); 1.1; 6; 20.1; 22.7
Ya. A. Eiduss	(a) **84**, K101 (1984)	1.2 (p. 18); 19; 21.7
H. M. Eissa	(a) **86**, 207 (1984)	10.1 (p. 103); 21.1
J. Ejimanya	(a) **83**, K87 (1984)	14.3.3 (p. 243); 11; 22.2.1
F. El Akkad	(a) **81**, 549 (1984)	13.4 (p. 174) and 20.3 (p. 346); 20.1; 22.4.4
A. Elango	(b) **122**, 715 (1984)	10.2 (p. 109) and 13.4 (p. 177); 20.1; 22.5.2
A. M. El-Dib	(b) **126**, 587 (1984)	13.1 (p. 165); 22.8
D. D. Eley	(a) **85**, 283 (1984)	14.3 (p. 229); 22.9
S. El-Halawany	(a) **84**, K89 (1984)	16 (p. 272); 14.3.1; 20.1; 22.8
M. M. El-Oker	(a) **83**, 263 (1984)	13.5.2 (p. 205) and 20.1 (p. 328); 13.1; 21
	(a) **83**, 613 (1984)	13.5.2 (p. 205) and 20.1 (p. 330); 21
	(a) **86**, 207 (1984)	10.1 (p. 103); 21.1
	(a) **86**, 805 (1984)	7 (p. 83) and 20.1 (p. 342); 21.1
A. El-Sadek	(a) **82**, 149 (1984)	11 (p. 124) and 14.3 (p. 223); 22.1.2
S. A. El-Sahhar	(a) **83**, 263 (1984)	13.5.2 (p. 205) and 20.1 (p. 328); 13.1; 21
A. A. El-Sharkawy	(a) **83**, 123 (1984)	8 (p. 86); 22.9
	(a) **85**, 429 (1984)	8 (p. 89); 22.5.2
P. V. Elyutin	(b) **124**, 279 (1984)	13 (p. 148); 2
G. A. Emelchenko	(b) **124**, K155 (1984)	13.5.1 (p. 202) and 20.1 (p. 334); 14.3.4; 22.4
V. V. Emtsev	(a) **85**, 575 (1984)	13.4 (p. 191); 14.3; 20.1; 22.1.2
A. Enache	(a) **81**, 191 (1984)	14.4 (p. 255) and 20.1 (p. 321); 22.9
R. Enderlein	(b) **126**, 575 (1984)	13.1 (p. 164); 22
P. Enders	(b) **121**, 39 (1984)	1.2 (p. 14); 4; 13; 22
	(b) **121**, 461 (1984)	1.2 (p. 15); 4; 6; 22
	(b) **124**, K21 (1984)	4 (p. 60); 22
	(b) **125**, K133 (1984)	13.1 (p. 164); 21.7
	(b) **126**, K177 (1984)	13.1 (p. 164); 21.7
S. Endo	(b) **124**, K101 (1984)	14.3 (p. 226); 13.1; 22.8
E. Endruschat	(a) **82**, K1 (1984)	1.2 (p. 15) and 1.4 (p. 28); 2; 11; 22.1.1
F. Engelke	(b) **125**, 483 (1984)	14.4.1 (p. 262); 19; 22.9
H. J. Engelmann	(a) **86**, 499 (1984)	1.3 (p. 26) and 13.3 (p. 171); 20; 21.1
C. Engler	(b) **122**, 745 (1984)	13.3 (p. 167); 4; 6; 8; 22
H. Englisch	(b) **123**, 711 (1984)	13 (p. 147)
	(b) **123**, K169 (1984)	13 (p. 148) and 14 (p. 210)
	(b) **124**, 373 (1984)	13 (p. 148)
	(b) **124**, K15 (1984)	6 (p. 74)
	(b) **125**, 393 (1984)	13 (p. 149); 20.3; 22
R. Englisch	(b) **123**, 711 (1984)	13 (p. 147)
	(b) **124**, 373 (1984)	13 (p. 148)
M. V. Entin	(b) **126**, 487 (1984)	6 (p. 79)
S. Enzo	(a) **84**, 371 (1984)	2 (p. 49); 21; 21.3
O. Erb	(a) **84**, 291 (1984)	16 (p. 272) and 20.1 (p. 333); 13.4; 22.8.1
J.-W. Erben	(a) **82**, K5 (1984)	2 (p. 42) and 3 (p. 55); 11; 22.1.2
	(a) **82**, K121 (1984)	1.5 (p. 33) and 3 (p. 56); 2; 11; 22.1.2
V. N. Ermakov	(b) **122**, K163 (1984)	13.4 (p. 178) and 14.3.4 (p. 250); 22.1.1
	(b) **125**, 815 (1984)	13.4 (p. 190) and 13.5.1 (p. 203); 13.3; 20.1; 22.4.1
T. G. Ermakova	(a) **81**, 377 (1984)	14.3 (p. 221); 22.9
A. N. Ermoshkin	(b) **121**, K59 (1984)	13.5.1 (p. 199) and 20.3 (p. 345); 20.1; 22.6
N. V. Ershov	(b) **121**, 451 (1984)	2 (p. 40); 21; 21.1.1
M. Sh. Erukhimov	(b) **123**, 105 (1984)	13.5.4 (p. 208); 18.2; 22
H. Eschrig	(b) **123**, K149 (1984)	13 (p. 147)
T. Eshita	(b) **122**, 489 (1984)	10.2 (p. 109) and 20.1 (p. 325); 20.3; 22.5.3
V. O. Esin	(a) **81**, 163 (1984)	3 (p. 55); 1.6
N. P. Esina	(a) **85**, 655 (1984)	14.3.3 (p. 246) and 20.3 (p. 356); 16; 22.2.1

V. F. Esipenko	(a) **82**, 405 (1984)	12.1 (p. 135); 21.1
C. Esnouf	(a) **85**, 463 (1984)	10.1 (p. 102) and 12.1 (p. 137); 21.3
G. P. Espinosa	(b) **122**, K11 (1984)	6 (p. 70) and 20.1 (p. 323); 2; 22
B. Esser	(b) **126**, 393 (1984)	20.3 (p. 357); 2; 13.4; 22
J. Estel	(b) **126**, 121 (1984)	6 (p. 79); 22.1.2
R. Ethiraj	(a) **85**, K27 (1984)	12.2 (p. 141) and 20 (p. 318); 22.5.4
L. R. Evangelista	(b) **126**, K29 (1984)	18 (p. 282)
F. Evangelisti	(b) **126**, 125 (1984)	2 (p. 52); 4; 20.1; 22.1.1; 22.1.2
R. A. Evarestov	(b) **122**, 231 (1984)	13.1 (p. 156)
	(b) **122**, 559 (1984)	13.1 (p. 156); 22
V. A. Evseev	(b) **122**, 349 (1984)	2 (p. 41) and 14.3 (p. 222); 14.3.4; 22.1.1
A. M. Evstigneev	(a) **83**, K31 (1984)	10.2 (p. 112); 14.3.4; 16; 20.1; 22.1.1
	(a) **86**, K83 (1984)	11 (p. 132) and 14.3.3 (p. 246); 22.2.1
A. A. Evtushenko	(a) **86**, K31 (1984)	10.2 (p. 120) and 14.4 (p. 259); 12.2; 22
K. Fabisiak	(a) **81**, 579 (1984)	2 (p. 41) and 14.3.1 (p. 233); 19; 22.1
W. R. Fahrner	(a) **86**, 789 (1984)	11 (p. 133) and 14.3.3 (p. 246); 14.3.2; 22.1.2
V. I. Fainberg	(a) **86**, K83 (1984)	11 (p. 132) and 14.3.3 (p. 246); 22.2.1
E. S. Falkevich	(a) **81**, 433 (1984)	1.4 (p. 27); 10.2; 22.1.2
G. Fantozzi	(a) **85**, 463 (1984)	10.1 (p. 102) and 12.1 (p. 137); 21.3
G. Farace	(b) **123**, 659 (1984)	13.4 (p. 181); 20.3; 22.6
B. Farag	(a) **81**, 549 (1984)	13.4 (p. 174) and 20.3 (p. 346); 20.1; 22.4.4
O. V. Farberovich	(b) **121**, 241 (1984)	13.1 (p. 153); 21.4
J. L. Farvacque	(b) **125**, 785 (1984)	13.4 (p. 190) and 20.1 (p. 338); 10.2; 22.2.3
	(b) **126**, 351 (1984)	13.1 (p. 164) and 20.1 (p. 340); 10.2; 22.2
D. Fasold	(a) **86**, 125 (1984)	20.1 (p. 340)
	(a) **86**, 485 (1984)	20.1 (p. 342); 11; 22.8.1
R. Fastow	(a) **84**, 49 (1984)	1.3 (p. 24) and 11 (p. 128); 1.5; 2; 21.1; 21.6
Fat Duen Ho	(b) **122**, 337 (1984)	10.2 (p. 107) and 13.4 (p. 176); 20.1; 22.8.1
C. Fau	(b) **125**, 831 (1984)	14.3 (p. 229); 13.1; 22.4.4
R. Fauquembergue	(a) **81**. 569 (1984)	14.3.2 (p. 238) and 14.3.3 (p. 241); 22.1.2
R. Fazio	(b) **124**, K177 (1984)	14.2 (p. 218)
	(b) **125**, K185 (1984)	14.2 (p. 218)
V. A. Fedorov	(a) **82**, K185 (1984)	20.2 (p. 343) and 20.3 (p. 349); 20.1; 22.8
V. E. Fedorov	(a) **83**, 147 (1984)	6 (p. 72) and 8 (p. 86); 22
	(a) **84**, K165 (1984)	14.1 (p. 213); 21
Yu. M. Fedorov	(a) **81**, K171 (1984)	20.3 (p. 347); 13.4; 22.5
V. V. Fedotova	(a) **84**, K63 (1984)	18.2 (p. 293); 22.8.2
H. Fehske	(b) **123**, 533 (1984)	18 (p. 280); 13; 21
	(b) **126**, 235 (1984)	13 (p. 151) and 18 (p. 282); 8
H. Feichtinger	(a) **81**, 343 (1984)	13.4 (p. 173); 14.3; 20.3; 22.8
V. A. Feigelman	(a) **86**, K51 (1984)	14.3 (p. 231) and 15 (p. 268); 22.1
M. S. Feldman	(a) **83**, K11 (1984)	6 (p. 72) and 20.1 (p. 328); 22.6
K. Feldmann	(b) **122**, K5 (1984)	1.3 (p. 22) and 1.4 (p. 28); 22.6; 22.8.1
I. Fellegvári	(a) **84**, 547 (1984)	20.1 (p. 334); 13.4; 22.8.2
P. Feltham	(a) **84**, K125 (1984)	10.1 (p. 100); 21.1
L. M. Feodorova	(a) **83**, 153 (1984)	10.2 (p. 111) and 14.3 (p. 225); 22.8.1
J. Rodriguez Fernandez	see under letter "R"	
M. Ferrari	(b) **123**, 659 (1984)	13.4 (p. 181); 20.3; 22.6
R. Ferrer	(b) **121**, K39 (1984)	8 (p. 84) and 18 (p. 278); 2
S. D. Ferris	(a) **84**, 363 (1984)	1.4 (p. 31) and 10.2 (p. 114); 14.3.3; 22.1.2
T. N. Fesenko	(a) **81**, 139 (1984)	2 (p. 39); 14.3; 15; 22.8
D. W. Field	(b) **123**, 479 (1984)	4 (p. 60); 6; 22
S. R. Figarova	(b) **121**. 11 (1984)	14.3.1 (p. 233); 6; 13.1; 15; 22
V. D. Fil	(b) **121**, 433 (1984)	7 (p. 81) and 14.1 (p. 211); 13.2; 21
G. M. Filaretova	(a) **85**, 655 (1984)	14.3.3 (p. 246) and 20.3 (p. 356); 16; 22.2.1

A. Ribeiro Filho	see under letter "R"	
S. A. Filin	(b) **123**, 605 (1984)	13.1 (p. 159) and 20 (p. 317); 21; 21.1
A. S. Filipchenko	(a) **84**, 541 (1984)	20.3 (p. 353); 22.2.3
J. Filipowicz	(a) **84**, K89 (1984)	16 (p. 272); 14.3.1; 20.1; 22.8
I. M. Filippov	(a) **82**, 511 (1984)	10.2 (p. 110) and 13.4 (p. 178); 11; 14.3; 22.1.2
N. G. Filonov	(a) **83**, 701 (1984)	14.3.3 (p. 243); 13.4; 22.2.1
G. Filoti	(a) **86**, 525 (1984)	5 (p. 66); 2; 21.1.1
H. J. Fischbeck	(b) **125**, 275 (1984)	13.1 (p. 162)
	(b) **125**, 665 (1984)	13.1 (p. 163)
A. Fischer	(a) **82**, 157 (1984)	9 (p. 92); 20.1; 22
F. Fischer	(b) **121**, 81 (1984)	6 (p. 67) and 20.1 (p. 320); 22.5.2
	(b) **124**, 61 (1984)	6 (p. 74) and 20.1 (p. 331); 10.2; 22.5.2
M. Fischer	(b) **121**, 59 (1984)	12.2 (p. 139); 22.3
A. Ya. Fishman	(b) **121**, 547 (1984)	18.3 (p. 298); 13.4; 22.8.2
H.-J. Fitting	(a) **81**, 323 (1984)	17 (p. 274); 22.6
	(a) **85**, 195 (1984)	17 (p. 276); 11; 21
I. P. Fittipaldi	(b) **121**, 197 (1984)	18.2 (p. 287); 13.3
U. Fleck	(b) **123**, 787 (1984)	13.1 (p. 160); 21.6
G. Fleury	(a) **83**, K103 (1984)	2 (p. 47) and 14.3.1 (p. 235); 22.1.3
C. Florea	(a) **81**, 191 (1984)	14.4 (p. 255) and 20.1 (p. 321); 22.9
V. Florescu	(a) **86**, 525 (1984)	5 (p. 66); 2; 21.1.1
R. Florian	(b) **124**, 741 (1984)	10.2 (p. 114) and 13.4 (p. 186); 20.3; 22.5.2
I. Földvári	(b) **121**, 117 (1984)	10.2 (p. 105) and 20.1 (p. 320); 22.5.2
	(a) **84**, 547 (1984)	20.1 (p. 334); 13.4; 22.8.2
B. Fornari	(b) **123**, K13 (1984)	6 (p. 72) and 20.1 (p. 327); 13.5; 22.8.1
A. Fortini	(b) **125**, 259 (1984)	13 (p. 149)
G. Fortunato	(a) **86**, 453 (1984)	14.3.3 (p. 246); 22.1.2
A. Fousková	(a) **81**, K117 (1984)	10.2 (p. 106); 14.4.1; 22.9
G. L. F. Fraga	(a) **83**, K187 (1984)	14.1 (p. 212); 21.1
Z. Frait	(b) **122**, 535 (1984)	19 (p. 307); 21
	(a) **85**, 179 (1984)	18.2.1 (p. 296); 19; 22.8.2
D. Fraitová	(b) **121**, 173 (1984)	19 (p. 305); 21.1; 21.1.1
	(b) **124**, 587 (1984)	18.2 (p. 293) and 19 (p. 311)
R. Francini	(b) **126**, 305 (1984)	13.5.1 (p. 204) and 20.1 (p. 339); 22.6
T. S. Frangulian	(b) **125**, K37 (1984)	10.2 (p. 116) and 20.1 (p. 336); 11; 22.5.2
W. Frank	(b) **124**, 69 (1984)	10.1 (p. 99); 21.1
H. Franke	(a) **83**, K73 (1984)	14.4 (p. 257); 22.8.1
R. Franz	(b) **125**, 779 (1984)	1.2 (p. 19); 13.4; 20.1; 22.1
Th. Frauenheim	(a) **86**, 735 (1984)	14.3.1 (p. 237); 1.2; 22.9
V. Frei	(b) **125**, 121 (1984)	6 (p. 76); 22.1; 22.1.1; 22.2.2; 22.5.2; 22.8
E. Zh. Freidenfeld	(a) **83**, K127 (1984)	14.4.1 (p. 261); 22.8.1
W. Frentrup	(a) **84**, 269 (1984)	17 (p. 275); 13.3; 22.1.2
H. Frey	(a) **83**, 437 (1984)	2 (p. 46) and 14.3.1 (p. 235); 3; 22.1.2
H.-G. Fritsche	(b) **121**, 705 (1984)	13 (p. 145); 21.6
H. P. Fritzer	(a) **86**, 749 (1984)	14.1 (p. 215); 1.2; 13.1; 18; 20.1; 22.6.1
D. Fröhlich	(b) **125**, 653 (1984)	13.5.1 (p. 203) and 20.1 (p. 337); 13.1; 22.4.3
	(b) **126**, 305 (1984)	13.5.1 (p. 204) and 20.1 (p. 339); 22.6
J. Fröhner	(b) **121**, K161 (1984)	13.5.2 (p. 205); 14.4; 22.9
	(a) **85**, 257 (1984)	19 (p. 312); 22.9
D. Fruchart	(a) **84**, 199 (1984)	18.2 (p. 292) and 18.4 (p. 302); 1.2; 14.1; 21.1
R. Fruchart	(a) **82**, K133 (1984)	1.1 (p. 11); 22.8
L. Fuentes	(b) **122**, K5 (1984)	1.3 (p. 22) and 1.4 (p. 28); 22.6; 22.8.1
R. Fuentes-Samaniego	(b) **121**, 101 (1984)	9 (p. 91) and 10 (p. 95)
	(b) **125**, 147 (1984)	9 (p. 93) and 10 (p. 95)
K. Fuji	(a) **83**, 567 (1984)	18.1 (p. 285) and 19 (p. 309); 18.4; 22.6.1
H. Fuji	(a) **86**, 75 (1984)	1.2 (p. 20); 22.5.4

Y. Fukuda	(a) **85**, K141 (1984)	17 (p. 276) and 20.3 (p. 356); 2; 22.6
S. A. Fysh	(a) **84**, 31 (1984)	1.3 (p. 24); 5; 22.6
M. Gaber	(a) **85**, 195 (1984)	17 (p. 276); 11; 21
R. G. Gabrielyan	(b) **123**, K97 (1984)	7 (p. 82)
	(b) **125**, K67 (1984)	5 (p. 66)
R. Ts. Gabrielyan	(a) **85**, 349 (1984)	1.1 (p. 13); 22.1.1
V. T. Gabrielyan	(a) **83**, 153 (1984)	10.2 (p. 111) and 14.3 (p. 225); 22.8.1
	(a) **86**, K105 (1984)	20.3 (p. 358); 22.8.1
J. C. Gâcon	(b) **124**, 325 (1984)	13.4 (p. 184) and 20.3 (p. 352); 18; 22.5.4
	(b) **124**, 731 (1984)	13.4 (p. 186) and 20.3 (p. 353); 18; 22.5.4
R. Gaethke	(b) **124**, 335 (1984)	13.5.1 (p. 201) and 20.3 (p. 352); 23
S. G. Gagarin	(b) **124**, 699 (1984)	13.1 (p. 161) and 13.3 (p. 168); 22.6
P. Gagliardelli	(b) **123**, 519 (1984)	10.2 (p. 112); 20.1; 20.3; 22.5.2
Yu. B. Gaididei	(b) **122**, 239 (1984)	13.5.1 (p. 199); 6; 22
G. N. Gaidukov	(a) **82**, 355 (1984)	1.3 (p. 23); 10.1; 21
R. P. Gairola	(b) **125**, 65 (1984)	6 (p. 76) and 8 (p. 88); 10.2; 22; 22.1.1
Z. Gałdecki	(a) **83**, K113 (1984)	1.1 (p. 12); 22.9
G. P. Gallerano	(b) **122**, K83 (1984)	10.2 (p. 108) and 13.4 (p. 176); 20.1; 22.5.2
H. Gamari-Seale	(b) **122**, 527 (1984)	18.2 (p. 290) and 18.4 (p. 301); 21.4
K. Oyama Gannon	see under letter "O"	
D. A. Garanin	(b) **124**, K133 (1984)	18.1 (p. 285) and 19 (p. 311)
H. Garem	(a) **84**, 481 (1984)	10.2 (p. 115); 22.1.2
K. B. Garg	(a) **83**, 279 (1984)	4 (p. 59) and 13 (p. 147); 20; 21.1; 21.1.1; 21.4
B. J. Garrard	(a) **86**, 75 (1984)	1.2 (p. 20); 22.5.4
R. K. Gartia	(a) **83**, 651 (1984)	20.3 (p. 351); 13.4; 22.8.1
L. S. Gasanov	(a) **82**, 307 (1984)	14.3.4 (p. 250); 22
F. M. Gashimzade	(b) **122**, 623 (1984)	13.1 (p. 157) and 20.1 (p. 325); 22
T. M. Gasymov	(b) **125**, 421 (1984)	15 (p. 267); 6; 13; 22
V. I. Gatalskaya	(a) **85**, K51 (1984)	11 (p. 130); 19; 22.1.2
K. Gatterer	(a) **86**, 749 (1984)	14.1 (p. 215); 1.2; 13.1; 18; 20.1; 22.6.1
M. Gatulle	(b) **121**, 59 (1984)	12.2 (p. 139); 22.3
A. Gavaix	(b) **122**, 255 (1984)	13.4 (p. 175) and 19 (p. 306); 22.8.1
B. Gaveau	(b) **121**, 47 (1984)	6 (p. 67)
F. F. Gavrilov	(b) **121**, K75 (1984)	19 (p. 305); 22
P. Gaworzewski	(a) **81**, K165 (1984)	10.2 (p. 107) and 13.4 (p. 175); 14.3; 22.1.2
	(a) **85**, 133 (1984)	1.4 (p. 32) and 10.2 (p. 116); 20.1; 22.1.2
	(a) **85**, 575 (1984)	13.4 (p. 191); 14.3; 20.1; 22.1.2
T. R. Gawron	(b) **125**, 341 (1984)	13.4 (p. 187); 20.1; 22
W. Gehlhoff	(a) **81**, 701 (1984)	13.4 (p. 175) and 19 (p. 306); 22.4.1; 22.4.2
R. Geick	(b) **125**, 131 (1984)	4 (p. 61); 1.1; 6; 20.1; 22.7
H.-D. Geiler	(a) **83**, K1 (1984)	3 (p. 56) and 11 (p. 126); 22.1.2
R. S. Gekht	(b) **122**, 509 (1984)	13 (p. 146) and 18.2 (p. 290)
B. R. Gelchinskii	(b) **125**, 81 (1984)	2 (p. 50) and 4 (p. 61); 21.2; 21.7
P. J. Gellings	(b) **122**, 317 (1984)	13.4 (p. 176) and 19 (p. 307); 22.8.1
S. De Gennaro	(b) **123**, 723 (1984)	14.2 (p. 217); 21.4
A. Geoffroy	(a) **83**, 709 (1984)	20.3 (p. 351); 22.4.1
A. N. Georgobiani	(a) **82**, 207 (1984)	13.4 (p. 176) and 16 (p. 269); 22.8
U. Geppert	(b) **124**, 55 (1984)	14.4.1 (p. 261); 19
N. N. Gerasimenko	(a) **86**, 185 (1984)	10.2 (p. 119) and 11 (p. 132); 22.1.2
	(a) **86**, 717 (1984)	11 (p. 133) and 14.3.3 (p. 246); 14.3.4; 22.1.2
W. Gericke	(a) **81**, 715 (1984)	20.3 (p. 347); 22.4.1
	(a) **81**. 733 (1984)	14.3.4 (p. 249); and 20.3 (p. 347); 22.4.2
H. Gerischer	(b) **122**, 651 (1984)	20.1 (p. 326); 13.1; 16; 22
E. Gerlach	(b) **121**, 757 (1984)	14.3.1 (p. 233); 22
D. Gerlich	(b) **123**, K5 (1984)	12.1 (p. 136) and 12.2 (p. 140); 21; 21.6; 22.5.2

A. P. Ges	(b) **122**, 47 (1984)	8 (p. 85) and 18.4 (p. 300); 7; 22.8.2
	(a) **84**, K63 (1984)	18.2 (p. 293); 22.8.2
T. Gessner	(a) **81**, 751 (1984)	11 (p. 123); 1.4; 22.1.2
J. Geurts	(b) **125**, 131 (1984)	4 (p. 61); 1.1; 6; 20.1; 22.7
G. Gevers	(a) **81** 105 (1984)	1.3 (p. 22); 14.3.1; 22.5
	(b) **123**, 377 (1984)	14.3.1 (p. 235); 13.4; 22.5
	(a) **84**, 273 (1984)	13.4 (p. 185) and 14.3.1 (p. 235); 22.5
R. Gevers	(a) **85**, 375 (1984)	1 (p. 10) and 10 (p. 96)
A. A. Ghani	(a) **84**, 337 (1984)	14.3 (p. 227) and 15 (p. 266); 22.8.2
A. A. Ghani Awad	see under letter "A"	
K. P. Ghatak	(b) **122**, K93 (1984)	14.3 (p. 223); 22
	(b) **123**, K143 (1984)	13.1 (p. 160) and 18 (p. 280); 22
	(b) **126**, K41 (1984)	13 (p. 151) and 18 (p. 283); 22.2.1
	(b) **126**, K47 (1984)	13.1 (p. 164); 22
S. K. Ghatak	(b) **125**, K125 (1984)	13.1 (p. 163); 22.8
M. N. Ghatikar	(b) **124**, K25 (1984)	13 (p. 148) and 18 (p. 280); 22.8
	(b) **125**, K125 (1984)	13.1 (p. 163); 22.8
	(b) **126**, 279 (1984)	20 (p. 318); 21.4
V. Ghiordănescu	(a) **84**, K139 (1984)	18.3 (p. 299) and 20.1 (p. 335); 22.8.2
M. H. Ghoniem	(a) **85**, 429 (1984)	8 (p. 89); 22.5.2
A. K. Ghosh	(b) **122**, 469 (1984)	1.3 (p. 22) and 4 (p. 59); 21.1; 21.1.1; 21.6
	(b) **125**, 97 (1984)	4 (p. 61) and 13.1 (p. 162); 21.1; 21.6
B. Ghosh	(a) **86**, 327 (1984)	20.3 (p. 357); 11; 13.4; 22.5.2
S. Ghosh	(b) **121**, 661 (1984)	13.5.2 (p. 205); 7; 22.2.3
	(b) **123**, 445 (1984)	1.2 (p. 16) and 6 (p. 72); 22.5
	(b) **124**, 395 (1984)	13.5.2 (p. 206); 7; 22.2.3
	(b) **124**, 489 (1984)	6 (p. 75) and 13.5.2 (p. 206); 7; 22.2.3
	(a) **85**, 535 (1984)	14 (p. 210) and 20 (p. 318); 22
M. H. Ben Ghozlen	(b) **125**, 87 (1984)	6 (p. 76) and 20.1 (p. 335); 22.9
G. Giaquinta	(b) **124**, K177 (1984)	14.2 (p. 218)
	(b) **125**, K185 (1984)	14.2 (p. 218)
M. Giehler	(a) **86**, 531 (1984)	2 (p. 54) and 20.1 (p. 342); 4; 6; 22.6.1
B. Giese	(b) **122**, 641 (1984)	13.1 (p. 157) and 13.5 (p. 196); 6; 22
	(b) **123**, 755 (1984)	13.1 (p. 159) and 14.3.4 (p. 251); 22
D. Gignoux	(a) **86**, 295 (1984)	18.4 (p. 303); 21.4
A. Gilewski	(a) **86**, 389 (1984)	14.2 (p. 218)
P. Gille	(a) **84**, K121 (1984)	10.2 (p. 115); 14.3; 22
C. Giovanella	(b) **126**, 125 (1984)	2 (p. 52); 4; 20.1; 22.1.1; 22.1.2
W. Giriat	(b) **124**, K53 (1984)	20.1 (p. 332); 13.1; 22.4.4
D. V. Gitsu	(b) **124**, K87 (1984)	14.3 (p. 226); 18; 22.7
B. A. Gizhevskii	(b) **121** 737 (1984)	15 (p. 264); 22.6
I. Gladkih	(b) **124**, 271 (1984)	13.4 (p. 183); 5; 21.1.1
H. Glaefeke	(a) **85**, K149 (1984)	13.4 (p. 192) and 17 (p. 277); 22
U. H. Gläser	(b) **121**, 673 (1984)	13.1 (p. 155); 20; 21.3
	(b) **124**, 423 (1984)	10.1 (p. 100) and 14.1 (p. 213); 21; 21.1
	(b) **126**, 741 (1984)	14.1 (p. 215); 13; 21; 21.1
K. D. Glinchuk	(a) **82**, 503 (1984)	13.4 (p. 178) and 20.3 (p. 349); 22.2.1
	(a) **83**, 625 (1984)	13.4 (p. 182) and 20.3 (p. 351); 11; 22.2.1
	(a) **84**, 237 (1984)	13.4 (p. 184); 22.1.2
	(a) **84**, 567 (1984)	13.4 (p. 186) and 20.3 (p. 353); 11; 22.2.1
	(a) **86**, 623 (1984)	13.4 (p. 195); 22.1.2
M. Gliński	(b) **125**, K85 (1984)	13.3 (p. 169) and 14.3 (p. 228); 22.2.3
M. Glogarová	(a) **82**, 47 (1984)	2 (p. 42) and 10 (p. 95)
E. Ya. Glushko	(b) **122**, 569 (1984)	13.5.1 (p. 200) and 20.3 (p. 349); 22.9
E. Gmelin	(b) **123**, 541 (1984)	8 (p. 87) and 18 (p. 280); 21.1
Yu. P. Gnatenko	(b) **123**, K63 (1984)	13.5.1 (p. 201) and 20.3 (p. 350); 22.3
N. K. Gobran	(a) **82**, 63 (1984)	1.2 (p. 15); 2; 21.1; 21.1.1

V. E. Godwin	(b) **122**, 295 (1984)	13.3 (p. 167) and 13.4 (p. 175)
O. Goede	(a) **81**, 695 (1984)	20.3 (p. 347); 22.4.1
	(b) **124**, 343 (1984)	13.4 (p. 184) and 20.3 (p. 352); 22.4.1
	(b) **126**, K159 (1984)	20.3 (p. 358); 22.4.1
F. Goedsche	(b) **123**, 143 (1984)	2 (p. 44) and 15 (p. 265); 14.1; 18.2; 21
V. B. Gohel	(b) **124**, 137 (1984)	13.1 (p. 160); 2; 14.1; 15; 21
M. Goiran	(b) **123**, K43 (1984)	13.1 (p. 158); 22.3
H. J. Goldsmid	(a) **81**, K127 (1984)	2 (p. 41) and 11 (p. 123); 8; 22.1.2
E. M. Goldys	(b) **124**, K79 (1984)	13.4 (p. 184) and 20.1 (p. 332); 18; 22.2.2
O. A. Golikova	(a) **86**, K51 (1984)	14.3 (p. 231) and 15 (p. 268); 22.1
H.-Chr. Golinsky	(a) **85**, 569 (1984)	13.4 (p. 191) and 14.3.1 (p. 236); 22.6
E. M. Gololobov	(a) **85**, K51 (1984)	11 (p. 130); 19; 22.1.2
W. Goltsche	(b) **121**, 471 (1984)	6 (p. 68) and 20 (p. 316); 13.5.2; 22
A. V. Goltsev	(b) **122**, 581 (1984)	6 (p. 70) and 20.1 (p. 325); 22
	(b) **126**, 537 (1984)	18.2 (p. 295) and 18.4 (p. 304)
Yu. V. Goltvianskii	(a) **82**, 221 (1984)	13.4 (p. 177) and 14.3.4 (p. 250); 14.4; 22
V. B. Golubev	(a) **83**, 589 (1984)	13.3 (p. 168) and 17 (p. 275); 11; 19; 22.9
A. Golzené	(b) **123**, K125 (1984)	11 (p. 126) and 19 (p. 309); 13.4; 22.2.1
A. M. Gómez	(a) **83**, K131 (1984)	10.1 (p. 99); 21.4
J. C. Gomez-Sal	(a) **86**, 295 (1984)	18.4 (p. 303); 21.4
L. L. Gonçalves	(b) **125**, 563 (1984)	18.4 (p. 303)
J. L. Gondar	(b) **125**, 221 (1984)	13.1 (p. 162); 22.2.1; 22.2.3
D. J. González	(b) **123**, 485 (1984)	4 (p. 60); 1.3; 21.2; 21.3
H. C. González	(a) **86**, 169 (1984)	10.1 (p. 102) and 11 (p. 132); 21.3
R. González	(a) **83**, 645 (1984)	20.3 (p. 351); 10.2; 13.4; 22.6
	(a) **84**, 179 (1984)	10.2 (p. 113) and 11 (p. 128); 20.1; 22.8.1
L. Gonzo	(b) **123**, 659 (1984)	13.4 (p. 181); 20.3; 22.6
E. P. Goodings	(a) **82**, 301 (1984)	1.1 (p. 11) and 14.3 (p. 223); 22.9
E. S. R. Gopal	(a) **85**, K165 (1984)	2 (p. 52) and 14.3 (p. 230); 22.1.3
I. K. Gopalakrishnan	(a) **85**, K89 (1984)	14.1 (p. 213); 21.1
B. S. V. Gopalam	(a) **83**, 693 (1984)	14.3.3 (p. 243) and 16 (p. 271); 22.1.2
R. V. Gopala Rao	see under letter "R"	
I. Gorczyca	(b) **125**, 229 (1984)	13.1 (p. 162); 12.2; 22.2.2; 22.2.3
P. S. Gordienko	(a) **84**, 387 (1984)	1.2 (p. 18) and 14.4.1 (p. 262); 8; 19; 22.8
W. S. Gordon	(a) **82**, 527 (1984)	14.3.3 (p. 242) and 20.3 (p. 349); 13.4; 22.4.1
E. A. Goremychkin	(b) **121**, 623 (1984)	13.4 (p. 174); 21.4
	(b) **124**, 355 (1984)	13.4 (p. 184); 19; 22.5
Yu. N. Gornostiriev	(b) **123**, 289 (1984)	13.3 (p. 168); 21
J. A. Gorri	(a) **82**, 85 (1984)	10.2 (p. 108); 22.5.2
I. Gościańska	(a) **82**, 491 (1984)	14.1 (p. 211) and 18.2.1 (p. 296); 21.4
B. N. Goshchitskii	(a) **85**, 435 (1984)	11 (p. 130); 2; 21.1
	(a) **86**, K59 (1984)	14.1 (p. 215) and 14.2 (p. 219); 1.2; 21; 21.1
L. Gosztonyi	(a) **84**, 547 (1984)	20.1 (p. 334); 13.4; 22.8.2
V. Gottschalch	(a) **82**, K137 (1984)	8 (p. 86); 1.1; 22.2.4
G. Götz	(a) **83**, 69 (1984)	1.2 (p. 16) and 3 (p. 56); 22.1.2
	(a) **83**, K1 (1984)	3 (p. 56) and 11 (p. 126); 22.1.2
	(a) **85**, K1 (1984)	1.6 (p. 38) and 11 (p. 129); 21
M. E. De Gouvéa	(b) **122**, K175 (1984)	19 (p. 308); 22.9
R. P. Goyal	(b) **123**, K109 (1984)	4 (p. 60); 22.5
A. I. Grachev	(a) **82**, K95 (1984)	16 (p. 270) and 20.1 (p. 324); 22.8
W. Graeff	(a) **82**, 35 (1984)	1 (p. 9) and 7 (p. 81); 14.4.1; 22.8.1
T. Grafe	(b) **123**, 663 (1984)	13.1 (p. 159) and 13.4 (p. 181); 22.6
G. Graiss	(a) **81**, 513 (1984)	18.2 (p. 289); 21.1.1
J. Grannec	(a) **81**, 105 (1984)	1.3 (p. 22); 14.3.1; 22.5
A. Gras-Marti	(a) **81**, 267 (1984)	11 (p. 123)
U. M. Grassano	(b) **122**, K83 (1984)	10.2 (p. 108) and 13.4 (p. 176); 20.1; 22.5.2
P. Grau	(a) **83**, 499 (1984)	12.2 (p. 141); 22.6

E. De Grave	(a) **83**, 93 (1984)	1 (p. 9); 5; 22
H. S. Greenside	(b) **126**, 511 (1984)	4 (p. 63) and 13 (p. 151); 12.2; 22.1; 22.1.1; 22.1.2
B. Greń	(a) **86**, 389 (1984)	14.2 (p. 218)
M. Grendel	(a) **82**, 519 (1984)	14.3.3 (p. 242); 14.3.4
G. Grenet	(b) **124**, 325 (1984)	13.4 (p. 184) and 20.3 (p. 352); 18; 22.5.4
	(b) **124**, 731 (1984)	13.4 (p. 186) and 20.3 (p. 353); 18; 22.5.4
V. P. Gribkovskii	(b) **126**, 415 (1984)	14.3.4 (p. 253)
V. V. Gribkovskii	(a) **86**, 573 (1984)	8 (p. 90); 11; 22.1; 22.2
S. A. Gridnev	(a) **83**, 131 (1984)	14.4.1 (p. 261); 10.2; 12.2; 22.8.1
V. I. Gridnev	(b) **126**, 565 (1984)	11 (p. 133); 13; 22.1.2
M. Griepentrog	(a) **84**, 269 (1984)	17 (p. 275); 13.3; 22.1.2
J. E. Griffiths	(b) **122**, K11 (1984)	6 (p. 70) and 20.1 (p. 323); 2; 22
J. Grigas	(a) **84**, 387 (1984)	1.2 (p. 18) and 14.4.1 (p. 262); 8; 19; 22.8
N. N. Grigorev	(b) **122**, K163 (1984)	13.4 (p. 178) and 14.3.4 (p. 250); 22.1.1
L. S. Grigoryan	(a) **84**, 597 (1984)	2 (p. 49) and 14.3.1 (p. 236); 19; 20.1; 22.9
G. R. Grigoryan	(b) **125**, K67 (1984)	5 (p. 66)
I. N. Grincheshen	(a) **85**, K85 (1984)	14.3 (p. 229) and 16 (p. 273); 22.8
A. U. Grinfelds	(a) **81**, K23 (1984)	20.3 (p. 345); 2; 13.5.3; 22.6
T. G. Grishchenko	(a) **86**, K63 (1984)	14.3 (p. 231) and 15 (p. 268); 22.1.2
P. Grochulski	(a) **83**, K113 (1984)	1.1 (p. 12); 22.9
	(a) **84**, K5 (1984)	1.1 (p. 12) and 8 (p. 87); 22.9
M. V. Grodetskii	(b) **123**, 169 (1984)	13.5.1 (p. 200); 20.1; 22.5
R. Groetzschel	(a) **85**, K35 (1984)	10.2 (p. 117) and 11 (p. 130); 22.1.2
V. Gröger	(a) **84**, 475 (1984)	10.1 (p. 100); 21.1
V. V. Gromov	(a) **83**, K153 (1984)	13.4 (p. 183) and 20.1 (p. 331); 22.8.2
J. Gronkowski	(a) **85**, 389 (1984)	1 (p. 10)
P. Grosbras	(a) **84**, 481 (1984)	10.2 (p. 115); 22.1.2
R. Grötzschel	(a) **84**, 171 (1984)	9 (p. 93); 11; 22.1.2
A. K. Grover	(a) **85**, K89 (1984)	14.1 (p. 213); 21.1
I. N. Groznov	(b) **123**, 183 (1984)	14.2 (p. 217); 5; 22.6.1
H. Gruber	(a) **86**, 749 (1984)	14.1 (p. 215); 1.2; 13.1; 18; 20.1; 22.6.1
R. Gründler	(b) **121**, K161 (1984)	13.5.2 (p. 205); 14.4; 22.9
	(b) **123**, K161 (1984)	13.5.2 (p. 205); 22.2.2
	(b) **126**, K149 (1984)	13.1 (p. 165) and 13.5.2 (p. 206); 14.4; 22.8
V. S. Grunin	(b) **123**, 353 (1984)	13.4 (p. 180) and 19 (p. 309); 22.6; 22.8.1
S. A. Grusha	(a) **86**, K83 (1984)	11 (p. 132) and 14.3.3 (p. 246); 22.2.1
S. L. Gruverman	(a) **84**, 423 (1984)	2 (p. 49) and 12.1 (p. 136); 1.2; 21; 21.4
A. N. Gruzintsev	(a) **82**, 207 (1984)	13.4 (p. 176) and 16 (p. 269); 22.8
V. G. Grygoriev	(a) **83**, 573 (1984)	11 (p. 127); 22.1.2
S. W. Gu	(b) **121**, K165 (1984)	13.5.3 (p. 207); 22
	(b) **122**, K41 (1984)	13.5.3 (p. 207); 22
Shi-wei Gu	see under letter "S"	
V. A. Gubanova	(b) **121**, K205 (1984)	19 (p. 306) and 20.1 (p. 322); 6; 13.4; 22.5.2
	(b) **124**, K75 (1984)	13.4 (p. 184) and 20.3 (p. 353); 10.2; 22.5.2
S. I. Gubarev	(b) **122**, K133 (1984)	13.4 (p. 178) and 18.2 (p. 290); 22.4.4
V. I. Gubskaya	(a) **85**, 585 (1984)	11 (p. 131) and 13.4 (p. 191); 22.1.2
V. V. Gudkov	(b) **121**, 433 (1984)	7 (p. 81) and 14.1 (p. 211); 13.2; 21
L. F. Gudymenko	(b) **124**, K139 (1984)	13.5 (p. 197) and 20.3 (p. 353); 22.4.1
S. Guha	(b) **122**, 799 (1984)	13 (p. 146) and 20 (p. 317); 18
	(a) **83**, 675 (1984)	7 (p. 82) and 14.3 (p. 226); 22.2.3
A. E. Guile	(a) **86**, 439 (1984)	14.3.2 (p. 239) and 14.3.4 (p. 253)
M. Guittard	(b) **124**, K105 (1984)	6 (p. 75) and 20.1 (p. 334); 22.3
F. Guiu	(a) **82**, 85 (1984)	10.2 (p. 108); 22.5.2
E. G. Gule	(b) **124**, K139 (1984)	13.5 (p. 197) and 20.3 (p. 353); 22.4.1
Z. Gumienny	(a) **83**, K51 (1984)	16 (p. 271) and 20.1 (p. 329); 22.3
H.-E. Gumlich	(b) **125**, 773 (1984)	13.4 (p. 190); 20.1; 22.4.1
H.-L. Günter	(b) **126**, 733 (1984)	6 (p. 79) and 15 (p. 268); 14.3; 22.1.1

P. Günter	(a) **84**, 103 (1984)	14.4.1 (p. 262) and 20.1 (p. 333); 22.8
A. C. Gupta	(b) **125**, 207 (1984)	13.2 (p. 166); 12.1; 15; 21.1
B. R. K. Gupta	(b) **123**, K109 (1984)	4 (p. 60); 22.5
K. C. Gupta	(b) **121**, 73 (1984)	4 (p. 58) and 6 (p. 67); 12.1; 21
S. K. Gupta	(b) **124**, K167 (1984)	19 (p. 311); 22.8
S. N. Gupta	(b) **125**, 211 (1984)	13 (p. 149); 20; 21.4
V. P. Gupta	(a) **81**, 281 (1984)	10.2 (p. 106); 5; 18.3; 18.4; 22.8
	(b) **125**, K1 (1984)	1 (p. 10) and 4 (p. 61); 22
P. Gürtler	(b) **124**, 335 (1984)	13.5.1 (p. 201) and 20.3 (p. 352); 23
V. E. Gusakov	(a) **81**, 469 (1984)	11 (p. 123); 22.1.1
A. N. Gusatinskii	(b) **125**, K129 (1984)	13 (p. 151) and 20 (p. 318); 22
D. A. Guseinov	(a) **86**, K19 (1984)	8 (p. 89); 22.8
G. D. Guseinov	(b) **124**, K69 (1984)	13.4 (p. 184) and 20.3 (p. 352); 22.8
R. R. Guseinov	(b) **125**, 237 (1984)	13.5.1 (p. 202)
T. M. Guseinov	(a) **82**, K87 (1984)	16 (p. 270); 22.1.3
D. A. Guseinova	(b) **122**, 623 (1984)	13.1 (p. 157) and 20.1 (p. 325); 22
A. I. Gusev	(a) **84**, 527 (1984)	1.3 (p. 25); 18.1; 22
	(a) **85**, 159 (1984)	10.1 (p. 101); 8; 12.1; 21.1
	(a) **86**, K11 (1984)	1.2 (p. 20) and 8 (p. 89); 6; 22
V. S. Gushchin	(a) **85**, 497 (1984)	18.1 (p. 286); 21.1.1
V. Gutan	(a) **81**, 715 (1984)	20.3 (p. 347); 22.4.1
G. L. Gutsev	(b) **121**, 595 (1984)	13.1 (p. 155); 17; 21.1
S. P. Gwet	(b) **123**, 503 (1984)	6 (p. 73) and 20.1 (p. 329); 22
Gyanesh Chandra	see under letter "C"	
Ha Vinh Tan	(b) **125**, 245 (1984)	13.5 (p. 198) and 20 (p. 318); 7
T. Hädrich	(a) **83**, 387 (1984)	14.2 (p. 217)
H. Haefner	(a) **86**, 833 (1984)	14.3.3 (p. 247) and 20.3 (p. 358); 22.2.2
W. E. Hagston	(a) **81**, 687 (1984)	14.3.4 (p. 249); and 20.3 (p. 347); 13.4; 22.4
V. A. Haisler	(b) **121**, K13 (1984)	6 (p. 68) and 20.1 (p. 320); 22.5
	(b) **125**, K103 (1984)	6 (p. 78) and 20.1 (p. 338); 22.5
D. Hajduković	(b) **123**, K35 (1984)	18.2 (p. 291)
	(b) **124**, K129 (1984)	18 (p. 281)
	(b) **125**, K51 (1984)	18.4 (p. 303)
	(b) **125**, K113 (1984)	13 (p. 150) and 18 (p. 282)
	(b) **126**, K33 (1984)	18.4 (p. 303)
	(b) **126**, K37 (1984)	18.4 (p. 303)
V. Hajko	(a) **82**, K159 (1984)	11 (p. 125) and 18.2 (p. 290); 2; 21.1.1
V. Hajko, Jr.	(a) **82**, K159 (1984)	11 (p. 125) and 18.2 (p. 290); 2; 21.1.1
N. A. Hakeem	(b) **123**, K1 (1984)	6 (p. 71) and 20.1 (p. 327); 22.8.2
S. El-Halawany	see under letter "E"	
L. E. Halliburton	(a) **81**, 253 (1984)	10.2 (p. 105); 11; 19; 20.1; 22.8.1
C. Hamann	(a) **82**, K5 (1984)	2 (p. 42) and 3 (p. 55); 11; 22.1.2
	(a) **82**, K111 (1984)	14.3.3 (p. 242) and 16 (p. 270); 22.9
	(a) **86**, 735 (1984)	14.3.1 (p. 237); 1.2; 22.9
A. Hamou	(a) **83**, K103 (1984)	2 (p. 47) and 14.3.1 (p. 235); 22.1.3
P. Hampl	(a) **86**, K117 (1984)	11 (p. 133) and 14.3 (p. 232); 22.2.1
P. H. Handel	(b) **121**, 429 (1984)	14.1 (p. 211); 21.6
M. Hanke	(b) **122**, K151 (1984)	13.1 (p. 157); 22
	(b) **123**, K39 (1984)	13.1 (p. 158)
O. P. Hansen	(b) **126**, 721 (1984)	14.1 (p. 215); 15; 21.7
Chu Hao	(a) **81**, 569 (1984)	14.3.2 (p. 238) and 14.3.3 (p. 241); 22.1.2
C. Hargitai	(b) **124**, 271 (1984)	13.4 (p. 183); 5; 21.1.1
P. Hargraves	(a) **85**, K101 (1984)	1.3 (p. 25); 2; 21.1
V. Hari Babu	see under letter "B"	
E. A. Harris	(b) **122**, 757 (1984)	13.4 (p. 178) and 19 (p. 308); 22.8.1

R. Harris	(b) **124**, K181 (1984)	15 (p. 267); 2; 21
E. Hartmann	(a) **86**, 401 (1984)	10.2 (p. 119) and 14.4 (p. 259); 22.8.2
H. Hartmann	(a) **81**, K199 (1984)	20.3 (p. 348); 22.4.1
U. Hartmann	(a) **82**, 481 (1984)	18.2 (p. 290)
F. Hashimoto	(a) **82**, 135 (1984)	10.2 (p. 108); 13.4; 22.1.1
	(a) **85**, 227 (1984)	10.2 (p. 117) and 13.4 (p. 189); 14.3; 22.1.1
	(a) **86**, K95 (1984)	10.2 (p. 121); 13.4; 14.3; 22.1.1
H. Hashimoto	(a) **85**, 335 (1984)	1.1 (p. 13)
	(a) **86**, 67 (1984)	1.4 (p. 32); 21.1; 21.6
I. Hashimoto	(a) **86**, 237 (1984)	10.1 (p. 103); 14.1; 21
H. F. Hassan	(b) **126**, 587 (1984)	13.1 (p. 165); 22.8
T. K. Hatwar	(b) **126**, 279 (1984)	20 (p. 318); 21.4
W. Hauffe	(a) **81**, 259 (1984)	1.6 (p. 36) and 11 (p. 122); 21.1
L. Havela	(a) **81**, 307 (1984)	18 (p. 278); 22.7
	(a) **82**, 191 (1984)	18 (p. 279); 22.7
R. Hayn	(b) **126**, 335 (1984)	2 (p. 53) and 13 (p. 151)
Th. Hayne	(b) **123**, 271 (1984)	13.5 (p. 196) and 20.1 (p. 327); 22.9
P. M. Hazzledine	(a) **81**, 473 (1984)	10.1 (p. 98); 12.1; 21.1; 21.6
G. Heber	(b) **122**, 133 (1984)	13 (p. 145) and 18 (p. 278)
M. Heber	(a) **82**, K5 (1984)	2 (p. 42) and 3 (p. 55); 11; 22.1.2
G. Heder	(b) **125**, 523 (1984)	10.2 (p. 117) and 19 (p. 313); 13.4; 22.5.2
Ok Hee Chung	see under letter "C"	
V. Heera	(b) **121**, 705 (1984)	13 (p. 145); 21.6
E. Hegenbarth	(a) **82**, K27 (1984)	8 (p. 85) and 14.4.1 (p. 260); 22.8.1
	(a) **83**, K23 (1984)	8 (p. 86) and 14.4.1 (p. 261); 22.8.1
K. Hehl	(a) **86**, 125 (1984)	20.1 (p. 340)
	(a) **86**, 485 (1984)	20.1 (p. 342); 11; 22.8.1
K. Heide	(a) **81**, K7 (1984)	6 (p. 68) and 19 (p. 305); 22.8.1
W. Heimbrodt	(a) **81**, 695 (1984)	20.3 (p. 347); 22.4.1
	(b) **126**, K159 (1984)	20.3 (p. 358); 22.4.1
M. Heinonen	(a) **83**, 279 (1984)	4 (p. 59) and 13 (p. 147); 20; 21.1; 21.1.1; 21.4
O. Heinrich	(b) **126**, 575 (1984)	13.1 (p. 164); 22
J. Hejtmanek	(a) **82**, 285 (1984)	14.3 (p. 223); 13.4; 15; 16; 22.8.1
H. W. Helberg	(a) **81**, 381 (1984)	10.2 (p. 106) and 14.3 (p. 221); 14.4; 22
K. Helming	(b) **126**, 43 (1984)	1.3 (p. 25) and 1.4 (p. 32); 21
Hemant Kumar	see under letter "K"	
M. Heming	(b) **124**, 315 (1984)	13.4 (p. 183) and 19 (p. 310); 22.8
F. Henneberger	(b) **121**, 685 (1984)	13.5.1 (p. 199) and 20.1 (p. 321); 22.4.1
	(b) **121**, K187 (1984)	20.1 (p. 322); 22.4.4
	(b) **121**, K191 (1984)	13.5.1 (p. 199) and 20.1 (p. 322); 22.4.1
K. Henneberger	(b) **124**, 213 (1984)	13.5.1 (p. 201) and 20.1 (p. 332); 18; 22.4.2
	(b) **125**, 693 (1984)	13.5.1 (p. 203); 22
D. Hennig	(b) **122**, K151 (1984)	13.1 (p. 157); 22
I. Henning	(a) **83**, K23 (1984)	8 (p. 86) and 14.4.1 (p. 261); 22.8.1
W. Henrion	(b) **123**, K69 (1984)	20.1 (p. 328); 13.1; 21
M. Henry	(b) **122**, 175 (1984)	13.4 (p. 175) and 13.5.3 (p. 207); 14.3; 19; 22.6.1
W. Hergert	(a) **85**, 641 (1984)	20.3 (p. 356); 10.2; 14.3; 22
J. van Herk	(b) **125**, 387 (1984)	20.3 (p. 354); 22.8
B. Hermoneit	(a) **85**, 553 (1984)	20.1 (p. 338) and 20.3 (p. 356); 13.4; 22.8
W. Herremans	(a) **85**, 375 (1984)	1 (p. 10) and 10 (p. 96)
K. Herrmann	(a) **86**, 433 (1984)	20.1 (p. 341); 20.2; 22
R. Herrmann	(b) **122**, K87 (1984)	14.1 (p. 211) and 14.3 (p. 223); 12.2; 22.2.3
	(b) **123**, 635 (1984)	7 (p. 82) and 13 (p. 147); 21.7
	(a) **83**, K207 (1984)	14.3 (p. 226); 13.1; 22.2.3
	(b) **125**, K85 (1984)	13.3 (p. 169) and 14.3 (p. 228); 22.2.3
	(a) **85**, K183 (1984)	12.2 (p. 142) and 14.3 (p. 230); 22.2.3
	(a) **86**, 427 (1984)	13.3 (p. 170) and 14.3.3 (p. 246); 22.2.3

A. Hertrampf	(b) **126**, 343 (1984)	13.4 (p. 192) and 19 (p. 313); 22.4.1
M. Hervieu	(a) **83**, 473 (1984)	1.1 (p. 12) and 1.4 (p. 30); 22.8.2
G. Hess	(b) **121**, 145 (1984)	18.4 (p. 300); 13.4; 18.3; 22.8.1
	(b) **123**, 123 (1984)	18.3 (p. 298) and 18.4 (p. 301); 22.8.1
H. Hess	(a) **85**, 543 (1984)	20.3 (p. 356); 22.8.1
I. Hevesi	(a) **81**, K67 (1984)	14.3 (p. 221); 22.6
	(a) **84**, 639 (1984)	14.3.4 (p. 252)
C. A. Hewett	(a) **84**, 49 (1984)	1.3 (p. 24) and 11 (p. 128); 1.5; 2; 21.1; 21.6
B. Hickey	(b) **124**, 279 (1984)	13 (p. 148); 2
P. Hicter	(b) **125**, 101 (1984)	4 (p. 61) and 13.1 (p. 162); 21.1; 21.4
M. Hidaka	(a) **81**, 177 (1984)	1.2 (p. 14); 6; 8; 22.5.4
	(a) **86**, 75 (1984)	1.2 (p. 20); 22.5.4
C. Hidalgo	(a) **83**, K93 (1984)	1.4 (p. 30) and 10.1 (p. 99); 21.1.1
N. T. Thuc Hien	(a) **81**, 353 (1984)	13.4 (p. 173) and 14.3.3 (p. 241); 22.2.2
	(a) **85**, 219 (1984)	13.4 (p. 188) and 14.3.3 (p. 245); 22.4.3
Nguyen Van Hieu	see under letter "N"	
A. S. Hilal	(a) **83**, 123 (1984)	8 (p. 86); 22.9
E. Hild	(a) **85**, 133 (1984)	1.4 (p. 32) and 10.2 (p. 116); 20.1; 22.1.2
D. Hildebrandt	(a) **85**, K35 (1984)	10.2 (p. 117) and 11 (p. 130); 22.1.2
V. Hinkel	(b) **122**, K45 (1984)	2 (p. 41) and 13 (p. 146); 19; 21.7; 22.1
B. Hintze	(a) **84**, 87 (1984)	6 (p. 74) and 8 (p. 87); 22.8
D. Hinz	(a) **84**, K133 (1984)	18 (p. 281); 21
C. Hirlimann	(b) **126**, K113 (1984)	6 (p. 80) and 20.1 (p. 341); 22.2.2
J. P. Hirth	(b) **121**, 101 (1984)	9 (p. 91) and 10 (p. 95)
	(a) **82**, 355 (1984)	1.3 (p. 23); 10.1; 21
	(b) **125**, 147 (1984)	9 (p. 93) and 10 (p. 95)
V. Hnatowicz	(a) **83**, 165 (1984)	11 (p. 126); 22.1.2; 22.6
Fat Duen Ho	see under letter "F"	
Hoang Ngoc Cam	(b) **126**, 247 (1984)	13.5 (p. 198)
Hoang Xuan Nguyen	see under letter "N"	
H.-R. Höche	(a) **86**, 229 (1984)	10.2 (p. 119); 22.2.1
H. D. Hochheimer	(b) **123**, 209 (1984)	13.1 (p. 158); 21.4
D. Hoehl	(a) **83**, 543 (1984)	1.4 (p. 30) and 11 (p. 127); 22.1.1; 22.1.2
H. Hoffmann	(b) **122**, K137 (1984)	18.2.1 (p. 296); 21.1.1; 21.4
J. Hoffmann	(b) **123**, 563 (1984)	18.3 (p. 298) and 18.4 (p. 302); 22.8.1
C. A. Hogarth	(b) **123**, 653 (1984)	20.1 (p. 330); 2; 13; 22.6
	(b) **126**, 293 (1984)	2 (p. 53) and 13.1 (p. 164); 19; 20.1; 22.6
J. H. C. Hogg	(a) **81**, 333 (1984)	13.4 (p. 173); 20.3; 22.4.1
W. Hohenau	(a) **82**, 119 (1984)	11 (p. 124) and 20.1 (p. 324); 22.6
B. Holland	(b) **126**, 511 (1984)	4 (p. 63) and 13 (p. 151); 12.2; 22.1; 22.1.1; 22.1.2
G. Hölzer	(a) **85**, K43 (1984)	10.2 (p. 117) and 11 (p. 130); 22.1.2
D. Hommel	(a) **81**, 695 (1984)	20.3 (p. 347); 22.4.1
	(a) **81**, K199 (1984)	20.3 (p. 348); 22.4.1
B. Hönerlage	(b) **121**, 649 (1984)	13.5.1 (p. 199); 20.1; 22.5
	(b) **123**, 201 (1984)	13.5.1 (p. 200); 20.1; 22.5
	(b) **124**, 221 (1984)	13.5.1 (p. 201) and 20.1 (p. 332); 22.5
Nguyen Hong Shon	see under letter "N"	
R. Honmura	(b) **121**, 197 (1984)	18.2 (p. 287); 13.3
H. Hora	(a) **81**, K127 (1984)	2 (p. 41) and 11 (p. 123); 8; 22.1.2
J. Horák	(a) **84**, K143 (1984)	20.1 (p. 335); 13.4; 14.3; 22.7.1
W. Hörig	(b) **121**, K55 (1984)	20.1 (p. 321); 13.1; 22.8
P. Höschl	(a) **82**, K199 (1984)	14.3 (p. 224); 13.4; 22.4.3
T. Hoshikawa	(a) **83**, 561 (1984)	14.1 (p. 212) and 18.1 (p. 285); 1.3; 18.2; 22.8
K. Hoshino	(a) **85**, K5 (1984)	1 (p. 10) and 2 (p. 50); 22.5.2
W. Hoyer	(a) **84**, 11 (1984)	2 (p. 47); 1.3; 22.7
	(a) **84**, K97 (1984)	1.3 (p. 25) and 2 (p. 49); 22.7

J. Hřebík	(a) **81**, 307 (1984)	18 (p. 278); 22.7
	(a) **82**, 191 (1984)	18 (p. 279); 22.7
K. Hricovíni	(b) **124**, K147 (1984)	13.1 (p. 162); 2; 17; 21.1.1
M. Hrobár	(a) **82**, 519 (1984)	14.3.3 (p. 242); 14.3.4
Tsing-Wong Hsu	(a) **81**, 439 (1984)	1.3 (p. 22) and 1.4 (p. 27)
Q. Huang	(a) **84**, 465 (1984)	10.1 (p. 100) and 12.1 (p. 136); 21
	(a) **86**, 593 (1984)	10.1 (p. 103) and 12.1 (p. 138); 21
S. Huant	(b) **121**, 289 (1984)	13.1 (p. 154); 22.3
	(b) **125**, 215 (1984)	13.1 (p. 162) and 19 (p. 312); 16; 22.2.3
G. Hüber	(b) **126**, K101 (1984)	14.1 (p. 214) and 19 (p. 314); 21.1
K. Hübner	(b) **121**, 505 (1984)	6 (p. 68) and 20.1 (p. 321); 2; 22.6
	(b) **124**, K171 (1984)	13.3 (p. 169); 22.1.2; 22.6
	(b) **125**, K79 (1984)	10.2 (p. 116) and 13.4 (p. 188); 22.6
	(b) **126**, K71 (1984)	13.4 (p. 193); 10.2; 22.6
W. Huhn	(b) **124**, 167 (1984)	13.1 (p. 161) and 13.5 (p. 197); 14.3; 20; 22
J. Humlíček	(b) **122**, 675 (1984)	13.1 (p. 157) and 20.1 (p. 326); 22
	(a) **83**, 581 (1984)	20.1 (p. 330); 22.8.2
Nguyen Dai Hung	see under letter "N"	
H.-J. Hunger	(a) **84**, K149 (1984)	13 (p. 149); 21.6
Nquy Huu Chi	see under letter "N"	
Young Hwan Choi	see under letter "Y"	
G. I. Ibaev	(a) **84**, K185 (1984)	14.3.1 (p. 236) and 14.3.3 (p. 245); 22.1.3; 22.4.2
N. M. Igonina	(a) **84**, 171 (1984)	9 (p. 93); 11; 22.1.2
S. Ihara	(a) **86**, 729 (1984)	10.2 (p. 121) and 14.4 (p. 259); 22.5.4
D. Ihle	(b) **121**, 217 (1984)	13.5.3 (p. 207); 22.6
T. Ikari	(b) **124**, K49 (1984)	20.3 (p. 352); 13.5.1; 22.3
	(a) **86**, K69 (1984)	14.3 (p. 231); 13.4; 22.3
Y. Ikeda	(a) **84**, 55 (1984)	1.1 (p. 12); 22.8.1
K. Ikezaki	(a) **85**, 615 (1984)	14.4 (p. 258); 22.9
V. Ya. Ilichev	(a) **81**, 507 (1984)	18.1 (p. 284) and 18.2 (p. 289); 14.1; 21.1; 21.1.1
I. A. Iliev	(b) **126**, 261 (1984)	20.1 (p. 339); 22.4.1
Z. P. Ilyukhina	(a) **82**, 207 (1984)	13.4 (p. 176) and 16 (p. 269); 22.8
R. M. Imamov	(a) **83**, K5 (1984)	1.6 (p. 37); 17; 22.1.2
	(a) **84**, 73 (1984)	1 (p. 9) and 17 (p. 275); 22.1.2
	(a) **86**, K1 (1984)	1.6 (p. 38) and 10.2 (p. 120); 22.2.1; 22.2.2
S. Imoto	(a) **83**, 567 (1984)	18.1 (p. 285) and 19 (p. 309); 18.4; 22.6.1
	(a) **83**, 729 (1984)	18.1 (p. 285) and 19 (p. 310); 18.4; 22.6.1
T. Imura	(a) **82**, 163 (1984)	10.1 (p. 98); 21.1
	(a) **83**, 529 (1984)	9 (p. 93); 2; 21.1; 21.1.1
	(a) **84**, 501 (1984)	10.1 (p. 100); 12.1; 21.1
V. L. Indenbom	(a) **83**, 195 (1984)	1.4 (p. 29) and 10.2 (p. 111); 22.1.2
M. P. Iñiguez	(b) **123**, 485 (1984)	4 (p. 60); 1.3; 21.2; 21.3
A. Ioanid	(a) **82**, K125 (1984)	1.2 (p. 16); 14.4.1; 22.8.1
S. P. Ionov	(b) **121**, 275 (1984)	13.1 (p. 154); 13.5.3
	(b) **123**, 183 (1984)	14.2 (p. 217); 5; 22.6.1
V. P. Iordatii	(b) **123**, K173 (1984)	14.2 (p. 217)
T. Irie	(b) **124**, K101 (1984)	14.3 (p. 226); 13.1; 22.8
	(b) **126**, K145 (1984)	13.1 (p. 165) and 20.1 (p. 341); 12.2; 22.8.2
A. I. Isaev	(a) **82**, K87 (1984)	16 (p. 270); 22.1.3
K. Ishigure	(a) **82**, K57 (1984)	5 (p. 64) and 18 (p. 279); 22.6
Y. Ishihara	(b) **121**, 407 (1984)	13.4 (p. 173) and 14.3 (p. 220); 22
M. Ishii	(b) **124**, K109 (1984)	6 (p. 75) and 20.1 (p. 334); 22
T. Ishii	(a) **85**, 615 (1984)	14.4 (p. 258); 22.9
S. Ishio	(a) **86**, 289 (1984)	18.1 (p. 286); 21.1; 21.1.1
M. S. Islam	(a) **81**, 197 (1984)	4 (p. 58); 20; 21

A. O. Ismailov	(a) **84**, K51 (1984)	18 (p. 281); 22.8
K. A. Ismailov	(a) **86**, K83 (1984)	11 (p. 132) and 14.3.3 (p. 246); 22.2.1
H. Isotalo	(b) **122**, 791 (1984)	14.3 (p. 223) and 15 (p. 265); 22.9
M. E. Itkis	(a) **81**, 367 (1984)	14 (p. 210); 22
N. Itoh	(b) **122**, 489 (1984)	10.2 (p. 109) and 20.1 (p. 325); 20.3; 22.5.3
T. Itoh	(b) **123**, 463 (1984)	6 (p. 72); 8; 21
	(b) **125**, 107 (1984)	6 (p. 76); 12.1; 21
	(b) **126**, 71 (1984)	6 (p. 78); 8; 21
Y. Itoh	(a) **83**, 529 (1984)	9 (p. 93); 2; 21.1; 21.1.1
P. G. Ivanitskii	(b) **121**, 623 (1984)	13.4 (p. 174); 21.4
A. V. Ivanov	(a) **84**, 417 (1984)	12.2 (p. 141); 22.9
M. A. Ivanov	(b) **121**, 547 (1984)	18.3 (p. 298); 13.4; 22.8.2
N. B. Ivanov	(b) **126**, K131 (1984)	13.5.3 (p. 207) and 18.4 (p. 304)
M. D. Ivanovich	(a) **82**, 275 (1984)	14.3.3 (p. 242); 22
V. I. Ivashchenko	(b) **121**, 583 (1984)	13.1 (p. 154); 22
V. M. Ivastchenko	(a) **84**, 669 (1984)	14.3.3 (p. 245) and 14.3.4 (p. 252); 22.2.1
E. L. Ivchenko	(b) **122**, 221 (1984)	18 (p. 279) and 20.1 (p. 322); 13.5.1; 22.4.2
Z. Ivić	(b) **123**, 135 (1984)	13 (p. 146) and 18.4 (p. 301); 20.1
K. Iwahana	(b) **126**, 693 (1984)	10.2 (p. 120); 18.1; 22.5.2
K. Iwasaki	(a) **81**, 485 (1984)	10.1 (p. 98) and 12.1 (p. 135); 21
	(a) **86**, 637 (1984)	10.1 (p. 104) and 12.1 (p. 138); 21
T. Iwashita	(b) **125**, 551 (1984)	18.2 (p. 294)
M. Iwata	(a) **85**, K105 (1984)	12.2 (p. 142); 22.9
K. Iwauchi	(a) **84**, 55 (1984)	1.1 (p. 12); 22.8.1
M. Izumi	(a) **82**, 163 (1984)	10.1 (p. 98); 21.1
A. Jabłoński	(a) **84**, 39 (1984)	1.3 (p. 24) and 1.6 (p. 37); 21; 21.1
A. S. Jacob	(b) **123**, K75 (1984)	13.4 (p. 180) and 20.1 (p. 328); 22.8.1
K. Jacobs	(a) **85**, 627 (1984)	14.3.3 (p. 245) and 20.3 (p. 356); 13.4; 22.2.2
P. W. M. Jacobs	(b) **124**, 261 (1984)	18 (p. 280) and 20.1 (p. 332); 22.5.2
M. A. Jacobson	(b) **122**, 183 (1984)	16 (p. 269) and 20.3 (p. 348); 22.4.1
B. Jacquier	(b) **124**, 325 (1984)	13.4 (p. 184) and 20.3 (p. 352); 18; 22.5.4
	(b) **124**, 731 (1984)	13.4 (p. 186) and 20.3 (p. 353); 18; 22.5.4
C. Jäger	(a) **81**, K7 (1984)	6 (p. 68) and 19 (p. 305); 22.8.1
Ch. Jäger	(a) **82**, K153 (1984)	9 (p. 92) and 19 (p. 308); 21.4
E. Jäger	(b) **121**, K119 (1984)	6 (p. 69) and 13.5 (p. 196); 22.1.2; 22.6
M. S. Jahan	(b) **126**, 687 (1984)	10.2 (p. 120) and 19 (p. 314); 11; 20.3; 22.5.3
L. Jahn	(a) **83**, K141 (1984)	18.2 (p. 292); 21.1; 21.4
	(a) **84**, K133 (1984)	18 (p. 281); 21
Mukesh Jain	(a) **82**, K181 (1984)	20.1 (p. 326); 13.1; 22.8
V. K. Jain	(b) **121**, K201 (1984)	19 (p. 306); 13.4; 22.8.1
S. K. Jaireth	(a) **81**, 281 (1984)	10.2 (p. 106); 5; 18.3; 18.4; 22.8
S. Jalaj	(b) **126**, 669 (1984)	13.3 (p. 170); 22.2.2
G. Jalbert	(b) **124**, 147 (1984)	13.1 (p. 160) and 20.1 (p. 331); 22
	(b) **125**, 605 (1984)	20.1 (p. 337); 20.2; 22.2.1; 22.4.1
M. Jałochowski	(a) **82**, 497 (1984)	14.3.1 (p. 234); 14.3.3; 21; 21.6
H. Jamshidi	(b) **122**, 461 (1984)	6 (p. 70) and 20.1 (p. 325); 22.2.1
Z. Janáček	(a) **85**, 77 (1984)	1.1 (p. 13); 22.1.2
A. R. Jani	(b) **124**, 137 (1984)	13.1 (p. 160); 2; 14.1; 15; 21
J. Janich	(a) **86**, 111 (1984)	1.2 (p. 20) and 14.4.1 (p. 263); 12.2; 22.8
R. Jankowiak	(b) **124**, 363 (1984)	13.5.1 (p. 202) and 20.3 (p. 352); 2; 22.9
L. Jansen	(b) **125**, 387 (1984)	20.3 (p. 354); 22.8
Z. Januškevičius	(a) **81**, K191 (1984)	11 (p. 123) and 14.3 (p. 222); 22.3
E. Jarocki	(b) **126**, 191 (1984)	13 (p. 151) and 18 (p. 282); 2; 21.1.1
R. Jašinskaitė	(a) **82**, K199 (1984)	14.3 (p. 224); 13.4; 22.4.3
M. Jergel	(a) **83**, 113 (1984)	2 (p. 44); 1.4; 21.1

S. Jetschke	(a) **86,** 125 (1984)	20.1 (p. 340)
	(a) **86,** 485 (1984)	20.1 (p. 342); 11; 22.8.1
A. Jeżowski	(a) **81,** 171 (1984)	8 (p. 84); 22.6
J. J. Jimenez-Rodriguez	(a) **81,** 267 (1984)	11 (p. 123)
Z. Jirák	(a) **83,** K117 (1984)	1.2 (p. 17); 22.8.1
M. Jirsa	(b) **124,** 609 (1984)	13.3 (p. 168) and 19 (p. 311)
	(b) **125,** 187 (1984)	13.3 (p. 169) and 19 (p. 312); 21
	(b) **126,** 547 (1984)	13.3 (p. 170) and 19 (p. 314)
R. N. Joarder	(b) **125,** K33 (1984)	2 (p. 50) and 8 (p. 88); 21.2
	(b) **126,** 443 (1984)	2 (p. 54) and 9 (p. 94); 21; 21.1
K. N. Jog	(a) **85,** 417 (1984)	1.2 (p. 19) and 12.2 (p. 142); 22.6
W. John	(b) **121,** 233 (1984)	13.1 (p. 153); 21
	(b) **121,** 641 (1984)	13.1 (p. 155); 17; 20; 22.8
	(b) **122,** 197 (1984)	13.1 (p. 155); 21
	(b) **126,** 335 (1984)	2 (p. 53) and 13 (p. 151)
S. G. Johnson	(a) **82,** 327 (1984)	1.6 (p. 36) and 14.3.2 (p. 238); 22.6
M. Jolivet	(a) **82,** 257 (1984)	10.2 (p. 109) and 14.3.1 (p. 234); 22.9
E. R. Jones, Jr.	(b) **121,** K125 (1984)	2 (p. 40) and 18 (p. 278); 21.1.1
K. S. Jones	(a) **82,** K101 (1984)	10.2 (p. 109) and 14.3 (p. 223); 22.1.1
P. L. Jones	(a) **81,** K51 (1984)	2 (p. 40) and 16 (p. 269); 22.1
D. P. Joshi	(a) **84,** 311 (1984)	14.3.3 (p. 244) and 16 (p. 272); 1.4; 22.1.2
S. K. Joshi	(b) **126,** 669 (1984)	13.3 (p. 170); 22.2.2
M. Jouanne	(b) **124,** K105 (1984)	6 (p. 75) and 20.1 (p. 334); 22.3
D. C. Joy	(a) **84,** 363 (1984)	1.4 (p. 31) and 10.2 (p. 114); 14.3.3; 22.1.2
Juh Tzeng Lue	(b) **126,** 713 (1984)	13.4 (p. 195) and 20.1 (p. 341); 2; 16; 22.1.2
C. Julien	(b) **124,** K105 (1984)	6 (p. 75) and 20.1 (p. 334); 22.3
T. Jung	(a) **81,** 463 (1984)	1.6 (p. 36) and 14.3.2 (p. 238); 22.2.3
	(a) **85,** K15 (1984)	1.5 (p. 34); 11; 22
G. Jungk	(b) **123,** 65 (1984)	6 (p. 71) and 20.1 (p. 327); 22.6
	(b) **124,** K1 (1984)	6 (p. 74) and 20.1 (p. 332); 2; 22.6
S. Jurga	(a) **81,** 77 (1984)	1.2 (p. 14); 19; 22.9
K. Jürgens	(a) **85,** K77 (1984)	20.3 (p. 355); 22.2.4
H. Käämbre	(a) **82,** K149 (1984)	10.2 (p. 110) and 17 (p. 274); 13.4; 22.5.2
	(a) **83,** K39 (1984)	10.2 (p. 112); 17; 20.3; 22.5.2
	(a) **83,** K171 (1984)	17 (p. 275); 11; 13.3; 13.4; 22.5.2
G. A. Kachurin	(a) **82,** 475 (1984)	11 (p. 125); 14.3; 22.1.2
S. Kačiulis	(a) **85,** K179 (1984)	14.3.3 (p. 246) and 14.3.4 (p. 253); 22.2.1
E. Kaden	(b) **126,** 733 (1984)	6 (p. 79) and 15 (p. 268); 14.3; 22.1.1
G. Kaden	(a) **83,** 375 (1984)	14.3.3 (p. 243) and 14.3.4 (p. 250); 13.3; 22.1.2
B. Kh. Kadirov	(a) **84,** K71 (1984)	13.1 (p. 161) and 14.3.3 (p. 244); 22.4.4
A. M. Kadomtseva	(a) **84,** 215 (1984)	18.4 (p. 302); 22.8.2
V. M. Kaganer	(a) **83,** 195 (1984)	1.4 (p. 29) and 10.2 (p. 111); 22.1.2
H.-M. Kagaya	(b) **121,** 501 (1984)	6 (p. 68); 8; 22.3
	(b) **121,** K1 (1984)	1.2 (p. 14); 22.2.1; 22.2.3
	(b) **121,** K113 (1984)	6 (p. 69); 22.2.2; 22.4.3
H.-Matsuo Kagaya	(b) **122,** K125 (1984)	6 (p. 71); 22.1.1; 22.1.2
	(b) **123,** 463 (1984)	6 (p. 72); 8; 21
	(b) **124,** 37 (1984)	6 (p. 74); 22.2; 22.4
	(b) **124,** 525 (1984)	6 (p. 75); 8; 21.2
	(b) **124,** K11 (1984)	1.2 (p. 17) and 4 (p. 60); 12.2; 22.2.1; 22.2.3
	(b) **125,** 107 (1984)	6 (p. 76); 12.1; 21
	(b) **125,** K99 (1984)	1.2 (p. 19) and 6 (p. 77); 8; 22.1.1; 22.1.2
	(b) **126,** 71 (1984)	6 (p. 78); 8; 21
	(b) **126,** K5 (1984)	6 (p. 79) and 8 (p. 89); 12.1; 21.2

	(b) **126**, K9 (1984)	12.2 (p. 142); 6; 22.2; 22.4
	(b) **126**, K109 (1984)	6 (p. 80); 8; 12.1; 21.2
M. Kahanda	(b) **125**, K181 (1984)	14.3 (p. 229) and 14.4 (p. 258); 22.5
H. G. Kahle	(b) **126**, 381 (1984)	1.2 (p. 19) and 13.4 (p. 193); 18.4; 20.1; 22.8.1
T. A. Kaidalova	(a) **84**, 387 (1984)	1.2 (p. 18) and 14.4.1 (p. 262); 8; 19; 22.8
N. Kaiser	(a) **81**, 99 (1984)	1.5 (p. 33) and 3 (p. 55); 2; 22
Zs. Kajcsos	(a) **82**, 143 (1984)	10.2 (p. 108); 22.6
	(a) **84**, 451 (1984)	10.2 (p. 115); 13.3; 22.6
S. L. Kakani	(b) **125**, 861 (1984)	14.2 (p. 218) and 18.2 (p. 294); 22.8
T. L. Kalabegishvili	(b) **122**, 95 (1984)	8 (p. 85) and 11 (p. 124); 20.1; 22.6
V. P. Kalashnikov	(b) **121**, 737 (1984)	15 (p. 264); 22.6
J. von Kalben	(a) **81**, 715 (1984)	20.3 (p. 347); 22.4.1
	(a) **81**, 733 (1984)	14.3.4 (p. 249) and 20.3 (p. 347); 22.4.2
V. V. Kalinin	(a) **86**, 185 (1984)	10.2 (p. 119) and 11 (p. 132); 22.1.2
V. T. Kalinnikov	(a) **81**, 139 (1984)	2 (p. 39); 14.3; 15; 22.8
M. Kalitzova	(a) **84**, K23 (1984)	1.4 (p. 31) and 11 (p. 128); 2; 22.1.1
P. Kálmán	(b) **122**, 501 (1984)	10.2 (p. 109); 22.5.2
J. Kalus	(b) **126**, 121 (1984)	6 (p. 79); 22.1.2
M. S. Kamara	(a) **84**, 613 (1984)	14.3.4 (p. 252); 22.2.1
K. Kambas	(b) **123**, K155 (1984)	13.1 (p. 160) and 20.1 (p. 330); 22.8
	(b) **124**, K105 (1984)	6 (p. 75) and 20.1 (p. 334); 22.3
V. Kamberský	(b) **126**, 547 (1984)	13.3 (p. 170) and 19 (p. 314)
U. Kambli	(b) **124**, 87 (1984)	13.5.1 (p. 201) and 13.5.4 (p. 208); 18.4; 20.1; 22.5
R. Kamel	(a) **82**, 63 (1984)	1.2 (p. 15); 2; 21.1; 21.1.1
N. Kameswari	(a) **86**, K15 (1984)	1.1 (p. 13) and 3 (p. 57); 22.6.1
A. A. Kaminskii	(a) **81**, K19 (1984)	20.2 (p. 343) and 20.3 (p. 345); 22.8.1
	(a) **82**, K185 (1984)	20.2 (p. 343) and 20.3 (p. 349); 20.1; 22.8
	(a) **83**, K65 (1984)	20.2 (p. 343) and 20.3 (p. 350); 13.4; 22.8
	(a) **83**, K159 (1984)	20.2 (p. 343) and 20.3 (p. 352); 20.1; 22.8.1
	(a) **83**, K165 (1984)	20.2 (p. 343) and 20.3 (p. 352); 20.1; 22.8.1
	(a) **84**, K81 (1984)	20.2 (p. 343) and 20.3 (p. 353); 13.4; 22.8
	(a) **85**, 553 (1984)	20.1 (p. 338) and 20.3 (p. 356); 13.4; 22.8
	(b) **126**, 659 (1984)	20.1 (p. 341) and 20.3 (p. 358); 13.4; 22.5
	(a) **86**, 345 (1984)	14.4 (p. 258) and 20.2 (p. 344); 1.1; 12.2; 20.1; 20.3; 22.8.1
V. M. Kaminskii	(b) **123**, K63 (1984)	13.5.1 (p. 201) and 20.3 (p. 350); 22.3
Y. Kamiura	(a) **82**, 135 (1984)	10.2 (p. 108); 13.4; 22.1.1
	(a) **85**, 227 (1984)	10.2 (p. 117) and 13.4 (p. 189); 14.3; 22.1.1
	(a) **86**, K95 (1984)	10.2 (p. 121); 13.4; 14.3; 22.1.1
L. N. Kamysheva	(a) **84**, K115 (1984)	14.4.1 (p. 262); 22.9
T. Kaneyoshi	(b) **121**, 197 (1984)	18.2 (p. 287); 13.3
	(b) **123**, 525 (1984)	2 (p. 45) and 18.2 (p. 291); 21.1.1
D. G. Kanhere	(b) **125**, K143 (1984)	13 (p. 151) and 17 (p. 276); 21
L. N. Kantorovich	(b) **123**, 325 (1984)	13.4 (p. 179) and 14.4 (256); 22.5.2
V. V. Kaplin	(b) **126**, 565 (1984)	11 (p. 133); 13; 22.1.2
D. V. Kapor	(b) **122**, 125 (1984)	18.2 (p. 289)
	(b) **123**, 135 (1984)	13 (p. 146) and 18.4 (p. 301); 20.1
	(b) **124**, 235 (1984)	13.5.1 (p. 201); 20.1
	(b) **125**, K45 (1984)	18.2 (p. 294)
	(b) **125**, K151 (1984)	6 (p. 78) and 13.5.1 (p. 204); 22.9
	(b) **126**, K175 (1984)	18.2 (p. 294)
V. A. Kapustin	(b) **122**, 549 (1984)	20.1 (p. 325); 13.5.1; 22.9
P. M. Karageorgy-Alkalaev	(a) **83**, K81 (1984)	2 (p. 45) and 14.3 (p. 225); 14.3.4; 22.1.2
S. M. Karagezian	(a) **83**, 153 (1984)	10.2 (p. 111) and 14.3 (p. 225); 22.8.1
Th. Karakostas	(a) **83**, 485 (1984)	1.4 (p. 30); 21
S. A. Karandashev	(a) **85**, 655 (1984)	14.3.3 (p. 246) and 20.3 (p. 356); 16; 22.2.1

W. Karaś	(b) **126**, 191 (1984)	13 (p. 151) and 18 (p. 282); 2; 21.1.1
L. G. Karaseva	(a) **83**, K153 (1984)	13.4 (p. 183) and 20.1 (p. 331); 22.8.2
J. P. Karins	(b) **126**, 157 (1984)	10.2 (p. 118) and 13.4 (p. 192); 22.1.2
A. E. Karkin	(a) **86**, K59 (1984)	14.1 (p. 215) and 14.2 (p. 219); 1.2; 21; 21.1
I. N. Karnaukhov	(b) **126**, 207 (1984)	10.1 (p. 102) and 18 (p. 282); 19
H. P. Karnthaler	(a) **81**, 473 (1984)	10.1 (p. 98); 12.1; 21.1; 21.6
P. Karnthaler	(a) **82**, 425 (1984)	1.4 (p. 29) and 10.2 (p. 110); 22.1.2
A. I. Karpechin	(a) **83**, 365 (1984)	11 (p. 126) and 14.3 (p. 225); 22.9
A. L. Karpei	(a) **82**, 365 (1984)	10.2 (p. 110); 22.1.2
E. Kartheuster	(b) **121**, 345 (1984)	13.4 (p. 172); 22.1.2; 22.2.2; 22.4.2
V. T. Karulkar	(a) **81**, 407 (1984)	14.3.3 (p. 241); 11; 14.4; 22.1.2
B. Karvaly	(a) **81**, K179 (1984)	14.3 (p. 222); 22.9
V. V. Karyagin	(b) **121**, 737 (1984)	15 (p. 264); 22.6
S. Kashchieva	(a) **83**, 411 (1984)	11 (p. 126) and 13.4 (p. 181); 14.3.3; 22.1.2
I. M. Kashirskii	(a) **82**, 537 (1984)	14.3.1 (p. 234) and 16 (p. 270); 13.1; 22.9
B. P. Kashnikov	(a) **85**, K39 (1984)	10.2 (p. 117) and 11 (p. 130); 22.1.2
	(a) **86**, 313 (1984)	13.4 (p. 193) and 19 (p. 314); 11; 22.1.2
V. A. Kasiyan	(b) **125**, 869 (1984)	13.4 (p. 190) and 14.3 (p. 229); 22.4.2
L. Kassamakova	(a) **84**, 561 (1984)	13.3 (p. 169) and 14.3.2 (p. 239); 22.1.2
A. Kasten	(b) **121**, 145 (1984)	18.4 (p. 300); 13.4; 18.3; 22.8.1
	(b) **123**, 123 (1984)	18.3 (p. 298) and 18.4 (p. 301); 22.8.1
	(b) **123**, 563 (1984)	18.3 (p. 298) and 18.4 (p. 302); 22.8.1
G. Kästner	(a) **83**, 543 (1984)	1.4 (p. 30) and 11 (p. 127); 22.1.1; 22.1.2
	(a) **84**, K23 (1984)	1.4 (p. 31) and 11 (p. 128); 2; 22.1.1
A. A. Katanov	(b) **125**, 421 (1984)	15 (p. 267); 6; 13; 22
S. P. Kathuria	(a) **81**, 533 (1984)	13.4 (p. 174) and 20.3 (p. 346); 22.5.2
	(a) **83**, 605 (1984)	20.3 (p. 351); 11; 13.4; 22.4
	(a) **86**, 327 (1984)	20.3 (p. 357); 11; 13.4; 22.5.2
M. Kato	(a) **81**, 439 (1984)	1.3 (p. 22) and 1.4 (p. 27)
	(b) **123**, 415 (1984)	1.2 (p. 16); 8; 12
	(a) **83**, 657 (1984)	14.1 (p. 212); 2; 8; 21; 21.6
Z. Katović	(a) **86**, 413 (1984)	14.4 (p. 259); 22.9
F. M. Katsapov	(a) **82**, 399 (1984)	1.4 (p. 28) and 1.6 (p. 37); 22.2.1
O. P. Katyal	(a) **86**, 759 (1984)	14.1 (p. 215); 1.3; 21
	(a) **86**, 765 (1984)	14.1 (p. 215); 21
B. Kaufmann	(a) **81**, K31 (1984)	13.4 (p. 173) and 20.3 (p. 345); 22
A. J. Kaur	(b) **123**, 71 (1984)	4 (p. 59); 6; 8; 22.5.2
W. Kautek	(b) **122**, 651 (1984)	20.1 (p. 326); 13.1; 16; 22
I. G. Kaverina	(a) **82**, 345 (1984)	1.5 (p. 33) and 3 (p. 55); 2; 22.1.2
P. de S. Kaviratne	(a) **86**, K121 (1984)	14 (p. 210); 22.8
H. Kawabe	(a) **83**, 207 (1984)	10.2 (p. 111); 22.1.2
I. Kawada	(b) **124**, K109 (1984)	6 (p. 75) and 20.1 (p. 334); 22
K. Kawamura	(b) **122**, 141 (1984)	10.2 (p. 107) and 18.4 (p. 300)
L. A. Kazakevich	(a) **82**, 511 (1984)	10.2 (p. 110) and 13.4 (p. 178); 11; 14.3; 22.1.2
A. A. Kazakov	(b) **121**, 177 (1984)	13.5.4 (p. 208); 18.2
	(b) **121**, 539 (1984)	18.2 (p. 288)
V. V. Kazmiruk	(a) **84**, K43 (1984)	1.4 (p. 31) and 10.2 (p. 114); 14.3.4; 22.1.2
A. V. Kazushchik	(a) **84**, K173 (1984)	14.3 (p. 228) and 15 (p. 267); 10.2; 22
T. S. Kê	(a) **84**, 157 (1984)	10.1 (p. 100) and 12.1 (p. 136); 21
	(a) **84**, 465 (1984)	10.1 (p. 100) and 12.1 (p. 136); 21
	(a) **86**, 593 (1984)	10.1 (p. 103) and 12.1 (p. 138); 21
N. Keddar	(a) **82**, 327 (1984)	1.6 (p. 36) and 14.3.2 (p. 238); 22.6
H. V. Keer	(a) **84**, K105 (1984)	1.2 (p. 18); 8; 14.1; 14.3; 18.1; 22.6
S. W. McKeever	see under letters "Mc"	
N. P. Kekelidze	(a) **83**, 139 (1984)	8 (p. 86); 22.1.1
M. G. Kekua	(a) **83**, 139 (1984)	8 (p. 86); 22.1.1
J. Kelber	(b) **122**, K99 (1984)	13.3 (p. 167); 22

G. KEMENY	(a) **81**, K179 (1984)	14.3 (p. 222); 22.9
S. KEMMLER-SACK	(a) **83**, 631 (1984)	13.4 (p. 182) and 20.3 (p. 351); 22.8
	(a) **86**, 685 (1984)	13.4 (p. 195) and 20.3 (p. 358); 22.8
W. T. MCKEOWN	see under letters "Mc"	
F. KERBE	(a) **84**, 451 (1984)	10.2 (p. 115); 13.3; 22.6
	(a) **82**, 143 (1984)	10.2 (p. 108); 22.6
H. KERKOW	(a) **82**, 125 (1984)	11 (p. 124); 3; 22.1.2
A. KESSLER	(b) **121**, 111 (1984)	10.2 (p. 105); 20.1; 22.5.2
	(b) **122**, 105 (1984)	10.2 (p. 107); 14.4; 22.5.2
E. KESSLER	(a) **81**, 585 (1984)	14.1 (p. 211) and 14.3.1 (p. 233); 8; 15; 21.7
F. R. KESSLER	(a) **85**, K77 (1984)	20.3 (p. 355); 22.2.4
T. KESZTHELYI	(b) **122**, 501 (1984)	10.2 (p. 109); 22.5.2
A. KEUPERS	(b) **122**, 781 (1984)	2 (p. 43) and 14.1 (p. 211); 21.6
I. B. KEVDINA	(b) **123**, 183 (1984)	14.2 (p. 217); 5; 22.6.1
P. K. KHABIBULLAEV	(b) **122**, 481 (1984)	10.2 (p. 109); 13.4; 22.5.2
	(b) **126**, 311 (1984)	13.1 (p. 164) and 13.4 (p. 192); 22.1.2
DAO KHAC AN	see under letters "D"	
R. M. KHACHATRYAN	(b) **122**, 725 (1984)	13.4 (p. 177); 19; 20.1; 22.8.1
Z. M. KHAKIMOV	(b) **126**, 311 (1984)	13.1 (p. 164) and 13.4 (p. 192); 22.1.2
A. O. KHALILOV	(a) **81**, K175 (1984)	14.3 (p. 222); 22.3.1
N. M. KHAMIDULLINA	(b) **126**, 253 (1984)	2 (p. 53); 20.1; 22.1.2
S. KHAN	(b) **124**, 489 (1984)	6 (p. 75) and 13.5.2 (p. 206); 7; 22.2.3
	(a) **85**, 535 (1984)	14 (p. 210) and 20 (p. 318); 22
R. KHANNA	(a) **84**, 95 (1984)	1.3 (p. 24); 6; 21
YU. P. KHAPACHEV	(b) **126**, 37 (1984)	1.5 (p. 35); 12
P. KHARE	(b) **124**, 29 (1984)	4 (p. 60) and 12.2 (p. 141); 22.5; 22.5.2
V. V. KHATKO	(a) **86**, 93 (1984)	1.5 (p. 35) and 2 (p. 53); 3; 4; 22.6
	(a) **86**, 477 (1984)	1.5 (p. 35) and 2 (p. 54); 3; 22.6
A. K. KHITRIN	(b) **126**, 481 (1984)	19 (p. 314)
E. P. KHLYBOV	(a) **86**, 389 (1984)	14.2 (p. 218)
H. E. KHODENKOV	(a) **84**, K135 (1984)	18.2 (p. 293); 22.8.2
I. I. KHODOS	(a) **84**, 79 (1984)	10.2 (p. 113); 1.4; 22.2.1
G. G. KHODZHABAGYAN	(a) **86**, 345 (1984)	14.4 (p. 258) and 20.2 (p. 344); 1.1; 12.2; 20.1; 20.3; 22.8.1
A. F. KHOKHLOV	(a) **81**, K15 (1984)	2 (p. 40) and 9 (p. 91); 22.1.2
L. E. KHOLODENKOV	(b) **125**, 365 (1984)	20.1 (p. 336) and 20.3 (p. 354); 13.4; 22.8.2
	(b) **126**, 659 (1984)	20.1 (p. 341) and 20.3 (p. 358); 13.4; 22.5
V. S. KHOMCHENKO	(a) **81**, 657 (1984)	14.3.4 (p. 249) and 20.3 (p. 346); 14.3.1; 22.6
A. KHOURY	(a) **83**, 403 (1984)	14.3.3 (p. 243) and 16 (p. 271); 22.1.2
L. G. KHVOSTANTSEV	(a) **82**, 389 (1984)	1.2 (p. 16); 14.1; 15; 21.7
	(a) **86**, K99 (1984)	13.1 (p. 165) and 15 (p. 268); 14.3; 22.1
E. KIBICKAS	(a) **85**, K179 (1984)	14.3.3 (p. 246) and 14.3.4 (p. 253); 22.2.1
M. KIBLER	(b) **124**, 325 (1984)	13.4 (p. 184) and 20.3 (p. 352); 18; 22.5.4
	(b) **124**, 731 (1984)	13.4 (p. 186) and 20.3 (p. 353); 18; 22.5.4
H. KIDO	(a) **83**, 561 (1984)	14.1 (p. 212) and 18.1 (p. 285); 1.3; 18.2; 22.8
W. KIENLE	(b) **124**, 69 (1984)	10.1 (p. 99); 21.1
J. KIERUL	(b) **124**, K121 (1984)	18.2 (p. 293); 21.1
M. KIKNADZE	(b) **125**, K63 (1984)	19 (p. 312); 22.5.3
J. KILMER	(b) **121**, 429 (1984)	14.1 (p. 211); 21.6
CHUL KOO KIM	see under letter "C"	
Q. KIM	(b) **122**, 263 (1984)	13.5.1 (p. 200) and 20.3 (p. 348); 22.4.3
TRAN KIM ANH	see under letter "T"	
L. C. KIMERLING	(a) **84**, 363 (1984)	1.4 (p. 31) and 10.2 (p. 114); 14.3.3; 22.1.2
P. KINDL	(a) **81**, 293 (1984)	11 (p. 123); 10.2; 22.9
R. KINK	(b) **124**, 335 (1984)	13.5.1 (p. 201) and 20.3 (p. 352); 23
G. G. KIRAKOSIAN	(b) **126**, K83 (1984)	10.2 (p. 119) and 14.3 (p. 230); 22.1.1
D. L. KIRK	(a) **82**, 327 (1984)	1.6 (p. 36) and 14.3.2 (p. 238); 22.6

V. V. Kirsanov	(a) **86**, 199 (1984)	9 (p. 94) and 10.1 (p. 102); 21.1.1
F.-G. Kirscht	(a) **81**, K165 (1984)	10.2 (p. 107) and 13.4 (p. 175); 14.3; 22.1.2
	(a) **85**, 133 (1984)	1.4 (p. 32) and 10.2 (p. 116); 20.1; 22.1.2
D. S. Kiselev	(b) **126**, 37 (1984)	1.5 (p. 35); 12
V. F. Kiselev	(a) **85**, 273 (1984)	13.4 (p. 189); 19; 20.3; 22.1.1; 22.1.2
	(a) **85**, K73 (1984)	13.3 (p. 169) and 20.3 (p. 355); 22.1.1; 22.6
Pran Kishan	(a) **84**, 535 (1984)	5 (p. 66); 22.8.2
S. B. Kislitsin	(a) **86**, 199 (1984)	9 (p. 94) and 10.1 (p. 102); 21.1.1
E. M. Kislitsina	(a) **86**, 199 (1984)	9 (p. 94) and 10.1 (p. 102); 21.1.1
L. B. Kiss	(a) **81**, K67 (1984)	14.3 (p. 221); 22.6
	(a) **84**, 639 (1984)	14.3.4 (p. 252)
W. Kissinger	(b) **123**, K119 (1984)	6 (p. 74) and 20.1 (p. 330); 13.1; 22.8
Y. Kitani	(b) **121**, 501 (1984)	6 (p. 68); 8; 22.3
	(b) **122**, K125 (1984)	6 (p. 71); 22.1.1; 22.1.2
	(b) **125**, K99 (1984)	1.2 (p. 19) and 6 (p. 77); 8; 22.1.1; 22.1.2
M. Kittler	(a) **81**, K131 (1984)	1.4 (p. 28) and 10.2 (p. 106); 14.3; 22.1.2
	(a) **84**, 143 (1984)	10.2 (p. 113); 14.3.3; 22.1.2
P. Kizler	(b) **121**, 111 (1984)	10.2 (p. 105); 20.1; 22.5.2
	(b) **122**, 105 (1984)	10.2 (p. 107); 14.4; 22.5.2
R. Klabes	(a) **82**, 475 (1984)	11 (p. 125); 14.3; 22.1.2
	(a) **82**, K5 (1984)	2 (p. 42) and 3 (p. 55); 11; 22.1.2
	(a) **82**, K121 (1984)	1.5 (p. 33) and 3 (p. 56); 2; 11; 22.1.2
	(a) **86**, 781 (1984)	11 (p. 133) and 14.3.1 (p. 237); 22.1.2
Th. Kleinefeld	(b) **125**, 293 (1984)	13.5.1 (p. 202) and 20.3 (p. 354); 22.5.1
P. Kleinert	(b) **122**, 81 (1984)	6 (p. 69); 20.1; 22.2.4
	(b) **124**, 559 (1984)	10.2 (p. 114); 11; 22.2.1
K. Kleinstück	(a) **86**, 117 (1984)	1.3 (p. 26); 18; 21.1; 21.1.1
J. J. Kleperis	(a) **81**, K121 (1984)	9 (p. 91); 21.6
	(a) **83**, 291 (1984)	13.4 (p. 180) and 20.1 (p. 328); 14.3.1; 19; 22.6
S. M. Kleshev	(b) **121**, 539 (1984)	18.2 (p. 288)
J. Klikorka	(a) **85**, K123 (1984)	2 (p. 52) and 20.1 (p. 339); 22.8
H. Klose	(a) **84**, 269 (1984)	17 (p. 275); 13.3; 22.1.2
G. Kluge	(a) **83**, 499 (1984)	12.2 (p. 141); 22.6
J. Kluge	(a) **81**, 675 (1984)	19 (p. 306) and 20.3 (p. 347); 13.4; 22.4.2
W. Knap	(b) **122**, 761 (1984)	13.1 (p. 157); 16; 22
G. Knežević	(b) **121**, 589 (1984)	13.5 (p. 196)
P. I. Knigin	(a) **81**, 11 (1984)	14.3.3 (p. 241) and 14.3.4 (p. 248); 16; 22.1.1; 22.1.2; 22.2.2
	(a) **84**, 301 (1984)	14.3.3 (p. 244) and 14.3.4 (p. 251); 16; 22
G. Knuyt	(b) **122**, 781 (1984)	2 (p. 43) and 14.1 (p. 211); 21.6
	(b) **123**, 763 (1984)	2 (p. 46) and 14.1 (p. 212); 21; 21.1
S. Kobayashi	(a) **83**, 219 (1984)	2 (p. 44) and 10.1 (p. 99); 21.1
Y. Kobayashi	(a) **84**, K29 (1984)	10.2 (p. 114) and 20 (p. 317); 22.9
B. V. Kobrin	(a) **83**, 159 (1984)	2 (p. 44) and 10.2 (p. 111); 22
S. Koc	(a) **83**, 418 (1984)	2 (p. 45) and 4 (p. 59); 17; 22.3
L. A. Kocharyan	(b) **125**, K67 (1984)	5 (p. 66)
N. M. L. Köche	(a) **85**, K65 (1984)	7 (p. 83) and 18 (p. 281); 19; 22.8.2
V. P. Kochereshko	(b) **122**, 221 (1984)	18 (p. 279) and 20.1 (p. 322); 13.5.1; 22.4.2
	(b) **124**, 213 (1984)	13.5.1 (p. 201) and 20.1 (p. 332); 18; 22.4.2
O. I. Kochetov	(b) **124**. 355 (1984)	13.4 (p. 184); 19; 22.5
J. Kočka	(a) **83**, 617 (1984)	13.1 (p. 160) and 20.1 (p. 331); 2; 22.1.2
	(b) **125**, K89 (1984)	2 (p. 50) and 14.3.1 (p. 236); 14.3.4; 22.1.2
É. Kocsárdy	(a) **84**, 165 (1984)	10.2 (p. 113); 19; 20.1; 22.6
Y. Koga	(a) **86**, K69 (1984)	14.3 (p. 231); 13.4; 22.3
E. M. Kogan	(b) **123**, 289 (1984)	13.3 (p. 168); 21
R. Kögler	(a) **82**, K5 (1984)	2 (p. 42) and 3 (p. 55); 11; 22.1.2
H.-J. Köhler	(a) **85**, K175 (1984)	14.2 (p. 218)

O. Kohmoto	(a) **85**, K155 (1984)	14.1 (p. 214); 2; 18.3; 21.1; 21.1.1
V. G. Kohn	(b) **123**, K149 (1984)	13 (p. 147)
	(a) **85**, 349 (1984)	1.1 (p. 13); 22.1.1
M. Koizumi	(a) **83**, 561 (1984)	14.1 (p. 212) and 18.1 (p. 285); 1.3; 18.2; 22.8
	(a) **85**, 523 (1984)	18.2 (p. 294); 14.1; 22.6.1
L. S. Kokhanchik	(a) **86**, 133 (1984)	14.4.1 (p. 263); 22.8.1
V. B. Kokshenev	(a) **81**, 171 (1984)	8 (p. 84); 22.6
K. Kolev	(a) **84**, K23 (1984)	1.4 (p. 31) and 11 (p. 128); 2; 22.1.1
L. G. Kolinova	(b) **124**, K155 (1984)	13.5.1 (p. 202) and 20.1 (p. 334); 14.3.4; 22.4
J. Kolk	(a) **86**, 219 (1984)	10.2 (p. 119); 19; 20.1; 20.3; 22.5.2
I. I. Kolkovskii	(a) **83**, 299 (1984)	11 (p. 126) and 13.4 (p. 180); 22.1.2
E. Kolley	(b) **122**, K27 (1984)	18.1 (p. 284) and 18.2 (p. 289); 21
	(b) **123**, 533 (1984)	18 (p. 280); 13; 21
W. Kolley	(b) **122**, K27 (1984)	18.1 (p. 284) and 18.2 (p. 289); 21
	(b) **123**, 533 (1984)	18 (p. 280); 13; 21
N. P. Kolmakova	(b) **121**, 185 (1984)	18.2 (p. 287) and 18.4 (p. 300)
B. T. Kolomiets	(a) **86**, 383 (1984)	16 (p. 273); 2; 14.3.1; 22; 22.1.2
V. V. Kolomoets	(b) **122**, K163 (1984)	13.4 (p. 178) and 14.3.4 (p. 250); 22.1.1
L. Z. Kon	(b) **123**, K173 (1984)	14.2 (p. 217)
R. V. Konakova	(a) **84**, 669 (1984)	14.3.3 (p. 245) and 14.3.4 (p. 252); 22.2.1
	(a) **86**, K83 (1984)	11 (p. 132) and 14.3.3 (p. 246); 22.2.1
L. Konczewicz	(a) **86**, 301 (1984)	13.4 (p. 193); 14.3; 22.2.3
A. M. Konin	(a) **86**, K55 (1984)	14.3 (p. 231); 22.1.1; 22.2.3
V. I. Kononenko	(a) **84**, 423 (1984)	2 (p. 49) and 12.1 (p. 136); 1.2; 21; 21.4
R. F. Konopleva	(b) **122**, 349 (1984)	2 (p. 41) and 14.3 (p. 222); 14.3.4; 22.1.1
N. Yu. Konstantinov	(a) **83**, K153 (1984)	13.4 (p. 183) and 20.1 (p. 331); 22.8.2
Chul Koo Kim	see under letter "C"	
H. J. A. Koopmans	(b) **122**, 317 (1984)	13.4 (p. 176) and 19 (p. 307); 22.8.1
Yu. V. Kopytko	(a) **81**, 661 (1984)	14.3.4 (p. 249) and 20.3 (p. 346); 22.4.1
D. V. Korbutyak	(b) **125**, 815 (1984)	13.4 (p. 190) and 13.5.1 (p. 203); 13.3; 20.1; 22.4.1
N. N. Koren	(a) **84**, K173 (1984)	14.3 (p. 228) and 15 (p. 267); 10.2; 22
I. Ya. Korenblit	(b) **124**, 79 (1984)	13.5.4 (p. 208) and 18.2 (p. 292); 22
M. Korkmaz	(b) **122**, 255 (1984)	13.4 (p. 175) and 19 (p. 306); 22.8.1
	(b) **126**, 323 (1984)	13.4 (p. 192) and 19 (p. 313); 22.8.1
D. Korn	(b) **126**, 425 (1984)	2 (p. 53) and 14.1 (p. 214); 21
A. Korner	(a) **81**, 473 (1984)	10.1 (p. 98); 12.1; 21.1; 21.6
A. Korneta	(a) **82**, K107 (1984)	14.4 (p. 256); 22.6
L. S. Kornienko	(a) **85**, 167 (1984)	11 (p. 129) and 20.1 (p. 337); 10.2; 22.5
V. V. Korobtsov	(a) **82**, 345 (1984)	1.5 (p. 33) and 3 (p. 55); 2; 22.1.2
Yu. S. Korolev	(a) **81**, 11 (1984)	14.3.3 (p. 241) and 14.3.4 (p. 248); 16; 22.1.1; 22.1.2; 22.2.2
O. V. Korolkova	(b) **122**, 599 (1984)	4 (p. 59); 20.1; 22; 22.5.2; 22.6; 22.8.1
A. D. Korotaev	(a) **82**, 405 (1984)	12.1 (p. 135); 21.1
E. R. Korutcheva	(b) **126**, K19 (1984)	1.2 (p. 20) and 10 (p. 96)
L. V. Korzunova	(a) **83**, K127 (1984)	14.4.1 (p. 261); 22.8.1
F. Kosek	(b) **121**, K43 (1984)	18 (p. 278); 22
J. Kossut	(b) **122**, 761 (1984)	13.1 (p. 157); 16; 22
M. E. Kost	(a) **83**, 461 (1984)	1.2 (p. 17); 21.4
P. Koštial	(a) **86**, K7 (1984)	1.5 (p. 35) and 7 (p. 83); 2; 22.1.3
E. Kotai	(a) **85**, K35 (1984)	10.2 (p. 117) and 11 (p. 130); 22.1.2
M. M. Kotlyarskii	(a) **83**, 245 (1984)	18.4 (p. 301); 1.4; 22.5
R. Kotzott	(b) **126**, 223 (1984)	13.5.4 (p. 209) and 20.1 (p. 339); 18.3
L. Koudelka	(a) **84**, K143 (1984)	20.1 (p. 335); 13.4; 14.3; 22.7.1
Surinder Koul	(a) **83**, 341 (1984)	14.3.1 (p. 235); 2; 22.1.3
L. Kovács	(a) **86**, 401 (1984)	10.2 (p. 119) and 14.4 (p. 259); 22.8.2
B. A. Koval	(a) **84**, 171 (1984)	9 (p. 93); 11; 22.1.2
V. Koval	(a) **81**, K191 (1984)	11 (p. 123) and 14.3 (p. 222); 22.3

M. V. Kovalchuk	(a) **81**, 427 (1984)	1.6 (p. 36) and 17 (p. 274); 22.1.1
V. B. Kovalchuk	(a) **82**, 533 (1984)	13.4 (p. 179); 14.3; 19; 22.1.2
	(a) **84**, 263 (1984)	13.4 (p. 185); 14.3; 22.1.2
	(a) **86**, 679 (1984)	13.4 (p. 195); 14.3; 19; 22.1.2
	(a) **86**, K91 (1984)	10.2 (p. 121); 22.1.2
N. P. Kovalenko	(b) **124**, 537 (1984)	2 (p. 48) and 12.1 (p. 136); 7; 21.2
A. V. Kovalshuk	(a) **82**, K111 (1984)	14.3.3 (p. 242) and 16 (p. 270); 22.9
Z. D. Kovalyuk	(b) **123**, K63 (1984)	13.5.1 (p. 201) and 20.3 (p. 350); 22.3
V. A. Kovarskii	(b) **123**, 345 (1984)	13.4 (p. 179); 16; 22
	(b) **123**, 671 (1984)	13.4 (p. 181); 20.1; 22.5
N. M. Kovtun	(a) **84**, 215 (1984)	18.4 (p. 302); 22.8.2
L. Kowalewski	(b) **124**, 599 (1984)	18.1 (p. 285)
G. Kowalski	(a) **84**, 119 (1984)	10.2 (p. 113); 22.1.2
L. Kozák	(a) **81**, 497 (1984)	9 (p. 91); 10.1; 18.2; 21.1.1
A. Kozanecki	(a) **86**, 263 (1984)	11 (p. 132) and 20.3 (p. 357); 10.2; 22.4.2
L. P. Kozeeva	(a) **83**, K65 (1984)	20.2 (p. 343) and 20.3 (p. 350); 13.4; 22.8
V. N. Kozhanov	(a) **86**, K59 (1984)	14.1 (p. 215) and 14.2 (p. 219); 1.2; 21; 21.1
M. J. Kozielski	(b) **122**, K23 (1984)	10.2 (p. 108) and 20.1 (p. 323); 6; 22.4.1
	(b) **123**, 99 (1984)	18.4 (p. 301) and 20.1 (p. 327); 13.5; 22.8
A. I. Kozlenkov	(b) **123**, 605 (1984)	13.1 (p. 159) and 20 (p. 317); 21; 21.1
A. I. Kozlov	(b) **126**, K59 (1984)	13.5.1 (p. 204) and 20.1 (p. 340); 22.3
G. V. Kozlov	(b) **125**, 493 (1984)	6 (p. 77) and 20.1 (p. 337); 1.2; 14.4.1; 22.8.1
M. A. Kozlov	(a) **86**, 117 (1984)	1.3 (p. 26); 18; 21.1; 21.1.1
S. N. Kozlov	(a) **83**, K149 (1984)	13.3 (p. 168); 22.1.1
G. Kozłowski	(b) **122**, 775 (1984)	14.2 (p. 216)
U. F. Kozmidis-Luburić	(b) **125**, K155 (1984)	6 (p. 78) and 13.5.1 (p. 204)
W. Kraak	(b) **122**, K87 (1984)	14.1 (p. 211) and 14.3 (p. 223); 12.2; 22.2.3
	(b) **123**, 635 (1984)	7 (p. 82) and 13 (p. 147); 21.7
	(a) **83**, K207 (1984)	14.3 (p. 226); 13.1; 22.2.3
	(b) **125**, K85 (1984)	13.3 (p. 169) and 14.3 (p. 228); 22.2.3
	(a) **85**, K183 (1984)	12.2 (p. 142) and 14.3 (p. 230); 22.2.3
G. A. Kraftmakher	(a) **81**, K159 (1984)	18.4 (p. 300); 22.8.2
G. P. Kramar	(a) **86**, 283 (1984)	18.3 (p. 299); 22.8.2
A. N. Krasiko	(a) **83**, K31 (1984)	10.2 (p. 112); 14.3.4; 16; 20.1; 22.1.1
	(a) **86**, K83 (1984)	11 (p. 132) and 14.3.3 (p. 246); 22.2.1
A. R. Krasnaya	(a) **85**, K145 (1984)	17 (p. 277); 10.2; 22.8.1
V. V. Krasnogolovets	(b) **123**, 365 (1984)	14.4 (p. 256); 6; 22.8.1
A. E. Krasovskii	(b) **123**, 787 (1984)	13.1 (p. 160); 21.6
H. Krause	(a) **85**, 569 (1984)	13.4 (p. 191) and 14.3.1 (p. 236); 22.6
E. Krautz	(a) **86**, 749 (1984)	14.1 (p. 215); 1.2; 13.1; 18; 20.1; 22.6.1
A. F. Kravchenko	(b) **121**, K13 (1984)	6 (p. 68) and 20.1 (p. 320); 22.5
	(b) **125**, K103 (1984)	6 (p. 78) and 20.1 (p. 338); 22.5
W. Krech	(a) **85**, K175 (1984)	14.2 (p. 218)
J. Kreissl	(a) **81**, 701 (1984)	13.4 (p. 175) and 19 (p. 306); 22.4.1; 22.4.2
J. Krempaský	(b) **124**, K147 (1984)	13.1 (p. 162); 2; 17; 21.1.1
U. Krey	(b) **122**, K137 (1984)	18.2.1 (p. 296); 21.1.1; 21.4
G. Kreysch	(a) **82**, 125 (1984)	11 (p. 124); 3; 22.1.2
	(a) **84**, 269 (1984)	17 (p. 275); 13.3; 22.1.2
A. Krier	(a) **81**, 681 (1984)	14.3.4 (p. 249) and 20.3 (p. 347); 22.4.1
	(a) **83**, 315 (1984)	20.3 (p. 350); 13.4; 22.4.1
P. Krishna	(a) **84**, 401 (1984)	1.2 (p. 18); 10.2; 22.3
V. G. Krishna Murthy	see under letter "M"	
R. Krishnan	(a) **82**, K173 (1984)	2 (p. 43) and 18.2 (p. 290); 21; 21.1
A. V. Krishna Rao	see under letter "R"	
P. Krispin	(a) **84**, 573 (1984)	13.4 (p. 187) and 14.3.3 (p. 244); 22.2.2
N. N. Kristoffel	(b) **122**, K183 (1984)	13.4 (p. 178); 10.2; 22.5.2
G. Krivaitė	(b) **122**, 623 (1984)	13.1 (p. 157) and 20.1 (p. 325); 22

M. A. KRIVOV	(a) **82**, K191 (1984)	11 (p. 125) and 14.3 (p. 224); 22.2.1; 22.8
H. J. KROKOSZINSKI	(b) **121**, 157 (1984)	18 (p. 278); 21.1
A. KRÓL	(b) **122**, K23 (1984)	10.2 (p. 108) and 20.1 (p. 323); 6; 22.4.1
	(b) **123**, K17 (1984)	6 (p. 72) and 10.2 (p. 111); 20.1; 22.4.3
H. KRONMÜLLER	(a) **81**, 227 (1984)	10.1 (p. 97) and 18.2 (p. 288); 11; 21.1; 21.1.1
V. T. KROTENKO	(b) **121**, 623 (1984)	13.4 (p. 174); 21.4
V. I. KRUGLOV	(b) **124**, 127 (1984)	13.5 (p. 197)
M. KRUSSANOVA	(a) **82**, 285 (1984)	14.3 (p. 223); 13.4; 15; 16; 22.8.1
I. V. KRYLOVA	(a) **83**, 589 (1984)	13.3 (p. 168) and 17 (p. 275); 11; 19; 22.9
YU. YU. KRYUCHKOV	(a) **81**, 217 (1984)	11 (p. 122); 14.3; 22.1.2; 22.2.1
M. KRZESIŃSKA	(a) **82**, 421 (1984)	7 (p. 82); 6; 12.2; 22.8.2
E. KUBALEK	(a) **83**, 307 (1984)	10.2 (p. 111) and 20.3 (p. 350); 13.4; 22.2
N. KUBICKI	(a) **86**, 427 (1984)	13.3 (p. 170) and 14.3.3 (p. 246); 22.2.3
J. KUČERA	(a) **81**, 497 (1984)	9 (p. 91); 10.1; 18.2; 21.1.1
P. V. KUCHINSKII	(a) **85**, 585 (1984)	11 (p. 131) and 13.4 (p. 191); 22.1.2
E. V. KUCHIS	(a) **82**, 569 (1984)	14.3.1 (p. 234); 1.5; 22.8.2
N. N. KUDELKIN	(a) **84**, K135 (1984)	18.2 (p. 293); 22.8.2
A. KUDIN	(a) **81**, K191 (1984)	11 (p. 123) and 14.3 (p. 222); 22.3
J. KUDRNOVSKÝ	(b) **124**, 179 (1984)	13.1 (p. 161); 17; 21
V. I. KUDRYASH	(a) **83**, 131 (1984)	14.4.1 (p. 261); 10.2; 12.2; 22.8.1
T. A. KUDYKINA	(b) **122**, K163 (1984)	13.4 (p. 178) and 14.3.4 (p. 250); 22.1.1
A. KUHN	(a) **84**, 411 (1984)	1.1 (p. 12); 19; 20.1; 22.9
G. KÜHN	(b) **121**, K55 (1984)	20.1 (p. 321); 13.1; 22.8
	(a) **84**, 87 (1984)	6 (p. 74) and 8 (p. 87); 22.8
W. KÜHN	(b) **123**, K39 (1984)	13.1 (p. 158)
A. KÜHNEL	(b) **124**, 55 (1984)	14.4.1 (p. 261); 19
	(a) **84**, 433 (1984)	14.4.1 (p. 262); 22.8.1
G. KÜHNEL	(a) **84**, 251 (1984)	13.4 (p. 185) and 14.3 (p. 227); 22.2.2
P. KUIVALAINEN	(b) **122**, 791 (1984)	14.3 (p. 223) and 15 (p. 265); 22.9
J. KULDA	(a) **82**, K141 (1984)	11 (p. 125); 22.1.2
	(a) **83**, 455 (1984)	1.1 (p. 12); 22.1.2
V. S. KULESHOV	(b) **121**, 561 (1984)	18.3 (p. 298) and 18.4 (p. 300)
B. I. KULIEV	(b) **121**, 11 (1984)	14.3.1 (p. 233); 6; 13.1; 15; 22
S. I. KULIEVA	(a) **84**, K185 (1984)	14.3.1 (p. 236) and 14.3.3 (p. 245); 22.1.3; 22.4.2
W. KULLMANN	(b) **125**, 131 (1984)	4 (p. 61); 1.1; 6; 20.1; 22.7
N. KULTSCHER	(a) **82**, 23 (1984)	1 (p. 9) and 7 (p. 81); 21.1
YU. A. KULYUPIN	(a) **85**, 45 (1984)	1.5 (p. 34) and 13.3 (p. 169); 21.7
A. KUMAR	(a) **85**, 297 (1984)	16 (p. 273); 2; 22.1.2
DEVENDRA KUMAR	(a) **82**, K11 (1984)	1.2 (p. 15) and 18 (p. 279); 22.8.2
G. S. KUMAR	(a) **85**, K27 (1984)	12.2 (p. 141) and 20 (p. 318); 22.5.4
HEMANT KUMAR	(a) **83**, 341 (1984)	14.3.1 (p. 235); 2; 22.1.3
S. KUMAR PAL	see under letter "P"	
SOBDEN KUMAR PAL	see under letter "P"	
V. KUMAR	(b) **121**, 487 (1984)	12.2 (p. 139); 14.4; 22.5
M. KUMARI	(b) **124**, 531 (1984)	8 (p. 88); 22.5.2
A. B. KUNZ	(b) **121**, K47 (1984)	20.1 (p. 320); 13.1; 23
J. V. KUNZLER	(a) **83**, K187 (1984)	14.1 (p. 212); 21.1
K. H. KUO	(a) **83**, 77 (1984)	1.3 (p. 23) and 1.4 (p. 29); 21.1
T. KURAISHI	(a) **82**, K169 (1984)	2 (p. 43) and 18.2 (p. 290); 21.1; 21.1.1
N. KURAMOCHI	(a) **85**, 121 (1984)	10.1 (p. 101) and 12.1 (p. 137); 21
N. I. KURDIANI	(b) **124**, 241 (1984)	13.4 (p. 183) and 20.1 (p. 332); 11; 22.2.1
M. V. KURIK	(a) **82**, K111 (1984)	14.3.3 (p. 242) and 16 (p. 270); 22.9
N. N. KURKIN	(b) **122**, K69 (1984)	19 (p. 307); 20.1; 22.6
A. N. KURMANTAEV	(a) **83**, 285 (1984)	13.4 (p. 180) and 14.3.1 (p. 235); 2; 20.1; 22.1.2
T. KUROBORI	(a) **83**, K179 (1984)	20.2 (p. 343); 10.2; 22.5.2
L. A. KUROCHKIN	(a) **85**, K51 (1984)	11 (p. 130); 19; 22.1.2

I. N. Kuropyatnik	(a) **84**, K165 (1984)	14.1 (p. 213); 21
H. D. Kürsten	(a) **81**, K19 (1984)	20.2 (p. 343) and 20.3 (p. 345); 22.8.1
F. V. Kusmartsev	(b) **121**, K87 (1984)	13.4 (p. 173)
W. Kużmicz	(a) **86**, K125 (1984)	14.3.3 (p. 247); 22.1.2
R. N. Kuzmin	(b) **125**, 461 (1984)	5 (p. 66); 1.6; 13.3; 21.1
L. M. Kuzmina	(b) **124**, 537 (1984)	2 (p. 48) and 12.1 (p. 136); 7; 21.2
A. I. Kuznetsov	(b) **121**, K59 (1984)	13.5.1 (p. 199) and 20.3 (p. 345); 20.1; 22.6
M. F. Kuznetsov	(a) **85**, K31 (1984)	11 (p. 130) and 20 (p. 318); 22.1.2; 22.2.1
V. I. Kuznetsov	(a) **82**, 467 (1984)	11 (p. 125) and 14.3.2 (p. 238); 13.4; 22.1.2
V. G. Kvachadze	(b) **122**, 95 (1984)	8 (p. 85) and 11 (p. 124); 20.1; 22.6
V. V. Kveder	(a) **84**, 149 (1984)	10.2 (p. 113) and 13.4 (p. 184); 22.1.2
J. Kvítek	(a) **83**, 165 (1984)	11 (p. 126); 22.1.2; 22.6
C. Y. Kwok	(a) **84**, K1 (1984)	1.5 (p. 34); 22
Le Vu Ky	(b) **124**, 111 (1984)	13 (p. 148) and 14.3.4 (p. 251)
Kyun Nahm	(b) **122**, K187 (1984)	14.3 (p. 223); 22.6
V. A. Labunov	(a) **86**, 573 (1984)	8 (p. 90); 11; 22.1; 22.2
R. Labusch	(a) **84**, 149 (1984)	10.2 (p. 113) and 13.4 (p. 184); 22.1.2
J. L. Lagzdons	(a) **84**, K197 (1984)	14.3.3 (p. 245) and 14.3.4 (p. 252); 14.4; 22.8.1
G. Lahajnar	(b) **123**, K83 (1984)	9 (p. 92) and 14.4 (p. 256); 19; 22.8.1
V. V. Laiko	(b) **126**, 481 (1984)	19 (p. 314)
S. V. J. Lakshman	(b) **123**, K75 (1984)	13.4 (p. 180) and 20.1 (p. 328); 22.8.1
M. Lakshmipathi Rao	see under letter "R"	
I. J. Lalov	(b) **125**, K147 (1984)	6 (p. 78) and 13.5.1 (p. 203); 20.1; 22.9
B. Lambert-Andron	(a) **82**, K133 (1984)	1.1 (p. 11); 22.8
P. T. Landsberg	(b) **126**, 141 (1984)	10 (p. 96)
J. van Landuyt	(a) **86**, 553 (1984)	1.3 (p. 26); 10.1; 21; 21.1
I. G. Lang	(b) **122**, 581 (1984)	6 (p. 70) and 20.1 (p. 325); 22
H. Lange	(a) **82**, 373 (1984)	1.6 (p. 37); 17; 21; 22.1.2
	(a) **82**, 459 (1984)	1.6 (p. 37) and 11 (p. 125); 21
	(b) **123**, K69 (1984)	20.1 (p. 328); 13.1; 21
H. Lange	(b) **126**, 261 (1984)	20.1 (p. 339); 22.4.1
O. L. de Lange	(a) **84**, 517 (1984)	10.1 (p. 100); 21
D. W. Langer	(b) **122**, 263 (1984)	13.5.1 (p. 200) and 20.3 (p. 348); 22.4.3
J. M. Langer	(a) **81**, K199 (1984)	20.3 (p. 348); 22.4.1
H. T. Langhammer	(b) **121**, 765 (1984)	13.1 (p. 155); 14.3; 15; 22
	(b) **123**, K47 (1984)	20.1 (p. 328); 13.1; 22.7
K. K. Laroia	(a) **84**, 535 (1984)	5 (p. 66); 22.8.2
A. Latuszynski	(b) **124**, 355 (1984)	13.4 (p. 184); 19; 22.5
S. S. Lau	(a) **81**, 123 (1984)	1.5 (p. 33) and 3 (p. 55); 21
	(a) **84**, 49 (1984)	1.3 (p. 24) and 11 (p. 128); 1.5; 2; 21.1; 21.6
N. Lauret	(b) **125**, 357 (1984)	13.4 (p. 188) and 20.3 (p. 354); 22.1.2
J. Lauzier	(a) **85**, 463 (1984)	10.1 (p. 102) and 12.1 (p. 137); 21.3
B. B. Lavrenčič	(b) **122**, 399 (1984)	1.2 (p. 15); 22.8.1
F. F. Lavrentev	(a) **86**, 177 (1984)	10.1 (p. 102); 21.1
K. Ławniczak-Jabłońska	(b) **123**, 627 (1984)	13.1 (p. 159); 20; 21.1
J. Ławrynowicz	(b) **121**, 47 (1984)	6 (p. 67)
M. Lazaridou	(b) **125**, K109 (1984)	10.2 (p. 117); 9; 22.5.3
	(b) **126**, K123 (1984)	10.2 (p. 120); 22.5.3
N. I. Lebedeva	(a) **84**, 171 (1984)	9 (p. 93); 11; 22.1.2
J. C. Le Bossé	see under letter "B"	
J. Lech	(b) **126**, 371 (1984)	13.4 (p. 192) and 19 (p. 314); 22.8
J. Leciejewicz	(a) **82**, 195 (1984)	18.4 (p. 301); 22.8
H. M. Ledbetter	(a) **85**, 89 (1984)	12.1 (p. 137); 21.1.1
J. Ledzion	(b) **124**, K121 (1984)	18.2 (p. 293); 21.1
B. S. Lee	(b) **124**, 593 (1984)	18.2 (p. 293)

S. J. Lee	(b) **122**, K73 (1984)	13.3 (p. 167); 14.3.2; 19; 22.1.2
A. Lehmann	(b) **121**, 505 (1984)	6 (p. 68) and 20.1 (p. 321); 2; 22.6
G. Lehmann	(b) **121**, K161 (1984)	13.5.2 (p. 205); 14.4; 22.9
	(a) **81**, 445 (1984)	2 (p. 40); 21.1.1
	(b) **124**, 315 (1984)	13.4 (p. 183) and 19 (p. 310); 22.8
A. Lehmeyer	(b) **126**, 223 (1984)	13.5.4 (p. 209) and 20.1 (p. 339); 18.3
N. Lehner	(b) **125**, 131 (1984)	4 (p. 61); 1.1; 6; 20.1; 22.7
A. Yu. Leiderman	(a) **83**, K81 (1984)	2 (p. 45) and 14.3 (p. 225); 14.3.4; 22.1.2
J. Leiro	(a) **83**, 279 (1984)	4 (p. 59) and 13 (p. 147); 20; 21.1; 21.1.1; 21.4
J. A. Leiro	(b) **125**, 305 (1984)	20 (p. 318); 13; 21
L. Lejček	(a) **82**, 47 (1984)	2 (p. 42) and 10 (p. 95)
A. A. Leksikov	(a) **81**, K171 (1984)	20.3 (p. 347); 13.4; 22.5
H. Lemke	(a) **83**, 637 (1984)	13.4 (p. 182); 14.3; 22.1.2
	(a) **85**, K133 (1984)	13.4 (p. 191); 14.3; 22.1.2
	(a) **85**, K137 (1984)	13.4 (p. 192); 14.3; 22.1.2
	(a) **86**, K39 (1984)	13.4 (p. 194); 14.3; 22.1.2
V. Lemos	(b) **122**, 53 (1984)	6 (p. 69) and 20.1 (p. 322); 22.8.1
E. I. Leonov	(a) **84**, 113 (1984)	6 (p. 74) and 20.1 (p. 333); 22.8
J. Leotin	(b) **123**, K43 (1984)	13.1 (p. 158); 22.3
Y. Lépine	(b) **122**, 151 (1984)	13.5.3 (p. 207)
M. C. Leroy	(b) **122**, 175 (1984)	13.4 (p. 175) and 13.5.3 (p. 207); 14.3; 19; 22.6.1
V. I. Levchenko	(a) **84**, K173 (1984)	14.3 (p. 228) and 15 (p. 267); 10.2; 22
A. A. Levin	(b) **126**, 311 (1984)	13.1 (p. 164) and 13.4 (p. 192); 22.1.2
S. D. Le Vine	see under letter "V"	
M. L. Levitan	(a) **85**, 51 (1984)	1.4 (p. 31)
N. L. Levshin	(a) **83**, K149 (1984)	13.3 (p. 168); 22.1.1
Le Vu Ky	see under letter "K"	
M. de Dios Leyva	see under letter "D"	
C. Lhermitte	(a) **83**, K103 (1984)	2 (p. 47) and 14.3.1 (p. 235); 22.1.3
R. Y. Li	(a) **86**, 813 (1984)	13.5.2 (p. 206) and 20 (p. 319); 21.6; 22.5.2
Zhen-ya Li	see under letter "Z"	
I. T. Liddell	(a) **84**, K75 (1984)	20.3 (p. 353); 22.4.1
T. Lideikis	(b) **122**, K179 (1984)	20.1 (p. 326); 13.4; 22
C.-D. Lien	(a) **81**, 123 (1984)	1.5 (p. 33) and 3 (p. 55); 21
A. Likforman	(b) **124**, K105 (1984)	6 (p. 75) and 20.1 (p. 334); 22.3
N. I. Likholit	(b) **123**, 691 (1984)	13.5 (p. 197) and 20.1 (p. 330); 6; 22
P. Lilley	(a) **85**, 235 (1984)	14.3.1 (p. 236) and 20.3 (p. 355); 13.4; 22.4.2
Young Soo Lim	see under letter "Y"	
J. H. Lind	(a) **81**, K51 (1984)	2 (p. 40) and 16 (p. 269); 22.1
G. Lindemann	(b) **123**, 319 (1984)	13.4 (p. 179); 16; 22.8
U. Lindner	(b) **124**, K39 (1984)	13.5.4 (p. 208); 18.2
	(b) **125**, 177 (1984)	13.5 (p. 197) and 18.4 (p. 303)
J. L. Lindström	(a) **85**, K109 (1984)	10.2 (p. 118); 22.1.2
C. H. Ling	(a) **84**, K1 (1984)	1.5 (p. 34); 22
L. F. Linnik	(a) **83**, K31 (1984)	10.2 (p. 112); 14.3.4; 16; 20.1; 22.1.1
L. G. Linnik	(a) **83**, K31 (1984)	10.2 (p. 112); 14.3.4; 16; 20.1; 22.1.1
B. Lippold	(b) **121**, 623 (1984)	13.4 (p. 174); 21.4
	(b) **122**, 377 (1984)	14 (p. 210); 21.4
Chen Li-Rong	(b) **124**, K113 (1984)	8 (p. 88); 21.2
K. Lischka	(b) **123**, 319 (1984)	13.4 (p. 179); 16; 22.8
A. A. Lisenko	(b) **121**, 583 (1984)	13.1 (p. 154); 22
M. P. Lisitsa	(b) **125**, 705 (1984)	13.5.1 (p. 203); 20.1; 22.3.1; 22.4.1
I. S. Lisitskii	(a) **85**, 167 (1984)	11 (p. 129) and 20.1 (p. 337); 10.2; 22.5
L. B. Litinskii	(b) **122**, K109 (1984)	1.2 (p. 16) and 13 (p. 146); 18.4; 23
A. S. Litovchenko	(a) **81**, K47 (1984)	14.4 (p. 255); 19; 22.8.1
	(a) **81**, K151 (1984)	18.1 (p. 284); 22.8.1

N. M. LITOVCHENKO	(a) **84**,	237 (1984)	13.4 (p. 184); 22.1.2
	(a) **86**,	623 (1984)	13.4 (p. 195); 22.1.2
V. G. LITOVCHENKO	(a) **84**,	285 (1984)	16 (p. 272); 13.4; 22.2.1
	(b) **125**,	815 (1984)	13.4 (p. 190) and 13.5.1 (p. 203); 13.3; 20.1; 22.4.1
D. C. LIU	(a) **85**,	69 (1984)	1.3 (p. 25) and 1.5 (p. 34); 22
Z. LIU	(a) **82**,	425 (1984)	1.4 (p. 29) and 10.2 (p. 110); 22.1.2
J. LLOPIS	(a) **82**,	111 (1984)	10.2 (p. 108) and 20.3 (p. 348); 20.1; 22.6
	(a) **83**,	645 (1984)	20.3 (p. 351); 10.2; 13.4; 22.6
C. LO	(a) **86**,	663 (1984)	5 (p. 66); 18.1; 20.1; 22.8.1
S. L. LOCK	(b) **124**,	593 (1984)	18.2 (p. 293)
H. LÖFFLER	(b) **126**,	741 (1984)	14.1 (p. 215); 13; 21; 21.1
	(a) **86**,	543 (1984)	1.3 (p. 26); 10.1; 21; 21.1
S. LOGOTHETIDIS	(b) **122**,	613 (1984)	20.1 (p. 325); 13.1; 22.8
K. LÖHNERT	(a) **83**,	307 (1984)	10.2 (p. 111) and 20.3 (p. 350); 13.4; 22.2
C. D. LOKHANDE	(a) **82**,	K195 (1984)	14.3.1 (p. 234); 20.1; 22.8
B. V. LOKSHIN	(a) **84**,	K101 (1984)	1.2 (p. 18); 19; 21.7
V. M. LOMAKO	(a) **85**,	585 (1984)	11 (p. 131) and 13.4 (p. 191); 22.1.2
	(a) **85**,	K57 (1984)	11 (p. 130) and 13.4 (p. 189); 22.2.1
B. A. LOMBOS	(b) **125**,	831 (1984)	14.3 (p. 229); 13.1; 22.4.4
V. A. LOMONOV	(a) **84**,	K81 (1984)	20.2 (p. 343) and 20.3 (p. 353); 13.4; 22.8
	(a) **85**,	553 (1984)	20.1 (p. 338) and 20.3 (p. 356); 13.4; 22.8
P. LONG	(a) **86**,	K45 (1984)	20.3 (p. 358); 22.8.1
J. LOOS	(b) **123**,	595 (1984)	6 (p. 73) and 18.3 (p. 299)
J. LOPEZ	(b) **124**,	381 (1984)	13.3 (p. 168); 21.1
J. M. LÓPEZ	(a) **85**,	423 (1984)	1.3 (p. 25) and 8 (p. 89); 21; 21.1
F. LÓPEZ-AGUILAR	(b) **123**,	219 (1984)	13.1 (p. 158); 22.6
V. A. LOPYREV	(a) **81**,	377 (1984)	14.3 (p. 221); 22.9
B. LORENZ	(b) **125**,	375 (1984)	4 (p. 61); 1.2; 22
	(b) **125**,	K193 (1984)	14.3 (p. 229); 22.6
	(b) **126**,	235 (1984)	13 (p. 151) and 18 (p. 282); 8
W. LORENZ	(b) **122**,	745 (1984)	13.3 (p. 167); 4; 6; 8; 22
P. LOŠŤÁK	(a) **84**,	K143 (1984)	20.1 (p. 335); 13.4; 14.3; 22.7.1
P. P. LOTTICI	(b) **122**,	431 (1984)	6 (p. 70) and 20.1 (p. 325); 2; 22.1
	(b) **125**,	K7 (1984)	6 (p. 77) and 20.1 (p. 336); 22.8
F. C. LOVEY	(a) **85**,	29 (1984)	1.4 (p. 31); 10.1; 21; 21.1
	(a) **86**,	553 (1984)	1.3 (p. 26); 10.1; 21; 21.1
P. L. LOYZANCE	(a) **85**,	359 (1984)	1.2 (p. 19); 22.8
L. LOZANO	(a) **81**,	105 (1984)	1.3 (p. 22); 14.3.1; 22.5
V. S. LUBIMOV	(b) **123**,	183 (1984)	14.2 (p. 217); 5; 22.6.1
A. LUBOMIRSKA-WITTLIN	(a) **84**,	K181 (1984)	14.3 (p. 228) and 16 (p. 272); 9; 13.4; 22.4.4
H. LUCHIAN	(b) **124**,	K125 (1984)	14.2 (p. 218) and 18 (p. 281); 21
T. LUCIŃSKI	(a) **84**,	607 (1984)	14.1 (p. 213); 2; 18.2.1; 21.1.1
JUH TZENG LUE	see under letter "J"		
P. F. LUGAKOV	(a) **82**,	467 (1984)	11 (p. 125) and 14.3.2 (p. 238); 13.4; 22.1.2
	(a) **82**,	511 (1984)	10.2 (p. 110) and 13.4 (p. 178); 11; 14.3; 22.1.2
	(a) **83**,	299 (1984)	11 (p. 126) and 13.4 (p. 180); 22.1.2
	(a) **83**,	521 (1984)	11 (p. 126); 13.4; 22.1.2
	(a) **84**,	457 (1984)	10.2 (p. 115) and 11 (p. 128); 14.3; 22.1.2
	(a) **85**,	441 (1984)	11 (p. 131); 13.4; 22.1.2
	(a) **86**,	309 (1984)	14.3.4 (p. 253); 13.4; 22.1.2
A. I. LUKANIN	(b) **121**,	525 (1984)	18.2 (p. 288); 21.4
	(b) **121**,	573 (1984)	18.2 (p. 288); 21.4
B. LUKASCH	(a) **82**,	125 (1984)	11 (p. 124); 3; 22.1.2
M. G. LUKASHEVICH	(a) **84**,	613 (1984)	14.3.4 (p. 252); 22.2.1
T. A. LUKASHEVICH	(a) **85**,	441 (1984)	11 (p. 131); 13.4; 22.1.2
F. LUKEŠ	(b) **122**,	675 (1984)	13.1 (p. 157) and 20.1 (p. 326); 22
	(a) **83**,	581 (1984)	20.1 (p. 330); 22.8.2

S. N. LUKIN	(a) **84**, 223 (1984)	18.4 (p. 302) and 19 (p. 310); 22.5
M. M. LUKINA	(a) **84**, 215 (1984)	18.4 (p. 302); 22.8.2
B. A. LUKIYANETS	(b) **124**, 693 (1984)	6 (p. 75) and 13 (p. 149); 22.3
V. V. LUKYANITSA	(a) **83**, 521 (1984)	11 (p. 126); 13.4; 22.1.2
	(a) **84**, 457 (1984)	10.2 (p. 115) and 11 (p. 128); 14.3; 22.1.2
L. LUNDGREN	(a) **84**, 199 (1984)	18.2 (p. 292) and 18.4 (p. 302); 1.2; 14.1; 21.1
C. W. LUNG	(a) **85**, K113 (1984)	10 (p. 96)
B. LUNN	(a) **86**, 821 (1984)	13.4 (p. 195) and 14.3 (p. 232); 20.1; 22.4.3
I. LUPSA	(a) **82**, K49 (1984)	18 (p. 279); 18.2; 21.1
A. LUSHCHIK	(a) **86**, 219 (1984)	10.2 (p. 119); 19; 20.1; 20.3; 22.5.2
CH. LUSHCHIK	(a) **86**, 219 (1984)	10.2 (p. 119); 19; 20.1; 20.3; 22.5.2
N. LUSHCHIK	(a) **86**, 219 (1984)	10.2 (p. 119); 19; 20.1; 20.3; 22.5.2
A. R. LUSIS	(a) **81**, K121 (1984)	9 (p. 91); 21.6
A. R. LŪSIS	(a) **83**, 291 (1984)	13.4 (p. 180) and 20.1 (p. 328); 14.3.1; 19; 22.6
	(a) **84**, K197 (1984)	14.3.3 (p. 245) and 14.3.4 (p. 252); 14.4; 22.8.1
V. S. LUTOVINOV	(b) **124**, K133 (1984)	18.1 (p. 285) and 19 (p. 311)
R. LUZZI	(b) **126**, 63 (1984)	8 (p. 89) and 11 (p. 131); 6; 22; 22.2.1
D. F. LYNCH	(b) **124**, 649 (1984)	1.6 (p. 38) and 13 (p. 148); 22
V. M. LYUBIN	(a) **86**, 383 (1984)	16 (p. 273); 2; 14.3.1; 22; 22.1.2
T. N. LYUBOVSKAYA	(a) **81**, 139 (1984)	2 (p. 39); 14.3; 15; 22.8
W. MAASS	(b) **122**, K137 (1984)	18.2.1 (p. 296); 21.1.1; 21.4
R. MACH	(a) **81**, 609 (1984)	14.3.4 (p. 249) and 20.3 (p. 346); 22.4.1
	(a) **81**, 723 (1984)	14.3.4 (p. 249) and 20.3 (p. 347); 14.3.1; 22.4.1
	(a) **81**, 733 (1984)	14.3.4 (p. 249) and 20.3 (p. 347); 22.4.2
T. A. MACHEKHINA	(a) **83**, 455 (1984)	1.1 (p. 12); 22.1.2
E. MACHO	(a) **82**, 111 (1984)	10.2 (p. 108) and 20.3 (p. 348); 20.1; 22.6
M.-P. MACHT	(a) **86**, 603 (1984)	9 (p. 94); 10.1; 21.1
R. MADAR	(a) **82**, K133 (1984)	1.1 (p. 11); 22.8
Y. MADHAVAN	(b) **125**, 513 (1984)	9 (p. 93); 22.4.3
P. MADHUSUDANA RAO	see under letter "R"	
K. MAEDA	(a) **83**, 219 (1984)	2 (p. 44) and 10.1 (p. 99); 21.1
T. MAEDA	(a) **82**, K169 (1984)	2 (p. 43) and 18.2 (p. 290); 21.1; 21.1.1
	(a) **83**, K195 (1984)	2 (p. 47) and 14.1 (p. 212); 21.1; 21.1.1
	(a) **84**, 55 (1984)	1.1 (p. 12); 22.8.1
J. MAEGE	(a) **84**, 573 (1984)	13.4 (p. 187) and 14.3.3 (p. 244); 22.2.2
O. I. MAEVA	(a) **84**, 285 (1984)	16 (p. 272); 13.4; 22.2.1
B. MAFFEO	(b) **124**, 757 (1984)	13.4 (p. 186); 20.1; 22.5.2
S. MAGER	(a) **86**, 117 (1984)	1.3 (p. 26); 18; 21.1; 21.1.1
A. MAGERL	(b) **123**, 85 (1984)	9 (p. 92); 21
T. G. MAGERRAMOV	(a) **84**, K185 (1994)	14.3.1 (p. 236) and 14.3.3 (p. 245); 22.1.3; 22.4.2
S. MAHAJAN	(b) **126**, 467 (1984)	8 (p. 90); 10.1; 21
S. A. MAHMOUD	(a) **84**, K67 (1984)	10.1 (p. 100) and 18.2 (p. 293); 21.1
J. MAIER	(b) **123**, K89 (1984)	14.4 (p. 256); 22
	(b) **124**, K187 (1984)	14.4 (p. 256); 22
H. S. MAITI	(a) **84**, 631 (1984)	14.3 (p. 227); 15; 22.6.1
S. N. MAITI	(b) **122**, 449 (1984)	4 (p. 59); 12.2; 22.5.4
B. MAJUMDAR	(a) **83**, K19 (1984)	12.2 (p. 140) and 20.3 (p. 350); 10.2; 22.5.2; 22.8.1
E. F. MAKAROV	(b) **123**, 183 (1984)	14.2 (p. 217); 5; 22.6.1
O. A. MAKAROV	(b) **121**, 293 (1984)	2 (p. 39) and 20.1 (p. 320); 13.1; 22
I. P. MAKAROVA	(b) **121**, 481 (1984)	1.2 (p. 15) and 6 (p. 68); 22.8
A. G. MAKHANEK	(b) **125**, 365 (1984)	20.1 (p. 336) and 20.3 (p. 354); 13.4; 22.8.2
	(b) **126**, 659 (1984)	20.1 (p. 341) and 20.3 (p. 358); 13.4; 22.5
A. SH. MAKHMUDOV	(b) **126**, 311 (1984)	13.1 (p. 164) and 13.4 (p. 192); 22.1.2
G. I. MAKOVETSKII	(a) **85**, K69 (1984)	14.1 (p. 213) and 18.2 (p. 294); 22.8
S. A. MAKSIMENKO	(a) **85**, K23 (1984)	1.6 (p. 38)

G. A. Maksimov	(a) **81**, K51 (1984)	2 (p. 40) and 9 (p. 91); 22.1.2
S. K. Maksimov	(a) **83**, 685 (1984)	10.2 (p. 112) and 14.3.2 (p. 239); 9; 22.2.4
	(a) **84**, 79 (1984)	10.2 (p. 113); 1.4; 22.2.1
	(a) **86**, 55 (1984)	1.4 (p. 32) and 10 (p. 96)
N. K. Maksimova	(a) **83**, 701 (1984)	14.3.3 (p. 243); 13.4; 22.2.1
A. Z. Maksymowicz	(b) **122**, 519 (1984)	2 (p. 43) and 18.2 (p. 290)
	(b) **126**, 191 (1984)	13 (p. 151) and 18 (p. 282); 2; 21.1.1
A. V. Malakhovskii	(b) **125**, 751 (1984)	18.2 (p. 294) and 20.1 (p. 337); 13.4; 22.8.1
J. Málek	(a) **85**, K123 (1984)	2 (p. 52) and 20.1 (p. 339); 22.8
P. Malétras	(a) **86**, 85 (1984)	1.3 (p. 25); 21; 21.1
C. Malgrange	(a) **85**, 389 (1984)	1 (p. 10)
S. J. Mali	(b) **125**, K143 (1984)	13 (p. 151) and 17 (p. 276); 21
B. Mallik	(a) **81**, K179 (1984)	14.3 (p. 222); 22.9
Yu. M. Malozovskii	(a) **81**, 391 (1984)	14.3.4 (p. 248) and 16 (p. 269); 22.2.3
O. L. Malta	(b) **124**, 147 (1984)	13.1 (p. 160) and 20.1 (p. 331); 22
V. I. Maltsev	(a) **85**, 529 (1984)	18.3 (p. 299); 22.8.2
V. Yu. Malyshev	(a) **81**, K185 (1984)	14.2 (p. 216); 1.2; 21.6
V. K. Malyutenko	(a) **81**, 391 (1984)	14.3.4 (p. 248) and 16 (p. 269); 22.2.3
B. R. Mamatkulov	(a) **81**, K85 (1984)	14.3.4 (p. 248); 22.1.2
	(a) **84**, 301 (1984)	14.3.3 (p. 244) and 14.3.4 (p. 251); 16; 22
A. N. Mamedov	(b) **124**, K155 (1984)	13.5.1 (p. 202) and 20.1 (p. 334); 14.3.4; 22.4
M. M. Mamedov	(b) **122**, 623 (1984)	13.1 (p. 157) and 20.1 (p. 325); 22
N. T. Mamedov	(a) **82**, K75 (1984)	20.3 (p. 349); 22.8
T. G. Mamedov	(b) **126**, K139 (1984)	13.1 (p. 165) and 20.1 (p. 341); 12.2; 22; 22.8
L. A. Manakova	(b) **121**, 275 (1984)	13.1 (p. 154); 13.5.3
J. Manam	(b) **121**, 393 (1984)	13.4 (p. 172) and 20.3 (p. 345); 10.2; 22.5.2
S. Mandache	(a) **86**, 525 (1984)	5 (p. 66); 2; 21.1.1
C. Mande	(a) **81**, 197 (1984)	4 (p. 58); 20; 21
	(b) **126**, 53 (1984)	1 (p. 10); 20; 21.1
A. Mandelis	(b) **122**, 687 (1984)	13.4 (p. 177); 22
M. Manfredi	(a) **86**, 333 (1984)	20.1 (p. 340) and 20.3 (p. 357); 13.4; 22.5.2
Nguyen Manh Son	see under letter "N"	
Yu. I. Mankov	(b) **125**, 197 (1984)	14.1 (p. 213); 18.2; 21
A. Manna	(b) **122**, 469 (1984)	1.3 (p. 22) and 4 (p. 59); 21.1; 21.1.1; 21.6
	(b) **125**, 97 (1984)	4 (p. 61) and 13.1 (p. 162); 21.1; 21.6
G. Manzke	(b) **125**, 693 (1984)	13.5.1 (p. 203); 22
O. V. Marchenko	(b) **123**, 183 (1984)	14.2 (p. 217); 5; 22.6.1
E. Marchese	(b) **125**, 721 (1984)	2 (p. 51) and 13 (p. 150)
M. J. Marcinkowski	(a) **83**, 59 (1984)	1.4 (p. 29) and 10 (p. 95)
	(b) **126**, 527 (1984)	10 (p. 96)
	(a) **86**, 460 (1984)	1.4 (p. 29) and 10 (p. 95)
P. P. Mardilovich	(b) **124**, 699 (1984)	13.1 (p. 161) and 13.3 (p. 168); 22.6
D. F. Mariani	(b) **124**, 465 (1984)	2 (p. 48); 21; 21.7
	(b) **126**, K1 (1984)	2 (p. 53); 21.7
M. M. Marinković	(b) **121**, 589 (1984)	13.5 (p. 196)
V. M. Markushev	(b) **124**, K69 (1984)	13.4 (p. 184) and 20.3 (p. 352); 22.8
R. Marquez	(a) **85**, 445 (1984)	10.2 (p. 117) and 12.2 (p. 142); 22.6.1
A. F. Marshall	(a) **82**, 55 (1984)	1.4 (p. 28); 21.1
A. Martinaitis	(a) **82**, K199 (1984)	14.3 (p. 224); 13.4; 22.4.3
G. Martínez	(b) **124**, 573 (1984)	13.4 (p. 185) and 18 (p. 281)
Cz. Marucha	(a) **81**, 185 (1984)	8 (p. 84); 21
S. B. Maslenkov	(b) **123**, 605 (1984)	13.1 (p. 159) and 20 (p. 317); 21; 21.1
A. V. Maslov	(a) **84**, 73 (1984)	1 (p. 9) and 17 (p. 275); 22.1.2
G. Massé	(a) **83**, K199 (1984)	14.3 (p. 226); 22.4.1
Sh. R. Mastov	(a) **85**, K31 (1984)	11 (p. 130) and 20 (p. 318); 22.1.2; 22.2.1
K. Masuda-Jindo	(b) **126**, 643 (1984)	13.4 (p. 194); 21; 21.1.1
Y. Masumoto	(b) **125**, 735 (1984)	13.5.1 (p. 203) and 20.3 (p. 355); 22.4.2

Y. Mathey	(b) **123**,	503 (1984)	6 (p. 73) and 20.1 (p. 329); 22
H. Mathieu	(a) **81**,	549 (1984)	13.4 (p. 174) and 20.3 (p. 346); 20.1; 22.4.4
P. C. Mathur	(a) **83**,	341 (1984)	14.3.1 (p. 235); 2; 22.1.3
V. K. Mathur	(b) **126**,	361 (1984)	13.4 (p. 192) and 20.3 (p. 357); 22
M. Matlak	(b) **123**,	155 (1984)	13 (p. 146) and 18 (p. 279)
M. Matos	(b) **124**,	757 (1984)	13.4 (p. 186); 20.1; 22.5.2
M. G. Matsko	(b) **125**,	613 (1984)	13.5.1 (p. 203) and 20.3 (p. 355); 13.3; 22.4.3
Y. Matsuoka	(b) **125**,	823 (1984)	10.1 (p. 101); 13.3; 21.6
G. Mattei	(b) **123**,	K13 (1984)	6 (p. 72) and 20.1 (p. 327); 13.5; 22.8.1
J. Matthäi	(a) **82**,	K5 (1984)	2 (p. 42) and 3 (p. 55); 11; 22.1.2
	(a) **82**,	K121 (1984)	1.5 (p. 33) and 3 (p. 56); 11; 22.1.2
	(a) **86**,	781 (1984)	11 (p. 133) and 14.3.1 (p. 237); 22.1.2
S. Matthies	(b) **126**,	43 (1984)	1.3 (p. 25) and 1.4 (p. 32); 21
L. Mattioli	(b) **123**,	K13 (1984)	6 (p. 72) and 20.1 (p. 327); 13.5; 22.8.1
A. Matulis	(b) **125**,	K71 (1984)	13.4 (p. 188)
T. A. Matveeva	(b) **124**,	K59 (1984)	13.1 (p. 161); 20; 21.4
E. E. Matyas	(a) **84**,	K193 (1984)	14.3 (p. 228); 22
Z. A. Matysina	(a) **82**,	K163 (1984)	1.2 (p. 16) and 18 (p. 279); 21; 21.1
W. Matz	(a) **84**,	11 (1984)	2 (p. 47); 1.3; 22.7
G. A. Maugin	(a) **81**,	519 (1984)	18.2 (p. 289); 12.2; 22.1.2; 22.6
C. Di Mauro	(b) **124**,	K177 (1984)	14.2 (p. 218)
	(b) **125**,	K185 (1984)	14.2 (p. 218)
V. May	(b) **125**,	693 (1984)	13.5.1 (p. 203); 22
A. P. Mayer	(b) **126**,	91 (1984)	6 (p. 78); 22.1.2
J. W. Mayer	(a) **84**,	49 (1984)	1.3 (p. 24) and 11 (p. 128); 1.5; 2; 21.1; 21.6
V. A. Mayer	(a) **82**,	475 (1984)	11 (p. 125); 14.3; 22.1.2
S. A. Mazen	(b) **123**,	K1 (1984)	6 (p. 71) and 20.1 (p. 327); 22.8.2
	(a) **84**,	337 (1984)	14.3 (p. 227) and 15 (p. 266); 22.8.2
	(a) **86**,	271 (1984)	18.3 (p. 299); 14.3; 22.8.2
V. V. Mazykin	(a) **81**,	K47 (1984)	14.4 (p. 255); 19; 22.8.1
	(a) **81**,	K151 (1984)	18.1 (p. 284); 22.8.1
P. Mazzetti	(b) **122**,	411 (1984)	2 (p. 42)
	(b) **123**,	403 (1984)	2 (p. 45)
S. W. McKeever	(b) **126**,	361 (1984)	13.4 (p. 192) and 20.3 (p. 357); 22
W. T. McKeown	(b) **126**,	K67 (1984)	13.3 (p. 170); 22.1
M. V. Medvedev	(b) **121**,	525 (1984)	18.2 (p. 288); 21.4
	(b) **121**,	573 (1984)	18.2 (p. 288); 21.4
A Meenakshi Sundaram	see under letter "S"		
R. M. Mehra	(a) **83**,	341 (1984)	14.3.1 (p. 235); 2; 22.1.3
H. Mehrer	(a) **81**,	497 (1984)	9 (p. 91); 10.1; 18.2; 21.1.1
A. O. O. Mekhrabov	(b) **126**,	453 (1984)	1.3 (p. 26) and 13 (p. 151); 21.1; 21.1.1
A. Sh. Mekhtiev	(a) **83**,	K215 (1984)	14.3.3 (p. 244); 22
S. D. Mekhtieva	(a) **82**,	K87 (1984)	16 (p. 270); 22.1.3
A. I. Melker	(b) **122**,	K17 (1984)	11 (p. 124)
	(a) **84**,	417 (1984)	12.2 (p. 141); 22.9
	(b) **126**,	133 (1984)	11 (p. 131)
	(b) **126**,	517 (1984)	11 (p. 132); 10.1; 21.1; 21.6
M. K. Melkonyan	(a) **81**,	47 (1984)	1.6 (p. 36) and 2 (p. 39); 10.2; 22.1.2
J. H. Mellor	(b) **122**,	757 (1984)	13.4 (p. 178) and 19 (p. 308); 22.8.1
F. E. A. Melo	(b) **122**,	53 (1984)	6 (p. 69) and 20.1 (p. 322); 22.8.1
B. A. Men	(a) **85**,	51 (1984)	1.4 (p. 31)
H. H. Mende	(a) **82**,	481 (1984)	18.2 (p. 290)
S. Y. Mensah	(b) **121**,	K209 (1984)	7 (p. 81) and 14.3 (p. 221); 22
M. Menyhárd	(a) **84**,	39 (1984)	1.3 (p. 24) and 1.6 (p. 37); 21; 21.1
	(a) **84**,	65 (1984)	1.3 (p. 24); 1.4; 9; 14.1; 15; 21; 21.1.1
V. V. Meriakri	(a) **81**,	K159 (1984)	18.4 (p. 300); 22.8.2

L. Merten	(b) **121**, 471 (1984)	6 (p. 68) and 20 (p. 316); 13.5.2; 22
	(b) **126**, 223 (1984)	13.5.4 (p. 209) and 20.1 (p. 339); 18.3
M. Messaoudi	(a) **84**, 411 (1984)	1.1 (p. 12); 19; 20.1; 22.9
	(a) **85**, 291 (1984)	14.3 (p. 229); 22.9
U. Messerschmidt	(a) **83**, 179 (1984)	10.2 (p. 111); 22.6
M. Meucci	(b) **122**, K83 (1984)	10.2 (p. 108) and 13.4 (p. 176); 20.1; 22.5.2
G. De Mey	(a) **82**, K91 (1984)	14 (p. 210)
B. Meyer	(b) **123**, K125 (1984)	11 (p. 126) and 19 (p. 309); 13.4; 22.2.1
H.-G. Meyer	(b) **122**, K105 (1984)	14.2 (p. 216); 21
	(a) **82**, 581 (1984)	14.2 (p. 217)
	(a) **85**, K175 (1984)	14.2 (p. 218)
O. Meyer	(a) **83**, 437 (1984)	2 (p. 46) and 14.3.1 (p. 235); 3; 22.1.2
P. Mialhe	(a) **83**, 403 (1984)	14.3.3 (p. 243) and 16 (p. 271); 22.1.2
J. L. Miane	(a) **81**, K71 (1984)	14.4 (p. 255); 22
R. Michalec	(a) **83**, 455 (1984)	1.1 (p. 12); 22.1.2
D. Michel	(b) **125**, 483 (1984)	14.4.1 (p. 262); 19; 22.9
P. Michel	(a) **85**, K1 (1984)	1.6 (p. 38) and 11 (p. 129); 21
P. Michon	(a) **85**, 399 (1984)	1.5 (p. 35) and 3 (p. 56); 2; 22.1.3
G. Micocci	(a) **82**, K79 (1984)	13.4 (p. 177); 14.4; 22.5
	(a) **85**, 609 (1984)	16 (p. 273); 2; 22.1.2
H. Miessner	(b) **125**, K93 (1984)	4 (p. 62); 10.1; 13.4; 21.1
L. Mihuţ	(b) **124**, 307 (1984)	20.3 (p. 352); 13.4; 22.5.2
M. Mijatović	(b) **126**, K33 (1984)	18.4 (p. 303)
M. P. Mikalkevičius	(a) **81**, K43 (1984)	14.3.1 (p. 233) and 16 (p. 269); 22
I. F. I. Mikhail	(b) **126**, 721 (1984)	14.1 (p. 215); 15; 21.7
G. V. Mikhailov	(b) **122**, 221 (1984)	18 (p. 279) and 20.1 (p. 322); 13.5.1; 22.4.2
M. Mikhailov	(a) **84**, K23 (1984)	1.4 (p. 31) and 11 (p. 128); 2; 22.1.1
N. N. Mikhailov	(b) **123**, 183 (1984)	14.2 (p. 217); 5; 22.6.1
A. V. Mikheenkov	(b) **122**, K109 (1984)	1.2 (p. 16) and 13 (p. 146); 18.4; 23
O. Yu. Mikityuk	(b) **125**, 815 (1984)	13.4 (p. 190) and 13.5.1 (p. 203); 13.3; 20.1; 22.4.1
P. Mikolajczak	(b) **124**, 355 (1984)	13.4 (p. 184); 19; 22.5
P. Mikula	(a) **82**, K141 (1984)	11 (p. 125); 22.1.2
	(a) **83**, 455 (1984)	1.1 (p. 12); 22.1.2
G. Milchberg	(b) **125**, 795 (1984)	13.4 (p. 190) and 20.3 (p. 356); 13.1; 22.4.3
A. Milewski	(a) **82**, K107 (1984)	14.4 (p. 256); 22.6
B. V. Mill	(a) **86**, 345 (1984)	14.4 (p. 258) and 20.2 (p. 344); 1.1; 12.2; 20.1; 20.3; 22.8.1
O. V. Miloslavskaya	(a) **83**, 245 (1984)	18.4 (p. 301); 1.4; 22.5
S. D. Milovidova	(a) **84**, K115 (1984)	14.4.1 (p. 262); 22.9
S. Kh. Milshtein	(a) **84**, 363 (1984)	1.4 (p. 31) and 10.2 (p. 114); 14.3.3; 22.1.2
N. S. Minaev	(a) **81**, 313 (1984)	10.2 (p. 106) and 13.4 (p. 173); 20.3; 22.1.2
S. Minagawa	(b) **124**, 565 (1984)	10.2 (p. 114) and 14.4 (p. 257)
M. R. Minev	(a) **81**, K141 (1984)	10.2 (p. 107); 20.1; 22.5.2
G. S. Mingaleev	(a) **83**, 365 (1984)	11 (p. 126) and 14.3 (p. 225); 22.9
	(a) **84**, 327 (1984)	14.3 (p. 227); 11; 22.9
J. Minge	(b) **123**, 27 (1984)	14.4.1 (p. 261); 4; 19; 22.8.1
C. Minier	(a) **85**, 463 (1984)	10.1 (p. 102) and 12.1 (p. 137); 21.3
E. Minni	(b) **123**, 627 (1984)	13.1 (p. 159); 20; 21.1
R. Mirianashvili	(b) **125**, K63 (1984)	19 (p. 312); 22.5.3
D. Mirjanić	(b) **124**, 235 (1984)	13.5.1 (p. 201); 20.1
	(b) **124**, K45 (1984)	13.5.1 (p. 202) and 13.5.4 (p. 208); 18.2
D. Lj. Mirjanić	(b) **121**, 589 (1984)	13.5 (p. 196)
A. V. Mirmelshtein	(a) **86**, K59 (1984)	14.1 (p. 215) and 14.2 (p. 219); 1.2; 21; 21.1
V. P. Miroshkin	(a) **84**, 645 (1984)	14.3 (p. 228); 22.8.2
N. Mirowska	(a) **83**, K51 (1984)	16 (p. 271) and 20.1 (p. 329); 22.3
E. B. Mirza	(a) **84**, K105 (1984)	1.2 (p. 18); 8; 14.1; 14.3; 18.1; 22.6
K. Misek	(b) **123**, 635 (1984)	7 (p. 82) and 13 (p. 147); 21.7

A. V. MISHCHENKO	(a) **84**, K165 (1984)	14.1 (p. 213); 21
A. V. MISHENKO	(a) **83**, 147 (1984)	6 (p. 72) and 8 (p. 86); 22
S. N. MISHRA	(b) **121**, K83 (1984)	5 (p. 64); 22.8
J. MISIEWICZ	(a) **83**, K51 (1984)	16 (p. 271) and 20.1 (p. 329); 22.3
D. S. MISRA	(a) **85**, 297 (1984)	16 (p. 273); 2; 22.1.2
S. V. MISYUL	(b) **121**, 481 (1984)	1.2 (p. 15) and 6 (p. 68); 22.8
A. V. MITIN	(a) **86**, 389 (1984)	14.2 (p. 218)
V. V. MITIN	(a) **84**, 669 (1984)	14.3.3 (p. 245) and 14.3.4 (p. 252); 22.2.1
N. CH. MITRA	(a) **84**, 493 (1984)	14.4 (p. 257); 22.6
V. YA. MITROFANOV	(b) **121**, 547 (1984)	18.3 (p. 298); 13.4; 22.8.2
R. MITROI	(a) **82**, 575 (1984)	16 (p. 270); 2; 22.9
A. I. MITSEK	(b) **121**, 185 (1984)	18.2 (p. 287) and 18.4 (p. 300)
T. MIURA	(a) **85**, 615 (1984)	14.4 (p. 258); 22.9
C. MIYAKE	(a) **83**, 567 (1984)	18.1 (p. 285) and 19 (p. 309); 18.4; 22.6.1
	(a) **83**, 729 (1984)	18.1 (p. 285) and 19 (p. 310); 18.4; 22.6.1
E. MIZERA	(a) **85**, 83 (1984)	1.4 (p. 31); 10.2; 22.4.2
	(a) **86**, K125 (1984)	14.3.3 (p. 247); 22.1.2
H. MIZUBAYASHI	(a) **85**, 121 (1984)	10.1 (p. 101) and 12.1 (p. 137); 21
A. R. MKRTCHYAN	(b) **125**, K67 (1984)	5 (p. 66)
Y. MLIK	(b) **125**, 87 (1984)	6 (p. 76) and 20.1 (p. 335); 22.9
K. MOCHIZUKI	(a) **85**, 249 (1984)	16 (p. 273) and 20.3 (p. 355); 22.4.4
S. MOCHIZUKI	(b) **126**, 105 (1984)	20.1 (p. 339); 6; 13.5; 22.6
M. MOCKER	(b) **123**, 247 (1984)	13.1 (p. 158) and 14.3 (p. 224); 20.3; 22.8
	(b) **123**, 755 (1984)	13.1 (p. 159) and 14.3.4 (p. 251); 22
A. MODY	(b) **121**, K33 (1984)	18.2 (p. 288)
	(b) **121**, K135 (1984)	18 (p. 278)
L. MOERNER	(a) **84**, 509 (1984)	10.1 (p. 100); 21
A. MOGUŠ-MILANKOVIĆ	(a) **86**, 413 (1984)	14.4 (p. 259); 22.9
M. T. MOHAMMAD	(a) **83**, 681 (1984)	20.1 (p. 331); 22.6
W. MÖHLING	(a) **83**, 195 (1984)	1.4 (p. 29) and 10.2 (p. 111); 22.1.2
J. MOLENDA	(b) **122**, 591 (1984)	13.1 (p. 156); 1.3; 21.1; 21.2
F. MOLLOT	(b) **123**, 99 (1984)	18.4 (p. 301) and 20.1 (p. 327); 13.5; 22.8
E. MOLVA	(b) **125**, 795 (1984)	13.4 (p. 190) and 20.3 (p. 356); 13.1; 22.4.3
F. MOMO	(a) **81**, K27 (1984)	2 (p. 40) and 19 (p. 306)
Y. MOMOSE	(a) **82**, K83 (1984)	14.4 (p. 256) and 17 (p. 274); 21; 22.9
M. MONDAL	(b) **122**, K93 (1984)	14.3 (p. 223); 22
	(b) **123**, K143 (1984)	13.1 (p. 160) and 18 (p. 280); 22
	(b) **126**, K41 (1984)	13 (p. 151) and 18 (p. 283); 22.2.1
	(b) **126**, K47 (1984)	13.1 (p. 164); 22
J. MONECKE	(b) **121**, 329 (1984)	13.1 (p. 154) and 14.3 (p. 220); 22
M. MONTAGNA	(b) **123**, 659 (1984)	13.4 (p. 181); 20.3; 22.6
A. MONTENERO	(a) **81**, K27 (1984)	2 (p. 40) and 19 (p. 306)
A. M. MONTI	(b) **121**, 513 (1984)	10.1 (p. 97); 9; 21.3
	(b) **126**, 163 (1984)	10.1 (p. 102); 21
D. B. DEMOOIJ	(a) **84**, 207 (1984)	18.2 (p. 293) and 18.4 (p. 302); 21; 21.4
	(a) **86**, 655 (1984)	1 (p. 10) and 18.2 (p. 295); 5; 22.8
P. C. MORAIS	(a) **85**, K65 (1984)	7 (p. 83) and 18 (p. 281); 19; 22.8.2
	(a) **86**, 163 (1984)	7 (p. 83); 19; 22.8.2
H. MORAWIEC	(a) **83**, K97 (1984)	1.2 (p. 17) and 10.1 (p. 99); 21; 21.1
M. MORENO	(b) **124**, K83 (1984)	10.2 (p. 113) and 13.4 (p. 184); 20.3; 22.5
P. MORETTI	(b) **124**, 297 (1984)	13.4 (p. 183); 22.5.2
D. V. MORGAN	(a) **83**, K87 (1984)	14.3.3 (p. 243); 11; 22.2.1
	(a) **86**, 439 (1984)	14.3.2 (p. 239) and 14.3.4 (p. 253)
G. J. MORGAN	(b) **124**, 279 (1984)	13 (p. 148); 2
J. F. MORHANGE	(b) **123**, 611 (1984)	2 (p. 46) and 14.4 (p. 257); 22.1.2
R. MORINEAU	(b) **122**, 175 (1984)	13.4 (p. 175) and 13.5.3 (p. 207); 14.3; 19; 22.6.1
TH. MORITZ	(b) **121**, K175 (1984)	13.1 (p. 155); 20.1; 22

N. V. Moseev	(a) **85**, 435 (1984)	11 (p. 130); 2; 21.1
V. A. Moskalenko	(b) **125**, 543 (1984)	18 (p. 282)
Yu. N. Moskvich	(a) **84**, K109 (1984)	1.1 (p. 12) and 14.4.1 (p. 262); 19; 22.8.1
C. Moţoc	(a) **82**, 575 (1984)	16 (p. 270); 2; 22.9
S. Motogi	(a) **81**, 519 (1984)	18.2 (p. 289); 12.2; 22.1.2; 22.6
M. A. de Moura	(b) **126**, 213 (1984)	18.4 (p. 303)
P. van Mourik	(a) **81**, 209 (1984)	10.1 (p. 97); 21
G. Mourou	(a) **83**, K203 (1984)	14.2 (p. 218)
P. Mrafko	(a) **83**, 113 (1984)	2 (p. 44); 1.4; 21.1
P. Mrozek	(a) **84**, 39 (1984)	1.3 (p. 24) and 1.6 (p. 37); 21; 21.1
A. Mrwa	(a) **82**, K111 (1984)	14.3.3 (p. 242) and 16 (p. 270); 22.9
E. N. S. Muccillo	(b) **122**, K127 (1984)	10.2 (p. 109) and 11 (p. 125); 14.4; 22.5.2
R. Muccilo	(b) **122**, K127 (1984)	10.2 (p. 109) and 11 (p. 125); 14.4; 22.5.2
A. V. Mudryi	(a) **81**, 313 (1984)	10.2 (p. 106) and 13.4 (p. 173); 20.3; 22.1.2
	(b) **125**, K75 (1984)	10.2 (p. 116) and 13.4 (p. 188); 22.1; 22.1.2
A. Mueller	(a) **84**, K97 (1984)	1.3 (p. 25) and 2 (p. 49); 22.7
E. Mugeński	(b) **125**, 381 (1984)	20.3 (p. 354); 10.2; 22.5.2
A. J. Muhammad	(a) **86**, 439 (1984)	14.3.2 (p. 239) and 14.3.4 (p. 253)
E. Mühle	(b) **121**, 623 (1984)	13.4 (p. 174); 21.4
	(b) **124**, 355 (1984)	13.4 (p. 184); 19; 22.5
Mukesch Jain	see under letter "J"	
E. Kh. Mukhamedzhanov	(a) **81**, 427 (1984)	1.6 (p. 36) and 17 (p. 274); 22.1.1
	(a) **83**, K5 (1984)	1.6 (p. 37); 17; 22.1.2
K. Mukherjee	(a) **81**, 439 (1984)	1.3 (p. 22) and 1.4 (p. 27)
	(a) **83**, 657 (1984)	14.1 (p. 212); 2; 8; 21; 21.6
M. L. Mukherjee	(a) **81**, 533 (1984)	13.4 (p. 174) and 20.3 (p. 346); 22.5.2
A. A. Mukhin	(a) **84**, 215 (1984)	18.4 (p. 302); 22.8.2
L. N. Mulay	(a) **86**, 663 (1984)	5 (p. 66); 18.1; 20.1; 22.8.1
B. G. Mulimani	(b) **124**, K181 (1984)	15 (p. 267); 2; 21
A. Müller	(a) **84**, 11 (1984)	2 (p. 47); 1.3; 22.7
B. Müller	(a) **83**, 499 (1984)	12.2 (p. 141); 22.6
G. O. Müller	(a) **81**, 597 (1984)	14.3.4 (p. 248) and 20.3 (p. 346); 22.4.1
	(a) **81**, 609 (1984)	14.3.4 (p. 249) and 20.3 (p. 346); 22.4.1
	(a) **81**, 723 (1984)	14.3.4 (p. 249) and 20.3 (p. 347); 14.3.1; 22.4.1
	(a) **81**, 733 (1984)	14.3.4 (p. 249) and 20.3 (p. 347); 22.4.2
	(b) **122**, 183 (1984)	16 (p. 269) and 20.3 (p. 348); 22.4.1
H. Müller	(b) **122**, 377 (1984)	14 (p. 210); 21.4
H. Müller	(b) **125**, K93 (1984)	4 (p. 62); 10.1; 13.4; 21.1
H.-J. Müller	(b) **124**, 747 (1984)	13.4 (p. 186) and 13.5.1 (p. 202); 20.3; 22.2.4
M. Müller	(a) **86**, 735 (1984)	14.3.1 (p. 237); 1.2; 22.9
W. Müller	(a) **83**, 437 (1984)	2 (p. 46) and 14.3.1 (p. 235); 3; 22.1.2
U. Müller-Jahreis	(a) **84**, 269 (1984)	17 (p. 275); 13.3; 22.1.2
A. I. Muminov	(b) **124**, 355 (1984)	13.4 (p. 184); 19; 22.5
I. Muminov	(a) **84**, 113 (1984)	6 (p. 74) and 20.1 (p. 333); 22.8
G. Munschy	(b) **126**, 329 (1984)	13.4 (p. 192) and 13.5.1 (p. 204); 22.2; 22.4
F. M. Muntyanu	(b) **124**, K87 (1984)	14.3 (p. 226); 18; 22.7
P. Munz	(a) **86**, 813 (1984)	13.5.2 (p. 206) and 20 (p. 319); 21.6; 22.5.2
L. A. Muradyan	(b) **121**, 481 (1984)	1.2 (p. 15) and 6 (p. 68); 22.8
V. V. Mürk	(b) **121**, K59 (1984)	13.5.1 (p. 199) and 20.3 (p. 345); 20.1; 22.6
V. G. Krishna Murthy	(a) **85**, K27 (1984)	12.2 (p. 141) and 20 (p. 318); 22.5.4
S. Ramana Murthy	(a) **81**, K137 (1984)	10.2 (p. 106) and 12.2 (p. 139); 22.8.2
	(a) **84**, 655 (1984)	14.3 (p. 228); 15; 22.8.2
V. Musilová	(a) **84**, 693 (1984)	14.3.3 (p. 245); 22.2.1
F. A. Mustafaev	(a) **82**, K75 (1984)	20.3 (p. 349); 22.8
N. B. Mustafaev	(b) **124**, K93 (1984)	15 (p. 266); 13.1; 21.7
	(b) **124**, K151 (1984)	13.1 (p. 162) and 20.1 (p. 334); 21.7
J. Muto	(a) **84**, K29 (1984)	10.2 (p. 114) and 20 (p. 317); 22.9

W. Mutscheller	(b) **122**, 73 (1984)	6 (p. 69) and 8 (p. 85); 22.8.1
G. F. Myachina	(a) **81**, 377 (1984)	14.3 (p. 221); 22.9
J. Mycielski	(b) **122**, 761 (1984)	13.1 (p. 157); 16; 22
	(b) **125**, 341 (1984)	13.4 (p. 187); 20.1; 22
A. M. Mysik	(a) **83**, 701 (1984)	14.3.3 (p. 243); 13.4; 22.2.1
S. N. Mysovskii	(b) **125**, K41 (1984)	10.2 (p. 116) and 13.4 (p. 188); 20.3; 22.5.2
A. V. Nabok	(a) **82**, 221 (1984)	13.4 (p. 177) and 14.3.4 (p. 250); 14.4; 22
G. Nachtwei	(a) **83**, K207 (1984)	14.3 (p. 226); 13.1; 22.2.3
F. Ya. Nad	(a) **81**, 367 (1984)	14 (p. 210); 22
E. M. Nadgornyi	(a) **83**, 179 (1984)	10.2 (p. 111); 22.6
G. S. Nadkarni	(a) **83**, 335 (1984)	14.3.1 (p. 235) and 14.3.4 (p. 250); 22.6.1
M. M. Nadoliisky	(a) **86**, K109 (1984)	14.4 (p. 259); 22.8.1
B. R. Nag	(a) **86**, 773 (1984)	14.3.4 (p. 254); 6; 13.1; 22.1.2
M. Nagabhooshanam	(a) **81**, K195 (1984)	14.3.1 (p. 234); 2; 22.2.3
V. M. Nagiev	(b) **125**, 75 (1984)	6 (p. 76) and 20.1 (p. 335); 22.8.1
	(b) **125**, 467 (1984)	6 (p. 77) and 20.1 (p. 337); 22.8.1
V. Nagirnyi	(b) **125**, 139 (1984)	10.2 (p. 115) and 20.3 (p. 354); 22.5.2
	(b) **126**, 407 (1984)	13.4 (p. 193) and 20.3 (p. 357); 22.5.4
	(b) **126**, 653 (1984)	13.4 (p. 194) and 20.3 (p. 358); 22.5.2
	(b) **126**, K63 (1984)	13.4 (p. 193) and 20.3 (p. 357); 22.5.2
D. L. Nagy	(b) **124**, 767 (1984)	13.4 (p. 186); 5; 22.8.1
Kyun Nahm	see under letter "K"	
K. Naito	(a) **82**, K57 (1984)	5 (p. 64) and 18 (p. 279); 22.6
I. Nakada	(b) **121**, 407 (1984)	13.4 (p. 173) and 14.3 (p. 220); 22
T. Nakamura	(b) **122**, 141 (1984)	10.2 (p. 107) and 18.4 (p. 300)
Y. Nakamura	(a) **81**, K109 (1984)	1.3 (p. 22); 5; 18.2; 21.1; 21.1.1; 21.6
	(a) **83**, K219 (1984)	1.3 (p. 22); 5; 18.2; 21.1; 21.1.1; 21.6
H. Nakanishi	(b) **124**, K101 (1984)	14.3 (p. 226); 13.1; 22.8
	(b) **126**, K145 (1984)	13.1 (p. 165) and 20.1 (p. 341); 12.2; 22.8.2
K. S. V. Nambi	(a) **82**, K71 (1984)	14.4.1 (p. 261) and 20.3 (p. 348); 22.8.2
	(a) **83**, K175 (1984)	17 (p. 275) and 20.3 (p. 352); 11; 13.4; 19; 22.8.1
L. Nánai	(a) **81**, K67 (1984)	14.3 (p. 221); 22.6
	(a) **84**, 639 (1984)	14.3.4 (p. 252)
N. M. Nancheva	(a) **81**, K141 (1984)	10.2 (p. 107); 20.1; 22.5.2
J. Narain	(b) **122**, 435 (1984)	4 (p. 59); 22.5.2
	(b) **123**, 497 (1984)	4 (p. 60); 6; 12.2; 22.6
	(b) **126**, 83 (1984)	4 (p. 62) and 12.2 (p. 142); 23
K. Narasimha Reddy	see under letter "R"	
F. S. Nasredinov	(a) **83**, 285 (1984)	13.4 (p. 180) and 14.3.1 (p. 235); 2; 20.1; 22.1.2
I. Natkaniec	(b) **123**, K115 (1984)	1.2 (p. 16) and 6 (p. 73); 22.9
J. Naudts	(b) **123**, 209 (1984)	13.1 (p. 158); 21.4
V. Naundorf	(a) **86**, 603 (1984)	9 (p. 94); 10.1; 21.1
M. Nawrocki	(b) **123**, 99 (1984)	18.4 (p. 301) and 20.1 (p. 327); 13.5; 22.8
W. Nazarewicz	(b) **122**, K23 (1984)	10.2 (p. 108) and 20.1 (p. 323); 6; 22.4.1
I. V. Nazarkin	(b) **122**, 349 (1984)	2 (p. 41) and 14.3 (p. 222); 14.3.4; 22.1.1
S. Z. Nazarova	(a) **83**, 11 (1984)	14.3 (p. 220) and 18 (p. 278); 13.1; 14.2; 15; 22
	(a) **86**, K11 (1984)	1.2 (p. 20) and 8 (p. 89); 6; 22
D. K. Neb	(b) **126**, K15 (1984)	4 (p. 62) and 6 (p. 79); 8; 23
E. Nebauer	(a) **84**, K39 (1984)	9 (p. 93); 14.3.2; 22.2.1
	(a) **85**, K169 (1984)	14.3.4 (p. 253); 9; 11; 22.2.1
D. D. Nedeoglo	(b) **125**, 869 (1984)	13.4 (p. 190) and 14.3 (p. 229); 22.4.2
Neeta Agnihotri	see under letter "A"	
B. B. S. Negi	(b) **123**, 491 (1984)	10.2 (p. 112); 6; 20.1; 22.5.2
A. Ya. Neiman	(a) **83**, 153 (1984)	10.2 (p. 111) and 14.3 (p. 225); 22.8.1
U. Neitzel	(b) **123**, 541 (1984)	8 (p. 87) and 18 (p. 280); 21.1

H. Nelkowski	(b) **122**, 285 (1984)	13.4 (p. 175) and 19 (p. 307); 22.4.2
E. Nembach	(b) **123**, K21 (1984)	10 (p. 95)
V. V. Nemoshkalenko	(b) **121**, 233 (1984)	13.1 (p. 153); 21
	(b) **123**, 787 (1984)	13.1 (p. 160); 21.6
C. Neogy	(b) **122**, 113 (1984)	10.2 (p. 107); 22.5.2
S. A. Nepijko	(a) **85**, 45 (1984)	1.5 (p. 34) and 13.3 (p. 169); 21.7
A. I. Nepomnyachikh	(b) **122**, 249 (1984)	10.2 (p. 107) and 13.4 (p. 175); 20.1; 20.3; 22.5.2
I. M. Nepomnyashi	(b) **126**, K53 (1984)	13.3 (p. 170) and 13.5.1 (p. 204)
D. Nesheva	(a) **82**, 243 (1984)	13.1 (p. 156) and 20.1 (p. 324); 2; 14.3.1; 22.4.1
F. G. Neshov	(a) **83**, 573 (1984)	11 (p. 127); 22.1.2
S. Nešpůrek	(a) **85**, 619 (1984)	14.3.4 (p. 252); 13.4; 22.1.2; 22.4.2
B. A. Nesterenko	(a) **82**, 221 (1984)	13.4 (p. 177) and 14.3.4 (p. 250); 14.4; 22
K. Skeff Neto	see under letter "S"	
H. Neumann	(b) **121**, 641 (1984)	13.1 (p. 155); 17; 20; 22.8
	(b) **121**, K55 (1984)	20.1 (p. 321); 13.1; 22.8
	(b) **123**, K119 (1984)	6 (p. 74) and 20.1 (p. 330); 13.1; 22.8
	(a) **84**, 87 (1984)	6 (p. 74) and 8 (p. 87); 22.8
	(b) **125**, K17 (1984)	6 (p. 77) and 20.1 (p. 336); 22.3
R. Neumann	(b) **123**, 641 (1984)	13.1 (p. 159); 21
M. Nevřiva	(a) **85**, 173 (1984)	18.2.1 (p. 296); 22.8.2
T. Ngoc	(a) **86**, K45 (1984)	20.3 (p. 358); 22.8.1
Hoang Ngoc Cam	see under letter "H"	
N. Nguyen	(a) **83**, 473 (1984)	1.1 (p. 12) and 1.4 (p. 30); 22.8.2
Nguyen Ai Viet	(b) **126**, 247 (1984)	13.5 (p. 198)
Nguyen Dai Hung	(a) **84**, K159 (1984)	20.1 (p. 335) and 20.3 (p. 354); 13.4; 22.8.1
Nguyen Hong Shon	(b) **122**, 361 (1984)	14.3 (p. 222); 18; 22
	(b) **126**, K105 (1984)	6 (p. 79) and 16 (p. 273); 7; 22.2.1
Hoang Xuan Nguyen	(b) **124**, 191 (1984)	13.5.1 (p. 201) and 20 (p. 317); 22.4.1
Nguyen Manh Son	(a) **84**, K159 (1984)	20.1 (p. 335) and 20.3 (p. 354); 13.4; 22.8.1
Nguyen Quoc Anh	(b) **121**, K97 (1984)	16 (p. 269); 22
	(b) **121**, K209 (1984)	7 (p. 81) and 14.3 (p. 221); 22
Nguyen Van Hieu	(b) **125**, 245 (1984)	13.5 (p. 198) and 20 (p. 318); 7
	(b) **126**, 247 (1984)	13.5 (p. 198)
J. E. Nicholls	(b) **121**, 339 (1984)	13.4 (p. 172) and 19 (p. 305); 22.4.3
	(a) **85**, 235 (1984)	14.3.1 (p. 236) and 20.3 (p. 355); 13.4; 22.4.2
	(a) **86**, 821 (1984)	13.4 (p. 195) and 14.3 (p. 232); 20.1; 22.4.3
G. Nicoară	(a) **84**, K139 (1984)	18.3 (p. 299) and 20.1 (p. 335); 22.8.2
M.-A. Nicolet	(a) **81**, 123 (1984)	1.3 (p. 33) and 3 (p. 55); 21
	(a) **86**, 471 (1984)	1.5 (p. 35); 2; 21
Al. Nicula	(a) **81**, K1 (1984)	1.4 (p. 27) and 19 (p. 305); 22.8.1
	(a) **83**, K15 (1984)	6 (p. 72) and 20.1 (p. 329); 2; 22.6.1
	(a) **84**, K17 (1984)	6 (p. 75) and 20.1 (p. 333); 2; 22.6.1
R. Niedzielski	(b) **123**, K29 (1984)	18.1 (p. 284); 21
B. Nieuwenhuijse	(b) **122**, 317 (1984)	13.4 (p. 176) and 19 (p. 307); 22.8.1
G. M. Niftiev	(b) **121**, K195 (1984)	20.3 (p. 345); 13.4; 22.3
	(a) **81**, K175 (1984)	14.3 (p. 222); 22.3.1
	(a) **83**, K61 (1984)	13.4 (p. 181); 16; 20.3; 22.3
	(a) **84**, K59 (1984)	5 (p. 65) and 18 (p. 281); 22.3
A. N. Nigam	(b) **125**, 207 (1984)	13.2 (p. 166); 12.1; 15; 21.1
R. R. Nigmatullin	(b) **123**, 739 (1984)	14.4 (p. 257)
	(b) **124**, 389 (1984)	14.4 (p. 257)
G. Nihoul	(b) **81**, 87 (1984)	1.1 (p. 11) and 1.4 (p. 27); 22.6
V. I. Nikitenko	(a) **84**, 443 (1984)	10.2 (p. 115) and 13.4 (p. 186); 14.4; 22.1.2
M. Nimis	(b) **123**, 519 (1984)	10.2 (p. 112); 20.1; 20.3; 22.5.2
V. N. Nishanov	(b) **126**, K53 (1984)	13.3 (p. 170) and 13.5.1 (p. 204)
T. Nishikawa	(a) **83**, 85 (1984)	1.5 (p. 33) and 17 (p. 275); 20.3; 22.5.2; 22.5.3
N. Nishimaki	(b) **122**, 269 (1984)	10.2 (p. 107) and 13.4 (p. 175); 20.1; 22.5.2

C. I. Nistor	(a) **86**. 525 (1984)	5 (p. 66); 2; 21.1.1
V. V. Nitsovich	(b) **125**, 815 (1984)	13.4 (p. 190) and 13.5.1 (p. 203); 13.3; 20.1; 22.4.1
G. P. Nizhnikova	(b) **121**, 241 (1984)	13.1 (p. 153); 21.4
T. Nobusada	(a) **82**, 135 (1984)	10.2 (p. 108); 13.4; 22.1.1
G. Nocke	(a) **83**, K43 (1984)	18.2 (p. 291); 21.1.1
M. Noguchi	(a) **82**, K83 (1984)	14.4 (p. 256) and 17 (p. 274); 21; 22.9
A. Noguera	(a) **82**, 553 (1984)	13.4 (p. 179) and 14.3 (p. 224); 8; 22.8
J. P. Noguier	(b) **125**, 357 (1984)	13.4 (p. 188) and 20.3 (p. 354); 22.1.2
A. Nohara	(a) **82**, 163 (1984)	10.1 (p. 98); 21.1
	(a) **84**, 501 (1984)	10.1 (p. 100); 12.1; 21.1
D. Nohavica	(a) **81**, K91 (1984)	14.3.3 (p. 241); 22.2
W. Nolting	(b) **123**, 155 (1984)	13 (p. 146) and 18 (p. 279)
H. M. van Noort	(a) **86**, 655 (1984)	1 (p. 10) and 18.2 (p. 295); 5; 22.8
V. Noreika	(b) **122**, K179 (1984)	20.1 (p. 326); 13.4; 22
D. O. Northwood	(a) **83**, 227 (1984)	10.1 (p. 99) and 12.1 (p. 136); 21
	(a) **84**, 509 (1984)	10.1 (p. 100); 21
	(a) **85**, 149 (1984)	10.1 (p. 101) and 12.1 (p. 137); 21.3
A. Nöthe	(b) **125**, 653 (1984)	13.5.1 (p. 203) and 20.1 (p. 337); 13.1; 22.4.3
P. Novák	(a) **85**, 173 (1984)	18.2.1 (p. 296); 22.8.2
R. Novák	(a) **85**, 173 (1984)	18.2.1 (p. 296); 22.8.2
V. A. Novikov	(a) **83**, K35 (1984)	10.2 (p. 112) and 11 (p. 126); 14.3; 22.8
M. Nowak	(a) **82**, 249 (1984)	16 (p. 269); 20.1; 22.4.4
I. W. Nowell	(a) **81**, K97 (1984)	1.1 (p. 11); 14.3; 22.9
	(a) **82**, 301 (1984)	1.1 (p. 11) and 14.3 (p. 223); 22.9
	(a) **83**, 39 (1984)	1.1 (p. 11); 14.3; 18.1; 22.9
Nquy Huu Chi	(a) **84**, K159 (1984)	20.1 (p. 335) and 20.3 (p. 354); 13.4; 22.8.1
S. B. Nurmagambetov	(b) **126**, 565 (1984)	11 (p. 133); 13; 22.1.2
P. V. Nutsubidze	(a) **82**, K53 (1984)	18 (p. 279); 22.8
M. Nylén	(b) **122**, 301 (1984)	13.1 (p. 156) and 13.3 (p. 167); 22.6

M. A. Ochando	(a) **83**, K93 (1984)	1.4 (p. 30) and 10.1 (p. 99); 21.1.1
H. L. Oczkowski	(a) **82**, 213 (1984)	13.4 (p. 176) and 20.3 (p. 348); 22.4.2
G. S. Oehrlein	(a) **85**, K109 (1984)	10.2 (p. 118); 22.1.2
G. Oelgart	(a) **81**, 541 (1984)	13.5.1 (p. 199) and 20.3 (p. 346); 22.2.4
	(a) **82**, 319 (1984)	14.3.3 (p. 242); 14.3.4; 22.2.4
	(a) **83**, 393 (1984)	13.4 (p. 180) and 14.3.3 (p. 243); 9; 22.2.4
	(a) **85**, 205 (1984)	11 (p. 129); 13; 22.2.4
	(a) **86**, 833 (1984)	14.3.3 (p. 247) and 20.3 (p. 358); 22.2.2
H. Oesterreicher	(a) **85**, K61 (1984)	18.2 (p. 294); 21.1
K. Oesterreicher	(a) **85**, K61 (1984)	18.2 (p. 294); 21.1
F. Ogiba	(a) **83**, K187 (1984)	14.1 (p. 212); 21.1
T. Ohkawa	(a) **85**, 335 (1984)	1.1 (p. 13)
Y. Ohno	(b) **121**, 407 (1984)	13.4 (p. 173) and 14.3 (p. 220); 22
K. Ohshima	(a) **81**, 129 (1984)	1.3 (p. 22); 21.1
H. Ohsugi	(b) **124**, 525 (1984)	6 (p. 75); 8; 21.2
S. Ohtaki	(a) **86**, 237 (1984)	10.1 (p. 103); 14.1; 21
Y. Ohtani	(a) **86**, 237 (1984)	10.1 (p. 103); 14.1; 21
H. Ohya-Nishiguchi	(a) **83**, 729 (1984)	18.1 (p. 285) and 19 (p. 310); 18.4; 22.6.1
V. K. Ojha	(b) **121**, 225 (1984)	13.1 (p. 153); 21.1
S. Okabe	(a) **83**, 85 (1984)	1.5 (p. 33) and 17 (p. 275); 20.3; 22.5.2; 22.5.3
Y. Okamoto	(a) **81**, 177 (1984)	1.2 (p. 14); 6; 8; 22.5.4
M. M. El-Oker	see under letter "E"	
Ok Hee Chung	see under letter "C"	
B. L. Oksengendler	(a) **81**, K145 (1984)	11 (p. 123) and 14.3.4 (p. 249); 22.1.2

S. Okuda	(a) **85**, 121 (1984)	10.1 (p. 101) and 12.1 (p. 137); 21
M. Okusawa	(b) **124**, 673 (1984)	13 (p. 149) and 17 (p. 276); 22.5
B. Olejníková	(a) **84**, 621 (1984)	13.4 (p. 187) and 14.3.4 (p. 252); 22.2.1
N. M. Olekhnovich	(a) **82**, 365 (1984)	10.2 (p. 110); 22.1.2
L. Oleari	(b) **124**, 707 (1984)	13.4 (p. 185) and 20.1 (p. 334); 22.8.1
M. C. Oliveros	(b) **124**, 457 (1984)	1.6 (p. 37); 1.2
B. Z. Olshanetskii	(a) **84**, K13 (1984)	1.6 (p. 37) and 2 (p. 48); 22.2.1; 22.2.4
A. Olszyna	(a) **86**, 453 (1984)	14.3.3 (p. 246); 22.1.2
M. S. Omar	(b) **123**, K119 (1984)	6 (p. 74) and 20.1 (p. 330); 13.1; 22.8
Om Prakash	see under letter "P"	
C. K. Ong	(a) **85**, 199 (1984)	20.1 (p. 337); 22.1.2
A. I. Onipko	(b) **123**, 37 (1984)	2 (p. 44) and 10 (p. 95)
N. A. Onishchenko	(b) **125**, 705 (1984)	13.5.1 (p. 203); 20.1; 22.3.1; 22.4.1
V. A. Onishchuk	(b) **123**, 183 (1984)	14.2 (p. 217); 5; 22.6.1
Z. Onyskiewicz	(b) **122**, 543 (1984)	18.1 (p. 284) and 18.4 (p. 301); 22.5
D. V. Oparin	(b) **121**, K75 (1984)	19 (p. 305); 22
Ch. Opitz	(b) **125**, K93 (1984)	4 (p. 62); 10.1; 13.4; 21.1
H. Opyrchał	(b) **121**, 117 (1984)	10.2 (p. 105) and 20.1 (p. 320); 22.5.2
P. Orlandi	(a) **82**, K129 (1984)	1.2 (p. 16) and 13.4 (p. 179); 5; 22.8
V. M. Orlov	(a) **84**, 113 (1984)	6 (p. 74) and 20.1 (p. 333); 22.8
B. A. Orłowski	(b) **126**, 285 (1984)	13.1 (p. 164); 22.5
C. Ortiz	(b) **123**, 79 (1984)	10.2 (p. 110) and 20.3 (p. 349); 22.5.2
S. Orzeszko	(a) **81**, 579 (1984)	2 (p. 41) and 14.3.1 (p. 233); 19; 22.1
K. Osamura	(a) **83**, 219 (1984)	2 (p. 44) and 10.1 (p. 99); 21.1
Z. Osetek	(b) **124**, K121 (1984)	18.2 (p. 293); 21.1
V. I. Osinskii	(a) **82**, 399 (1984)	1.4 (p. 28) and 1.6 (p. 37); 22.2.1
V. E. Oskin	(a) **86**, 363 (1984)	11 (p. 132) and 14.4 (p. 259); 22.9
M. Oskooie-Tabrizi	(a) **86**, 663 (1984)	5 (p. 66); 18.1; 20.1; 22.8.1
Yu. A. Ossipyan	(a) **84**, 149 (1984)	10.2 (p. 113) and 13.4 (p. 184); 22.1.2
W. von der Osten	(b) **122**, 203 (1984)	13.5.1 (p. 199); 6; 20.1; 22.5.1
	(b) **125**, 293 (1984)	13.5.1 (p. 202) and 20.3 (p. 354); 22.5.1
L. N. Oster	(a) **85**, K145 (1984)	17 (p. 277); 10.2; 22.8.1
J. Oswald	(a) **81**, 343 (1984)	13.4 (p. 173); 14.3; 20.3; 22.8
L. C. Otero-Diaz	(a) **82**, 379 (1984)	1.3 (p. 23) and 10.2 (p. 110); 19; 22.6.1
K. Ots	(a) **83**, K109 (1984)	1.5 (p. 33) and 2 (p. 47); 20.1; 22.1.2; 22.6
J. Otto	(b) **121**, 81 (1984)	6 (p. 67) and 20.1 (p. 320); 22.5.2
K. L. Ovanesyan	(a) **82**, K185 (1984)	20.2 (p. 343) and 20.3 (p. 349); 20.1; 22.8
	(a) **83**, K159 (1984)	20.2 (p. 343) and 20.3 (p. 352); 20.1; 22.8.1
S. G. Ovchinnikov	(b) **123**, 105 (1984)	13.5.4 (p. 208); 18.2; 22
K. Oyama Gannon	(b) **124**, 261 (1984)	18 (p. 280) and 20.1 (p. 332); 22.5.2
C. Oytana	(a) **83**, 235 (1984)	10.1 (p. 99); 12.1; 21; 21.1.1
R. P. Ozerov	(b) **122**, 599 (1984)	4 (p. 59); 20.1; 22; 22.5.2; 22.6; 22.8.1
G. Paasch	(b) **122**, K99 (1984)	13.3 (p. 167); 22
B. D. Padalia	(b) **121**, K65 (1984)	20 (p. 316); 21; 21.4
	(b) **122**, K59 (1984)	20 (p. 316); 22.8.2
	(b) **126**, 279 (1984)	20 (p. 318); 21.4
M. Pagannone	(b) **123**, K13 (1984)	6 (p. 72) and 20.1 (p. 327); 13.5; 22.8.1
J. Paitz	(a) **84**, 547 (1984)	20.1 (p. 334); 13.4; 22.8.2
	(a) **86**, 401 (1984)	10.2 (p. 119) and 14.4 (p. 259); 22.8.2
A. Paja	(b) **125**, K171 (1984)	14.1 (p. 214); 2; 21.1.1
P. Pajas	(b) **121**, 293 (1984)	2 (p. 39) and 20.1 (p. 320); 13.1; 22
L. Pajasová	(b) **121**, 293 (1984)	2 (p. 39) and 20.1 (p. 320); 13.1; 22
H-R. Pak	(b) **123**, 415 (1984)	1.2 (p. 16); 8; 12
S. Pak	(b) **126**, 565 (1984)	11 (p. 133); 13; 22.1.2
S. Kumar Pal	(a) **86**, 43 (1984)	2 (p. 53); 4; 22.5.2

SOBHEN KUMAR PAL	(a) **83**,	429 (1984)	2 (p. 46) and 4 (p. 60); 22.5.2
S. PALCHAUDHURI	(b) **125**,	K33 (1984)	2 (p. 50) and 8 (p. 88); 21.2
	(b) **126**,	443 (1984)	2 (p. 54) and 9 (p. 94); 21; 21.1
T. PALEWSKI	(a) **84**,	K47 (1984)	18.2 (p. 293); 21.5
B. R. PAMPLIN	(b) **123**,	K119 (1984)	6 (p. 74) and 20.1 (p. 330); 13.1; 22.8
E. T-S. PAN	(a) **86**,	471 (1984)	1.5 (p. 35); 2; 21
T. J. PANEK	(a) **83**,	K97 (1984)	1.2 (p. 17) and 10.1 (p. 99); 21; 21.1
V. E. PANIN	(b) **121**,	265 (1984)	13 (p. 144); 21; 22
D. PANKNIN	(a) **82**,	171 (1984)	11 (p. 124); 14.3; 22.1.2
T. PANNAPARAYIL	(a) **86**,	663 (1984)	5 (p. 66); 18.1; 20.1; 22.8.1
YA. I. PANOVA	(a) **84**,	645 (1984)	14.3 (p. 228); 22.8.2
	(a) **86**,	283 (1984)	18.3 (p. 299); 22.8.2
V. A. PANTELEEV	(a) **81**,	K15 (1984)	2 (p. 40) and 9 (p. 91); 22.1.2
C. PAPATRIANTAFILLOU	(b) **122**,	527 (1984)	18.2 (p. 290) and 18.4 (p. 301); 21.4
G. V. PAPAYAN	(b) **122**,	549 (1984)	20.1 (p. 325); 13.5.1; 22.9
K. M. PARASKEVOPOULOS	(b) **124**,	229 (1984)	13.5.2 (p. 205) and 20.1 (p. 332); 22.8
M. PARDAVI-HORVÁTH	(a) **84**,	547 (1984)	20.1 (p. 334); 13.4; 22.8.2
R. PAREJA	(a) **84**,	179 (1984)	10.2 (p. 113) and 11 (p. 128); 20.1; 22.8.1
I. A. PARFIANOVICH	(b) **121**,	401 (1984)	13.4 (p. 172) and 20.1 (p. 320); 10.2; 22.5.2
	(b) **122**,	K183 (1984)	13.4 (p. 178); 10.2; 22.5.2
	(b) **125**,	K41 (1984)	10.2 (p. 116) and 13.4 (p. 188); 20.3; 22.5.2
A. PARISINI	(b) **125**,	K7 (1984)	6 (p. 77) and 20.1 (p. 336); 22.8
H. PARK	(a) **81**,	K51 (1984)	2 (p. 40) and 16 (p. 269); 22.1
S. PARKE	(b) **122**,	757 (1984)	13.4 (p. 178) and 19 (p. 308); 22.8.1
T. J. PARKER	(b) **122**,	461 (1984)	6 (p. 70) and 20.1 (p. 325); 22.2.1
J. PARŠELIŪNAS	(a) **85**,	K179 (1984)	14.3.3 (p. 246) and 14.3.4 (p. 253); 22.2.1
G. PARTHASARATHY	(a) **85**,	K165 (1984)	2 (p. 52) and 14.3 (p. 230); 22.1.3
A. V. PARYGIN	(a) **84**,	K101 (1984)	1.2 (p. 18); 19; 21.7
M. V. PASECHNIK	(b) **121**,	623 (1984)	13.4 (p. 174); 21.4
L. PASEMANN	(a) **84**,	133 (1984)	1.4 (p. 30) and 10.2 (p. 113); 14.3.3; 22.1.2
	(a) **85**,	641 (1984)	20.3 (p. 356); 10.2; 14.3; 22
M. PASEMANN	(a) **81**,	273 (1984)	1.4 (p. 27) and 10.2 (p. 106); 5; 22.1.2
	(a) **81**,	751 (1984)	11 (p. 123); 1.4; 22.1.2
E. M. PASHAEV	(a) **84**,	73 (1984)	1 (p. 9) and 17 (p. 275); 22.1.2
	(a) **86**,	K1 (1984)	1.6 (p. 38) and 10.2 (p. 120); 22.2.1; 22.2.2
N. PASHOV	(a) **84**,	K23 (1984)	1.4 (p. 31) and 11 (p. 128); 2; 22.1.1
A. PASTUREL	(b) **125**,	101 (1984)	4 (p. 61) and 13.1 (p. 162); 21.1; 21.4
W. PASTUSIAK	(b) **122**,	K195 (1984)	19 (p. 308); 22.5.3
	(b) **126**,	K77 (1984)	13.4 (p. 193) and 19 (p. 314); 22.5.3
F. PASZTI	(a) **85**,	K35 (1984)	10.2 (p. 117) and 11 (p. 130); 22.1.2
S. K. PATAPIS	(a) **81**,	K57 (1984)	15 (p. 264); 22.6
C. PATEL	(b) **122**,	461 (1984)	6 (p. 70) and 20.1 (p. 325); 22.2.1
A. P. PATHAK	(b) **122**,	171 (1984)	11 (p. 124) and 13 (p. 145); 22.1.2
M. S. PATHAK	(b) **124**,	489 (1984)	6 (p. 75) and 13.5.2 (p. 206); 7; 22.2.3
I. B. PATRINA	(b) **123**,	353 (1984)	13.4 (p. 180) and 19 (p. 309); 22.6; 22.8.1
G. L. PAUL	(a) **81**,	K127 (1984)	2 (p. 41) and 11 (p. 123); 8; 22.1.2
J. PAVEL	(a) **82**,	47 (1984)	2 (p. 42) and 10 (p. 95)
M. PAVEL	(a) **81**,	K117 (1984)	10.2 (p. 106); 14.4.1; 22.9
L. I. PAVLOV	(a) **84**,	555 (1984)	13.4 (p. 186); 20.2; 22.4.1
S. T. PAVLOV	(b) **122**,	581 (1984)	6 (p. 70) and 20.1 (p. 325); 22
V. A. PAVLOV	(a) **85**,	11 (1984)	10.1 (p. 97); 9; 21
	(a) **85**,	315 (1984)	10.1 (p. 97); 11; 21
A. A. PAVLYUK	(a) **83**,	K65 (1984)	20.2 (p. 343) and 20.3 (p. 350); 13.4; 22.8
S. H. PAWAR	(a) **82**,	K195 (1984)	14.3.1 (p. 234); 20.1; 22.8
A. PAWEŁEK	(a) **85**,	K117 (1984)	10.1 (p. 102) and 12.1 (p. 137); 21
	(a) **86**,	K27 (1984)	10.1 (p. 103) and 12.1 (p. 138)
Z. PAWŁOWSKA	(b) **121**,	695 (1984)	13.1 (p. 155); 21.1

A. Paz	(a) **82**, 85 (1984)	10.2 (p. 108); 22.5.2
S. J. Pearton	(a) **82**, K101 (1984)	10.2 (p. 109) and 14.3 (p. 223); 22.1.1
V. I. Pecherskaya	(a) **81**, 507 (1984)	18.1 (p. 284) and 18.2 (p. 289); 14.1; 21.1; 21.1.1
M. A. Pedrosa	(a) **84**, 179 (1984)	10.2 (p. 113) and 11 (p. 128); 20.1; 22.8.1
V. S. Pekar	(a) **81**, 661 (1984)	14.3.4 (p. 249) and 20.3 (p. 346); 22.4.1
	(a) **81**, 709 (1984)	20.3 (p. 347); 22
J. Pełka	(b) **123**, 627 (1984)	13.1 (p. 159); 20; 21.1
A. Percheron-Guegan	(b) **125**, 101 (1984)	4 (p. 61) and 13.1 (p. 162); 21.1; 21.4
R. Pérez	(a) **83**, 445 (1984)	1.1 (p. 11); 21.1
	(a) **85**, 113 (1984)	1.4 (p. 31); 10.1
M. M. A. Perik	(b) **122**, 317 (1984)	13.4 (p. 176) and 19 (p. 307); 22.8.1
Yu. E. Perlin	(a) **85**, 553 (1984)	20.1 (p. 338) and 20.3 (p. 356); 13.4; 22.8
P. Petelenz	(b) **125**, 765 (1984)	13.4 (p. 189); 20.1; 22.9
J. W. Petersen	(a) **81**, 239 (1984)	10.2 (p. 105); 5; 22.1.2
A. Petraitiene	(b) **122**, K179 (1984)	20.1 (p. 326); 13.4; 22
V. F. Petrenko	(b) **121**, 421 (1984)	14.3.4 (p. 248) and 14.4 (p. 255); 22.6
	(a) **86**, K31 (1984)	10.2 (p. 120) and 14.4 (p. 259); 12.2; 22
A. G. Petrosyan	(a) **82**, K185 (1984)	20.2 (p. 343) and 20.3 (p. 349); 20.1; 22.8
	(a) **83**, K159 (1984)	20.2 (p. 343) and 20.3 (p. 352); 20.1; 22.8.1
A. K. Petrosyan	(b) **122**, 725 (1984)	13.4 (p. 177); 19; 20.1; 22.8.1
H.-D. Pfannes	(b) **124**, 767 (1984)	13.4 (p. 186); 5; 22.8.1
H. K. Phan	(a) **81**, K81 (1984)	14.3.3 (p. 241); 14.3.4; 22.1.2
J. C. Phillips	(b) **122**, K11 (1984)	6 (p. 70) and 20.1 (p. 323); 2; 22
L. H. Phu	(a) **81**, K81 (1984)	14.3.3 (p. 241); 14.3.4; 22.1.2
Phuong An	(b) **122**, 331 (1984)	13.4 (p. 176) and 20.1 (p. 323); 22.4.2
R. Pickenhain	(a) **85**, 627 (1984)	14.3.3 (p. 245) and 20.3 (p. 356); 13.4; 22.2.2
	(a) **86**, 833 (1984)	14.3.3 (p. 247) and 20.3 (p. 358); 22.2.2
U. Pietsch	(a) **82**, K137 (1984)	8 (p. 86); 1.1; 22.2.4
	(b) **126**, 595 (1984)	4 (p. 63); 13; 22.1.2
G. I. Pilipenko	(b) **121**, K75 (1984)	19 (p. 305); 22
O. Pilla	(b) **123**, 659 (1984)	13.4 (p. 181); 20.3; 22.6
R. G. Pillay	(b) **121**, K141 (1984)	18.2 (p. 288); 22.7
F. Pille	(b) **125**, 483 (1984)	14.4.1 (p. 262); 19; 22.9
A. N. Pilyankevich	(a) **82**, 449 (1984)	1.4 (p. 29) and 10.2 (p. 110); 22.3
E. Pinčík	(a) **86**, 795 (1984)	13.3 (p. 171) and 14.3.3 (p. 246); 22.2.1
L. Pintschovius	(b) **126**, 121 (1984)	6 (p. 79); 22.1.2
R. Piotrzkowski	(b) **124**, 411 (1984)	13.4 (p. 184) and 16 (p. 271); 12.2; 22.2.3
P. Pipinys	(a) **83**, K69 (1984)	13.4 (p. 181) and 14.3.4 (p. 251); 22.6
E. Pippel	(a) **81**, K93 (1984)	1.4 (p. 28); 10.2; 22.8.1
J. Piqueras	(a) **82**, 111 (1984)	10.2 (p. 108) and 20.3 (p. 348); 20.1; 22.6
	(a) **83**, 645 (1984)	20.3 (p. 351); 10.2; 13.4; 22.6
A. S. T. Pires	(b) **122**, K175 (1984)	19 (p. 308); 22.9
E. Pirouz	(a) **82**, 425 (1984)	1.4 (p. 29) and 10.2 (p. 110); 22.1.2
J. M. Pirrung	(a) **83**, 437 (1984)	2 (p. 46) and 14.3.1 (p. 235); 3; 22.1.2
C. Pisani	(b) **122**, 211 (1984)	13.1 (p. 156); 13.2; 21.2; 21.3
	(b) **125**, 721 (1984)	2 (p. 51) and 13 (p. 150)
Yu. V. Pisarevskii	(a) **86**, 345 (1984)	14.4 (p. 258) and 20.2 (p. 344); 1.1; 12.2; 20.1; 20.3; 22.8.1
D. I. Piskunov	(a) **86**, 55 (1984)	1.4 (p. 32) and 10 (p. 96)
S. K. Plachkova	(b) **122**, K53 (1984)	13.1 (p. 156); 14.3; 15; 22
	(a) **83**, 349 (1984)	14.3 (p. 225) and 15 (p. 266); 8; 22.8
	(b) **125**, 659 (1984)	1.2 (p. 19) and 14.4.1 (p. 263); 13.1; 22.8
N. M. Plakida	(b) **124**, 45 (1984)	7 (p. 82) and 14.4.1 (p. 261); 22.8.1
R. Planel	(b) **123**, 99 (1984)	18.4 (p. 301) and 20.1 (p. 327); 13.5; 22.8
J. A. Plascak	(b) **124**, K35 (1984)	18.2 (p. 292)
A. E. Plaudius	(a) **84**, K153 (1984)	20.3 (p. 354); 2; 10.2; 13.4; 20.1; 22.6
F. Pleiter	(b) **121**, K141 (1984)	18.2 (p. 288); 22.7

V. G. Plekhanov	(a) **81**,	K23 (1984)	20.3 (p. 345); 2; 13.5.3; 22.6
P. Plenkiewicz	(b) **126**,	285 (1984)	13.1 (p. 164); 22.5
B. Ploss	(b) **123**,	K155 (1984)	13.1 (p. 160) and 20.1 (p. 330); 22.8
G. S. Plotnikov	(a) **85**,	273 (1984)	13.4 (p. 189); 19; 20.3; 22.1.1; 22.1.2
	(a) **85**,	K73 (1984)	13.3 (p. 169) and 20.3 (p. 355); 22.1.1; 22.6
V. N. Plotnikov	(a) **86**,	93 (1984)	1.5 (p. 35) and 2 (p. 53); 3; 4; 22.6
D. Ploumbidis	(b) **122**,	K45 (1984)	2 (p. 41) and 13 (p. 146); 19; 21.7; 22.1
	(b) **122**,	K49 (1984)	13 (p. 146) and 19 (p. 307); 21.1
	(b) **126**,	K101 (1984)	14.1 (p. 214) and 19 (p. 314); 21.1
R. Plumier	(b) **123**,	553 (1984)	18.2 (p. 291) and 18.4 (p. 301)
R. B. Pode	(b) **122**,	K79 (1984)	20.3 (p. 348); 20.1; 22.5.2
A. A. Podrezov	(a) **82**,	355 (1984)	1.3 (p. 23); 10.1; 21
A. D. Pogrebnyak	(a) **81**,	217 (1984)	11 (p. 122); 14.3; 22.1.2; 22.2.1
	(a) **83**,	K35 (1984)	10.2 (p. 112) and 11 (p. 126); 14.3; 22.8
	(a) **85**,	K31 (1984)	11 (p. 130) and 20 (p. 318); 22.1.2; 22.2.1
	(a) **86**,	191 (1984)	11 (p. 132); 22.1.2; 22.2.1
U. W. Pohl	(b) **125**,	773 (1984)	13.4 (p. 190); 20.1; 22.4.1
A. L. Poimanov	(a) **83**,	K215 (1984)	14.3.3 (p. 244); 22
G. P. Pokhil	(a) **85**,	K39 (1984)	10.2 (p. 117) and 11 (p. 130); 22.1.2
K. Polgár	(b) **125**,	K13 (1984)	6 (p. 77) and 20.1 (p. 336); 22.8.1
V. Z. Polinger	(b) **125**,	401 (1984)	6 (p. 76) and 13.4 (p. 188); 10.2; 22.5.2
M. Połomska	(a) **82**,	K33 (1984)	1.2 (p. 15) and 14.4.1 (p. 260); 22.8.1
V. M. Polyanovskii	(b) **122**,	K115 (1984)	6 (p. 70) and 7 (p. 81); 19; 22
W. Pompe	(b) **122**,	35 (1984)	2 (p. 41); 10.2; 12.2
H.-E. Ponath	(b) **123**,	691 (1984)	13.5 (p. 197) and 20.1 (p. 330); 6; 22
E. G. Ponyatovskii	(a) **83**,	461 (1984)	1.2 (p. 17); 21.4
I. Pop	(b) **123**,	K53 (1984)	13.1 (p. 159) and 18.3 (p. 298); 21.1; 21.4
M. Pope	(b) **125**,	573 (1984)	13 (p. 150) and 13.5.1 (p. 202); 22.9
M. Popescu	(a) **82**,	K125 (1984)	1.2 (p. 16); 14.4.1; 22.8.1
A. Popitsch	(a) **86**,	749 (1984)	14.1 (p. 215); 1.2; 13.1; 18; 20.1; 22.6.1
A. S. Popov	(a) **81**,	669 (1984)	13.4 (p. 174) and 14.3.3 (p. 242); 20.3; 22.2.2
D. E. Popov	(b) **125**,	251 (1984)	11 (p. 129) and 20 (p. 318); 13; 22.1.3; 22.5.2
D. N. Popov	(a) **81**,	K141 (1984)	10.2 (p. 107); 20.1; 22.5.2
V. A. Popov	(b) **121**,	561 (1984)	18.3 (p. 298) and 18.4 (p. 300)
V. P. Popov	(a) **85**,	K39 (1984)	10.2 (p. 117) and 11 (p. 130); 22.1.2
M. N. Popova	(b) **124**,	K69 (1984)	13.4 (p. 184) and 20.3 (p. 352); 22.8
N. S. Popovich	(a) **85**,	K85 (1984)	14.3 (p. 229) and 16 (p. 273); 22.8
S. Porowski	(a) **86**,	301 (1984)	13.4 (p. 193); 14.3; 22.2.3
V. N. Porozkov	(a) **81**,	163 (1984)	3 (p. 55); 1.6
H. Port	(b) **124**,	741 (1984)	10.2 (p. 114) and 13.4 (p. 186); 20.3; 22.5.2
J. C. Portal	(b) **121**,	289 (1984)	13.1 (p. 154); 22.3
W. Possart	(a) **84**,	319 (1984)	14.3.2 (p. 239); 22.9
S. Yu. Potapenko	(b) **123**,	191 (1984)	13.1 (p. 158); 13.3; 14.3.1; 22
A. I. Potapov	(a) **82**,	K191 (1984)	11 (p. 125) and 14.3 (p. 224); 22.2.1; 22.8
U. Potzel	(b) **123**,	85 (1984)	9 (p. 92); 21
M. Pouchard	(a) **82**,	285 (1984)	14.3 (p. 223); 13.4; 15; 16; 22.8.1
E. D. Pozhidaev	(a) **85**,	591 (1984)	14.3 (p. 230); 11; 22.9
	(a) **86**,	709 (1984)	14.3.2 (p. 240); 11; 22.9
J. Pozo	(b) **121**,	K39 (1984)	8 (p. 84) and 18 (p. 278); 2
V. Prabhawalkar	(b) **121**,	K65 (1984)	20 (p. 316); 21; 21.4
R. B. Prabhu	(b) **124**,	K25 (1984)	13 (p. 148) and 18 (p. 280); 22.8
A. Pradel	(a) **82**,	295 (1984)	13.1 (p. 156) and 14.3 (p. 223); 8; 15; 20.1; 22.8
H. Pradella	(b) **122**,	285 (1984)	13.4 (p. 175) and 19 (p. 307); 22.4.2
J. Prakash	(b) **121**,	127 (1984)	10.2 (p. 105) and 20.1 (p. 320); 22.5.2
Om Prakash	(b) **122**,	K59 (1984)	20 (p. 316); 22.8.2
Chandra Prakash	(a) **84**,	535 (1984)	5 (p. 66); 22.8.2
S. Prakash	(b) **126**,	467 (1984)	8 (p. 90); 10.1; 21

PRAN KISHAN	see under letter "K"	
B. PRASAD	(b) **122**, 91 (1984)	8 (p. 85); 21.2
	(b) **123**, K9 (1984)	6 (p. 71); 4; 21.3
	(b) **124**, 497 (1984)	6 (p. 75) and 8 (p. 87); 21.2
C. PRASAD	(b) **125**, 629 (1984)	13.1 (p. 163); 22.2
K. PRASAD	(a) **84**, K1 (1984)	1.5 (p. 34); 22
B. N. PRASOLOV	(a) **83**, 131 (1984)	14.4.1 (p. 261); 10.2; 12.2; 22.8.1
H. PRESTING	(b) **126**, 11 (1984)	6 (p. 67) and 13.1 (p. 153); 22.1; 22.2; 22.4
M. T. PRIMATAROWA	(b) **121**, 357 (1984)	13.4 (p. 172) and 13.5.1 (p. 199)
	(b) **123**, 573 (1984)	13.5.4 (p. 208) and 18.2 (p. 291)
A. PRIEMUTH	(a) **82**, 561 (1984)	14.3 (p. 224) and 15 (p. 265); 22.7
U. PRÖHL	(a) **82**, 561 (1984)	14.3 (p. 224) and 15 (p. 265); 22.7
F. PROKERT	(b) **124**, 503 (1984)	1.2 (p. 17) and 6 (p. 75); 22.5
A. V. PROKHOROVICH	(a) **82**, 503 (1984)	13.4 (p. 178) and 20.3 (p. 349); 22.2.1
	(a) **83**, 625 (1984)	13.4 (p. 182) and 20.3 (p. 351); 11; 22.2.1
	(a) **84**, 567 (1984)	13.4 (p. 186) and 20.3 (p. 353); 11; 22.2.1
V. I. PROTASOV	(a) **85**, 105 (1984)	1.2 (p. 18); 10.1; 21.1
	(a) **85**, 435 (1984)	11 (p. 130); 2; 21.1
B. N. PROVOTOROV	(b) **126**, 481 (1984)	19 (p. 314)
Z. PRZEŁOŻNY	(b) **122**, 775 (1984)	14.2 (p. 216)
J. PULS	(b) **121**, K187 (1984)	20.1 (p. 322); 22.4.4
	(b) **121**, K191 (1984)	13.5.1 (p. 199) and 20.1 (p. 322); 22.4.1
J. PUŁTORAK	(a) **82**, 11 (1984)	14.3.3 (p. 241); 22.1.1
	(a) **85**, K93 (1984)	14.3.3 (p. 245); 22
K. PURANDAR	(b) **121**, K79 (1984)	13.4 (p. 173) and 20.1 (p. 321); 22.9
A. L. PUSHKARCHUK	(b) **124**, 699 (1984)	13.1 (p. 161) and 13.3 (p. 168); 22.6
	(b) **125**, K75 (1984)	10.2 (p. 116) and 13.4 (p. 188); 22.1; 22.1.2
KH. I. PUSHKAROV	(b) **123**, 573 (1984)	13.5.4 (p. 208) and 18.2 (p. 291)
L. PŮST	(b) **121**, 173 (1984)	19 (p. 305); 21.1; 21.1.1
	(b) **122**, 535 (1984)	19 (p. 307); 21
	(b) **123**, 115 (1984)	19 (p. 309); 18.2; 21
	(a) **85**, 179 (1984)	18.2.1 (p. 296); 19; 22.8.2
A. PUTZKA	(b) **124**, 767 (1984)	13.4 (p. 186); 5; 22.8.1
A. A. PUZANOV	(a) **83**, 573 (1984)	11 (p. 127); 22.1.2

DANG TRAN QUAN	see under letter "D"
NGUYEN QUOC ANH	see under letter "N"

S. I. RADAUTSAN	(a) **84**, K169 (1984)	14.3.3 (p. 245); 20.3; 22.2.2; 22.4.3
S. RADHAKRISHNA	(b) **124**, 481 (1984)	6 (p. 75) and 20.1 (p. 333); 22.5.4
S. RADHAKRISHNAN	(a) **83**, 335 (1984)	14.3.1 (p. 235) and 14.3.4 (p. 250); 22.6.1
J. B. RADUNOVIĆ	(b) **122**, 383 (1984)	13 (p. 145) and 14 (p. 210)
E. RADZHABOV	(b) **123**, K79 (1984)	13.4 (p. 180) and 20.3 (p. 350); 10.2; 22.5.2
J. RAFAŁOWICZ	(a) **81**, 185 (1984)	8 (p. 84); 21
R. RAI	(b) **124**, K167 (1984)	19 (p. 311); 22.8
A. K. RAJAURIA	(b) **122**, 435 (1984)	4 (p. 59); 22.5.2
S. RAJENDRAN	(a) **84**, 631 (1984)	14.3 (p. 227); 15; 22.6.1
K. S. RAJU	(a) **81**, K93 (1984)	1.4 (p. 28); 10.2; 22.8.1
V. K. RAJU	(a) **81**, K37 (1984)	14.3.4 (p. 248); 22.9
F. S. RAKHMENKULOV	(a) **83**, 147 (1984)	6 (p. 72) and 8 (p. 86); 22
	(a) **84**, K165 (1984)	14.1 (p. 213); 21
S. V. RAKITIN	(a) **85**, K31 (1984)	11 (p. 130) and 20 (p. 318); 22.1.2; 22.2.1
P. N. RAM	(b) **123**, 491 (1984)	10.2 (p. 112); 6; 20.1; 22.5.2
K. RAMACHANDRAN	(b) **125**, 513 (1984)	9 (p. 93); 22.4.3
A. RAMANAND	(b) **122**, 11 (1984)	8 (p. 84) and 12.1 (p. 135); 6; 21.4

S. RAMANA MURTHY	see under letter "M"	
K. V. RAMANATHAN	(a) **81**, 407 (1984)	14.3.3 (p. 241); 11; 14.4; 22.1.2
K. V. RAMANUJACHARI	(a) **86**, K15 (1984)	1.1 (p. 13) and 3 (p. 57); 22.6.1
M. G. RAMAZANZADE	(a) **83**, 323 (1984)	14.3.4 (p. 250); 16; 22.3
R. RAMÍREZ	(b) **124**, 573 (1984)	13.4 (p. 185) and 18 (p. 281)
D. RAMOUL	(b) **124**, 403 (1984)	14.3 (p. 226) and 15 (p. 266); 22
J. P. S. RANA	(b) **121**, 493 (1984)	4 (p. 58); 12.2; 22.5.2
	(b) **126**, 495 (1984)	4 (p. 63); 12.2; 22.5.2
V. V. RANDOSHKIN	(a) **84**, K135 (1984)	18.2 (p. 293); 22.8.2
A. RANFAGNI	(b) **124**, 297 (1984)	13.4 (p. 183); 22.5.2
A. RANGWALA	(b) **121**, K135 (1984)	18 (p. 278)
A. V. KRISHNA RAO	(b) **122**, 709 (1984)	10.2 (p. 109) and 20.1 (p. 326); 22.5.2
B. M. RAO	(a) **81**, 209 (1984)	10.1 (p. 97); 21
K. V. RAO	(b) **122**, 709 (1984)	10.2 (p. 109) and 20.1 (p. 326); 22.5.2
M. LAKSHMIPATHI RAO	(a) **82**, 229 (1984)	10.2 (p. 108) and 20.3 (p. 348); 22.5.2
P. MADHUSUDANA RAO	(a) **83**, K145 (1984)	20.3 (p. 351); 22.8
R. RAMJI RAO	(b) **122**, 11 (1984)	8 (p. 84) and 12.1 (p. 135); 6; 21.4
R. V. GOPALA RAO	(a) **83**, 429 (1984)	2 (p. 46) and 4 (p. 60); 22.5.2
	(b) **125**, K33 (1984)	2 (p. 50) and 8 (p. 88); 21.2
	(b) **126**, 443 (1984)	2 (p. 54) and 9 (p. 94); 21; 21.1
R. V. G. RAO	(a) **86**, 43 (1984)	2 (p. 53); 4; 22.5.2
S. R. RAO	(a) **81**, K195 (1984)	14.3.1 (p. 234); 2; 22.2.3
T. S. RAO	(a) **82**, K37 (1984)	12.2 (p. 140); 22.8.2
T. SESHAGIRI RAO	(a) **81**, K137 (1984)	10.2 (p. 106) and 12.2 (p. 139); 22.8.2
V. SITAKARA RAO	(a) **84**, 631 (1984)	14.3 (p. 227); 15; 22.6.1
U. V. SUBBA RAO	(a) **82**, 229 (1984)	10.2 (p. 108) and 20.3 (p. 348); 22.5.2
E. I. RASHBA	(b) **121**, K87 (1984)	13.4 (p. 173)
I. H. RASHED	(a) **85**, 429 (1984)	8 (p. 89); 22.5.2
V. I. RASHUPKIN	(a) **81**, K185 (1984)	14.2 (p. 216); 1.2; 21.6
E. M. RASPOPOVA	(a) **81**, 139 (1984)	2 (p. 39); 14.3; 15; 22.8
D. T. RASULOV	(a) **83**, K211 (1984)	14.3.3 (p. 244) and 16 (p. 271); 22.4.1; 22.4.3; 22.4.4
H. RATAJCZAK	(a) **82**, 491 (1984)	14.1 (p. 211) and 18.2.1 (p. 296); 21.4
R. P. S. RATHORE	(b) **121**, 73 (1984)	4 (p. 58) and 6 (p. 67); 12.1; 21
	(b) **122**, K119 (1984)	6 (p. 70); 21
	(b) **126**, 87 (1984)	6 (p. 78); 12.1; 21.3
I. G. RATISHVILI	(b) **125**, 533 (1984)	10.1 (p. 101); 21.1.1
V. V. RATNAM	(b) **121**, 393 (1984)	13.4 (p. 172) and 20.3 (p. 345); 10.2; 22.5.2
A. M. RATNER	(b) **124**, 249 (1984)	13.4 (p. 183); 13.5.1; 23
R. RAUCH	(b) **123**, 93 (1984)	10.2 (p. 110); 20.1; 20.3; 22.5.3
H. RAUH	(b) **125**, 131 (1984)	4 (p. 61); 1.1; 6; 20.1; 22.7
B. RAUSCHENBACH	(a) **85**, 473 (1984)	1.3 (p. 25) and 11 (p. 131); 21; 21.1.1
L. P. RAUTIAN	(b) **122**, 549 (1984)	20.1 (p. 325); 13.5.1; 22.9
B. RAVEAU	(a) **83**, 473 (1984)	1.1 (p. 12) and 1.4 (p. 30); 22.8.2
N. M. RAVINDRA	(b) **123**, 611 (1984)	2 (p. 46) and 14.4 (p. 257); 22.1.2
A. K. RAY	(b) **121**, K47 (1984)	20.1 (p. 320); 13.1; 23
B. RAY	(a) **84**, K75 (1984)	20.3 (p. 353); 22.4.1
D. RAY CHAUDHURI	see under letter "C"	
A. RAYMOND	(a) **83**, K199 (1984)	14.3 (p. 226); 22.4.1
S. M. RAZA	(a) **84**, K125 (1984)	10.1 (p. 100); 21.1
T. M. RAZYKOV	(a) **84**, K71 (1984)	13.1 (p. 161) and 14.3.3 (p. 244); 22.4.4
C. RAZZETTI	(b) **125**, K7 (1984)	6 (p. 77) and 20.1 (p. 336); 22.8
S. A. REBROV	(a) **84**, K169 (1984)	14.3.3 (p. 245); 20.3; 22.2.2; 22.4.3
K. NARASIMHA REDDY	(a) **82**, 229 (1984)	10.2 (p. 108) and 20.3 (p. 348); 22.5.2
K. V. REDDY	(a) **83**, K145 (1984)	20.3 (p. 351); 22.8
P. J. REDDY	(a) **81**, K37 (1984)	14.3.4 (p. 248); 22.9
P. V. REDDY	(a) **82**, K37 (1984)	12.2 (p. 140); 22.8.2
Y. P. REDDY	(b) **121**, K79 (1984)	13.4 (p. 173) and 20.1 (p. 321); 22.9

L. Redlich	(a) **81**,	715 (1984)	20.3 (p. 347); 22.4.1
A. R. Regel	(a) **83**,	285 (1984)	13.4 (p. 180) and 14.3.1 (p. 235); 2; 20.1; 22.1.2
P. Reiche	(a) **85**,	553 (1984)	20.1 (p. 338) and 20.3 (p. 356); 13.4; 22.8
K. Reimann	(b) **125**,	653 (1984)	13.5.1 (p. 203) and 20.1 (p. 337); 13.1; 22.4.3
H. Reimer	(b) **124**,	61 (1984)	6 (p. 74) and 20.1 (p. 331); 10.2; 22.5.2
T. L. Reinecke	(b) **122**,	K65 (1984)	13.3 (p. 167)
P. Reineker	(b) **124**,	657 (1984)	13.5.1 (p. 202); 22.9
	(b) **125**,	593 (1984)	13.5.1 (p. 203); 22.9
J. Reinhardt	(a) **81**,	323 (1984)	17 (p. 274); 22.6
U. Reinsperger	(a) **81**,	715 (1984)	20.3 (p. 347); 22.4.1
J. P. Remeika	(b) **122**,	K11 (1984)	6 (p. 70) and 20.1 (p. 323); 2; 22
A. Remón	(a) **82**,	111 (1984)	10.2 (p. 108) and 20.3 (p. 348); 20.1; 22.6
A. A. Rempel	(a) **84**,	527 (1984)	1.3 (p. 25); 18.1; 22
	(a) **86**,	K11 (1984)	1.2 (p. 20) and 8 (p. 89); 6; 22
P. Rennert	(b) **121**,	673 (1984)	13.1 (p. 155); 20; 21.3
P. J. Reuters	(a) **83**,	719 (1984)	14.3.3 (p. 243); 14.4; 22; 22.1.2
H. Reuther	(a) **83**,	173 (1984)	2 (p. 44) and 9 (p. 92); 22.6.1
F. Reynaud	(a) **84**,	393 (1984)	1.2 (p. 18) and 1.3 (p. 24); 21
I. S. Rez	(b) **122**,	599 (1984)	4 (p. 59); 20.1; 22; 22.5.2; 22.6; 22.8.1
B. Rheinländer	(a) **86**,	833 (1984)	14.3.3 (p. 247) and 20.3 (p. 358); 22.2.2
J. F. Rhodes	(b) **126**,	361 (1984)	13.4 (p. 192) and 20.3 (p. 357); 22
L. A. Amarante Ribeiro	see under letter „A"		
A. Ribeiro Filho	(b) **124**,	457 (1984)	1.6 (p. 37); 1.2
	(b) **125**,	45 (1984)	1.2 (p. 18) and 20.1 (p. 335); 1.5
M. Richardson	(a) **83**,	279 (1984)	4 (p. 59) and 13 (p. 147); 20; 21.1; 21.1.1; 21.4
W. Richert	(b) **121**,	609 (1984)	13.5 (p. 196); 22.1.1; 22.2.1
	(b) **121**,	633 (1984)	13 (p. 145); 13.5; 22
A. Richter	(b) **122**,	35 (1984)	2 (p. 41); 10.2; 12.2
H. Richter	(a) **84**,	143 (1984)	10.2 (p. 113); 14.3.3; 22.1.2
J. Richter	(b) **125**,	839 (1984)	1.3 (p. 25) and 14.1 (p. 214); 21
R. Richter	(b) **123**,	143 (1984)	2 (p. 44) and 15 (p. 265); 14.1; 18.2; 21
W. Richter	(b) **125**,	131 (1984)	4 (p. 61); 1.1; 6; 20.1; 22.7
M. Rick	(a) **83**,	K183 (1984)	14.3 (p. 226); 22
V. Riede	(b) **123**,	K47 (1984)	20.1 (p. 328); 13.1; 22.7
	(b) **123**,	K119 (1984)	6 (p. 74) and 20.1 (p. 330); 13.1; 22.8
	(b) **125**,	K17 (1984)	6 (p. 77) and 20.1 (p. 336); 22.3
M. Riedel	(a) **85**,	K175 (1984)	14.2 (p. 218)
B. Ries	(b) **124**,	363 (1984)	13.5.1 (p. 202) and 20.3 (p. 352); 2; 22.9
A. Rigamonti	(a) **82**,	K129 (1984)	1.2 (p. 16) and 13.4 (p. 179); 5; 22.8
A. Rimeika	(a) **83**,	K69 (1984)	13.4 (p. 181) and 14.3.4 (p. 251); 22.6
C. Rincón	(a) **81**,	K77 (1984)	13.4 (p. 173) and 14.3 (p. 221); 22.8
G. Riontino	(a) **84**,	371 (1984)	2 (p. 49); 21; 21.3
P. S. Riseborough	(b) **122**,	161 (1984)	13 (p. 145); 18.1; 21; 21.4
Lj. M. Ristovski	(b) **123**,	309 (1984)	13.4 (p. 179) and 13.5 (p. 196); 22.9
G. Ritter	(b) **124**,	767 (1984)	13.4 (p. 186); 5; 22.8.1
D. R. Rizdvyanetskii	(a) **82**,	K163 (1984)	1.2 (p. 16) and 18 (p. 279); 21; 21.1
A. Rizk	(a) **81**,	513 (1984)	18.2 (p. 289); 21.1.1
A. Rizzo	(a) **85**,	609 (1984)	16 (p. 273); 2; 22.1.2
Th. Roatsch	(a) **82**,	581 (1984)	14.2 (p. 217)
A. Röder	(a) **84**,	319 (1984)	14.3.2 (p. 239); 22.9
J. Rodriguez Fernandez	(a) **86**,	295 (1984)	18.4 (p. 303); 21.4
H. Roemer	(a) **84**,	K177 (1984)	14.4 (p. 258); 22.5
C. Roetti	(b) **122**,	211 (1984)	13.1 (p. 156); 13.2; 21.2; 21.3
E. Roick	(b) **124**,	335 (1984)	13.5.1 (p. 201) and 20.3 (p. 352); 23
M. Rolland	(a) **82**,	257 (1984)	10.2 (p. 109) and 14.3.1 (p. 234); 22.9
A. I. Romanenko	(a) **84**,	K165 (1984)	14.1 (p. 213); 21
A. E. Romanov	(b) **122**,	35 (1984)	2 (p. 41); 10.2; 12.2

E. P. Romanov	(a) **86**, K59 (1984)	14.1 (p. 215) and 14.2 (p. 219); 1.2; 21; 21.1
S. I. Romanov	(a) **82**, 475 (1984)	11 (p. 125); 14.3; 22.1.2
S. N. Romanov	(b) **122**, K17 (1984)	11 (p. 124)
	(b) **126**, 133 (1984)	11 (p. 131)
	(b) **126**, 517 (1984)	11 (p. 132); 10.1; 21.1; 21.6
I. D. Romanova	(a) **83**, 701 (1984)	14.3.3 (p. 243); 13.4; 22.2.1
J. Rosa	(a) **86**, 319 (1984)	13.4 (p. 194); 20.1; 20.3; 22.2
A. Röseler	(b) **123**, 65 (1984)	6 (p. 71) and 20.1 (p. 327); 22.6
	(b) **124**, K1 (1984)	6 (p. 74) and 20.1 (p. 332); 2; 22.6
J. Röseler	(b) **124**, 213 (1984)	13.5.1 (p. 201) and 20.1 (p. 332); 18; 22.4.2
J. Rosenzweig	(a) **83**, 357 (1984)	14.3 (p. 225); 15; 22
	(a) **83**, K183 (1984)	14.3 (p. 226); 22
	(a) **85**, K81 (1984)	7 (p. 83) and 16 (p. 273); 14.3; 22.4.3
M. Rösler	(b) **121**, 609 (1984)	13.5 (p. 196); 22.1.1; 22.2.1
	(b) **126**, 629 (1984)	17 (p. 277); 13.5.2; 21
H. Rossmann	(b) **121**, 685 (1984)	13.5.1 (p. 199) and 20.1 (p. 321); 22.4.1
	(b) **121**, K191 (1984)	13.5.1 (p. 199) and 20.1 (p. 322); 22.4.1
C. Roşu	(a) **82**, 575 (1984)	16 (p. 270); 2; 22.9
L. Roth	(a) **85**, K15 (1984)	1.5 (p. 34); 11; 22
M. Rothbauer	(a) **84**, 693 (1984)	14.3.3 (p. 245); 22.2.1
M. Rouzeyre	(a) **81**, 549 (1984)	13.4 (p. 174) and 20.3 (p. 346); 20.1; 22.4.4
C. L. Roy	(b) **121**, 717 (1984)	13.3 (p. 167)
D. Roy	(b) **122**, 449 (1984)	4 (p. 59); 12.2; 22.5.4
	(a) **82**, K11 (1984)	1.2 (p. 15) and 18 (p. 279); 22.8.2
G. Roy	(b) **121**, 717 (1984)	13.3 (p. 167)
J. B. Roy	(b) **121**, 743 (1984)	14.3.2 (p. 238) and 14.3.4 (p. 248); 22.1.2
	(a) **86**, K79 (1984)	14.3.3 (p. 246); 14.3.2; 22.2.4
S. K. Roy	(b) **123**, 45 (1984)	6 (p. 71); 5; 10.1; 21
M. Deb Roy	see under letter "D"	
O. V. Rozanov	(a) **84**, K109 (1984)	1.1 (p. 12) and 14.4.1 (p. 262); 19; 22.8.1
F. Rozpłoch	(a) **81**, 579 (1984)	2 (p. 41) and 14.3.1 (p. 233); 19; 22.1
E. I. Rozum	(b) **125**, 251 (1984)	11 (p. 129) and 20 (p. 318); 13; 22.1.3; 22.5.2
	(b) **126**, 565 (1984)	11 (p. 133); 13; 22.1.2
A. Rubaszek	(b) **124**, 159 (1984)	13 (p. 148); 13.2; 21
S. V. Rubin	(a) **83**, K81 (1984)	2 (p. 45) and 14.3 (p. 225); 14.3.4; 22.1.2
V. M. Rubinov	(a) **85**, K159 (1984)	14.3.1 (p. 236); 11; 22.2.2
V. V. Rubinov	(a) **81**, 11 (1984)	14.3.3 (p. 241) and 14.3.4 (p. 248); 16; 22.1.1; 22.1.2; 22.2.2
H. Rubio	(a) **85**, 511 (1984)	18.2.1 (p. 297); 2; 21.1.1
I. Rückmann	(b) **126**, K135 (1984)	13.5.1 (p. 204) and 20.1 (p. 341); 22.4.1; 22.4.2
Yu. V. Rud	(a) **82**, K191 (1984)	11 (p. 125) and 14.3 (p. 224); 22.2.1; 22.8
	(b) **125**, 759 (1984)	13.4 (p. 189) and 20.3 (p. 355); 22.4.4
F. G. Rüdenauer	(a) **81**, 247 (1984)	9 (p. 91); 11; 22.1.2
A. I. Rudenko	(a) **83**, 365 (1984)	11 (p. 126) and 14.3 (p. 225); 22.9
	(a) **84**, 327 (1984)	14.3 (p. 227); 11; 22.9
V. N. Rudko	(b) **126**, 703 (1984)	14.2 (p. 219)
G. Rudlof	(a) **85**, K149 (1984)	13.4 (p. 192) and 17 (p. 277); 22
V. S. Rukhlov	(b) **125**, K121 (1984)	13 (p. 150) and 18 (p. 282)
N. V. Rumak	(a) **86**, 93 (1984)	1.5 (p. 35) and 2 (p. 53); 3; 4; 22.6
	(a) **86**, 477 (1984)	1.5 (p. 35) and 2 (p. 54); 3; 22.6
R. Rünger	(b) **122**, K49 (1984)	13 (p. 146) and 19 (p. 307); 21.1
P. G. Rustamov	(a) **86**, K113 (1984)	14.3 (p. 232) and 15 (p. 268); 22.8
N. Rusu	(b) **123**, K53 (1984)	13.1 (p. 159) and 18.3 (p. 298); 21.1; 21.4
V. V. Ryabchenkov	(a) **85**, 553 (1984)	20.1 (p. 338) and 20.3 (p. 356); 13.4; 22.8
A. O. Rybaltovskii	(a) **85**, 167 (1984)	11 (p. 129) and 20.1 (p. 337); 10.2; 22.5
V. Rybka	(a) **83**, 165 (1984)	11 (p. 126); 22.1.2; 22.6
	(a) **86**, K117 (1984)	11 (p. 133) and 14.3 (p. 232); 22.2.1

R. M. Ryskulbekova	(b) **125**, 81 (1984)	2 (p. 50) and 4 (p. 61); 21.2; 21.7
I. A. Ryzhkin	(b) **121**, 421 (1984)	14.3.4 (p. 248) and 14.4 (p. 255); 22.6
	(a) **86**, K31 (1984)	10.2 (p. 120) and 14.4 (p. 259); 12.2; 22
	(a) **86**, K87 (1984)	7 (p. 83); 22
E. Rzepka	(b) **125**, K13 (1984)	6 (p. 77) and 20.1 (p. 336); 22.8.1
H. Rzewuski	(a) **86**, 263 (1984)	11 (p. 132) and 20.3 (p. 357); 10.2; 22.4.2

S. C. Sabharwal	(a) **86**, 327 (1984)	20.3 (p. 357); 11; 13.4; 22.5.2
R. Kh. Sabirov	(b) **125**, K117 (1984)	18 (p. 282) and 19 (p. 313)
B. A. Sabrah	(b) **123**, K1 (1984)	6 (p. 71) and 20.1 (p. 327); 22.8.2
F. C. Sá Barreto	(b) **121**, K151 (1984)	13 (p. 145) and 18.2 (p. 288)
A. El-Sadek	see under letter "E"	
F. M. Sadigov	(a) **86**, K113 (1984)	14.3 (p. 232) and 15 (p. 268); 22.8
R. Z. Sadykhov	(a) **84**, K51 (1984)	18 (p. 281); 22.8
E. K. Sadykov	(b) **123**, 703 (1984)	5 (p. 65) and 19 (p. 309); 18.1
M. Saeki	(b) **124**, K109 (1984)	6 (p. 75) and 20.1 (p. 334); 22
V. S. Saenko	(a) **83**, 365 (1984)	11 (p. 126) and 14.3 (p. 225); 22.9
	(a) **84**, 327 (1984)	14.3 (p. 227); 11; 22.9
	(a) **84**, 585 (1984)	11 (p. 129) and 14.3 (p. 227); 22.9
	(a) **85**, 591 (1984)	14.3 (p. 230); 11; 22.9
	(a) **86**, 363 (1984)	11 (p. 132) and 14.4 (p. 259); 22.9
	(a) **86**, 709 (1984)	14.3.2 (p. 240); 11; 22.9
V. L. Safonov	(b) **126**, 197 (1984)	13.5.4 (p. 209) and 18.4 (p. 303); 22.5
S. A. El-Sahhar	see under letter „E"	
S. N. Sahu	(b) **122**, 661 (1984)	13.1 (p. 157); 22.1; 22.2
	(b) **126**, 157 (1984)	10.2 (p. 118) and 13.4 (p. 192); 22.1.2
G. Said	(a) **84**, K67 (1984)	10.1 (p. 100) and 18.2 (p. 293); 21.1
A. G. Saif	(b) **121**, 749 (1984)	14.2 (p. 216)
S. Saito	(a) **82**, K169 (1984)	2 (p. 43) and 18.2 (p. 290); 21.1; 21.1.1
	(a) **83**, K195 (1984)	2 (p. 47) and 14.1 (p. 212); 21.1; 21.1.1
H. Saka	(a) **82**, 163 (1984)	10.1 (p. 98); 21.1
A. Sakalas	(a) **82**, K199 (1984)	14.3 (p. 224); 13.4; 22.4.3
D. Della Sala	see under letter "D"	
E. Yu. Salaev	(a) **82**, K75 (1984)	20.3 (p. 349); 22.8
	(b) **126**, K139 (1984)	13.1 (p. 165) and 20.1 (p. 341); 12.2; 22; 22.8
A. M. Saletskii	(a) **85**, K73 (1984)	13.3 (p. 169) and 20.3 (p. 355); 22.1.1; 22.6
O. P. Salita	(a) **86**, 177 (1984)	10.1 (p. 102); 21.1
A. V. Salker	(a) **84**, K105 (1984)	1.2 (p. 18); 8; 14.1; 14.3; 18.1; 22.6
E. A. Salkov	(a) **83**, K31 (1984)	10.2 (p. 112); 14.3.4; 16; 20.1; 22.1.1
J. Salm	(a) **84**, 379 (1984)	1.5 (p. 34) and 2 (p. 49); 22.6
Z. A. Salnik	(a) **84**, 237 (1984)	13.4 (p. 184); 22.1.2
	(a) **86**, 623 (1984)	13.4 (p. 195); 22.1.2
V. N. Salomatov	(b) **121**, 401 (1984)	13.4 (p. 172) and 20.1 (p. 320); 10.2; 22.5.2
	(b) **122**, K183 (1984)	13.4 (p. 178); 10.2; 22.5.2
	(b) **125**, K41 (1984)	10.2 (p. 116) and 13.4 (p. 188); 20.3; 22.5.2
K. Saminadayar	(b) **125**, 795 (1984)	13.4 (p. 190) and 20.3 (p. 356); 13.1; 22.4.3
T. Sanadze	(b) **125**, K63 (1984)	19 (p. 312); 22.5.3
C. Sanchez	(b) **122**, 175 (1984)	13.4 (p. 175) and 13.5.3 (p. 207); 14.3; 19; 22.6.1
R. Sanctuary	(a) **84**, 103 (1984)	14.4.1 (p. 262) and 20.1 (p. 333); 22.8
E. Sandow	(a) **85**, K169 (1984)	14.3.4 (p. 253); 9; 11; 22.2.1
Sang Don Choi	see under letter "C"	
N. Sankarraman	(a) **83**, 335 (1984)	14.3.1 (p. 235) and 14.3.4 (p. 250); 22.6.1
C. Santandrea	(b) **123**, 541 (1984)	8 (p. 87) and 18 (p. 280); 21.1
S. Santhanam	(a) **83**, K77 (1984)	14.3.1 (p. 235); 22.7
M. P. Dos Santos	see under letter "D"	

S. P. Sanyal	(b) **123**, 453 (1984)	4 (p. 60); 12.2; 22.5.4
	(b) **125**, 113 (1984)	4 (p. 61); 6; 12.2; 22; 22.5.3
	(a) **85**, 417 (1984)	1.2 (p. 19) and 12.2 (p. 142); 22.6
S. Sapru	(b) **125**, K161 (1984)	20.3 (p. 356); 11; 13.4; 22.5.2
R. S. Saraswat	(b) **122**, K169 (1984)	19 (p. 308); 18.1; 22.8.1
O. G. Sarbei	(b) **126**, 607 (1984)	13 (p. 152) and 20 (p. 319); 22.2.1
R. M. Sardarly	(b) **124**, K69 (1984)	13.4 (p. 184) and 20.3 (p. 352); 22.8
S. E. Sarkisov	(a) **82**, K185 (1984)	20.2 (p. 343) and 20.3 (p. 349); 20.1; 22.8
	(a) **83**, K165 (1984)	20.2 (p. 343) and 20.3 (p. 352); 20.1; 22.8.1
	(a) **84**, K81 (1984)	20.2 (p. 343) and 20.3 (p. 353); 13.4; 22.8
	(a) **85**, 553 (1984)	20.1 (p. 338) and 20.3 (p. 356); 13.4; 22.8
	(a) **86**, 345 (1984)	14.4 (p. 258) and 20.2 (p. 344); 1.1; 12.2; 20.1; 20.3; 22.8.1
J. Sárközi	(b) **122**, 501 (1984)	10.2 (p. 109); 22.5.2
E. F. Sarmento	(b) **121**, 197 (1984)	18.2 (p. 287); 13.3
A. K. Sarychev	(b) **121**, K215 (1984)	14.3 (p. 221); 22.3
A. Sasaki	(a) **85**, K105 (1984)	12.2 (p. 142); 22.9
Y. Sasaki	(a) **86**, 289 (1984)	18.1 (p. 286); 21.1; 21.1.1
A. P. Saščiuk	(a) **86**, K55 (1984)	14.3 (p. 231); 22.1.1; 22.2.3
A. M. Satanin	(b) **123**, 191 (1984)	13.1 (p. 158); 13.3; 14.3.1; 22
T. Satow	(a) **85**, K5 (1984)	1 (p. 10) and 2 (p. 50); 22.5.2
R. Satyanarayana	(a) **84**, 655 (1984)	14.3 (p. 228); 15; 22.8.2
W. Sauer	(a) **81**, 541 (1984)	13.5.1 (p. 199) and 20.3 (p. 346); 22.2.4
M. A. Savchenko	(b) **122**, 367 (1984)	14.2 (p. 216); 21.4
B. N. Savenko	(a) **83**, K117 (1984)	1.2 (p. 17); 22.8.1
I. M. Savić	(b) **123**, K35 (1984)	18.2 (p. 291)
	(b) **124**, K129 (1984)	18 (p. 281)
	(b) **125**, K51 (1984)	18.4 (p. 303)
	(b) **125**, K113 (1984)	13 (p. 150) and 18 (p. 282)
	(b) **126**, K37 (1984)	18.4 (p. 303)
M. M. Savin	(a) **81**, 415 (1984)	10.1 (p. 97); 21
E. J. Savino	(b) **121**, 513 (1984)	10.1 (p. 97); 9; 21.3
E. M. Savitskii	(b) **121**, 623 (1984)	13.4 (p. 174); 21.4
V. V. Savyak	(a) **86**, K63 (1984)	14.3 (p. 231) and 15 (p. 268); 22.1.2
H. C. Saxena	(b) **122**, K119 (1984)	6 (p. 70); 21
N. N. Saxena	(b) **126**, 669 (1984)	13 (p. 146) and 20 (p. 317); 22.6
R. B. Saxena	(b) **121**, 661 (1984)	13.5.2 (p. 205); 7; 22.2.3
V. K. Saxena	(b) **126**, K29 (1984)	18 (p. 282)
A. Scacco	(b) **122**, K83 (1984)	10.2 (p. 108) and 13.4 (p. 176); 20.1; 22.5.2
E. Schachinger	(b) **123**, 773 (1984)	14.2 (p. 217)
H. Schaeben	(b) **123**, 425 (1984)	1.3 (p. 23) and 1.4 (p. 29)
J. P. Schaffer	(a) **81**, K51 (1984)	2 (p. 40) and 16 (p. 269); 22.1
U. Schaller	(b) **122**, K87 (1984)	14.1 (p. 211) and 14.3 (p. 223); 12.2; 22.2.3
	(a) **85**, K183 (1984)	12.2 (p. 142) and 14.3 (p. 230); 22.2.3
W. Scharff	(a) **82**, K5 (1984)	2 (p. 42) and 3 (p. 55); 11; 22.1.2
	(a) **82**, K121 (1984)	1.5 (p. 33) and 3 (p. 56); 2; 11; 22.1.2
A. Scharmann	(a) **83**, 537 (1984)	10.2 (p. 112); 14.4; 22.8.1
	(a) **84**, 291 (1984)	16 (p. 272) and 20.1 (p. 333); 13.4; 22.8.1
F. Schauer	(a) **85**, 619 (1984)	14.3.4 (p. 252); 13.4; 22.1.2; 22.4.2
A. Scheifler	(a) **81**, 541 (1984)	13.5.1 (p. 199) and 20.3 (p. 346); 22.2.4
M. Schell	(a) **85**, 569 (1984)	13.4 (p. 191) and 14.3.1 (p. 236); 22.6
H. Schelle	(b) **121**, 673 (1984)	13.1 (p. 155); 20; 21.3
M. Schenk	(a) **84**, K121 (1984)	10.2 (p. 115); 14.3; 22
S. Scherf	(b) **125**, 177 (1984)	13.5 (p. 197) and 18.4 (p. 303)
U. Scherz	(b) **124**, 287 (1984)	6 (p. 74) and 20.1 (p. 332); 13.4; 22.2.2; 22.4.1
W. Schiller	(b) **126**, K91 (1984)	14.1 (p. 214) and 15 (p. 267); 21
M. Schlüter	(b) **126**, 511 (1984)	4 (p. 63) and 13 (p. 151); 12.2; 22.1; 22.1.1; 22.1.2

K. SCHMALZ	(a) **81**, K165 (1984)	10.2 (p. 107) and 13.4 (p. 175); 14.3; 22.1.2
	(a) **85**, 575 (1984)	13.4 (p. 191); 14.3; 20.1; 22.1.2
B. J. SCHMID	(b) **124**, 657 (1984)	13.5.1 (p. 202); 22.9
	(b) **125**, 593 (1984)	13.5.1 (p. 203); 22.9
D. SCHMID	(b) **124**, 741 (1984)	10.2 (p. 114) and 13.4 (p. 186); 20.3; 22.5.2
B. SCHMIDT	(a) **81**, 751 (1984)	11 (p. 123); 1.4; 22.1.2
	(a) **83**, 165 (1984)	11 (p. 126); 22.1.2; 22.6
E. SCHMIDT	(b) **122**, 675 (1984)	13.1 (p. 157) and 20.1 (p. 326); 22
	(a) **83**, 581 (1984)	20.1 (p. 330); 22.8.2
G. K. SCHMIDT	(a) **82**, 75 (1984)	10.1 (p. 98) and 12.1 (p. 135); 1.4; 21
	(a) **82**, 413 (1984)	10.1 (p. 98) and 12.1 (p. 135); 1.4; 21
M. SCHMIDT	(a) **85**, K19 (1984)	1.5 (p. 35) and 13.3 (p. 169); 17; 21.1.1
W. SCHMIDT	(b) **123**, 641 (1984)	13.1 (p. 159); 21
W. F. SCHMIDT	(b) **126**, K165 (1984)	14.3.4 (p. 253) and 20.1 (p. 342); 23
B. SCHNABEL	(a) **82**, K153 (1984)	9 (p. 92) and 19 (p. 308); 21.4
F. SCHNEIDER	(a) **85**, 455 (1984)	10.2 (p. 118) and 11 (p. 131); 5; 22.2.2
H. A. SCHNEIDER	(b) **123**, K161 (1984)	13.5.2 (p. 205); 22.2.2
	(b) **126**, K149 (1984)	13.1 (p. 165) and 13.5.2 (p. 206); 14.4; 22.8
CH. SCHNITTLER	(b) **122**, K99 (1984)	13.3 (p. 167); 22
M. SCHNÜRER	(a) **81**, 463 (1984)	1.6 (p. 36) and 14.3.2 (p. 238); 22.2.3
	(a) **86**, 427 (1984)	13.3 (p. 170) and 14.3.3 (p. 246); 22.2.3
G. SCHOLZ	(b) **123**, 641 (1984)	13.1 (p. 159); 21
A. SCHÖNHALS	(b) **124**, 515 (1984)	2 (p. 48); 12.2
J. SCHREIBER	(a) **86**, 229 (1984)	10.2 (p. 119); 22.2.1
W. SCHRITTENLACHER	(b) **122**, 285 (1984)	13.4 (p. 175) and 19 (p. 307); 22.4.2
	(b) **122**, 735 (1984)	13.4 (p. 178); 19; 22.4.2
K.-W. SCHRÖDER	(a) **81**, K131 (1984)	1.4 (p. 28) and 10.2 (p. 106); 14.3; 22.1.2
W. SCHROF	(b) **124**, 741 (1984)	10.2 (p. 114) and 13.4 (p. 186); 20.3; 22.5.2
W. SCHRÖTER	(a) **81**, 273 (1984)	1.4 (p. 27) and 10.2 (p. 106); 5; 22.1.2
G. SCHUBERT	(b) **125**, 839 (1984)	1.3 (p. 25) and 14.1 (p. 214); 21
R. SCHUCHARDT	(b) **122**, K151 (1984)	13.1 (p. 157); 22
E. M. SCHULSON	(a) **85**, 481 (1984)	10.1 (p. 102); 1.4; 21; 21.1
D. SCHULTZE	(a) **81**, K19 (1984)	20.2 (p. 343) and 20.3 (p. 345); 22.8.1
	(a) **85**, 553 (1984)	20.1 (p. 338) and 20.3 (p. 356); 13.4; 22.8
G. SCHULZ	(a) **81**, 723 (1984)	14.3.4 (p. 249); and 20.3 (p. 347); 14.3.1; 22.4.1
	(a) **81**, 733 (1984)	14.3.4 (p. 249) and 20.3 (p. 347); 22.4.2
J. SCHULZE	(a) **81**, 463 (1984)	1.6 (p. 36) and 14.3.2 (p. 238); 22.2.3
L. SCHUMANN	(b) **121**, 505 (1984)	6 (p. 68) and 20.1 (p. 321); 2; 22.6
C. SCHWAB	(b) **123**, K125 (1984)	11 (p. 126) and 19 (p. 309); 13.4; 22.2.1
R. SCHWABE	(b) **122**, 629 (1984)	13.5.1 (p. 200) and 20.3 (p. 349); 22.2.2
L. O. SCHWAN	(b) **124**, 741 (1984)	10.2 (p. 114) and 13.4 (p. 186); 20.3; 22.5.2
D. SCHWARZ	(a) **84**, K177 (1984)	14.4 (p. 258); 22.5
C. F. SCHWERDTFEGER	(b) **125**, 333 (1984)	13.4 (p. 187) and 19 (p. 312); 22.6.1
S. J. SCHWERDTFEGER	(b) **125**, 333 (1984)	13.4 (p. 187) and 19 (p. 312); 22.6.1
M. SCROCCO	(b) **125**, 713 (1984)	13.1 (p. 163) and 17 (p. 276); 21; 21.6
M. T. SEBASTIAN	(a) **84**, 401 (1984)	1.2 (p. 18); 10.2; 22.3
L. N. SEDLÁKOVÁ	(a) **83**, 455 (1984)	1.1 (p. 12); 22.1.2
N. N. SEDOV	(a) **85**, 45 (1984)	1.5 (p. 34) and 13.3 (p. 169); 21.7
V. D. SEDYKH	(a) **82**, K63 (1984)	2 (p. 42) and 18.2 (p. 289); 5; 21.1.1
D. SEGERS	(a) **81**, 209 (1984)	10.1 (p. 97); 21
H. K. SEHGAL	(a) **82**, K181 (1984)	20.1 (p. 326); 13.1; 22.8
M. SEIDEL	(a) **83**, K43 (1984)	18.2 (p. 291); 21.1.1
P. SEIDEL	(b) **122**, K105 (1984)	14.2 (p. 216); 21
	(a) **85**, K175 (1984)	14.2 (p. 218)
G. SEIFERT	(b) **121**, 705 (1984)	13 (p. 145); 21.6
M. SEIFERT	(a) **86**, 691 (1984)	1.4 (p. 32) and 13.4 (p. 195); 20.1; 22.2.1
W. SEIFERT	(a) **85**, 627 (1984)	14.3.3 (p. 245) and 20.3 (p. 356); 13.4; 22.2.2

M. Seilenthal	(a) **83**, K109 (1984)	1.5 (p. 33) and 2 (p. 47); 20.1; 22.1.2; 22.6
S. Sekimoto	(a) **84**, 55 (1984)	1.1 (p. 12); 22.8.1
G. Sekretarczyk	(b) **122**, 549 (1984)	20.1 (p. 325); 13.5.1; 22.9
B. Selle	(a) **81**, 715 (1984)	20.3 (p. 347); 22.4.1
G. K. Semin	(a) **84**, K101 (1984)	1.2 (p. 18); 19; 21.7
V. N. Semkin	(b) **122**, 549 (1984)	20.1 (p. 325); 13.5.1; 22.9
K. Sen	(a) **85**, 603 (1984)	14.3.4 (p. 252); 13.4; 22.1.2
P. K. Sen	(b) **124**, 117 (1984)	13.5.1 (p. 201) and 20.1 (p. 331); 22.2
J. P. Senateur	(a) **84**, 199 (1984)	18.2 (p. 292) and 18.4 (p. 302); 1.2; 14.1; 21.1
D. Senulienė	(a) **84**, 113 (1984)	6 (p. 74) and 20.1 (p. 333); 22.8
A. V. Serebryakov	(a) **82**, K63 (1984)	2 (p. 42) and 18.2 (p. 289); 5; 21.1.1
P. P. Seregin	(a) **83**, 285 (1984)	13.4 (p. 180) and 14.3.1 (p. 235); 2; 20.1; 22.1.2
A. V. Serikov	(b) **121**, 451 (1984)	2 (p. 40); 21; 21.1.1
J. Serna	(a) **83**, K131 (1984)	10.1 (p. 99); 21.4
T. Seshagiri Rao	see under letter "R"	
V. P. Seth	(b) **121**, K201 (1984)	19 (p. 306); 13.4; 22.8.1
J. P. Šetrajčić	(b) **122**, 125 (1984)	18.2 (p. 289)
	(b) **123**, 135 (1984)	13 (p. 146) and 18.4 (p. 301); 20.1
	(b) **124**, 235 (1984)	13.5.1 (p. 201); 20.1
	(b) **124**, K45 (1984)	13.5.1 (p. 202) and 13.5.4 (p. 208); 18.2
	(b) **125**, 743 (1984)	6 (p. 77) and 13.5.1 (p. 203)
	(b) **125**, K45 (1984)	18.2 (p. 294)
	(b) **125**, K151 (1984)	6 (p. 78) and 13.5.1 (p. 204); 22.9
	(b) **125**, K155 (1984)	6 (p. 78) and 13.5.1 (p. 204)
	(b) **126**, K175 (1984)	18.2 (p. 294)
S. B. Sevastianov	(a) **86**, 717 (1984)	11 (p. 133) and 14.3.3 (p. 246); 14.3.4; 22.1.2
A. L. Shabalov	(a) **83**, K11 (1984)	6 (p. 72) and 20.1 (p. 328); 22.6
M. B. Shah	(a) **86**, 275 (1984)	10.1 (p. 103) and 18.2 (p. 295); 21
S. Shah	(b) **125**, 211 (1984)	13 (p. 149); 20; 21.4
M. G. Shakhtakhtinskii	(b) **124**, K93 (1984)	15 (p. 266); 13.1; 21.7
	(b) **124**, K151 (1984)	13.1 (p. 162) and 20.1 (p. 334); 21.7
V. I. Shalaev	(a) **85**, 11 (1984)	10.1 (p. 97); 9; 21
	(a) **85**, 315 (1984)	10.1 (p. 97); 11; 21
J. Shanker	(b) **121**, 487 (1984)	12.2 (p. 139); 14.4; 22.5
	(b) **122**, 65 (1984)	12.2 (p. 139); 22.5.2
	(b) **122**, 435 (1984)	4 (p. 59); 22.5.2
	(b) **123**, 11 (1984)	4 (p. 58); 6; 12.2; 22.5.2
	(b) **123**, 71 (1984)	4 (p. 59); 6; 8; 22.5.2
	(b) **123**, 497 (1984)	4 (p. 60); 6; 12.2; 22.6
	(b) **125**, K21 (1984)	4 (p. 61); 12.2; 22.5.2
	(b) **126**, 77 (1984)	12.2 (p. 142); 22.5.2
	(b) **126**, 83 (1984)	4 (p. 62) and 12.2 (p. 142); 23
V. P. Shantarovich	(b) **123**, 183 (1984)	14.2 (p. 217); 5; 22.6.1
	(a) **83**, 159 (1984)	2 (p. 44) and 10.2 (p. 111); 22
A. A. El-Sharkawy	see under letter "E"	
A. K. Sharma	(a) **84**, 439 (1984)	7 (p. 82); 22.4.1
K. C. Sharma	(b) **122**, 423 (1984)	2 (p. 42); 22.1.1; 22.1.2
	(b) **125**, 505 (1984)	2 (p. 51) and 7 (p. 83); 21
	(b) **126**, 59 (1984)	2 (p. 52); 8; 22.5.2
R. V. Sharma	(a) **85**, 39 (1984)	2 (p. 50); 22.5
E. G. Sharoyan	(b) **122**, 725 (1984)	13.4 (p. 177); 19; 20.1; 22.8.1
	(a) **84**, 597 (1984)	2 (p. 49) and 14.3.1 (p. 236); 19; 20.1; 22.9
D. Shaw	(a) **86**, 629 (1984)	9 (p. 94); 10.2; 22.2.1
N. M. Shchagina	(a) **86**, 111 (1984)	1.2 (p. 20) and 14.4.1 (p. 263); 12.2; 22.8
L. A. Shebanov	(a) **83**, K127 (1984)	14.4.1 (p. 261); 22.8.1
L. N. Shehata	(b) **121**, 749 (1984)	14.2 (p. 216)
E. G. Sheikhet	(a) **81**, 433 (1984)	1.4 (p. 27); 10.2; 22.1.2

S. S. Sheinin	(a)	**83**,	445 (1984)	1.1 (p. 11); 21.1
M. K. Sheinkman	(b)	**123**.	295 (1984)	13.4 (p. 179) and 14.3 (p. 224); 22.2.1
R. M. Shelby	(b)	**123**,	79 (1984)	10.2 (p. 110) and 20.3 (p. 349); 22.5.2
A. U. Sheleg	(a)	**86**,	517 (1984)	1.2 (p. 20); 22.3
Y. Q. Sheng	(a)	**86**,	813 (1984)	13.5.2 (p. 206) and 20 (p. 319); 21.6; 22.5.2
M. El Sherif	(b)	**125**,	493 (1984)	6 (p. 77) and 20.1 (p. 337); 1.2; 14.4.1; 22.8.1
W. F. Sherman	(b)	**122**,	461 (1984)	6 (p. 70) and 20.1 (p. 325); 22.2.1
M. A. Sheromov	(a)	**86**,	117 (1984)	1.3 (p. 26); 18; 21.1; 21.1.1
V. A. Shershel	(a)	**84**,	263 (1984)	13.4 (p. 185); 14.3; 22.1.2
	(a)	**86**,	679 (1984)	13.4 (p. 195); 14.3; 19; 22.1.2
S. Shigetomi	(b)	**124**,	K49 (1984)	20.3 (p. 352); 13.5.1; 22.3
	(a)	**86**,	K69 (1984)	14.3 (p. 231); 13.4; 22.3
S. Shigetomi	(a)	**86**,	K69 (1984)	14.3 (p. 231); 13.4; 22.3
M. Sh. Shikhsaidov	(a)	**84**,	79 (1984)	10.2 (p. 113); 1.4; 22.2.1
M. Shimada	(a)	**83**,	561 (1984)	14.1 (p. 212) and 18.1 (p. 285); 1.3; 18.2; 22.8
	(a)	**85**,	523 (1984)	18.2 (p. 294); 14.1; 22.6.1
Y. Shimada	(a)	**83**,	255 (1984)	18.2.1 (p. 296); 2; 21.1
	(a)	**84**,	K55 (1984)	2 (p. 48) and 18.2.1 (p. 296); 21
S. Shimanuki	(b)	**122**,	269 (1984)	10.2 (p. 107) and 13.4 (p. 175); 20.1; 22.5.2
S. Shindo	(b)	**125**,	161 (1984)	11 (p. 129)
M. Shiojiri	(a)	**84**,	55 (1984)	1.1 (p. 12); 22.8.1
S. Shionoya	(b)	**125**,	735 (1984)	13.5.1 (p. 203) and 20.3 (p. 355); 22.4.2
G. O. Shironyan	(a)	**82**,	K185 (1984)	20.2 (p. 343) and 20.3 (p. 349); 20.1; 22.8
Yu. M. Shirshov	(a)	**82**,	221 (1984)	13.4 (p. 177) and 14.3.4 (p. 250); 14.4; 22
Shi-wei Gu	(b)	**125**,	323 (1984)	6 (p. 76) and 13.5.1 (p. 202)
A. P. Shkadarevich	(b)	**125**,	K41 (1984)	10.2 (p. 116) and 13.4 (p. 188); 20.3; 22.5.2
G. M. Shmelev	(b)	**121**,	K97 (1984)	16 (p. 269); 22
Nguyen Hong Shon	see under letter "N"			
	(b)	**121**,	K209 (1984)	7 (p. 81) and 14.3 (p. 221); 22
	(b)	**122**,	361 (1984)	14.3 (p. 222); 18; 22
	(b)	**126**,	K105 (1984)	6 (p. 79) and 16 (p. 273); 7; 22.2.1
K. N. Shrivastava	(b)	**125**,	11 (1984)	6 (p. 67) and 18.2 (p. 287); 13; 22.5
	(b)	**125**,	441 (1984)	6 (p. 67) and 18.2 (p. 287); 13; 22.5
S. B. Shrivastava	(b)	**121**,	225 (1984)	13.1 (p. 153); 21.1
A. A. Shtanov	(a)	**85**,	K85 (1984)	14.3 (p. 229) and 16 (p. 273); 22.8
Yu. M. Shulga	(b)	**121**,	595 (1984)	13.1 (p. 155); 17; 21.1
A. I. Shulgin	(b)	**123**,	605 (1984)	13.1 (p. 159) and 20 (p. 317); 21; 21.1
V. V. Shusha	(a)	**83**,	299 (19 4)	11 (p. 126) and 13.4 (p. 180); 22.1.2
	(a)	**86**,	309 (1984)	14.3.4 (p. 253); 13.4; 22.1.2
S. D. Shutov	(a)	**84**,	343 (1984)	14.3.4 (p. 251); 2; 16; 22
L. A. Shuvalov	(b)	**123**,	K83 (1984)	9 (p. 92) and 14.4 (p. 256); 19; 22.8.1
	(a)	**83**,	131 (1984)	14.4.1 (p. 261); 10.2; 12.2; 22.8.1
	(a)	**85**,	409 (1984)	1.2 (p. 19); 14.4.1; 19; 22.8.1
	(a)	**86**,	111 (1984)	1.2 (p. 20) and 14.4.1 (p. 263); 12.2; 22.8
Shu Zhen	(a)	**86**,	155 (1984)	1.3 (p. 26) and 4 (p. 62); 22.7
D. Siapkas	(b)	**124**,	229 (1984)	13.5.2 (p. 205) and 20.1 (p. 332); 22.8
A. S. Sidorkin	(a)	**84**,	K115 (1984)	14.4.1 (p. 262); 22.9
V. A. Sidorov	(a)	**81**,	K15 (1984)	2 (p. 40) and 9 (p. 91); 22.1.2
	(a)	**82**,	389 (1984)	1.2 (p. 16); 14.1; 15; 21.7
	(a)	**86**,	K99 (1984)	13.1 (p. 165) and 15 (p. 268); 14.3; 22.1
V. E. Sidorov	(a)	**85**,	497 (1984)	18.1 (p. 286); 21.1.1
P. Siebert	(a)	**86**,	543 (1984)	1.3 (p. 26); 10.1; 21; 21.1
W. Siegel	(a)	**84**,	251 (1984)	13.4 (p. 185) and 14.3 (p. 227); 22.2.2
M. A. Sigal	(b)	**121**,	161 (1984)	18.3 (p. 298) and 19 (p. 305)
S. Sikorski	(a)	**82**,	265 (1984)	14.3.2 (p. 238); 14.3.3
J. Sikula	(a)	**84**,	693 (1984)	14.3.3 (p. 245); 22.2.1

A. Šileika	(b) **122**, 623 (1984)	13.1 (p. 157) and 20.1 (p. 325); 22
	(b) **125**, 645 (1984)	14.4.1 (p. 263); 20.1; 22.8
A. R. Silin	(a) **81**, K23 (1984)	20.3 (p. 345); 2; 13.5.3; 22.6
I. M. Silvestrova	(a) **86**, 345 (1984)	14.4 (p. 258) and 20.2 (p. 344); 1.1; 12.2; 20.1; 20.3; 22.8.1
P. R. Silva	(b) **121**, K19 (1984)	13 (p. 144) and 18.2 (p. 287)
	(b) **121**, K25 (1984)	13 (p. 144) and 18.2 (p. 287)
	(b) **121**, K151 (1984)	13 (p. 145) and 18.2 (p. 288)
V. Šima	(b) **121**, K131 (1984)	18.4 (p. 300); 21.4
A. A. Simashkevich	(a) **84**, 343 (1984)	14.3.4 (p. 251); 2; 16; 22
A. V. Simashkevich	(b) **125**, 869 (1984)	13.4 (p. 190) and 14.3 (p. 229); 22.4.2
S. S. Simeonov	(a) **82**, 275 (1984)	14.3.3 (p. 242); 22
I. Šimkiene	(b) **122**, K179 (1984)	20.1 (p. 326); 13.4; 22
D. J. Simkin	(b) **124**, 261 (1984)	18 (p. 280) and 20.1 (p. 332); 22.5.2
O. Simmich	(a) **82**, K21 (1984)	1.4 (p. 28) and 9 (p. 92); 3; 21
G. Simon	(b) **124**, 799 (1984)	14.3 (p. 227); 13.1; 15; 22.7
S. Simon	(a) **81**, K1 (1984)	1.4 (p. 27) and 19 (p. 305); 22.8.1
Yu. V. Simonenko	(a) **86**, K63 (1984)	14.3 (p. 231) and 15 (p. 268); 22.1.2
V. I. Simonov	(b) **121**, 481 (1984)	1.2 (p. 15) and 6 (p. 68); 22.8
S. Simons	(b) **121**, K109 (1984)	6 (p. 69) and 8 (p. 84)
M. V. Simonyan	(a) **84**, 597 (1984)	2 (p. 49) and 14.3.1 (p. 236); 19; 20.1; 22.9
S. Simov	(a) **84**, K23 (1984)	1.4 (p. 31) and 11 (p. 128); 2; 22.1.1
S. Simozar	(a) **81**, 55 (1984)	1.3 (p. 22) and 2 (p. 39); 8; 21; 21.1
E. H. Sin	(a) **85**, 199 (1984)	20.1 (p. 337); 22.1.2
L. Sincan	(a) **85**, K129 (1984)	20.1 (p. 339); 2; 22.6.1
A. K. Singh	(a) **81**, 281 (1984)	10.2 (p. 106); 5; 18.3; 18.4; 22.8
I. Singh	(b) **126**, 669 (1984)	13.3 (p. 170); 22.2.2
O. P. Singh	(b) **125**, K1 (1984)	1 (p. 10) and 4 (p. 61); 22
R. Singh	(a) **84**, 659 (1984)	14.3.3 (p. 244); 13.4; 22.1.2
R. K. Singh	(b) **123**, 453 (1984)	4 (p. 60); 12.2; 22.5.4
	(b) **124**, 29 (1984)	4 (p. 60) and 12.2 (p. 141); 22.5; 22.5.2
	(b) **125**, 113 (1984)	4 (p. 61); 6; 12.2; 22; 22.5.3
	(a) **85**, 417 (1984)	1.2 (p. 19) and 12.2 (p. 142); 22.6
	(b) **126**, K15 (1984)	4 (p. 62) and 6 (p. 79); 8; 23
R. P. Singh	(b) **121**, 493 (1984)	4 (p. 58); 12.2; 22.5.2
	(b) **126**, 495 (1984)	4 (p. 63); 12.2; 22.5.2
S. P. Singh	(b) **126**, 87 (1984)	6 (p. 78); 12.1; 21.3
T. J. Singh	(b) **124**, K5 (1984)	8 (p. 87)
V. A. Singh	(b) **122**, 661 (1984)	13.1 (p. 157); 22.1; 22.2
Y. Singh	(b) **123**, 733 (1984)	14.2 (p. 217)
R. Singha	(a) **83**, 651 (1984)	20.3 (p. 351); 13.4; 22.8.1
I. K. Sinishchuk	(a) **86**, K83 (1984)	11 (p. 132) and 14.3.3 (p. 246); 22.2.1
V. V. Sinitskii	(a) **82**, 537 (1984)	14.3.1 (p. 234) and 16 (p. 270); 13.1; 22.9
E. P. Sinyavskii	(b) **123**, 671 (1984)	13.4 (p. 181); 20.1; 22.5
M. Siradjuddin	(a) **81**, K37 (1984)	14.3.4 (p. 248); 22.9
P. Sircar	(a) **85**, 649 (1984)	14.3.3 (p. 246); 14.3.4; 22.2.1
P. Široký	(a) **83**, 581 (1984)	20.1 (p. 330); 22.8.2
D. I. Sirota	(b) **121**, 185 (1984)	18.2 (p. 287) and 18.4 (p. 300)
V. Sitakara Rao	see under letter "R"	
A. A. Sitnikova	(a) **81**, 433 (1984)	1.4 (p. 27); 10.2; 22.1.2
V. V. Skafaru	(b) **125**, 543 (1984)	18 (p. 282)
K. Skeef Neto	(a) **85**, K65 (1984)	7 (p. 83) and 18 (p. 281); 19; 22.8.2
	(a) **86**, 163 (1984)	7 (p. 83); 19; 22.8.2
I. Yu. Sklyadneva	(b) **121**, 265 (1984)	13 (p. 144); 21; 22
Ž. M. Škrbić	(b) **125**, K151 (1984)	6 (p. 78) and 13.5.1 (p. 204); 22.9
	(b) **125**, K155 (1984)	6 (p. 78) and 13.5.1 (p. 204)
M. J. Škrinjar	(b) **122**, 125 (1984)	18.2 (p. 289)

	(b) **123**, 135 (1984)	13 (p. 146) and 18.4 (p. 301); 20.1
	(b) **124**, K45 (1984)	13.5.1 (p. 202) and 13.5.4 (p. 208); 18.2
	(b) **125**, K45 (1984)	18.2 (p. 294)
	(b) **125**, K151 (1984)	6 (p. 78) and 13.5.1 (p. 204); 22.9
	(b) **126**, K175 (1984)	18.2 (p. 294)
A. V. Skripov	(b) **126**, 557 (1984)	13.1 (p. 164) and 19 (p. 314); 21.1
P. A. Skubenko	(b) **123**, K63 (1984)	13.5.1 (p. 201) and 20.3 (p. 350); 22.3
L. N. Skuja	(a) **84**, K153 (1984)	20.3 (p. 354); 2; 10.2; 13.4; 20.1; 22.6
L. N. Skuya	(a) **81**, K23 (1984)	20.3 (p. 345); 2; 13.5.3; 22.6
M. M. Slavinskii	(b) **125**, K67 (1984)	5 (p. 66)
J. Šlechta	(b) **123**, K133 (1984)	8 (p. 87) and 18 (p. 280); 2
A. Ślęzak	(b) **126**, 371 (1984)	13.4 (p. 192) and 19 (p. 314); 22.8
V. V. Slisenko	(b) **121**, 623 (1984)	13.4 (p. 174); 21.4
U. Sliwczuk	(b) **122**, 203 (1984)	13.5.1 (p. 199); 6; 20.1; 22.5.1
V. K. Slovyanskikh	(a) **82**, K53 (1984)	18 (p. 279); 22.8
T. Słupkowski	(a) **83**, 329 (1984)	14.3 (p. 225); 22.9
L. Smardz	(a) **84**, K129 (1984)	1.5 (p. 34) and 18.2.1 (p. 296); 19; 21.1
Z. Smetana	(b) **121**, K131 (1984)	18.4 (p. 300); 21.4
B. I. Smirnov	(a) **82**, 441 (1984)	10.2 (p. 110); 22.5.2
V. P. Smirnov	(b) **122**, 231 (1984)	13.1 (p. 156)
	(b) **122**, 559 (1984)	13.1 (p. 156); 22
I. S. Smirnova	(b) **126**, 177 (1984)	10.2 (p. 118) and 14.3 (p. 230); 22.2; 22.4
T. V. Smirnova	(b) **122**, 47 (1984)	8 (p. 85) and 18.4 (p. 300); 7; 22.8.2
A. E. Smith	(b) **123**, 619 (1984)	1.6 (p. 37) and 13 (p. 147); 22; 22.6
	(b) **124**, 649 (1984)	1.6 (p. 38) and 13 (p. 148); 22
I. O. Smith	(a) **83**, 227 (1984)	10.1 (p. 99) and 12.1 (p. 136); 21
	(a) **84**, 509 (1984)	10.1 (p. 100); 21
	(a) **85**, 149 (1984)	10.1 (p. 101) and 12.1 (p. 137); 21.3
F. Smutný	(a) **82**, K33 (1984)	1.2 (p. 15) and 14.4.1 (p. 260); 22.8.1
	(b) **126**, 459 (1984)	1.2 (p. 20) and 14.4.1 (p. 263); 22.8.1
T. N. Snezhkova	(a) **82**, 441 (1984)	10.2 (p. 110); 22.5.2
I. I. Snighiryova	(a) **84**, 79 (1984)	10.2 (p. 113); 1.4; 22.2.1
O. V. Snitko	(a) **83**, K31 (1984)	10.2 (p. 112); 14.3.4; 16; 20.1; 22.1.1
F. C. T. So	(a) **86**, 471 (1984)	1.5 (p. 35); 2; 21
M. Šob	(b) **123**, 649 (1984)	13 (p. 147); 18.2; 21.1.1
D. Sobczynska	(a) **86**, 453 (1984)	14.3.3 (p. 246); 22.1.2
Sobden Kumar Pal	see under letter "P"	
V. V. Sobolev	(b) **126**, K59 (1984)	13.5.1 (p. 204) and 20.1 (p. 340); 22.3
R. Sobolewski	(a) **83**, K203 (1984)	14.2 (p. 218)
H. Sobotta	(b) **123**, K47 (1984)	20.1 (p. 328); 13.1; 22.7
	(b) **123**, K119 (1984)	6 (p. 74) and 20.1 (p. 330); 13.1; 22.8
	(b) **125**, K17 (1984)	6 (p. 77) and 20.1 (p. 336); 22.3
A. Sodeika	(a) **81**, K191 (1984)	11 (p. 123) and 14.3 (p. 222); 22.3
H. Sodolski	(a) **86**, 703 (1984)	14.4. (p. 259)
V. I. Sokolov	(b) **122**, 703 (1984)	13.5.1 (p. 200); 16; 20.1; 22.4.4
	(b) **124**, K155 (1984)	13.5.1 (p. 202) and 20.1 (p. 334); 14.3.4; 22.4
S. V. Sokolskii	(a) **86**, 177 (1984)	10.1 (p. 102); 21.1
H. Solbrig	(b) **123**, 237 (1984)	2 (p. 44) and 13 (p. 146)
	(b) **124**, K63 (1984)	13.3 (p. 168); 21.1
A. V. Soldatov	(b) **125**, K129 (1984)	13 (p. 151) and 20 (p. 318); 22
N. K. Solovarov	(b) **122**, K69 (1984)	19 (p. 307); 20.1; 22.6
N. E. Solovev	(a) **86**, K51 (1984)	14.3 (p. 231) and 15 (p. 268); 22.1
V. N. Solovev	(a) **83**, 553 (1984)	2 (p. 46) and 9 (p. 93); 22.1.2
T. Soma	(b) **121**, 501 (1984)	6 (p. 68); 8; 22.3
	(b) **121**, K1 (1984)	1.2 (p. 14); 22.2.1; 22.2.3
	(b) **121**, K113 (1984)	6 (p. 69); 22.2.2; 22.4.3
	(b) **122**, K125 (1984)	6 (p. 71); 22.1.1; 22.1.2

	(b) **123**, 463 (1984)	6 (p. 72); 8; 21
	(b) **124**, 37 (1984)	6 (p. 74); 22.2; 22.4
	(b) **124**, 525 (1984)	6 (p. 75); 8; 21.2
	(b) **124**, K11 (1984)	1.2 (p. 17) and 4 (p. 60); 12.2; 22.2.1; 22.2.3
	(b) **125**, 107 (1984)	6 (p. 76); 12.1; 21
	(b) **125**, K99 (1984)	1.2 (p. 19) and 6 (p. 77); 8; 22.1.1; 22.1.2
	(b) **126**, 71 (1984)	6 (p. 78); 8; 21
	(b) **126**, K5 (1984)	6 (p. 79) and 8 (p. 89); 12.1; 21.2
	(b) **126**, K9 (1984)	12.2 (p. 142); 6; 22.2; 22.4
	(b) **126**, K109 (1984)	6 (p. 80); 8; 12.1; 21.2
K. Somaiah	(a) **82**, 201 (1984)	13.4 (p. 176) and 20.3 (p. 348); 20.1; 22.5.4
	(a) **85**, 265 (1984)	10.2 (p. 117); 20.1; 20.3; 22.5.4
F. Somma	(b) **122**, K83 (1984)	10.2 (p. 108) and 13.4 (p. 176); 20.1; 22.5.2
H. Sommer	(b) **121**, 641 (1984)	13.1 (p. 155); 17; 20; 22.8
Nguyen Manh Son	see under letter "N"	
Youn Soo Lim	see under letter "Y"	
T. Soovik	(b) **125**, 139 (1984)	10.2 (p. 115) and 20.3 (p. 354); 22.5.2
	(b) **126**, 407 (1984)	13.4 (p. 193) and 20.3 (p. 357); 22.5.4
	(b) **126**, 653 (1984)	13.4 (p. 194) and 20.3 (p. 358); 22.5.2
	(b) **126**, K63 (1984)	13.4 (p. 193) and 20.3 (p. 357); 22.5.2
G. Sorge	(a) **86**, 111 (1984)	1.2 (p. 20) and 14.4.1 (p. 263); 12.2; 22.8
J. Soria	(a) **82**, 379 (1984)	1.3 (p. 23) and 10.2 (p. 110); 19; 22.6.1
B. Sorkin	(a) **82**, K149 (1984)	10.2 (p. 110) and 17 (p. 274); 13.4; 22.5.2
	(a) **83**, K39 (1984)	10.2 (p. 112); 17; 20.3; 22.5.2
	(a) **83**, K171 (1984)	17 (p. 275); 11; 13.3; 13.4; 22.5.2
L. M. Sorokin	(a) **81**, 433 (1984)	1.4 (p. 27); 10.2; 22.1.2
L. E. Soshnikov	(b) **122**, 47 (1984)	8 (p. 85) and 18.4 (p. 300); 7; 22.8.2
A. Sotgiu	(a) **81**, K27 (1984)	2 (p. 40) and 19 (p. 306)
T. Soumura	(a) **82**, K169 (1984)	2 (p. 43) and 18.2 (p. 290); 21.1; 21.1.1
	(a) **83**, K195 (1984)	2 (p. 47) and 14.1 (p. 212); 21.1; 21.1.1
C. Sourisseau	(b) **123**, 503 (1984)	6 (p. 73) and 20.1 (p. 329); 22
J. M. Spaeth	(b) **125**, 523 (1984)	10.2 (p. 117) and 19 (p. 313); 13.4; 22.5.2
V. Spânu	(a) **86**, 525 (1984)	5 (p. 66); 2; 21.1.1
H. N. Spector	(b) **123**, 747 (1984)	14.1 (p. 212) and 14.3 (p. 225); 21.7
	(b) **126**, K171 (1984)	14.3.2 (p. 239)
G. Spörl	(a) **82**, K27 (1984)	8 (p. 85) and 14.4.1 (p. 260); 22.8.1
M. J. Springis	(b) **123**, 335 (1984)	10.2 (p. 110) and 20.3 (p. 350); 22.6
	(b) **125**, K165 (1984)	20.3 (p. 356); 22.6
J. Spyridelis	(b) **122**, 613 (1984)	20.1 (p. 325); 13.1; 22.8
T. K. K. Srinivasan	(b) **124**, 481 (1984)	6 (p. 75) and 20.1 (p. 333); 22.5.4
K. K. P. Srivastava	(b) **121**, K83 (1984)	5 (p. 64); 22.8
R. S. Srivastava	(a) **84**, 311 (1984)	14.3.3 (p. 244) and 16 (p. 272); 1.4; 22.1.2
H. Stachowiak	(b) **121**, 307 (1984)	11 (p. 122) and 13.4 (p. 172); 21
	(b) **121**, K69 (1984)	13 (p. 145); 21
	(b) **124**, 159 (1984)	13 (p. 148); 13.2; 21
J. L. Staehli	(b) **122**, 629 (1984)	13.5.1 (p. 200) and 20.3 (p. 349); 22.2.2
A. Stahl	(b) **124**, 167 (1984)	13.1 (p. 161) and 13.5 (p. 197); 14.3; 20; 22
K. V. Stamenov	(a) **84**, 555 (1984)	13.4 (p. 186); 20.2; 22.4.1
C. V. Stancampiano	(a) **83**, K203 (1984)	14.2 (p. 218)
F. Stangler	(a) **84**, 475 (1984)	10.1 (p. 100); 21.1
J. Stankiewicz	(b) **124**, K53 (1984)	20.1 (p. 332); 13.1; 22.4.4
P. Ya. Starostin	(a) **85**, K57 (1984)	11 (p. 130) and 13.4 (p. 189); 22.2.1
V. I. Startsev	(a) **81**, K11 (1984)	2 (p. 40) and 12.1 (p. 135); 10.1; 21
H. Staudinger	(b) **123**, 319 (1984)	13.4 (p. 179); 16; 22.8
B. Stébé	(b) **126**, 329 (1984)	13.4 (p. 192) and 13.5.1 (p. 204); 22.2; 22.4
A. V. Stefanovich	(b) **122**, 367 (1984)	14.2 (p. 216); 21.4
R. Stegmann	(a) **81**, 541 (1984)	13.5.1 (p. 199) and 20.3 (p. 346); 22.2.4

U. Steigenberger	(b) **125**,	131 (1984)	4 (p. 61); 1.1; 6; 20.1; 22.7
W. Steiger	(a) **81**,	247 (1984)	9 (p. 91); 11; 22.1.2
E. Steinbeiss	(a) **84**,	379 (1984)	1.5 (p. 34) and 2 (p. 49); 22.6
V. F. Stelmakh	(a) **84**,	613 (1984)	14.3.4 (p. 252); 22.2.1
Yu. N. Stelmakhov	(a) **83**,	245 (1984)	18.4 (p. 301); 1.4; 22.5
S. I. Stenin	(a) **84**,	K13 (1984)	1.6 (p. 37) and 2 (p. 48); 22.2.1; 22.2.4
A. Stepanescu	(b) **122**,	411 (1984)	2 (p. 42)
	(b) **123**,	403 (1984)	2 (p. 45)
A. P. Stepanov	(b) **126**,	557 (1984)	13.1 (p. 164) and 19 (p. 314); 21.1
S. A. Stepanov	(a) **81**,	47 (1984)	1.6 (p. 36) and 2 (p. 39); 10.2; 22.1.2
	(a) **86**,	143 (1984)	1.1 (p. 13); 1.6; 22.1.2
J. D. Stephenson	(a) **84**,	19 (1984)	1.4 (p. 30) and 1.6 (p. 37); 14.4.1; 22.8.1
J. Stępie ń-Damm	(a) **83**,	K113 (1984)	1.1 (p. 12); 22.9
A. Stern	(b) **124**,	K171 (1984)	13.3 (p. 169); 22.1.2; 22.6
O. Štika	(a) **83**,	617 (1984)	13.1 (p. 160) and 20.1 (p. 331); 2; 22.1.2
M. J. Stillman	(b) **124**,	261 (1984)	18 (p. 280) and 20.1 (p. 332); 22.5.2
F. Stobiecki	(a) **82**,	K177 (1984)	2 (p. 43) and 18.2.1 (p. 296); 21.1.1
	(a) **84**,	K129 (1984)	1.5 (p. 34) and 18.2.1 (p. 296); 19; 21.1
T. Stobiecki	(b) **125**,	K171 (1984)	14.1 (p. 214); 2; 21.1.1
	(b) **126**,	191 (1984)	13 (p. 151) and 18 (p. 282); 2; 21.1.1
T. Stöckert	(b) **123**,	K161 (1984)	13.5.2 (p. 205); 22.2.2
	(b) **126**,	K149 (1984)	13.1 (p. 165) and 13.5.2 (p. 206); 14.4; 22.8
V. F. Stolarenko	(b) **125**,	K37 (1984)	10.2 (p. 116) and 20.1 (p. 336); 11; 22.5.2
A. V. Stolyarenko	(b) **125**,	705 (1984)	13.5.1 (p. 203); 20.1; 22.3.1; 22.4.1
H. Stolz	(b) **122**,	203 (1984)	13.5.1 (p. 199); 6; 20.1; 22.5.1
	(b) **124**,	201 (1984)	13.5 (p. 197); 13.5.1; 22.2.1
F. Storbeck	(a) **81**,	259 (1984)	1.6 (p. 36) and 11 (p. 122); 21.1
	(a) **86**,	499 (1984)	1.3 (p. 26) and 13.3 (p. 171); 20; 21.1
M. Stordeur	(b) **121**,	765 (1984)	13.1 (p. 155); 14.3; 15; 22
	(b) **123**,	K47 (1984)	20.1 (p. 328); 13.1; 22.7
	(b) **124**,	439 (1984)	14.3 (p. 226); 15; 22.7
	(b) **124**,	799 (1984)	14.3 (p. 227); 13.1; 15; 22.7
J. P. Stradins	(a) **81**,	K121 (1984)	9 (p. 91); 21.6
H. Stramska	(b) **124**,	431 (1984)	14.3.4 (p. 251) and 18 (p. 280); 22
P. Středa	(b) **124**,	K97 (1984)	14.3.2 (p. 239) and 14.3.4 (p. 251); 22.2.1
	(b) **125**,	849 (1984)	13.3 (p. 170); 8; 14.1; 15; 18; 21
R. Strehlow	(b) **123**,	663 (1984)	13.1 (p. 159) and 13.4 (p. 181); 22.6
	(b) **123**,	K39 (1984)	13.1 (p. 158)
I. Streit	(b) **122**,	629 (1984)	13.5.1 (p. 200) and 20.3 (p. 349); 22.2.2
P. Streitenberger	(b) **125**,	681 (1984)	13 (p. 150); 14.4; 21
V. L. Strizhevskii	(b) **123**,	691 (1984)	13.5 (p. 197) and 20.1 (p. 330); 6; 22
H. Strusny	(a) **85**,	K35 (1984)	10.2 (p. 117) and 11 (p. 130); 22.1.2
H. Stubb	(b) **122**,	791 (1984)	14.3 (p. 223) and 15 (p. 265); 22.9
J. Stuchlík	(a) **83**,	617 (1984)	13.1 (p. 160) and 20.1 (p. 331); 2; 22.1.2
C. M. Su	(a) **84**,	157 (1984)	10.1 (p. 100) and 12.1 (p. 136); 21
U. V. Subba Rao	see under letter "R"		
M. Subotowicz	(b) **124**,	355 (1984)	13.4 (p. 184); 19; 22.5
A. Subrahmanyam	(a) **85**,	K93 (1984)	14.4 (p. 257); 11; 22.5.2
S. Sudhakar	(a) **85**,	K27 (1984)	12.2 (p. 141) and 20 (p. 318); 22.5.4
R. Sudharsanan	(b) **124**,	481 (1984)	6 (p. 75) and 20.1 (p. 333); 22.5.4
M. Suezawa	(a) **82**,	235 (1984)	13.4 (p. 177); 20.1; 22.1.2
	(a) **85**,	469 (1984)	10.2 (p. 118) and 13.4 (p. 191); 20.1; 22.1.2
V. I. Sugakov	(b) **126**,	703 (1984)	14.2 (p. 219)
T. Sugiura	(a) **84**,	K9 (1984)	3 (p. 56); 10.2; 22.2.2
M. Suhrke	(b) **126**,	393 (1984)	20.3 (p. 357); 2; 13.4; 22
T. Sukegawa	(a) **84**,	K9 (1984)	3 (p. 56); 10.2; 22.2.2
A. L. Sukhman	(a) **84**,	423 (1984)	2 (p. 49) and 12.1 (p. 136); 1.2; 21; 21.4

A. A. SUKHOVSKII	(a)	**84**, K109 (1984)	1.1 (p. 12) and 14.4.1 (p. 262); 19; 22.8.1
S. V. SUKHVALO	(a)	**82**, 569 (1984)	14.3.1 (p. 234); 1.5; 22.8.2
A. SUKIENNICKI	(a)	**85**, 189 (1984)	18.2.1 (p. 297); 22.8.2
C. SUŁKOWSKI	(a)	**83**, K191 (1984)	14.2 (p. 218)
F. A. SULTAN	(a)	**83**, 263 (1984)	13.5.2 (p. 205) and 20.1 (p. 328); 13.1; 21
G. D. SULTANOV	(a)	**84**, K59 (1984)	5 (p. 65) and 18 (p. 281); 22.3
K. SUMINO	(a)	**82**, 235 (1984)	13.4 (p. 177); 20.1; 22.1.2
	(a)	**85**, 469 (1984)	10.2 (p. 118) and 13.4 (p. 191); 20.1; 22.1.2
K. SUMIYAMA	(a)	**81**, K109 (1984)	1.3 (p. 22); 5; 18.2; 21.1; 21.1.1; 21.6
	(a)	**83**, K219 (1984)	1.3 (p. 22); 5; 18.2; 21.1; 21.1.1; 21.6
A. MEENAKSHI SUNDARAM	(a)	**86**, K15 (1984)	1.1 (p. 13) and 3 (p. 57); 22.6.1
M. SUNDBERG	(a)	**85**, 83 (1984)	1.4 (p. 31); 10.2; 22.4.2
C. M. SUNTA	(a)	**81**, 533 (1984)	13.4 (p. 174) and 20.3 (p. 346); 22.5.2
E. SUONINEN	(b)	**123**, 627 (1984)	13.1 (p. 159); 20; 21.1
V. V. SUPRUNCHIK	(a)	**86**, 313 (1984)	13.4 (p. 193) and 19 (p. 314); 11; 22.1.2
P. SÜPTITZ	(a)	**82**, 157 (1984)	9 (p. 92); 20.1; 22
K. SURENDRANATH	(b)	**122**, 345 (1984)	14 (p. 210) and 15 (p. 265); 22.8
D. K. SURI	(b)	**124**, K167 (1984)	19 (p. 311); 22.8
SURINDER KOUL	see under letter "K"		
V. SURIYANARAYANAN	(b)	**125**, 513 (1984)	9 (p. 93); 22.4.3
T. P. SURKOVA	(b)	**124**, K155 (1984)	13.5.1 (p. 202) and 20.1 (p. 334); 14.3.4; 22.4
M. SURMA	(b)	**121**, 209 (1984)	14.2 (p. 216)
YU. P. SUROV	(a)	**83**, K35 (1984)	10.2 (p. 112) and 11 (p. 126); 14.3; 22.8
S. V. SURYANARAYANA	(b)	**123**, 453 (1984)	4 (p. 60); 12.2; 22.5.4
H. SÜSSMANN	(a)	**82**, 561 (1984)	14.3 (p. 224) and 15 (p. 265); 22.7
E. V. SUVOROV	(a)	**83**, 195 (1984)	1.4 (p. 29) and 10.2 (p. 111); 22.1.2
K. SUZUKI	(a)	**85**, 249 (1984)	16 (p. 273) and 20.3 (p. 355); 22.4.4
	(a)	**86**, 729 (1984)	10.2 (p. 121) and 14.4 (p. 259); 22.5.4
B. G. SVENSSON	(a)	**85**, K109 (1984)	10.2 (p. 118); 22.1.2
C. S. SWAMY	(a)	**86**, K15 (1984)	1.1 (p. 13) and 3 (p. 57); 22.6.1
K. L. SWEENEY	(a)	**81**, 253 (1984)	10.2 (p. 105); 11; 19; 20.1; 22.8.1
M. A. S. SWEET	(a)	**84**, K75 (1984)	20.3 (p. 353); 22.4.1
P. ŚWIDERSKI	(b)	**122**, K147 (1984)	13.5.1 (p. 200); 22.4.4
V. K. SYAL	(b)	**126**, 59 (1984)	2 (p. 52); 8; 22.5.2
H. SYHRE	(a)	**81**, 247 (1984)	9 (p. 91); 11; 22.1.2
N. N. SYRBU	(b)	**125**, K17 (1984)	6 (p. 77) and 20.1 (p. 336); 22.3
S. R. SYRTSOV	(b)	**123**, K165 (1984)	13.1 (p. 160) and 14.4.1 (p. 261); 22
A. J. SZADKOWSKI	(a)	**84**, K181 (1984)	14.3 (p. 228) and 16 (p. 272); 9; 13.4; 22.4.4
A. SZEKERES	(a)	**84**, 561 (1984)	13.3 (p. 169) and 14.3.2 (p. 239); 22.1.2
J. SZNAJD	(b)	**124**, 103 (1984)	18 (p. 280)
V. SZÖCS	(b)	**125**, K137 (1984)	13.5.1 (p. 203); 22.9
E. SZONTAGH	(a)	**84**, 165 (1984)	10.2 (p. 113); 19; 20.1; 22.6
M. SZUSKIEWICZ	(b)	**123**, 649 (1984)	13 (p. 147); 18.2; 21.1.1
S. SZUSZKIEWICZ	(b)	**123**, 649 (1984)	13 (p. 147); 18.2; 21.1.1
T. SZUTA-BUCHACZ	(a)	**82**, 421 (1984)	7 (p. 82); 6; 12.2; 22.8.2
J. SZYMASZEK	(a)	**83**, K191 (1984)	14.2 (p. 218)
H. SZYMCZAK	(b)	**122**, K37 (1984)	7 (p. 81) and 13.5.3 (p. 207); 22.6
A. SZYTUŁA	(a)	**82**, 195 (1984)	18.4 (p. 301); 22.8
	(a)	**84**, 229 (1984)	18.2 (p. 293) and 18.4 (p. 303); 1.2; 22.8
E. D. TABACHNIKOVA	(a)	**81**, K11 (1984)	2 (p. 40) and 12.1 (p. 135); 10.1; 21
V. P. TABAKAR	(b)	**123**, K173 (1984)	14.2 (p. 217)
B. G. TAGIEV	(b)	**121**, K195 (1984)	20.3 (p. 345); 13.4; 22.3
	(a)	**81**, K175 (1984)	14.3 (p. 222); 22.3.1
	(a)	**83**, K61 (1984)	13.4 (p. 181); 16; 20.3; 22.3
	(a)	**84**, K59 (1984)	5 (p. 65) and 18 (p. 281); 22.3

H. TAGUCHI	(a) **85**, 523 (1984)	18.2 (p. 294); 14.1; 22.6.1
M. A. TAGUYRDZHANOV	(a) **86**, 383 (1984)	16 (p. 273); 2; 14.3.1; 22; 22.1.2
M. TAKAHASHI	(a) **86**, 289 (1984)	18.1 (p. 286); 21.1; 21.1.1
Y. TAKAHASHI	(b) **124**, K11 (1984)	1.2 (p. 17) and 4 (p. 60); 12.2; 22.2.1; 22.2.3
M. TAKEDA	(a) **86**, 67 (1984)	1.4 (p. 32); 21.1; 21.6
H. TAKEI	(a) **83**, 467 (1984)	1.2 (p. 17); 12.2; 22.8.1
	(a) **86**, 841 (1984)	1.2 (p. 17); 12.2; 22.8.1
S. J. TAKERIS	(a) **81**, K121 (1984)	9 (p. 91); 21.6
H. TAKEUCHI	(a) **83**, 567 (1984)	18.1 (p. 285) and 19 (p. 309); 18.4; 22.6.1
	(a) **83**, 729 (1984)	18.1 (p. 285) and 19 (p. 310); 18.4; 22.6.1
N. TAKEUCHI	(a) **83**, K179 (1984)	20.2 (p. 343); 10.2; 22.5.2
	(a) **85**, K141 (1984)	17 (p. 276) and 20.3 (p. 356); 2; 22.6
S. TAKEUCHI	(a) **83**, 219 (1984)	2 (p. 44) and 10.1 (p. 99); 21.1
I. E. TALANIN	(a) **81**, 433 (1984)	1.4 (p. 27); 10.2; 22.1.2
I. TALE	(a) **86**, 319 (1984)	13.4 (p. 194); 20.1; 20.3; 22.2
S. P. TAMHANKAR	(a) **82**, K195 (1984)	14.3.1 (p. 234); 20.1; 22.8
G. TAMULAITIS	(b) **126**, K135 (1984)	13.5.1 (p. 204) and 20.1 (p. 341); 22.4.1; 22.4.2
I. TAMURA	(b) **123**, 525 (1984)	2 (p. 45) and 18.2 (p. 291); 21.1.1
H. S. TAN	(a) **85**, 199 (1984)	20.1 (p. 337); 22.1.2
K. L. TAN	(b) **125**, K161 (1984)	20.3 (p. 356); 11; 13.4; 22.5.2
HA VINH TAN	see under letter "H"	
A. TANAKA	(a) **84**, K9 (1984)	3 (p. 56); 10.2; 22.2.2
N. TANAKA	(a) **81**, 129 (1984)	1.3 (p. 22); 21.1
S. TANIGAWA	(b) **125**, 823 (1984)	10.1 (p. 101); 13.3; 21.6
K. TANIMURA	(b) **122**, 489 (1984)	10.2 (p. 109) and 20.1 (p. 325); 20.3; 22.5.3
B. G. TANKHILEVICH	(b) **124**, 79 (1984)	13.5.4 (p. 208) and 18.2 (p. 292); 22
S. B. TARASENKO	(b) **123**, K25 (1984)	18.4 (p. 301); 19
E. I. TARASOVA	(b) **124**, 249 (1984)	13.4 (p. 183); 13.5.1; 23
A. M. TARATIN	(b) **122**, 29 (1984)	11 (p. 123); 21
	(b) **124**, 641 (1984)	11 (p. 128); 13; 22.1.2
V. TARTACHNIK	(a) **81**, K191 (1984)	11 (p. 123) and 14.3 (p. 222); 22.3
J. C. TEDENAC	(a) **82**, 295 (1984)	13.1 (p. 156) and 14.3 (p. 223); 8; 15; 20.1; 22.8
G. B. TEITELBAUM	(b) **125**, K121 (1984)	13 (p. 150) and 18 (p. 282)
P. TEKIEL	(b) **121**, K91 (1984)	14.2 (p. 216)
V. T. TELEPA	(b) **123**, K25 (1984)	18.4 (p. 301); 19
	(a) **83**, K135 (1984)	18.1 (p. 285) and 18.4 (p. 302); 1.2; 22.8
	(a) **84**, 223 (1984)	18.4 (p. 302) and 19 (p. 310); 22.5
G. VAN TENDELOO	(a) **82**, 55 (1984)	1.4 (p. 28); 21.1
	(a) **84**, 185 (1984)	1.1 (p. 12) and 1.4 (p. 30); 21.1; 21.6
	(a) **85**, 29 (1984)	1.4 (p. 31); 10.1; 21; 21.1
	(a) **86**, 553 (1984)	1.3 (p. 26); 10.1; 21; 21.1
K. TENNAKONE	(b) **125**, K181 (1984)	14.3 (p. 229) and 14.4 (p. 258); 22.5
	(a) **86**, K121 (1984)	14 (p. 210); 22.8
A. TEPORE	(a) **82**, K79 (1984)	13.4 (p. 177); 14.4; 22.5
	(a) **85**, 609 (1984)	16 (p. 273); 2; 22.1.2
S. TERAKADO	(b) **125**, 823 (1984)	10.1 (p. 101); 13.3; 21.6
S. F. TEREKHOVA	(b) **125**, 705 (1984)	13.5.1 (p. 203); 20.1; 22.3.1; 22.4.1
H. TERLETZKI	(a) **86**, 789 (1984)	11 (p. 133) and 14.3.3 (p. 246); 14.3.2; 22.1.2
L. A. TERSIEV	(b) **125**, 659 (1984)	1.2 (p. 19) and 14.4.1 (p. 263); 13.1; 22.8
G. I. TESLENKO	(a) **81**, 391 (1984)	14.3.4 (p. 248) and 16 (p. 269); 22.2.3
J. TETERIS	(a) **83**, K47 (1984)	20.1 (p. 329); 2; 22
D. P. TEWARI	(a) **84**, 439 (1984)	7 (p. 82); 22.4.1
S. A. TEYS	(a) **84**, K13 (1984)	1.6 (p. 37) and 2 (p. 48); 22.2.1; 22.2.4
V. E. TEZLEVAN	(a) **82**, 207 (1984)	13.4 (p. 176) and 16 (p. 269); 22.8
L. DO THANH	see under letter "D"	
D. THÉBAULT	(b) **125**, 357 (1984)	13.4 (p. 188) and 20.3 (p. 354); 22.1.2
D. THEIS	(a) **81**, 647 (1984)	1.5 (p. 33) and 14.3.4 (p. 249); 20.3; 22.4.1

E. Thomas	(a) **84**, K97 (1984)	1.3 (p. 25) and 2 (p. 49); 22.7
P. A. Thomas	(b) **123**, 611 (1984)	2 (p. 46) and 14.4 (p. 257); 22.1.2
Dang Dinh Thong	see under letters "Da"	
D. Thornberry	(b) **121**, K125 (1984)	2 (p. 40) and 18 (p. 278); 21.1.1
G. Thörner	(b) **126**, 617 (1984)	17 (p. 277); 13.3
N. T. Thuc Hien	see under letter "H"	
I. Thurzo	(a) **82**, 519 (1984)	14.3.3 (p. 242); 14.3.4
	(a) **86**, 795 (1984)	13.3 (p. 171) and 14.3.3 (p. 246); 22.2.1
F. Thuselt	(b) **122**, 629 (1984)	13.5.1 (p. 200) and 20.3 (p. 349); 22.2.2
	(b) **123**, 257 (1984)	13.5.2 (p. 205); 22.1.1; 22.1.2; 22.2.1; 22.2.2
	(a) **84**, 677 (1984)	13.5 (p. 197); 14.3.3; 22.2.2
L. Tichý	(a) **85**, K123 (1984)	2 (p. 52) and 20.1 (p. 339); 22.8
I. M. Tiginyanu	(a) **82**, 207 (1984)	13.4 (p. 176) and 16 (p. 269); 22.8
E. A. Tikhonova	(a) **81**, 69 (1984)	1 (p. 9); 10
D. R. Tilley	(b) **124**, 457 (1984)	1.6 (p. 37); 1.2
	(b) **125**, 45 (1984)	1.2 (p. 18) and 20.1 (p. 335); 1.5
I. N. Timchenko	(b) **125**, 869 (1984)	13.4 (p. 190) and 14.3 (p. 229); 22.4.2
M. Timko	(a) **82**, K159 (1984)	11 (p. 125) and 18.2 (p. 290); 2; 21.1.1
V. B. Timofeev	(b) **122**, K133 (1984)	13.4 (p. 178) and 18.2 (p. 290); 22.4.4
V. A. Timofeeva	(a) **83**, K165 (1984)	20.2 (p. 343) and 20.3 (p. 352); 20.1; 22.8.1
A. Tiourine	(b) **122**, K83 (1984)	10.2 (p. 108) and 13.4 (p. 176); 20.1; 22.5.2
N. P. Tishchenko	(a) **83**, 513 (1984)	9 (p. 92); 10.2; 23
V. D. Tkachenko	(a) **81**, K151 (1984)	18.1 (p. 284); 22.8.1
V. D. Tkachev	(a) **81**, 313 (1984)	10.2 (p. 106) and 13.4 (p. 173); 20.3; 22.1.2
	(b) **125**, K75 (1984)	10.2 (p. 116) and 13.4 (p. 188); 22.1; 22.1.2
	(a) **85**, K43 (1984)	10.2 (p. 117) and 11 (p. 130); 22.1.2
Yu. A. Tkhorik	(a) **86**, K83 (1984)	11 (p. 132) and 14.3.3 (p. 246); 22.2.1
R. Tognato	(a) **82**, K17 (1984)	1.6 (p. 36); 10.1; 21
	(a) **85**, K47 (1984)	10.1 (p. 101); 21
	(a) **86**, K35 (1984)	10.2 (p. 120); 22
N. Toichev	(b) **121**, K205 (1984)	19 (p. 306) and 20.1 (p. 322); 6; 13.4; 22.5.2
E. I. Tolpygo	(b) **123**, 295 (1984)	13.4 (p. 179) and 14.3 (p. 224); 22.2.1
M. Tomak	(b) **122**, 295 (1984)	13.3 (p. 167) and 13.4 (p. 175)
	(b) **126**, 677 (1984)	13.4 (p. 195); 22.1.2
	(b) **126**, 729 (1984)	2 (p. 54) and 14.1 (p. 215); 21.1; 21.6
P. M. Tomchuk	(b) **123**, 365 (1984)	14.4 (p. 256); 6; 22.8.1
A. Tomita	(a) **85**, K141 (1984)	17 (p. 276) and 20.3 (p. 356); 2; 22.6
R. D. Tomlinson	(b) **121**, 641 (1984)	13.1 (p. 155); 17; 20; 22.8
	(a) **84**, 87 (1984)	6 (p. 74) and 8 (p. 87); 22.8
J. W. Tomm	(a) **86**, 433 (1984)	20.1 (p. 341); 20.2; 22
M. Tonelli	(b) **122**, K83 (1984)	10.2 (p. 108) and 13.4 (p. 176); 20.1; 22.5.2
A. H. Toneyan	(a) **85**, 349 (1984)	1.1 (p. 13); 22.1.1
B. Topič	(a) **85**, 409 (1984)	1.2 (p. 19); 14.4.1; 19; 22.8.1
M. Topić	(a) **86**, 413 (1984)	14.4 (p. 259); 22.9
W. Torbicz	(a) **86**, 453 (1984)	14.3.3 (p. 246); 22.1.2
V. I. Torgashev	(b) **126**, 459 (1984)	1.2 (p. 20) and 14.4.1 (p. 263); 22.8.1
M. I. Török	(a) **81**, K67 (1984)	14.3 (p. 221); 22.6
	(a) **84**, 639 (1984)	14.3.4 (p. 252)
	(b) **126**, K97 (1984)	14.3.4 (p. 253)
V. V. Torokin	(a) **84**, 423 (1984)	2 (p. 49) and 12.1 (p. 136); 1.2; 21; 21.4
V. Toşa	(b) **123**, K53 (1984)	13.1 (p. 159) and 18.3 (p. 298); 21.1; 21.4
B. S. Tošić	(b) **121**, 589 (1984)	13.5 (p. 196)
	(b) **125**, 743 (1984)	6 (p. 77) and 13.5.1 (p. 203)
	(b) **125**, K155 (1984)	6 (p. 78) and 13.5.1 (p. 204)
S. D. Toshev	(a) **86**, K109 (1984)	14.4 (p. 259); 22.8.1
A. Tóth	(b) **122**, 501 (1984)	10.2 (p. 109); 22.5.2
D. H. J. Totterdell	(a) **86**, 821 (1984)	13.4 (p. 195) and 14.3 (p. 232); 20.1; 22.4.3

J. Toulouse	(a) **84**, 411 (1984)	1.1 (p. 12); 19; 20.1; 22.9
K. D. Tovstyuk	(b) **124**, 693 (1984)	6 (p. 75) and 13 (p. 149); 22.3
Tran Chot	(a) **86**, K73 (1984)	14.3.4 (p. 253); 22.1.2
Tran Kim Anh	(a) **84**, K159 (1984)	20.1 (p. 335) and 20.3 (p. 354); 13.4; 22.8.1
Dang Tran Quan	see under letter "D"	
M. Trapp	(a) **81**, 541 (1984)	13.5.1 (p. 199) and 20.3 (p. 346); 22.2.4
	(a) **84**, K39 (1984)	9 (p. 93); 14.3.2; 22.2.1
S. B. Trickey	(b) **121**, K47 (1984)	20.1 (p. 320); 13.1; 23
S. Trimper	(b) **126**, K23 (1984)	18.2 (p. 295); 1.2
S. Trivedi	(a) **81**, K179 (1984)	14.3 (p. 222); 22.9
R. Troć	(b) **123**, K29 (1984)	18.1 (p. 284); 21
A. I. Trokhimets	(b) **124**, 699 (1984)	13.1 (p. 161) and 13.3 (p. 168); 22.6
A. L. Tronconi	(a) **86**, 163 (1984)	7 (p. 83); 19; 22.8.2
A. C. Tropper	(b) **123**, 79 (1984)	10.2 (p. 110) and 20.3 (p. 349); 22.5.2
A. N. Trukhin	(a) **84**, K153 (1984)	20.3 (p. 354); 2; 10.2; 13.4; 20.1; 22.6
R. Trykozko	(a) **84**, K89 (1984)	16 (p. 272); 14.3.1; 20.1; 22.8
H.-U. Tschirner	(b) **123**, K177 (1984)	2 (p. 46) and 14.3 (p. 225); 13.4; 22
W. S. Tse	(a) **86**, K23 (1984)	6 (p. 79) and 20.1 (p. 341); 22
A. V. Tsikunov	(a) **82**, 467 (1984)	11 (p. 125) and 14.3.2 (p. 238); 13.4; 22.1.2
Tsing-Wong Hsu	see under letter "H"	
O. B. Tsiok	(a) **86**, K99 (1984)	13.1 (p. 165) and 15 (p. 268); 14.3; 22.1
V. G. Tsirelson	(b) **122**, 599 (1984)	4 (p. 59); 20.1; 22; 22.5.2; 22.6; 22.8.1
S. V. Tsivinskii	(a) **81**, 299 (1984)	10.1 (p. 97); 1.4; 21; 21.1; 21.1.1; 22.1.1; 22.1.2
T. Tsuchida	(a) **86**, 237 (1984)	10.1 (p. 103); 14.1; 21
T. Tsuji	(a) **82**, K57 (1984)	5 (p. 64) and 18 (p. 279); 22.6
S. Tsunekawa	(a) **83**, 467 (1984)	1.2 (p. 17); 12.2; 22.8.1
	(a) **86**, 841 (1984)	1.2 (p. 17); 12.2; 22.8.1
A. E. Tsurkan	(a) **84**, K169 (1984)	14.3.3 (p. 245); 20.3; 22.2.2; 22.4.3
G. I. Tsurkan	(b) **121**, K97 (1984)	16 (p. 269); 22
	(b) **121**, K209 (1984)	7 (p. 81) and 14.3 (p. 221); 22
	(b) **122**, 361 (1984)	14.3 (p. 222); 18; 22
	(b) **126**, K105 (1984)	6 (p. 79) and 16 (p. 273); 7; 22.2.1
M. Tuichiev	(a) **85**, K159 (1984)	14.3.1 (p. 236); 11; 22.2.2
A. V. Tulinov	(a) **85**, K39 (1984)	10.2 (p. 117) and 11 (p. 130); 22.1.2
K. K. Tumkevičius	(a) **81**, K43 (1984)	14.3.1 (p. 233) and 16 (p. 269); 22
C. Tuncay	(b) **126**, 677 (1984)	13.4 (p. 195); 22.1.2
Yen Tzu-Tung	(b) **124**, K113 (1984)	8 (p. 88); 21.2
A. A. Turinge	(a) **85**, K39 (1984)	10.2 (p. 117) and 11 (p. 130); 22.1.2
A. Turos	(a) **83**, 437 (1984)	2 (p. 46) and 14.3.1 (p. 235); 3; 22.1.2
A. Twardowski	(b) **122**, K147 (1984)	13.5.1 (p. 200); 22.4.4
B. P. Tyagi	(a) **85**, 603 (1984)	14.3.4 (p. 252); 13.4; 22.1.2
E. A. Tyavlovskaya	(a) **82**, 399 (1984)	1.4 (p. 28) and 1.6 (p. 37); 22.2.1
T. Tyc	(a) **81**, 171 (1984)	8 (p. 84); 22.6
I. Tychina	(a) **81**, K191 (1984)	11 (p. 123) and 14.3 (p. 222); 22.3
A. P. Tyutnev	(a) **83**, 365 (1984)	11 (p. 126) and 14.3 (p. 225); 22.9
	(a) **84**, 327 (1984)	14.3 (p. 227); 11; 22.9
	(a) **84**, 585 (1984)	11 (p. 129) and 14.3 (p. 227); 22.9
	(a) **85**, 591 (1984)	14.3 (p. 230); 11; 22.9
	(a) **86**, 363 (1984)	11 (p. 132) and 14.4 (p. 259); 22.9
	(a) **86**, 709 (1984)	14.3.2 (p. 240); 11; 22.9
V. I. Tyutyunnik	(b) **121**, K75 (1984)	19 (p. 305); 22
Juh Tzeng Lue	see under letter "J"	

O. Uemura	(a) **85**, K5 (1984)	1 (p. 10) and 2 (p. 50); 22.5.2
Ya. A. Ugai	(a) **86**, K51 (1984)	14.3 (p. 231) and 15 (p. 268); 22.1

S. R. UGLOV	(b) **125**, 251 (1984)	11 (p. 129) and 20 (p. 318); 13; 22.1.3; 22.5.2
	(b) **126**, 565 (1984)	11 (p. 133); 13; 22.1.2
K. UHLIG	(b) **123**, K177 (1984)	2 (p. 46) and 14.3 (p. 225); 13.4; 22
K. UHLMANN	(a) **85**, K19 (1984)	1.5 (p. 35) and 13.3 (p. 169); 17; 21.1.1
W. ULRICI	(a) **84**, 243 (1984)	13.4 (p. 185) and 20.1 (p. 333); 22.2.1
	(a) **86**, 691 (1984)	1.4 (p. 32) and 13.4 (p. 195); 20.1; 22.2.1
A. G. ULYASHIN	(b) **125**, K75 (1985)	10.2 (p. 116) and 13.4 (p. 188); 22.1; 22.1.2
CHUNG-IN UM	see under letter "C"	
M. UMENO	(a) **83**, 207 (1984)	10.2 (p. 111); 22.1.2
B. S. UMURZAKOV	(b) **124**, K75 (1984)	13.4 (p. 184) and 20.3 (p. 353); 10.2; 22.5.2
H.-J. UNGER	(b) **121**, 371 (1984)	2 (p. 39) and 13.4 (p. 172); 22
K. UNGER	(b) **123**, 229 (1984)	13.1 (p. 158)
	(a) **84**, 677 (1984)	13.5 (p. 197); 14.3.3; 22.2.2
D. UNGUR	(b) **124**, K117 (1984)	2 (p. 49) and 19 (p. 311); 22.8
H.-G. UNRUH	(a) **84**, K177 (1984)	14.4 (p. 258); 22.5
	(b) **126**, 115 (1984)	1.2 (p. 19) and 14.4 (p. 258); 6; 22.8.1
S. UNTERRICKER	(a) **85**, 455 (1984)	10.2 (p. 118) and 11 (p. 131); 5; 22.2.2
Y. UNUMA	(b) **125**, 735 (1984)	13.5.1 (p. 203) and 20.3 (p. 355); 22.4.2
U. N. UPADHYAYA	(b) **125**, 861 (1984)	14.2 (p. 218) and 18.2 (p. 294); 22.8
G. C. UPRETI	(b) **122**, K169 (1984)	19 (p. 308); 18.1; 22.8.1
I. N. URALTSEV	(b) **121**, K191 (1984)	13.5.1 (p. 199) and 20.1 (p. 322); 22.4.1
	(b) **122**, 221 (1984)	18 (p. 279) and 20.1 (p. 322); 13.5.1; 22.4.2
	(b) **124**, 213 (1984)	13.5.1 (p. 201) and 20.1 (p. 332); 18; 22.4.2
L. URAY	(a) **82**, 545 (1984)	1.3 (p. 23) and 14.1 (p. 212); 15; 21
	(a) **84**, 65 (1984)	1.3 (p. 24); 1.4; 9; 14.1; 15; 21; 21.1.1
K. URBAN	(a) **81**, 145 (1984)	1.2 (p. 14); 1.3; 11; 21.1
V. URBONAVIČIUS	(a) **84**, 387 (1984)	1.2 (p. 18) and 14.4.1 (p. 262); 8; 19; 22.8
A. R. URMANOV	(a) **83**, 573 (1984)	11 (p. 127); 22.1.2
N. URYÛ	(b) **125**, 551 (1984)	18.2 (p. 294)
V. V. USTINOV	(b) **123**, 289 (1984)	13.3 (p. 168); 21
D. I. UZUNOV	(b) **126**, K19 (1984)	1.2 (p. 20) and 10 (p. 96)

J. VACÍK	(a) **83**, 165 (1984)	11 (p. 126); 22.1.2; 22.6
K. S. VAHVASELKÄ	(a) **83**, 103 (1984)	1.3 (p. 23) and 2 (p. 44); 14.1; 21; 21.1
B. A. VAID	(b) **122**, 423 (1984)	2 (p. 42); 22.1.1; 22.1.2
	(b) **125**, 505 (1984)	2 (p. 51) and 7 (p. 83); 21
	(b) **126**, 59 (1984)	2 (p. 52); 8; 22.5.2
S. N. VAIDYA	(a) **86**, 565 (1984)	1.2 (p. 21) and 6 (p. 80); 8; 23
A. A. VAKHNENKO	(b) **122**, 239 (1984)	13.5.1 (p. 199); 6; 22
K. VALACKA	(b) **122**, K179 (1984)	20.1 (p. 326); 13.4; 22
J. A. VALBIS	(b) **123**, 335 (1984)	10.2 (p. 110) and 20.3 (p. 350); 22.6
	(b) **125**, K165 (1984)	20.3 (p. 356); 22.6
L. M. VALIEV	(a) **84**, K51 (1984)	18 (p. 281); 22.8
G. VALIUKONIS	(b) **122**, 623 (1984)	13.1 (p. 157) and 20.1 (p. 325); 22
S. N. VALKOVSKII	(a) **83**, 179 (1984)	10.2 (p. 111); 22.6
V. VALVODA	(a) **81**, 203 (1984)	4 (p. 58); 21.1
	(a) **81**, K113 (1984)	1.6 (p. 36); 21.1
	(a) **83**, K123 (1984)	6 (p. 74) and 8 (p. 87); 21
D. M. VANDERWALKER	(a) **82**, 93 (1984)	10.2 (p. 108); 22.1.2
	(a) **85**, 61 (1984)	1.4 (p. 31); 22.1.2
	(a) **86**, 507 (1984)	1.6 (p. 38) and 10.2 (p. 120); 1.2; 21.1; 22.1.2
M. VANĚČEK	(a) **83**, 617 (1984)	13.1 (p. 160) and 20.1 (p. 331); 2; 22.1.2
NGUYEN VAN HIEU	see under letter "N"	
A. V. VANNIKOV	(a) **83**, 365 (1984)	11 (p. 126) and 14.3 (p. 225); 22.9
	(a) **84**, 327 (1984)	14.3 (p. 227); 11; 22.9
	(a) **84**, 585 (1984)	11 (p. 129) and 14.3 (p. 227); 22.9

	(a) **85**,	591 (1984)	14.3 (p. 230); 11; 22.9
	(a) **86**,	363 (1984)	11 (p. 132) and 14.4 (p. 259); 22.9
	(a) **86**,	709 (1984)	14.3.2 (p. 240); 11; 22.9
D. Varchon	(a) **83**,	235 (1984)	10.1 (p. 99); 12.1; 21; 21.1.1
R. A. Vardanian	(b) **126**,	K83 (1984)	10.2 (p. 119) and 14.3 (p. 230); 22.1.1
R. S. Vardanian	(a) **84**,	K7 (1984)	1 (p. 9)
P. Varotsos	(b) **125**,	K109 (1984)	10.2 (p. 117); 9; 22.5.3
G. B. Varshney	(b) **121**,	493 (1984)	4 (p. 58); 12.2; 22.5.2
	(b) **126**,	495 (1984)	4 (p. 63); 12.2; 22.5.2
R. K. Varshney	(b) **122**,	65 (1984)	12.2 (p. 139); 22.5.2
	(b) **126**,	77 (1984)	12.2 (p. 142); 22.5.2
A. R. Vasconcellos	(b) **126**,	63 (1984)	8 (p. 89) and 11 (p. 131); 6; 22; 22.2.1
A. N. Vasilev	(b) **125**,	477 (1984)	6 (p. 77); 13.4
T. K. Vasileva	(a) **86**,	K109 (1984)	14.4 (p. 259); 22.8.1
A. A. Vasilkevich	(b) **121**,	623 (1984)	13.4 (p. 174); 21.4
V. V. Vasin	(b) **121**,	451 (1984)	2 (p. 40); 21; 21.1.1
P. Văsina	(a) **84**,	693 (1984)	14.3.3 (p. 245); 22.2.1
L. V. Vassilev	(b) **121**,	203 (1984)	13.1 (p. 153); 1.2; 22.8
	(b) **125**,	K177 (1984)	14.3 (p. 229) and 15 (p. 267); 1.2; 22.3.1
B. Vasvári	(b) **124**,	683 (1984)	13 (p. 149); 2; 21.1
V. N. Vasyukov	(a) **84**,	223 (1984)	18.4 (p. 302) and 19 (p. 310); 22.5
Yu. D. Vaulin	(a) **84**,	K13 (1984)	1.6 (p. 37) and 2 (p. 48); 22.2.1; 22.2.4
C. Vautier	(a) **85**,	399 (1984)	1.5 (p. 35) and 3 (p. 56); 2; 22.1.3
J. Vávra	(a) **83**,	455 (1984)	1.1 (p. 12); 22.1.2
P. R. Vaya	(a) **83**,	693 (1984)	14.3.3 (p. 243) and 16 (p. 271); 22.1.2
L. Vecsernyés	(a) **85**,	133 (1984)	1.4 (p. 32) and 10.2 (p. 116); 20.1; 22.1.2
H. Vega	(a) **86**,	117 (1984)	1.3 (p. 26); 18; 21.1; 21.1.1
A. S. Vengurlekar	(a) **81**,	407 (1984)	14.3.3 (p. 241); 11; 14.4; 22.1.2
R. K. Veliev	(a) **86**,	K19 (1984)	8 (p. 89); 22.8
T. F. Veremeichik	(b) **124**,	719 (1984)	13.4 (p. 186) and 20.1 (p. 334); 22.6
G. S. Verma	(b) **124**,	K5 (1984)	8 (p. 87)
R. D. Verma	(a) **83**,	K19 (1984)	12.2 (p. 140) and 20.3 (p. 350); 10.2; 22.5.2; 22.8.1
T. S. Verma	(b) **121**,	487 (1984)	12.2 (p. 139); 14.4; 22.5
N. V. Vershinina	(a) **86**,	717 (1984)	11 (p. 133) and 14.3.3 (p. 246); 14.3.4; 22.1.2
A. Vervoitte	(b) **124**,	87 (1984)	13.5.1 (p. 201) and 13.5.4 (p. 208); 18.4; 20.1; 22.5
Ya. I. Verzimacha	(a) **82**,	K111 (1984)	14.3.3 (p. 242) and 16 (p. 270); 22.9
S. Ves	(b) **122**,	613 (1984)	20.1 (p. 325); 13.1; 22.8
Nguyen Ai Viet	see under letter "N"		
C. Viger	(a) **83**,	K103 (1984)	2 (p. 47) and 14.3.1 (p. 235); 22.1.3
D. Vignaud	(b) **125**,	785 (1984)	13.4 (p. 190) and 20.1 (p. 338); 10.2; 22.2.3
	(b) **126**,	351 (1984)	13.1 (p. 164) and 20.1 (p. 340); 10.2; 22.2
J. Vignolo	(b) **124**,	465 (1984)	2 (p. 48); 21; 21.7
	(b) **126**,	K1 (1984)	2 (p. 53); 21.7
G. Viliani	(b) **123**,	659 (1984)	13.4 (p. 181); 20.3; 22.6
S. D. Le Vine	(b) **121**,	K125 (1984)	2 (p. 40) and 18 (p. 278); 21.1.1
Ha Vinh Tan	see under letter "H"		
A. P. Vinogradov	(b) **121**,	K215 (1984)	14.3 (p. 221); 22.3
S. V. Vintsents	(a) **85**,	273 (1984)	13.4 (p. 189); 19; 20.3; 22.1.1; 22.1.2
M. I. Vladimir	(b) **125**,	543 (1984)	18 (p. 282)
V. I. Vladimirov	(b) **122**,	35 (1984)	2 (p. 41); 10.2; 12.2
Yu. L. Vladislavskii	(b) **125**,	155 (1984)	10.2 (p. 116); 22.5.2
N. Vlahovici	(a) **82**,	K125 (1984)	1.2 (p. 16); 14.4.1; 22.8.1
N. A. Vlasenko	(a) **81**,	657 (1984)	14.3.4 (p. 249) and 20.3 (p. 346); 14.3.1; 22.6
	(a) **81**,	661 (1984)	14.3.4 (p. 249) and 20.3 (p. 346); 22.4.1
S. V. Vlasov	(b) **121**,	241 (1984)	13.1 (p. 153); 21.4
R. M. Vlasova	(b) **122**,	549 (1984)	20.1 (p. 325); 13.5.1; 22.9
C. M. van Vliet	(b) **121**,	429 (1984)	14.1 (p. 211); 21.6

M. Voelskow	(a)	**82**,	475 (1984)	11 (p. 125); 14.3; 22.1.2
	(a)	**82**,	K5 (1984)	2 (p. 42) and 3 (p. 55); 11; 22.1.2
	(a)	**82**,	K121 (1984)	1.5 (p. 33) and 3 (p. 56); 2; 11; 22.1.2
	(a)	**86**,	781 (1984)	11 (p. 133) and 14.3.1 (p. 237); 22.1.2
S. V. Voitikov	(b)	**126**,	415 (1984)	14.3.4 (p. 253)
V. V. Voitsekhovskii	(a)	**85**,	167 (1984)	11 (p. 129) and 20.1 (p. 337); 10.2; 22.5
G. Völkel	(a)	**85**,	257 (1984)	19 (p. 312); 22.9
J. Völkl	(b)	**123**,	85 (1984)	9 (p. 92); 21
F. Völklein	(a)	**81**,	585 (1984)	14.1 (p. 211) and 14.3.1 (p. 233); 8; 15; 21.7
O. L. Volkova	(a)	**82**,	441 (1984)	10.2 (p. 110); 22.5.2
V. G. Vologin	(a)	**85**,	529 (1984)	18.3 (p. 299); 22.8.2
G. Vones	(b)	**123**,	773 (1984)	14.2 (p. 217)
S. A. Vorobev	(a)	**81**,	217 (1984)	11 (p. 122); 14.3; 22.1.2; 22.2.1
	(b)	**124**,	641 (1984)	11 (p. 128); 13; 22.1.2
Yu. V. Vorobev	(b)	**123**,	295 (1984)	13.4 (p. 179) and 14.3 (p. 224); 22.2.1
S. A. Vorobiev	(b)	**122**,	29 (1984)	11 (p. 123); 21
	(b)	**125**,	251 (1984)	11 (p. 129) and 20 (p. 318); 13; 22.1.3; 22.5.2
	(a)	**85**,	K31 (1984)	11 (p. 130) and 20 (p. 318); 22.1.2; 22.2.1
	(b)	**126**,	565 (1984)	11 (p. 133); 13; 22.1.2
I. N. Voronin	(b)	**124**,	793 (1984)	14.3 (p. 227)
V. I. Voronin	(a)	**86**,	K59 (1984)	14.1 (p. 215) and 14.2 (p. 219); 1.2; 21; 21.1
Yu. I. Voronovskii	(a)	**86**,	133 (1984)	14.4.1 (p. 263); 22.8.1
M. Vrána	(a)	**83**,	455 (1984)	1.1 (p. 12); 22.1.2
S. Vratislav	(a)	**83**,	K117 (1984)	1.2 (p. 17); 22.8.1
G. M. Vujičić	(b)	**123**,	K93 (1984)	14.2 (p. 217)
Le Vu Ky	see under letter "K"			
A. P. Vyatkin	(a)	**83**,	701 (1984)	14.3.3 (p. 243); 13.4; 22.2.1
E. G. Vyatkin	(b)	**122**,	29 (1984)	11 (p. 123); 21
A. A. Wafik	(a)	**86**,	271 (1984)	18.3 (p. 299); 14.3; 22.8.2
A. H. Wafik	(a)	**81**,	513 (1984)	18.2 (p. 289); 21.1.1
C. Wagner	(a)	**82**,	149 (1984)	11 (p. 124) and 14.3 (p. 223); 22.1.2
M. Wagner	(b)	**122**,	73 (1984)	6 (p. 69) and 8 (p. 85); 22.8.1
	(a)	**83**,	69 (1984)	1.2 (p. 16) and 3 (p. 56); 22.1.2
	(a)	**83**,	K1 (1984)	3 (p. 56) and 11 (p. 126); 22.1.2
J. Walachova	(a)	**81**,	K91 (1984)	14.3.3 (p. 241); 22.2
B. Walker	(b)	**124**,	741 (1984)	10.2 (p. 114) and 13.4 (p. 186); 20.3; 22.5.2
R. G. Walmsley	(a)	**82**,	55 (1984)	1.4 (p. 28); 21.1
F. Walz	(a)	**81**,	227 (1984)	10.1 (p. 97) and 18.2 (p. 288); 11; 21.1; 21.1.1
	(a)	**82**,	179 (1984)	18.2 (p. 289); 18.3; 21.1.1
	(a)	**85**,	503 (1984)	18.2 (p. 294); 2; 10.1; 21.1; 21.1.1
U. Walzer	(b)	**125**,	55 (1984)	4 (p. 61); 12.1; 13.1; 21.1
	(b)	**126**,	752 (1984)	4 (p. 61); 12.1; 13.1; 21.1
K. Wandel	(a)	**81**,	739 (1984)	20.3 (p. 347); 22.2.4
	(a)	**82**,	K205 (1984)	14.3.3 (p. 242); 22.2.4
B. M. Wanklyn	(a)	**86**,	75 (1984)	1.2 (p. 20); 22.5.4
G. H. Wannier	(b)	**121**,	603 (1984)	13 (p.145)
S. Waplak	(b)	**123**,	27 (1984)	14.4.1 (p. 261); 4; 19; 22.8.1
Y. Warita	(a)	**86**,	729 (1984)	10.2 (p. 121) and 14.4 (p. 259); 22.5.4
F. Warkusz	(a)	**84**,	K85 (1984)	14.1 (p. 213)
A. V. R. Warrier	(a)	**82**,	K181 (1984)	20.1 (p. 326); 13.1; 22.8
J. Waşicki	(b)	**123**,	K115 (1984)	1.2 (p. 16) and 6 (p. 73); 22.9
S. M. Wasim	(a)	**81**,	K77 (1984)	13.4 (p. 173) and 14.3 (p. 221); 22.8
	(a)	**82**,	553 (1984)	13.4 (p. 179) and 14.3 (p. 224); 8; 22.8
H. F. J. Watanabe	(a)	**83**,	467 (1984)	1.2 (p. 17); 12.2; 22.8.1
	(a)	**86**,	841 (1984)	1.2 (p. 17); 12.2; 22.8.1

T. Watanabe	(a) **85**,	K5 (1984)	1 (p. 10) and 2 (p. 50); 22.5.2
Y. Watanabe	(b) **122**,	309 (1984)	13.4 (p. 176); 10.2; 18.1; 19; 22.8
D. N. Waters	(b) **123**,	653 (1984)	20.1 (p. 330); 2; 13; 22.6
A. Watterich	(b) **121**,	117 (1984)	10.2 (p. 105) and 20.1 (p. 320); 22.5.2
	(a) **84**,	165 (1984)	10.2 (p. 113); 19; 20.1; 22.6
Z. Wawrzak	(a) **83**,	K113 (1984)	1.1 (p. 12); 22.9
	(a) **84**,	K5 (1984)	1.1 (p. 12) and 8 (p. 87); 22.9
B. Weber	(a) **85**,	K1 (1984)	1.6 (p. 38) and 11 (p. 129); 21
H. H. Weber	(b) **122**,	183 (1984)	16 (p. 269) and 20.3 (p. 348); 22.4.1
H.-J. Weber	(b) **126**,	305 (1984)	13.5.1 (p. 204) and 20.1 (p. 339); 22.6
D. Wegener	(b) **121**,	413 (1984)	2 (p. 39) and 14.3 (p. 220); 6; 22
H. Wegener	(b) **121**,	K103 (1984)	14.1 (p. 211); 21
R. K. Wehner	(b) **126**,	91 (1984)	6 (p. 78); 22.1.2
C. M. Weinert	(b) **124**,	287 (1984)	6 (p. 74) and 20.1 (p. 332); 13.4; 22.2.2; 22.4.1
H. Weinert	(a) **81**,	739 (1984)	20.3 (p. 347); 22.2.4
	(a) **82**,	K205 (1984)	14.3.3 (p. 242); 22.2.4
G. F. Weir	(b) **124**,	279 (1984)	13 (p. 148); 2
G. Weiser	(b) **123**,	271 (1984)	13.5 (p. 196) and 20.1 (p. 327); 22.9
P. Weissbrodt	(a) **81**,	259 (1984)	1.6 (p. 36) and 11 (p. 122); 21.1
C. Weissmantel	(a) **82**,	K5 (1984)	2 (p. 42) and 3 (p. 55); 11; 22.1.2
	(a) **82**,	K121 (1984)	1.5 (p. 33) and 3 (p. 56); 2; 11; 22.1.2
L. Wendler	(b) **121**,	K119 (1984)	6 (p. 69) and 13.5 (p. 196); 22.1.2; 22.6
	(b) **123**,	469 (1984)	6 (p. 72) and 20.1 (p. 329); 13.5; 22.2.1; 22.2.2; 22.5.3
	(b) **124**,	K161 (1984)	13.3 (p. 169) and 13.5.2 (p. 206)
	(b) **125**,	K27 (1984)	6 (p. 77) and 20 (p. 318); 13.5.2; 21.6; 22.2.2
H. Werheit	(b) **125**,	779 (1984)	1.2 (p. 19); 13.4; 20.1; 22.1
A. Werner	(b) **123**,	209 (1984)	13.1 (p. 158); 21.4
M. Werner	(a) **86**,	543 (1984)	1.3 (p. 26); 10.1; 21; 21.1
P. Werner	(a) **85**,	83 (1984)	1.4 (p. 31); 10.2; 22.4.2
U. Werner	(a) **85**,	205 (1984)	11 (p. 129); 13; 22.2.4
J. Wernisch	(a) **84**,	39 (1984)	1.3 (p. 24) and 1.6 (p. 37); 21; 21.1
W. Wesch	(a) **85**,	K169 (1984)	14.3.4 (p. 253); 9; 11; 22.2.1
D. Weschenfelder	(b) **121**,	145 (1984)	18.4 (p. 300); 13.4; 18.3; 22.8.1
C. Wesołowska	(a) **82**,	K67 (1984)	13.4 (p. 177); 22.6
J. M. Wesselinowa	(b) **121**,	317 (1984)	13 (p. 144) and 14.4.1 (p. 260); 22.8.1
	(b) **121**,	K147 (1984)	18.2 (p. 288); 13.5.4; 22
	(b) **121**,	K157 (1984)	13 (p. 145) and 14.4.1 (p. 260)
	(b) **123**,	585 (1984)	18.2 (p. 291); 6; 13; 22
	(b) **123**,	K129 (1984)	13 (p. 147) and 18.2 (p. 291); 22
	(b) **126**,	229 (1984)	18.2 (p. 294); 22.8
K. Wesser	(b) **126**,	157 (1984)	10.2 (p. 118) and 13.4 (p. 192); 22.1.2
G. Weyer	(a) **81**,	239 (1984)	10.2 (p. 105); 5; 22.1.2
E. Wieser	(a) **81**,	247 (1984)	9 (p. 91); 11; 22.1.2
	(a) **82**,	171 (1984)	11 (p. 124); 14.3; 22.1.2
	(a) **82**,	475 (1984)	11 (p. 125); 14.3; 22.1.2
	(a) **82**,	K5 (1984)	2 (p. 42) and 3 (p. 55); 11; 22.1.2
	(a) **82**,	K121 (1984)	1.5 (p. 33) and 3 (p. 56); 2; 11; 22.1.2
M. H. van Wijngaarden	(a) **81**,	209 (1984)	10.1 (p. 97); 21
T. Wiktorczyk	(a) **82**,	K67 (1984)	13.4 (p. 177); 22.6
R. Wille	(b) **126**,	305 (1984)	13.5.1 (p. 204) and 20.1 (p. 339); 22.6
M. R. Willis	(a) **85**,	283 (1984)	14.3 (p. 229); 22.9
K. Winer	(b) **124**,	473 (1984)	2 (p. 48) and 6 (p. 75); 22.1.2
W. Winkler	(a) **84**,	193 (1984)	18.2 (p. 292); 5; 22.6
H. Wipf	(b) **123**,	85 (1984)	9 (p. 92); 21
Th. Wirth	(a) **82**,	459 (1984)	1.6 (p. 37) and 11 (p. 125); 21
W. Wischert	(a) **83**,	631 (1984)	13.4 (p. 182) and 20.3 (p. 351); 22.8

M. Wobst	(a) **84**, 11 (1984)	2 (p. 47); 1.3; 22.7
	(a) **84**, K97 (1984)	1.3 (p. 25) and 2 (p. 49); 22.7
U. Woggon	(a) **81**, 739 (1984)	20.3 (p. 347); 22.2.4
	(a) **82**, K205 (1984)	14.3.3 (p. 242); 22.2.4
R. Wojciechowski	(b) **124**, 599 (1984)	18.1 (p. 285)
W. Wojdowski	(b) **123**, K101 (1984)	6 (p. 73) and 20.1 (p. 330); 22.8
L. Wojtczak	(b) **121**, 47 (1984)	6 (p. 67)
A. Wolf	(a) **82**, K5 (1984)	2 (p. 42) and 3 (p. 55); 11; 22.1.2
D. Wolf	(a) **86**, 685 (1984)	13.4 (p. 195) and 20.3 (p. 358); 22.8
H. Wonn	(b) **123**, 787 (1984)	13.1 (p. 160); 21.6
J. A. Woollam	(a) **85**, 69 (1984)	1.3 (p. 25) and 1.5 (p. 34); 22
F. Wooten	(b) **124**, 473 (1984)	2 (p. 48) and 6 (p. 75); 22.1.2
J. Wosik	(a) **82**, K177 (1984)	2 (p. 43) and 18.2.1 (p. 296); 21.1.1
D. Wruck	(a) **86**, 691 (1984)	1.4 (p. 32) and 13.4 (p. 195); 20.1; 22.2.1
W. Wüchner	(b) **126**, 381 (1984)	1.2 (p. 19) and 13.4 (p. 193); 18.4; 20.1; 22.8.1
F. Wulf	(a) **86**, 789 (1984)	11 (p. 133) and 14.3.3 (p. 246); 14.3.2; 22.1.2
H.-J. Wünsche	(b) **124**, 747 (1984)	13.4 (p. 186) and 13.5.1 (p. 202); 20.3; 22.2.4
B. Wyncke	(b) **125**, 493 (1984)	6 (p. 77) and 20.1 (p. 337); 1.2; 14.4.1; 22.8.1
S. Wysocki	(b) **121**, 731 (1984)	13.4 (p. 174); 14.3.2; 22.6
Xian-Wu Zou	(b) **121**, K171 (1984)	13.1 (p. 155); 21.2
Hoang Xuan Nguyen	see under letter "N"	
J. V. Yakhmi	(a) **85**, K89 (1984)	14.1 (p. 213); 21.1
E. B. Yakimov	(a) **84**, 443 (1984)	10.2 (p. 115) and 13.4 (p. 186); 14.4; 22.1.2
	(a) **84**, K43 (1984)	1.4 (p. 31) and 10.2 (p. 114); 14.3.4; 22.1.2
A. V. Yakovlev	(a) **83**, K153 (1984)	13.4 (p. 183) and 20.1 (p. 331); 22.8.2
H. Yamaguchi	(a) **86**, 237 (1984)	10.1 (p. 103); 14.1; 21
S. Yamashita	(a) **81**, 177 (1984)	1.2 (p. 14); 6; 8; 22.5.4
S. C. Yan	(a) **86**, 593 (1984)	10.1 (p. 103) and 12.1 (p. 138); 21
Chuan-zhang Yang	see under letter "C"	
V. K. Yanovskii	(a) **81**, 399 (1984)	1.2 (p. 14) and 14.4.1 (p. 260); 14.3; 15; 20.1; 22.8.1
V. M. Yartsev	(b) **126**, 501 (1984)	14.3 (p. 231); 6; 20.1; 22.9
N. A. Yarykin	(a) **84**, 443 (1984)	10.2 (p. 115) and 13.4 (p. 186); 14.4; 22.1.2
G. Y. Yashin	(b) **121**, K13 (1984)	6 (p. 68) and 20.1 (p. 320); 22.5
Yu. N. Yashkir	(b) **123**, 691 (1984)	13.5 (p. 197) and 20.1 (p. 330); 6; 22
V. Ya. Yaskolko	(a) **85**, K145 (1984)	17 (p. 277); 10.2; 22.8.1
K. Yasutake	(a) **83**, 207 (1984)	10.2 (p. 111); 22.1.2
V. Yu. Yavid	(a) **81**, 469 (1984)	11 (p. 123); 22.1.1
H. Q. Ye	(a) **83**, 77 (1984)	1.3 (p. 23) and 1.4 (p. 29); 21.1
Yen Tzu-Tung	see under letter "T"	
Do Yi-Jing	(b) **124**, K113 (1984)	8 (p. 88); 21.2
S. Yoneyama	(a) **82**, 135 (1984)	10.2 (p. 108); 13.4; 22.1.1
Youn Soo Lim	(b) **122**, K187 (1984)	14.3 (p. 223); 22.6
C. J. Young	(a) **84**, 517 (1984)	10.1 (p. 100); 21
Young Hwan Choi	(b) **122**, K187 (1984)	14.3 (p. 223); 22.6
S. A. Yousef	(a) **83**, 263 (1984)	13.5.2 (p. 205) and 20.1 (p. 328); 13.1; 21
S. G. Yudin	(a) **86**, 573 (1984)	8 (p. 90); 11; 22.1; 22.2
Yu-kui Zhou	(b) **125**, 323 (1984)	6 (p. 76) and 13.5.1 (p. 202)
M. S. Yunusov	(a) **81**, K145 (1984)	11 (p. 123) and 14.3.4 (p. 249); 22.1.2
	(b) **126**, 311 (1984)	13.1 (p. 164) and 13.4 (p. 192); 22.1.2
V. P. Yurev	(a) **83**, 245 (1984)	18.4 (p. 301); 1.4; 22.5
V. N. Zabluda	(b) **125**, 751 (1984)	18.2 (p. 294) and 20.1 (p. 337); 13.4; 22.8.1
R. Zach	(a) **84**, 229 (1984)	18.2 (p. 293) and 18.4 (p. 303); 1.2; 22.8

M. S. Zaghloul	(a) **85**, 429 (1984)	8 (p. 89); 22.5.2
S. Y. Zaginaichenko	(a) **82**, K163 (1984)	1.2 (p. 16) and 18 (p. 279); 21; 21.1
W. Zagozdzon-Wosik	(a) **86**, K125 (1984)	14.3.3 (p. 247); 22.1.2
M. A. Zaikovskaya	(a) **81**, K145 (1984)	11 (p. 123) and 14.3.4 (p. 249); 22.1.2
V. M. Zakharov	(b) **121**, 293 (1984)	2 (p. 39) and 20.1 (p. 320); 13.1; 22
Yu. V. Zakharov	(b) **125**, 197 (1984)	14.1 (p. 213); 18.2; 21
T. Zaleski	(b) **123**, K115 (1984)	1.2 (p. 16) and 6 (p. 73); 22.9
V. M. Zaletin	(b) **121**, K13 (1984)	6 (p. 68) and 20.1 (p. 320); 22.5
	(b) **125**, K103 (1984)	6 (p. 78) and 20.1 (p. 338); 22.5
M. Załużny	(b) **123**, K57 (1984)	13.1 (p. 159) and 20.1 (p. 328); 22; 22.2.1
V. I. Zametin	(b) **124**, 625 (1984)	20.1 (p. 333); 1.2; 14.4.1; 22; 22.5.2; 22.8.1; 22.9
M. B. Zapart	(a) **82**, 67 (1984)	1.2 (p. 15) and 19 (p. 307); 13.4; 22.8.1
W. Zapart	(a) **82**, 67 (1984)	1.2 (p. 15) and 19 (p. 307); 13.4; 22.8.1
D. Zappe	(a) **86**, 407 (1984)	14.3.4 (p. 253); 13.4; 22.5
V. V. Zaretskii	(a) **86**, 517 (1984)	1.2 (p. 20); 22.3
V. G. Zavodinskii	(a) **82**, 345 (1984)	1.5 (p. 33) and 3 (p. 55); 2; 22.1.2
N. S. Zayats	(a) **82**, 503 (1984)	13.4 (p. 178) and 20.3 (p. 349); 22.2.1
	(a) **83**, 625 (1984)	13.4 (p. 182) and 20.3 (p. 351); 11; 22.2.1
	(a) **84**, 567 (1984)	13.4 (p. 186) and 20.3 (p. 353); 11; 22.2.1
S. Zazubovich	(b) **125**, 139 (1984)	10.2 (p. 115) and 20.3 (p. 354); 22.5.2
	(b) **126**, 407 (1984)	13.4 (p. 193) and 20.3 (p. 357); 22.5.4
	(b) **126**, 653 (1984)	13.4 (p. 194) and 20.3 (p. 358); 22.5.2
	(b) **126**, K63 (1984)	13.4 (p. 193) and 20.3 (p. 357); 22.5.2
K. Žďánský	(a) **81**, 353 (1984)	13.4 (p. 173) and 14.3.3 (p. 241); 22.2.2
	(a) **85**, 219 (1984)	13.4 (p. 188) and 14.3.3 (p. 245); 22.4.3
J. J. Żebrowski	(a) **85**, 189 (1984)	18.2.1 (p. 297); 22.8.2
E. Zedler	(a) **81**, 445 (1984)	2 (p. 40); 21.1.1
B. Žekš	(b) **122**, 399 (1984)	1.2 (p. 15); 22.8.1
M. Zelený	(a) **81**, 307 (1984)	18 (p. 278); 22.7
	(a) **82**, 191 (1984)	18 (p. 279); 22.7
H. R. Zelsmann	(b) **125**, 795 (1984)	13.4 (p. 190) and 20.3 (p. 356); 13.1; 22.4.3
N. Zeman	(b) **124**, 767 (1984)	13.4 (p. 186); 5; 22.8.1
L. A. Zemnukhova	(a) **84**, 387 (1984)	1.2 (p. 18) and 14.4.1 (p. 262); 8; 19; 22.8
G. Zentai	(a) **86**, 383 (1984)	16 (p. 273); 2; 14.3.1; 22; 22.1.2
A. Zentko	(a) **82**, K159 (1984)	11 (p. 125) and 18.2 (p. 290); 2; 21.1.1
B. S. Zhang	(a) **84**, 465 (1984)	10.1 (p. 100) and 12.1 (p. 136); 21
J. Zhang	(b) **121**, K165 (1984)	13.5.3 (p. 207); 22
	(a) **83**, 357 (1984)	14.3 (p. 225); 15; 22
L. D. Zhang	(a) **84**, 465 (1984)	10.1 (p. 100) and 12.1 (p. 136); 21
Z. Zhang	(a) **83**, 77 (1984)	1.3 (p. 23) and 1.4 (p. 29); 21.1
M. Zhanibekov	(b) **121**, K205 (1984)	19 (p. 306) and 20.1 (p. 322); 6; 13.4; 22.5.2
K. R. Zhdanov	(a) **83**, 147 (1984)	6 (p. 72) and 8 (p. 86); 22
Shu Zhen	see under letter "S"	
Zhen-ya Li	(b) **126**, K125 (1984)	18.2 (p. 295); 18.4
G. M. Zhidomirov	(b) **124**, 699 (1984)	13.1 (p. 161) and 13.3 (p. 168); 22.6
S. K. Zhilinskii	(a) **81**, 367 (1984)	14 (p. 210); 22
Z. I. Zhmurova	(a) **84**, K81 (1984)	20.2 (p. 343) and 20.3 (p. 353); 13.4; 22.8
Yu-kui Zhou	see under letter "Y"	
M. F. Zhu	(a) **86**, 471 (1984)	1.5 (p. 35); 2; 21
E. A. Zhurakovskii	(b) **121**, 583 (1984)	13.1 (p. 154); 22
A. F. Zhuravlev	(b) **126**, 207 (1984)	10.1 (p. 102) and 18 (p. 282); 19
A. V. Zhuravlev	(a) **83**, K135 (1984)	18.1 (p. 285) and 18.4 (p. 302); 1.2; 22.8
K. Žičkus	(b) **121**, K51 (1984)	20.1 (p. 320); 13.1; 22.8
	(b) **125**, 645 (1984)	14.4.1 (p. 263); 20.1; 22.8
M. Ziegler	(a) **84**, 79 (1984)	10.2 (p. 113); 1.4; 22.2.1
A. van der Ziel	(b) **121**, 429 (1984)	14.1 (p. 211); 21.6
R. Zielińska	(b) **123**, K17 (1984)	6 (p. 72) and 10.2 (p. 111); 20.1; 22.4.3

O. Ziep	(b) **123**,	247 (1984)	13.1 (p. 158) and 14.3 (p. 224); 20.3; 22.8
P. Ziesche	(b) **121**,	705 (1984)	13 (p. 145); 21.6
	(b) **123**,	787 (1984)	13.1 (p. 160); 21.6
G. Zimmerer	(b) **124**,	335 (1984)	13.5.1 (p. 201) and 20.3 (p. 352); 23
J. Zimmermann	(a) **81**,	569 (1984)	14.3.2 (p. 238) and 14.3.3 (p. 241); 22.1.2
R. Zimmermann	(b) **121**,	609 (1984)	13.5 (p. 196); 22.1.1; 22.2.1
	(b) **124**,	191 (1984)	13.5.1 (p. 201) and 20 (p. 317); 22.4.1
	(b) **124**,	201 (1984)	13.5 (p. 197); 13.5.1; 22.2.1
O. Zmeškal	(a) **85**,	619 (1984)	14.3.4 (p. 252); 13.4; 22.1.2; 22.4.2
M. Zoaeter	(b) **124**,	403 (1984)	14.3 (p. 226) and 15 (p. 266); 22
O. J. Żogał	(a) **82**,	K153 (1984)	9 (p. 92) and 19 (p. 308); 21.4
V. F. Zolin	(b) **124**,	K69 (1984)	13.4 (p. 184) and 20.3 (p. 352); 22.8
I. Zólomy	(a) **82**,	K209 (1984)	14.3.3 (p. 243); 14.3.4; 22.1.2
A. B. Zolotovitskii	(b) **121**,	737 (1984)	15 (p. 264); 22.6
A. V. Zotov	(a) **82**,	345 (1984)	1.5 (p. 33) and 3 (p. 55); 2; 22.1.2
N. V. Zotova	(a) **85**,	655 (1984)	14.3.3 (p. 246) and 20.3 (p. 356); 16; 22.2.1
Xian-Wu Zou	see under letter "X"		
A. Zoul	(a) **86**,	673 (1984)	14.3.2 (p. 240); 14.3.4; 22.2.1
E. Zschech	(a) **86**,	117 (1984)	1.3 (p. 26); 18; 21.1; 21.1.1
H. Zscheile	(b) **121**,	K161 (1984)	13.5.2 (p. 205); 14.4; 22.9
E. M. Zub	(a) **82**,	569 (1984)	14.3.1 (p. 234); 1.5; 22.8.2
R. Żuberek	(a) **82**,	K177 (1984)	2 (p. 43) and 18.2.1 (p. 296); 21.1.1
A. Žukauskas	(b) **126**,	K135 (1984)	13.5.1 (p. 204) and 20.1 (p. 341); 22.4.1; 22.4.2
I. Zupančič	(b) **123**,	K83 (1984)	9 (p. 92) and 14.4 (p. 256); 19; 22.8.1
S. Žurauskas	(b) **121**,	385 (1984)	13.4 (p. 172); 14.3.3 22.1.2
R. Zürcher	(a) **84**,	475 (1984)	10.1 (p. 100); 21.1
A. K. Zvezdin	(a) **84**,	215 (1984)	18.4 (p. 302); 22.8.2
A. I. Zvyagin	(a) **82**,	67 (1984)	1.2 (p. 15) and 19 (p. 307); 13.4; 22.8.1
I. P. Zvyagin	(a) **86**,	375 (1984)	2 (p. 53) and 14.3.2 (p. 239); 22.1.2
I. I. Zyatkov	(a) **84**,	645 (1984)	14.3 (p. 228); 22.8.2

19. Magnetic Resonance

20. Optical Properties
    20.1 Spectra. Optical Constants (X-Ray Spectra See 20)
    20.2 Lasers
    20.3 Luminescence (X-Ray Spectra See 20)

21. Metals. Alloys
    21.1 Transition Metals (Group Sc to Zn) and Their Alloys
        21.1.1 Fe and Fe Alloys
    21.2 Alkali Metals
    21.3 Alkali-Earth Metals
    21.4 Rare-Earth Metals
    21.5 Actinides
    21.6 Noble Metals
    21.7 Semi-Metals (Semiconducting Alloys See 22.7)

22. Semiconductors and Ionic Crystals
    22.1 Elements
        22.1.1 Germanium
        22.1.2 Silicon
        22.1.3 Group VI Elements
    22.2 III–V Compounds
        22.2.1 Arsenides
        22.2.2 Phosphides
        22.2.3 Antimonides
        22.2.4 Mixed Crystals
    22.3 IV–IV, II–V, and III–VI Compounds
        22.3.1 Mixed Crystals
    22.4 II–VI Compounds
        22.4.1 Sulphides
        22.4.2 Selenides
        22.4.3 Tellurides
        22.4.4 Mixed Crystals
    22.5 Halides
        22.5.1 Silver Halides
        22.5.2 Alkali Halides
        22.5.3 Alkali-Earth Halides
        22.5.4 Mixed Crystals
    22.6 Simple Oxides (CdO, ZnO See 22.4)
        22.6.1 Mixed Crystals
    22.7 Semiconducting Intermetallic Compounds
        22.7.1 Mixed Crystals
    22.8 Tertiary and Higher Compounds
        22.8.1 Salts of Oxy-Acids (Carbonates, Phosphates, Silicates Including Mica etc.)
        22.8.2 Spinels, Garnets, Ferrites
    22.9 Organic Semiconductors

23. Solidified Gases

For comments on the use of the classification scheme see p. 6.

82913/82914

Manuscripts and letters for
physica status solidi (b) — basic research and
physica status solidi (a) — applied research

should be sent to

DDR-1086 Berlin, Leipziger Str. 3—4, P. B. 1233

or

Internationaler Buchversand GmbH, 1000 Berlin 19, Eschenallee 21.

The object of "physica status solidi (b)" and "physica status solidi (a)" is to publish rapidly important and topical results concerning fundamental research in the field of solid state physics and aspects of application and experimental techniques of fundamental solid state research, respectively. The issues of "physica status solidi (b)" and "physica status solidi (a)" will be published on the 1st and the 16th day of each month, respectively. Either series will contain the following categories:

1. Review articles written by arrangement with the editors. They will be remunerated.
2. Original papers selected from those which have not been submitted for publication elsewhere, after being reviewed. (The latest date for acceptance of contributions is the 8th and the 23rd day of each month for "physica status solidi (b)" and "physica status solidi (a)", respectively, 50 days before the delivery of the journal.)
3. Short notes published in the off-set part of the journal. They are, if necessary after being reviewed, selected by the editorial staff. (The latest date for acceptance of manuscripts for this part is the 12th and the 27th day of each month for "physica status solidi (b)" and "physica status solidi (a)", respectively, 16 days before the delivery of the journal.)
4. Pre-printed titles of papers to be published in the next issue of both series.

Editorial Information

Manuscripts should be submitted preferentially in English. Besides, papers in German, French, and Russian are admitted. The final decision on the incorporation of the papers in the series (b) or (a) is taken by the editorial staff after consultation of the referee. Authors are asked to observe the following rules:

1. Manuscripts have to be written as concisely as possible. At most, 20 and 4 pages (28 lines per page; abstracts, tables, references, figures, and captions of figures included, where 3 figures are counted as about one page) are admitted for original papers and short notes, respectively. All manuscripts should be type-written on one side only. Two copies are required. Badly arranged manuscripts are returned.
2. The author's name, initials, and institution should be given.
3. One abstract of about ten lines in English and one in German, French, or Russian are required for original papers.
4. Because of lack of space conclusions (if any) must not have the character of a summary (i. e. substantially a repetition of the contents of the abstract).
5. Formulae should not be written too small and not with pencil. Separate lines for formulae are desirable. Indistinct writing will lead to a return of the manuscript.
6. Drawings should be enclosed on separate sheets of paper, not exceeding the maximum and minimum sizes of 20 × 30 and 10 × 15 cm², respectively. Photographs should be submitted, on white paper, only for review articles and original papers. Duplicates are required.
7. Captions for the figures should be submitted on separate sheets of paper, and special instructions with regard to the size of illustrations should be given on these.
8. References should be numbered in order of appearance and should be given, on a separate sheet, in the form used in physica status solidi.
9. SI-units should be used.
10. Each paper should contain a proposal for the subject classification according to the classification scheme appearing on the second and third cover of each issue.

The authors will be sent galley proofs of review articles and original papers. Proof reading of short notes and pre-printed titles will be done by the editorial staff.
For each original paper, and short note, a total of 75 reprints is available free of charge. The number of reprints free of charge per review article is 25. Requests for additional reprints (against payment) should be given together with the submitted manuscript. **The deadline for orders of additional reprints is the 15th and the 30th day of each month** for "physica status solidi (b)" and "physica status solidi (a)", respectively, two weeks before the delivery of the journal.

Bei Fragen zur Produktsicherheit wenden Sie sich bitte an:
If you have any questions regarding product safety,
please contact:

Walter de Gruyter GmbH
Genthiner Straße 13
10785 Berlin
productsafety@degruyterbrill.com